AF327642

The series NanoScience and Technology is focused on the fascinating nano-world, mesoscopic physics, analysis with atomic resolution, nano and quantum-effect devices, nanomechanics and atomic-scale processes. All the basic aspects and technology-oriented developments in this emerging discipline are covered by comprehensive and timely books. The series constitutes a survey of the relevant special topics, which are presented by leading experts in the field. These books will appeal to researchers, engineers, and advanced students.

NanoScience and Technology

For further volumes:
www.springer.com/series/3705

Vladimir M. Fomin

Editor

Physics
of Quantum
Rings

 Springer

Editor
Vladimir M. Fomin
Leibniz Institute for Solid State
  and Materials Research
Dresden, Germany

ISSN 1434-4904   NanoScience and Technology
ISBN 978-3-642-39196-5   ISBN 978-3-642-39197-2 (eBook)
DOI 10.1007/978-3-642-39197-2
Springer Heidelberg New York Dordrecht London

Library of Congress Control Number: 2013948471

Foreword

Physics of Solids developed into an independent discipline at the end of the 30ies of last century with the formulation of the theory of electronic band structure. Boundary condition for the validity of this theory is the assumption of an infinitely extended crystal showing no defects, interfaces or surfaces. Almost at the same time the quantum mechanical problem of a particle in a one-dimensional potential well was solved for the first time.

Semiconductors were recognized as an important class of solids only a decade later, although the first roots go back to the 19th century when Ferdinand Braun, well known as the inventor of the cathode ray tube, wrote in 1874 his thesis on "Current Conduction through Sulfur-Metals". The subject of this thesis got much later the name "Schottky Diode". The developments of the transistor by William Shockley and his coworkers starting 1947 and of III–V-compounds by Heinrich Welker already in 1951 present landmarks decisive for the advancement of modern multi-targeted technologies enabling today solar cells, microprocessors or semiconductor lasers, to mention a few device groups having diffused into our daily life. Indeed it is unthinkable to live without such devices enabling, in particular, modern communication technologies.

Heterostructures, layered semiconductor/semiconductor or semiconductor/insulator structures like Si/SiO_2, were essential parts of devices like transistors from the very beginning. With the advent of III–V-based heterostructures, presenting the basis for light emitting, but also highly efficient light harvesting devices, the materials basis for a wealth of devices and systems broadened enormously and the scientific community embarked to explore "chemical engineering" in a very systematic way. Twice Nobel prizes were awarded for the physics of Si- and III–V-based heterostructures in 1985 and 2000. The limits of combining materials of varying chemical composition on top of each other were discovered to be controlled by the variation of lattice constants between different materials. If this difference is too large, defects like dislocations develop and the device properties degrade. Thus, the original enthusiasm on "chemical engineering" was fast decaying at the end of the 80ies of last century and almost entirely "lattice-matched heterostructures" were thought to be useful, restricting enormously the range of structures being available for III–V-based applications or fundamental physics investigations.

At the end of the 60ies the first nanostructures caught very rapidly increasing interest of the community. Dingle and coauthors fabricated the first "particle-in-a-box" structure, which is called today a "quantum well" or a "two-dimensional structure". The fundamental band gap of a thin layer of a narrow-band-gap material, with a thickness below the de Broglie wavelength of a charge carrier, inserted between two barriers of larger-gap materials, was discovered to be thickness dependent, thus confirming the theoretical prediction. The emission wavelength of a laser based on quantum wells is consequently tunable via the thickness of the active layer. This discovery marks the advent of modern nanostructure physics. Soon later in the 70ties and 80ties, research moved to structures of still lower dimensionality, like one-dimensional and zero-dimensional structures, quantum wires and quantum dots (QDs). Efficient technologies for easy fabrication of defect-free nanostructures were missing, however, and the interest faded away until the beginning of the 90ties. Then the Stranski-Krastanow mode of self-organized growth of strained zero-dimensional nanostructures was discovered [1], theoretically founded by modern theory of surface physics and demonstrated to present the basis of active layers for e.g. lasers with lower threshold current density than ever thought of [2]. Surprisingly, two paradigms of modern semiconductor physics had to be given up at the same time by these discoveries: the "lattice match paradigm for heterostructures" and the "fabrication paradigm" that lithography based method must be employed to create quantum wires and QDs. A minimum amount of strain induced by lattice mismatch of the heterostructures is the driving source for QD formation. Zero-dimensional structures, from the point of view of their electronic properties, do not resemble any more classical semiconductors with their continuous dispersion of energy as a function of momentum. They behave like giant hydrogen atoms in a dielectric cage and show a very simple twofold degenerate energy level system [2] thus presenting a potential source of qubits and entangled photons.

In the 21st century, the hallmarks of modern solid state physics, far beyond just semiconductors, are design, fabrication, study and applications of the now existing great variety of nanostructures. Among them, quantum rings, which are the subject of the present book, take an outstanding place, because they are not simply zero-dimensional coherent clusters of atoms or molecules on a surface. Quantum rings combine sizes at the nanoscale with a non-trivial topology: doubly-connectedness of a ring or even more complicated topological properties like one-sidedness of a Möbius strip. This combination leads again to the occurrence of unique physical properties, in particular, persistent currents. Quantum rings present a unique playground for quantum mechanical paradigms. Their physical properties are designed by controlling the geometry of a ring and the magnetic flux threading it, as well as by creating assemblies of quantum rings.

The present book gives an exhaustive and clear overview of this vigorously developing field, starting with a comprehensive pedagogical introduction of the fundamentals, via a profound presentation of the key technologies for their fabrication, characterization tools, discoveries and findings, to a discussion of the most recent advancements and current research activities. The style of the book is highly motivating for both experienced and young scientists: it finally leads a reader straightforwardly towards still open problems in this fascinating field.

The book is written by a group of the world's leading scientists of this field, who have provided fundamental contributions to the fabrication, characterization and theoretical analysis of quantum rings. Hence a reader receives a unique access to their "scientific laboratory", in particular, about state-of-the art methods of growth (MBE, droplet epitaxy, lithographic patterning, ...), characterization (Scanning-probe imaging like STM, SEM, XSTM, ...) and theoretical analysis of nanostructures and metamaterials.

Based on their unprecedented tunability, quantum rings are highly prospective as elemental base for various applications: photonic detectors and sources, including single-photon emitters, nanoflash memories, qubits for spintronic quantum computing, magnetic random access memory, recording medium and other spintronic devices ... The book contains road maps for the implementation of quantum rings into such real-world devices.

This book will be the required reading for all those who are active in nanoscience, nanotechnology and the applications of quantum rings.

References

1. V. Shchukin, D. Bimberg, Rev. Mod. Phys. **71**, 1125 (1999)
2. D. Bimberg, M. Grundmann, N.N. Ledentsov, *Quantum Dot Heterostructures* (Wiley, Chichester, 1999)

Berlin, Germany Dieter Bimberg

Preface

For the first time in monographic literature, the present book provides a broad panorama of the physics of quantum rings with emphasis on *modern advancements* in theoretical and experimental investigations of semiconductor quantum rings. It is written in a style which makes these issues accessible to theoretical physicists, experimental researchers, and technologists with different levels of experience: from graduate and PhD students to experts. The book is also intended to convey the fascination of quantum rings to specialists in other disciplines: mathematics, chemistry, electronic and optical engineering, and information technologies. Our goal is that this book will succeed in invigorating research interests towards the further development of fundamental insight in and applications of quantum rings.

It starts with an introduction into the fundamental physics of quantum rings as a heuristically unique playground for the quantum-mechanical paradigm and a concise overview of the state-of-the-art in the field, with a particular emphasis on the quantum interference phenomena like the Aharonov-Bohm effect in quantum rings (Chap. 1). The book consists of three main parts, though the borders between them are conventional: Part I. Fabrication, characterization and physical properties, Part II. Aharonov-Bohm effect for excitons and Part III. Theory.

The *first part* represents three advanced methods of fabrication of quantum rings: self-organized growth, droplet epitaxy and lithographic patterning, as well as their characterization based on scanning-probe-microscopy. It opens with Chap. 2 (by Wen Lei and Axel Lorke) and Chap. 3 (by Jorge M. García, Benito Alén, Juan Pedro Silveira and Daniel Granados) representing fundamentals of the self-organized growth and optical properties of semiconductor quantum rings. In Chap. 4 (by myself, Vladimir N. Gladilin, Jozef T. Devreese and Paul M. Koenraad) we discuss how the modern characterization of self-assembled InGaAs/GaAs quantum rings using X-STM has allowed for a development of an adequate model of their shape, which quantitatively explains the Aharonov-Bohm effect observed in the magnetization. Self-organized formation of highly distinct GaSb/GaAs quantum-ring structures and their X-STM characterization are presented in Chap. 6 (by Andrea Lenz and Holger Eisele).

Scanning-probe electronic imaging of lithographically patterned quantum rings, which is discussed in Chap. 5 (by Frederico R. Martins, Hermann Sellier, Marco G. Pala, Benoit Hackens, Vincent Bayot and Serge Huant), can access to the intimate properties of buried electronic systems. Another promising way of controllable self-assembled fabrication of quantum rings—by droplet epitaxy—is overviewed in Chap. 7 (by Jiang Wu and Zhiming M. Wang) with emphasis on ordered arrays and in Chap. 8 (by Stefano Sanguinetti, Takaaki Mano and Takashi Kuroda), where the focus is on semiconductor quantum-ring complexes.

The *second part* deals with the Aharonov-Bohm effect for multi-electron systems, in particular, for excitons and plasmons. In Chap. 9, Alexander V. Chaplik and Vadim M. Kovalev review theoretical investigations on novel versions of the Aharonov-Bohm effect in quantum rings, including that for electronic Wigner molecules, polarized neutral and charged excitons, and polarons, as well as its manifestations in the longitudinal magnetoresistance. Also, the role of the spin-orbit interaction in the electronic properties of quantum rings is revealed. Theory meets experiment on the Aharonov-Bohm effect for neutral excitons in quantum rings in Chap. 10 (by Marcio D. Teodoro, Vivaldo L. Campo, Jr., Victor Lopez-Richard, Euclydes Marega, Jr., Gilmar E. Marques and Gregory J. Salamo). Remarkably robust optical Aharonov-Bohm effect occurs in type-II quantum dots presented in Chap. 11 (by Ian R. Sellers, Igor L. Kuskovsky, Alexander O. Govorov and Bruce D. McCombe). Chapter 12 (by Fei Ding, Bin Li, François M. Peeters, Val Zwiller, Armando Rastelli and Oliver G. Schmidt) describes the observation and manipulation of Aharonov-Bohm-type oscillations in a single quantum ring.

The *third part* represents advancements in theory of quantum rings. The effects of a tensile-strained insertion layer on strain and the electronic structure of quantum rings are analyzed in Chap. 13 (by Pilkyung Moon, Euijoon Yoon, Won Jun Choi, Jae Dong Lee and Jean-Pierre Leburton) using the model advanced in Chap. 4. The basic approaches to theoretical modeling of electronic and optical properties of semiconductor quantum rings are overviewed by Oliver Marquardt in Chap. 14; it can also serve as a tutorial for students. A survey on Coulomb interaction in finite-width quantum rings is provided in Chap. 15 (by Benjamin Baxevanis and Daniela Pfannkuche). Booming studies on general topological aspects of quantum rings are illuminated by Benny Lassen, Morten Willatzen and Jens Gravesen in Chap. 16 on differential-geometry methods applied to rings and Möbius nanostructures. In Chap. 17, Carlos Segarra, Josep Planelles and Juan I. Climente discuss effects of hole mixing in semiconductor quantum rings and show that the strong strain potential may compete against the band-offset potential in quantum rings. Engineering of electron states and spin relaxation in quantum rings and quantum dot-ring nanostructures is reviewed in Chap. 18 (by Marcin Kurpas, Elżbieta Zipper and Maciej M. Maśka).

The *main message* of the present book is that the front-line methods of fabrication and characterization of quantum rings together with the sophisticated cutting-edge theoretical research have allowed for accumulation of a significant thesaurus of fundamental information on their behavior. This highly diversified knowledge underpins numerous suggestions for prospective applications of quantum rings as a highly tunable elemental base for future device design and optimization, in particular, in optoelectronics and spintronics, magnetic memory devices, photonic sources and detectors, and information storage and processing.

Acknowledgements

I am indebted to my teacher and friend Evghenii Petrovich Pokatilov of blessed memory, who introduced me into the fascinating world of theoretical physics. My special thanks are due to Lutz Wendler: we started together the investigations in the field of physics of quantum rings. This was made possible by awarding me a Humboldt Fellowship, for which I am deeply grateful to the Alexander von Humboldt Foundation.

I want to thank very warmly my colleagues and friends for fruitful and motivating collaborations related to various problems of physics of quantum rings: Janneke H. Blokland, Iris M.A. Bominaar-Silkens, Murat Bozkurt, Alexander V. Chaplik, Liviu F. Chibotaru, Peter C.M. Christianen, Jozef T. Devreese, Jorge M. García, Vladimir N. Gladilin, Alexander O. Govorov, Daniel Granados, Suwit Kiravittaya, Niek A.J.M. Kleemans, Serghei N. Klimin, Paul M. Koenraad, Arkady A. Krokhin, Jan Kees Maan, Vyacheslav R. Misko, Victor V. Moshchalkov, Peter Offermans, Oliver G. Schmidt, Alfonso G. Taboada, Jacques Tempere, Eric C.M. van Genuchten, Joachim H. Wolter, Uli Zeitler and Hu Zhao.

I am extremely grateful to many colleagues for insightful discussions and stimulating interactions: Dieter Bimberg, Markus Büttiker, Manuel Cardona, Peter Cendula, Venkat Chandrasekhar, Valeri G. Grigoryan, Yoseph Imry, John C. Inkson, Peter Kratzer, Jörg P. Kotthaus, the late Rolf Landauer, Axel Lorke, Dominiqe Mailly, Felix von Oppen, Carmine Ortix, Pierre M. Petroff, Jeroen van den Brink, Achim Wixforth, Roger Wördenweber and Vladimir I. Yudson.

I am very thankful to the State University of Moldova, my *alma mater*, where my research career was launched. Various stages of my research activities were performed in a number of institutions, to which I am very grateful: Martin-Luther University of Halle-Wittenberg, University of Antwerp, Eindhoven University of Technology, Catholic University of Leuven, Research Center Jülich, University of Duisburg-Essen, and, most recently, Institute for Integrative Nanosciences (IIN)—Leibniz Institute for Solid State and Materials Research (IFW) Dresden.

I highly appreciate great enthusiasm, time and effort of all contributors to this book. I would like to express my deep gratitude to Claus E. Ascheron, Senior Editor at Springer, for a vigorous support of the idea to publish it, to Donatas Akmanavičius for its thorough production at VTeX and to Alexander A. Balandin, Jorge M. García, Vladimir N. Gladilin, Alexander O. Govorov, Serge Huant, Paul M. Koenraad, Axel Lorke, Ian R. Sellers, Bartłomiej Szafran and Zhiming M. Wang for their valuable support while I was preparing this book for publication.

With profound gratitude I keep the memory of my parents, who nurtured my aspiration to comprehend the world. Special thanks are due to my wife and children for their understanding and patience in the course of my work on this book.

Dresden, Germany   Vladimir M. Fomin

Contents

Part III Theory

List of Contributors

Benito Alén Instituto de Microelectrónica de Madrid, Tres Cantos, Madrid, Spain

Benjamin Baxevanis I. Institute for Theoretical Physics, University of Hamburg, Hamburg, Germany

V. Bayot IMCN/NAPS, Université catholique de Louvain, Louvain-la-Neuve, Belgium

V.L. Campo Jr. Departamento de Física, Universidade Federal de São Carlos, São Carlos, São Paulo, Brazil

A.V. Chaplik Institute of Semiconductor Physics, Novosibirsk, Russia; Novosibirsk State University, Novosibirsk, Russia

Won Jun Choi Center for OptoElectronic Convergence System, Korea Institute of Science and Technology, Seoul, Korea

Juan I. Climente Departament de Química Física i Analítica, Universitat Jaume I, Castelló, Spain

J.T. Devreese Theory of Quantum and Complex Systems, University of Antwerp, Antwerp, Belgium

F. Ding Institute for Integrative Nanosciences, IFW Dresden, Dresden, Germany

Holger Eisele Institut für Festkörperphysik, Technische Universität Berlin, Berlin, Germany

Vladimir M. Fomin Institute for Integrative Nanosciences, IFW Dresden, Dresden, Germany

Jorge M. García Instituto de Microelectrónica de Madrid, Tres Cantos, Madrid, Spain

V.N. Gladilin Theory of Quantum and Complex Systems, University of Antwerp, Antwerp, Belgium

A.O. Govorov Department of Physics & Astronomy, Clippinger Research Laboratory, Ohio University, Athens, OH, USA

Daniel Granados Instituto de Microelectrónica de Madrid, Tres Cantos, Madrid, Spain; IMDEA Nanociencia, Madrid, Spain

Jens Gravesen Department of Mathematics, Technical University of Denmark, Kgs. Lyngby, Denmark

B. Hackens IMCN/NAPS, Université catholique de Louvain, Louvain-la-Neuve, Belgium

S. Huant Institut Néel, CNRS & Université Joseph Fourier, Grenoble, France

P.M. Koenraad Photonics and Semiconductor Nanophysics, Eindhoven University of Technology, Eindhoven, The Netherlands

V.M. Kovalev Institute of Semiconductor Physics, Novosibirsk, Russia; Novosibirsk State University, Novosibirsk, Russia

Takashi Kuroda National Institute for Materials Science, Tsukuba, Japan

Marcin Kurpas Department of Theoretical Physics, University of Silesia, Katowice, Poland

I.L. Kuskovsky Department of Physics, Queens College, City University of New York, Flushing, NY, USA

Benny Lassen Mads Clausen Institute, University of Southern Denmark, Sønderborg, Denmark

Jean-Pierre Leburton Department of Electrical and Computer Engineering, University of Illinois at Urbana-Champaign, Urbana, IL, USA

JaeDong Lee Department of Emerging Materials Science, Daegu Gyeongbuk Institute of Science and Technology, Daegu, Korea

Wen Lei School of Electrical, Electronic and Computer Engineering, The University of Western Australia, Crawley, WA, Australia

Andrea Lenz Institut für Festkörperphysik, Technische Universität Berlin, Berlin, Germany

B. Li Departement Fysica, Universiteit Antwerpen, Antwerpen, Belgium

V. Lopez-Richard Departamento de Física, Universidade Federal de São Carlos, São Carlos, São Paulo, Brazil

Axel Lorke Faculty of Physics and CENIDE, Universität Duisburg-Essen, Duisburg, Germany

Takaaki Mano National Institute for Materials Science, Tsukuba, Japan

E. Marega Jr. Instituto de Física de São Carlos, Universidade de São Paulo, São Carlos, São Paulo, Brazil

Oliver Marquardt Tyndall National Institute, Cork, Ireland

G.E. Marques Departamento de Física, Universidade Federal de São Carlos, São Carlos, São Paulo, Brazil

F. Martins IMCN/NAPS, Université catholique de Louvain, Louvain-la-Neuve, Belgium

Maciej M. Maśka Department of Theoretical Physics, University of Silesia, Katowice, Poland

B.D. McCombe Department of Physics, Fronczak Hall, University at Buffalo, The State University of New York, Buffalo, NY, USA

Pilkyung Moon Department of Physics, Tohoku University, Sendai, Japan

M.G. Pala IMEP-LAHC, UMR 5130, CNRS/INPG/UJF/UdS, Grenoble, France

F.M. Peeters Departement Fysica, Universiteit Antwerpen, Antwerpen, Belgium

Daniela Pfannkuche I. Institute for Theoretical Physics, University of Hamburg, Hamburg, Germany

Josep Planelles Departament de Química Física i Analítica, Universitat Jaume I, Castelló, Spain

A. Rastelli Institute of Semiconductor and Solid State Physics, Johannes Kepler University Linz, Linz, Austria

G.J. Salamo Arkansas Institute for Nanoscale Materials Science and Engineering, University of Arkansas, Fayetteville, AR, USA

Stefano Sanguinetti LNESS and Dipartimento di Scienza dei Materiali, Universitá di Milano Bicocca, Milano, Italy

O.G. Schmidt Institute for Integrative Nanosciences, IFW Dresden, Dresden, Germany

Carlos Segarra Departament de Química Física i Analítica, Universitat Jaume I, Castelló, Spain

I.R. Sellers Department of Physics & Astronomy, University of Oklahoma, Norman, OK, USA

H. Sellier Institut Néel, CNRS & Université Joseph Fourier, Grenoble, France

Juan Pedro Silveira Instituto de Microelectrónica de Madrid, Tres Cantos, Madrid, Spain

M.D. Teodoro Departamento de Física, Universidade Federal de São Carlos, São Carlos, São Paulo, Brazil

Zhiming M. Wang State Key Laboratory of Electronic Thin Film and Integrated Devices, University of Electronic Science and Technology of China, Chengdu, P.R. China

Morten Willatzen Mads Clausen Institute, University of Southern Denmark, Sønderborg, Denmark; Department of Photonics Engineering, Technical University of Denmark, Kgs. Lyngby, Denmark

Jiang Wu State Key Laboratory of Electronic Thin Film and Integrated Devices, University of Electronic Science and Technology of China, Chengdu, P.R. China

Euijoon Yoon Department of Materials Science and Engineering, Seoul National University, Seoul, Korea

Elżbieta Zipper Department of Theoretical Physics, University of Silesia, Katowice, Poland

V. Zwiller Kavli Institute of Nanoscience, Delft University of Technology, Delft, The Netherlands

Chapter 1
Quantum Ring: A Unique Playground for the Quantum-Mechanical Paradigm

Vladimir M. Fomin

"The Ring *is… not just a story…: it's a cosmos."*
(R. Lepage about the tetralogy The Ring of the Nibelung
of R. Wagner)
http://wagnersdream.metoperafamily.org/robert-interview.html

Abstract The physics of quantum rings is reviewed from basic concepts rooted in the quantum-mechanical paradigm—via unprecedented challenges brilliantly overcome by both theory and experiment—to promising application perspectives.

1.1 Prologue

Doubly-connected (ring-like) structures at the scale of nanometers (nanoscale) are generally termed *Quantum Rings* (QRs). They exhibit a unique density of states for charge carriers and quantum fields and hence a vast variety of physical properties, which are cardinally different from those of singly-connected structures (like quantum dots).

Circular electric currents prophetically introduced by Ampère [1, 2] to explain the origin of magnetism: "…un aimant doit être considéré comme un assemblage de courans électriques qui ont lieu dans des plans perpendiculaires à son axe…"[1] were an essential precursor of persistent currents in the modern physics of QRs. A magnetic field was related to the currents circulating along concentric paths: "…à chacun des pôles d'un aimant, les courants électriques dont il se compose sont dirigés suivant des courbes fermées concentriques…"[2] Quantum mechanics predicts that small enough ring-like structures threaded by a magnetic flux, in the

---

[1]"… a magnet should be considered as an assembly of electric currents that occur in planes perpendicular to its axis…" (Translation by V. M. F.)

[2]"… at each of the poles of a magnet, the electrical currents, of which it consists, are directed along concentric closed curves…" (Translation by V. M. F.)

V.M. Fomin (✉)
Institute for Integrative Nanosciences, IFW Dresden, Helmholtzstraße 20, 01069 Dresden, Germany
e-mail: v.fomin@ifw-dresden.de

V.M. Fomin (ed.), *Physics of Quantum Rings*, NanoScience and Technology,
DOI 10.1007/978-3-642-39197-2_1, © Springer-Verlag Berlin Heidelberg 2014

equilibrium state, carry *persistent* (dissipationless) circulating electron currents that do not require an external power source. A prerequisite is that the electron state keeps quantum coherence over the whole doubly-connected system.

There have been a number of reviews representing various aspects of physics of QRs, for example, effects of a finite width of the QRs [3], mesoscopic phenomena in QRs with strongly coupled polarons [4], possible types of III–V semiconductor QRs [5], equilibrium properties of mesoscopic metal rings [6], ring-like nanostructures as a leitmotif in plasmonics and nanophotonics [7], theoretical modeling of the self-organized QRs on the basis of the modern characterization of those nanostructures [8], theoretical analysis and experimental observations of persistent currents by virtue of the magnetic flux quantization phenomenon [9], and advancements in experimental and theoretical physics of QRs [10]. In the present Chapter, we discuss a number of contributions to the physics of QRs, essential for the topics of the present book—(i) fundamentals of physics of QRs and (ii) semiconductor QRs—without any claim for an exhaustive presentation of the extensive literature in this vigorously developing field.

1.2 At Dawn

The following studies, commenced already at the very early stage of the quantum physics, unraveled the key properties of persistent currents in ring-like quantum structures.

For calculating the magnetically induced current densities of aromatic hydrocarbon ring molecules, Pauling [11] advanced a hypothesis that the external electrons in the benzene molecule can *circulate freely* and provide a very large contribution to the diamagnetic susceptibility with the magnetic field normal to the plane of the carbon hexagon: "We may well expect that in these regions the potential function representing the interaction of an electron with the nuclei and other electrons in the molecule would be approximately cylindrically symmetrical with respect to the hexagonal axis of the molecule, the electron, some distance above or below the plane of the nuclei, passing almost imperceptibly from the field of one carbon atom to that of the next."

Within the framework of a quantum-mechanical derivation, London [12] demonstrated that the diamagnetic susceptibility of aromatic ring molecules was related to a *current circulating around the opening* induced by the magnetic field: "La susceptibilité… correspond à des courants induits qui circulent d'un atome à l'autre autour de la chaîne cyclique."[3] This current belonged to the *ground state*, in analogue with superconducting currents: "Nous pouvons… disant que les combinaisons aromatiques se comportent comme des supraconducteurs."[4]

---

[3]"The susceptibility… corresponds to the induced currents that flow from one atom to another around the cyclic chain." (Translation by V.M.F.)

[4]"We can… say that the aromatic combinations behave as superconductors." (Translation by V.M.F.)

Calculating the magnetic response of ultrasmall magnetic ring-shaped particles on the basis of the Schrödinger equation, Hund [13] showed that both at zero temperature and in thermodynamic equilibrium at temperature $T > 0$, there existed a *total current circulating around the annulus*, which was *dissipationless*: "...ein wesentlicher Teil des der diamagnetischen Magnetisierung entsprechenden Stromes um das Loch herumfließt; dieser Strom hat keine Joulesche Warme, da die Besetzung der Zustande dem Temperaturgleichgewicht entspricht."[5] Further, it was demonstrated that a set of eigenstates found for an electron in a ring in a magnetic field B led to jumps in the magnetization from negative (diamagnetic) to positive (paramagnetic) at certain values of the applied magnetic field: "...es tritt zu der negativen (diamagnetischen) Magnetisierung plötzlich eine konstante positive Magnetisierung hinzu, und dies wiederholt sich nach einem gewissen Zuwachs von B."[6] As a result, the current circulating around the opening of the ring acquires *a zigzag form* as a function of the applied magnetic field (shown in Fig. 4 of [13]).

Systematically developing the earlier ideas, Dingle found [14] that the equilibrium properties calculated for small *free-electron systems* in a perfect ring and in a perfect infinite cylinder were sensitive to the *magnetic flux Φ* threading the system, the magnetic permittivity consisting of a steady part and periodic in the magnetic flux terms. The fundamental dimensionless quantity, which determined the *periodic* dependence, was (in the modern notation) the ratio Φ/Φ_0, where the magnetic flux quantum $\Phi_0 = h/e$ was determined by universal constants: the Planck constant h and the elementary charge e. Dingle already noticed the challenges in observing those periodic terms: "...a single cylinder would possess only a very small magnetic moment, whilst it would be difficult to ensure a uniform radius for a bundle of cylinders"—a conclusion that has remained very urgent for experimentalists ever since then. The key challenges in detecting persistent currents experimentally are twofold: they produce exceptionally small signals and they are very sensitive to the environment [15].

1.3 Fundamentals of Topological Effects

A fundamental role of the ring-topology for the quantum-mechanical paradigm was unraveled within the theory of a geometric phase [16–19]. Berry [19] provided a simple but intuitively appealing derivation of the geometric (Berry) phase, which will be recalled below. A system is considered whose Hamiltonian H depends on a set of varying parameters $\mathbf{R} \equiv \mathbf{R}(t)$ forming a closed path C between the instant $t = 0$ and the instant $t = T$ such that $\mathbf{R}(0) = \mathbf{R}(T)$.

---

[5]"...an essential part of the current corresponding to the diamagnetic magnetization flows around the annulus; this current produces no Joule heat, as the population of states corresponds to the thermal equilibrium." (Translation by V.M.F.)

[6]"...a constant positive magnetization occurs to be suddenly added to the negative (diamagnetic) magnetization, and this is repeated after a certain increase in B." (Translation by V.M.F.)

The evolution of the state of the system is governed by the Schrödinger equation:

$$i\hbar\big|\dot\psi(t)\big\rangle = H\big(\mathbf{R}(t)\big)\big|\psi(t)\big\rangle. \tag{1.1}$$

At any instant t, the eigenstates satisfy the stationary Schrödinger equation

$$H(\mathbf{R})\big|n(\mathbf{R})\big\rangle = E_n(\mathbf{R})\big|n(\mathbf{R})\big\rangle, \tag{1.2}$$

where $|n(\mathbf{R})\rangle$ is single-valued in the region that includes C. Within the *adiabatic* approximation [20], the system prepared in one of these states $|n(\mathbf{R}(0))\rangle$ will evolve with the Hamiltonian $H(\mathbf{R}(t))$ and be in the state $|n(\mathbf{R}(t))\rangle$ at the instant t. A gauge-invariant generalization to the phase-coherence phenomena in *nonadiabatically* evolving quantum systems was proposed by Aharonov and Anandan [21]. The solution to the Schrödinger equation (1.1) is sought in the form

$$\big|\psi(t)\big\rangle = \exp\!\left[-\frac{i}{\hbar}\int_0^t d\tau\, E_n\big(\mathbf{R}(\tau)\big)\right]\exp\!\big[i\gamma_n(t)\big]\big|n\big(\mathbf{R}(t)\big)\big\rangle. \tag{1.3}$$

Substituting (1.3) into the Schrödinger equation (1.1) and taking into account (1.2), we find the equation for the *geometric phase* $\gamma_n(t)$:

$$\dot\gamma_n(t) = i\big\langle n\big(\mathbf{R}(t)\big)\big|\nabla_{\mathbf{R}} n\big(\mathbf{R}(t)\big)\big\rangle \cdot \dot{\mathbf{R}}(t).$$

(A gauge can be chosen so that the Aharonov-Bohm phase is included in the dynamical phase instead of the geometric phase, see, e.g., [22, 23].) The total phase change of the state of (1.3) on the path C

$$\big|\psi(T)\big\rangle = \exp\!\left[-\frac{i}{\hbar}\int_0^T d\tau\, E_n\big(\mathbf{R}(\tau)\big)\right]\exp\!\big[i\gamma_n(C)\big]\big|\psi(0)\big\rangle \tag{1.4}$$

is then determined by the *geometric phase change*

$$\gamma_n(C) = i\oint_C \big\langle n(\mathbf{R})\big|\nabla_{\mathbf{R}} n(\mathbf{R})\big\rangle \cdot d\mathbf{R}. \tag{1.5}$$

A generalization of the phase factor $i\gamma_n(C)$ in (1.4) (which was initially derived for a non-degenerate Hamiltonian) to the Hamiltonian with degenerate energy levels, was provided in terms of the path-ordered integrals involving non-Abelian gauge fields [24].

Magnetic field is an important tool revealing physical effects due to the doubly-connected topology. Consider a magnetic flux line [25] (or tube) carrying a flux Φ. For positions $\mathbf{R}$ outside of the flux line (tube), the magnetic field is zero, but there exists a set of gauge-equivalent vector potentials $\mathbf{A}(\mathbf{R})$ such that for any closed path C threaded by the magnetic flux line (tube)

$$\oint_C \mathbf{A}(\mathbf{R}) \cdot d\mathbf{R} = \Phi.$$

Further, let a particle carrying a charge q be confined to a box at $\mathbf{R}$, which is not penetrated by the flux line (tube). Without a flux, the Hamiltonian of the particle $H(\mathbf{p}, \mathbf{r} - \mathbf{R})$ depends on the momentum $\mathbf{p}$ and the relative position $\mathbf{r} - \mathbf{R}$ and possesses the eigenfunctions $\psi_n(\mathbf{r} - \mathbf{R})$ that satisfy (1.2) with eigenenergies independent of $\mathbf{R}$. With non-zero flux, the states $|n(\mathbf{R})\rangle$ satisfy

$$H\big(\mathbf{p} - q\mathbf{A}(\mathbf{r}), \mathbf{r} - \mathbf{R}\big)\big|n(\mathbf{R})\big\rangle = E_n\big|n(\mathbf{R})\big\rangle \tag{1.6}$$

with the eigenenergies unaffected by the vector potential. Solutions of (1.6) are obtained in terms of the Dirac phase factor [26]

$$\big\langle\mathbf{r}\big|n(\mathbf{R})\big\rangle = \exp\left[i\frac{q}{\hbar}\int_{\mathbf{R}}^{\mathbf{r}} d\boldsymbol{\rho}\cdot\mathbf{A}(\boldsymbol{\rho})\right]\psi_n(\mathbf{r} - \mathbf{R}). \tag{1.7}$$

Within a thought experiment, the box is transported round a closed doubly-connected path C threaded by the flux line (tube). Any such path is topologically equivalent to a *ring*. The integrand in the geometric phase change of (1.5) is then

$$\big\langle n(\mathbf{R})\big|\nabla_{\mathbf{R}}n(\mathbf{R})\big\rangle = \int d^3r\,\psi_n^*(\mathbf{r} - \mathbf{R})\left[-i\frac{q}{\hbar}\mathbf{A}(\mathbf{R})\psi(\mathbf{r} - \mathbf{R}) + \nabla_{\mathbf{R}}\psi_n(\mathbf{r} - \mathbf{R})\right]$$

$$= -i\frac{q}{\hbar}\mathbf{A}(\mathbf{R}).$$

The integral of the second term in the integrand vanishes because of the wave function normalization. Consequently, the geometric phase change

$$\gamma_n(C) = \frac{q}{\hbar}\oint_C \mathbf{A}(\mathbf{R})\cdot d\mathbf{R} = \frac{q\varPhi}{\hbar} \tag{1.8}$$

is independent of n. Thus, a charged particle *gains a phase* as it moves over a closed path about the flux line (tube):

$$\big|\psi(\varPhi)\big\rangle = \exp\left[i\frac{q\varPhi}{\hbar}\right]\big|\psi(\varPhi = 0)\big\rangle. \tag{1.9}$$

The geometric phase occurring in (1.9) leads to a *quantum interference* between the states of the particles in the transported box and those in a box that was not moved about the flux line (tube). There are numerous manifestations of this quantum interference, which is known as the Aharonov-Bohm effect [25, 27]. They are revealed in the electronic spectra, magnetization, optical and transport properties of QRs and, in particular, represented in the present book. Observation of the Aharonov-Bohm effect was significantly facilitated by nano-scale fabrication and low-temperature detection techniques, which minimize dephasing, as demonstrated in the beautiful experiment on dephasing in electron interference by a 'which-path' detector [28]. The phase acquired by a particle with nonzero spin can also follow from spin-orbit-coupling instead of a magnetic field (Aharonov-Casher effect) [29].

Oscillating persistent currents were extensively investigated in superconductor QRs, which are beyond the scope of the present book; see [30–32] for references.

1.4 Renaissance

In their works dealing with the flux quantization in *superconducting* rings, Byers and Yang [22] and Bloch [23] showed that "the magnetic flux through any surface whose boundary loop lies entirely in superconductors is quantized in units" $\Phi_0^{\text{sup}} = h/(2e)$, where $2e$ is the charge of a Cooper pair [22]. As a result, a general theorem follows: all physical properties of a doubly-connected system are periodic in the magnetic flux through the opening Φ with the period Φ_0^{sup}. The experimental detection of the periodicity of the magnetization as a function of magnetic flux ("magnetic flux quantization") in superconducting rings [33] and cylinders [34] was used to demonstrate that charge in superconductors was carried in units of $2e$ [34].

Gunther and Imry [35] analyzed persistent currents in a hollow, cylindrically shaped superconductor taking into account that the magnetic flux consists of two parts: that due to the external magnetic field and that due to the current. The flux quantization was shown to be exhibited when the cylinder was thick enough as compared to the penetration depth of the superconductor and thin enough as compared to the temperature-dependent coherence length as to exclude the off-diagonal long-range order.

Kulik [36, 37] discussed the persistent currents and the flux quantization in a hollow thin-walled normal metallic cylinder and ring threaded by a tube of magnetic flux-lines that were confined within an inner cylinder (a magnetic coil) with a radius smaller than the radius of the outer cylinder, in which no electric or magnetic field was present.

In cylindrical bismuth single-crystal whiskers 200–800 nm thick, oscillations in the longitudinal magnetoresistance with the period $\Phi_0/\cos\theta$ (θ was the angle of the tilt of the magnetic field with respect to the cylinder axis) observed by Brandt et al. [38, 39] were interpreted as a possible manifestation of the Aharonov-Bohm effect.

Büttiker, Imry and Landauer [40] were the first to consider persistent currents in a strictly one-dimensional *normal-metal* ring with disorder. It was concluded that "Small and strictly one-dimensional rings of normal metal, driven by an external magnetic flux, act like superconducting rings with a Josephson junction, except that $2e$ is replaced by e." These authors noticed a fundamental analogy between the energy spectrum of an electron traversing the ring and that of an electron in a periodic potential: it consisted of bands of width V with band gaps Δ. Such band states carried persistent currents. The heuristic value of the possibility to conduct a persistent-current calculation using the widely developed solid-state band-structure theory could be hardly overestimated. It was pointed out, that the band energy in a ring oscillated periodically as a function of the enclosed flux: $E_n(\Phi) = E_n(\Phi + \Phi_0)$ and carried the (*single-band*) *persistent current* $I_n = -\mathrm{d}E_n(\Phi)/\mathrm{d}\Phi$. For a geometrically perfect ring, the persistent currents carried by consecutive bands had opposite signs.

The key criteria for a possible observation of the persistent current are represented in Chap. 4 of [41]. Firstly, the electron level width (determined as $\hbar/\tau_\phi$ through the inelastic scattering time τ_ϕ) must be much smaller than the typical values of the band gap Δ and the bandwidth V. The latter condition is equivalent to the requirement that *phase coherence be maintained along the whole ring*, i.e., the phase-coherence length l is larger than the mean circumference L of the ring (ballistic regime). With increasing disorder, when the electron free path is smaller than the ring circumference (diffusive regime) the period of the Aharonov-Bohm effect becomes $\Phi_0/2$ [42]. Secondly, the *temperature must be low* enough: $k_B T \leq \Delta$. Otherwise, the sum of the persistent currents (with alternating signs) carried by the occupied levels would lead to a strong reduction of the overall persistent current.

The seminal work by Büttiker et al. [40] initiated a tremendous interest in the persistent current problem, starting with the papers on the resistance of small one-dimensional rings of normal metal driven by an external time-dependent magnetic flux [43] and the persistent currents and the absorption of power in the ring that was coupled via a single current lead to a dissipative electron reservoir [44].

The first evidence for persistent currents in mesoscopic rings was provided in the following three pioneering experiments. The persistent current in the *diffusive regime*, where the *elastic* (non-dephasing) *mean free path l* was much smaller than the mean circumference L of the ring, was measured in an ensemble of 10^7 copper rings [45] with a SQUID magnetometer and for a single (isolated) gold ring [46] using a highly sensitive thin-film miniature dc-SQUID magnetometer. In semiconductors, the persistent current was first detected for a lithographically prepared single GaAs ring in the *ballistic regime*, i.e., for $L < l$, [47] using a special technique, where the sample and the SQUID were made on the same chip. Further measurements of persistent currents were made on arrays of gold QRs [48, 49] and an ensemble of 10^5 disconnected silver rings [50].

The problem of matching theoretical predictions with the emerging experimental evidence stimulated the further intensive research aimed at a development of more realistic models of QRs, taking into account effects due to the finite size, disorder of different nature, and the electron-electron interaction.

For the metallic QRs (in the diffusive regime) the magnitudes of the persistent currents occurred much larger (by two orders of magnitude) than those predicted using the model of non-interacting electrons [51, 52], while for the semiconductor QRs in the ballistic regime this simple theory seemed to agree with experiment. This stimulated investigations (see [3, 41] for details) of the following issues: (i) the role of the choice of the statistical ensemble (canonical versus grand canonical) to calculate average values of persistent currents [53–55], (ii) the role of spin in producing the fractional Aharonov-Bohm effect [56–58], (iii) the role of the electron-electron interaction [59–61], (iv) the role of correlations due to the electron-electron interaction beyond the first-order perturbation approach [62–65].

Important conceptual ingredients to resolve the discrepancy between the measured and observed values of the magnitude of persistent currents in metallic QRs were (i) the argument of local charge neutrality in volume elements larger than the screening length [60, 66] and (ii) the fact that the effect of disorder may be strongly reduced by the electron-electron interaction [67–69]. Another interesting way to get agreement was based on the diamagnetic sign of the persistent currents observed in metal QRs, e.g., by [45], which suggested that the materials were weak superconductors [62] with a very low critical temperature [65]. Attractive electron-electron interaction may enhance the magnetic response of a QR due to the contribution of high energy levels [62]. Resolving the contradiction between experiment and theory in what concerns the magnitude of persistent currents in metallic QRs has been recognized as a major open challenge in mesoscopic physics [49, 63, 64, 70].

A rigorous quantum-mechanical theory of persistent currents developed for QRs in the ballistic regime revealed that the coupling between the different channels of the electron motion caused the occurrence of higher harmonics of Φ_0 in the persistent current. In particular, the halving of the fundamental period of the persistent current may occur in a single finite-width [71] or finite-height [72] QR due to the coupling of the azimuthal and, correspondingly, radial or paraxial electron motions by virtue of the impurity scattering.

If the magnetic field penetrated the conducting region of the finite-width QR, the Aharonov-Bohm-type oscillations due to the magnetic flux threading the opening coexisted with the diamagnetic shift of energy levels due to the magnetic field in the QR and were aperiodic [73, 74].

The role of the electron-electron interaction in a finite-width QR for a sufficiently low density, at which the correlation energy is much larger than the Fermi energy, consisted in formation of an N-electron Wigner molecule with relative angular motions of the electrons in the form of harmonic oscillations and radial motions depending on the shape of the confining potential [75–77]. The results for highly correlated electrons were, generally speaking, distinct from those for free electrons, except for the case of low temperatures, when a high-symmetry equilibrium configuration of electrons occurred by virtue of the strong repulsion between them. A Wigner molecule determined the ring-specific rich spectra of absorption, photoluminescence (PL), and Raman scattering [78, 79], which were significantly distinct from those of free electrons in a QR. Study of electronic transitions in QRs caused by a high-frequency inhomogeneous piezoelectric field accompanying a surface acoustic wave unveiled another possibility to distinguish the Wigner-molecule-regime from that of the free electrons by virtue of a different mechanism of the electronic absorption: for free electrons the dipole matrix element was other than zero, while in the Wigner molecule the absorption occurred due to quadruple and higher multipolar transitions [80].

Effect of the spin-orbit interaction was shown to dramatically change persistent currents in QRs as a function of the magnetic flux as compared to the case without the spin-orbit coupling; in particular, it may suppress the first Fourier harmonic in the persistent current and thus simulate the $\Phi_0/2$-periodicity [81].

The presence of magnetic impurities in a QR may induce bistability of the persistent current of two interacting electrons and a hysteresis in its dependence on the magnetic flux [82].

Interesting effects were unveiled in systems, where quantum rings were coupled to quantum dots. For a mesoscopic ring with a quantum dot inserted in one of its arms, it was shown that the phase of Aharonov-Bohm oscillations was not related to the dot charge alone but instead to the total charge of the system [83]. In the presence of the Aharonov-Bohm flux, a charge response of a mesoscopic ring coupled to a side-branch quantum dot revealed a sequence of plateaus of diamagnetic and paramagnetic states, while a mesoscopic ring containing an embedded quantum dot with leads exhibited a number of sharp peaks in the persistent current depending on the parity of the total number of electrons in the system [84].

Emergence of novel materials, e.g., carbon nanotubes, provided a new playground for observation and investigation of the Aharonov-Bohm effect [85].

1.5 Florescence

As the cornerstone of high-tech industry of the twenty-first century, nanostructures [86] are known as the cradle of new fabrication technologies [87], new characterization instruments [88, 89], and new theoretical insights [41].

A remarkable breakthrough in the physics of QRs was related to the discovery of the self-organized formation of QRs of a few tens of nm in diameter in 1997 by García, Medeiros-Ribeiro, Schmidt, Ngo, Feng, Lorke, Kotthaus and Petroff for the InAs/GaAs system [90]. It opened unprecedented perspectives to fabricate, characterize and investigate large arrays of semiconductor QRs as well as to control their size and shape.

1.5.1 Self-assembly Through Partial Overgrowth

It was demonstrated that by using a partial capping process the shape and size of InAs self-assembled quantum dots grown by Molecular Beam Epitaxy (MBE) may be modified in a way that led to the fabrication of self-assembled QRs [90]. The fabrication process was monitored using the *in situ* Reflection High-Energy Electron Diffraction (RHEED) technique [91], cross-section Transmission Electron Microscopy (TEM) [90, 92] and Atomic Force Microscopy (AFM) [90, 93, 94] measurements.

Two mechanisms were revealed, which mainly contributed to the self-assembled formation of QRs via partial overgrowth technique. One of them was *kinetic diffusion*: the In atoms, due to their higher diffusion mobility at the interface as compared to the diffusion mobility of the Ga atoms, could diffuse out of the partially capped quantum dots outwards onto the surface of the surrounding GaAs forming a ring-shaped InGaAs island [95]. Another mechanism was based on the thermodynamically driven *dewetting*: the imbalance of surface and interface forces acting upon the partially capped islands InAs/GaP [93]. Formation of liquid In droplets on the top of the InAs quantum dots under partial capping due to the *stress-induced melting* effect was established experimentally [91] and theoretically [96].

Unlike mesoscopic QRs defined lithographically, the self-assembled QRs, embedded in a GaAs matrix, could function in the quantum limit, *free of decoherence*, owing to scattering processes [97, 98]. The energy spectra of self-assembled QRs were thoroughly studied through their peculiar optical properties using PL [94, 99],

Time-Resolved PL and PL Excitation [100], as well as Photoemission Microscopy [101] in both single QRs and QR-arrays.

Being embedded in a heterostructure, the QRs can be electrically tuned by an electric field, and carriers can be injected with single-electron/single-hole precision. The detailed energy structure of electrons (holes) in QRs was obtained using the following three spectroscopic techniques. The PL (optical emission) of a single QR changed as electrons were added one-by-one. The emission energy changed abruptly whenever an electron was added, the sizes of the jumps revealing a shell structure [98]. Capacitance-voltage measurements allowed for probing the single-particle and many-particle ground states as a function of the applied electric field. Far-infrared absorption spectra demonstrated the effect of flux quantization on the intraband transitions.

1.5.2 Characterization

Cross-Sectional Scanning Tunneling Microscopy (X-STM) and Scanning-Gate Microscopy (SGM) belong to the advanced characterization methods that can access the intimate behavior of buried electronic systems and have been successfully exploited to get insight into the geometric structure of QRs.

X-STM of self-assembled InGaAs/GaAs QRs revealed the remaining quantum dot material whereas the Atomic Force Microscopy (AFM) represented the erupted QD material [102]. Based on this structural information from the X-STM measurements, a model of a self-assembled QR as a *singly-connected* "quantum volcano" (with a strong dip rather than opening in the center) was substantiated [102–104]. The electron magnetization was calculated as a function of the applied magnetic field for single-electron [74, 103] and two-electron [105] QRs. Quite surprisingly, even though those nanostructures were *singly-connected* and *anisotropic*, they exhibited the Aharonov-Bohm behavior, which was generally considered to be restricted to *doubly-connected* topologies. This was due to the fact that the electron wave functions in a "quantum volcano" were decaying towards the center so rapidly (exponentially) that they were *topologically identical* to the electron wave functions in doubly-connected QRs. The theory allowed for a quantitative explanation of the Aharonov-Bohm oscillations in the magnetization observed using the torsion magnetometry on those ring-like structures [106]. For measurements of the persistent currents in metal QRs, a micromechanical detector based on cantilever torsion magnetometry was proved to provide orders of magnitude greater sensitivity than SQUID-based detectors [15].

The Aharonov-Bohm oscillations of conductance in a mesoscopic ring defined by dry etching in a two-dimensional electron gas below the surface of an $Al_xGa_{1-x}As/GaAs$ heterostructure and interrupted by two tunnel barriers were modified by a perpendicular magnetic field and a bias voltage [107]. As a result, the nonequilibrium electron dephasing time was found to be significantly shortened at high voltages and magnetic fields.

Studies of the lithographically patterned InGaAs-based QRs by means of SGM provided unique imaging of Aharonov-Bohm interferences in real space and the electronic local density-of-states at low magnetic fields [108] and Coulomb islands in the quantum Hall regime at high magnetic fields and very low temperatures [109]. This allowed for unveiling the spatial structure of transport inside a quantum Hall interferometer and, subsequently, for deciphering the high-magnetic field magnetoresistance oscillations. Scanning-probe technique unraveled, also, a counter-intuitive behavior of a two-path network patterned from a GaInAs heterojunction in the form of a rectangular QR-structure connected to a source and a drain via two openings [110]. The antidot in the initial rectangular QR-structure could then be bypassed by a third path for the electrons. Partially blocking the electron transport through this additional branch by using SGM resulted in an *increased* current through the whole device. This counter-intuitive effect was interpreted as a mesoscopic analog of the Braess paradox known for classical networks.

1.5.3 *Various Materials Systems*

The self-assembly was proved to be an efficient method of QR formation also in diverse materials systems, for instance, InAs/InP [111], Ge/Si [112] and GaSb/GaAs [113]. Capping the InAs or InGaAs quantum dots by a GaAs/AlAs layer before annealing allowed for impeding the inward diffusion of the Ga and Al atoms and resulted in nicely shaped self-assembled QR-structures [114].

In contrast to the InGaAs/GaAs materials system, where capping of quantum dots followed by a growth interruption was necessary to initiate the quantum-dot to QR transformation, GaSb QRs occurred just after the deposition of GaSb on GaAs(001). Ring-shaped GaSb/GaAs quantum dots, grown by MBE, were characterized using X-STM [115]. These QRs, as distinct from the self-assembled InGaAs/GaAs QRs, possessed *a clear central opening* extending over about 40 % of the outer base length and were therefore truly doubly-connected objects.

A distinct series of quantized modes in the vortex state observed in the spin excitations of ferromagnetic rings at the micrometer scale, fabricated using electron beam lithography, was attributed to spin waves that circulate around the ring and interfere constructively [116]. This is a representative example of the vigorously developing spin-wave physics in devices with topologically nontrivial magnetization profile.

1.5.4 *Droplet Epitaxy and Lithography*

Besides the above-described partial overgrowth technique within the Stranski-Krastanov growth mode, another technique of the QRs fabrication was developed, which allowed for preparation of strain free GaAs/AlGaAs QRs and QR-complexes—droplet epitaxy [117–119]. This method started with formation of group-III liquid-metal droplets within a Volmer-Weber growth mode on the substrate surface by supplying their pure molecular beam. The nanostructures were subsequently formed by being exposed to a group-V element (As, Sb, P). The temporal evolution of these nanometer-scale objects was tracked in situ during the growth process using the RHEED technique [120].

Droplet epitaxy has demonstrated a unique ability to assemble QR nanostructures of complex morphologies ranging from single QRs [121], concentric double QRs [118, 122] and double-QR complexes [123] to concentric higher-order multiple QRs [124] and coupled QR/disks [125]. In GaAs double QRs formed by the droplet epitaxy, the size and height of the QRs were shown to depend on the supplied As flux. At a low As flux, larger and flatter rings were obtained. The formation of outer and inner rings was attributed to crystallization of out-diffused Ga and nanodrilling of Ga on the GaAs surface, correspondingly [118].

There has been a continuing insightful analysis of lithographically determined QRs. Magnetotransport experiments in the Coulomb blockade regime [126] and magnetoresistance measurements [127] on closed rings, fabricated with AFM oxidation lithography, confirmed that a microscopic understanding of energy levels of band charge carriers in QRs with the spin-orbit interaction could be extended to a many-electron system.

1.5.5 Novel Manifestations of the Aharonov-Bohm Effect

New conditions for manifestation of the Aharonov-Bohm effect through *neutral composite entities* consisting of charged particles in QRs were actively sought for starting with the seminal paper by Chaplik [81]. For an exciton in a one-dimensional QR placed in a perpendicular magnetic field, he found a Φ_0-periodic dependence of the exciton binding energy on the magnetic flux. In [128] the same result for the exciton ground-state energy was obtained within another analytical approach.

Extending the Berry's analysis of the phase evolution for a charge carrier, given in (1.9), consider the case when a particle (exciton) composed of an electron ($q = -e$) and a hole ($q = e$) is confined to a box that is transported around an opening of a doubly-connected system threaded by a magnetic flux line (tube). The wave function of the exciton

$$\left| \Psi(\Phi_h, \Phi_e) \right\rangle = \exp\left[i \frac{e(\Phi_h - \Phi_e)}{\hbar} \right] \left| \Psi(\Phi_h = 0, \Phi_e = 0) \right\rangle \qquad (1.10)$$

gains a phase, which is determined by a difference between the magnetic fluxes through the paths C_h and C_e encircled by the hole and the electron, respectively:

$$\Phi_h = \frac{e}{\hbar} \oint_{C_h} \mathbf{A}(\mathbf{R}) \cdot d\mathbf{R}, \qquad \Phi_e = \frac{e}{\hbar} \oint_{C_e} \mathbf{A}(\mathbf{R}) \cdot d\mathbf{R}. \qquad (1.11)$$

If the exciton is *polarized*, the paths C_h and C_e are different from each other (cp. [129]), the quantum interference, according to (1.10), is caused by the magnetic flux $(\Phi_h - \Phi_e)$ through the area between the two paths. Being manifested mainly through optical response of QRs, it is called *excitonic (or optical) Aharonov-Bohm effect* [129–132].

An example of the occurrence of the excitonic Aharonov-Bohm effect in transport phenomena was provided by the following feature of the vertical transport through a QR, which was immersed in a dielectric matrix: the tunnel current, as a function of magnetic flux for a given voltage across the structure, had the form of modulated oscillations with a characteristic period Φ_0 [133]. The optical Aharonov-Bohm effect became more prominent if the dc electric field was applied in the plane containing a QR [134] or in the vertical direction [135], because of the enhanced polarization of the exciton. Control over the Aharonov-Bohm oscillations in the energy spectrum of a QR could be realized also using low-frequency electromagnetic radiation [136].

The first experimental verification of the excitonic Aharonov-Bohm effect in self-assembled QRs was obtained by tracing patterns of the PL intensity under increasing magnetic field at different temperatures [137]. The role of the built-in piezoelectric fields in strained QR-systems consisted in changing the sequence of maxima and minima of the Aharonov-Bohm oscillations. For those observations, a correlation between the electron and hole due to the Coulomb interaction was shown to be a necessary condition.

The existence of the optical Aharonov-Bohm effect was first demonstrated through PL for type-II InP/GaAs quantum dots [138] and Zn(SeTe) quantum dots in ZnTe/ZnSe superlattice [139, 140]. Large and persistent oscillations in both the energy and the intensity of the PL unveiled the presence of coherently rotating exciton states. These remarkably robust Aharonov-Bohm oscillations were shown to persist until 180 K. The magnitude of the observed effects was attributed to the geometry of the columnar type-II structures investigated, which created a ring-like topology of the electron state. The advantage of this geometry to favor the optical Aharonov-Bohm effect was demonstrated in magnetic (ZnMn)Te quantum dot structures, where the strength of the Aharonov-Bohm interference effect could be controlled by the spin disorder in the system [140, 141].

The first MPL study of single neutral excitons was performed in single self-assembled InGaAs/GaAs QRs, which were fabricated by MBE combined with AsBr$_3$ *in situ* etching [135]. Oscillations in the neutral exciton radiative recombination energy and in the emission intensity were detected as a function of the applied magnetic field. Effective control over the period of the oscillations was achieved through a gate potential that modified the exciton confinement [135, 142]. Strain was shown to play a crucial role to govern the localization of electrons and holes in type-I semiconductor QRs, eventually leading to spatially separated charge carriers [143].

Singly charged excitons (trions) and multiply charged excitons in QRs were extensively studied, both theoretically and experimentally. The period of oscillations of the binding energy of charged complexes in magnetic flux was shown to differ from Φ_0, being determined by the number of electrons and the ratio of effective masses of the electron and the hole [144]. The diamagnetic shift of the exciton

PL line was found to be positive for a neutral exciton and negative for a trion and other negatively charged complexes [145]. Circularly polarized magnetophotoluminescence (MPL) spectra of a single QR, fabricated with the modulated-barrier approach, were dominated by two features: a high-energy line due to neutral exciton recombination and a low-energy line owing to emission from charged excitons [146]. Measured photon energy from charged exciton recombination as a function of the magnetic field clearly revealed the oscillatory behavior, which was in antiphase with the calculated electron's energy [147].

The excitonic Aharonov-Bohm effect, originally considered for a one-dimensional model, was shown to remain essentially unchanged in QRs of finite width [148]. Though the Aharonov-Bohm oscillations of the oscillator strength as a function of the magnetic flux for the ground state of the exciton decreased with increasing the QR width, their amplitude remained finite down to radius-to-width ratios less than unity due to the non-simply-connectedness of the confinement potential. That implied that the key condition needed for the observation of the excitonic Aharonov-Bohm effect was the avoidance of the QR center.

The exciton energy spectra and optical transitions spectra calculated for excitons with the realistic confinement potential of self-assembled QRs [102] taking into account the strain revealed a very high sensitivity to the size, anisotropic shape, and composition of a QR [149]. Photoluminescence spectroscopy of a large ensemble of InAs/GaAs QRs in magnetic fields up to 30 T for different excitation densities unveiled that the confinement of an electron and a hole along with the Coulomb interaction suppressed the excitonic Aharonov-Bohm effect in these QRs [99, 150]. This suppression was confirmed also by MPL studies of type-II self-assembled GaSb/GaAs QRs [151].

Another manifestation of the Aharonov-Bohm effect in neutral formations was related to magnetoplasmon oscillations in QRs [152]. The plasmon frequency in a finite-width QR was constituted of a monotonous part superposed with Aharonov-Bohm oscillations. Their period and amplitude were found to vary with the magnetic field.

Polaron shift in QRs revealed the non-monochromaticity of the Aharonov-Bohm oscillations, which was attributed to the difference in the magnetic fluxes that are encircled by different electron trajectories [153]. When an exciton was generated, the contributions of the electron and the hole to the polarization of the medium had opposite signs, and it was therefore important to take the finiteness of the ring into account when calculating the net effect determined by the wave functions of the particles.

1.5.6 Advancements of Theory

Embedding QRs in various multilayer structures is an important tool to control their physical properties. The non-trivial role of strain in QR multilayer systems was theoretically revealed in Ref. [154]. In GaAs-capped $InAs/In_{0.53}Ga_{0.47}As$ QRs, there occurred an anomalous strain relaxation: GaAs embedded in the $In_{0.53}Ga_{0.47}As$ matrix considerably weakened each strain component and biaxial strain by providing enough room for the atomic relaxation of InAs. GaAs embedded in $In_{0.53}Ga_{0.47}As$ acted as a potential barrier for both electrons and heavy holes and as a potential well for light holes. The weak positive biaxial strain of InAs along with the strong negative biaxial strain of GaAs in a QR led to an enhancement of the light-hole character of the states in the valence band of a QR as compared to those in a quantum dot.

Calculating the strain profile as well as the charge carriers energy and other properties (shell filling, spin polarization, exciton fine structure, magnetization...) in QRs requires the extensive use of a great variety of the advanced tools of the modern theoretical physics, of which we name below only a few.

Exact diagonalization method revealed the fractional Aharonov-Bohm effect of a few-electron system in a one-dimensional QR taking into account spin, disorder and the Coulomb interaction [155]. A great challenge for the theory—to find analytical solutions for quantum states in QRs—was addressed for two electrons on a one-dimensional QR for particular values of the radius [156]. Many-electron QRs were studied using a number of versions of the Density Functional Theory [157, 158].

After calculating the strain with the atomistic Valence Force Field method, the electronic properties were derived in the framework of the Empirical Pseudopotential method [159] or the Empirical Tight-Binding method [81, 160–162]. A continuum description of the QR system was assumed within the single-band Effective-Mass approximation [163, 164] and its diversified generalizations onto multiband $k \cdot p$ approaches [165]. Of importance for modeling the self-assembled QRs was the finding [166], that the 14-band $k \cdot p$ model can accommodate for the correct symmetry of the underlying GaAs zincblende lattice, which was not reflected in the standard 8-band model. The ground-state energy of the few-particle systems in QRs was calculated within the Configuration Interaction method [135, 159].

Transfer-Matrix method was employed to account for the mutual influence of the radial and azimuthal motions in the presence of impurities in the finite-width QR [71]. The Landauer-Büttiker formalism was used to analyze transport properties of QRs [167, 168]. Using the Keldysh Green's function formalism enabled unveiling two contributions, thermodynamic and kinetic, to the disorder-averaged magnetization of QRs [169].

Path-Integral Quantum-Monte-Carlo method was applied for investigation of the energy spectra of few-electron systems in QRs as a function of their geometry [58]. The interplay between the confinement geometry and the Coulomb interaction was pronouncedly manifested through the electronic properties of a QR. The ground state of a perfect QR containing a small number of interacting electrons was analyzed as a function of its geometric parameters: ring radius, radial confinement, and eccentricity. A Path-Integral Quantum-Monte-Carlo calculation demonstrated a strong dependence of the total spin of the ground state on the structure geometry. For instance, for a three-electron QR, changing the radius produced a spin polarization of the ground state, while an elliptical deformation resulted in a spin-depolarized ground state [170].

A finite mixing of the heavy-hole subband with the light-hole subband in self-assembled InAs/GaAs QRs was shown to be larger than in quantum dots and critical in determining the hole spin properties. The large light-hole component in QRs underpinned their perspectives for applications requiring enhanced tunneling rates [171] and spin-orbit mediated control [172].

1.6 Multi-Faceted Horizons

In view of the emerging high-tech realizations, finding and exploiting novel phenomena in QR-structures will be the key issues for the future development in the theoretical and experimental physics of QRs. At the time of writing this chapter, the perspective research directions in the field range from non-trivial topologies, new materials, alignment and assembly of QRs arrays—through engineering QR-based metamaterials—to device design, manufacturing, and application.

1.6.1 Novel Topological Structures

Nanostructure fabrication techniques have allowed for generating topologically nontrivial manifolds at the micro- and nanoscale with manmade space metrics, which determine the energy spectrum and other physical properties of electrons confined in such objects. For instance, when spooling a single crystalline $NbSe_3$ ribbon on a selenium droplet, surface tension produces a twist in the ribbon, leading to the formation of a one-sided Möbius ring [173]. Analytical and computational differential geometry methods have been developed to examine particle quantum eigenstates and eigenenergies in curved and strained nanostructures [174, 175]. Significant changes in eigenstate symmetry and eigenenergy are revealed due to the interplay between curvature and strain effects for bending radii of a few nanometers. Curvature effects become negligible at bending radii above ~ 50 nm.

Symbiosis of the geometric potential and an inhomogeneous twist renders an observation of the topology effect on the electron ground-state energy in microscale Möbius rings into the realm of experimental verification. A *"delocalization-to-localization"* transition for the electron ground state is unveiled in inhomogeneous Möbius rings. This transition can be quantified through the Aharonov-Bohm effect on the ground-state persistent current as a function of the magnetic flux threading the Möbius ring [176]. The theoretical analysis of such topologically nontrivial manifolds at the nanoscale will have practical relevance, as any pertinent fabrication techniques are likely to generate geometric and structural inhomogeneities.

1.6.2 Graphene QRs

Electronic quantum interference in QR-structures based on graphene has been investigated with a focus on the interplay between the Aharonov-Bohm effect and the peculiar electronic and transport properties of this material [168]. The first experimental realization of a graphene ring structure has been provided by [177]. In this work, the authors investigate the Aharonov-Bohm oscillations in diffusive single-layer graphene as a function of the magnetic field, which is applied perpendicularly to the graphene plane in a two-terminal setup. They find clear magnetoconductance oscillations with the expected period of Φ_0 on the top of a low-frequency background signal due to universal conductance fluctuations. A significant increase in the oscillation amplitude at strong magnetic fields close to the onset of the quantum Hall regime is strong enough to make the second harmonic (oscillations of period $\Phi_0/2$) visible in the frequency spectrum. Such a behavior, observed in smaller rings using a two-terminal as well as four-terminal geometry is attributed to scattering on magnetic impurities [178].

Additional tunability is introduced into the graphene ring device by applying a side gate potential to one of the ring arms. Investigation of the influence of such side gates on a four-terminal geometry in the diffusive regime reveals phase shifts of the Aharonov-Bohm oscillations as a function of the gate voltage as well as phase jumps of π at zero magnetic field—direct consequences of the electrostatic Aharonov-Bohm effect as well as the generalized Onsager relations [179]. The electrostatic Aharonov-Bohm effect appears to be more feasible in graphene QRs than in metal QRs due to the low screening of this material [180, 181]. Voltage-driven charge-carrier states ranging from metallic to semiconductor ones are theoretically revealed for QRs determined by a set of concentric circular gates over a graphene sheet placed on a substrate [182].

1.6.3 Ordering of QRs. Metamaterials

Great efforts have been devoted to achieve vertical and lateral alignment of QRs. Stacking of three InGaAs/GaAs QRs is demonstrated to provide a broad-area laser [183]. In QR complexes and stacks, novel correlations occur, which allow for control over their electronic and magnetic properties.

One-dimensional ordered QR-chains have been fabricated on a quantum-dot superlattice template by MBE. The quantum-dot superlattice template is prepared by stacking multiple quantum-dot layers. The lateral ordering is introduced by engineering the strain field of a multi-layer InGaAs quantum-dot superlattice. QR chains are then formed by partially capping InAs quantum dots with a thin layer of GaAs which introduces a morphological change from quantum dots to QRs [184]. It is shown that two-dimensional periodically aligned QR-arrays can be fabricated on GaAs high-index [(311)B and (511)B] surfaces [185].

An alternative approach to self-assembly of aligned QRs is to create an artificially ordered template by pre-patterning. Nanosphere lithography is used to create ordered GeSi quantum dots, and ordered GeSi QRs are subsequently formed by capping the quantum dots with a thin Si capping layer [186]. When the Si capping layer is more than 3 nm thick, most quantum dots are converted into QRs. Additional fabrication techniques, such as Ar^+ sputter redeposition using porous alumina templates [187] and laser-interference lithography in conjunction with electrochemical deposition [188], are promising, low-cost, and scalable tools for producing ordered QR-arrays.

QRs are a very promising building block for metamaterials. High-density assemblies of QRs may contain clusters of close or even partially overlapping QRs [114]. A moderate coupling between adjacent QRs in a cluster is shown to significantly influence the energy spectrum: while its lowest part may preserve the single-QR behavior, the high-energy part is strongly modified [189]. Metamaterials consisting of split nanosized gold QRs are found to possess unusual electromagnetic response properties like a negative index of refraction for wavelengths in the micrometer region [190, 191], where the resonance wavelength scales linearly with the size of the circuit. A possible control over the electromagnetic response in QR composite metamaterials made from metals and semiconductors is an attractive goal for further investigations.

1.6.4 Photonic Sources and Detectors

Using QRs as photon sources and detectors is based on their unique optical properties associated with the excitonic Aharonov-Bohm effect [192]. Using QRs, a technique is theoretically devised to completely freeze and release individual photons at will by tuning magnetic and electric fields that enable QRs *to trap and store light* [192]. Application of these QRs as light capacitors or buffers is expected in the fields of photonic computing and communications technologies [192, 193]. The shallow bound-state energy levels of the InGaAs QRs are shown to be feasible for detecting photons in the terahertz regime [194].

Single InAs/GaAs QRs embedded in a photonic crystal lattice are demonstrated to allow for single-photon emission and photon antibunching between the exciton and biexciton emissions [195]. This extends the realm of the QRs investigations towards quantum electrodynamics. The antibunching of photons observed in a double QR [196] is a clear signature of a single photon emitter.

1.6.5 Spintronics. Magnetic Memory

An effective confinement-governed wave function engineering is explored theoretically in systems with QRs: two-dimensional complex nanostructures in the form of double concentric QRs and dot-QR nanostructures that consist of a QR with a quantum dot inside. The higher spin stability in a QR than in a quantum dot makes QRs attractive for the realization of spin qubits, because the relaxation and decoherence processes take place in the time scale that is sufficiently long for spin manipulations and readout [197]. The dot-QR nanostructure allows, by changing the potential barrier separating the dot from the QR and the potential well offset between the dot and the QR, for a significant alteration of coherent, optical and transport properties of the structure. In particular, the spin relaxation time of dot-QR nanostructures, used as spin qubits or spin memory devices, can be modified by orders of magnitude [198]. A crossover from the ballistic to the resonant tunneling transport, unveiled for an ideal one-dimensional QR with spin-orbit interaction, underpins the suggestion to use QRs for fabricating one-qubit spintronic quantum gate and thus, for quantum information processing [167]. An Aharonov-Bohm interferometer consisting of a QR with two quantum dots embedded in its arms reveals sensitive spin-polarized electron transmission that might be useful for spintronics applications [199].

QR arrays offer superior prospects in high density magnetic memory applications as magnetic random access memory, recording medium, and other spintronic devices [200].

In conclusion, a great variety of semiconductor QR-systems, in particular, single and multiple QRs, ordered arrays of QRs, complexes of QRs in combination with other nanostructures, Möbius QRs, have been fabricated with advanced high-tech methods, characterized with cutting-edge technologies, and analyzed with innovative theoretical tools. Their unique *doubly-connected topology* and the *ring-like density of states* for charge carriers, spins, plasmon and photon fields provide a veritable cornucopia of fascinating properties and possibilities to boost development of the strategic domains of technology: quantum computing based on photon and spin manipulations, photonic emitters and detectors, magnetic memory, and engineering of novel metamaterials.

Acknowledgements I wish to express my greatest gratitude to Axel Lorke, Ian R. Sellers, Paul M. Koenraad, and Alexander O. Govorov, who have read the manuscript of the present chapter, for their critical comments and valuable suggestions.

References

1. A.-M. Ampère, Ann. Chim. Phys. **15**, 59 (1820)
2. A.-M. Ampère, Ann. Chim. Phys. **15**, 170 (1820)
3. L. Wendler, V.M. Fomin, Phys. Status Solidi (b) **191**, 409 (1995)
4. M. Bayindir, I.O. Kulik, in *Quantum Mesoscopic Phenomena and Mesoscopic Devices in Microelectronics*, ed. by I.O. Kulik, R. Ellialtıoğlu. NATO Science Series, Series C: Mathematical and Physical Sciences, vol. 559 (Kluwer, Dordrecht, 2000), pp. 283–292
5. B.C. Lee, O. Voskoboynikov, C.P. Lee, Physica E **24**, 87 (2004)
6. L. Saminadayar, C. Bäuerle, D. Mailly, in *Encyclopedia of Nanoscience and Nanotechnology*, vol. 3, ed. by H.S. Nalwa (American Scientific, Valencia, 2004), pp. 267–285
7. P. Nordlander, ACS Nano **3**, 488 (2009)
8. V.M. Fomin, L.F. Chibotaru, J. Nanoelectron. Optoelectron. **4**, 3 (2009)
9. I.O. Kulik, Low Temp. Phys. **36**, 1063 (2010)
10. V.M. Fomin (ed.), *A special issue on Modern Advancements in Experimental and Theoretical Physics of Quantum Rings*. J. Nanoelectron. Optoelectron. **6**, pp. 1–86 (2011)
11. L. Pauling, J. Chem. Phys. **4**, 673 (1936)
12. F.J. London, J. Phys. Radium **8**, 397 (1937)
13. F. Hund, Ann. Phys. (Leipz.), 5. Folge **32**, 102 (1938)
14. R.B. Dingle, Proc. R. Soc. Lond. A **212**, 47 (1952)

15. A.C. Bleszynski-Jayich, W.E. Shanks, B. Peaudecerf, E. Ginossar, F. von Oppen, L. Glazman, J.G.E. Harris, Science **326**, 272 (2009)
16. S.M. Rytov, Dokl. Akad. Nauk SSSR **18**, 263 (1938) [English translation in: B. Markovsky and S. I. Vinitsky, Eds., *Topological Phases in Quantum Theory* (World Scientific, Singapore, 1989), pp. 6–10]
17. V.V. Vladimirskii, Dokl. Akad. Nauk SSSR **21**, 222 (1941) [English translation in: B. Markovsky and S. I. Vinitsky, Eds., *Topological Phases in Quantum Theory* (World Scientific, Singapore, 1989), pp. 11–16]
18. S. Pancharatnam, Proc. Indian Acad. Sci. A **44**, 247 (1956)
19. M.V. Berry, Proc. R. Soc. Lond. A **392**, 45 (1984)
20. M. Born, V. Fock, Z. Phys. **51**, 165 (1928)
21. Y. Aharonov, J. Anandan, Phys. Rev. Lett. **58**, 1593 (1987)
22. N. Byers, C.N. Yang, Phys. Rev. Lett. **7**, 46 (1961)
23. F. Bloch, Phys. Rev. **137**, A787 (1965)
24. F. Wilczek, A. Zee, Phys. Rev. Lett. **52**, 2111 (1984)
25. Y. Aharonov, D. Bohm, Phys. Rev. **115**, 485 (1959)
26. P.A.M. Dirac, Proc. R. Soc. Lond. A **133**, 60 (1931)
27. W. Ehrenbug, R.E. Siday, Proc. R. Soc. Lond. B **62**, 8 (1949)
28. E. Buks, R. Schuster, M. Heiblum, D. Mahalu, V. Umansky, Nature **391**, 871 (1998)
29. Y. Aharonov, A. Casher, Phys. Rev. Lett. **53**, 319 (1984)
30. V.M. Fomin, V.R. Misko, J.T. Devreese, V.V. Moshchalkov, Phys. Rev. B **58**, 11703 (1998)
31. H. Zhao, V.M. Fomin, J.T. Devreese, V.V. Moshchalkov, Solid State Commun. **125**, 59 (2003)
32. J.T. Devreese, V.M. Fomin, V.N. Gladilin, J. Tempere, Physica C **470**, 848 (2010)
33. R. Doll, M. Näbauer, Phys. Rev. Lett. **7**, 51 (1961)
34. B.S. Deaver Jr., W.M. Fairbank, Phys. Rev. Lett. **7**, 43 (1961)
35. L. Gunther, Y. Imry, Solid State Commun. **7**, 1391 (1969)
36. I.O. Kulik, JETP Lett. **11**, 275 (1970)
37. I.O. Kulik, Sov. Phys. JETP **31**, 1172 (1970)
38. N.B. Brandt, D.V. Gitsu, A.A. Nikolaeva, Ya.G. Ponomarev, JETP Lett. **24**, 272 (1976)
39. N.B. Brandt, E.N. Bogachek, D.V. Gitsu, G.A. Gogadze, I.O. Kulik, A.A. Nikolaeva, Ya.G. Ponomarev, Sov. J. Low Temp. Phys. **8**, 358 (1982)
40. M. Büttiker, Y. Imry, R. Landauer, Phys. Lett. A **96**, 365 (1983)
41. Y. Imry, *Introduction to Mesoscopic Physics* (Oxford University Press, Oxford, 1997). 234 pp.
42. J.P. Carini, K.A. Muttalib, S.R. Nagel, Phys. Rev. Lett. **53**, 102 (1984)
43. R. Landauer, M. Büttiker, Phys. Rev. Lett. **54**, 2049 (1985)
44. M. Büttiker, Phys. Rev. B **32**, 1846 (1985)
45. L.P. Lévy, G. Dolan, J. Dunsmuir, H. Bouchiat, Phys. Rev. Lett. **64**, 2074 (1990)
46. V. Chandreasekhar, R.A. Webb, M.J. Brandy, M.B. Ketchen, W.J. Gallagher, A. Kleinsasser, Phys. Rev. Lett. **67**, 3578 (1991)
47. D. Mailly, C. Chapelier, A. Benoit, Phys. Rev. Lett. **70**, 2020 (1993)
48. E.M.Q. Jariwala, P. Mohanty, M.B. Ketchen, R.A. Webb, Phys. Rev. Lett. **86**, 1594 (2001)
49. H. Bluhm, N.C. Koshnick, J.A. Bert, M.E. Huber, K.A. Moler, Phys. Rev. Lett. **102**, 136802 (2009)
50. R. Deblock, R. Bel, B. Reulet, H. Bouchiat, D. Mailly, Phys. Rev. Lett. **89**, 206803 (2002)
51. H.-F. Cheung, Y. Gefen, E.K. Riedel, W.-H. Shih, Phys. Rev. B **37**, 6050 (1988)
52. F. von Oppen, E.K. Riedel, Phys. Rev. Lett. **66**, 84 (1991)
53. B.L. Altshuler, Y. Gefen, Y. Imry, Phys. Rev. Lett. **66**, 88 (1991)
54. H. Bouchiat, G. Montambaux, J. Phys. (Paris) **50**, 2695 (1989)
55. H.-F. Cheung, E.K. Riedel, Y. Gefen, Phys. Rev. Lett. **62**, 587 (1989)
56. D. Loss, P. Goldbart, Phys. Rev. B **43**, 13762 (1991)
57. J.F. Weisz, R. Kishore, F.V. Kusmartsev, Phys. Rev. B **49**, 8126 (1994)
58. A. Emperador, F. Pederiva, E. Lipparini, Phys. Rev. B **68**, 115312 (2003)

59. V. Ambegaokar, U. Eckern, Phys. Rev. Lett. **65**, 381 (1990)
60. A. Schmid, Phys. Rev. Lett. **66**, 80 (1991)
61. U. Eckern, A. Schmid, Europhys. Lett. **18**, 457 (1992)
62. M. Schechter, Y. Oreg, Y. Imry, Y. Levinson, Phys. Rev. Lett. **90**, 026805 (2003)
63. U. Eckern, P. Schwab, V. Ambegaokar, Phys. Rev. Lett. **93**, 209701 (2004)
64. M. Schechter, Y. Oreg, Y. Imry, Y. Levinson, Phys. Rev. Lett. **93**, 209702 (2004)
65. H. Bary-Soroker, O. Entin-Wohlman, Y. Imry, Phys. Rev. Lett. **101**, 057001 (2008)
66. N. Argaman, Y. Imry, Phys. Scr. T **49**, 333 (1993)
67. A. Altland, S. Iida, A. Müller-Groeling, H.A. Weidenmüller, Ann. Phys. **219**, 148 (1992)
68. A. Müller-Groeling, H.A. Weidenmüller, C.H. Lewenkopf, Europhys. Lett. **22**, 193 (1993)
69. A. Müller-Groeling, H.A. Weidenmuller, Phys. Rev. B **49**, 4752 (1994)
70. Y. Imry, *Introduction to Mesoscopic Physics*, 2nd edn. (Oxford University Press, Oxford, 2005). 236 pp.
71. L. Wendler, V.M. Fomin, A.A. Krokhin, Phys. Rev. B **50**, 4642 (1994)
72. L. Wendler, V.M. Fomin, Z. Phys. B **96**, 373 (1995)
73. W.C. Tan, J.C. Inkson, Phys. Rev. B **53**, 6947 (1996)
74. V.M. Fomin, V.N. Gladilin, S.N. Klimin, J.T. Devreese, N.A.J.M. Kleemans, P.M. Koenraad, Phys. Rev. B **76**, 235320 (2007)
75. A.O. Govorov, A.V. Chaplik, L. Wendler, V.M. Fomin, JETP Lett. **60**, 643 (1994)
76. L. Wendler, V.M. Fomin, A.V. Chaplik, Solid State Commun. **96**, 809 (1995)
77. L. Wendler, V.M. Fomin, A.V. Chaplik, A.O. Govorov, Z. Phys. B **100**, 211 (1996)
78. L. Wendler, V.M. Fomin, A.V. Chaplik, A.O. Govorov, Physica B **227**, 397 (1996)
79. L. Wendler, V.M. Fomin, A.V. Chaplik, A.O. Govorov, Phys. Rev. B **54**, 4794 (1996)
80. V.M. Kovalev, A.V. Chaplik, Semiconductors **37**, 1195 (2003)
81. A.V. Chaplik, JETP Lett. **62**, 900 (1995)
82. L. Wendler, V.M. Fomin, Phys. Rev. B **51**, 17814 (1995)
83. A. Levy Yeyati, M. Büttiker, Phys. Rev. B **52**, R14360 (1995)
84. M. Büttiker, C.A. Stafford, Phys. Rev. Lett. **76**, 495 (1996)
85. C. Schönenberger, A. Bachtold, Ch. Strunk, J.-P. Salvetat, J.-M. Bonard, L. Forró, T. Nussbaumer, Nature **397**, 673 (1999)
86. D. Bimberg (ed.), *Semiconductor Nanostructures* (Springer, Berlin, 2008). 357 pp.
87. D. Bimberg, M. Grundmann, N.N. Ledentsov, *Quantum Dot Heterostructures* (Wiley, Chichester, 1999). 338 pp.
88. G. Binnig, H. Rohrer, IBM J. Res. Dev. **30**, 4 (1986)
89. R. Wiesendanger, *Scanning Probe Microscopy: Analytical Methods* (Springer, Berlin, 1998). 216 pp.
90. J.M. García, G. Medeiros-Ribeiro, K. Schmidt, T. Ngo, J.L. Feng, A. Lorke, J. Kotthaus, P.M. Petroff, Appl. Phys. Lett. **71**, 2014 (1997)
91. J.M. García, J. Silveira, F. Briones, Appl. Phys. Lett. **77**, 409 (2000)
92. D. Granados, J.M. García, T. Ben, S.I. Molina, Appl. Phys. Lett. **86**, 071918 (2005)
93. R. Blossey, A. Lorke, Phys. Rev. E **65**, 021603 (2002)
94. D. Granados, J.M. García, Appl. Phys. Lett. **82**, 2401 (2003)
95. A. Lorke, R.J. Luyken, J.M. García, P.M. Petroff, Jpn. J. Appl. Phys. **40**, 1857 (2001)
96. D.J. Bottomley, Appl. Phys. Lett. **80**, 4747 (2002)
97. A. Lorke, R.J. Luyken, A.O. Govorov, J. Kotthaus, J.M. García, P.M. Petroff, Phys. Rev. Lett. **84**, 2223 (2000)
98. R.J. Warburton, C. Schaflein, D. Haft, F. Bickel, A. Lorke, K. Karrai, J. Garcia, W. Schoenfeld, P. Petroff, Nature **405**, 926 (2000)
99. N.A.J.M. Kleemans, J.H. Blokland, A.G. Taboada, H.C.M. van Genuchten, M. Bozkurt, V.M. Fomin, V.N. Gladilin, D. Granados, J.M. García, P.C.M. Christianen, J.C. Maan, J.T. Devreese, P.M. Koenraad, Phys. Rev. B **80**, 155318 (2009)
100. B. Alén, J. Martínez-Pastor, D. Granados, J.M. García, Phys. Rev. B **72**, 155331 (2005)
101. G. Biasiol, S. Heun, L. Sorba, J. Nanoelectron. Optoelectron. **6**, 20–33 (2011)

102. P. Offermans, P.M. Koenraad, J.H. Wolter, D. Granados, J.M. García, V.M. Fomin, V.N. Gladilin, J.T. Devreese, Appl. Phys. Lett. **87**, 131902 (2005)
103. V.M. Fomin, V.N. Gladilin, J.T. Devreese, P. Offermans, P.M. Koenraad, J.H. Wolter, J.M. García, D. Granados, AIP Conf. Proc. **772**, 803 (2005)
104. P. Offermans, P.M. Koenraad, J.H. Wolter, D. Granados, J.M. García, V.M. Fomin, V.N. Gladilin, J.T. Devreese, Physica E **32**, 41 (2006)
105. V.M. Fomin, V.N. Gladilin, J.T. Devreese, N.A.J.M. Kleemans, P.M. Koenraad, Phys. Rev. B **77**, 205326 (2008)
106. N.A.J.M. Kleemans, I.M.A. Bominaar-Silkens, V.M. Fomin, V.N. Gladilin, D. Granados, A.G. Taboada, J.M. García, P. Offermans, U. Zeitler, P.C.M. Christianen, J.C. Maan, J.T. Devreese, P.M. Koenraad, Phys. Rev. Lett. **99**, 146808 (2007)
107. W.G. van der Wiel, Yu.V. Nazarov, S. De Franceschi, T. Fujisawa, J.M. Elzerman, E.W.G.M. Huizeling, S. Tarucha, L.P. Kouwenhoven, Phys. Rev. B **67**, 033307 (2003)
108. B. Hackens, F. Martins, T. Ouisse, H. Sellier, S. Bollaert, X. Wallart, A. Cappy, J. Chevrier, V. Bayot, S. Huant, Nat. Phys. **2**, 826 (2006)
109. B. Hackens, F. Martins, S. Faniel, C.A. Dutu, H. Sellier, S. Huant, M. Pala, L. Desplanque, X. Wallart, V. Bayot, Nat. Commun. **1**, 39 (2010)
110. M.G. Pala, S. Baltazar, P. Liu, H. Sellier, B. Hackens, F. Martins, V. Bayot, X. Wallart, L. Desplanque, S. Huant, Phys. Rev. Lett. **108**, 076802 (2012)
111. T. Raz, D. Ritter, G. Bahir, Appl. Phys. Lett. **82**, 1706 (2003)
112. J. Cui, Q. He, X.M. Jiang, Y.L. Fan, X.J. Yang, F. Xue, Z.M. Jiang, Appl. Phys. Lett. **83**, 2907 (2003)
113. S. Kobayashi, C. Jiang, T. Kawazu, H. Sakaki, Jpn. J. Appl. Phys. **43**, L662 (2004)
114. B.C. Lee, C.P. Lee, Nanotechnology **15**, 848 (2004)
115. R. Timm, H. Eisele, A. Lenz, L. Ivanova, G. Balakrishnan, D. Huffaker, M. Dähne, Phys. Rev. Lett. **101**, 256101 (2008)
116. J. Podbielski, F. Giesen, D. Grundler, Phys. Rev. Lett. **96**, 167207 (2006)
117. S. Sanguinetti, N. Koguchi, T. Mano, T. Kuroda, J. Nanoelectron. Optoelectron. **6**, 34 (2011)
118. J. Wu, Z.M. Wang, A.Z. Li, Z. Zeng, S. Li, G. Chen, G.J. Salamo, J. Nanoelectron. Optoelectron. **6**, 58 (2011)
119. Ch. Heyn, A. Stemmann, Ch. Strelow, T. Köppen, D. Sonnenberg, A. Graf, S. Mendach, W. Hansen, J. Nanoelectron. Optoelectron. **6**, 62 (2011)
120. Á. Nemcsics, Ch. Heyn, A. Stemmann, A. Schramm, H. Welsch, W. Hansen, Mater. Sci. Eng. B **165**, 118 (2009)
121. T. Kuroda, T. Mano, T. Ochiai, S. Sanguinetti, K. Sakoda, G. Kido, N. Koguchi, Phys. Rev. B **72**, 205301 (2005)
122. T. Mano, T. Kuroda, S. Sanguinetti, T. Ochiai, T. Tateno, J. Kim, T. Noda, M. Kawabe, K. Sakoda, G. Kido, N. Koguchi, Nano Lett. **5**, 425 (2005)
123. S. Huang, Z. Niu, Z. Fang, H. Ni, Z. Gong, J. Xia, Appl. Phys. Lett. **89**, 031921 (2006)
124. C. Somaschini, S. Bietti, N. Koguchi, S. Sanguinetti, Nano Lett. **9**, 3419 (2009)
125. C. Somaschini, S. Bietti, S. Sanguinetti, N. Koguchi, A. Fedorov, Nanotechnology **21**, 125601 (2010)
126. A. Fuhrer, S. Lüscher, T. Ihn, T. Heinzel, K. Ensslin, W. Wegscheider, M. Bichler, Nature **413**, 822 (2001)
127. B. Grbić, R. Leturcq, T. Ihn, K. Ensslin, D. Reuter, A.D. Wieck, Physica E **40**, 1273 (2008)
128. R.A. Römer, M.E. Raikh, Phys. Rev. B **62**, 7045 (2000)
129. A.O. Govorov, S.E. Ulloa, K. Karrai, R.J. Warburton, Phys. Rev. B **66**, 081309 (2002)
130. A.V. Kalameitsev, A.O. Govorov, V.M. Kovalev, JETP Lett. **68**, 669 (1998)
131. A.V. Chaplik, JETP Lett. **75**, 292 (2002)
132. M. Grochol, F. Grosse, R. Zimmermann, Phys. Rev. B **74**, 115416 (2006)
133. V.M. Kovalev, A.V. Chaplik, J. Exp. Theor. Phys. **95**, 912 (2002)
134. A.V. Maslov, D.S. Citrin, Phys. Rev. B **67**, 121304(R) (2003)
135. F. Ding, N. Akopian, B. Li, U. Perinetti, A. Govorov, F.M. Peeters, C.C. Bof Bufon, C. Deneke, Y.H. Chen, A. Rastelli, O.G. Schmidt, V. Zwiller, Phys. Rev. B **82**, 075309 (2010)

136. V.M. Kovalev, A.V. Chaplik, Europhys. Lett. **77**, 47003 (2007)
137. M.D. Teodoro, V.L. Campo Jr., V. Lopez-Richard, E. Marega Jr., G.E. Marques, Y. Galvao Gobato, F. Iikawa, M.J.S.P. Brasil, Z.Y. AbuWaar, V.G. Dorogan, Yu.I. Mazur, M. Benamara, G.J. Salamo, Phys. Rev. Lett. **104**, 086401 (2010)
138. E. Ribeiro, A.O. Govorov, W. Carvalho Jr., G. Medeiros-Ribeiro, Phys. Rev. Lett. **92**, 126402 (2004)
139. I.R. Sellers, V.R. Whiteside, I.L. Kuskovsky, A.O. Govorov, B.D. McCombe, Phys. Rev. Lett. **100**, 136405 (2008)
140. I.R. Sellers, A.O. Govorov, B.D. McCombe, J. Nanoelectron. Optoelectron. **6**, 4 (2011)
141. I.R. Sellers, V.R. Whiteside, A.O. Govorov, W.C. Fan, W.-C. Chou, I. Khan, A. Petrou, B.D. McCombe, Phys. Rev. B **77**, 241302 (2008)
142. F. Ding, B. Li, N. Akopian, U. Perinetti, Y.H. Chen, F.M. Peeters, A. Rastelli, V. Zwiller, O.G. Schmidt, J. Nanoelectron. Optoelectron. **6**, 51 (2011)
143. M. Tadić, N. Čukarić, V. Arsoski, F.M. Peeters, Phys. Rev. B **84**, 125307 (2011)
144. A.O. Govorov, A.V. Chaplik, JETP Lett. **66**, 454 (1997)
145. A.V. Chaplik, J. Exp. Theor. Phys. **92**, 169 (2001)
146. M. Bayer, M. Korkusinski, P. Hawrylak, T. Gutbrod, M. Michel, A. Forchel, Phys. Rev. Lett. **90**, 186801 (2003)
147. M. Korkusiński, P. Hawrylak, M. Bayer, Phys. Status Solidi (b) **234**, 273 (2002)
148. C. González-Santander, F. Domínguez-Adame, R.A. Römer, Phys. Rev. B **84**, 235103 (2011)
149. V.M. Fomin, V.N. Gladilin, J.T. Devreese, N.A.J.M. Kleemans, M. Bozkurt, P.M. Koenraad, Phys. Status Solidi (b) **245**, 2657 (2008)
150. V.M. Fomin, V.N. Gladilin, J.T. Devreese, J.H. Blokland, P.C.M. Christianen, J.C. Maan, A.G. Taboada, D. Granados, J.M. García, N.A.J.M. Kleemans, H.C.M. van Genuchten, M. Bozkurt, P.M. Koenraad, Proc. SPIE **7364**, 736402 (2009)
151. M. Ahmad Kamarudin, M. Hayne, R.J. Young, Q.D. Zhuang, T. Ben, S.I. Molina, Phys. Rev. B **83**, 115311 (2011)
152. V.M. Kovalev, A.V. Chaplik, JETP Lett. **90**, 679 (2009)
153. V.M. Kovalev, A.V. Chaplik, J. Exp. Theor. Phys. **101**, 686 (2005)
154. P. Moon, K. Park, E. Yoon, J.P. Leburton, Phys. Status Solidi RRL **3**, 76 (2009)
155. K. Niemela, P. Pietilainen, P. Hyvonen, T. Chakraborty, Europhys. Lett. **36**, 533 (1996)
156. P.-F. Loos, P.M.W. Gill, Phys. Rev. Lett. **108**, 083002 (2012)
157. S. Viefers, P. Singha Deo, S.M. Reimann, M. Manninen, M. Koskinen, Phys. Rev. B **62**, 106683 (2000)
158. E. Räsänen, S. Pittalis, C.R. Proetto, E.K.U. Gross, Phys. Rev. B **79**, 121305 (2009)
159. W. Zhang, Z. Su, M. Gong, C.-F. Li, G.-C. Guo, L. He, Europhys. Lett. **83**, 67004 (2008)
160. Y. Meir, Y. Gefen, O. Entin-Wohlman, Phys. Rev. Lett. **63**, 798 (1989)
161. T. Meier, P. Thomas, S.W. Koch, K. Maschke, Phys. Status Solidi (b) **234**, 283 (2002)
162. P. Potasz, A.D. Güçlü, P. Hawrylak, Phys. Rev. B **82**, 075425 (2010)
163. T. Chakraborty, P. Pietilainen, Phys. Rev. B **50**, 8460 (1994)
164. J.A. Barker, R.J. Warburton, E.P. O'Reilly, Phys. Rev. B **69**, 035327 (2004)
165. J.I. Climente, J. Planelles, J. Nanoelectron. Optoelectron. **6**, 81 (2011)
166. O. Marquardt, S. Schulz, Ch. Freysoldt, S. Boeck, T. Hickel, E.P. O'Reilly, J. Neugebauer, Opt. Quantum Electron. **44**, 183 (2012)
167. S. Bellucci, P. Onorato, Phys. Rev. B **78**, 235312 (2008)
168. J. Schelter, P. Recher, B. Trauzettel, Solid State Commun. **52**, 1411 (2012)
169. O.L. Chalaev, V.E. Kravtsov, Phys. Rev. Lett. **89**, 176601 (2002)
170. B. Baxevanis, D. Pfannkuche, J. Nanoelectron. Optoelectron. **6**, 76 (2011)
171. M.F. Doty, J.I. Climente, M. Korkusinski, M. Scheibner, A.S. Bracker, P. Hawrylak, D. Gammon, Phys. Rev. Lett. **102**, 047401 (2009)
172. S.E. Economou, J.I. Climente, A. Badolato, A.S. Bracker, D. Gammon, M.F. Doty, Phys. Rev. B **86**, 085319 (2012)
173. S. Tanda, T. Tsuneta, Y. Okajima, K. Inagaki, K. Yamaya, N. Hatakenaka, Nature **417**, 397 (2002)

174. J. Gravesen, M. Willatzen, Phys. Rev. A **72**, 032108 (2005)
175. B. Lassen, M. Willatzen, J. Gravesen, J. Nanoelectron. Optoelectron. **6**, 68 (2011)
176. V.M. Fomin, S. Kiravittaya, O.G. Schmidt, Phys. Rev. B **86**, 195421 (2012)
177. S. Russo, J.B. Oostinga, D. Wehenkel, H.B. Heersche, S.S. Sobhani, L.M.K. Vandersypen, A.F. Morpurgo, Phys. Rev. B **77**, 085413 (2008)
178. M. Huefner, F. Molitor, A. Jacobsen, A. Pioda, C. Stampfer, K. Ensslin, T. Ihn, Phys. Status Solidi (b) **246**, 2756 (2009)
179. M. Huefner, F. Molitor, A. Jacobsen, A. Pioda, C. Stampfer, K. Ensslin, T. Ihn, New J. Phys. **12**, 043054 (2010)
180. B. Szafran, Phys. Rev. B **77**, 205313 (2008)
181. B. Szafran, Phys. Rev. B **77**, 235314 (2008)
182. L. Villegas-Lelovsky, C. Trallero-Giner, V. Lopez-Richard, G.E. Marques, C.E.P. Villegas, M.R.S. Tavares, Nanotechnology **23**, 385201 (2012)
183. F. Suárez, D. Granados, M.L. Dotor, J.M. García, Nanotechnology **15**, S126 (2004)
184. J. Wu, Z.M. Wang, K. Holmes, E. Marega, Y. Mazur, G. Salamo, J. Nanopart. Res. **14**, 919 (2012)
185. J. Wu, Z.M. Wang, K. Holmes, E. Marega Jr., Z. Zhou, H. Li, Y.I. Mazur, G.J. Salamo, Appl. Phys. Lett. **100**, 203117 (2012)
186. Y. Ma, J. Cui, Y. Fan, Z. Zhong, Z. Jiang, Nanoscale Res. Lett. **6**, 205 (2011)
187. K.L. Hobbs, P.R. Larson, G.D. Lian, J.C. Keay, M.B. Johnson, Nano Lett. **4**, 167 (2004)
188. R. Ji, W. Lee, R. Scholz, U. Gösele, K. Nielsch, Adv. Mater. **18**, 2593 (2006)
189. J. Planelles, J.I. Climente, F. Rajadell, Physica E **33**, 370 (2006)
190. R.A. Shelby, D.R. Smith, S. Schultz, Science **292**, 77 (2001)
191. S. Linden, C. Enrich, M. Wegener, J. Zhou, T. Koschny, C.M. Soukoulis, Science **306**, 1351 (2004)
192. A.M. Fischer, V.L. Campo Jr., M.E. Portnoi, R.A. Römer, Phys. Rev. Lett. **102**, 096405 (2009)
193. M.F. Borunda, X. Liu, A.A. Kovalev, X.-J. Liu, T. Jungwirth, J. Sinova, Phys. Rev. B **78**, 245315 (2008)
194. H. Ling, S. Wang, C. Lee, M. Lo, J. Appl. Phys. **105**, 034504 (2009)
195. E. Gallardo, L.J. Martínez, A.K. Nowak, D. Sarkar, D. Sanvitto, H.P. van der Meulen, J.M. Calleja, I. Prieto, D. Granados, A.G. Taboada, J.M. García, P.A. Postigo, J. Opt. Soc. Am. B **27**, A21 (2010)
196. M. Abbarchi, C. Mastrandrea, A. Vinattieri, S. Sanguinetti, T. Mano, T. Kuroda, N. Koguchi, K. Sakoda, M. Gurioli, Phys. Rev. B **79**, 085308 (2009)
197. E. Zipper, M. Kurpas, J. Sadowski, M.M. Maska, J. Phys. Condens. Matter **23**, 115302 (2011)
198. E. Zipper, M. Kurpas, M. Maska, New J. Phys. **14**, 093029 (2012)
199. E.R. Hedin, Y.S. Joe, J. Appl. Phys. **110**, 026107 (2011)
200. Z.C. Wen, H.X. Wei, X.F. Han, Appl. Phys. Lett. **91**, 122511 (2007)

Part I
Fabrication, Characterization and Physical Properties

Chapter 2
Growth and Spectroscopy of Semiconductor Quantum Rings

Wen Lei and Axel Lorke

Abstract Quantum rings are unique nanostructures as they are topologically not simply connected and therefore different from most other low-dimensional systems such as quantum dots, quantum wires or quantum wells. This topology gives rise to an intriguing energy structure, in particular, when a magnetic field is applied such that a flux can penetrate through the ring's interior. Flux quantization will lead to a ground state, which has a non-vanishing angular momentum, and the intraband transitions are affected by the corresponding change in dipole-allowed transitions. Quantum rings, which are of the order of 10 nm in size are of particular interest, because they make it possible to study these systems in the true quantum limit.

In this chapter, we will review the growth techniques, which lead to the self-organized formation of quantum rings of a few tens of nanometers in diameter. The mechanisms will be discussed, which 'invert' the geometry of InAs islands, grown in the Stranski-Krastanov mode on GaAs, when they are partially capped with GaAs. When the thus formed nanorings are embedded in a suitable heterostructure, they can be electrically tuned and carriers can be injected with single-electron/single-hole precision. We will discuss, how the number of carriers and the strength of the applied field influence the single-particle and many-particle ground states, which can be probed by capacitance-voltage measurements. Also, far-infrared absorption spectra will be presented, which show the influence of flux quantization on the intraband transitions. These spectroscopic techniques, together with photoluminescence data obtained on single rings as well as on ring ensembles, make it possible to obtain an in-depth view into the detailed energetic structure of nanoscopic rings.

W. Lei (✉)
School of Electrical, Electronic and Computer Engineering, The University of Western Australia, Crawley 6009, WA, Australia
e-mail: wen.lei@uwa.edu.au

A. Lorke
Faculty of Physics and CENIDE, Universität Duisburg-Essen, Lotharstr. 1, 47058 Duisburg, Germany
e-mail: axel.lorke@uni-due.de

V.M. Fomin (ed.), *Physics of Quantum Rings*, NanoScience and Technology,
DOI 10.1007/978-3-642-39197-2_2, © Springer-Verlag Berlin Heidelberg 2014

2.1 Introduction

In the last two decades, much attention has been devoted to semiconductor nanostructures such as quantum dots (QDs), quantum wires (QWRs), and quantum rings (QRs) due to the interest in fundamental physics study and potential device applications [1, 2]. These nanostructures have a size around a few or a few tens of nanometers, and thus exhibit strong size confinement in one or more dimensions, resulting in fascinating physical properties and possibly enhanced performance for optical and optoelectronic devices [1, 2]. At the same time, the rapid development of micro- and nano-processing technology like electron beam lithography, and ion bean milling make it possible to construct semiconductor nanostructures with various geometries. However, the micro- and nano-processing can induce defects and contamination to these nanostructures and thus affect their physical properties and device performance [3, 4]. These problems are overcome with the development and maturing of self-assembled growth of semiconductor nanostructures using advanced epitaxial techniques such as molecular beam epitaxy (MBE) and metal organic chemical vapor deposition (MOCVD). In self-assembled growth of lattice mismatched semiconductor nanostructures, coherent nanostructures such as QDs are spontaneously formed by relaxing the strain energy (caused by lattice mismatch) accumulated in the system once the epitaxial layer thickness exceeds a critical value, which is usually labeled as Stranski-Krastanov (S-K) growth [5]. These coherent nanostructures, unlike lithography defined ones, are defect-free and thus exhibit excellent optical and electrical properties. As a result, these self-assembled nanostructures are widely studied for applications in optoelectronics devices, quantum cryptography, and quantum computing, etc. [1, 2], and indeed high performance QD devices have been demonstrated recently [6].

In addition to nanostructures with simple geometry like QDs, self-assembled growth technique can also be utilized to construct nanostructures with complex geometries such as QRs. Physically, ring structures have holes in their geometrical center, which theoretically defines the capability of trapping a single magnetic flux in their interior and thus offers an exciting opportunity to study electronic wavefunction phases in magneto-optical experiments [7]. Though the trapping of a single flux quantum in a small molecule such as benzene was predicted many years ago, it was never observed in nowadays laboratories considering the huge magnetic fields (50,000 T) required. But, QRs, a kind of nanoscale ring structures provide us a unique opportunity to study rings in true quantum limit. So, three-dimensional topological quantum effects such as oscillation in magnetization, unusual magnetic susceptibility behaviour, etc can be observed with the magnetic fields available in today's laboratories. Therefore, defect-free QR structures are especially interesting for fundamental physics studies some of which were never explored before. Since the first experimental demonstration of self-assembled QRs in 1999 [8], a huge amount of attention has been devoted to these artificial ring structures for their fundamental physics study and device applications, and significant advance has been made in both material growth and spectroscopy study. For example, not only randomly distributed QRs but also ordered QRs were obtained by engineering the strain distribution in the system [9, 10]. New epitaxial technique such as droplet epitaxy [11] was developed for fabricating QRs. Various spectroscopic techniques were developed to investigate the optical, electrical and magnetic properties of these ring structures [12–17].

In this chapter, we will mainly review the recent research progress of semiconductor QRs, mainly InGaAs QRs, including both material growth and spectroscopy study. The content will be organized as follows: first, we will review the self-assembled growth of semiconductor QRs *via* partial overgrowth technique; then we will discuss the related theoretical model for the formation of QRs; and then we will briefly introduce the droplet epitaxy of QRs; after that we will review the interband spectroscopy study of InGaAs QRs; finally, we will discuss the intraband spectroscopy study of InGaAs QRs.

2.2 Epitaxial Growth of III–V Semiconductor Quantum Rings

Apart from micro- and nano-processing technology (lithography and etching), there are generally two kinds of epitaxial techniques to fabricate semiconductor QRs: (1) partial overgrowth; (2) droplet epitaxy. These two growth techniques will be discussed in the following:

2.2.1 Partial Overgrowth

The partial overgrowth technique is based on a capping and annealing process of self-assembled QDs. First, self-assembled QDs are grown *via* standard S-K growth mode. In this S-K growth, the epitaxial growth starts with two-dimensional layer-by-layer growth, and the strain energy (caused by lattice mismatch) accumulated in the system increases as the epilayer thickness increases. When the epitaxial layer thickness exceeds a critical value, the system becomes unstable and tries to relax its strain energy by forming three dimensional islands, where the transition from two-dimensional growth (wetting layer (WL)) to three-dimensional growth (island) takes place. The critical thickness is determined by the lattice mismatch between the epitaxial layer and substrate. For example, self-assembled InAs/GaAs QDs are formed when the deposited InAs layer is thicker than 2 monolayers (MLs) typically. Then, self-assembled QDs are ***partially*** covered with a thin capping layer, e.g. GaAs layer. After that, the partially covered quantum dots are ***annealed*** for a short time under a certain high temperature (typically >450 °C). During the annealing process, self-assembled QDs changes into QRs due to outward migration of atoms or dewetting process. It must be stressed that *to make sure the formation of quantum rings the self-assembled quantum dots must be **partially not completely** capped and must be **annealed for a short time***.

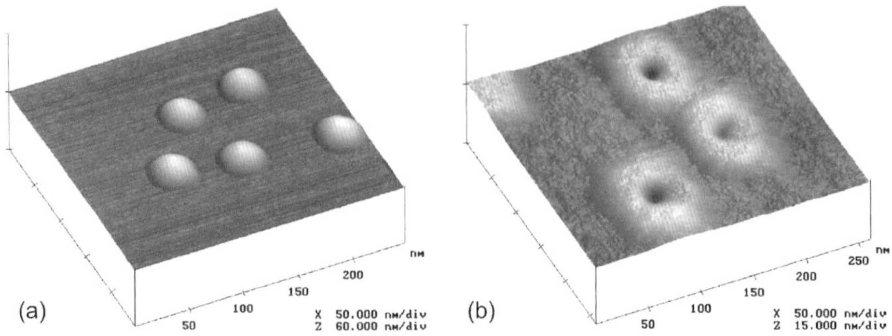

Fig. 2.1 Representative AFM images of InAs QDs (**a**) and InGaAs QRs (**b**) grown on GaAs substrates

2.2.1.1 InGaAs Quantum Rings—Material Growth and Structural Characterization

Based on our previous work, we will mainly focus on InGaAs QRs here. Other QRs will be discussed in Sect. 2.2.1.2. The InGaAs QRs are usually grown on GaAs substrates using MBE. A typical growth process is as follows [18–20]: first, several hundreds of nanometers of GaAs are deposited on GaAs (001) substrate at a growth temperature of 600 °C, in order to smoothen the growth front and achieve better growth quality. Then, the growth temperature is reduced to 530 °C, and 1.7 MLs of InAs is deposited, which is just a little over the critical thickness necessary for the transition from two-dimensional to three-dimensional growth (S-K growth). This results in the formation of a random array of self-assembled QDs. Then, the QDs are immediately covered partially with 4 nm GaAs, and then the growth is interrupted for 30 seconds. Note that the 30 seconds growth interruption at a growth temperature of 530 °C serves as an annealing process, which triggers the morphological transformation from dots to rings. After that, the growth is stopped and the sample is cooled down to room temperature for characterization.

Figure 2.1 shows the atomic force microscopy (AFM) images of InAs QDs obtained before partial capping and that of InGaAs QRs obtained after partial overgrowth and annealing. As shown in Fig. 2.1(a), the InAs QDs have an average height of ∼6 nm and an average diameter of ∼20 nm, which is quite normal for self-assembled InAs QDs formed by the S-K process. However, after partial overgrowth and annealing, the shape of the nano-islands drastically changes from lens-shaped dots into ring-like islands, as shown in Fig. 2.1(b). These ring structures, compared with InAs QDs, present an increased lateral size (between 60 and 140 nm in outer diameter), a reduced height (about 2 nm) and a well-defined center hole of about 20 nm in diameter. Therefore, these islands are a kind of nanoscale rings, usually called QRs. It is observed that the diameter of the center hole is comparable to the QD diameter, indicating the evolution relationship between the QDs and the QRs. The formation of InGaAs QRs here might be caused by the outward diffusion of In atoms from the QDs and their subsequent alloying with surrounding GaAs matrix

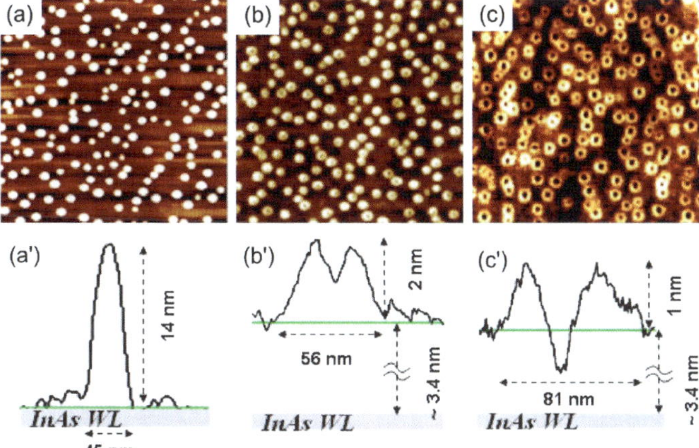

Fig. 2.2 AFM images and representative surface profiles (′) of InAs QDs grown under the same conditions but capped and annealed at different temperatures: (a) without capping and annealing; (b) capping and annealing at 450 °C; (c) capping and annealing at 470 °C. The AFM image size: 1 μm × 1 μm. (Figure adapted from Ref. [24])

material. Usually, it is assumed that during the overgrowth the heteroepitaxial system is in equilibrium, and thus the morphology and composition of self-assembled QDs are only little affected by overgrowth [21]. For example, the lens-like QD shape might be changed to flat-lens-like QDs. No such dot-to-ring transition is expected for usual full overgrowth process. However, the results here suggest that the epitaxial system during the *partial* overgrowth process is not in a complete equilibrium, and thus the morphology of the QDs can be modified dramatically if suitable growth conditions are met. Therefore, by controlling the growth parameters which affect the equilibrium of the system such as migration and alloying of In atoms, the morphology and composition of the InAs QDs can be engineered in a wide range.

It is well known that during the self-assembled growth of QDs, the growth temperature has a significant influence on the growth kinetic of the system and thus the morphology of the QDs [22, 23]. Similarly, during the partial overgrowth of QDs the temperature of the annealing process will also have a significant effect on the migration of In atoms and thus the morphology and composition of the QRs. Figure 2.2 shows the morphology changes of InAs QDs prepared under the same conditions but capped and annealed at different temperatures [24]. For all the three samples shown in Fig. 2.2, the InAs QDs were obtained by depositing 2.6 MLs of InAs on GaAs at 520 °C under As_2 atmosphere. For the sample shown in Fig. 2.2(a), after the QD growth no additional capping and annealing steps were introduced and the sample was cooled down to room temperature for characterization. For the samples shown in Fig. 2.2(b) and (c), after the QD growth the growth temperature was reduced to 450 and 470 °C, respectively, for the following capping and annealing processes: a thin GaAs layer of 2.5 nm was deposited to partially cover the QDs, and then a 50 s annealing under As_2 flux was introduced at the same temperature. It is observed that after the capping and annealing processes, the morphology of QDs

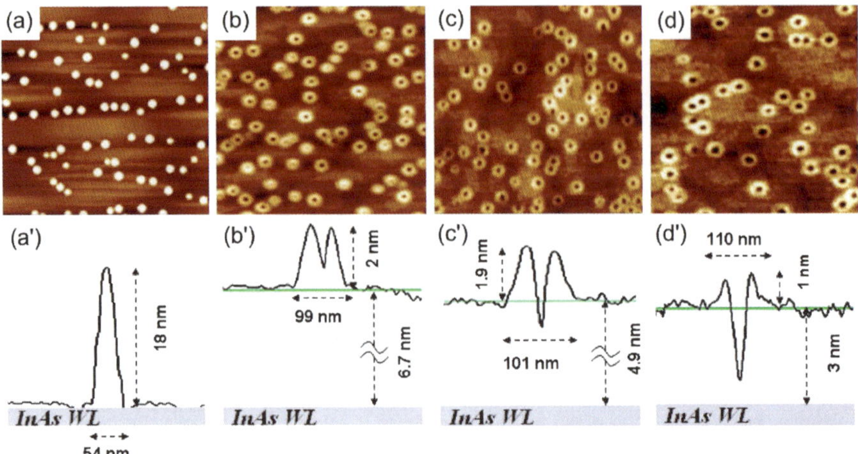

Fig. 2.3 AFM images and representative surface profiles (′) of the InAs QDs capped with 0 (**a**), 5.5 (**b**), 4 (**c**), 2.5 (**d**) nm GaAs layers and annealed at 505 °C for 50 seconds. The AFM image size: 1 μm × 1 μm. (Figure adapted from Ref. [24])

changes dramatically. The original QDs, shown in Fig. 2.2(a), change into volcano-like structures, as shown in Fig. 2.2(b), or a well defined ring structure, as shown in Fig. 2.2(c), depending on the annealing temperature. This suggests that the capping and annealing temperature plays an important role in the formation of QRs. At a higher capping and annealing temperature like 470 °C, the In atoms are very mobile and the outward migration of In atoms from the uncovered top of each QD is significant, leaving behind a deep hole in the center depleted of InAs. Consequently, a complete ring structures is formed. In contrast, at lower capping and annealing temperatures like 450 °C, the In atoms are not as mobile as they are at higher temperatures, and much less In atoms migrate out of the QDs, resulting in volcano-like structures (no deep hole in the center).

In addition to the migration kinetics of In atoms, the amount of InAs materials available for outward migration is another factor affecting the formation of QRs, providing us another approach to engineer the morphology of InGaAs QRs—changing the thickness of GaAs capping layer. To study the effect of GaAs capping layer's thickness on the formation of InGaAs QRs, Ling and Lee have capped InAs QDs with GaAs layers of different thicknesses and then annealed them to observe the change of their morphology [24]. Figure 2.3 shows the AFM images and the representative surface profiles of the InAs QDs capped with GaAs layers of different thicknesses and annealed at 505 °C for 50 seconds. For all the samples shown in Fig. 2.3, InAs QDs were formed by depositing 3.2 MLs of InAs on a GaAs substrate at a growth temperature of 540 °C. For the sample shown in Fig. 2.3(a), the InAs QDs were cooled down directly without capping and annealing after their growth. For the samples shown in Figs. 2.3(b), (c) and (d), the partial capping and annealing processes were used to form the QRs. After the deposition of InAs QDs, the growth temperature was reduced to 505 °C and GaAs capping layers of different thicknesses were deposited. After that, 50 seconds growth interruption at 505 °C was introduced, and then the samples were cooled down to room temperature for characterization. The GaAs capping layers have a thickness of 5.5, 4, and 2.5 nm for the samples shown in Figs. 2.3(b), (c), and (d), respectively. As shown in these images, the well-formed QDs (Fig. 2.3(a)) evolve into volcano-like dots (Fig. 2.3(b)), into rings with shallow node in the center (Fig. 2.3(c)) and into rings with deep node in the center (Fig. 2.3(d)) after the partial overgrowth. Obviously, the thickness of GaAs capping layer plays an important role in determining the final shape of islands. The amount of In atoms that can migrate out from the center portion of the QDs depends on the amount of InAs QD surface left without capping, which is determined by the thickness of GaAs capping layer. For the sample shown in Fig. 2.3(a), which doesn't have a GaAs capping layer, InAs QDs are obtained with a height around 18 nm and a diameter around 54 nm. In the sample shown in Fig. 2.3(b), which has a thicker GaAs capping layer of 5.5 nm, the QDs turn into volcano-like structures with a center crater about 1.4 nm deep. And, these volcano-like islands present a height around 8.7 nm and a lateral size around 99 nm. Furthermore, for the sample shown in Fig. 2.3(d), which has a thinner GaAs cap layer of 2.5 nm, well-defined QRs are formed with a center hole of 3 nm. These well-defined QRs show a height around 4 nm and a lateral diameter around 110 nm. Compared with the InAs QDs obtained, the InGaAs QRs show a much reduced height and an increased lateral size, which is due to the fact the InAs material that moves from the top of the QDs spreads out to the adjacent area above the GaAs capping layer. Since the QD sizes are the same for all the samples studied here, the InAs QDs with a thicker cap layer will have less exposed area that allows In atoms to move outward upon annealing, leading to the formation of only volcano-like dot structures. In contrast, the QDs with a thinner cap layer will have more exposed In atoms that can move outward upon annealing, leading to the formation of ring structures. Not only that, the imbalance of the surface tension at the interface between the cap layer and the QD droplet (dewetting process) also plays an important role in the morphology change, which will be discussed in detail in Sect. 2.2.1.3.

As demonstrated above, the outward diffusion of In atoms from InAs QDs and their subsequent alloying with GaAs matrix play an important role in the formation of InGaAs QR structures. Experimentally, both the outward diffusion of In atoms from InAs QDs and their alloying with GaAs matrix can be evidenced by the redistribution of In composition in the QRs and their surrounding areas, which can be measured using some techniques such as cross-sectional scanning tunneling microscopy (X-STM), and X-ray Photoemission Electron Microscopy (XPEEM). A more detailed discussion of the In distribution taken from cross-sectional STM measurements can be found in Chap. 4. In the following, we will introduce the XPEEM technique to probe the distribution of In concentration in the QR samples [25].

For this XPEEM study, InAs QDs were grown on GaAs substrate by depositing about 2 MLs of InAs QDs at 540 °C. After the growth of InAs QDs, the growth temperature was reduced to 490 °C, and then 2 nm GaAs was deposited to partially cover the QDs. Then, a 30 sec annealing was introduced at the same temperature to

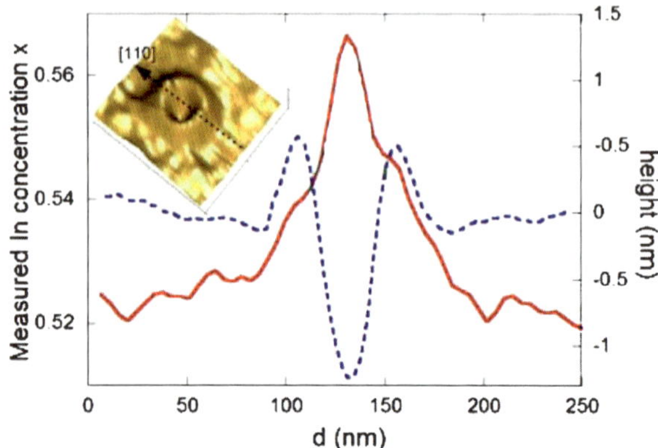

Fig. 2.4 AFM height profile (*short dashed line*) and XPEEM In composition profile (*solid line*) of a QR along the [110] direction. The inset presents the AFM image of a representative surface QR. (Figure adapted from Ref. [25])

facilitate the formation of InGaAs QRs. After that, the sample was rapidly cooled down and capped with a protective layer of As_4. The morphological and chemical composition mapping of the sample surface (after thermal removal of the protective As_4 capping) were performed at room temperature by XPEEM, which allows obtaining photoelectron spectra with energy resolution of 0.3 eV and lateral resolution down to 25 nm. Figure 2.4 presents the In concentration profile of a QR along the [110] direction measured with XPEEM. The inset of Fig. 2.4 shows an AFM image of a QR from the same sample after exposing to air. The dashed line in Fig. 2.4 is an average of 10 AFM line scans through the QR centers along the [110] direction. It is observed that the QR has a rim height around 0.5 nm, and a central hole depth around 1.2 nm, with respect to the surrounding surface. And, the average outer and inner diameters of the QR are around 69 nm and 26 nm, respectively. As shown by the solid line in Fig. 2.4, the Indium concentration decreases from about 0.57 at the center of the QR to 0.52 above the WL, suggesting the outward diffusion of In atoms from InAs QDs and their subsequent alloying with GaAs matrix. The In concentration profile presents a double structure, with a center narrow peak corresponding to the center hole of the QR, and a shoulder structure to the surrounding rim. And the In concentration profiles obtained along the [110] and [11-0] are virtually identical, in contrast to the X-STM results on embedded InGaAs QRs discussed in Chap. 4, which might be caused by the different growth conditions.

As discussed above, the InGaAs QRs basically can be considered to arise from the InAs QDs. Therefore, the spatial distribution of these InGaAs QRs is determined by that of InAs QDs. However, the S-K growth of InAs QDs is a random self-organization process, instead of an ordered growth process, leading to a random distribution of InGaAs QRs. For their device applications even fundamental physics study, it is very important to achieve the ordered growth of these InGaAs QRs. For InAs QDs, their ordered growth have been studied for many years and significant advanced has been made in this aspect [26–29]. By engineering the strain field in the system, the preferred nucleation sites for InAs QDs can be well controlled and both lateral and vertical ordered QD array can be achieved. Similarly, this strain field technique can also be applied to achieve the ordered growth of InGaAs QRs.

In the following, we will briefly introduce how to achieve the ordered growth (both lateral and vertical) of InGaAs QRs.

Lateral Ordered Growth Both one-dimensional and two-dimensional ordered lateral growth of InGaAs QRs were achieved recently by engineering the strain field distribution in the system, where one-dimensional ordered growth was observed on GaAs (100) surface while two-dimensional ordered growth was observed on high index GaAs surfaces [8]. For this study, InGaAs QRs were grown on GaAs substrates with various orientation such as (100), (311)B, and (511)B surfaces. Figures 2.5(a) and (b) show the schematic illustration of the sample structure and the mechanism of ordered growth. After the growth of 200 nm GaAs buffer layer, 8 periods $In_{0.6}Ga_{0.4}As$/GaAs QD superlattices were grown to create a strong strain field in the system, which led to the formation of laterally aligned QDs in the top layers. The aligned QDs served as a self-assembled template for the ordered growth of subsequent InGaAs QRs. After the growth of the InGaAs/GaAs QD superlattices, the growth temperature was reduced to 540 °C and 2.1 ML InAs was deposited to grow InAs QDs. After that, a growth interruption of several tens of seconds was introduced to promote the lateral ordering of InAs QDs. Then, a thin GaAs layer (4 nm) was deposited to partially cap the InAs QDs and form InGaAs QRs. The growth details can be found in Ref. [8]. Figures 2.5(c), (d) and (e) show the AFM images of the InGaAs QRs obtained. From left to right, the AFM images correspond to InGaAs QRs grown on (100), (311)B, and (511)B GaAs substrates, respectively. Obviously, as indicated by Fig. 2.5(c), self-organized InGaAs QRs are formed and aligned along the [1-10] direction on GaAs (100) surface. The ordered QRs are evolved from the top layer of InAs QDs formed on the template created by the InGaAs/GaAs QD superlattices. Therefore, the lateral alignment of InGaAs QRs is mainly caused by the elastic interactions among the strained self-assembled QDs [30, 31]. Because the QDs prefer to nucleate at local strain minima, the QD arrangement in the top layers, both QD size and spacing, becomes progressively uniform after stacking multiple layers of QDs. In the meantime, due to the high anisotropy of In migration, QDs on the (100) surface are elongated along the [1-10] direction. Furthermore, the elongated or oval-shaped QDs introduce elongated strain field along the [1-10] direction, leading to an alignment of strain minima along the [1-10] direction after stacking multiple layers of QDs. However, the alignment of QDs is much less pronounced along the [110] direction due to the weaker strain correlation in this direction. Therefore, the InGaAs QRs grown on (001) GaAs surface only show one-dimensional alignment even though the InGaAs/GaAs QD superlattice template is used. More interestingly, the InGaAs QRs on (311)B and (511)B surfaces present two-dimensional ordering as shown in Figs. 2.5(d) and (e). This two-dimensional ordering might be related to the preferred strain relaxation direction and the nominal surface step separation [8].

Vertical Ordered Growth Vertical ordered growth of InGaAs QRs can also be achieved by modifying the strain fields in the stacked structures. As suggested by the previous study on vertical ordered growth of InAs QDs/QWRs [27–29, 32], the

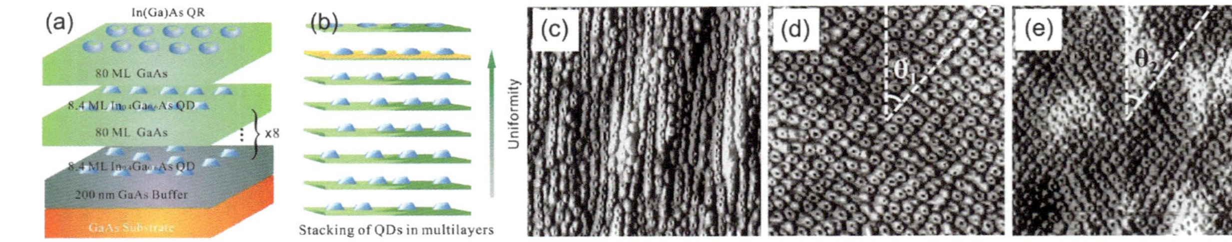

Fig. 2.5 Schematic sample structure (**a**) for ordered growth of InGaAs QRs and illustration (**b**) showing lateral alignment of QRs by using QD superlattices; AFM images of InGaAs QRs grown on GaAs (001) (**c**), GaAs (311) (**d**) and GaAs (511) (**e**) surfaces using QD superlattice template. The AFM image size: 2.5 μm × 2.5 μm. (Figure adapted from Ref. [8])

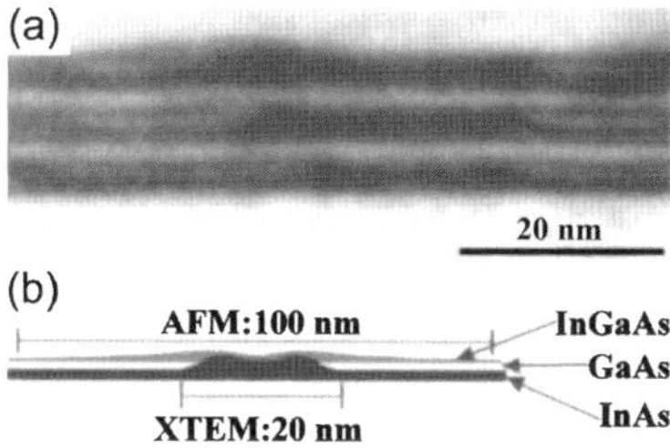

Fig. 2.6 X-TEM image of three layers of InGaAs QRs grown with 3 nm GaAs spacer (**a**), and illustration of material distribution of a InGaAs QRs determined from the X-TEM image. (Figure adapted from Ref. [10])

strain field generated in the bottom layers can be transferred to the top layer and thus provides preferential nucleation sites for the islands in the top layers when appropriate spacer layer (composition and thickness) is used. Therefore, by choosing spacer layers with appropriate thickness and composition both vertical correlation and anti-correlation of the QRs can be achieved during the stacked growth. For this study, three layers of InGaAs QRs were grown on GaAs (001) substrates and the InGaAs QRs were obtained by partially covering InAs QDs with 2 nm GaAs [10]. The top layer of InGaAs QRs were left uncovered for AFM characterization. To study the vertical ordered growth, GaAs spacers with different thicknesses (1.5, 3, 4.5, 6, 10, and 14 nm) were used for the stacked growth. The AFM results showed that when the spacer thickness is thicker than 6 nm, the aspect ratio and density of the top layer QRs are similar to those of the single layer QRs grown under the same conditions, indicating that a thin GaAs layer (≤ 6 nm) should be used to ensure the transfer of the strain field from the bottom QR layers to the top QR layers. Cross-sectional Transmission Electron Microscopy (X-TEM) measurements were performed on the samples with 3 and 6 nm GaAs spacer. Figure 2.6 shows (002) dark field X-TEM images of the QR superlattices with 3 nm GaAs spacer. Obviously, the InGaAs QR superlattices with a 3 nm GaAs spacer present vertical ordering of the QRs. Vertical ordering of the QRs was also observed in the samples with 6 nm GaAs spacer (images not shown here). As shown in Fig. 2.6, the InAs WL is observed as a dark-gray region at the bottom of each layer, while the GaAs capping layer is observed as a light-gray GaAs region surrounding each QR. On top of the GaAs capping layer there is a second gray region which is observed as the InGaAs alloy coming from the InAs migrated out of the center of each QD during the QR formation. Therefore, the location of each QR can be determined by looking for the dark-gray InAs-rich regions surrounded by these tri-layers. The details of this X-TEM study can be found in Ref. [10].

2.2.1.2 Other III–V Quantum Rings

In the last few years some other III–V semiconductor QRs such as InAs/InP quantum rings were also obtained *via* partial overgrowth technique. It must be noted that different from InGaAs/GaAs QRs, InAs and InP share the same group III element—In. Therefore, the formation of InAs/InP QRs cannot be explained by the outward diffusion of group III elements from the QDs and subsequent alloying

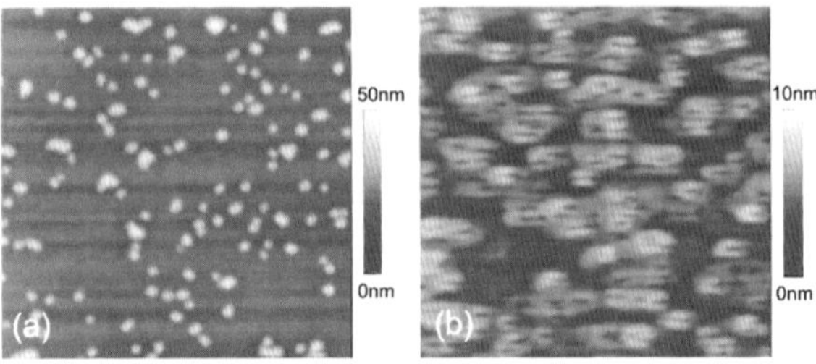

Fig. 2.7 AFM images of InAs/InP QDs (**a**) and QRs (**b**) obtained. The AFM image size: 2 μm × 2 μm. (Images adapted from Ref. [33])

with the surrounding capping layers. In this case, the formation of InAs/InP QRs is mainly caused by the *dewetting process* in the system, which will be discussed in Sect. 2.2.1.3.

The InAs/InP QRs were reported by Raz et al. [33] and were grown on InP(100) at 495 °C by a metal-organic MBE system. Trimethylindium, triethylgallium, arsine, and phosphine were used as group III and V sources, respectively. A 200-nm-thick InP buffer layer was first deposited. After a 40 s growth interruption, 2.1 ML of InAs was deposited to form InAs QDs. After another 40 s growth interruption the InAs QDs were partially capped by 1-nm-thick InP layer. The samples were then cooled down to room temperature for characterization. For comparison, self-assembled InAs/InP QDs were also grown under the same conditions but without the 1-nm-thick InP cap layer. Figure 2.7 shows the AFM images of both the InAs QDs and InAs QRs obtained. The InAs QDs have an average density of 3×10^9 cm^{-2}, and present a lens-like shape with an average base diameter of 75 nm and an average height of 15 nm. However, with 1 nm InP capping layer, ring-like InAs structures are formed which are elongated along the [110] direction. And, the QRs have an average outer diameters of $\sim$220 nm along the [110] direction, and $\sim$ 110 nm along the [1-10] direction, and an average height of $\sim$2.5 nm. The growth details of these InAs/InP QRs can be found in Ref. [33]. The interesting point about the formation of InAs/InP QRs is that both InAs and InP have the same group III element, and there is no driven force for the diffusion of group III atoms. Therefore, the kinetic diffusion model for the formation of InGaAs quantum rings is not applicable here. The formation of InAs/InP quantum rings is related to the dewetting process in the system, which will be discussed in the Sect. 2.2.1.3.

2.2.1.3 Formation Mechanism of InGaAs Quantum Rings

It can be seen from the above discussion that both kinetic diffusion and dewetting process contribute to the formation of QRs via partial overgrowth technique. Depending on the material system and growth conditions, either kinetic diffusion or dewetting process may dominate the formation process of the QRs. In the following, we will introduce these two processes:

As for the kinetic diffusion model, it is based on a kinetic diffusion process caused by the different mobility of different atoms at high temperatures. Figure 2.8 shows the illustration about how a QD evolves into a QR with a kinetic diffusion

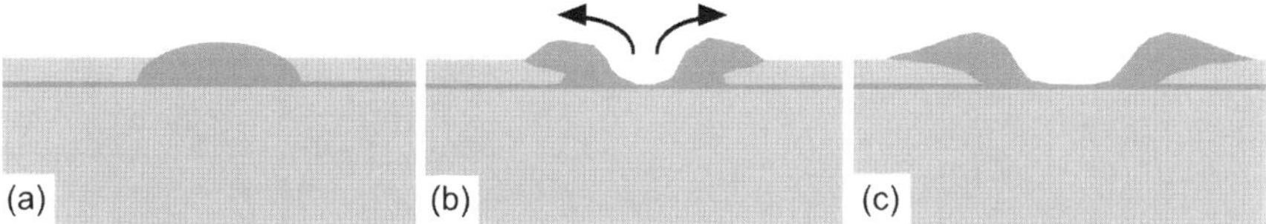

Fig. 2.8 Illustrations about the formation of InGaAs QRs *via* the kinetic diffusion model. (**a**) InAs QDs partially capped with GaAs layer; (**b**) outward diffusion of In atoms upon annealing; (**c**) formation of InGaAs QRs. *Dark gray colour* represents indium-rich material, while *light gray colour* represents GaAs material

process. Figure 2.8(a) shows the configuration of an InAs QD structure, immediately after the deposition of a thin GaAs capping layer. It is assumed that in this case the GaAs capping layer does not completely cover the QDs and the top of the InAs QD is left uncovered. This is supported by the fact that the strain relaxation at the top of InAs QDs makes them unfavorable locations for the growth of GaAs [34]. Furthermore, the previous experiments about the effect of different capping layer thicknesses on the QR formation also support this assumption. As discussed before, at a growth temperature above 450 °C the In atoms are quite mobile on the surface [35], while the Ga atoms diffuse only little after they have been incorporated into the crystal lattice. Therefore, the In atoms can diffuse out of the partially capped islands onto the surface of the surrounding GaAs, as shown in Fig. 2.8(b). As schematically illustrated in Fig. 2.8(c), after an appropriate annealing time the outward diffused In atoms will alloy with the surrounding GaAs and form an In-rich $In_xGa_{1-x}As$ rim around the void at the former location of the InAs QD. The rim and void formed can explain the observed volcano shape of the islands. Furthermore, the important role played by the kinetic diffusion can be reflected by the elongated shape of QRs which is due to the fact that Indium atoms diffuse much faster along the [1-10] direction than along the [110] direction [35]. An extreme example is given in Fig. 2.9 [20], where the length of a InGaAs QR along the [1-10] direction increases up to 600 nm. In this extreme example, a substrate holder with a large thermal mass was used during the growth, and thus the sample cooled down slowly, which gave much more time for the outward diffusion of In atoms and resulted in a strongly elongated island shape. In Fig. 2.9, the hole in the center is hardly discernible (see the arrow), but it clearly reflect the strong anisotropy of the outer edge of the island. This agrees with the assumption that the GaAs capping layer hardly diffuses during the formation of the QRs.

Though the above kinetic diffusion model can well explain a number of experimental observations discussed before, it cannot account for the formation of InAs/InP QRs. Even for InGaAs QRs, there are still some experimental phenomena which cannot be explained with this kinetic diffusion model. First, as shown in Figs. 2.2(b) and 2.3(b), under some growth conditions both the islands with center holes and those without center holes can co-exist which suggests an abrupt formation process instead of a continuous hole formation caused by atom diffusion. Second, the sharp outer island edge seen in Figs. 2.2 and 2.3 is not compatible with a diffusive picture. With the kinetic diffusion model, the island edge should be a

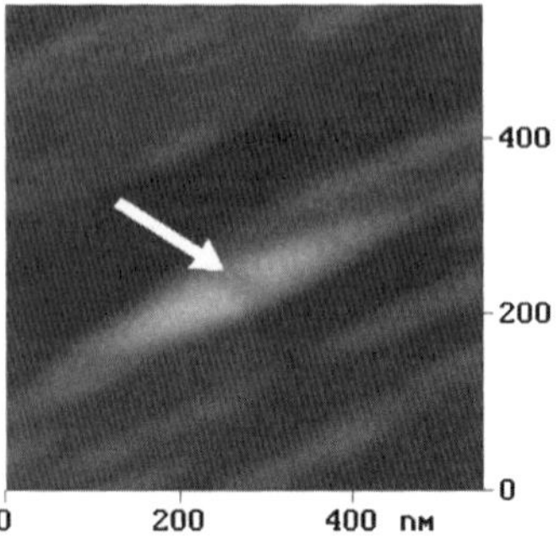

Fig. 2.9 AFM image of an InGaAs QR grown with larger thermal mass substrate holder. The arrow refers to the central hole of the QR

slope instead of sharp edge. Finally, the strongly different aspect ratios of the QRs seen in Figs. 2.5 and 2.7 do not agree with the kinetic diffusion model. For a purely diffusive process, the aspect ratio of the islands is determined by the different diffusion lengths along the [110] and [1-10] directions and is thus constant over time. Therefore, it can be concluded that apart from the kinetic diffusion model, other formation mechanisms must be taken into account.

In the meantime, if we look into the literature [36], we find there are striking similarities between the ring-like islands and the structures observed in dewetting processes. This leads us to reckon that surface and interface forces also play a role in the transformation of QDs into QRs during the partial overgrowth process. In the following, the dewetting process promoting the formation of ring structures will be discussed [37]. Figure 2.10 shows the schematic dewetting process for the formation of InGaAs QRs. In a solid-on-liquid case, the wetting angle θ of uncapped InAs QDs is given by the balance of forces at the foot of the QD,

$$\gamma_{ac} = \gamma_{bc} \cos(\theta) + \gamma_{ab} \tag{2.1}$$

Here, γ_{ij} is the force at the interface between materials i and j, which in the present InAs/GaAs system are $a = \text{GaAs}$, $b = \text{InAs}$, and $c = \text{As-vapor}$. For partially covered QDs, the corresponding relation would be

$$\gamma_{ac} = \gamma_{bc} \cos(\theta) - \gamma_{ab} \cos(\theta), \tag{2.2}$$

which is obviously incompatible with (2.1). The configuration in Fig. 2.10(b) therefore leaves an unbalanced net outward force $\Delta F = \gamma_{ab}(1 + \cos(\theta))$ (Fig. 2.10(c)), which promotes the disintegration of the partially capped islands and the formation of the central hole. It is possible that under the circumstance of partial capping some In droplets might be formed on the top of the InAs QDs due to the stress-induced melting effect, the detailed discussion of which can be found in Chap. 3. In addition to the stress measurements in Chap. 3, the cross-sectional STM study by J.M. Ulloa et al. [38] also showed that during the partial capping of InAs WLs with GaAs layers some In atoms will segregate to the surface of capping layer, which might migrate to the top of InAs QDs and form liquid In droplets. The formation of In droplets on the top of InAs QDs will facilitate the dewetting process and lead to the formation of QRs. Figure 2.10(d) schematically illustrates that the balance of forces which might indeed be better met in volcano-shaped islands.

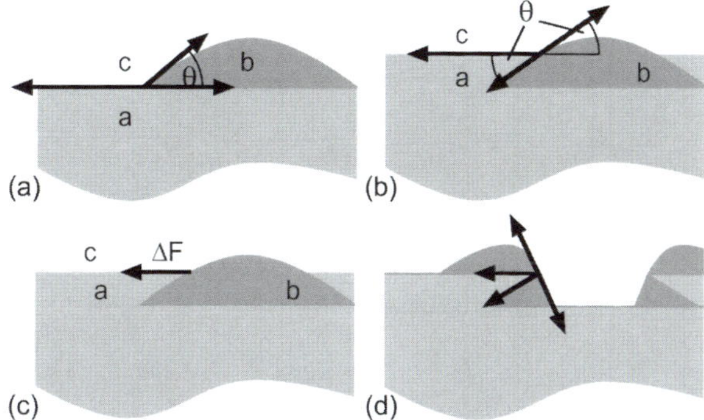

Fig. 2.10 Schematic dewetting process for the formation of a ring structure: (**a**) InAs QD on GaAs buffer; (**b**) InAs QD covered partially with GaAs capping layer; (**c**) unbalanced net outward force formed on the QD; (**d**) formation of volcano-shaped island. The phases are indicated as solid/GaAs (**a**, *light gray colour*), liquid/InAs (**b**, *dark gray colour*), and vapor (**c**). The *arrows* represent the interfacial tensions

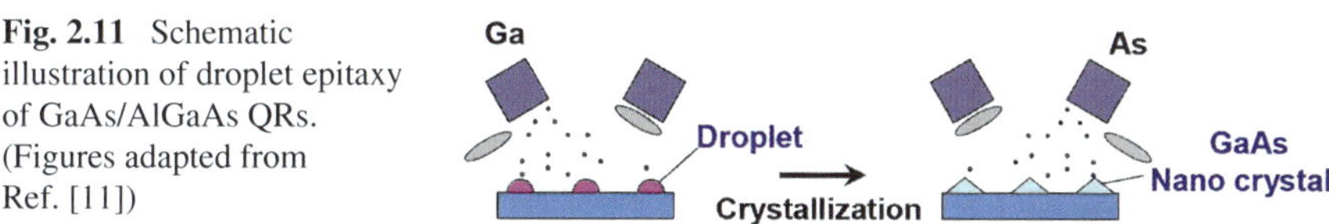

Fig. 2.11 Schematic illustration of droplet epitaxy of GaAs/AlGaAs QRs. (Figures adapted from Ref. [11])

As discussed above, the formation of the QRs *via* partial overgrowth process is promoted by at least two different mechanisms. One is the diffusion of In atoms out of the InAs QDs and their subsequent alloying with surrounding matrix, which is at least a prominent effect for the formation of InGaAs QRs. The other is a dewetting process, promoted by the imbalance of surface and interface forces acting upon the partially capped islands, which is more prominent for the material system with common group III element such as InAs/InP QRs.

2.2.2 Droplet Epitaxy

Droplet epitaxy is another kind of self-assembled growth technique for semiconductor nanostructures, which was developed in recent years [11, 39]. Compared with S-K growth, where group III and V atoms are deposited simultaneously during the growth of nanostructures, for droplet epitaxy the group III and group V atoms are deposited consecutively. As shown in Fig. 2.11, droplet epitaxy generally consists of two main steps [39, 40]: the first step is to deposit ultra-fine liquid droplets of group-III metals with low melting points such as Ga and In on substrates by supplying their molecular beams in the absence of group V elements; the second step is to form semiconductor nanostructures by exposing metal droplets to molecular beams of group-V elements such as As to crystallize the metal droplets. Obviously,

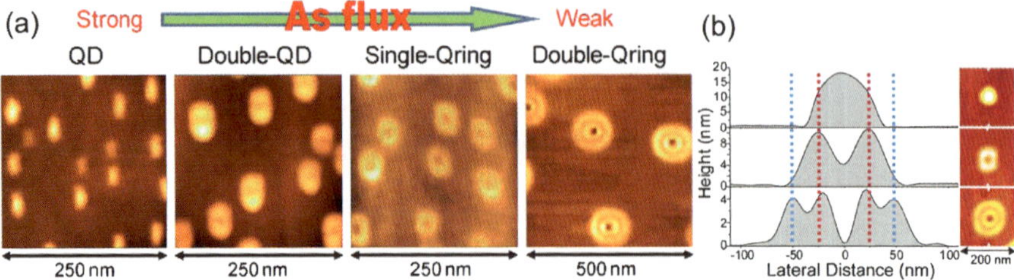

Fig. 2.12 (**a**) AFM images of the GaAs/AlGaAs nanostructures obtained with different As flux intensities during the crystallization process, and (**b**) the lateral profile of a typical island of QD, single QR and double QR obtained. (Figures adapted from Ref. [11])

droplet epitaxy, different from S-K growth, does not rely on the lattice mismatch in the system to form nanostructures. So, both lattice matched and lattice mismatched nanostructures can be achieved with this novel growth technique. In the following, we will briefly introduce the droplet epitaxy of both lattice matched and lattice mismatched semiconductor QRs.

2.2.2.1 MBE Droplet Epitaxy of GaAs/AlGaAs Quantum Rings

By tuning the growth parameters during the crystallization process of Ga droplets, high optical quality GaAs/AlGaAs quantum structures can be obtained, including QDs, single QRs, and concentric double QRs. The typical droplet epitaxy process for GaAs/AlGaAs nanostructures are as follows [11, 39]: first, 0.1 μm-thick AlGaAs barrier layer is grown on a GaAs (001) substrate at 580 °C. Then the substrate temperature is ramped down to 200–350 °C with the pressure of the residual As atmosphere below 1.3×10^{-6} Pa. And then, liquid Ga droplets are deposited on the substrate by supplying ~4 equivalent MLs of Ga atoms. The average diameter of the resulting Ga droplets is ca. 10–50 nm. After that, As molecular flux is introduced into the growth chamber at 200 °C to crystallize the Ga droplets and form GaAs nanostructures. GaAs nanostructures with different shapes can be obtained by varying the growth conditions. Figure 2.12 shows AFM images of the GaAs nanostructures obtained by crystallizing with various As flux intensities. When a As flux of 2.7×10^{-2} Pa (beam equivalent pressure (BEP)) is used, standard GaAs QDs are formed due to the suppression of Ga diffusion on the substrate surface [40, 41]. With reducing the As flux intensity during the crystallization process, central holes become visible gradually. When the As flux is reduced to 1.3×10^{-3} Pa BEP, the central holes of the islands become very clear and well-defined single-rings are obtained [41, 42]. It is also noted that when an intermediate As flux of 8×10^{-3} Pa BEP is used, the shape of the nanostructures becomes highly anisotropic, and laterally coupled double-QDs are obtained [42]. With further reducing the As flux to 1.3×10^{-4} Pa BEP, double-ring structures are obtained with a well-defined central hole [39]. Hence, GaAs nanostructures with various shapes can be obtained by just modifying the growth conditions during the droplet epitaxy process. Figure 2.12(b) shows the cross-sectional profiles of some typical nanostructures measured by AFM. It is observed that the single-rings and inner rings of double-rings present the same lateral size as the base size of the initial droplets, indicating the efficient crystallization at the edges of the droplets. However the outer ring of double-QRs shows a larger diameter than the initial Ga droplets, suggesting that the Ga atoms migrate away from the droplets, and the As atoms migrate toward the droplets. The details of these novel GaAs/AlGaAs QRs can be found in Refs. [11] and [39]. In addition to the migration of Ga atoms, the formation of GaAs QRs might also be related to the dewetting of Ga droplets during the crystallization process. The direct deposition of liquid Ga droplets on the surface will facilitate the dewetting of Ga droplets and the formation of QRs during the process of As crystallization, as discussed before.

2.2.2.2 MOCVD Droplet Epitaxy of InAsSb/InP Quantum Rings

Apart from MBE growth technique, MOCVD growth technique can also be used for droplet epitaxy, which is seldom explored. For the MOCVD epitaxial growth of semiconductor films, it is usually assumed that the thin film growth is accompanied by chemical reactions between different metal-organic precursors [43], instead of alloying of different atoms in the MBE droplet epitaxy. So—it was assumed—MOCVD cannot be used for droplet epitaxy. However, in reality MOCVD can indeed be used for droplet epitaxy. And, both lattice matched GaAs/AlGaAs and lattice mismatched InAsSb/InP QRs have been achieved in our group. In the following, we will briefly introduce our most recent results on the droplet epitaxy of InAsSb/InP QRs.

The InAsSb QRs were grown on semi-insulating InP (001) substrates in a horizontal flow MOCVD reactor (AIX200/4) at a pressure of 180 mbar. Trimethylindium (TMIn), trimethylgallium (TMGa), trimethylantimony (TMSb), phosphine (PH_3), and arsine (AsH_3) were used as the precursors and ultra-high purity H_2 as the carrier gas. The InAsSb nanostructures were grown using the following layer sequence: first, a 50 nm InP layer and 100 nm $In_{0.53}Ga_{0.47}As$ matrix layer were deposited at 650 °C, then the temperature was dropped to 480 °C and the growth was interrupted for 10 seconds with all sources removed from the reactor (for eliminating the influence of AsH_3 source on the later deposition of In droplets). After that at 480 °C, 8 equivalent MLs of liquid In droplets were deposited, which were immediately exposed to a (AsH_3 + TMSb) flow with various TMSb/(AsH_3 + TMSb) flow ratios for several tens of seconds. After that, the sample was cooled down with the protection of (AsH_3 + TMSb) flow. The morphology of these InAsSb nanostructures was characterized with AFM in tapping mode and scanning electron microscopy (SEM). Figure 2.13 shows the representative AFM and SEM images of the InAsSb/InP QRs obtained. It is observed that InAsSb rings with dilute density are formed on the sample surface. The rings have a height of 3–7 nm, and a lateral length of 110 nm along the [110] direction and 170 nm along the [1-10] direction. Since MOCVD droplet epitaxy is a quite new topic, the theoretical model for the formation of these QRs is still missing and remains to be explored further.

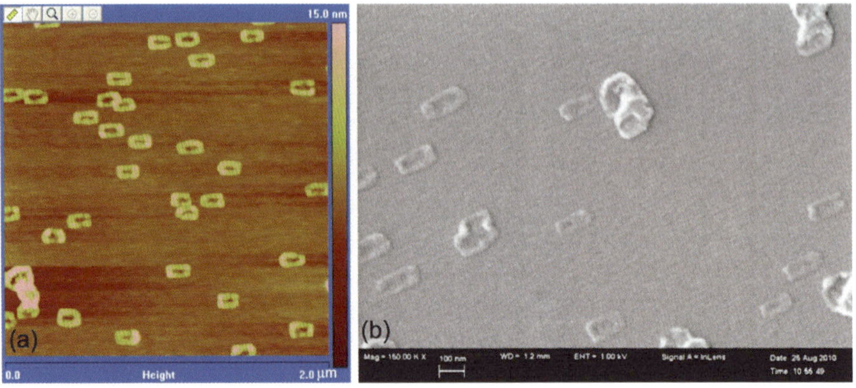

Fig. 2.13 AFM image (**a**) and SEM image (**b**) of InAsSb/InP QRs obtained *via* MOCVD droplet epitaxy. The AFM scan size of (**a**) is 2 μm × 2 μm

2.3 Spectroscopy Study of InGaAs Quantum Rings

Because of their ring geometry, QRs, theoretically, can trap single magnetic flux and thus have many exciting physical properties such as Aharonov-Bohm (AB) effect, which are not observed in QD systems. In the following, we will review the recent work on the spectroscopy study of InGaAs QRs, including interband and intraband spectroscopies.

2.3.1 Interband Spectroscopy

Interband spectroscopy involves the transition of carriers between the valence and conduction bands in the QRs, including ground state and excited state transitions. By studying interband transitions, we can obtain information about the QD bandgap energy, carrier excitation, carrier occupancy, carrier recombination, etc., which are important for both fundamental physics study and device applications. Also, the effect of external fields (electrical, optical and magnetic) on the interband transitions is also interesting. In the following, we will first discuss the fundamental interband transition in the rings *via* transmission spectroscopy, then the effect of external fields on the interband transitions including electron injection and magnetic field, and last the carrier lifetime associated with the interband transitions in the rings.

To study the fundamental interband transitions and the effect of electron injection on these transitions, Pettersson et al. [12] and Warburton et al. [13] used InGaAs QRs embedded in a capacitor-type MISFET (metal-insulator-semiconductor field effect transistor) structure, which allows tuning the electron occupancy in the QRs. This MISFET structure consisted of a heavily Si-doped GaAs back contact layer as electron reservoir, a GaAs spacer separating the rings from the back contact layer but allowing for tunneling of electrons between the back contact and the QRs, a InGaAs

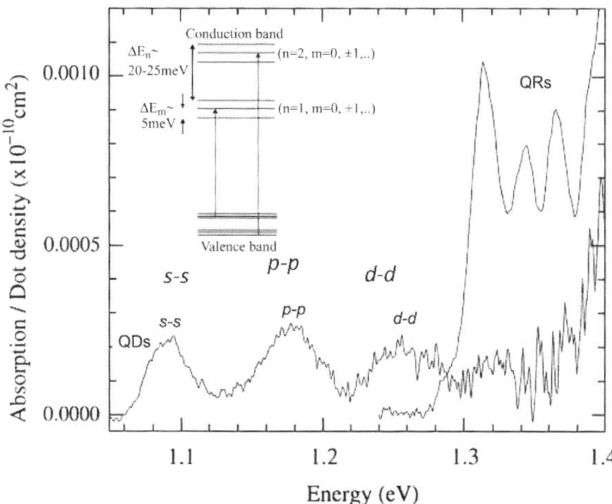

Fig. 2.14 Absorption per density of absorbers for the QDs and QRs at 4.2 K. The absorption was determined by subtracting the measured transmission from 1. The *inset* shows schematically the relevant electronic structure for the fundamental and the first excited transitions. (Figure adapted from Ref. [12])

QR layer, a GaAs spacer and AlAs/GaAs superlattice blocking layer to suppress the surface leakage current or tunneling between QRs and top gate electrode. The In-GaAs QRs were grown using the partial overgrowth technique discussed before. With this capacitor structure, an external voltage will cause a linear electrostatic potential across the structure, and the applied gate voltage V_g can be converted into an energy shift of the energy levels in the QRs with respect to the Fermi energy level in the back contact by using a certain "lever arm" [14]. Single electrons are loaded into the QRs when the energy level of the corresponding state is shifted below the Fermi energy in the back contact. By measuring the AC capacitance spectra of this capacitor, the gate voltage needed to load electrons into certain energy level can be determined as the tunneling of electrons between the back contact and the QRs will increase the capacitance and lead to a capacitance peak in the capacitance-voltage (C-V) spectra. Therefore, with this C-V spectroscopy, the number of electrons injected into the QRs can be well controlled, providing a good platform for studying interband transitions in the rings. The details of this C-V spectroscopy will be discussed in Sect. 2.3.2.2; here we only utilize this C-V spectroscopy to control the number of electrons injected into the QRs.

Generally, there are two kinds of spectroscopies about interband transitions in semiconductors: one is absorption/transmission; the other is emission. Different from emission spectra such as photoluminescence (PL), transmission spectra can not only measure the bandgap energy but also the absorption oscillator strength. To measure the transmission spectra of neutral rings (empty rings), Pettersson et al. tuned the gate voltage to be more negative than the onset of electron tunneling. Figure 2.14 shows the resulting interband absorption spectrum for the InGaAs QRs [12]. For comparison, a InAs QD spectrum is also displayed. As shown in Fig. 2.14, three interband transitions are observed for both InAs QDs and InGaAs QRs. As for the InAs QDs, the three interband transitions are labeled as s–s, p–p, and d–d transitions, representing the interband transitions from the s, p and d states in the valence band to the s, p and d states in the conduction band, respectively [43].

For InGaAs QRs, three interband transition peaks are also observed, but they show quite different properties, which will be discussed in the following.

As for the ground state interband transition, the QRs have an energy of 1.31 eV, instead of the 1.1 eV for the QDs, which indicates that the vertical confinement potential is stronger in the QRs. This can be attributed to the fact that the QRs are thinner than the QDs in the samples studied here (the details about the QR and QD morphology can be found in Ref. [12]). Apart from the transition energy, the absorption oscillator strengths are also quite different for the QRs and QDs. To calculate the absorption oscillator strengths, four Gaussian curves are used to fit the absorption curves: three for the dot or ring resonances and one for the absorption of the WL. Because the area of each Gaussian curve is linearly related to the density of absorbers (extracted from their C-V spectra) the oscillator strengths of each transition can be determined by evaluating the Gaussian fits. For the ground state transition, the absorption oscillator strength is determined to be 10.9 and 31 for the QDs and QRs. Since the absorption oscillator strength represents the overlap between the electron and hole wave functions, the larger oscillator strength value for QRs, compared with that of the QDs indicates that the lateral confinement in the QRs is much weaker than that in the QDs, which can be explained by their larger lateral sizes. As for the excited state transitions, the QRs also exhibit different characteristics compared with the QDs. The first excited state transition of the QRs only shows smaller absorption oscillator strength (18) than the ground state transition. In contrast, the absorption oscillator strength of the first excited state transition of the QDs is almost twice as large as that of the ground state transition simply because the degeneracy is a factor of two higher. And, the first excited state transition of the QRs exhibits only a diamagnetic shift when applying a magnetic field, i.e. s state character, while the first excited state transition of the QDs splits into two in magnetic field, i.e. p-state character. Furthermore, the QRs also show different behavior in the electron occupancy compared with the QDs. The quantum states of the QRs can be labeled by two quantum numbers, m and n, where m is the quantum number of angular momentum projected along the growth direction, while n is the quantum number describing the eigenstates in the radial confinement potential. The energy separation of the states with different m but the same n can be estimated to be a few meV from the geometry of the QRs. In contrast, the separation of states with the same m but different n is ca. 20 meV for the electrons. In the interband transmission study here, transitions between different sets of radial states are observed here (see the inset in Fig. 2.14).

Since the interband transition involves electrons in the conduction band and holes in the valence band, the electron/hole occupancy in the conduction/valence band will have significant influence on the interband transitions. As discussed before, the number of electrons injected into the QRs can be well controlled by tuning the gate voltage applied to the sample, which makes it possible to study the effect of electron occupancy on the interband transitions. In the following, we will discuss how the electron occupancy affects the interband transition in a single InGaAs QR which was characterized with micro-PL spectroscopy [13]. In this study, the partial capping was used to shift the interband transitions into an energy range >1.2 eV, where sufficiently sensitive detectors are available [44] (cf. also Fig. 2.14). The micro-PL spectra of the QR was measured as a function of gate voltage V_g, which controlled the number of electrons loaded into the QR. Figure 2.15 shows the colour-scale plot of the PL data measured. At $V_g = -0.7$ V, a single, sharp peak is observed at 1.266 eV which is the emission from a single electron-hole pair in the QR. With increasing gate voltage, at $V_g = -0.6$ V the PL energy jumps to 1.26 eV, due to the trapping of an additional electron in the QR. At higher V_g, further steps can be observed in the PL energy, where each step denotes that an additional electron is added to the QR. It should be mentioned that the first jump at -0.6 V (6.0 meV in energy) represents the binding energy of a singly charged exciton (X^{1-}). The second jump represents the energy required for adding an additional electron to the X^{1-} to form X^{2-}, and so on. The first jump in energy is large, the second small, the third also small, and the fourth reasonably large. These changes in PL energy are an optical demonstration of Hund's rules for electron charging. When an electron is added with the benefit of exchange energy, such as the second and third steps, the charging energy is small. When an electron is added to complete a sub-shell, the charging energy is large.

Figure 2.15(b) shows the micro-PL spectra of the QR measured under different electron occupancy configurations. For the jump from X^{1-} to X^{2-}, a satellite appears on the low-energy side of the main PL peak. The satellite can be observed in the colour-scale plot, and is a clear feature in the individual spectra. The X^{2-} has two lines because it has two different final states [45], as shown in Fig. 2.15(b). For X^{2-}, the two possible final states present either parallel spins (a triplet state) or anti-parallel spins (a singlet state). Analogous to the excited states of the helium atom, these two states are separated by twice the exchange energy, $2X_{sp}$, and have degeneracies of 3 and 1, respectively. The splitting is of 3.6 meV, which denotes $X_{sp} = 1.8$ meV. For X^{3-} the final state is analogous to the excited lithium atom. Theoretically, the splitting is of $3X_{sp}$ and experimentally, the splitting is 1.44 times larger than that of the X^{2-}, close to the predicted 1.5. The X^{4-} state is one electron short of a filled p shell and should behave like X^{2-}. Experimentally, the splitting for X^{4-} returns almost to the value for X^{2-} but the PL peak becomes very broad, as shown in Fig. 2.15(b). At large and negative V_g, the PL signal disappears, which can be attributed to the field ionization of the excitons.

Apart from electron occupancy, magnetic field also has significant influence on the interband transition in the QRs. The InGaAs QRs for this study are grown with partial overgrowth technique, and embedded in GaAs matrix for micro-PL study under different magnetic fields, where the growth details and sample structures can be found in Ref. [15]. For surface morphology characterization, surface QRs are also grown under the same growth conditions as those of embedded QRs. Figure 2.16(a) shows the AFM image of a representative surface QR. The QR has a rim diameter of 35 nm, a height of 1.3 nm, and a center dip of ~ 2 nm. As shown by its lateral height profile in Fig. 2.16(b), the QR is anisotropic and the rim of the surface QR is higher along the [1-10] direction. This anisotropic shape will have significant influence on the diamagnetic shift of the QR.

Figure 2.16(c) shows the micro-PL spectra of a single QR measured under different excitation powers. Clearly, two PL peaks are observed, which can be attributed

Fig. 2.15 (**a**) Colour-scale plot of the micro-PL data *vs* gate voltage at 4.2 K. *Dark blue, green* and *yellow* correspond to low, medium and high signals, respectively. The main features are from a single QR. The PL signal at 1.265 eV, pronounced around 0.0 V, is from a second QR; (**b**) micro-PL spectra from a single QR for different charge states. Spectra showing at $V_g = -0.76, -0.16, -0.10, 0.40, 0.22$ and 0.50 V, correspond to the emission from the X, X^{1-}, X^{2-}, X^{3-}, X^{4-} and X^{5-} excitons, respectively. (Figures adapted from Ref. [13])

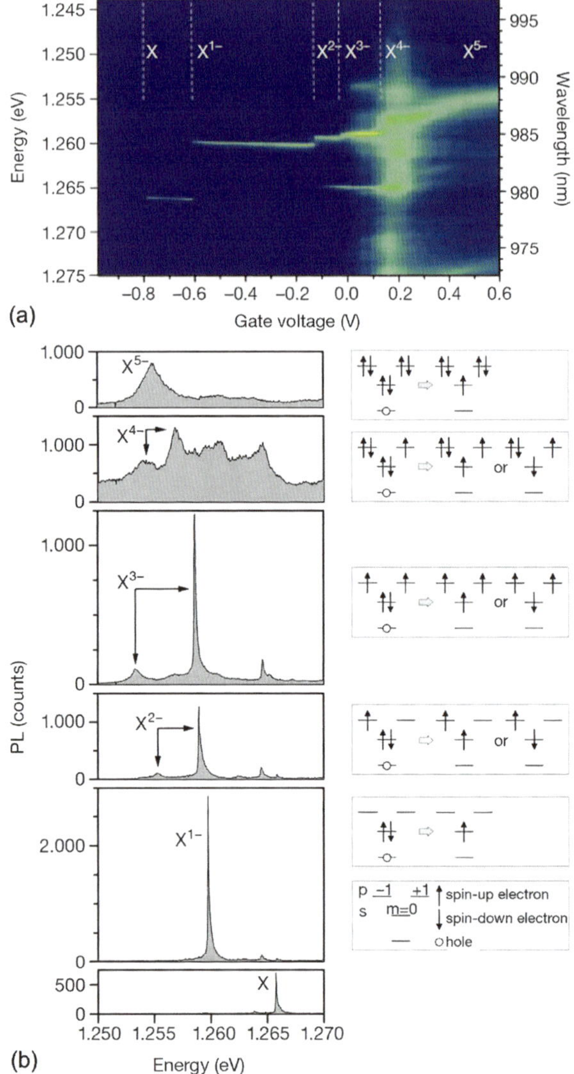

to the recombination of exciton (X) and biexciton (XX) states according to their linear and quadratic power dependence of the intensity as shown in the inset of Fig. 2.16(c). Figure 2.17 shows the micro-PL spectra of a QR measured under different magnetic fields. Because of the spin Zeeman effect both the X and XX lines split into a cross circularly polarized doublet when an external magnetic field is applied on the QR along the growth direction. Furthermore, considering the fact that the exciton state is the final state of the spin-singlet biexciton state both X and XX states have an identical energy splitting of 131 μeV/T, which corresponds to an excitonic g factor of $|g| = 2.3$. As shown in Fig. 2.17(b), the average energy of each Zeeman

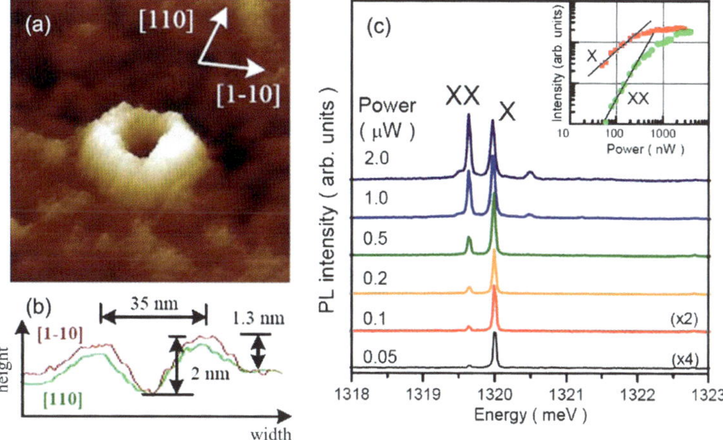

Fig. 2.16 (**a**) AFM image of the surface QR; (**b**) topographical line scans along the [110] and the [1-10] directions; (**c**) power-dependent micro-PL spectra of a single QR. The *inset* shows the integrated intensity of X and XX lines *vs* excitation power. (Figures adapted from Ref. [15])

Fig. 2.17 (**a**) Micro-PL spectra of a QR measured under different magnetic fields. (**b**) Emission peak energies of the X and XX Zeeman doublets *vs* magnetic field. *Dash lines* are quadratic fits to the averages of the Zeeman doublets. (Figures adapted from Ref. [15])

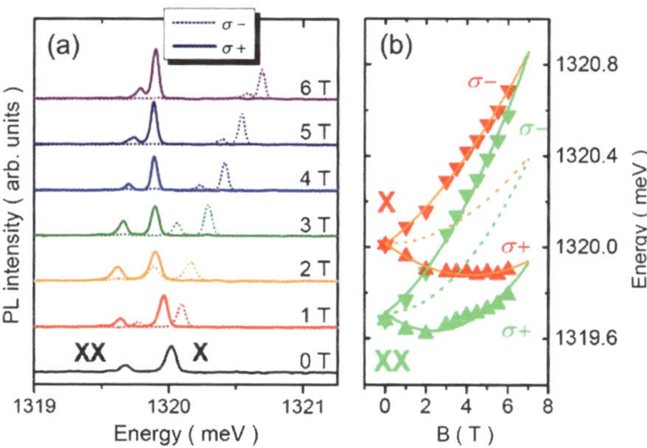

doublet presents a quadratic dependence on magnetic field B (diamagnetic shift), which can be fitted to γB^2, where γ is the diamagnetic coefficient. For the investigated QR the fitting gives an average diamagnetic coefficient of $\gamma_X = 6.8\ \mu\text{eV/T}^2$ for the X state and $\gamma_{XX} = 14.8\ \mu\text{eV/T}^2$ for the XX state.

It is interesting to observe that for the QR studied the diamagnetic coefficient of the XX state is larger than that of the X state, which is different from the case of QDs, where the diamagnetic coefficient of the XX state is usually similar to, or even smaller than, that of the X state. Since the diamagnetic coefficient γ is proportional to the area of the excitonic wave function, the large γ value for the XX state indicates that the wave function of the XX state is more extended than that of the X state in the QR. The more sensitive diamagnetic response of the XX state might be due to the fact that the QRs studied here do not have perfect azimuthal

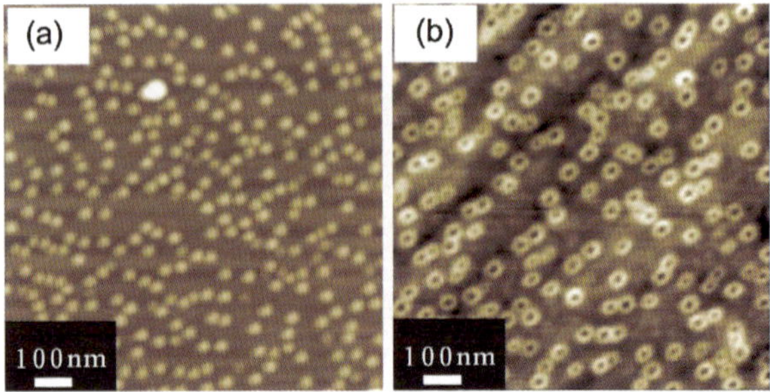

Fig. 2.18 AFM images of surface QDs (**a**) and surface InGaAs QRs (**b**) fabricated. (Figures adapted from Ref. [16])

symmetry as suggested by the anisotropic shape of the representative surface QR measured with AFM (see Figs. 2.16(a) and (b)).

In the following, another important physical characteristic associated with interband transition will be discussed—carrier lifetime. For a corresponding investigation by Lin et al., InGaAs QRs were grown *via* partial overgrowth technique and embedded in GaAs matrix for time-resolved PL study. The details of the sample growth and time-resolved PL can be found in Ref. [16]. For comparison, an InAs QD sample was also grown without the partial overgrowth process. Figure 2.18 shows the AFM images of surface InAs QDs and InGaAs QRs. The QDs have an average base diameter of about 20 nm and a height of 2 nm. The final QR shape has a base width of 60 nm, a height of 1 nm, and an inner diameter of 30 nm.

Figure 2.19 shows the micro-PL spectra and the time-resolved PL spectra of the QD and QR samples measured at different temperatures. It is observed that the ground state interband transition energies of the QD and QR are 1.21 and 1.25 eV at 15 K, respectively. The GaAs matrix and WL emission are observed at around 1.51 and 1.43 eV for both samples. Excited state transitions ($\sim$1.30 eV at 15 K) are observed for the QR, as shown in Fig. 2.19(c). Figures 2.19(b) and (d) present the PL transient spectra of the QD and QR samples measured at different temperatures. The detection energies were fixed at the PL peaks associated with the ground state interband transition at different temperatures, as indicated by dots shown in Figs. 2.19(a) and (c). The fast rise time of the order of instrumental resolution denotes that there is no phonon bottleneck effect in both the QD and QR samples studied. For the QD sample, the decay time of the ground state is about 1.1 ns at a low temperature, which drops to less than 0.5 ns when the temperature increases to room temperature. In contrast, the decay lifetime of the QR becomes longer with increasing temperature and reaches 10.5 ns at room temperature. The temperature dependence of the decay time for both samples is shown in Fig. 2.19(e). It is observed that at temperatures above 150 K, the exciton dynamics change dramatically in the QRs, and the decay lifetime increases significantly. For the excited states of QRs, they also present a similar temperature characteristic of the PL transient behavior, as shown in Fig. 2.19(f). The long lifetime of the excitons at a high temperature might be due to the thermal population of dark states, which can neither be accessed by absorbing photons nor relax to other energy states nonradiatively, competing with the exciton radiative recombination ground state. It has been reported that the overlap between the electron and hole wave functions in the QRs is reduced due to the presence of

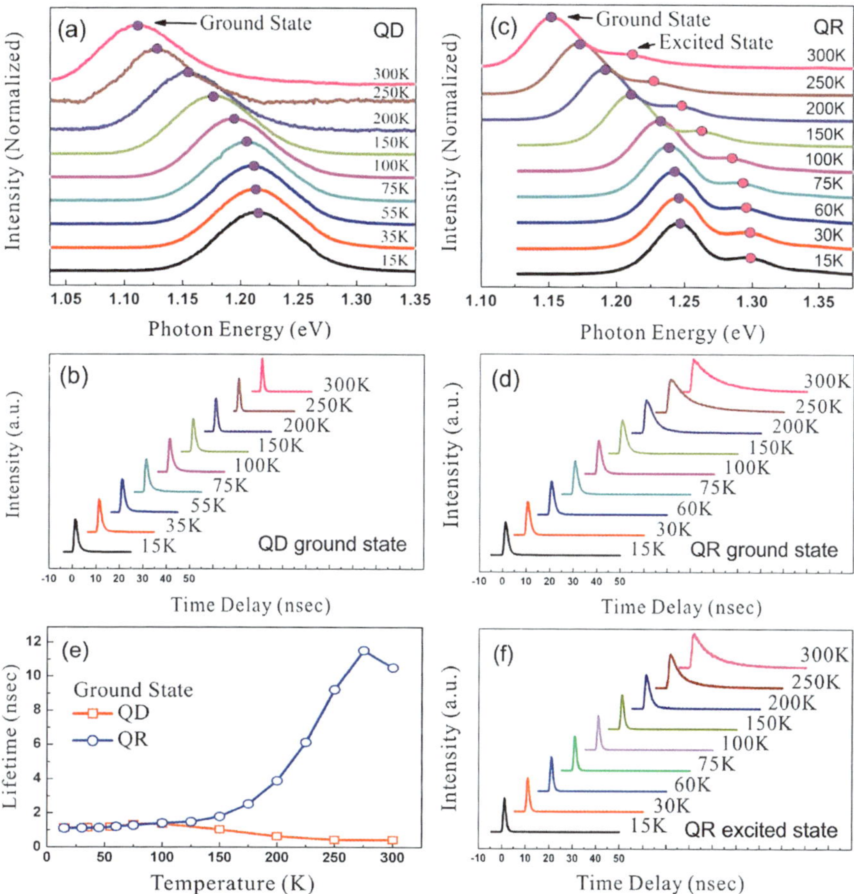

Fig. 2.19 Micro-PL spectra (**a**) and PL transient spectra (@ ground state energy) (**b**) of QDs at different temperatures; micro-PL (**c**) and PL transient spectra (@ ground state energy) (**d**) of QRs; temperature dependence of the PL decay time of the ground states in the QD and QR (**e**); PL transient spectra at first excited state energy of QRs (**f**). (Figures adapted from Ref. [16])

piezoelectric/strain potential and a large asymmetry in the QR profiles [46, 47]. This will cause a reduction in exciton oscillator strength, and thus a longer decay time is expected for QR structures. It is possible that at a temperature above 150 K, the electrons and holes tend to occupy states with reduced wave function overlap, which leads to an increase of the PL decay time in the QRs.

2.3.2 Intraband Spectroscopy

Apart from their interband spectroscopy, QRs are expected to also have interesting intraband physical properties based on their unique ring shape. In the following, we

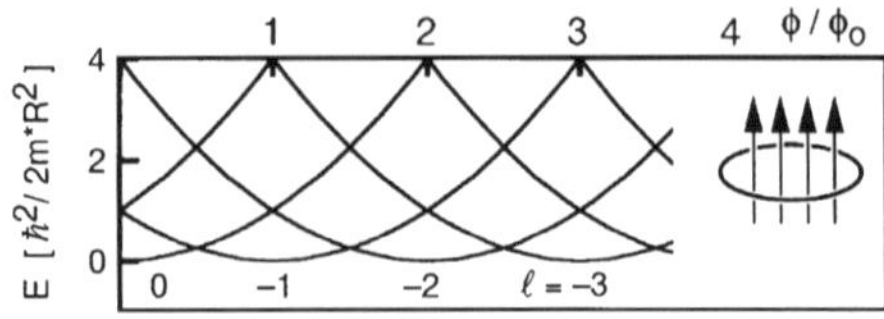

Fig. 2.20 Energy levels of an ideal one-dimensional QR as a function of the magnetic flux Ø penetrating the QR area

will review intraband spectroscopy studies of InGaAs QRs, mainly focusing on their C-V spectra and far-infrared (FIR) spectra.

2.3.2.1 Flux Quantization

Before discussing the experimental results, a simple model describing the electron states in QRs is introduced here [17]. More in-depth treatments can be found in Chaps. 4 and 7–14. The electronic states in QRs can be described using a simple model of a circular, one-dimensional wire, bent into a circle of radius R. Then, the energy levels of the QRs follow from the periodic boundary conditions to $E_l = \frac{\hbar^2}{2m^*} k_l^2$ with $k_l = l \frac{1}{R}$. When a magnetic flux $\phi = \pi R^2 B$ penetrates the interior of the QR, an additional phase is applied on the electron on its way around the QR, which leads to

$$E_l = \frac{\hbar^2}{2m^* R^2} \left(l + \frac{\phi}{\phi_0} \right)^2, \quad l = 0, \pm 1, \pm 2, \ldots, \tag{2.3}$$

with ϕ_0 being the flux quantum. Thus, with increasing magnetic field B, the ground state will change from angular momentum $l = 0$ to one with higher and higher negative l [see Fig. 2.20], a fact related to the persistent currents in mesoscopic rings [7, 48–50]. Furthermore, a periodic, Aharonov-Bohm-type oscillation in the ground state energy will occur, as shown in Fig. 2.20. Obviously, each change in the ground state will lead to a pronounced change in the possible transitions, which can be reflected in the C-V and FIR spectra, which will be discussed later.

2.3.2.2 Transport Spectroscopy (C-V)

Capacitance-voltage spectroscopy can be used to investigate the many-particle ground state energies in the conduction or valence band of QRs. For the studies discussed in the following [17–19, 51], InGaAs QRs were obtained with the partial overgrowth technique and embedded in a MISFET structure, which is schematically depicted in Fig. 2.21(a). The InAs islands (QRs/QDs) were embedded in nominally undoped GaAs, which was again sandwiched between a highly doped GaAs layer (serving as a back contact) and a surface Schottky gate. For the growth of InGaAs QRs, approximately 1.5–1.7 MLs of InAs were deposited at 520 °C to form InAs QDs which were then capped with GaAs layer of various thicknesses (1 nm and 5 nm) and annealed for 1 min to form InGaAs QRs. A typical AFM image of the InGaAs QRs obtained can be seen in Fig. 2.1(b). The InGaAs QRs have an outer

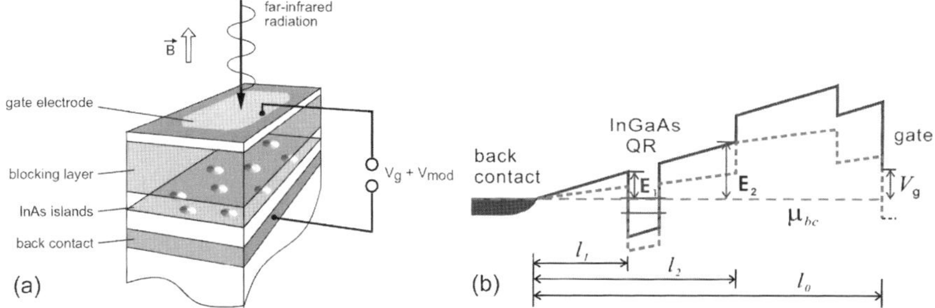

Fig. 2.21 Schematic sample structure (**a**) and conduction band structure (**b**) of the MISFET sample used for C-V spectroscopy study. The structure parameters l_1, l_2, and l_0 are the distance from the back contact to the QR layer, 2DEG (formed at the interface between GaAs spacer and AlGaAs blocking layer), and sample surface, respectively. E_1 and E_2 represent the energy difference between the Fermi energy and the GaAs conduction band edge at the position of the QR layer and the 2DEG, respectively

diameter between 60 and 140 nm, a height of 2 nm and an inner diameter of 20 nm for the well-defined center hole. The distance between the InGaAs islands and the back contact is small (25 nm here) enough to allow electron tunneling, so that in general, the islands are in equilibrium with the back contact. Because of their large distance from the Schottky gate and an inserted AlAs/GaAs superlattice blocking barrier, no charge transfer between the islands and the top gate will occur in the range of voltages investigated here. Figure 2.21(b) shows the schematic conduction band of the MISFET structure. The electric field between the back contact and the gate is given by the Schottky barrier and the applied gate voltage V_g. By changing V_g, the energy levels of the islands with respect to the back contact can be shifted. Assuming that the charge in the island layer is negligible and will not pin the conduction band, the energy shift is directly proportional to the gate voltage V_g. The proportionality factor is given by the 'lever arm' determined by the sample geometry, i.e. the ratio between the tunneling barrier thickness l_1 and total thickness l_0, $\Delta E = eV_g l_1 / l_0$ (see Fig. 2.21(b)) [14, 18]. It should be mentioned that this lever arm is not valid after the charging of InAs WL, see Ref. [14]. For appropriately chosen sample dimensions, the energy levels of the islands can be shifted so that for moderately negative voltages, the islands are void of electrons and can be charged with electrons one by one with increasing bias. The exact voltages at which additional electrons can tunnel into the islands can be monitored by superimposing upon V_g a small modulation voltage V_{mod}, which will cause the electron to oscillate between the back contact and the islands. This alternating current is detected in the outside circuit as an additional capacitive signal.

As discussed above, the electronic structure of QRs is different from that of QDs. Therefore, it is interesting to know the difference between the C-V spectrum of QRs and that of QDs. Vice versa, the unique characteristics of the CV spectra of QRs may be used for confirming the formation of QR structures inside the embedded structure. Figure 2.22 shows the C-V spectra of InGaAs nanostructures obtained with

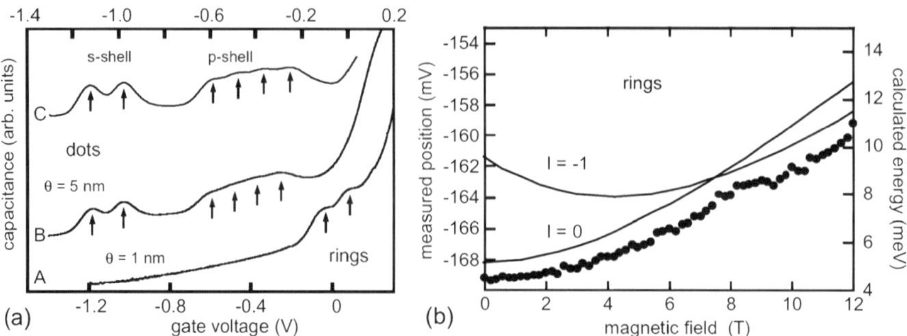

Fig. 2.22 (**a**) *C-V* spectra of InGaAs nanostructures. Curves A and B correspond to the InGaAs islands capped with 1 and 5 nm GaAs, while curve C corresponds to the InAs QDs; (**b**) magnetic field induced shift of the single electron charging peak. The *solid line* in (**b**) shows the results of single particle calculation of the ground state energy in a parabolic ring potential

GaAs capping layers of different thicknesses θ. Curve A presents the *C-V* spectrum of the QR sample with $\theta = 1$ nm. Two single-electron charging peaks (arrows) are observed just before the onset of a large increase in capacitance, caused by the formation of a two dimensional electron gas in the MISFET structure [52, 53]. For comparison, a QD sample is also studied and the *C-V* characteristic of the QD sample is also presented as curve C in Fig. 2.22 [52, 53]. For the QD sample, six single electron charging events are observed, which are grouped as s-shell and p-shell charging. It is observed that the *C-V* spectrum of the sample with $\theta = 5$ nm (curve B), that was otherwise grown identical to the $\theta = 1$ nm sample, exhibits no change in the electronic properties compared to the QD sample (curve C). This indicates that 5 nm GaAs is sufficient to completely cover the InAs QDs, so that no out-diffusion and ring formation will occur. This information is of great importance, as it is a further indication of the structural change that takes place when the growth is interrupted. The shift of the first charging peak from -1.2 V (trace B) to about 0 V (trace C) can be explained by an upward shift of the ground state energy caused by the reduced height of the QRs compared to the QDs. Figure 2.22(a) only intuitively shows that the electronic states in the $\theta = 1$ nm QR sample are *different* from those in the QD sample. The most convincing evidence for the formation of QR structure will be the unique magnetic characteristic of ring shape—Aharonov-Bohm-effect. One manifestation of the Aharonov-Bohm effect is the stepwise increase of the ground-states' angular momentum with increasing magnetic field as more and more flux quanta penetrate the interior of the QR [49–51], see Fig. 2.20.

Here, the ground-state energy of the single electron state in the QR sample can be obtained by plotting the position of the lowest charging peak as a function of the magnetic field. For comparison, the energy levels of single-particle states in the InGaAs QRs is calculated using the model developed by Chakraborty et al. [49] and a parabolic ring potential $U(r) = \frac{1}{2}m^*\omega_0^2(r - R_0)^2$ is used [17–19], where ω_0 is the characteristic frequency of the radial confinement and R_0 is the effective radius of the ring. Parameters $R_0 = 14$ nm and $\hbar\omega_0 = 12$ meV are used in the calculation to achieve the best fitting with experimental data, while the electron effective mass is assumed to be $m^* = 0.07m_e$, a value obtained from spectroscopic investigations of InAs QDs embedded in GaAs matrix [52, 54]. Figure 2.22(b) shows the charging peak energies and calculated energy levels of the QRs as a function of magnetic fields. As indicated by the arrow, at $B = 8.2$ T, corresponding to a flux of $\pi R_0^2 B \approx \phi_0$, a change in the slope can be identified in the data which is attributed to the magnetic-field-induced ground state transition from $l = 0$ to $l = -1$. Note that the shift of the charging peak is given in mV [right-hand scale in Fig. 2.22(b)]. From a comparison with the calculations (left-hand scale) a voltage-to-energy conversion factor of $f = e\Delta V_g / \Delta E = 1.8$ can be obtained. Using the lever arm model, which is quite accurate in the case of QDs [14, 17, 18], the present layer structure gives $f = 7$. This discrepancy is mainly due to the fact that the slopes at which the $l = 0$ and $l = -1$ states intersect strongly depend on the detailed choice of the confining potential. Converting the separation between the lowest charging peaks in Fig. 2.22(b) into energy, we obtain a Coulomb interaction energy of roughly 20 meV for both QRs and QDs. This similarity is somewhat surprising, considering the larger lateral size of the QRs. The missing central part, which will decrease the effective area, may partly be responsible for the large Coulomb interaction in the QRs. Furthermore, the small electronic ring diameter of about 28 nm also seems surprising, considering the large outer diameter of the islands as shown in Fig. 2.1(b). Analysis of the island height profile shows that, as in a real volcano, the highest elevation is close to the inner hole, which is roughly 20 nm in diameter. Since the electronic states are expected to be confined to the parts where the InGaAs island is the thickest, the height profile and the electronic determination of the QR diameter are in good agreement.

This C-V spectroscopy, combined with self-consistent numerical solution of the one-dimensional Schrödinger/Poisson equations, allows it to determine the electron energy levels of the QRs with respect to the GaAs conduction band edge [14]. The QR sample for this study was grown on semi-insulating GaAs (001) substrate. The active region of the sample was grown using the following layer sequence: first, a 60 nm thick, heavily Si-doped GaAs back contact layer, then 25 nm undoped GaAs as a tunneling barrier, then 1.4–1.7 ML InGaAs QR layer (first 1.4–1.7 ML InAs was deposited at 585 °C to form QDs, then the QDs were capped by 2 nm GaAs at 545 °C, after that a 30 s growth interruption was introduced to form QRs), then 30 nm GaAs spacer layer and 34 periods of a AlAs (3 nm)/GaAs (1 nm) superlattice as the blocking layer, followed by a 5 nm thick GaAs cap layer. As shown in the inset of Fig. 2.23(a), the InGaAs QRs have a average height of about 2 nm, an average outer diameter of about 49 nm, and an average inner diameter of about 20 nm. The density of these QRs is around 1.56×10^{10} cm^{-2}. The typical C-V spectrum of the QR sample studied is shown in Fig. 2.23(a). Clearly, at least four charging features can be observed, which are labeled as s_1 (0.489 V), s_2 (0.696), WL (0.821 V), and 2DEG (1.719 V). The low QR density in the present sample makes it reasonable to assume that the charge accumulated in the QRs is not sufficient to pin the conduction band. So, the linear lever arm approximation $\Delta E_1 = eV_g l_1 / l_0$ can be used up to the onset of the charging of the InAs WL to convert gate voltages into charging energies. By using this linear lever arm approximation, the energetic distance from

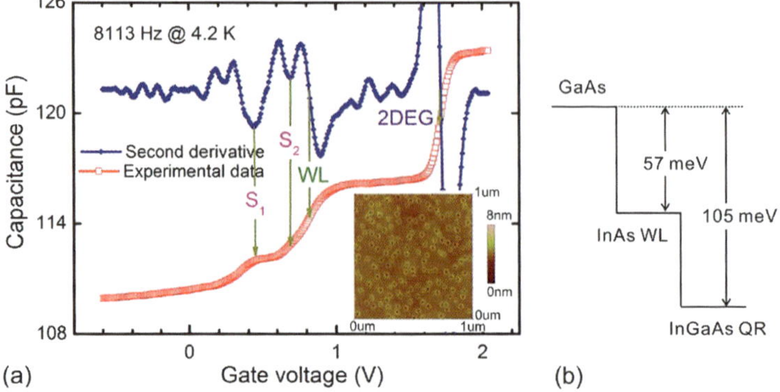

Fig. 2.23 (**a**) Typical C-V spectrum of the QRs and its second derivative. The *arrows* mark the positions of the charging events. The *inset* shows the typical AFM image of the sample surface (scan size: 1×1 μm^2 and height contrast: 8 nm); (**b**) conduction band structure of the sample determined from C-V spectra and 1D numerical simulations

the electron ground state (s_1) in the QRs to the electron ground state in InAs WL can be extracted to be 48.5 meV. Moreover, from the voltage difference between s_1 and s_2 charging peaks, the electron Coulomb blockade energy E_{ss}^C is found to be 26 meV. Because of its high density of states, the InAs WL can hold a large amount of charge, which pins the conduction band and makes the lever arm approximation invalid. By using the "1D Poisson/Schrödinger" simulation as described in Ref. [14], the energetic distance from the electron ground state of WL to GaAs conduction band edge can be found to be 57 meV with the help of the linear lever arm approximation $E_2/E_1 = l_2/l_1$ (valid up to the charging of WL). Combined with the energetic distance from the ground state of QRs to the ground state of WL extracted, the complete picture of the energy levels in conduction band of the QR sample can be obtained, as shown in Fig. 2.23(b).

2.3.2.3 Far-Infrared Absorption

The FIR absorption experiments were performed on the same sample used for the C-V measurements in Sect. 2.3.2.2. Figure 2.24(a) shows the normalized FIR transmission spectra of the sample with $\theta = 1$ nm at $V_g = 0.143$ V (upward arrow in Fig. 2.22(a)) measured with two different magnetic fields B (0 and 10 T), applied perpendicular to the plane of the QRs. Comparing the carrier density obtained from either C-V or FIR spectroscopy with the QR density determined by AFM, it is found that, at this gate voltage, each QR is filled with approximately $n_e = 2$ electrons, which means that each capacitance maximum corresponds to the filling of one electron per ring. Figure 2.24(b) shows the FIR response as a function of the magnetic field. As indicated by the different symbols, the resonances in Fig. 2.24(b) can be grouped into the following modes: two resonances (o), which degenerate at $B = 0$

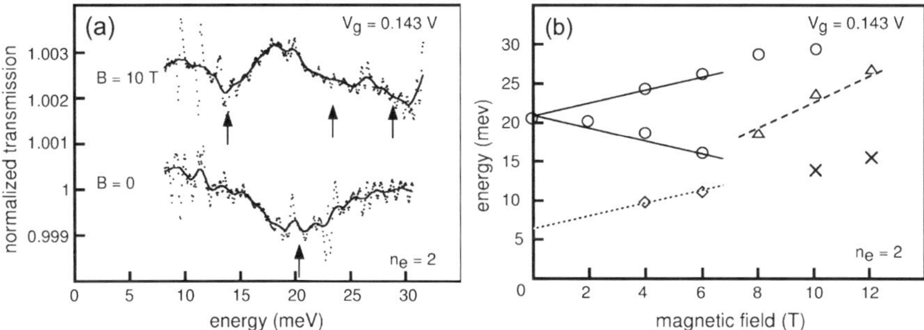

Fig. 2.24 (**a**) Normalized FIR transmission spectra of QRs (filled with $n_e = 2$ electrons each) at magnetic fields $B = 0$ and $B = 10$ T. Minima in the transmission (*arrows*) correspond to electronic excitations. The *solid lines* are a smoothed representation of the data points. *Curves* are offset for clarity. (**b**) Resonance positions *vs* magnetic field

and exhibit orbital Zeeman splitting when a magnetic field is applied and a low-lying mode ($\Diamond$) which, due to an insufficient signal-to-noise ratio at very low energies, can only be detected above 10 meV, but extrapolates to ≈ 7 meV at $B = 0$. This mode disappears at about $B = 7$ T, when also the lower o mode vanishes and a new mode ($\triangle$) appears. The resonances summarized in Fig. 2.24(b) differ quite strongly from those observed in QDs [54]. For QDs, two resonances are usually observed: one increases with increasing field, while the other decreases. More interestingly, Fig. 2.24(b) can be directly compared to the excitation spectrum of QRs, as calculated, e.g., by Halonen et al. [55]. Even though these calculations were performed for a ring with much larger dimensions, all of the above experimental features are in good qualitative agreement with the calculated energy dispersion. Furthermore, when using the effective mass of $m^* = 0.07 m_e$, the slopes of the o modes ($\pm \frac{1}{2}\hbar\omega_c$, solid lines), the $\Diamond$ mode ($\frac{1}{2}\hbar\omega_c$, dotted line), and the ($\triangle$) mode ($\hbar\omega_c$, dashed line) are all in agreement with the calculations. Therefore, it can be concluded that indeed the QR structure is formed and preserved in the structures. Combined with the C-V spectroscopy study discussed above, the change in the spectrum at about $B = 8$ T is caused by the magnetic-field induced change in the ground state. The $\times$ mode is not found in the calculated ring excitations in Ref. [54]. No sound explanation is found for this $\times$ mode, but it is possible that these resonances arise from the presence of a few large QDs which have not developed into QRs and therefore do not change their ground state at $B = 8$ T.

2.4 Summary

In this chapter, we have reviewed the recent progress of the growth and spectroscopy study of semiconductor QRs, mainly InGaAs QRs. With partial overgrowth technique, high quality self-assembled QRs can be obtained by choosing appropriate growth conditions. Also, ordered growth (both lateral and vertical) can be achieved for the QRs by engineering the strain field and thus their preferential nucleation sites in the system. Apart from partial overgrowth technique, droplet epitaxy appears as another promising growth technique for fabricating QR structures. These QR structures demonstrate some unique physical properties such as oscillation in magnetization due to their ring-like shape, in addition to the quantum confinement effect caused by their nanoscale sizes. The electron occupancy of these QRs is demonstrated to follow the rules of electron occupancy in atoms (Hund's rule). Though substantial progress has been achieved in the growth and spectroscopy study of these QRs, more work is needed to further explore the unique physical properties associated with their special ring-shape, and their applications in devices, which will be the focus of the future work.

Acknowledgement This work was supported by the Deutsche Forschungsgemeinschaft, Bundesministerium für Bildung und Forschung and the Australian Research Council.

References

1. D. Bimberg, M. Grundmann, N.N. Ledentsov, *Quantum Dot Heterostructures* (Wiley, New York, 1998) and references therein
2. P.M. Petroff, A. Lorke, A. Imamoglu, Phys. Today **54**, 46 (2001)
3. L. Wang, A. Rastelli, S. Kiravittaya, M. Benyoucef, O.G. Schmidt, Adv. Mater. **21**, 2601 (2009)
4. R. Songmuang, S. Kiravittaya, O.G. Schmidt, Appl. Phys. Lett. **82**, 2892 (2003)
5. I.N. Stranski, L. Krastanow, Akad. Wiss. Wien **146**, 797 (1938)
6. Z.M. Wang (ed.), *Quantum Dot Devices*. Lecture Notes in Nanoscale Science and Technology, vol. 13 (Springer, Berlin, 2012)
7. M. Büttiker, Y. Imry, R. Landauer, Phys. Lett. A **96**, 365 (1983)
8. J. Wu, Z.M. Wang, K. Holmes, E. Marega Jr., Z. Zhou, H. Li, Y.I. Mazur, G.J. Salamo, Appl. Phys. Lett. **100**, 203117 (2012)
9. J.M. Garcia, G. Medeiros-Ribeiro, K. Schmit, T. Ngo, J.L. Feng, A. Lorke, J.P. Kotthaus, P.M. Petroff, Appl. Phys. Lett. **71**, 2014 (1997)
10. D. Granados, J.M. García, T. Ben, S.I. Molina, Appl. Phys. Lett. **86**, 071918 (2005)
11. T. Mano, T. Kuroda, K. Kuroda, K. Sakoda, J. Nanophotonics **3**, 031605 (2009)
12. H. Pettersson, R.J. Warburton, A. Lorke, K. Karrai, J.P. Kotthaus, J.M. Garcia, P.M. Petroff, Physica E **6**, 510 (2000)
13. R.J. Warburton, C. Schäflein, D. Haft, F. Bickel, A. Lorke, K. Karrai, J.M. Garcia, W. Schoenfeld, P.M. Petroff, Nature **405**, 926 (2000)
14. W. Lei, C. Notthoff, A. Lorke, D. Reuter, A.D. Wieck, Appl. Phys. Lett. **96**, 033111 (2010)
15. T.C. Lin, C.H. Lin, H.S. Ling, Y.J. Fu, W.H. Chang, S.D. Lin, C.P. Lee, Phys. Rev. B **80**, 081304R (2009)
16. C.H. Lin, H.S. Lin, C.C. Huang, S.K. Su, S.D. Lin, K.W. Sun, C.P. Lee, Y.K. Liu, M.D. Yang, J.L. Shen, Appl. Phys. Lett. **94**, 183101 (2009)
17. A. Lorke, R.J. Luyken, A.O. Govorov, J.P. Kotthaus, J.M. Garcia, P.M. Petroff, Phys. Rev. Lett. **84**, 2223 (2000)
18. A. Lorke, J.M. Garcia, R. Blossey, R.J. Luyken, P.M. Petroff, Adv. Solid State Phys. **43**, 125 (2003)
19. A. Lorke, R.J. Luyken, J.M. Garcia, P.M. Petroff, Jpn. J. Appl. Phys. **40**, 1857 (2001)
20. A. Lorke, R. Blossey, J.M. Garcia, M. Bichler, G. Abstreiter, Mater. Sci. Eng. B **88**, 225 (2002)

21. H. Eisele, A. Lenz, R. Heitz, R. Timm, M. Dähne, Y. Temko, T. Suzuki, K. Jacobi, J. Appl. Phys. **104**, 124301 (2008)
22. W. Lei, H.H. Tan, C. Jagadish, Appl. Phys. Lett. **95**, 013108 (2009)
23. W. Lei, J. Nanopart. Res. **13**, 1647 (2011)
24. H.S. Ling, C.P. Lee, J. Appl. Phys. **102**, 024314 (2007)
25. R. Magri, S. Heun, G. Biasiol, A. Locatelli, T.O. Mentes, L. Sorba, in *Physics of Semiconductors: 29th International Conference on the Physics of Semiconductors*. AIP Conference Proceedings, vol. 1199 (2010), p. 3
26. C. Zhao, Y.H. Chen, C.X. Cui, B. Xu, J. Sun, W. Lei, L.K. Lu, Z.G. Wang, J. Chem. Phys. **123**, 094708 (2005)
27. N. Sritirawisarn, F.W.M. van Otten, R. Nötzel, J. Phys. Conf. Ser. **245**, 012004 (2010)
28. Q. Xie, A. Madhukar, P. Chen, N.P. Kobayashi, Phys. Rev. Lett. **75**, 2542 (1995)
29. W. Lei, Y.H. Chen, P. Jin, X.L. Ye, Y.L. Wang, B. Xu, Z.G. Wang, Appl. Phys. Lett. **88**, 063114 (2006)
30. G. Springholz, M. Pinczolits, V. Holy, S. Zerlauth, I. Vavra, G. Bauer, Physica E **9**, 149 (2001)
31. J. Tersoff, C. Teichert, M.G. Lagally, Phys. Rev. Lett. **76**, 1675 (1996)
32. H.X. Li, T. Daniels-Race, M.A. Hasan, Appl. Phys. Lett. **80**, 1367 (2002)
33. T. Raz, D. Ritter, G. Bahir, Appl. Phys. Lett. **82**, 1706 (2003)
34. I. Kegel, T.H. Metzger, A. Lorke, J. Peisl, J. Stangl, G. Bauer, J.M. García, P.M. Petroff, Phys. Rev. Lett. **85**, 1694 (2000)
35. V. Bressler-Hill et al., Phys. Rev. B **50**, 8479 (1994) and references therein
36. S. Herminghaus, K. Jacobs, K. Mecke, J. Bischof, A. Fery, M. Ibn-Elhaj, S. Schlagowski, Science **282**, 5390 (1998)
37. R. Blossey, A. Lorke, Phys. Rev. E **65**, 021603 (2002)
38. J.M. Ulloa, P. Offermans, P.M. Koenraad, in *Handbook of Self Assembled Semiconductor Nanostructures for Novel Devices in Photonics and Electronics*, ed. by M. Henini (Elsevier, Oxford, 2008), pp. 165–200
39. T. Mano, T. Kuroda, S. Sanguinetti, T. Ochiai, T. Tateno, J. Kim, T. Noda, M. Kawabe, K. Sakoda, G. Kido, N. Koguchi, Nano Lett. **5**, 425 (2005)
40. T. Mano, N. Koguchi, J. Cryst. Growth **278**, 108 (2005)
41. K. Watanabe, N. Koguchi, Y. Gotoh, Jpn. J. Appl. Phys. **39**, L79 (2000)
42. M. Yamagiwa, T. Mano, T. Kuroda, T. Takeno, K. Sakoda, G. Kido, N. Koguchi, F. Minami, Appl. Phys. Lett. **89**, 113115 (2006)
43. R.J. Warburton, B.T. Miller, C.S. Dürr, C. Bödefeld, K. Karrai, J.P. Kotthaus, G. Medeiros-Ribeiro, P.M. Petroff, S. Huant, Phys. Rev. B **58**, 16221 (1998)
44. R.J. Warburton, C. Schäflein, D. Haft, F. Bickel, A. Lorke, K. Karrai, J.M. Garcia, W. Schoenfeld, P.M. Petroff, Physica E **9**, 124 (2001)
45. A. Wojs, P. Hawrylak, Phys. Rev. B **55**, 13066 (1997)
46. A.O. Govorov, S.E. Ulloa, K. Karrai, R.J. Warburton, Phys. Rev. B **66**, 081309 (2002)
47. J.A. Barker, R.J. Warburton, E.P. O'Reilly, Phys. Rev. B **69**, 035327 (2004)
48. A.G. Aronov, Yu.V. Sharvin, Rev. Mod. Phys. **59**, 755 (1987)
49. T. Chakraborty, P. Pietiläinen, Phys. Rev. B **50**, 8460 (1994)
50. L. Wendler, V.M. Fomin, Phys. Status Solidi (b) **191**, 409 (1995)
51. A. Lorke, R.J. Luyken, Physica B **256–258**, 424 (1998)
52. B.T. Miller, W. Hansen, S. Manus, R.J. Luyken, A. Lorke, J.P. Kotthaus, S. Huant, G. Medeiros-Ribeiro, P.M. Petroff, Phys. Rev. B **56**, 6764 (1997)
53. H. Drexler, D. Leonard, W. Hansen, J.P. Kotthaus, P.M. Petroff, Phys. Rev. Lett. **73**, 2252 (1994)
54. M. Fricke, A. Lorke, J.P. Kotthaus, G. Medeiros-Ribeiro, P.M. Petroff, Europhys. Lett. **36**, 197 (1996)
55. V. Halonen, P. Pietiläinen, T. Chakraborty, Europhys. Lett. **33**, 377 (1996)

Chapter 3
0D Band Gap Engineering by MBE Quantum Rings: Fabrication and Optical Properties

Jorge M. García, Benito Alén, Juan Pedro Silveira, and Daniel Granados

Abstract In this chapter we show how it is possible to modify the shape and size of InAs on GaAs self assembled quantum dots grown by Molecular Beam Epitaxy (MBE) by introducing a pause during the capping process, also known as partial overgrowth technique (García et al. in Appl. Phys. Lett. 71:2014, 1997; Appl. Phys. Lett. 72:3172, 1998; Granados and García in Appl. Phys. Lett. 82:2401, 2003). Under certain growth-pause capping conditions it is possible to obtain self-assembled quantum rings. The changes in shape and size lead to a modification of the quantum confinement potential and enables the control over fundamental physical properties, such as the optical emission energy from ground or excited states, the magnitude of its fine structure splitting or the sign of its permanent electric dipole moment.

The partial capping technique has played a key role in the engineering of 0D nanostructures with tailor made properties (Michler et al. in Science 290(5500): 2282, 2000; Kiravittaya et al. in Rep. Prog. Phys. 72(4):046502, 2009). For example, it has allowed to fabricate a single-photon source that is based on a single self assembled quantum nanostructure embedded in a high-quality factor microcavity structure (Michler et al. in Science 290(5500):2282, 2000). Another example is the possibility to engineer what has been called "the smallest rings of electricity" (see Chap. 2), which unveil novel magnetic properties associated to non-trivial topologies at the nanoscale (Fomin (ed.) in J. Nanoelectron. Optoelectron., vol. 6. American Scientific, 2011) (see Chaps. 2, 4, 14, 17, and 18).

J.M. García (✉) · B. Alén · J.P. Silveira · D. Granados
Instituto de Microelectrónica de Madrid, Calle Isaac Newton, 8, 28760 Tres Cantos, Madrid, Spain
e-mail: jm.garcia@csic.es

B. Alén
e-mail: benito@imm.cnm.csic.es

J.P. Silveira
e-mail: juanpi.silveira@gmail.com

D. Granados
IMDEA Nanociencia, Calle Faraday, 9, Ciudad Universitaria de Cantoblanco, 28049 Madrid, Spain
e-mail: daniel.granados@imdea.org

V.M. Fomin (ed.), *Physics of Quantum Rings*, NanoScience and Technology,
DOI 10.1007/978-3-642-39197-2_3, © Springer-Verlag Berlin Heidelberg 2014

Typically, the formation of uniform and high quality self-assembled quantum dots, requires fixed growing parameters that does not allow to control independently the size, shape and overall density of the ensemble. The partial capping technique allows to have two separate sets of growing parameters, or "control knobs": one for optimum QD nucleation and another employed for tuning size and shape during partial capping.

It is well known that the accumulated stress during growth of heteroepitaxial materials systems can lead to the self-assembly of quantum dots. But these driving forces are the very same ones responsible for a disassembling process that takes places during the capping process. Embedding an ensemble of elastically relaxed islands into a matrix material with a smaller lattice parameter, puts into play forces that compete dynamically with the capping process. Some examples of these processes are: atomic segregation, material interchange, surface reconstructions changes, stress-induced melting and de-wetting. The islands on the surface (either pyramid-, dome- or lens-shaped), will not preserve intact their structure after being capped. That is why it is so crucial to understand and control in detail the growth and embedding mechanisms of stressed materials. A way to achieve atomic level control of these dissociation mechanisms is to introduce a growth pause during capping to let the system relax.

This chapter focuses on the understanding and control of the capping process of quantum dots, and how under certain capping conditions, it is possible to obtain quantum rings and other nanostructures with quantum properties engineered at will.

We show *in situ*, real time, accumulated stress and Reflection High Energy Electron Diffraction (RHEED) measurements during InAs on GaAs(001) growth that shed light on the complicated processes that take place during growth and capping of lattice mismatched, and therefore strained, nanostructures. The experiments show that a large amount of indium melts due to the stress accumulation. This highly mobile material plays a key role on the transformations of self assembled nanostructures. For example, this liquid indium strongly segregates during the capping with GaAs, resulting in asymmetrical final soft barrier potentials.

We present a model to explain the formation of quantum rings under certain partial capping conditions which takes into account the competition processes between de-wetting [see Chap. 2], stress-induced melting of InAs and In/Ga exchange/alloying. We present Atomic Force Microscope (AFM) results of nanostructures with different shapes obtained under various partial capping conditions. The role of As_2 in the final formation of rings is also discussed.

Changes in the size and shape of the self-assembled nanostructures induced during the partial capping process allows additional 0D band gap engineering that will change accordingly their optical properties. This is based on measurements of continuous wave photoluminescence (PL), time-resolved PL (TRPL) and photoluminescence excitation (PLE) obtained both, in QRs ensembles, and in single quantum rings. Nanostructures capped under different conditions and with different emission energies are compared. As the diameter to height ratio increases, the radiative lifetime and the splitting between ground and excited states decreases leading to sizeable effects in the recombination dynamics as a function of the temperature.

We show that a smaller height and a strong In/Ga exchange induced by the partial capping process are responsible of the reduction of the electrical polarizability and the inversion of the permanent dipole moment in these QRs. Finally, the voltage dependent micro-PL and micro-PLE spectra of charge tunable QRs are presented and analyzed in the framework of a central parabolic confinement model including coulomb interactions in the strong confinement regime.

3.1 Introduction

The fabrication and physics of self-assembled quantum dots (QD) [7–9] has been intensively studied over the last three decades. Their discrete density of states and rather large confinement and Coulomb interaction energies have made them a realm of wealth for solid state quantum mechanics and for applied physics. Actually, the number of applications is growing continuously in many fields like: quantum information technologies and new generation of photovoltaic devices.

Quite early it was demonstrated that it is possible to modify the shape and size of self assembled quantum dots grown by Molecular Beam Epitaxy (MBE) (MBE) by using a partial capping process [1–3]. The changes in shape and size lead to the fabrication of self assembled quantum rings(QRs). Like QD, self-assembled quantum rings confine the electrical carriers in all three spatial directions and therefore are similarly characterized by a discrete density of states [10]. Such modifications of the quantum confinement potential of the nanostructures enable the control over fundamental physical properties, such as the optical emission of the ground and excited states [11], the magnitude of its fine structure splitting and its oscillator strength or the sign of its permanent electric dipole moment [12, 13].

The partial capping technique has played a key role in the engineering of 0D nanostructures with tailor made properties [5]. For example, it has allowed to fabricate a single-photon source that is based on a single self assembled quantum nanostructure embedded in a high-quality factor microcavity structure [4]. Another example is the possibility to engineer what has been called "the smallest rings of electricity" (see Chap. 2), which unveil magnetic properties associated to non-trivial topologies at the nanoscale [6, 11, 14] (see Chaps. 2, 4, 10, 14 and 18). The experimental observation of the Aharonov and Bohm (AB) [15] effect demonstrate the true quantum nature of these rings.

In this chapter we show that a strong segregation of stress-induced melting of InAs during growth of InAs on GaAs is responsible for the formation of quantum rings under certain partial capping conditions. We focus the discussion of the fabrication process based on accumulated stress measurements, Reflection High Energy Electron Diffraction (RHEED) observation and Atomic Force Microscope (AFM) measurements. Changes in the size and shape of the self-assembled nanostructures induced during the partial capping process will change their optical properties accordingly. To show that, we present results on continuous wave photoluminescence (PL), time-resolved PL (TRPL) and photoluminescence excitation (PLE) obtained both, in QRs ensembles, and in single quantum rings.

3.2 Fabrication of Quantum Dots and Quantum Rings

Typically, the formation of uniform and high quality self-assembled quantum dot requires fixed parameters that do not allow to control independently the size, shape and overall density of the ensemble [16]. The partial capping technique allows to have two separate sets of growing parameters, or "control knobs": one for optimum uncapped islands (QD) nucleation and another employed for tuning size and shape during partial capping.

It is well known that the accumulated stress during growth of heteroepitaxial materials systems can lead to the self-assembly of uniform size distribution of In-GaAs islands that after further capping behave as quantum dots. But these driving forces are the same ones responsible for a disassembling process that dramatically takes places during capping. Embedding an ensemble of elastically relaxed islands into a matrix material with a lattice mismatch, sets into play some non-trivial disassembling mechanisms like: atomic segregation [17], surface exchange between column-III atoms [18], surface reconstructions changes [19], stress-induced melting [20] and de-wetting [21].

The islands on the surface (either pyramid-, dome- or lens-shaped) [22], will not preserve their structure after being capped. That is why it is so crucial to understand and control in detail the growth and embedding mechanisms of stressed materials. A way to achieve atomic level control of these dissociation mechanisms is to introduce a growth pause during capping and let the system reorganize its configuration [2, 3].

We take a unique point of view to study the formation of 0D self assembled nanostructures. As the driving force for their formation is the accumulation of stress and its subsequent partial relaxation, we quantify by direct, *in situ*, and real-time measurements the accumulated stress during formation of InAs QD on GaAs(001). As it usually happens, a new technique sheds new light on the understanding of these processes and led to the discovery some unexpected mechanisms in action during growth of InAs on GaAs[001]. The experiments show that a large amount of indium is in a strain-free state, melted, due to the stress accumulation [20, 23].

This highly mobile material plays a key role on the transformations of nanostructures during capping. Two examples show how relevant it is this process. In one case, during growth of single layer InAs quantum well on GaAs, liquid indium/InAs strongly segregates during capping with GaAs, resulting in vertically asymmetrical soft barrier potentials [20, 24]. The second case is a stress-induced melting of InAs islands grown on GaAs that allows the formation of self assembled quantum rings(QRs) [3].

3.2.1 Experimental Set Up

An *in situ* characterization technique commonly found in MBE growth equipment is Reflection High Energy Electron Diffraction (RHEED) (see Fig. 3.1 and inset to

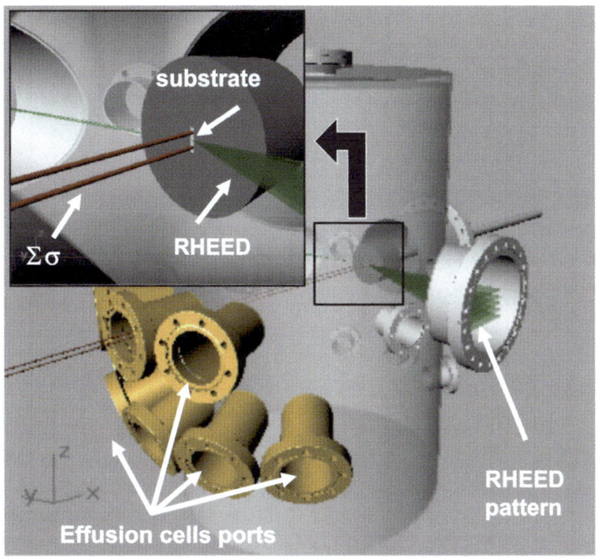

Fig. 3.1 Experimental configuration of the MBE at IMM. During growth, two parallel laser beams (*red lines*) impinge normally to the surface of a 100–150 µm-thick cantilever-shaped substrate. Simultaneously, the RHEED pattern can be monitored (*green lines*) and be used for accurate control of substrate temperature

Fig. 3.1). Somewhat rarer is a MBE with other *in situ* characterization technique as Differential Reflectance (DR). But what is really unique of the MBE at the Institute of Microelectronics of Madrid, (IMM) is that it is equipped with RHEED, DR and *in situ accumulated stress ($\sum \sigma$)* measurement that allows a wide range of *in situ* studies.

In situ, real time measurement of $\sum \sigma$ during heteroepitaxial MBE growth is performed by direct determination of strain induced substrate curvature using a laser deflection technique. This technique consists of obtaining the stress accumulated on the sample by measuring its bending, using to that aim the deflection of two laser parallel beams impinging on the sample surface (Fig. 3.1). The sample must be lever-shaped, with a sufficient small thickness (100–150 µm) to allow its bending during growth of strained layers. We use two separate substrates oriented along the principal crystallographic directions ([110] and [1$\bar{1}$0]) to get information about the anisotropic stress effect [25]. The sample is clamped to one of its end to a special sample holder (Fig. 3.2a) with an aperture in the center that allows its bending. The separation of the reflected spots at a certain distance from the sample surface is measured with a photodiode or a CCD camera [26] (Fig. 3.2b, c, d), and then it can be related to the changes in total stress accumulated ($\sum \sigma$) through Stoney's equation [27, 28]. This method is easy to implement in a molecular beam epitaxy (MBE) reactor if there is optical access normal to the sample, since all the optics and detection apparatus are outside the vacuum chamber.

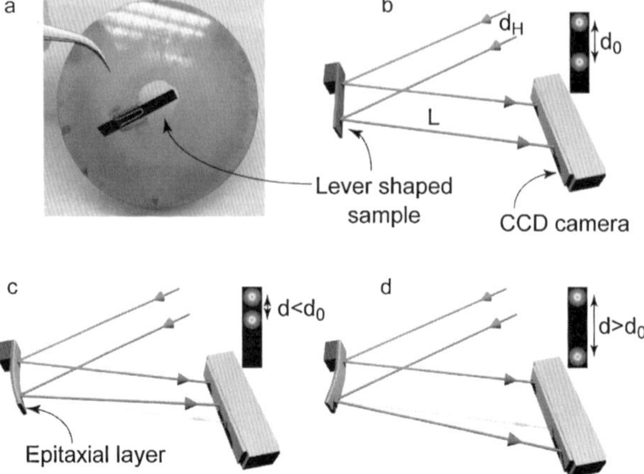

Fig. 3.2 (**a**) The sample holder has a hole in the center that allows a cantilever-shaped thin substrate to bend freely during growth. (**b**) Two parallel laser beams impinging on the substrate allow the measurement of changes of the radius of curvature (**c**, **d**), and therefore, a direct quantification of the accumulated stress

3.2.2 Segregation of Indium During Growth of One Monolayer of InAs on GaAs

Segregation of In during growth by MBE of $In_xGa_{1-x}As$ quantum wells embedded in a GaAs matrix is an issue that has been known for a long time [29] and that has a tremendous impact in the final quality of the quantum wells.

We have performed *in situ* measurements of $\sum \sigma$ during the growth of the simplest material system: one monolayer of InAs on GaAs. For these experiments, a dose of In atoms sufficient to grow one nominal monolayer of InAs (1 In ML) was supplied at different $T_{\text{substrate}}$ from 170 °C to 520 °C [23]. During deposition of 1 equivalent ML of In, an approximately linear and isotropic increase of accumulated stress is observed, as expected for an isotropic mismatched layer. Figure 3.4 shows accumulated stress evolution at $T_{\text{substrate}} = 470\,°C$.

Furthermore, and very surprisingly when it was observed for the first time, during capping of InAs with GaAs, it is measured a progressive increase of $\Sigma\sigma$ until a final steady stress state is reached (see Fig. 3.3). Interpretation of this observation is clear if we consider that our experimental method is sensitive to any material that induces strain/stress during deposition: only a fraction of the delivered In is incorporated epitaxially into the lattice as InAs. There is some indium material on the surface that is not contributing to the total accumulated stress. As some works based on thermodynamical calculation of Gibbs free energy [30] has shown, it can be assumed that this material is in a liquid state.

As schematically depicted in Fig. 3.4(a), there is an accumulation of a liquid phase [l] of InAs (or In) on the surface, not contributing to the increase of stress. During subsequent GaAs capping this liquid indium is progressively incorporated

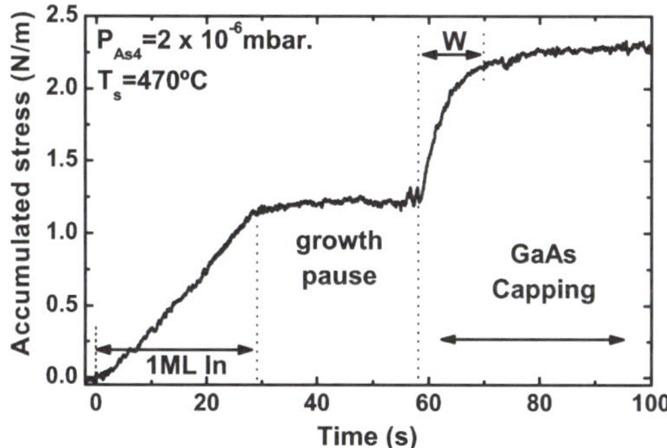

Fig. 3.3 Accumulated stress evolution ($\sum \sigma$) during 1 InAs ML deposition, growth interruption and subsequent GaAs capping. The presence of a liquid [s] and a solid phase of InAs [s] (see Fig. 3.4) is necessary to account for the fact that the cantilever GaAs(001) substrate bends during InAs deposition, and bends further during GaAs capping [30]. Segregation length (W) is defined in the figure and can be easily obtained from the time evolution during GaAs capping

Fig. 3.4 (a) During deposition of 1 ML of InAs on GaAs a large fraction of supplied indium stays in a stress-free liquid phase [l], in coexistence with patches of pseudomorphic solid InAs [s]. (b) During subsequent GaAs capping this liquid indium is progressively incorporated into the GaAs matrix until total In exhaustion. Segregation length (W) is also depicted in the figure

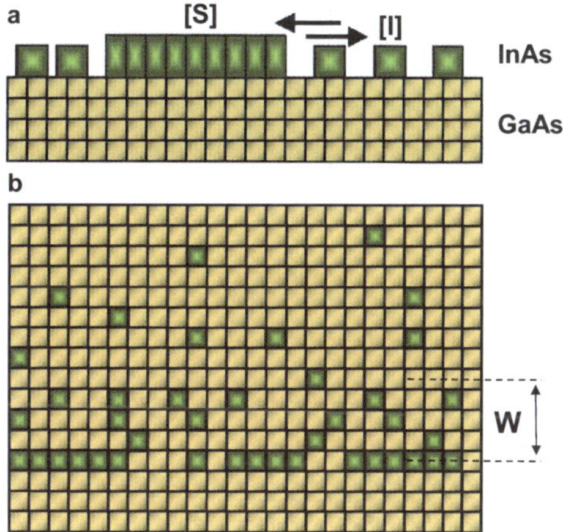

into the GaAs matrix and increases stress until total In exhaustion. Actually, this method provides a simple way to measure directly the segregation length (W) (see Figs. 3.3 and 3.4) of In during fabrication of In$_x$Ga$_{1-x}$As Quantum Wells.

The fraction of stress-free In can be evaluated from the ratio between partial accumulated stress before and after capping with a thick GaAs layer. We find that at typical QD growth temperatures ($T_{substrate} > 450\,°C$), as much as $\sim$50 % of the supplied In stays in a liquid phase [20, 23]. This large amount of stress-induced segregated liquid In strongly modifies the commonly assumed picture for QD self-assembling process, and it has tremendous impact on the final size and shape of the confinement potential after the nanostructures are capped, as is discussed is Sect. 3.2.4.

Liquid indium on a substrate surface is not such a "rara avis" as it may look. Actually, another fabrication method of self assembled quantum nanostructures, and in particular rings [31], is droplet epitaxy and it relies in the presence of liquid III

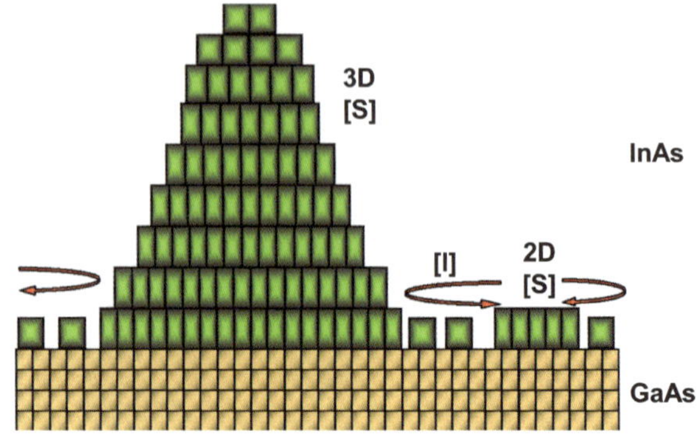

Fig. 3.5 During InAs QD growth a large fraction of supplied indium stays in a stress-free liquid phase ([l]), in coexistence with QD (3D[s]) and patches of pseudomorphic solid InAs (2D[s]). The tip of the dots is largely relaxed

element on the surface. This method, (as described in Chap. 8) consists of forming group III liquid metal droplets on the substrate surface by supplying their pure molecular beam. Droplets form on the surface based on a Volmer-Weber growth mode. The nanostructures are then formed by being exposed to a group V element (arsenization, phosphorization, antimonization).

3.2.3 Formation of Quantum Dots

The nucleation of 3D InAs islands on GaAs is certainly a complex scenario and it is schematically depicted in Fig. 3.5. When the QD are formed, the surface must be in a quasi-equilibrium of three phases: 2D InAs solid islands (2D[s]), InAs islands (3D[s]) and liquid/stress-free In[l]. Due to the large lattice mismatch between InAs and GaAs, strain energy competes very efficiently with chemical bonding energy, which is controlled by surface In supersaturation. This competition determines the equilibrium ratio between pseudomorphic InAs and unbonded In.

The appearance of QD (with smaller accumulation of stress due to tip relaxation) displaces previous equilibrium and reduces the observed accumulated stress rate as a function of delivered In (see inflection point labeled as "QD formation 3D RHEED" in Fig. 3.6). However, quantitative analysis of stress relaxation due to QD formation should take into account large mass transfer process between migrating liquid In, InAs 2D solid islands and solid InAs 3D islands and has not been fully addressed yet.

The large increase of stress observed during GaAs capping (Fig. 3.6) is partially due to the incorporation of a large amount of liquid In, as has been discussed in Sect. 3.2.2. During capping of the islands, there is a strong increase of stress on the basal plane of the InAs islands that will induce a QD melting. During further capping, Indium incorporates in the form of an $In_xGa_{1-x}As$ alloy with a gradually decreasing x composition. Strong In-Ga intermixing has been identified to be responsible for QD size change during capping [1, 32].

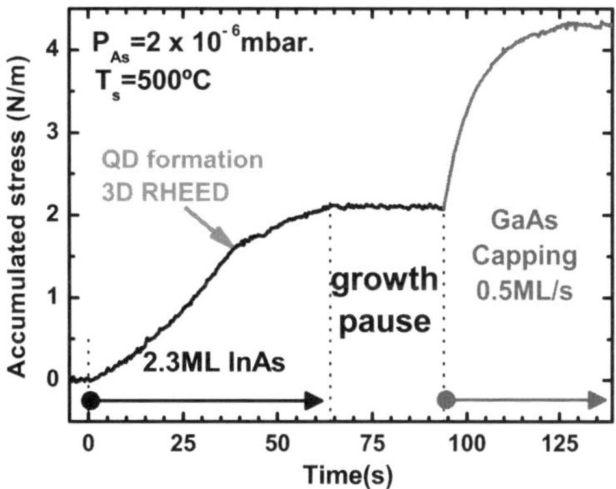

Fig. 3.6 Accumulated stress evolution during growth of 2.3 InAs ML deposition, growth interruption and subsequent GaAs capping. A clear reduction of stress increase rate is observed when QDs start to nucleate, in coincidence with the appearance of a 3D RHEED pattern

3.2.4 Formation of Quantum Rings

Control over these processes allows manipulation of shapes and sizes of ensembles of nanostructures, and in particular, the creation of self assembled quantum rings.

Figure 3.7(a) shows an AFM image of QDs with a full width at half maximum (FWHM) vertical size distribution of 9 % centered at 10 nm. When a 2 nm cap is deposited at 540 °C under 4×10^{-6} mbar. As$_4$ beam equivalent pressure (BEP) each QD is transformed into an elongated dash-like nanostructure (Fig. 3.7(b)). Cap deposition at lower substrate temperatures, $T_{cap} = 500$ °C, produces two camel hump-like nanostructures (Fig. 3.7(c)). Figure 3.7(d) shows quantum rings formed at $T_{cap} = 500$ °C under As$_2$ flux. It is worth mentioning that we never obtained QRs for ensembles of QDs with vertical dimensions measured by AFM less than 7 nm. Next we propose a possible mechanism for ring formation (see Fig. 3.8).

Some authors have suggested [21, 33] that a change in the balance of surface free energy due to the thin capping layer creates an outward pointing de-wetting force. This force brings about material redistribution and results in a ring shaped structure. There are theoretical results [30] showing that a biaxial epitaxial stress of InAs on GaAs [001] leads to a mixture of stress-free matter including a liquid phases of In and/or InAs. Moreover, this effect has been observed [23] to be independent of T in the 170–520 °C range. The driving force for this melting is the elastic energy. It is expected that when a relaxed QD is partially covered, and consequently compressed, it undergoes a phase transition into liquid phase, and allows a de-wetting process [21] that expels highly mobile InAs from the QD.

In order to explain the different morphologies observed in Fig. 3.7, we have to consider that there is also a simultaneous Ga-In alloying process during partial capping of the dots. In Fig. 3.8 a scheme of the proposed mechanisms involved is depicted. The presence of liquid InAs is mainly a T independent process, whereas Ga–In alloying is strongly T dependent [34] and quenches the mobility of In. The competition between these two processes at different capping temperatures can explain our experimental observations. Although liquid InAs must have very high mobility, at large temperatures ($T_{cap} = 540$ °C) a strong In-Ga alloying process results in the

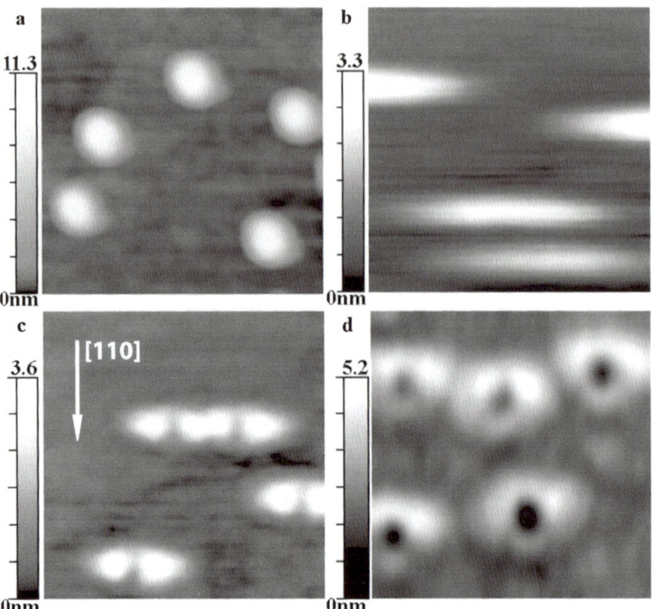

Fig. 3.7 250 nm × 250 nm AFM images of ensembles of (**a**) QDs grown at $T_{\mathrm{substrate}} = 540\,^\circ\mathrm{C}$ under As$_2$. (**b**) Dash-like nanostructures obtained when the QD are capped with 2 nm of GaAs at $T_{\mathrm{cap}} = 540\,^\circ\mathrm{C}$ under As$_4$. (**c**) Camel hump-like nanostructures obtained using a 2 nm capping of GaAs at $T_{\mathrm{cap}} = 500\,^\circ\mathrm{C}$ under As$_4$ and (**d**) quantum rings obtained after capping with 2 nm of GaAs at $T_{\mathrm{cap}} = 500\,^\circ\mathrm{C}$ employing As$_2$. Crystallographic orientation is the same for all images

Fig. 3.8 Scheme of various process at play during self-assembled quantum dot (QD) and quantum ring (QR) fabrication. (**a**) Tall QDs have a stressed basal plane and a relaxed tip. A thin cap induces an increase in the stress of both the base and the tip. The tip will suffer a stress-induced melting. (**b**) At $T_{\mathrm{substrate}} = 500\,^\circ\mathrm{C}$, under As$_2$ pressure, QRs are formed when there is an enhanced isotropic de-wetting of the melted tip of the island and a reduction of alloying at the basal plane of the dot

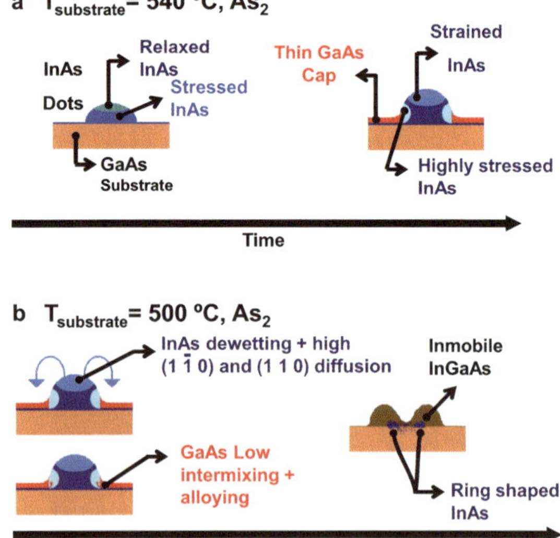

formation of an immobile InGaAs solid phase at the base of the dot, resulting in a dash-like nanostructure (Fig. 3.7(b)). For lower cap temperatures ($T_{\mathrm{cap}} = 500\ ^\circ$C), GaInAs alloying is reduced in the base of the dot, so there is large fraction of InAs that melts under compression and allows de-wetting of center region. This leads to the formation of a depleted region surrounded by immobile InGaAs. This growing conditions produces camel hump-like nanostructures (Fig. 3.7(c)). These elongated nanostructures (Fig. 3.7(b) and (c)) clearly shows a large anisot.opy that can be related to strain anisotropy, to chemical anisotropy or to anisotropic In-diffusion. Our measurement along [110] and [1$\bar{1}$0] crystallographic directions suggest that there are no strain anisotropy effects.

The observation of quantum ring formation under As_2 is related to an enhancement of the reactivity along the [110] direction [35]. For example, during GaAs homoepitaxy, it is well known that the migration distance of both principal directions is different due to the anisotropic chemical reactivity of the steps [35]. Arsenic terminated B steps (perpendicular to the [1$\bar{1}$0] direction) provide highly active sites for gallium adsorption whereas gallium terminated A steps are less reactive to group III migrating atoms. The structures elongated along the [1$\bar{1}$0] direction usually found when growing by MBE and step flow growth mode on B type vicinal surfaces are attributed to this chemical reactivity anisotropy. The step flow of A-type facets can be achieved for higher As pressures [35]. If we assume similar chemical behavior of the original InAs island steps, an increase of As pressure will produce enhancement of the reactivity of A steps to Ga atoms. The use of As_2 then has similar effects to employing a larger As_4 flux [36]. The chemical reactivity of the surface during ring formation has been monitored by RHEED. At the temperature at which the partial capping is deposited over the QD layer, the GaAs surface shows a C(4×4) RHEED. In this reconstruction, A and B steps present similar chemical reactivity, and therefore an isotropically diffusion for Ga and In atoms is expected. This isotropic reactivity is responsible of the final ring morphology. Immediately after the partial capping, it is observed a (2×3) RHEED pattern, characteristic of an InGaAs alloy.

Once the material redistribution has taken place during the growth pause, these nanostructures preserve their structure during further capping, as demonstrated by cross section transmission electron microscope (TEM) [37] and scanning tunneling microscope (STM) cross section measurements [38] (see Chap. 4). The growth pause during partial capping enhances the formation of InGaAs alloy, and reduces concentration of large amounts on In in some parts of the InAs islands. This two facts contribute to preservation of the quantum ring shape during subsequent capping.

In spite of the tremendous morphological transformation that is taking place during ring formation; accumulated stress measurements (Fig. 3.9) show no changes in the absolute value during ring formation. It may be possible that the reorganization process of the material is taking place faster that the temporal resolution of $\sum \sigma$ measurement technique. Although our data show no changes during material reorganization, this growth pause has drastic consequences in the final confinement potential, as it is thoroughly discussed in Sect. 3.3.

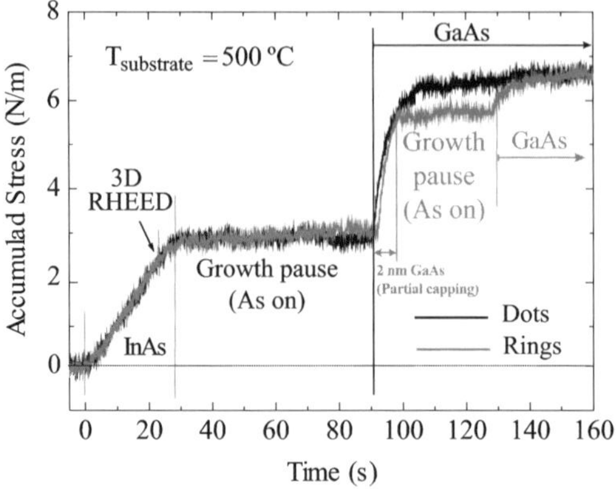

Fig. 3.9 Accumulated stress $\Sigma\sigma$ measurement during growth of InAs for Island formation and its subsequent QD (*blue line*) and QR (*red line*) formation. The amount of deposited InAs is the same in both growths. The total accumulated stress measurement is the same in both cases. The rings are formed during the second growth pause (introduced only for the ring formation) and no significant changes in $\sum\sigma$ are observed

3.3 Optical Properties

The electronic and optical properties of self-assembled quantum dots (QD) have been intensively studied over the last years using both, spatially averaged (QD ensemble), and spatially resolved (single QD) spectroscopies. Their discrete density of states and rather large confinement and Coulomb interaction energies have made of QD a joyful playground for solid state quantum mechanics. This in turn have made the number of applications in quantum information technologies, and in other fields, like new generation photovoltaics, to grow very rapidly.

Like QD, self-assembled quantum rings (QRs) confine the electrical carriers in all three spatial directions and therefore are similarly characterized by a discrete density of states. However, QRs have attracted the interest of the scientific community mainly because, as established by Aharonov and Bohm (AB), their quantum mechanical properties are periodic in the magnetic field flux penetrating it [15]. Semiconductor QRs made of III–V compound materials benefit of strong interband optical transitions which can be exploited to observe and manipulate such purely quantum interference effect. To this respect, it has been found that the existence or absence of a different radial confinement for electrons and holes, or the possibility to control the QR charge are of paramount importance for the observation of the AB oscillations in magnetic field (see Chaps. 9–12 and references therein) [6, 39–49].

In the previous section, the growth mechanisms that allow us to tailor the shape of quantum nanostructures from QD to QR have been discussed. In the following, some of their optical properties at zero magnetic field will be presented and compared at the ensemble and single nanostructure level. When no magnetic field is applied, the shape and composition determine the electronic structure of the nanostructures in the single particle picture. Realistic models have been drawn taking into account the

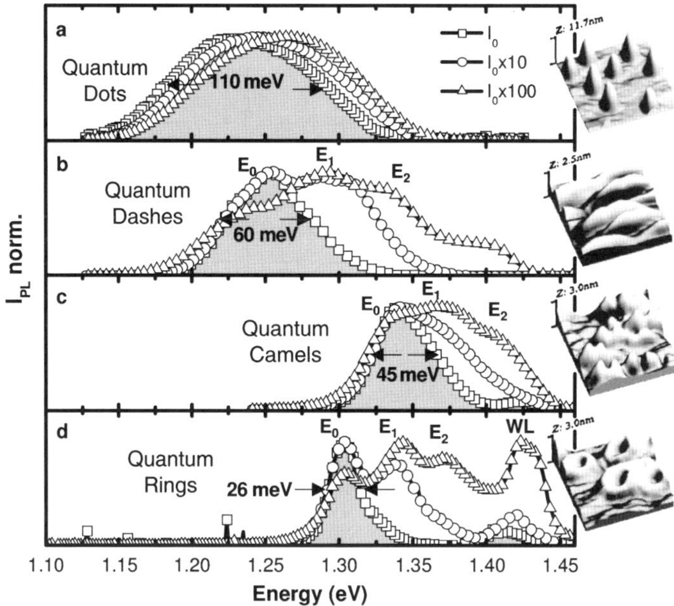

Fig. 3.10 Normalized photoluminescence spectra obtained at 20 K and different excitation powers

true confinement geometry [50, 51], and refinements like strain [52, 53], piezoelectric fields [52], or valence band mixing contributions [54] (see Chaps. 13–18 and references therein). This notwithstanding, several authors have achieved considerable success catching only the overall ring potential geometry with 2D parabolic-like models, but at the expense of some fitting of the model dependent parameters to the experimental data [6, 11, 40, 42, 55–57].

3.3.1 Shape Dependent PL and TRPL

To establish a complete picture linking QDs and QRs, it seems desirable to compare the optical properties of nanostructures which exhibit different shapes. As explained in Sect. 3.2.4, the growth conditions can dramatically affect the morphology of nanostructures which are otherwise made of the same constituent materials, namely InAs over GaAs(001). Figure 3.10 shows ensemble photoluminescence (PL) spectra measured for increasing excitation densities in four different samples. Each sample contains nanostructures with a different morphology, as shown in the nearby AFM images.

At low excitation densities, the emission finds its maximum, E_0, at 1.23 eV for QD, 1.25 eV for quantum dashes (QDh), 1.34 eV for quantum camels (QC), and 1.30 eV for QR. The large inhomogeneous broadening of the QD sample must be associated to the dispersion of QD heights observed by AFM (7 ± 3 nm and aspect

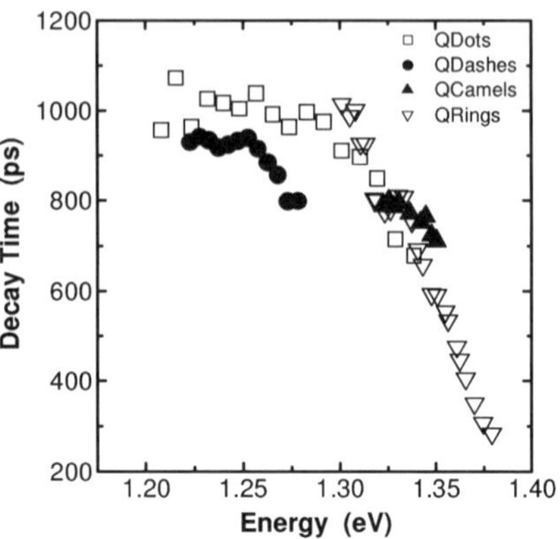

Fig. 3.11 Decay lifetime measured at 20 K in different samples

ratio $\sim$3—diameter to height ratio). A large inhomogeneous broadening is undesirable in most applications. The partial capping and annealing method provides a venue to equalize this broadening. During this process, large QDs in the ensemble are trimmed leaving behind smaller nanostructures with smaller height and better homogeneity. The best example being the QR sample with a full width at half maximum $FWHM \sim 26$ meV, roughly one fifth of the QD value as shown in Fig. 3.10.

Due to the strong In-Ga exchange that takes place during the annealing step, knowing the height of the nanostructure is not enough to determine its ground state energy. Composition and shape play also an important role. Quantum dashes and camels have similar heights according to AFM, 2.0 ± 0.5 nm [3]. Yet, their emission bands are 90 meV apart as a result of the different annealing temperature, and thus different alloying degree and shape (Sect. 3.2.4). On the other hand, as we demonstrated for quantum camels and rings, the shape can be controlled independently of the annealing temperature by changing the As flux. Altogether, the partial capping method provides three adjustable parameters to fabricate QR samples with different emission energy between 1.30–1.36 eV at low temperatures [3, 13, 58, 59].

In Fig. 3.10, the excited states present in each sample are identified according to results found in PL and PLE characterization [13, 58, 59]. The corresponding excess energies measured from the ground state are $E_1 = 55$ meV $E_2 = 110$ meV for QDh, $E_1 = 35$ meV $E_2 = 68$ meV for QC, and $E_1 = 34$ meV $E_2 = 68$ meV for QR. The combination of large aspect ratios (≥ 20) and strong alloying produces a soft confinement potential in the growth plane. For many purposes, such potential can be described using a two dimensional parabolic potential whose most relevant characteristic is an evenly spaced electronic shell structure as observed here. In this framework, the larger the energy splitting among excited states, $E_{e(h)} = \dfrac{\hbar^2}{m^*_{e(h)} l^2_{e(h)}}$, the more confined in the plane are the electron and hole wavefunctions, where $l_{e(h)}$ and $m^*_{e(h)}$ are the carrier confinement lengths and carrier effective masses, respectively [60].

Figure 3.11 summarizes the decay times obtained from time resolved PL experiments (TRPL) performed in the same four samples [58, 59]. For each sample, the ground state decay lifetime was measured as a function of energy. The decay times

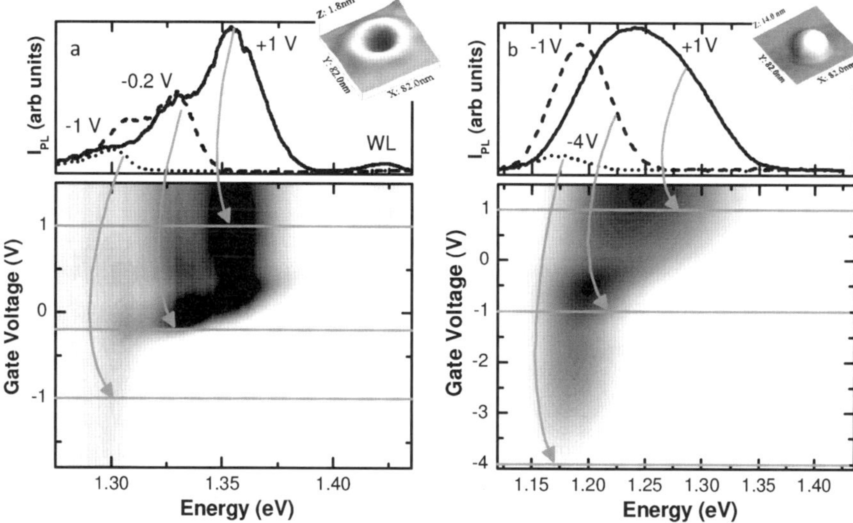

Fig. 3.12 PL spectra measured at 15 K for three external bias applied to (**a**) the QR sample and (**b**) the QD sample. The *lower panels* show contour plots of the photoluminescence covering the full voltage range

are around 1 ns, a typical value found in the literature for self-assembled dots [61]. As observed in Fig. 3.11, ground state energy and decay lifetime are inversely correlated in the different samples. In large nanostructures, strain and piezoelectric fields separate the electron and hole in opposite directions along the QD vertical axis [62]. In nanostructures with smaller height and large diameter, this effect is less intense making the electron-hole overlap larger and thus the decay time shorter [63]. For similar reasons, quantum dashes emitting at the same energy as QD have slightly shorter decay times thanks to their smaller height.

In the QR sample, the experiment is extended beyond the fundamental transition into the excited states region. Above 1.35 eV the TRPL signal correspond entirely to excited shell recombination and the small measured lifetimes ($\leq 600\,$ps) are due the additional contribution of carrier relaxation down to the ground state in parallel with radiative recombination.

3.3.2 Bias Dependent PL and TRPL

Embedding the nanostructures in a p-i-n diode or a Schottky diode allows us to study their response to a vertical electric field. We have investigated two samples containing self-assembled QDs or QRs embedded in the intrinsic region of otherwise identical Schottky diodes [59].

Under forward bias, the QR PL spectrum consists of three narrow emission bands at $\sim 1.300\,$eV, $\sim 1.326\,$eV, $\sim 1.354\,$eV, as shown in Fig. 3.12(a). As discussed previ-

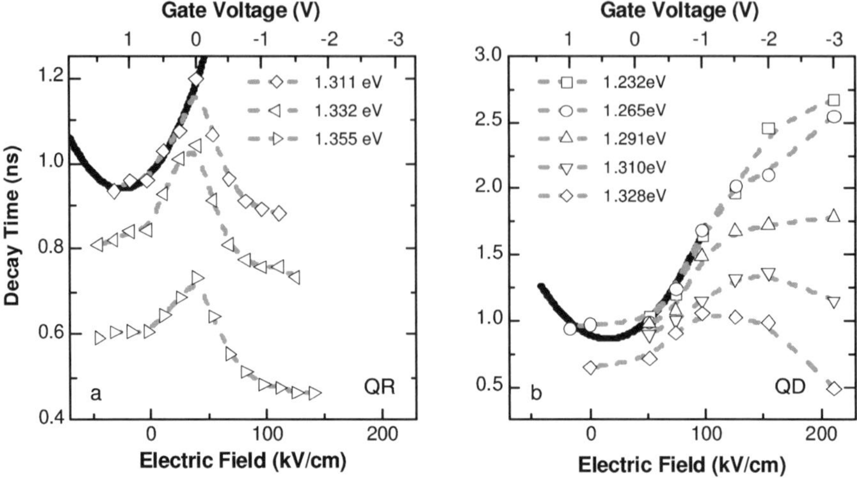

Fig. 3.13 (a) Evolution of the decay lifetime measured for different energies as a function of the electric field applied to the QR sample. The region of initial lifetime increase is fitted to a parabolic dispersion [*thick line*]. (b) Same as (a) but for the QD sample

ously, they shall be associated with the ground state, first and second excited state, respectively. The wetting layer (WL) contribution is also visible at 1.42 eV. When a reverse bias is applied to the device, electron tunneling out of the QRs takes place quenching the ensemble averaged PL bands. The high energy excited states quench faster thanks to their narrower tunneling barriers, and, at -1 V, only the emission from the QRs ground state remains visible.

Figure 3.12(b) shows the outcome of the same experiment performed in the QD sample. The large inhomogeneous broadening of this sample allows to compare QRs with QDs of varying ground state energy. As it can be observed, large QD emitting at $\sim$1.17 eV are still visible at $V_g = -4$ V, while those emitting around 1.30 eV are already quenched at ~ -1 V. To this respect, small QDs behave as QRs.

TRPL decay curves have been also recorded in the energy and bias range of interest for both samples [59]. Figure 3.13 shows the resulting decay lifetimes, τ, as a function of the applied electric field. Quantum dots emitting at 1.23 eV are characterized by a monotonic increase of τ with the reverse bias. The decay time quickly raises from 1.02 ± 0.02 ns at 0 V to 2.65 ± 0.15 ns at $V_g = -3$ V at this energy. Smaller QDs, emitting above 1.3 eV, show in comparison a less pronounced increase of the exciton lifetime in the same bias region, followed by a voltage region beyond -2 V where the lifetime decreases.

Quantum rings emitting at 1.31 eV show a trend similar to small QDs, with a region of decay time increase between 1 V (935 ± 20 ps) and 0 V (1196 ± 30 ps), and then a region of decrease when the device is reverse biased to -1 V (886 ± 30 ps) [Fig. 3.13(b)]. The tunneling of electrons out of the nanostructures is responsible for both, the quenching of the luminescence, and the decrease of the exciton lifetime observed at negative voltages. Meanwhile, the region of lifetime increase observed in both samples can be explained by the separation of the electron-hole pair, and therefore the reduction of the oscillator strength, predicted by the quantum confined Stark effect [64]. The differences between large and small QDs must be attributed to the larger polarizability of the exciton wavefunction in the former and the increasing importance of electron tunneling as the confinement energy increases in the latter [59]. Moreover, the analysis of the oscillator strength reduction can offer more insight in the differences between quantum dots and rings, as discussed below.

Due to strain and piezoelectric field dependence of the valence and conduction band edges, the electron and hole are spatially separated along the growth direction even at zero bias [52, 62, 65]. When the electron lies above the hole, a negative permanent dipole moment is created and a positive electric field has to be applied along z to cancel the initial carrier separation. At this critical field, the radiative lifetime finds its minimum and the exciton energy finds its maximum. While the situation just described is the typical for pure InAs QD, the opposite alignment, hole above the electron, occurs when the In/Ga ratio varies within the nanostructure (In increasing from the base to the top) [62]. Figure 3.13 shows that the radiative lifetime reaches a minimum at ~ -25 kV/cm (~ 16 kV/cm) for QRs (QDs) emitting at the center of their inhomogeneous emission band. Within the limits of the perturbation theory [66], this means that both nanostructures have permanent dipole moments with opposite signs. In particular, our result implies a constant composition profile for the studied QD, and the presence of a noticeable In–Ga composition grading in the QR. Just as expected from the partial capping and annealing steps introduced during the growth of the latter [3, 38].

3.3.3 Single QR μPL and μPLE

Figure 3.14(a) shows the band diagram of the n-type Schottky heterostructure which has been used to investigate single QR in a vertical electrical field [10, 13]. Under reverse bias, electrons and holes photogenerated by the excitation laser can be trapped in the QR and generate luminescence, or can be swept by the electric field and generate photocurrent. In addition, a short tunneling barrier (~ 25 nm) allows to control the number of electrons confined in the QR by changing bias [11].

When the number of electrons in the QR varies, new Coulomb interactions come into play which change the ground state energy. Such energy shifts are small compared with the inhomogeneous broadening of the ensemble emission, but can be readily observed in the luminescence spectrum of a single QR, as shown in Fig. 3.14(b). At slightly negative voltages, the neutral exciton emission line, X^0, dominates the spectrum. Applying small positive voltages, one electron tunnels from the back contact into the QR shifting the emission to lower energies by 3.2 meV. This shifted line corresponds to radiative recombination of a negatively charged exciton or negative trion, X^{1-}. At 0.65 V, a new shift of the emission peak by 1.3 meV indicates a new tunneling event and the formation of a doubly charged exciton [10].

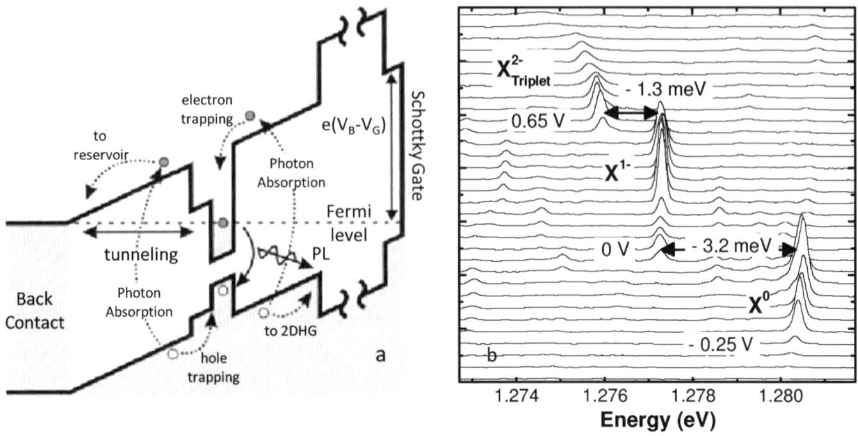

Fig. 3.14 (**a**) Energy band diagram in the vicinity of a QR layer embedded in a Schottky diode under reverse bias. (**b**) Voltage dependent μ-PL spectra of a single QR measured at 4.2 K

Using a two-dimensional parabolic potential, analytical expressions can be derived for $E_e + E_h$, $\Delta E_{X^0,X^{1-}}$ and $\Delta E_{X^{1-},X^{2-}}$ which depend on the carrier confinement lengths, l_e and l_h, as well as on the In$_{1-x}$Ga$_x$As composition entering linearly in the dielectric constant, $\epsilon_r(x)$, and effective masses, $m_e^*(x)$ and $m_e^*(x)$ [60]:

$$E_e + E_h = \frac{\hbar^2}{l_e^2 m_e^*(x)} + \frac{\hbar^2}{l_h^2 m_h^*(x)} \tag{3.1}$$

$$\Delta E_{X^0,X^{1-}} = \frac{e^2\sqrt{\pi}}{4\pi\epsilon_0\epsilon_r(x)}\left[\frac{1}{\sqrt{2}l_e} - \frac{1}{\sqrt{l_e^2 + l_h^2}}\right] \tag{3.2}$$

$$\Delta E_{X^{1-},X^{2-}} = \frac{e^2\sqrt{\pi}}{4\pi\epsilon_0\epsilon_r(x)}\left[\frac{3}{4\sqrt{2}l_e} - \frac{2l_e^2 + l_h^2}{2(l_e^2 + l_h^2)^{\frac{3}{2}}} + \frac{1}{4\sqrt{2}l_e}\right] \tag{3.3}$$

Introducing $\Delta E_{X^0,X^{1-}} = -3.2$ meV, $\Delta E_{X^{1-},X^{2-}} = -1.3$ meV and the splitting among excited states measured in as–grown samples, $E_e + E_h = 34$ meV (Fig. 3.10), we obtain for our QR: $x = 0.24$, $l_e = 9.1$ nm, $l_h = 5.0$ nm and therefore, $E_e = 27$ meV and $E_h = 7$ meV.

With these values, the emission bands of the different QR samples can be compared and labeled according to the model. As it can be seen in Fig. 3.15(a), as-grown QRs emit light from ground and excited states with equal angular momentum (s–s, p–p, d–d, …) which are split by 34 meV. However, when embedded in a forward biased n-type Schottky diode, QRs are populated with excess electrons and the observed band splitting is just 27 meV. We shall conclude that, in this situation, the emission bands come from unpaired shell recombination with the hole in its ground state and the electron in an excited state (s–s, s–p, s–d, …). Radiative recombination from unpaired shells is in principle forbidden by dipolar selection rules and, nevertheless, has been observed before in the ensemble absorption spectrum of similar QR samples [67] and also in μ-PLE spectra of InGaAs lens-shaped QDs [68].

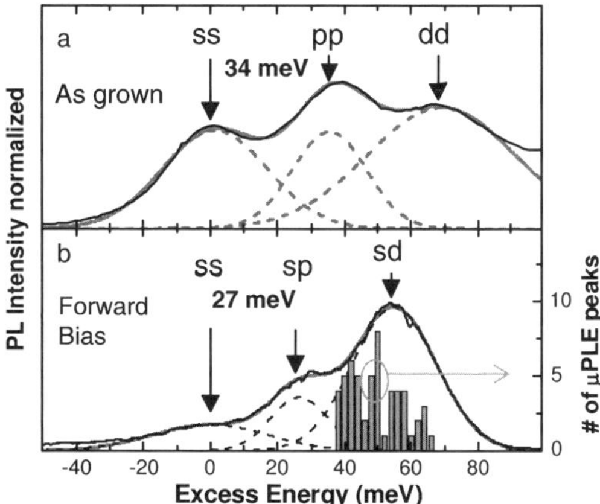

Fig. 3.15 PL spectrum measured at 15 K for (**a**) the as-grown QR sample and (**b**) for a Schottky heterostructure containing QR under forward bias. The histogram shows the number of μ-PLE peaks detected in the same energy region

Their relative dominance in our experiment might be tentatively attributed to orbital mixing caused by coulomb interaction which is enhanced in our case by resonant injection of electrons into excited states.

We conclude our study of the inter-band optical properties of self-assembled QRs discussing the results obtained in excitation μ-PL of single QR (μ-PLE) [13]. Figure 3.16 shows examples of μ-PLE spectra measured for different QR resonances. Apart of the WL and GaAs edge, most sharp resonances appear in the range $\Delta E \sim 35$–60 meV due to laser absorption in p and d excited shells, as shown in Fig. 3.15(b). Sub-shell degeneracies are lifted by the Coulomb interactions, and also by the crystal lattice and/or confining potential in-plane asymmetry. QRs are nearly round objects with a small elongation along the [1$\bar{1}$0] crystal direction [3]. Therefore, we do not expect a large splitting between the p_x and p_y states. Doublet structures with energy splitting of <2 meV are often observed between 39 an 46 meV, while, in other cases, the doublet structures are not present and the number of peaks is also smaller indicating even more symmetric QRs. This is in contrast with InGaAs QDs where $p_x - p_y$ excited state energy splittings larger than 5 meV have been measured by μ-PLE [69].

In conclusion, it is possible to modify the shape and size of InAs on GaAs self assembled quantum dots grown by MBE by introducing a pause during the capping process, also known as partial overgrowth technique. Under certain growth-pause capping conditions it is possible to obtain self-assembled quantum rings. The changes in shape and size lead to a modification of the quantum confinement potential and enables the control over fundamental physical properties. We have shown that a large fraction of stress-induced liquid indium plays a key role in the transformation of the nanostructures. We have analyzed the steady state and time-dependent photoluminescence spectrum of ensemble and single QR both, as-grown and embedded in Schottky diodes. Our study reveals that self-assembled QRs are nearly round and thinner than QD nanostructures with a graded InGaAs composition along the growth direction and average $x_{Ga} = 0.24$. In the growth plane, the energy barrier potential can be described with a two-dimensional harmonic oscillator or parabolic

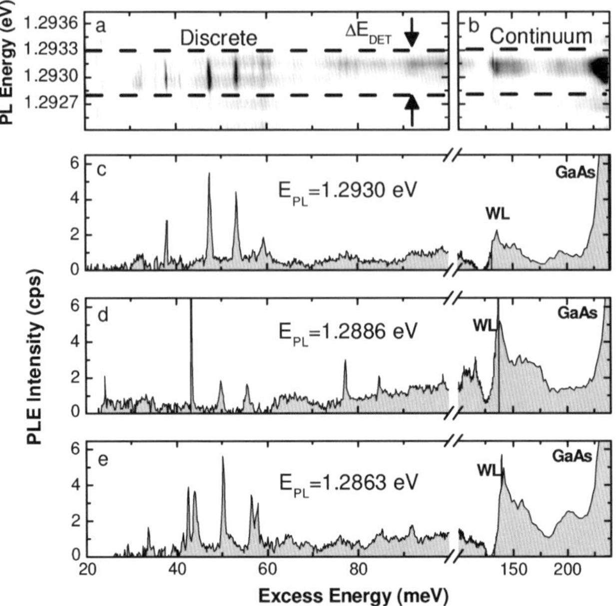

Fig. 3.16 μ-PLE image of a single QR resonance measured varying the laser excitation energy (**a**) below and (**b**) above the WL. (**c**), (**d**) and (**e**) Typical μ-PLE spectra obtained averaging different μ-PLE image over ΔE_{DET}

model. To this respect, we find that holes are more spatially confined than electrons, $l_h = 5.0$ nm vs $l_e = 9.1$ nm, and also have smaller confinement energy, $E_h = 7$ meV vs $E_e = 27$ meV. The control of the charge in a single QR has been also demonstrated opening the path for the demonstration of quantum interference Aharanov-Bohm oscillations in these systems.

Acknowledgements This work would not be possible without the support from Spanish projects EPIC-NANOTICS (MINECO TEC2011-29120-C05-04), Q&C Light (CAM S2009ESP-1503), NANOSELF project (TIC2002-04096), Comunidad Autónoma de Madrid (07T/0062/2000), and CICYT (TIC99-1035-C02); and EU project NANOMAT (G5RD-CT-2001-00545).

References

1. J.M. García, G. Medeiros-Ribeiro, K. Schmidt, T. Ngo, J.L. Feng, A. Lorke, J. Kotthaus, P. Petroff, Appl. Phys. Lett. **71**, 2014 (1997)
2. J.M. García, T. Mankad, P.O. Holtz, J. Wellman, P.M. Petroff, Appl. Phys. Lett. **72**, 3172 (1998)
3. D. Granados, J.M. García, Appl. Phys. Lett. **82**, 2401 (2003)
4. P. Michler, A. Kiraz, C. Becher, W.V. Schoenfeld, P.M. Petroff, L. Zhang, E. Hu, A. Imamoglu, Science **290**(5500), 2282 (2000)
5. S. Kiravittaya, A. Rastelli, O.G. Schmidt, Rep. Prog. Phys. **72**(4), 046502 (2009)

6. V.M. Fomin (ed.), *A special issue on Modern Advancements in Experimental and Theoretical Physics of Quantum Rings*. J. Nanoelectron. Optoelectron. **6** (American Scientific, 2011)
7. D.J. Eaglesham, M. Cerullo, Phys. Rev. Lett. **64**, 1943 (1990)
8. D. Leonard, M. Krishnamurthy, C.M. Reaves, S. Denbaars, P. Petroff, Appl. Phys. Lett. **63**, 3203 (1993)
9. J.Y. Marzin, J. Gerald, A. Izrael, D. Barrier, G. Gastard, Phys. Rev. Lett. **713**, 716 (1994)
10. R.J. Warburton, C. Schäflein, D. Haft, F. Bickel, A. Lorke, K. Karrai, J.M. García, W. Schoenfeld, P.M. Petroff, Nature **405**, 926 (2000)
11. A. Lorke, R.J. Luyken, A.O. Govorov, J.P. Kotthaus, J.M. García, P.M. Petroff, Phys. Rev. Lett. **84**, 2223 (2000)
12. R.J. Warburton, C. Schulhauser, D. Haft, C. Schäflein, K. Karrai, J.M. García, W. Schoenfeld, P.M. Petroff, Phys. Rev. B **65**, 113303 (2002)
13. B. Alén, J. Martínez-Pastor, D. Granados, J.M. García, Phys. Rev. B **72**, 155331 (2005)
14. N.A.J.M. Kleemans, I.M.A. Bominaar-Silkens, V.M. Fomin, V.N. Gladilin, D. Granados, A.G. Taboada, J.M. García, P. Offermans, U. Zeitler, P.C.M. Christianen, J.C. Maan, J.T. Devreese, P.M. Koenraad, Phys. Rev. Lett. **99**, 146808 (2007)
15. Y. Aharonov, D. Bohm, Phys. Rev. B **115**, 485 (1959)
16. P.B. Joyce, T.J. Krzyzewski, G.R. Bell, T.S. Jones, S. Malik, D. Childs, R. Murray, Phys. Rev. B **62**, 10891 (2000)
17. T. Kawai, H. Yonezu, Y. Ogasawara, D. Saito, K. Pak, J. Cryst. Growth **201**, 1146 (1999)
18. J. Moison, C. Guille, F. Houzay, F. Barthe, M.V. Rompay, Phys. Rev. B **40**, 6149 (1989)
19. J. Silveira, F. Briones, J. Cryst. Growth **201–202**, 113 (1999)
20. J.M. García, J. Silveira, F. Briones, Appl. Phys. Lett. **77**, 409 (2000)
21. R. Blossey, A. Lorke, Phys. Rev. E **65**, 021603 (2002)
22. M. Grundmann, O. Stier, D. Bimberg, Phys. Rev. B **52**, 11969 (1995)
23. J.P. Silveira, J.M. García, F. Briones, J. Cryst. Growth **227/228**, 995 (2001)
24. O.B.E. Tournier, K. Ploog, Appl. Phys. Lett. **60**, 287 (1992)
25. J.M. García, L. González, M.U. González, J.P. Silveira, Y. González, F. Briones, J. Cryst. Growth **227–228**, 975 (2001)
26. D. Alonso-Álvarez, J.M. Ripalda, B. Alén, J.M. Llorens, A. Rivera, F. Briones, Adv. Mater. **23**, 5256 (2011)
27. J. Floro, E. Chason, S. Lee, R. Twesten, R. Hwang, L. Freund, J. Electron. Mater. **26**, 969 (1997)
28. M.U. González, L. González, J.M. García, Y. González, J.P. Silveira, F. Briones, Microelectron. J. **35**(1), 13 (2004)
29. J. Massies, F. Turco, A. Saletes, J. Contour, J. Cryst. Growth **80**, 307 (1987)
30. D.J. Bottomley, Appl. Phys. Lett. **80**, 4747 (2002)
31. T. Mano, T. Kuroda, S. Sanguinetti, T. Ochiai, T. Tateno, J. Kim, T. Noda, M. Kawabe, K. Sakoda, G. Kido, N. Koguchi, Nano Lett. **5**, 425 (2005)
32. Q. Xie, P. Chen, A. Madhukar, Appl. Phys. Lett. **65**, 2051 (1994)
33. A. Lorke, R. Blossey, J.M. García, M. Bichler, G. Abstreiter, Mater. Sci. Eng. B **88**, 225 (2002)
34. P.B. Joyce, T.J. Krzyzewski, G.R. Bell, B.A. Joyce, T.S. Jones, Phys. Rev. B **58**, R15981 (1998)
35. Y. Horikoshi, H. Yamaguchi, F. Briones, M. Kawashima, J. Cryst. Growth **105**, 326 (1990)
36. T. Ogura, D. Kishimoto, T. Nishinaga, J. Cryst. Growth **226**, 179 (2001)
37. D. Granados, J.M. García, T. Ben, S.I. Molina, Appl. Phys. Lett. **86**, 071918 (2005)
38. P. Offermans, P.M. Koenraad, J.H. Wolter, D. Granados, J.M. García, V.M. Fomin, V.N. Gladilin, J.T. Devreese, Appl. Phys. Lett. **87**, 131902 (2005)
39. R.A. Römer, R.E. Raikh, Phys. Rev. B **62**, 7045 (2000)
40. J. Song, S.E. Ulloa, Phys. Rev. B **63**, 125302 (2001)
41. H. Hu, J.L. Zhu, D.J. Li, J.J. Xiong, Phys. Rev. B **63**, 195307 (2001)
42. I. Galbraith, F.J. Braid, R.J. Warburton, Phys. Status Solidi A **190**, 781 (2002)
43. A.O. Govorov, S.E. Ulloa, K. Karrai, R.J. Warburton, Phys. Rev. B **66**, 081309 (2002)

44. M. Bayer, M. Korkusinski, P. Hawrylak, T. Gutbrod, M. Michel, A. Forchel, Phys. Rev. Lett. **90**, 186801 (2003)
45. E. Ribeiro, A.O. Govorov, J.G.M.R.W. Carvalho, Phys. Rev. Lett. **92**, 126402 (2004)
46. I.R. Sellers, V.R. Whiteside, L. Kuskovsky, A.O. Govorov, B.D. McCombe, Phys. Rev. Lett. **100**, 136405 (2008)
47. A.M. Fischer, J.V.L. Campo, M.E. Portnoi, R.A. Römer, Phys. Rev. Lett. **102**, 096405 (2009)
48. M.D. Teodoro, J.V.L.R.V.L. Campo, J.G.E.M.E. Marega, Y.G. Gobato, F. Iikawa, M.J.S.P. Brasil, Z.Y. AbuWaar, V.G. Dorogan, Y.I. Mazur, M. Benamara, G.J. Salamo, Phys. Rev. Lett. **104**, 086401 (2010)
49. F. Ding, N. Akopian, B. Li, U. Perinetti, A. Govorov, F.M. Peeters, C.C.B. Bufon, C. Deneke, Y.H. Chen, A. Rastelli, O.G. Schmidt, V. Zwiller, Phys. Rev. B **82**, 075309 (2010)
50. J.M. Llorens, C. Trallero-Giner, A. García-Cristobal, A. Cantarero, Phys. Rev. B **64**, 035309 (2001)
51. O. Voskoboynikov, Y. Li, H.M. Lu, C.F. Shih, C.P. Lee, Phys. Rev. B **66**, 155306 (2002)
52. J.A. Barker, R.J. Warburton, E.P. O'Reilly, Phys. Rev. B **69**, 035327 (2004)
53. J.I. Climente, J. Planelles, F. Rajadell, J. Phys. Condens. Matter **17**, 1573 (2005)
54. J.I. Climente, J. Planelles, W. Jaskólski, Phys. Rev. B **68**, 075307 (2003)
55. A. Emperador, M. Pi, M. Barranco, A. Lorke, Phys. Rev. B **62**, 4573 (2000)
56. A. Puente, L. Serra, Phys. Rev. B **63**, 125334 (2001)
57. H. Hu, G.M. Zhang, J.L. Zhu, J.J. Xiong, Phys. Rev. B **63**, 045320 (2001)
58. J. Gomis, J. Martínez-Pastor, B. Alén, D. Granados, J. García, P. Roussignol, Eur. Phys. J. B **54**, 471 (2006)
59. B. Alén, J. Bosch, J. Martínez-Pastor, D. Granados, J.M. García, L. González, Phys. Rev. B **75**, 045319 (2007)
60. R.J. Warburton, B.T. Miller, C.S. Dürr, C. Bödefeld, K. Karrai, J.P. Kotthaus, G. Medeiros-Ribeiro, P.M. Petroff, S. Huant, Phys. Rev. B **58**, 16221 (1998)
61. R. Heitz, A. Kalburge, Q. Xie, M. Grundmann, P. Chen, A. Hoffmann, A. Madhukar, D. Bimberg, Phys. Rev. B **57**, 9050 (1998)
62. J.A. Barker, E.P. O'Reilly, Phys. Rev. B **61**, 13840 (2000)
63. A. Greilich, M. Schwab, T. Berstermann, T. Auer, R. Oulton, D. Yakovlev, M. Bayer, V. Stavarache, D. Reuter, A. Wieck, Phys. Rev. B **73**, 045323 (2006)
64. E.E. Mendez, G. Bastard, L.L. Chang, L. Esaki, H. Morkoc, R. Fischer, Phys. Rev. B **26**, 7101 (1982)
65. O. Stier, M. Grundmann, D. Bimberg, Phys. Rev. B **59**, 5688 (1999)
66. W. Sheng, J.P. Leburton, Phys. Rev. B **67**, 125308 (2003)
67. H. Pettersson, R. Warburton, A. Lorke, K. Karrai, J. Kotthaus, J. García, P. Petroff, Physica E **6**, 510 (2000)
68. R. Oulton, J.J. Finley, A.I. Tartakovskii, D.J. Mowbray, M.S. Skolnick, M. Hopkinson, A. Vasanelli, R. Ferreira, G. Bastard, Phys. Rev. B **68**, 235301 (2003)
69. J.J. Finley, P.W. Fry, A.D. Ashmore, A. Lemaitre, A.I. Tartakovskii, R. Oulton, D.J. Mowbray, M.S. Skolnick, M. Hopkinson, P.D. Buckle, P.A. Maksym, Phys. Rev. B **63**, R161305 (2001)

Chapter 4
Self-organized Quantum Rings: Physical Characterization and Theoretical Modeling

V.M. Fomin, V.N. Gladilin, J.T. Devreese, and P.M. Koenraad

Abstract An adequate modeling of the self-organized quantum rings is possible only on the basis of the modern characterization of those nanostructures. We discuss an atomic-scale analysis of the indium distribution of self-organized InGaAs quantum rings (QRs). The analysis of the shape, size and composition of self-organized InGaAs QRs at the atomic scale reveals that AFM only shows the material coming out of the QDs during the QR formation. The remaining QD material, as observed by Cross-Sectional Scanning Tunneling Microscopy (X-STM), shows an asymmetric indium-rich crater-like shape with a depression rather than an opening at the center and determines the observed ring-like electronic properties of QR structures. A theoretical model of the geometry and materials properties of the self-organized QRs is developed on that basis and the magnetization is calculated as a function of the applied magnetic field. Although the real QR shape differs strongly from an idealized circular-symmetric open-ring structure, Aharonov-Bohm-type oscillations in the magnetization have been predicted to survive. They have been observed using the torsion magnetometry on InGaAs QRs. Examples of prospective applications of QRs are presented that do and do not utilize the topological properties of QRs.

V.M. Fomin (✉)
Institute for Integrative Nanosciences, IFW Dresden, Dresden, Germany
e-mail: v.fomin@ifw-dresden.de

V.N. Gladilin · J.T. Devreese
Theory of Quantum and Complex Systems, University of Antwerp, Antwerp, Belgium

V.N. Gladilin
e-mail: vladimir.gladilin@ua.ac.be

J.T. Devreese
e-mail: jozef.devreese@ua.ac.be

P.M. Koenraad
Photonics and Semiconductor Nanophysics, Eindhoven University of Technology, Eindhoven, The Netherlands
e-mail: P.M.Koenraad@tue.nl

V.M. Fomin (ed.), *Physics of Quantum Rings*, NanoScience and Technology,
DOI 10.1007/978-3-642-39197-2_4, © Springer-Verlag Berlin Heidelberg 2014

4.1 Introduction

Remarkable advancements in the field of self-assembled quantum rings (QRs) are a result of mutually stimulating developments of novel fabrication technologies, powerful characterization instruments and adequate theoretical approaches. A striking example is the rapidly progressing production and investigation of $In_x Ga_{1-x} As$ self-assembled QRs, which are represented in detail in Chaps. 2 and 3 and Refs. [1–4]. A deep insight into the structure of QRs and quantum dots is obtained on the base of the Cross-Sectional Scanning Tunneling Microscopy (X-STM), see, for instance, Chaps. 5 and 6 and reviews [5, 6]. In the present chapter, we demonstrate that a high level of complexity is needed for a dedicated theoretical model to adequately represent the specific features of QRs as determined by X-STM. The model of a self-assembled QR, which is *singly connected* and *anisotropic,* suggested in Refs. [7, 8] has found a vast number of applications and developments. In particular, it was used to demonstrate that Aharonov-Bohm oscillations in the magnetization and the magnetic susceptibility peak persist even though self-assembled QRs are singly connected and show a pronounced shape anisotropy, to interpret the magnetization behavior of a self-assembled QR in a nice quantitative agreement with the experiment [9], to demonstrate that inherent structural asymmetry combined with the interparticle Coulomb interactions, plays a crucial role in the diamagnetic shift of excitons and biexcitons in self-assembled QRs [10], to analyze effects due to a tensile-strained insertion layer on strain and electronic structure of self-assembled QRs (Chap. 13 and Refs. [11, 12]). Using significantly simplified models of the QR shape, only qualitative conclusions could be drawn, e.g., on the energy spectra [13] and magnetic response [14].

$In_x Ga_{1-x} As$ QRs are formed by capping self-organized quantum dots (QDs) grown by Stranski-Krastanov (SK) mode with a layer thinner than the dot height and subsequent annealing [1]. During this process, the QD material suffers an anisotropic redistribution, resulting in elongated ring-shaped islands on the surface, with crater-like holes in their centers as was shown with Atomic Force Microscopy (AFM) measurements [1]. The dot-to-ring transition has been attributed to a dewetting process which expels indium from the QD [15] and a simultaneous temperature dependent Ga–In alloying process [16]. Capacitance and far-infrared spectroscopy measurements on buried QR structures have provided evidence of the Aharonov-Bohm oscillation [2]. Although no direct measurements were available, an electronic radius of about 14 nm was deduced. This led to the conclusion that the QR shape as determined from AFM topography is preserved when buried [2]. Until Ref. [17], however, no structural measurements of buried QRs have been available. Evolution of InAs/GaAs QDs partially capped with GaAs has been experimentally studied as an annealing process transforms them first into QRs and later into holes penetrating the whole cap layer [18], shape, composition, and optical emission being monitored by using AFM, X-ray photoemission microscopy, and photoluminescence (PL), respectively. The sizes of self-assembled QRs determined by Cross-Sectional Transmission Electron Microscopy (X-TEM) [19] matched the AFM results. In-plane mapping by grazing-incidence X-ray diffraction [20] has revealed the lateral extent of strained regions in the buried $In_x Ga_{1-x} As$ QRs; a comparison between strain and composition maps of QDs and QRs allows for inferring on how the QR configuration affects its optical characteristics. Development of conductive scanning microscopy-based characterization tools allows for the investigation of the electrical properties of individual self-assembled GeSi QRs at the nanoscale by combining the scanning Kelvin microscopy, conductive AFM and scanning capacitance microscopy [21]. On this base, the electrical properties of a QR have been explained by the fact that the QR rim has a lower barrier height with respect to the tip and a higher carrier density, giving rise to a higher conductivity at the rim compared to the QR opening and the wetting layer.

4.2 X-STM Characterization

Measurements of the vertical Stark effect of excitons confined to individual QRs [4] showed rather large dipole moments with opposite sign as compared to QDs [22]. Theoretical calculations indicated that both the observed electronic radius and dipole moment of the QRs were inconsistent with the geometry as determined by AFM. In order to unambiguously resolve this discrepancy, twenty layers of QRs separated from each other by 18 nm were grown by using solid source molecular-beam epitaxy (MBE) on a Si-doped n-type GaAs (001) substrate, and the shape, size, and composition of buried QRs were analyzed at the atomic scale by cross-sectional scanning tunneling microscopy (X-STM) [7, 8]. The X-STM measurements were performed in an ultra-high-vacuum chamber on the orthogonal (110) and ($1\bar{1}0$) cross-sectional surfaces.

Figure 4.1 shows an AFM image of the surface layer of the structure. The ring-shaped islands (density 10^{10} cm^{-2}) are elongated along the [$1\bar{1}0$] direction, with an outer size of about 100 by 70 nm and an average height of about 1 nm. The holes in the center of the islands are asymmetric as well and have a size of 30 by 20 nm and a depth of about 0.5–1.5 nm. Figure 4.2(a) shows a large scale X-STM image of three of the QR layers. The bright spots correspond to In atoms in the top layer of the cleaved surface. In the image the cross-sections of two flat indium-rich nanostructures can be seen. The averaged height profiles, taken across the middle QR layer in the growth direction between the points A and B, are displayed in Fig. 4.2(b). The profiles clearly indicate two peaks in the indium concentration. The highest peak can be attributed to the wetting layer on which the QDs are formed during growth. The presence of the second indium layer is attributed to the accumulation of segregated indium from the wetting layer at the surface of the partial capping layer and to surface migration of indium atoms that have been expelled from the quantum dots during QR formation [24–27]. The separation between the wetting layer and the second indium layer increases with about 2 bilayers towards the nanostructure. This change in separation is in agreement with the height and diameter of the uncapped QRs as measured by AFM. Thus at least to some extent the shape of the ring-shaped islands as observed by AFM is preserved after capping.

Fig. 4.1 1×1 μm^2 atomic force microscopy image showing the anisotropic distribution of InAs/GaAs dot material after 2 nm GaAs capping and 1 min annealing at growth temperature under As$_2$ flux. The height scale is 0 (*dark*) to 2.5 nm (*bright*). The height of the ring-shaped islands is about 1 nm. Atomic steps can be seen in the image. After Ref. [7]

Enlarged views of the nanostructures can be seen in Figs. 4.2(c) and 4.2(d) where we show filled states topography images of the orthogonal (1̄10) and (110) cleavage planes, which correspond to the short and long axis of the ring-shaped islands observed by AFM, respectively. The nanostructures have a crater-like shape which can be attributed to the remainder of the quantum dots after the QR formation process. It is clearly seen that these quantum craters do not have an opening at the center. Furthermore, in the [110] direction, the rim of the quantum crater appears brighter and higher (8 BLs) compared to the [1̄10] direction where the rim is less pronounced. This asymmetry is attributed to the preferential diffusion of the dot material in the [1̄10] direction [28] as can be seen from the elongation of the ring-shaped islands in Fig. 4.1. The indium-rich asymmetric crater-like shapes, observed in the X-STM measurements, differ substantially from the ring-shaped islands on the surface of uncapped QR structures. The observation that buried QRs have a smaller size and a larger height, than the ring-shaped islands as revealed by AFM, offers an explanation for the observed discrepancies between the measured and theoretical values for the electronic radius and dipole moment of the QRs [23].

Figure 4.3 schematically shows a QR sample used to study persistent currents in QRs. The sample was grown by molecular beam epitaxy and contains 29 mutually decoupled periods [17]. Each period consists of a nanostructured InAs layer, between two 24 nm GaAs layers, and a 2 nm doped (7×10^{16} cm^{-3} Si) GaAs layer, that provides electrons to the InAs nanostructures. Using a one-dimensional Poisson solver, the average number of electrons per nanostructure was estimated to be less than 1.5. The sample is capped by a final nanostructured layer. By performing AFM on this layer, the nanostructure density was determined to be 9×10^{9} cm^{-2} per layer. PL experiments showed a single peak, indicating a unimodal size distribution, at 1.3 eV, which is typical for these nanostructures [16]. The FWHM of the PL peak is 40 meV, from which the estimated size dispersion is about 5 %.

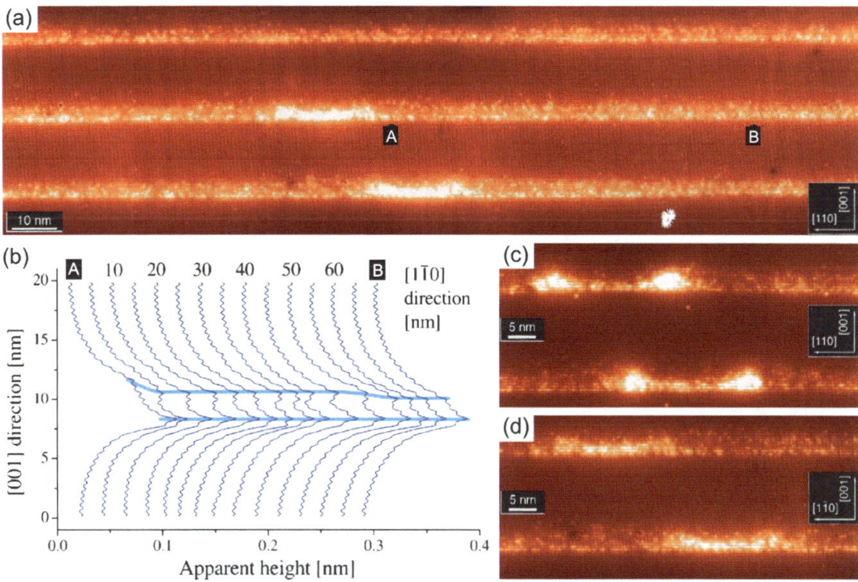

Fig. 4.2 (**a**) Empty states X-STM image showing two buried QRs, $V_{\text{sample}} = 1.45$ V. (**b**) Averaged apparent height profiles taken in the growth ([001]) direction across the InAs layer between points A and B of panel (**a**). The height profiles are averaged over a distance of 10 nm and show two peaks in the indium concentration. Panels (**c**) and (**d**) show the filled states topography images of a cleaved QR in the $(1\bar{1}0)$ plane and (110) plane, respectively, $V_{\text{sample}} = -3$ V. The height scale is 0 (*dark*) to 0.25 nm (*bright*). After Ref. [7]

Fig. 4.3 Layer structure of the InAs/GaAs sample for magnetization, consisting of 29 layers of self-assembled InAs nanostructures. The nanostructures are located between two 24 nm GaAs layers. The repeated sequence contains a modulation doping layer, which provides electrons to the nanostructured layers. Two additional doping layers are inserted to accommodate for the depletion toward the capping layer and the undoped substrate. After Ref. [9]

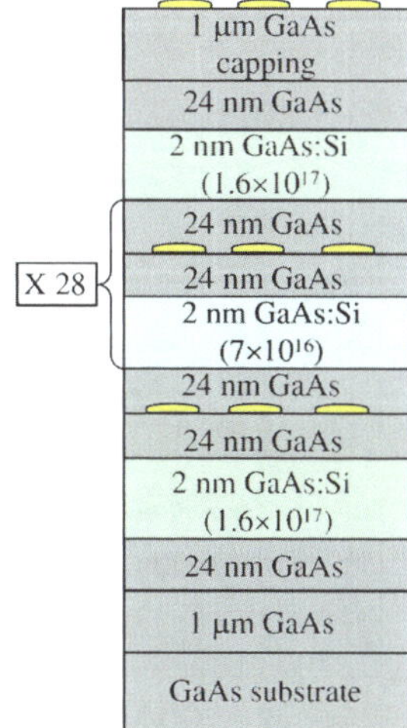

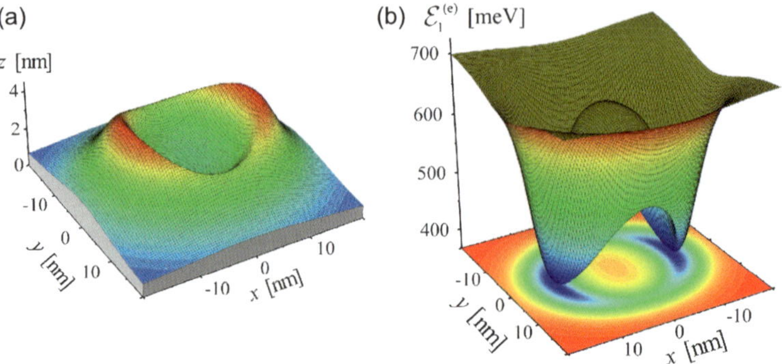

Fig. 4.4 (**a**) Height of a QR as a function of the radial and the azimuthal coordinates as modeled by (4.1) with $R = 11.5$ nm, $h_0 = 1.6$ nm, $h_M = 3.6$ nm, $h_\infty = 0.4$ nm, $\gamma_0 = 3$ nm, $\gamma_\infty = 5$ nm, $\xi_h = 0.2$, $\xi_\gamma = \xi_R = 0$. (**b**) The adiabatic potential governing the electron motion in an unstrained $\mathrm{In_{0.6}Ga_{0.4}As}$ QR shown in panel (**a**). After Ref. [31]

4.3 Modeling of Shape and Materials Properties

By imaging QRs at a high voltage [$V_{\text{sample}} = -3$ V; see Figs. 4.2(c) and 4.2(d)], electronic contributions to the contrast in the image are minimized and only the true outward surface relaxation due to the lattice mismatch (7 %) between the InAs and surrounding GaAs is imaged. The outward relaxation of the cleaved surface is used to determine the indium composition of the quantum craters [29, 30]. The quantum craters are modeled with a varying-thickness InGaAs layer embedded in an infinite GaAs medium. The bottom of the InGaAs layer is assumed to be perfectly flat and parallel to the xy-plane. The height of the $\mathrm{In}_x\mathrm{Ga}_{1-x}\mathrm{As}$ layer as a function of the radial coordinate ρ and the angular coordinate φ is modeled by the expression

$$h(\rho,\varphi) = h_0 + \frac{[\tilde{h}_M(\varphi) - h_0]\{1 - [\rho/\tilde{R}(\varphi) - 1]^2\}}{\{[\rho - \tilde{R}(\varphi)]/\tilde{\gamma}_0(\varphi)\}^2 + 1}, \quad \rho \leq \tilde{R}(\varphi),$$

$$h(\rho,\varphi) = h_\infty + \frac{\tilde{h}_M(\varphi) - h_\infty}{\{[\rho - \tilde{R}(\varphi)]/\tilde{\gamma}_\infty(\varphi)\}^2 + 1}, \quad \rho > \tilde{R}(\varphi) \tag{4.1}$$

with

$$\tilde{h}_M(\varphi) = h_M(1 + \xi_h \cos 2\varphi), \qquad \tilde{\gamma}_0(\varphi) = \gamma_0(1 + \xi_\gamma \cos 2\varphi),$$

$$\tilde{\gamma}_\infty(\varphi) = \gamma_\infty(1 + \xi_\gamma \cos 2\varphi), \qquad \tilde{R}(\varphi) = R(1 + \xi_R \cos 2\varphi). \tag{4.2}$$

Here, h_0 corresponds to the thickness at the center of the crater, h_M to the rim height and h_∞ to the thickness of the $\mathrm{In}_x\mathrm{Ga}_{1-x}\mathrm{As}$ layer far away from the ring-like structure. The γ_0 and γ_∞ parameters define the inner and outer slopes of the rim, respectively. The parameters ξ_h, ξ_γ, ξ_R describe the ring-shape anisotropy. An example of the QR shape, described by (4.1), is shown in Fig. 4.4(a).

A three-dimensional finite-element calculation based on elasticity theory has been applied to determine the relaxation of the cleaved surface of the modeled QR.

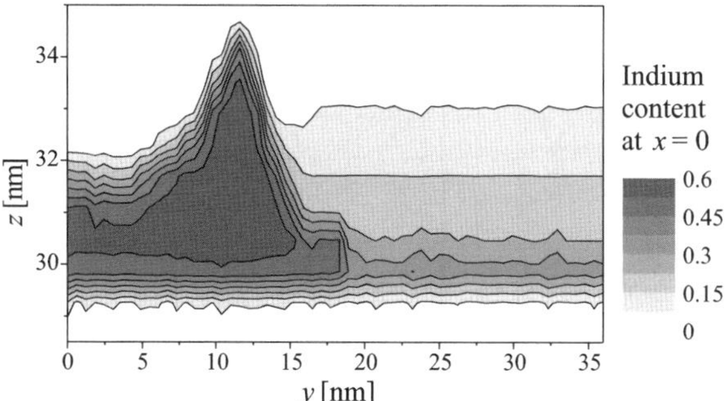

Fig. 4.5 Indium distribution in the yz plane in the QR. After Ref. [31]

With $R = 10.75$ nm, $h_0 = 1.6$ nm, $h_M = 3.6$ nm, $h_\infty = 0.4$ nm, $\gamma_0 = \gamma_\infty = 3$ nm, $\xi_h = 0.2$, $\xi_\gamma = -0.25$ and $\xi_R = 0.07$, an indium concentration of 55 % (see Fig. 4.5) results in a calculated surface relaxation that matches the measured relaxation of the cleaved surface, as shown in [8]. This set of geometric parameters of a QR is selected as standard for the calculations discussed below, unless stated otherwise.

4.4 Theory of Electronic Properties of One-Electron Rings, Including Magnetization

The single-particle Hamiltonians in a strained ring have the form [32, 33]

$$H_e = -\frac{\hbar^2}{2}\left(\nabla + \frac{ie}{\hbar}\mathbf{A}\right)\frac{1}{m_e(\mathbf{r}_e)}\left(\nabla + \frac{ie}{\hbar}\mathbf{A}\right) + V_e(\mathbf{r}_e) + \delta E_e(\mathbf{r}_e) - eV_P(\mathbf{r}_e),$$
(4.3)

$$H_h = -\frac{\hbar^2}{2}\left(\nabla - \frac{ie}{\hbar}\mathbf{A}\right)\frac{1}{m_h(\mathbf{r}_h)}\left(\nabla - \frac{ie}{\hbar}\mathbf{A}\right) - V_h(\mathbf{r}_h) - \delta E_h(\mathbf{r}_h) + eV_P(\mathbf{r}_h),$$
(4.4)

where $m_e(\mathbf{r}_e)$ $[m_h(\mathbf{r}_h)]$ is the conduction-electron [heavy-hole] mass and $\mathbf{A} = \mathbf{e}_\varphi H\rho/2$ is the vector potential of the uniform magnetic field $\mathbf{B} = \mathbf{e}_z B$. $V_e(\mathbf{r}_e)$ $[V_h(\mathbf{r}_h)]$ is the bottom of the conduction band [top of the valence band], determined by the In content x, in the absence of strain. The strain-induced shift of the conduction band

$$\delta E_\beta = a_\beta(\varepsilon_{xx} + \varepsilon_{yy} + \varepsilon_{zz})$$
(4.5)

depends on the hydrostatic component of the strain tensor ε_{jk}. Here, $\beta = e, h$.

The shear strains give rise to the piezoelectric potential

$$V_P(\mathbf{r}) = -\frac{1}{4\pi\varepsilon_0\varepsilon_r}\int\frac{\text{div}\mathbf{P}}{|\mathbf{r} - \mathbf{r}'|}d^3\mathbf{r}'$$
(4.6)

determined by the piezoelectric polarization $P_i = e_{ijk}\varepsilon_{jk}$, where for InAs and GaAs only the piezoelectric moduli $e_{123} = e_{213} = e_{312}$ are other than zero; ε_r is the relative dielectric constant. The relevant materials parameters are: $e_{123}^{\text{InAs}} = 0.045$ C/m^2,

$e_{123}^{GaAs} = 0.16$ C/m^2, $a_e^{InAs} = -5.08$ eV, $a_e^{GaAs} = -7.17$ eV, $a_h^{InAs} = 1.00$ eV, $a_h^{GaAs} = 1.16$ eV [34]. The band gap and the effective masses as well as the parameters e_{123} and a_β for In$_x$Ga$_{1-x}$As are taken from a linear interpolation between the corresponding values for InAs and GaAs. We further assume that for the conduction and valence bands in In$_x$Ga$_{1-x}$As the band edge variations with x in the absence of strain are to each other as 7:3.

The single-particle Schrödinger equations are solved within the adiabatic approximation, using the Ansatz:

$$\Psi^{(\beta)}(\mathbf{r}) = \psi_k^{(\beta)}(z; \rho, \varphi)\Phi_{kj}^{(\beta)}(\rho, \varphi), \tag{4.7}$$

where the index k numbers subbands due to the size quantization along the z-axis:

$$\left[-\frac{\hbar^2}{2} \frac{\partial}{\partial z} \frac{1}{m_\beta(\rho, \varphi, z)} \frac{\partial}{\partial z} \pm V_\beta(\rho, \varphi, z) \pm \delta E_\beta(\rho, \varphi, z) \mp eV_P(\rho, \varphi, z) \right]$$

$$\times \psi_k^{(\beta)}(z; \rho, \varphi) = \mathcal{E}_k^{(\beta)}(\rho, \varphi)\psi_k^{(\beta)}(z; \rho, \varphi). \tag{4.8}$$

In (4.8), the upper (lower) sign stands for conduction electrons (heavy holes), i.e. for $\beta = e$ ($\beta = h$). The Schrödinger equation (4.8) for the "fast" degree of freedom (along the z-axis) is solved numerically, obtaining, in particular, the adiabatic potentials $\mathcal{E}_k^{(\beta)}(\rho, \varphi)$.

An example of the calculated adiabatic potential $\mathcal{E}_1^{(e)}(\rho, \varphi)$, which corresponds to the lowest state of the size quantization along the z-axis, is shown in Fig. 4.4(b). A QR—though it reveals a potential hill near its axis—is a singly connected structure. So, it is not evident whether or not electronic states in it resemble those in a doubly connected (ideal ring-like) geometry. Moreover, the adiabatic potential possesses two pronounced minima, which can be regarded as the potential profile of two quantum dots. If the potential minima are deep enough, then the electron is localized in one of those quantum dots, and no persistent current occurs at all.

Figure 4.6 shows the adiabatic potentials $\mathcal{E}_1^{(e)}(\rho, \varphi)$ corresponding to the indium distribution (see Fig. 4.5) and strain data for a realistic QR as found using the finite-element numerical calculation package ABAQUS [35], which is based on the elasticity theory. Due to strain, the depth of a potential well for an electron significantly decreases (cp. panels b and a in Fig. 4.6). The influence of the piezoelectric potential on the shape of the adiabatic potential $\mathcal{E}_1^{(e)}$ along the x- and y-axes is almost negligible. For the direction $x = y$, the effect of the piezoelectric potential on $\mathcal{E}_1^{(e)}$ is more pronounced, but still does not seem to be crucial in governing the electron in-plane motion.

The Schrödinger equations for the "slow" degrees of freedom,

$$\left[-\frac{\hbar^2}{2}\left(\nabla_{\rho,\varphi} \mp \frac{e}{\hbar}\mathbf{A} \right) \frac{1}{m_k^{(\beta)}(\rho, \varphi)} \left(\nabla_{\rho,\varphi} \mp \frac{e}{\hbar}\mathbf{A} \right) + \mathcal{E}_k^{(\beta)}(\rho, \varphi) \right]\Phi_{kj}^{(\beta)}(\rho, \varphi)$$

$$= E_{kj}^{(\beta)}\Phi_{kj}^{(\beta)}(\rho, \varphi), \tag{4.9}$$

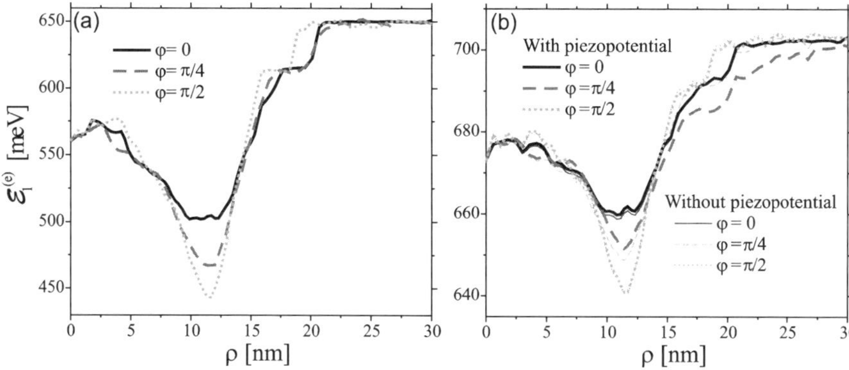

Fig. 4.6 Adiabatic potential, calculated in the absence (**a**) and in the presence (**b**) of strain, as a function of the radial coordinate ρ for three different in-plane directions, determined by the angular coordinate φ. In panel (**b**), *heavy* (*thin*) *curves* are obtained with (without) the piezoelectric potential. After Ref. [31]

Fig. 4.7 Magnetic moment induced by the ground-state persistent current as a function of the applied magnetic field for unstrained $In_{0.6}Ga_{0.4}As$ QRs with $R = 10.75$ nm, $h_0 = 1.6$ nm, $h_M = 3.6$ nm, $\gamma_0 = \gamma_\infty = 3$ nm, $\xi_R = 0$ at different values of the anisotropy parameters ξ_h and ξ_γ. After Ref. [31]

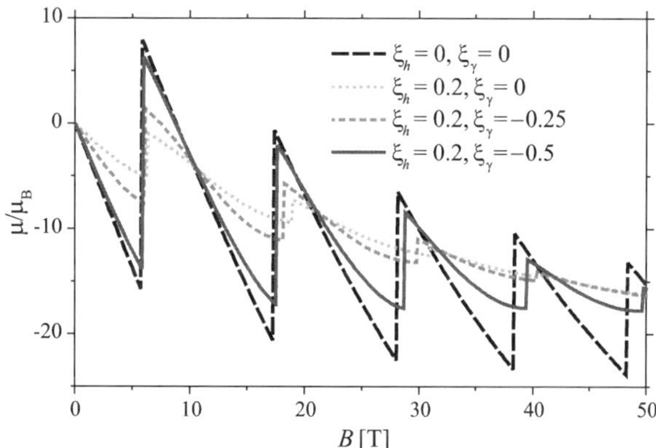

with the upper (lower) sign for $\beta = e$ ($\beta = h$) and the effective masses

$$m_k^{(\beta)}(\rho, \varphi) = \int dz \left| \psi_k^{(\beta)}(z; \rho, \varphi) \right|^2 m_\beta(\rho, \varphi, z) \qquad (4.10)$$

determine the eigenstates of the in-plane motion, which are labeled by the index j. We are interested in the lowest states of an electron and a hole in the QR. Therefore, we restrict our calculations to the states in the lowest subband of the (strong) size-quantization along the z-axis (i.e., we consider states with $k = 1$). For each value of the applied magnetic field, the electron and hole eigenstates in the QR are found by numerical diagonalization of the adiabatic Hamiltonian, which enters the lhs of (4.9).

The effect of the ring-height anisotropy on the oscillations of the calculated zero-temperature electron magnetic moment μ as a function of magnetic field B [36] is illustrated in Fig. 4.7, where we compare the magnetic moment for QRs with nonzero

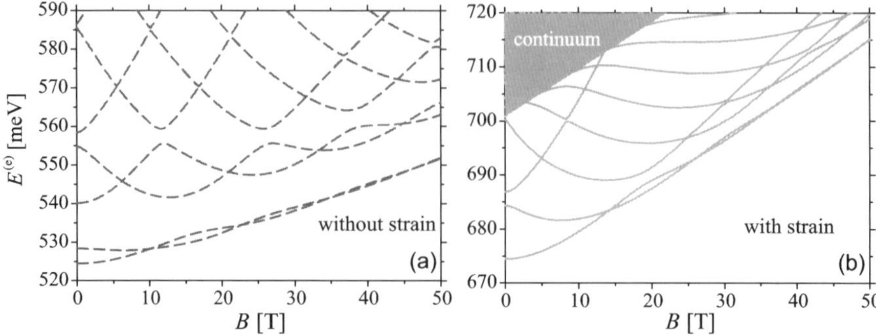

Fig. 4.8 Electron energy spectra in a QR, calculated without (**a**) and with (**b**) taking into account effects due to strain. Energies are counted from the bottom of the conduction band in unstrained InAs. The region of continuum as assessed by our numerical simulation is shadowed. After Ref. [31]

ξ_h and ξ_γ to the results for a perfectly symmetric ring. The magnetic moment of an electron in the applied magnetic field B is calculated as

$$\mu = -\frac{\mu_B}{Z} \sum_j \exp\left(-\frac{E_{1j}^{(e)}}{k_B T}\right) \frac{\partial E_{1j}^{(e)}}{\partial B}, \tag{4.11}$$

where μ_B is the Bohr magneton, T is the temperature and $Z = \sum_j \exp(-E_{1j}^{(e)}/(k_B T))$. Shape anisotropy of the QR results in a mixing of electron states with different magnetic quantum numbers. Variations of the height of the rim $\tilde{h}_M(\varphi)$ with φ tend to suppress oscillations of μ versus B. However, well-pronounced oscillations of $\mu(B)$ can be expected even for QRs with a strong shape anisotropy, provided that the width of the rim changes as a function of φ in antiphase with the rim height. Remarkably, this condition is satisfied for realistic SAQRs as characterized by X-STM [7] (see also Fig. 4.6).

In Fig. 4.8, the lowest electron energy levels, calculated with and without effects due to strain, are shown as a function of the applied magnetic field B. Since the potential well for an electron in the strained QR is relatively shallow [see Fig. 4.6(b)], there are only few discrete energy levels below the continuum of states in the GaAs barrier. Due to a reduced potential barrier at the center of a strained ring, the effective electron radius decreases when taking into account strain. Correspondingly, the transition magnetic fields, where the ground and first excited electron states interchange, are higher in a strained ring than in an unstrained one. Strain-induced effects reduce the magnitude of variations of the adiabatic potential as a function of the azimuthal angle. As a result, the mixing of electron states with different magnetic quantum numbers, which occurs due to shape anisotropy of a QR, is weakened. Consequently, at relatively weak fields B the energy spacing between the lowest electron state (which arises from the state with $L = 0$ in a circularly symmetric ring) and the first excited state (which arises from the state with $L = 1$ in a circularly symmetric ring) is strongly enhanced due to strain, while the zero-field splitting between

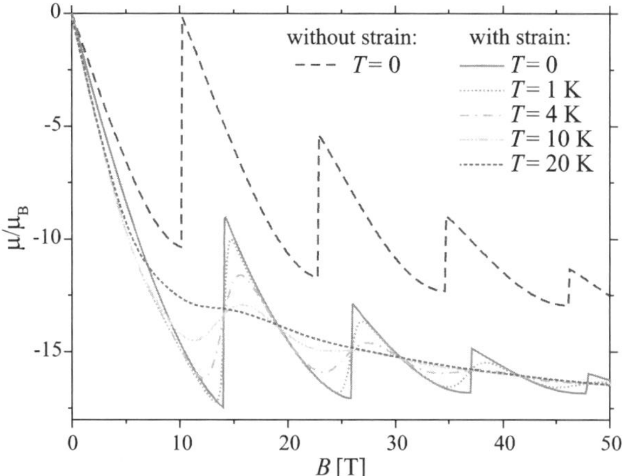

Fig. 4.9 Magnetic moment of an electron, calculated with and without taking into account effects due to strain. In the case of a strained QR, the results for different temperatures are shown. After Ref. [31]

the first and second excited states (which correspond to $L = 1$ and $L = -1$ in a circularly symmetric ring) is significantly reduced.

As shown in Fig. 4.9, the main effect of strain on the behavior of the electron magnetic moment $\mu(B)$ is a shift of transition fields towards higher B. This shift, already noticed when discussing the electron energy spectra, appears because strain makes shallower the potential well in the rim. When decreasing the depth of this potential well, the electron states tend to those in a flat disk. This leads to an overall shift of the curve $\mu(B)$ (at nonzero B) towards larger negative values. The oscillation amplitudes for $\mu(B)$ are not significantly influenced by strain. There are two competitive effects of strain on these amplitudes. On the one hand, due to strain-induced reduction of the potential-well depth, there is an increasing penetration of the electron wave function into barriers, so that the effective width of the ring increases. Such an increase of the ring width tends to decrease the oscillation amplitude. On the other hand, the strain-induced reduction of the potential-well depth weakens the influence of shape anisotropy on electron states. Correspondingly, the suppression effect of shape anisotropy on the oscillations of $\mu(B)$ is weakened, too. As further illustrated in Fig. 4.9, an increase of the temperature tends to smooth out the Aharonov-Bohm oscillations of $\mu(B)$. Importantly, suppression of the first Aharonov-Bohm oscillations (at relatively low B) is not dramatic for liquid-helium temperatures.

4.5 Observation of the AB Effect Through Magnetization

In Ref. [9], the AB oscillations of persistent currents were detected via the magnetic moment of electrons in a highly homogeneous ensemble of InAs self-assembled QRs. The magnetic moment of the nanostructures was obtained from magnetization experiments using a torque magnetometer [37]. These measurements were performed at temperatures of $T = 1.2$ K and $T = 4.2$ K in magnetic fields up to

Fig. 4.10 (**a**) Oscillation in the magnetic moment per electron, obtained at 1.2 K and at 4.2 K, after subtracting the linear background from the measured signal, dividing by the total number of electrons, and averaging over several measurements. The *inset* shows the first derivative of the experimental magnetic moment with respect to B at $T = 1.2$ K. (**b**) Calculated magnetic moment, and its derivative (*inset*), of a single electron in a nanostructure at different temperatures. After Ref. [9]

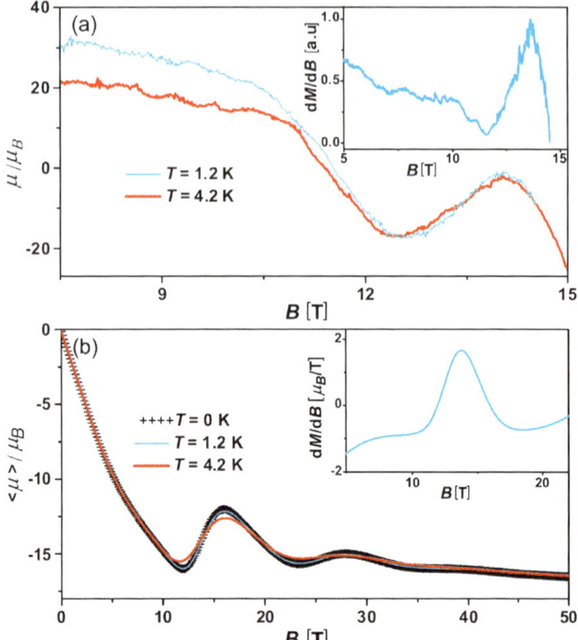

15 T. The sensitivity of the magnetometer was 2.8×10^{-12} J/T, i.e. $3 \times 10^{11} \mu_B$ at $B = 14$ T, and it was limited by mechanical noise, which was about 8 % of the experimentally observed oscillation magnitude. The total magnetization of the sample was due to about 1.5×10^{11} nanostructures with a total number of electrons $N \sim 2.2 \times 10^{11}$. Over the entire magnetic field range, a relatively large background signal was observed due to the substrate and to dia- and paramagnetic materials close to the sample. To enhance the visibility of the AB oscillations, this linear background was subtracted and the signal was normalized to the total number of electrons N in the sample, resulting in the magnetic moment per electron $\mu = M/N$ [Fig. 4.10(a)]. To prove that the observed oscillation is not an artefact due to the background subtraction, we also plot the first derivative of the signal [inset to Fig. 4.10(a)], which is much less sensitive to the monotonous background. This procedure reveals an oscillation around 14 T as a fingerprint of the AB effect.

Figure 4.10(b) shows the calculated magnetization curve, which results from averaging over a QR ensemble with a size dispersion of 5 %, consistent with the measured width of the PL peak. As seen from a comparison between Figs. 4.10(a) and 4.10(b), the model described in the previous section accurately explains the position of the observed AB oscillation around 14 T, rather than at 5 T expected for an ideal 1D ring of the same radius. This difference in the position of the AB oscillations is due to the influence of strain in the self-assembled "volcano-like" QRs as well as to the singly-connectedness of these QRs. Figure 4.10(b) also shows the calculated magnetic moment for higher magnetic fields that are not yet accessible by magnetization experiments. The higher order AB oscillations are strongly damped. This is a consequence of the presence of the magnetic field in the rim of the QRs, which enhances the electron localization close to the minima of the adiabatic potential. In Fig. 4.10(b) the calculated results are plotted for three temperatures. Without including size variations of the QRs, the calculated amplitude of the AB oscillations decreases with increasing temperature (see Fig. 4.9). A negligible temperature effect on the electron magnetic moment in Fig. 4.10(b) is due to the QR ensemble averaging.

The shown results confirm the existence of an oscillatory persistent current in self-assembled QRs containing only a single electron. Even though the investigated nanostructures are singly connected and anisotropic, they show the AB behavior that is generally considered to be restricted to ideal (doubly connected) topologies.

4.6 Theory of Two-Electron Systems and Excitons in Quantum Rings

The purpose of the first part of this section is to consider the contribution from QRs with two electrons to the magnetization. The Hamiltonian of the two electrons in a QR is represented as

$$H_{ee}(\mathbf{r}_1, \mathbf{r}_2) = H_e(\mathbf{r}_1) + H_e(\mathbf{r}_2) + V_{\text{Coul}}(\mathbf{r}_1, \mathbf{r}_2), \tag{4.12}$$

where $H_e(\mathbf{r}_1)$ is the single-electron Hamiltonian [see (4.3)], $V_{\text{Coul}}(\mathbf{r}_1, \mathbf{r}_2)$ describes the Coulomb interaction between the electrons with radius-vectors $\mathbf{r}_1$ and $\mathbf{r}_2$. In order to obey the Pauli exclusion principle, the spin-singlet (spin-triplet) states in the two-electron rings must possess orbital wave functions which are symmetric (antisymmetric) with respect to the permutation of the coordinates of electrons. Aimed at finding two-electron eigenstates, we start with constructing the basis functions, which describe the orbital wave functions of spin-singlet and spin-triplet states in the absence of the electron-electron interaction:

$$\Psi_{j_1 j_2}^{(ee,0)}(\mathbf{r}_1, \mathbf{r}_2) = c_{j_1 j_2} \left[\Psi_{1 j_1}^{(e)}(\mathbf{r}_1) \Psi_{1 j_2}^{(e)}(\mathbf{r}_2) + \Psi_{1 j_1}^{(e)}(\mathbf{r}_2) \Psi_{1 j_2}^{(e)}(\mathbf{r}_1) \right], \tag{4.13}$$

$$\Psi_{j_1 j_2}^{(ee,1)}(\mathbf{r}_1, \mathbf{r}_2) = c_{j_1 j_2} \left[\Psi_{1 j_1}^{(e)}(\mathbf{r}_1) \Psi_{1 j_2}^{(e)}(\mathbf{r}_2) - \Psi_{1 j_1}^{(e)}(\mathbf{r}_2) \Psi_{1 j_2}^{(e)}(\mathbf{r}_1) \right], \quad j_1 \neq j_2, \tag{4.14}$$

where $c_{j_1 j_2} = 1/\sqrt{2}$ for $j_1 \neq j_2$ and $c_{j_1 j_1} = 1/2$. We numerically diagonalize the Hamiltonian (4.12) in the above basis, looking for the wave functions of the two interacting electrons in the form

$$\tilde{\Psi}_J^{(ee,S)}(\mathbf{r}_1, \mathbf{r}_2) = \sum_{j_1=1}^{j_{max}} \sum_{j_2=1}^{j_1-S} A_{J j_1 j_2} \Psi_{j_1 j_2}^{(ee,S)}(\mathbf{r}_1, \mathbf{r}_2), \tag{4.15}$$

where $S = 0$ ($S = 1$) in the case of spin-singlet (spin-triplet) states (see Ref. [38] for more details).

For short, the states $\Psi_{j_1 j_2}^{(ee,S)}$ for $S = 0, 1$ are labeled below as $(j_1, j_2)^S$, where the numbers j_1 and j_2 correspond to the order of the single-electron energy levels

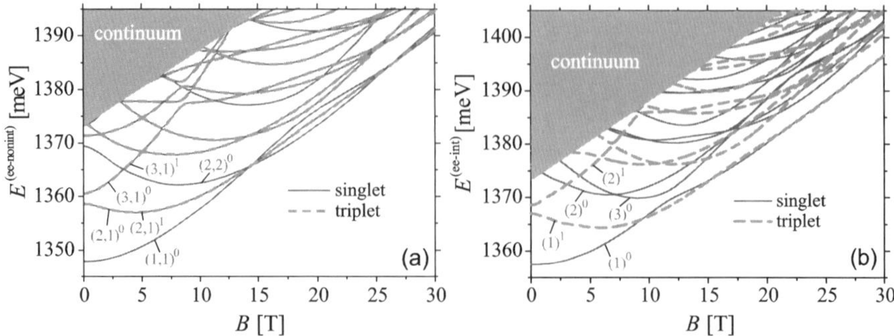

Fig. 4.11 Energy spectrum of two noninteracting (**a**) and interacting (**b**) electrons in a strained QR as a function of the applied magnetic field. For $S = 0, 1$ the states $\Psi_{j_1,j_2}^{ee,S}$ of noninteracting electrons are labeled as $(j_1, j_2)^S$, where the numbers j_1 and j_2 correspond to the order of the single-electron energy levels at $B = 0$. All triplet energy levels $(j_1, j_2)^1$ for $j_1 \neq j_2$ overlap with singlet energy levels $(j_1, j_2)^0$. The states $\tilde{\Psi}_J^{ee,S}$ of interacting electrons are labeled as $(J)^S$ for $S = 0, 1$. The region of the energy continuum as obtained from our numerical simulation is shadowed. After Ref. [38]

at $B = 0$. The states $\tilde{\Psi}_J^{(ee,S)}$ are labeled as $(J)^S$ for $S = 0, 1$. In Fig. 4.11 the calculated two-electron energy spectra are plotted for the cases of no electron-electron interaction and with the Coulomb interaction taken into account. It is worth recalling here that the index j, rather than the electron angular momentum L, is specific for single-electron eigenstates in an anisotropic QR. Approximately, the correspondence between the states of two electrons in a QR with account for the Coulomb interaction and the states of two non-interacting electrons in a QR is as follows: $(1, 1)^0 \rightarrow (1)^0$; $(2, 1)^0 \rightarrow (2)^0$; $(2, 2)^0 \rightarrow (3)^0$; $(2, 1)^1 \rightarrow (1)^1$; $(3, 1)^1 \rightarrow (2)^1$. Due to the Coulomb interaction, the degeneracy between the spin-singlet and spin-triplet states is lifted. For example, there is a significant splitting between the spin-singlet $(2)^0$ and the spin-triplet $(1)^1$ states, although the corresponding states of two non-interacting electrons $(2, 1)^0$ and $(2, 1)^1$ are degenerate. This fact indicates a strong exchange interaction in the QR under consideration.

While at relatively low magnetic fields the ground state in Fig. 4.11(b) corresponds to the lowest spin-singlet energy level, at $B \approx 10$ T the ground state becomes spin-triplet. With further increasing magnetic field, the lowest spin-singlet and spin-triplet states sequentially replace each other as the ground state. This behavior is reminiscent of the Aharonov-Bohm effect in a single-electron QR. At $B \approx 12$ T the state originating from $(1, 1)^0$ reveals anticrossing from those states, which originate from $(2, 1)^0$ and $(2, 2)^0$.

In Fig. 4.12, the calculated magnetic moment μ of the two non-interacting and interacting electrons is plotted as a function of the applied magnetic field. The Coulomb interaction leads to a more complicated oscillating structure of μ versus B as compared to the case when the electron-electron interaction is absent. In particular, the first oscillation of the magnetic moment shifts due to the Coulomb interaction towards weaker magnetic fields. One of the reasons for this shift is an

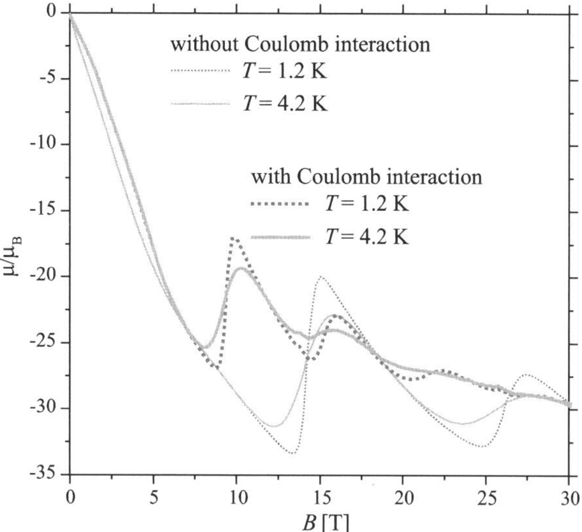

Fig. 4.12 The calculated magnetic moment of two noninteracting (interacting) electrons in a strained quantum ring is shown by the *thin* (*heavy*) *lines* for two different temperatures. After Ref. [38]

increase of the effective electronic radius of the ring due to the mutual Coulomb repulsion of the two electrons. At $B > 15$ T, the Aharonov-Bohm oscillations of the magnetic moment are still present, but substantially smoothed out.

As implied by Fig. 4.12, for two-electron anisotropic QRs with radial sizes $\sim$10 nm, the Aharonov-Bohm-effect-related phenomena appear at magnetic fields $\sim$10 T. In the experiment on magnetization [9], no appreciable oscillations are detected in the above region. Therefore one may assume that the observed Aharonov-Bohm effect [9] is mainly due to the single-electron QRs in the ensemble of rings under investigation.

In Refs. [39, 40], the method presented in Ref. [38] for two-electron QRs, has been extended to excitons in QRs and applied to calculate the exciton energy spectra and the optical transition probabilities in QRs with a realistic anisotropic singly connected shape. The Hamiltonian of an exciton in a strained quantum ring is

$$H_{\text{ex}} = H_e + H_h + V_{\text{Coul}}(\mathbf{r}_e, \mathbf{r}_h), \tag{4.16}$$

where H_e [H_h] is the single-particle Hamiltonian of an electron [a hole] determined by (4.3) [(4.4)], $V_{\text{Coul}}(\mathbf{r}_e, \mathbf{r}_h)$ describes the Coulomb interaction between an electron and a hole with radius-vectors $\mathbf{r}_e$ and $\mathbf{r}_h$, respectively. We consider here only heavy holes and treat them within the one-band model.

In order to find exciton eigenstates, we start with constructing the basis functions, which describe a non-interacting eh-pair:

$$\Psi^{(eh)}_{j_e j_h}(\mathbf{r}_e, \mathbf{r}_h) = \Psi^{(e)}_{1 j_e}(\mathbf{r}_e) \Psi^{(h)}_{1 j_h}(\mathbf{r}_h). \tag{4.17}$$

Then we diagonalize the Hamiltonian (4.16) in the above basis, looking for the exciton wavefunctions in the form

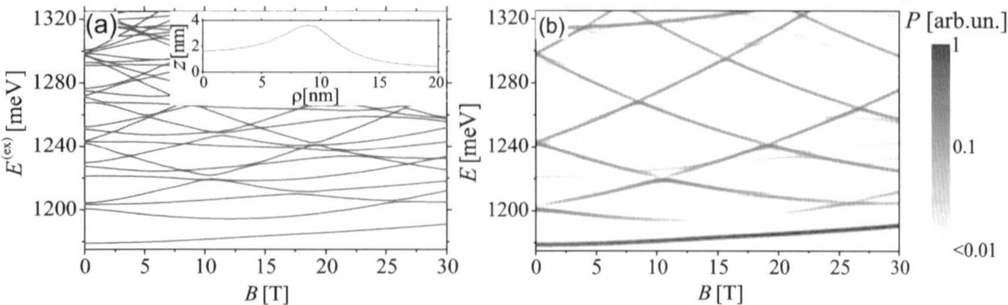

Fig. 4.13 (**a**) Energies of the lowest dipole active exciton states in an unstrained axially symmetric In$_{0.55}$Ga$_{0.45}$As/GaAs ring-like structure as a function of the applied magnetic field. *Inset*: Shape of the structure. (**b**) Spectral distribution of the optical transition probabilities P for an exciton as a function of the applied magnetic field ($\Gamma = 1$ meV). After Ref. [40]

$$\Psi_J^{(ex)}(\mathbf{r}_e, \mathbf{r}_h) = \sum_{j_e, j_h=1}^{j_{max}} A_{Jj_e j_h} \Psi_{j_e j_h}^{(eh)}(\mathbf{r}_e, \mathbf{r}_h). \tag{4.18}$$

Then for each value of the applied magnetic field B the lowest exciton states are found by numerical diagonalization of the Hamiltonian (4.16) in the basis (4.17).

The effect of the electron-hole Coulomb interaction on the energy spectrum and transition probabilities is illustrated in Fig. 4.13 for an unstrained azimuthally isotropic In$_{0.55}$Ga$_{0.45}$As/GaAs ring-like structure with the shape shown in the inset to Fig. 4.13(a). Due to strong selection rules, applicable for the highly symmetric structure under consideration, only a small fraction of the states of the non-interacting eh-pair, namely the states with zero envelope angular momentum $L_e + L_h = 0$, are dipole active. As seen from Fig. 4.13(a), the electron-hole Coulomb interaction, which mixes eh-states with the same $L_e + L_h$ and different $L_e(=-L_h)$, n_e, n_h, leads to anticrossing for the lowest two dipole-active energy levels (for a non-interacting eh-pair, the corresponding levels cross each other). This mixing of states due to the electron-hole Coulomb interaction is accompanied by a redistribution of oscillator strengths (usually, in favor of lower-lying states). This is illustrated in Fig. 4.13(b), where we plot the spectral distributions of the calculated probabilities P for optical transitions to different states of an exciton. In order to enhance visualization, a small Gaussian broadening (~ 1 meV) is introduced for all the energy levels. In the energy range, which corresponds to anticrossing of the lowest two dipole active energy levels, the upper of these two levels becomes "dark", while the transition probability P for the lower level increases.

The electron energy spectrum, the magnetization, and the optical-transition probabilities were analyzed for excitons in strained axially symmetric ring-like structures [see the inset in Fig. 4.14(a)] with different In concentration in the In-rich region. The strain-induced flattening of the adiabatic potential makes the structure under consideration "disk-like" rather than "ring-like". However, when increasing the In content x, the depth of the adiabatic potential well in the In-rich region significantly increases both for an electron and for a hole. This results, in particular, in a lowering of the electron ground-state energy [see Fig. 4.14(a)]. In Fig. 4.14(b), we

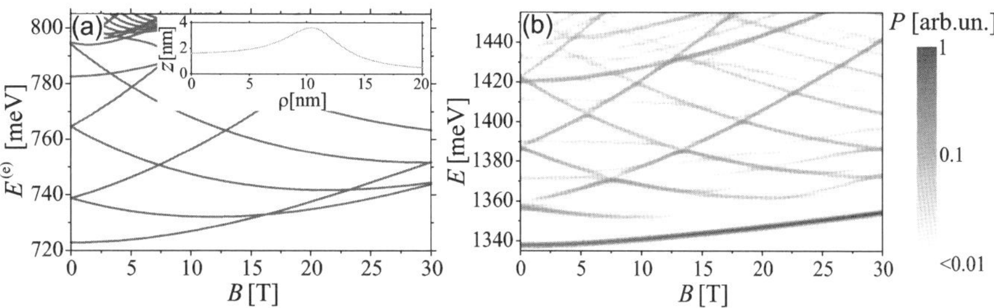

Fig. 4.14 (**a**) Energies of the lowest electron states in a strained axially symmetric $In_{0.65}Ga_{0.35}As/GaAs$ ring-like structure as a function of the applied magnetic field. *Inset*: Shape of the structure. (**b**) Spectral distribution of the optical transition probabilities P for an exciton as a function of the applied magnetic field ($\Gamma = 1$ meV). After Ref. [40]

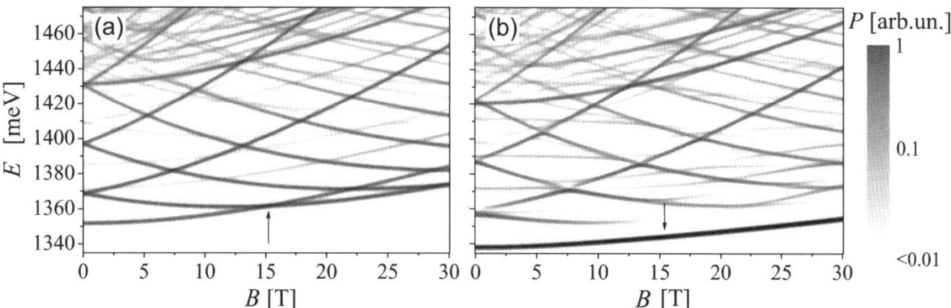

Fig. 4.15 Calculated optical transition probabilities for an anisotropic strained $In_{0.7}Ga_{0.3}As/GaAs$ QR with $R = 9$ nm, $h_0 = 1.6$ nm, $h_M = 3.6$ nm, $h_\infty = 0.4$ nm, $\gamma_0 = \gamma_\infty = 3$ nm, $\xi_R = \xi_\gamma = 0$, $\xi_h = 0.2$, $\Gamma = 1$ meV, in the case of (**a**) a noninteracting electron-hole pair and (**b**) an interacting electron-hole pair. The *arrows* correspond to the first excitonic AB resonance in the ground state. After Ref. [41]

show spectral distributions of the calculated probabilities P for optical transitions to different states of an exciton in a strained $In_{0.65}Ga_{0.35}As/GaAs$ ring-like structure. At relatively low magnetic fields B, a complicated pattern of P starts for energies ~ 100 meV above the lowest exciton state. This energy region actually corresponds to continuum states.

In Figs. 4.15 and 4.16 we show the calculated probabilities P of optical transitions for ring-like structures with $x = 0.7$ and with realistic shape anisotropy. As seen from Fig. 4.15(a), for a noninteracting electron-hole pair there is a crossover in the ground-state energy around $B = 15$ T, in agreement with magnetization experiments [9]. The Coulomb interaction results into a smooth behavior of the ground-state energy as function of B, as shown in Fig. 4.15(b). Due to the Coulomb interaction the oscillator strengths of the two lowest "bright" *eh*-states are significantly redistributed in favor of the lowest level. Therefore, despite a relatively large splitting between these two exciton energy levels, the upper one is practically unresolvable in the case of an appreciable energy-level broadening, $\Gamma = 10$ meV (see Fig. 4.16).

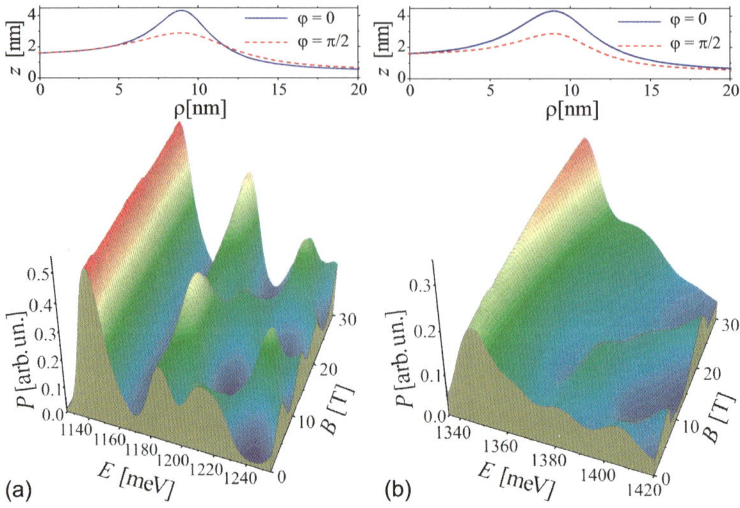

Fig. 4.16 (**a**) Shape of the structure (*top*) and broadened spectral distributions of P for an exciton in the anisotropic unstrained $In_{0.7}Ga_{0.3}As/GaAs$ QR with $R = 9$ nm, $h_0 = 1.6$ nm, $h_M = 3.6$ nm, $h_\infty = 0.4$ nm, $\gamma_0 = \gamma_\infty = 3$ nm, $\xi_R = 0$, $\xi_h = 0.2$, $\xi_\gamma = -0.25$, and with $\Gamma = 10$ meV. (**b**) Same as in panel (**a**) but for a strained ring with $\xi_\gamma = 0$. After Ref. [39]

4.7 Experiments on Excitonic Properties of Quantum Rings

In Ref. [41], the exciton energy level structure of a large ensemble of InAs/GaAs quantum rings was experimentally investigated by PL spectroscopy in magnetic fields up to 30 T for different excitation densities. For the PL studies, a sample containing a single layer of QRs [16] was mounted in a liquid-helium bath cryostat at $T = 4.2$ K. Static magnetic fields up to 30 T were applied parallel to the growth direction and the PL was detected in the Faraday configuration. The dependence of the QR emission energy on the excitation density is shown in Fig. 4.17(a). The ground-state emission energy of the QRs is centered at around 1.308 eV, typical for these nanostructures [16]. The ground-state emission has an inhomogeneous broadening with a full width at half maximum of 20 meV. With increasing excitation density, two additional peaks can be resolved. These peaks have an energy of 39 and 63 meV above the ground-state energy.

To investigate the influence of the ring-like geometry on the excitonic behavior in the excited states of the QRs, the magneto-PL of these structures for higher excitation intensities was measured. With increasing B, both resolvable excited states split up in two separate peaks. Each of the PL peaks splits further with a smaller energy separation into two peaks of opposite circular polarization [see Fig. 4.17(b)]. Figure 4.18(a) shows the higher excitation data in σ^--polarization as a function of B at intervals of 1 T. The calculated transition probabilities are plotted in Fig. 4.18(b). The calculated and measured spectra show a qualitative resemblance. Within the theoretical model, which was successfully applied to explain the magnetization behavior of QRs on similar samples [9, 31], one finds that for all realistic ring param-

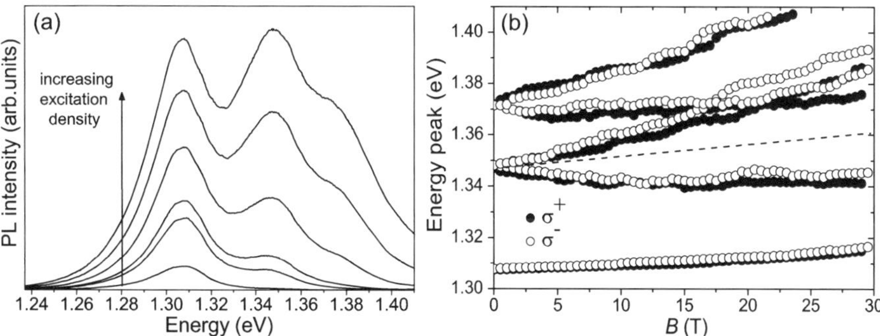

Fig. 4.17 (**a**) PL as a function of excitation density, for which the lowest (highest) excitation density is 102 W cm^{-2} (105 W cm^{-2}). Two excited states can be distinguished for higher excitation density located 38 and 63 meV above the ground-state emission energy. (**b**) The energy diagram showing the peak position in B in both σ^{-}- (*empty circles*) and σ^{+}- (*filled circles*) polarization. The QRs exhibit splittings into two branches corresponding to the different excited states, in contrast to QDs, where a third peak (indicated by the *dashed line*) is observed. After Ref. [41]

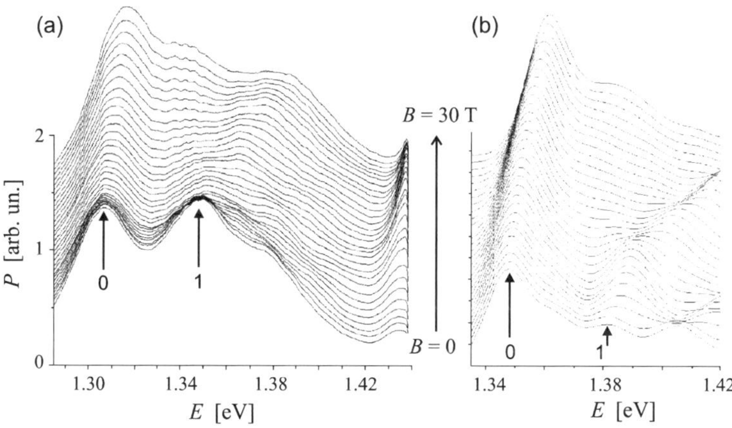

Fig. 4.18 (**a**) Experimental σ^{+}-PL spectra of self-assembled InGaAs/GaAs QRs in magnetic fields from 0 T (the *lowest curve*) to 30 T (the *highest curve*) with the step 1 T. The two lowest peaks at $B = 0$ are indicated with labels "0" (at 1.30735 eV) and "1" (at 1.34945 eV), corresponding to the peaks "0" and "1" in panel (**b**), where the theoretical spectra of the optical transition probabilities are shown [cf. Fig. 4.16(b)]. After Ref. [40]

eters the PL of the first-excited state is concealed by the ground-state luminescence if an inhomogeneous broadening of about 20 meV is included. The ring character of the nanostructures under consideration results in non-equidistant energy level splittings in the exciton diagram and in a magnetic-field induced splitting of each excited state into two states. This is in contrast to what has been observed in the measurements on QDs [42–44], which show a magnetic-field induced splitting of the d-state into three states and equidistant energy levels.

In contrast to the AB effect of single electrons in QRs under consideration, an excitonic AB effect in an ensemble of those QRs was not observed nor expected based on the theoretical model [40, 41]. The absence of prominent oscillations in the ground-state energy of the calculated exciton spectra as compared to the case of a noninteracting electron-hole pair is a consequence of the Coulomb interaction. As implied by the results of magneto-PL measurements on single InGaAs/GaAs QRs [45], the amplitude of the AB oscillations in the ground-state energy of a QR is as small as about few μeV, so that in a PL spectrum of an ensemble of QRs these oscillations are fully hidden due to the size and shape dispersion of QRs.

4.8 Applications of QRs

From a fundamental point of view, semiconductor QRs are fascinating topological structures, whose unique features are represented in the present book. From an applied point of view it is less obvious, where the applications of QRs can be found, if one wants to utilize the unique topological properties of QRs. The suggested applications of QRs that make use of the topological properties are, not surprisingly, primarily found in the field of spintronics and involve spin manipulation of carriers captured in or passing through the ring structure. In order to exploit the topological properties of QRs, it is crucial that the phase scattering and spin scattering rates of carriers in the QR, which lead to decoherence of QR states, are small. This unfortunately limits any by now suggested application to low temperatures and makes the chance of real applications in the near future very dim. However, additional applications have been suggested that do not depend on the topological properties of QRs, but use, for instance, the special charge distribution or the frequently observed anisotropy in the shape of self-assembled rings. Below, we will first discuss some examples of applications that do not utilize the topological properties of QRs and then a number of examples that are based on topologically induced features.

In 2003, Granados and García [16] reported a strong dependence of the PL spectrum on the actual shape of InAs/GaAs self-assembled quantum structures. Depending on the growth conditions during capping and annealing of InAs self-assembled dots they observe QDs, QRs and more complicated structures. The luminescence for these structures varied from 1.1 eV (QDs) to 1.35 eV (QRs) depending on the shape of the nanostructure. This shift is of course due to the shallower confinement of the carriers in a QR compared to those captured in a QD. This feature does not, however, make them attractive, because typically more effort is put in getting the emission of semiconductor nanostructures at longer (rather than shorter) wavelengths that are accessible by standard InAs/GaAs QDs. Nevertheless, InAs/GaAs QR structures have been analyzed for their use in THz detectors, because they give rise to a very low dark current [46] and GaAs/AlGaAs QRs have been proposed in solar cells because these strain-free QRs might provide a better quantum efficiency [47], see also Chap. 7. In these applications, neither the topology, nor the ring-shaped charge distribution is essential for the operation of the device.

The following applications make use of the special geometry of the ring without exploiting the topological aspects. Due to the fact, discussed in detail in the present chapter, that InAs/GaAs QRs are often asymmetric [7], they can give rise to a strong optical anisotropy [48, 49]. This property can be useful in situations, where optical polarization detection or manipulation is needed. The ring-shaped potential, independent of any in-plane anisotropy, is important in type-II QRs that have been shown to work as memory devices with a long storage time because oi the spatial separation of the electron and hole, which prevents carrier recombination [50]. In these devices the hole is confined in a dot potential, whereas the electron is confined in a ring potential that runs around the dot. Wavefunction engineering in a similar quantum dot-ring nanostructure was proposed in a theoretical study by Zipper et al. [51], see also Chap. 18. They proposed electrical gating to control exciton relaxation and absorption by manipulating the confinement parameters. The envisioned device can be actively tuned from being highly absorbing to almost transparent in a specific part of the infrared or microwave spectrum.

More sophisticated is the idea to control the transition probability of an electron-hole pair captured in a ring by transforming a bright exciton state into a dark exciton state and vice versa [52]. In this device, the combination of a perpendicular magnetic and in-plane electric field is used to control the exciton state and this should make it possible to store and to release an optically excited exciton in a QR, which can also be read out by an additional approach. Interestingly, GaAs/AlGaAs double QRs obtained by droplet epitaxy [53] have been proven to show photon antibunching, which is essential for single photon devices, see also Chap. 8. Only the inner ring showed a strong photon antibunching, whereas this is relaxed in the outer ring probably due to a stronger inhomogeneity in the potential of the outer ring. The authors suggest that these single photon emitting ring structures can be of relevance for quantum computational devices.

The most exciting applications are of course those that are based on the topological properties of the ring geometry. Not without surprise, these applications lie in the field of quantum computing, where spin manipulation is of essence. For instance, sequential spin flips were observed in lithographic GaAs/AlGaAs QRs [54]. In these QRs, operating in the Coulomb blockade regime, experimental results were obtained that showed the ability to probe individual spin flips, which is an important step towards accurate spin control. The detection of spin flips was possible due to the characteristic electronic spectrum of the QR structure. Spin relaxation in QRs and QDs was explored theoretically in detail by Zipper et al. [55], see also Chap. 18. The authors report higher spin stability in a QR than in a QD, supporting the use of QRs as a spin qubit. They also present a discussion on quantum state manipulation in a QR in relation to its use as a spin qubit [55].

In summary, the ongoing studies revealed great potentialities of QRs as basis elements for a broad spectrum of applications, starting from THz detectors, efficient solar cells and memory devices, through electrically tunable optical valves and single photon emitters, and further to spin qubits for quantum computing.

References

1. J.M. García, G. Medeiros-Ribeiro, K. Schmidt, T. Ngo, J.L. Feng, A. Lorke, J. Kotthaus, P.M. Petroff, Appl. Phys. Lett. **71**, 2014 (1997)
2. A. Lorke, R.J. Luyken, A.O. Govorov, J.P. Kotthaus, J.M. García, P.M. Petroff, Phys. Rev. Lett. **84**, 2223 (2000)
3. R.J. Warburton, C. Schäflein, D. Haft, F. Bickel, A. Lorke, K. Karrai, J.M. García, W. Schoenfeld, P.M. Petroff, Nature **405**, 926 (2000)
4. R.J. Warburton, C. Schulhauser, D. Haft, C. Schäflein, K. Karrai, J.M. García, W. Schoenfeld, P.M. Petroff, Phys. Rev. B **65**, 113303 (2002)
5. V.M. Fomin, J. Nanoelectron. Optoelectron. **6**, 1 (2011)
6. G. Biasiol, S. Heun, Phys. Rep. **500**, 117 (2011)
7. P. Offermans, P.M. Koenraad, J.H. Wolter, D. Granados, J.M. García, V.M. Fomin, V.N. Gladilin, J.T. Devreese, Appl. Phys. Lett. **87**, 131902 (2005)
8. P. Offermans, P.M. Koenraad, J.H. Wolter, D. Granados, J.M. García, V.M. Fomin, V.N. Gladilin, J.T. Devreese, Physica E **32**, 41 (2006)
9. N.A.J.M. Kleemans, I.M.A. Bominaar-Silkens, V.M. Fomin, V.N. Gladilin, D. Granados, A.G. Taboada, J.M. García, P. Offermans, U. Zeitler, P.C.M. Christianen, J.C. Maan, J.T. Devreese, P.M. Koenraad, Phys. Rev. Lett. **99**, 146808 (2007)
10. T.-C. Lin, C.-H. Lin, H.-S. Ling, Y.-J. Fu, W.-H. Chang, S.-D. Lin, C.-P. Lee, Phys. Rev. B **80**, 081304(R) (2009)
11. P. Moon, K. Park, E. Yoon, J.-P. Leburton, Phys. Status Solidi RRL **3**, 76 (2009)
12. P. Moon, W.J. Choi, K. Park, E. Yoon, J.D. Lee, J. Appl. Phys. **109**, 103701 (2011)
13. V. Arsoski, N. Čukarić, M. Tadić, F.M. Peeters, Phys. Scr. T **149**, 014054 (2012)
14. L.M. Thu, W.T. Chiu, O. Voskoboynikov, Phys. Rev. B **85**, 205419 (2012)
15. R. Blossey, A. Lorke, Phys. Rev. E **65**, 021603 (2002)
16. D. Granados, J.M. García, Appl. Phys. Lett. **82**, 2401 (2003)
17. D. Granados, J.M. García, T. Ben, S.I. Molina, Appl. Phys. Lett. **86**, 071918 (2005)
18. V. Baranwal, G. Biassol, S. Heun, A. Locatelli, T.O. Mentes, M.N. Orti, L. Sorba, Phys. Rev. B **80**, 155328 (2009)
19. C.H. Lin, H.S. Lin, C.C. Huang, S.K. Su, S.D. Lin, K.W. Sun, C.P. Lee, Y.K. Liu, M.D. Yang, J.L. Shen, Appl. Phys. Lett. **94**, 183101 (2009)
20. M.D. Teodoro, A. Malachias, V. Lopes-Oliveira, D.F. Cesar, V. Lopez-Richard, G.E. Marques, E. Marega Jr., M. Benamara, Yu.I. Mazur, G.J. Salamo, J. Appl. Phys. **112**, 014319 (2012)
21. Y. Lv, J. Cui, Z.M. Jiang, X. Yang, Nanoscale Res. Lett. **7**, 659 (2012)
22. P.W. Fry, I.E. Itskevich, D.J. Mowbray, M.S. Skolnick, J.J. Finley, J.A. Barker, E.P. O'Reilly, L.R. Wilson, I.A. Larkin, P.A. Maksym, M. Hopkinson, M. Al-Khafaji, J.P.R. David, A.G. Cullis, G. Hill, J.C. Clark, Phys. Rev. Lett. **84**, 733 (2000)
23. J.A. Barker, R.J. Warburton, E.P. O'Reilly, Phys. Rev. B **69**, 035327 (2004)
24. L.G. Wang, P. Kratzer, M. Scheffler, Q.K.K. Liu, Appl. Phys. A, Mater. Sci. Process. **73**, 161 (2001)
25. Z.R. Wasilewski, S. Fafard, J.P. McCaffrey, J. Cryst. Growth **201/202**, 1131 (1999)
26. E. Steimetz, T. Wehnert, H. Kirmse, F. Poser, J.-T. Zettler, W. Neumann, W. Richter, J. Cryst. Growth **221**, 592 (2000)
27. A. Lenz, H. Eisele, R. Timm, S.K. Becker, R.L. Sellin, U.W. Pohl, D. Bimberg, M. Dähne, Appl. Phys. Lett. **85**, 3848 (2004)
28. K. Shiraishi, Appl. Phys. Lett. **60**, 1363 (1992)
29. P. Offermans, P. Koenraad, J. Wolter, K. Pierz, M. Roy, P. Maksym, Physica E **26**, 236 (2005)
30. D.M. Bruls, J.W.A.M. Vugs, P.M. Koenraad, H.W.M. Salemink, J.H. Wolter, M. Hopkinson, M.S. Skolnick, F. Long, S.P.A. Gill, Appl. Phys. Lett. **81**, 1708 (2002)
31. V.M. Fomin, V.N. Gladilin, S.N. Klimin, J.T. Devreese, N.A.J.M. Kleemans, P.M. Koenraad, Phys. Rev. B **76**, 235320 (2007)
32. M. Grundmann, O. Stier, D. Bimberg, Phys. Rev. B **52**, 11969 (1995)
33. J.A. Barker, E.P. O'Reilly, Phys. Rev. B **61**, 13840 (2000)

34. C.G. Van de Walle, Phys. Rev. B **39**, 1871 (1989)
35. http://www.simulia.com/products/abaqus_fea.html
36. V.M. Fomin, V.N. Gladilin, J.T. Devreese, P. Offermans, P.M. Koenraad, J.H. Wolter, J.M. García, D. Granados, AIP Conf. Proc. **772**, 803 (2005)
37. M.R. Schaapman, P.C.M. Christianen, J.C. Maan, D. Reuter, A.D. Wieck, Appl. Phys. Lett. **81**, 1041 (2002)
38. V.M. Fomin, V.N. Gladilin, J.T. Devreese, N.A.J.M. Kleemans, P.M. Koenraad, Phys. Rev. B **77**, 205326 (2008)
39. V.M. Fomin, V.N. Gladilin, J.T. Devreese, N.A.J.M. Kleemans, M. Bozkurt, P.M. Koenraad, Phys. Status Solidi (b) **245**, 2657 (2008)
40. V.M. Fomin, V.N. Gladilin, J.T. Devreese, J.H. Blokland, P.C.M. Christianen, J.C. Maan, A.G. Taboada, D. Granados, J.M. García, N.A.J.M. Kleemans, H.C.M. van Genuchten, M. Bozkurt, P.M. Koenraad, Proc. SPIE **7364**, 736402 (2009)
41. N.A.J.M. Kleemans, J.H. Blokland, A.G. Taboada, H.C.M. van Genuchten, M. Bozkurt, V.M. Fomin, V.N. Gladilin, D. Granados, J.M. García, P.C.M. Christianen, J.C. Maan, J.T. Devreese, P.M. Koenraad, Phys. Rev. B **80**, 155318 (2009)
42. S. Raymond, S. Studenikin, A. Sachrajda, Z. Wasilewski, S.J. Cheng, W. Sheng, P. Hawrylak, A. Babinski, M. Potemski, G. Ortner, M. Bayer, Phys. Rev. Lett. **92**, 187402 (2004)
43. A. Babinski, M. Potemski, S. Raymond, J. Lapointe, Z.R. Wasilewski, Phys. Rev. B **74**, 155301 (2006)
44. S. Awirothananon, S. Raymond, S. Studenikin, M. Vachon, W. Render, A. Sachrajda, X. Wu, A. Babinski, M. Potemski, S. Fafard, S.J. Cheng, M. Korkusinski, P. Hawrylak, Phys. Rev. B **78**, 235313 (2008)
45. F. Ding, N. Akopian, B. Li, U. Perinetti, A. Govorov, F.M. Peeters, C.C. Bof Bufon, C. Deneke, Y.H. Chen, A. Rastelli, O.G. Schmidt, V. Zwiller, Phys. Rev. B **82**, 075309 (2010)
46. G. Huang, W. Guo, P. Bhattacharya, G. Ariyawansa, A.G.U. Perera, Appl. Phys. Lett. **94**, 101115 (2009)
47. J. Wu, Z.M. Wang, V.G. Dorogan, S. Li, Z. Zhou, H. Li, J. Lee, E.S. Kim, G.J. Salamo, Appl. Phys. Lett. **101**, 043904 (2012)
48. D. Granados, J.M. García, J. Cryst. Growth **251**, 213 (2007)
49. W. Zhang, Z. Su, M. Gong, C.-F. Li, G.-C. Guo, L. He, Europhys. Lett. **83**, 67004 (2008)
50. R.J. Young, E.P. Smakman, A.M. Sanchez, P. Hodgson, P.M. Koenraad, M. Hayne, Appl. Phys. Lett. **100**, 082104 (2012)
51. E. Zipper, M. Kurpas, M.M. Maska, New J. Phys. **14**, 093029 (2012)
52. A.M. Fisher, V.L. Campo, M.E. Portnoi, R. Rudolf, Phys. Rev. Lett. **102**, 096405 (2009)
53. M. Abbarchi, C.A. Mastrandrea, A. Vinattieri, S. Sanguinetti, T. Mano, T. Kuroda, N. Koguchi, K. Sakoda, M. Gurioli, Phys. Rev. B **79**, 085308 (2009)
54. E. Rasanen, A. Muhle, M. Aichinger, R.J. Haug, Phys. Rev. B **84**, 165320 (2011)
55. E. Zipper, M. Kurpas, J. Sadowski, M.M. Maska, J. Phys. Condens. Matter **23**, 115302 (2011)

Chapter 5
Scanning Probe Electronic Imaging of Lithographically Patterned Quantum Rings

F. Martins, H. Sellier, M.G. Pala, B. Hackens, V. Bayot, and S. Huant

Abstract Quantum rings patterned from two-dimensional semiconductor heterostructures exhibit a wealth of quantum transport phenomena at low temperature and in a magnetic field that can be mapped in real space thanks to dedicated scanning probe techniques. Here, we summarize our studies of GaInAs-based quantum rings by means of scanning gate microscopy both at low magnetic field, where Aharonov-Bohm interferences and the electronic local density of states are imaged, and at high magnetic field and very low temperatures, where the scanning probe can image Coulomb islands in the quantum Hall regime. This allows decrypting the apparent complexity of the magneto-resistance of a mesoscopic system in this regime. Beyond imaging and beyond a strict annular shape of the nanostructure, we show that this scanning-probe technique can also be used to unravel a new counter-intuitive behavior of branched-out rectangular quantum rings, which turns out to be a mesoscopic analog of the Braess paradox, previously known for road or other classical networks only.

F. Martins · B. Hackens · V. Bayot
IMCN/NAPS, Université catholique de Louvain, Louvain-la-Neuve, Belgium

F. Martins
e-mail: frederico.rodrigues@uclouvain.be

B. Hackens
e-mail: benoit.hackens@uclouvain.be

V. Bayot
e-mail: vincent.bayot@uclouvain.be

H. Sellier · S. Huant (✉)
Institut Néel, CNRS & Université Joseph Fourier, BP 166, 38042 Grenoble, France
e-mail: serge.huant@grenoble.cnrs.fr

H. Sellier
e-mail: hermann.sellier@grenoble.cnrs.fr

M.G. Pala
IMEP-LAHC, UMR 5130, CNRS/INPG/UJF/UdS, Minatec, Grenoble, France
e-mail: pala@minatec.inpg.fr

V.M. Fomin (ed.), *Physics of Quantum Rings*, NanoScience and Technology,
DOI 10.1007/978-3-642-39197-2_5, © Springer-Verlag Berlin Heidelberg 2014

5.1 Introduction

Electron systems confined in mesoscopic quantum rings (QRs) patterned from semiconductor heterostructures exhibit a wealth of quantum transport phenomena at low temperature and in a magnetic field such as the Aharonov-Bohm effect (AB) or the quantum Hall effect (QH). These effects have usually been observed thanks to measurements of the device electrical resistance vs the magnetic field. Such data yield "global" information on the phenomenon, i.e. on the scale of the device. During the last fifteen years, numerous attempts have been made to obtain real-space information on these mesoscopic phenomena down to the nanometer scale (i.e. on a smaller scale than the device size), thanks to dedicated scanning probe techniques. Mapping locally these phenomena give new insights, which allow for a more in-depth comparison with simulations. This chapter focuses on GaInAs-based open QRs that are imaged by scanning gate microscopy (SGM), a variant of electric atomic force microscopy (AFM), which can access to the intimate behavior of buried electronic systems, not accessible to the tip of scanning tunneling microscopy (STM).

After a brief introduction to SGM in Sect. 5.2, Sect. 5.3 focuses on the low-magnetic field range where the conductance modulations of a ring device induced by the scanning probe provide rich patterns that are either concentric or radial with the ring topology. The concentric patterns, primarily seen when the tip scans outside the ring area [1], image in real space the AB interferences taking birth in the ring device as a consequence of its ability to capture a magnetic flux or to differentially probe in its two arms a remote electrostatic potential. Radial patterns, that are seen when the probe scans directly over the ring, indirectly map the electronic local density of states (LDOS) at the Fermi energy [2, 3], as does STM in a direct way for surface electron systems [4]. Quantum simulations give a limit to the range of validity for the correspondence between conductance and LDOS maps and show how robust this correspondence can be against, for example, the introduction of impurities in the ring materials [5].

At high magnetic field, which is the focus of Sect. 5.4, the electron system enters into the QH regime, where electrons should only be transmitted through spatially separated edge states (ESs) near integer filling of the Landau levels. In contrast to extended 2D systems that exhibit vanishing longitudinal resistance concomitant to Hall plateaus in the QH regime, mesoscopic devices lead to surprising observations [6], such as pseudo-AB oscillations with "sub-periods" and "super-periods" compared to the orthodox AB oscillations seen at low field. To explain these observations, recent models [7] put forward the theory that, as ES come close to each other, electrons can hop between ES, or tunnel through quantum Hall islands. Such islands, also called localized states, are made of ES rotating around hills or dips in the potential landscape, or around the central antidot in a ring geometry. SGM reveals to be very powerful in locating QH islands in GaInAs QRs, and in revealing the spatial structure of transport inside the QH interferometer that they form [8]. Locations of QH islands are found by modulating, with the scanning tip, the tunneling between ESs and confined electron orbits. Tuning the magnetic field, SGM unveils a continuous evolution of active QH islands [9]. This allows decrypting the complexity of high-magnetic field magnetoresistance oscillations, and opens the way to further local scale manipulations of QH localized states.

In Sect. 5.5 we consider the possibility to control the electron transport through the buried semiconductor nanostructure by means of the SGM tip. In doing so, we find evidence for a counterintuitive behavior of mesoscopic networks [9] that presents a striking similarity with the Braess paradox encountered in traffic or classical networks [10]. A simulation of quantum transport in a two-branch mesoscopic network of rectangular shape reveals that adding a third branch can paradoxically reduce transport efficiency. This manifests itself in a sizable conductance drop of the network. A SGM experiment using the tip to modulate the transmission of one branch in the network reveals the occurrence of this paradox by mapping the conductance variation as a function of the tip voltage and position [9].

5.2 A Brief Introduction to the Technique of Scanning-Gate Microscopy

Unlike common AFM-based imaging techniques, SGM does not rely on a measurement of the cantilever property (i.e. its deflection angle, or resonance frequency shift), but rather of the device electrical characteristics. The principle of SGM [11–13] is sketched in Fig. 5.1. A voltage-biased (V_{tip}) metal-coated AFM tip is laterally scanned at an altitude of a few tens of nm over the device surface to perturb locally its electrical conductance G (or resistance). The changes in the device conductance ΔG are mapped as a function of the relative tip-sample position (x, y) to draw a $\Delta G(x, y)$ SGM map. Depending on the device impedance, a current-biased device (I) or voltage-biased (V) configuration can be used. The whole setup is immersed into a cryostat to operate down to below 100 mK for the coldest SGM setups [8, 14]. Optionally, an external magnetic field can be applied. The combined low-temperature and magnetic field environment requires the use of cryogenic magnetic-free displacement units, such as for example titanium-made inertial step motors [15, 16] for the *in situ* coarse positioning of the tip relative to the nanostructure over a few millimeters. In our setup, fine positioning over a few micrometers for image acquisition is ensured by commercially available piezoelectric scanner elements.

Using an AFM environment allows for locating the active device by measuring the sample topography, e.g. by using the dynamic mode of the AFM. Instead of using an optical method to measure the AFM cantilever deflection, as commonly done in AFM, it is advantageous to use a light-free setup [17], so that photosensitive devices remain under dark conditions during the entire experiment. One solution consists in gluing the AFM cantilever on a piezoelectric tuning fork, and monitoring the shift Δf of its resonance frequency observed when the tip approaches the surface. Sample topography is performed by using a feedback loop on Δf while scanning the tip over the device surface. Once the device topography has been mapped,

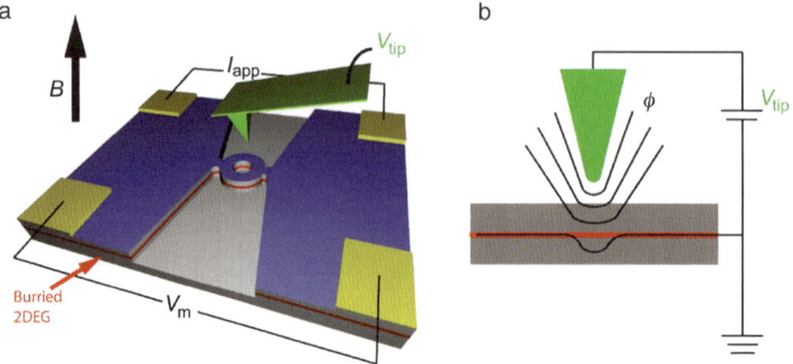

Fig. 5.1 Principle of scanning gate microscopy. Frame **a** depicts a sketch of a SGM experiment on a ring-shaped device that hosts a buried 2DEG. A current I_{app} is applied to the device and the voltage drop V_m across it is measured, while the biased tip (V_{tip}) is raster scanned over the device. A perpendicular magnetic field **B** is applied. In frame **b**, *curved lines* sketch the tip-induced isopotential lines Φ felt by the buried electron system

SGM is performed by lifting the tip at some tens of nanometers (typically 20 to 50 nm) above the surface and scanning it along a plane parallel to the 2D electron gas (2DEG), with a voltage applied to the tip. SGM measurements are carried out without contact between tip and sample, so that there is no electrical current through the tip, which acts indeed as a flying nanogate.

SGM has been used to image a broad range of transport-related phenomena in various nanostructures, such as for example the branching of conductance channels transmitted through quantum point contacts (QPCs) [12, 13], magnetic steering in a series of connected QPCs [18], Coulomb blockade in quantum dots [19–21], scarred wavefunctions in quantum billiards [22], and various graphene-based nanodevices [23–28], including in the quantum Hall regime [29]. We refer the reader to Ref. [30] for a more extensive review. In the rest of this chapter, we focus on our own work devoted to GaInAs-based QRs.

5.3 Imaging of Quantum Rings in the Low-Field Aharonov-Bohm Regime

If electrons maintain their phase coherence over sufficiently long distances, an open QR sees its conductance peaking when electron waves interfere constructively at the output contact, and decreasing to a minimum for destructive interferences. Varying either the electrostatic potential in one arm, e.g., by approaching the SGM biased tip as shown schematically in Fig. 5.2, or the magnetic flux captured by the QR allows the interference to be tuned. This gives rise to the electrostatic [31] and magnetic [32] AB oscillations in the ring conductance.

These interference phenomena can be imaged in real space by SGM [1]. An example of such imaging is shown in Fig. 5.3. Here, the QR is patterned from a

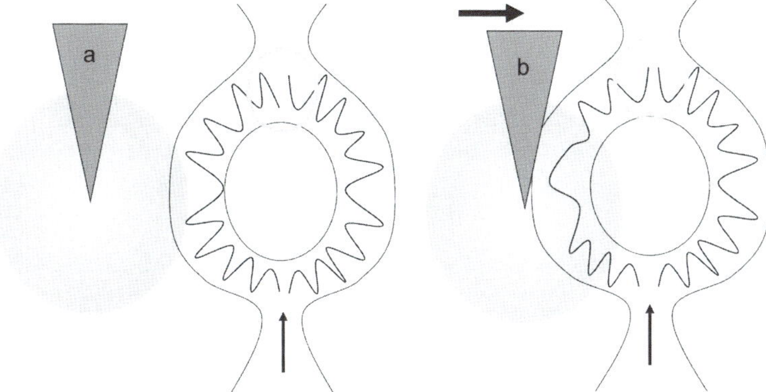

Fig. 5.2 Sketch of the electrostatic AB effect in a QR due to the tip bias. The cone and sphere stand for the tip and potential range, respectively. Approaching the tip to the QR from the left modifies the electrostatic potential felt by both arms, tuning the wavefunction interference at the output of the device from destructive (**a**) to constructive (**b**)

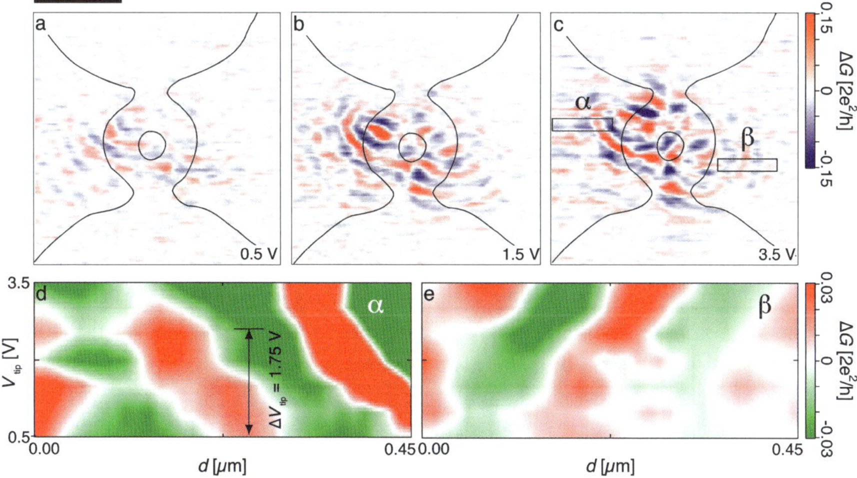

Fig. 5.3 SGM images (**a–c**) of a GaInAs QR whose geometry is shown schematically by *full lines* (image size: 2 µm × 2 µm; $T = 4.2$ K; zero magnetic field). The inner and outer ring diameters are 210 nm and 610 nm, respectively. The images are acquired for different tip voltages V_{tip} indicated on the figures and are all filtered [1] to compensate for a slowly varying strong background, which masks part of the interference pattern. Frames **d** and **e** are two sequences of profile plots as function of the tip bias. Each *horizontal line* corresponds to a vertical average of the conductance map in regions (α) and (β) shown in **c**, respectively. Adapted from [33]

Ga$_{0.3}$In$_{0.7}$As-based heterojunction with carrier concentration and mobility at 4.2 K of 2.0×10^{16} cm^{-2} and 100.000 cm^2 V^{-1} s^{-1}, respectively. The QR is connected to the 2D electron reservoir, which is buried 25 nm below the free surface, by two upper and lower narrow constrictions. The mean-free path and coherence length in

$Ga_{0.3}In_{0.7}As$ at 4.2 K are 2 μm and 1 μm, respectively, so that the electron transport is in the ballistic and (partly) coherent regimes. The coherent nature is confirmed by the observation of AB oscillations in the magneto-conductance when the QR is subjected to a perpendicular magnetic field [1].

One also observes an electrostatic AB effect which gives rise, at low magnetic field, to a well-developed fringe pattern in the SGM conductance image of the QR when the tip scans outside the QR. This outer pattern is mainly concentric with the ring geometry, as can be seen in the sequence of images shown in Fig. 5.3a–c obtained at different voltages applied to the tip. The interference pattern is here best seen on the left part of the ring, possibly due to a ring asymmetry. The qualitative interpretation in terms of a scanning-gate-induced electrostatic AB effect is that as the tip approaches the QR, either from the left or right, the electrical potential mainly increases on the corresponding side of the QR (see Fig. 5.2). This induces a phase difference between electron wavefunctions traveling through the two arms of the ring, and/or bends the electron trajectories, tuning the interference alternatively from constructive to destructive, thereby producing the observed pattern. Figure 5.3d, e shows how the interference pattern evolves for increasing tip voltages when the tip scans over the left hand-side and right hand-side regions of the QR, respectively [33]. It is clear from this figure that for increasing tip voltages the interference fringes shift away from the QR to the left in d and to the right in e, respectively. This is a direct manifestation of the tip-induced electrostatic AB effect. From Fig. 5.3d, we find that a phase shift of π is obtained for a tip bias variation $\Delta_{\mathrm{tip}} = 1.75$ V [33].

Now, modifying the magnetic field strength, another phase term contributes through the magnetic AB effect, i.e. the capture of the magnetic flux threading the QR area. The flux periodicity of such oscillations correspond to the flux quantum $\varphi_0 = h/e$. This displaces the whole fringe pattern with respect to the QR. This displacement is periodic in magnetic field strength with the same periodicity (here 13 mT, in nice agreement with the average area of the QR) than the AB oscillations seen in the magneto-conductance [1, 33], which gives further support to the interpretation in terms of AB effects. This interpretation was confirmed by density functional theory [34].

In Fig. 5.3a–c, it is clear that the conductance images also exhibit a complex pattern when the tip scans directly over the QR region. These inner fringes are linked to the local electron-probability density in the QR [2, 3, 5], provided that the tip potential is weak enough not to distort the QR electron density (see also Ref. [34]). A detailed analysis based on quantum mechanical simulations of the electron probability density, including a model tip potential, the magnetic field, and randomly distributed impurities, reproduces the main experimental features and demonstrates the relationship between SGM conductance maps and electron probability density, i.e. LDOS, at the Fermi energy. An example of such a relationship is shown in Fig. 5.4 in the case of a realistic QR perturbed by positively charged impurities. Although impurities distort the LDOS, this distortion is reflected back in the conductance image in such a way that the conductance map can still be seen as a mirror

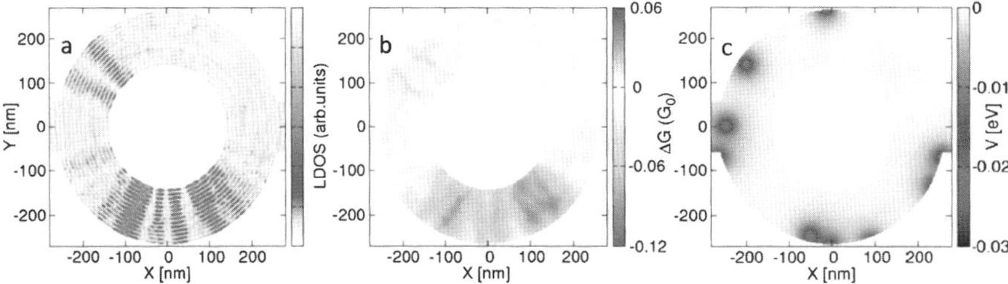

Fig. 5.4 Quantum simulation of a SGM experiment on a QR in the presence of positively charged impurities. The outer diameter, inner diameter and opening width are 530 nm, 280 nm, and 120 nm, respectively. The effective mass is 0.04 m_0 as in GaInAs (m_0 is the free electron mass). The Fermi energy is $E_F = 107.4$ meV. Frames **a** and **b** are simulated images of the LDOS and conductance changes (in units of G_0, the quantum of conductance), respectively, calculated for the random distribution of positively-charged impurities shown in frame **c**. In the simulation, the tip potential has a Lorentzian shape with 10 nm range and amplitude $E_F/50$. Adapted from [5]

of the electronic LDOS. As shown in Fig. 5.4, both the LDOS and conductance images tend to develop radial fringes, which are mostly, but not entirely, anchored to the impurity locations.

The discussion above suggests that SGM can be viewed as the analog of STM [4] for imaging the electronic LDOS in open mesoscopic systems buried under an insulating layer. It can also be seen as the counterpart of the near-field scanning optical microscope that can image photonic [35, 36] LDOS in confined nanostructures, provided that the excitation light source can be considered as point-like such as in active tips based on fluorescent nano-objects [37].

5.4 Imaging Quantum Rings in the Quantum Hall Regime

In the previous sections we focused on transport at low magnetic field (B) through a QR. The wave-like nature of electrons could be revealed by periodic AB oscillations in the magneto-resistance of the device. They originate from the different phases that electrons acquire along both arms of a QR when B is applied perpendicular to the 2D electron system. In this section, we will discuss another type of periodic magneto-resistance oscillations, which show up at high B in QRs.

At high magnetic field the electron transport picture changes drastically as the cyclotron radius shrinks, highly degenerate Landau levels (LLs) form and the 2DEG enters in the "QH regime". When the Fermi energy lies between two Landau levels (i.e. around integer LL filling factor ν), the bulk of the electron system becomes insulating and current flows through counter-propagating one-dimensional channels, the so-called ESs, confined along the borders of the device, where the Fermi energy crosses LLs. In macroscopic devices, scattering between opposite ESs vanishes and the electron mean free path becomes of the order of several millimeters. Moreover, QH islands (QHIs), i.e. electrons trapped in closed ESs pinned around potential inhomogeneities, remain electrically isolated and do not contribute to electron trans-

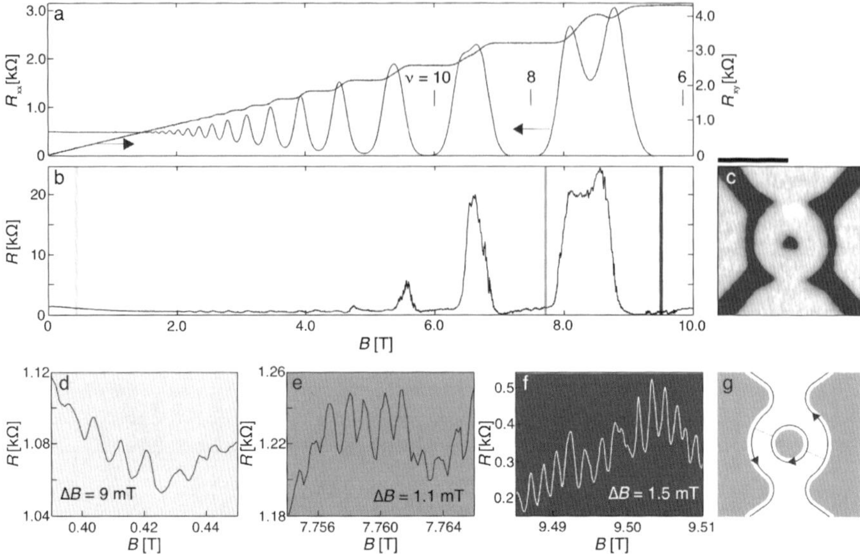

Fig. 5.5 Quantum Hall effect on a mesoscopic QR. In **a** the step-like and oscillating curves are the longitudinal (R_{xx}) and the transverse (R_{xy}) resistances of a Hall bar, respectively, as function of the magnetic field (B). At high B, R_{xy} is quantized in $e^2/(N \cdot h)$, where N is the number of fully occupied Landau levels, while R_{xx} becomes zero. The AFM micrograph of the QR is shown in **b** where the *black bar* represents 1 μm. In **c** we display the magnetoresistance measured across the QR in the same range of B and at a temperature $T = 100$ mK. Close-ups of **b** are shown in **d**–**f**. At low-B (**d**), the periodicity of the oscillations is $\Delta B_{AB} = 9$ mT which, given the geometry presented in **c**, is consistent with AB interferences. At high magnetic field, for $N = 8$, shown in **e** the periodicity is $\Delta B = 1.1$ mT whereas, for $N = 6$, displayed in **f**, we find $\Delta B = 1.5$ mT. The latter oscillations are explained, as sketched in **g**, as tunneling between edge states through a Coulomb island located around the central anti-dot of the QR. This gives rise to periodic oscillations with $\Delta B \cdot N = \Delta B_{AB}$

port. This gives rise, as shown in Fig. 5.5a, to plateaus in the transverse resistance R_{xy}, and to a vanishing longitudinal resistance R_{xx}. measured in a macroscopic Hall bar (for review see e.g. Ref. [38]).

However, when the size of the device is reduced and becomes comparable with the size of QHIs, this picture is no longer valid. In such conditions, several experiments have reported "sub-periodic" and "super-periodic" magnetoresistance oscillations, i.e. with a flux period corresponding to a fraction, or a multiple of the usual AB period observed at low magnetic field (see sections above) [6, 39–44]. Driven by these intriguing results, theoretical efforts have explained these oscillations within a model where Coulomb interactions dominate [7]. In small devices, the counter-propagating ESs are indeed brought close to each other so that electrons may tunnel between them, either directly, and/or through a QHI that mediates electron transmission [45, 46]. Rather than coherent effects, the discrete nature of electrons has naturally been put forward.

The experiments discussed in this section are performed at a temperature $T = 100$ mK, inside a ^{3}He/^{4}He dilution refrigerator, equipped with a superconducting coil that can provide a magnetic field up to 15/17 T. The 2DEG in which the QR is patterned is located 25 nm below the surface and, at low-T, the electron density and mobility are 1.4×10^{16} m^{-2} and 4 m^2/V s, respectively. Figure 5.5c shows the device topography. The QR has an average outer diameter of 1 µm, two apertures and a central antidot diameter of approx. 300 nm. The magneto-resistance of the quantum ring (R *versus* B), measured simultaneously with R_{xx} and R_{xy} in the bulk, is shown in Figs. 5.5d–f. Figure 5.5b shows R *versus* B over the full magnetic field range, from 0 to 10 T. At low magnetic field, Fig. 5.5d, periodic AB oscillations with $\Delta B_{AB} = 9$ mT are observed. This is consistent with an average radius of 380 nm for the QR. In Figs. 5.5e, f, two B-ranges are zoomed, around $\nu = 8$ and 6, respectively. In these ranges the magnetoresistance of the device displays oscillations with two different "sub-periods": $\Delta B = 1.1$ mT around $\nu = 8$ and, around $\nu = 6$, $\Delta B = 1.5$ mT. As sketched in Fig. 5.5g, one can understand these oscillations within a Coulomb-dominated model where electrons tunnel between propagating ESs through a QHI with discrete energy levels, located around the central anti-dot of the QR [7]. The basis of this model is that a change in magnetic field induces a periodic change in the QHI energy with respect to the ES energy. For each flux quantum added to the QHI, one electron has to be added to each populated ES in the QHI, which means that, in this case, Coulomb blockade oscillations are observed, with a period:

$$\Delta B = (\varphi_0/A)/N;$$

where N is the number of filled ESs around the QHI of area A. Indeed, the periods measured in Fig. 5.5e and f are consistent with this relation. Nevertheless, shifting the magnetic field range to $B = [9.67\ \text{T}, 9.74\ \text{T}]$ (still around $\nu = 6$), the magnetoresistance, as displayed in Fig. 5.6a, reveals "super-period" oscillations with $\Delta B = 17$ mT. Using the previous model, we conclude that they correspond to ES loops with a radius of $\sim$65 nm, which would not fit around the QR antidot. However, it could well be located, as drawn in Fig. 5.6b, somewhere in the 300-nm-wide arms of the QR, or near its openings, and be connected to the propagating ES through tunnel junctions.

In order to precisely locate such a QHI, we now use SGM since this technique is particularly well adapted to image electronic transport through a buried 2DEG in the quantum Hall regime [8, 14, 30, 33, 47–50]. Here, we measure the resistance of the device as the electrically polarized AFM tip scans over the surface. Within the Coulomb-dominated model, as the negatively biased tip approaches a QHI, it progressively changes the electrostatic potential experienced by electrons trapped in the QHI and modifies the QHI surface which, in turn, induces an energy imbalance between the QHI and the ESs [7, 8]. As with the magnetic field, this imbalance then allows electrons to tunnel between ESs and the QHI in the Coulomb blockade regime, whenever a QHI energy level lies in the energy window defined by the propagating ES potential. In a SGM map, one therefore expects to observe sets of concentric resistance fringes, each one corresponding to a Coulomb blockade peak, encircling each active QHI.

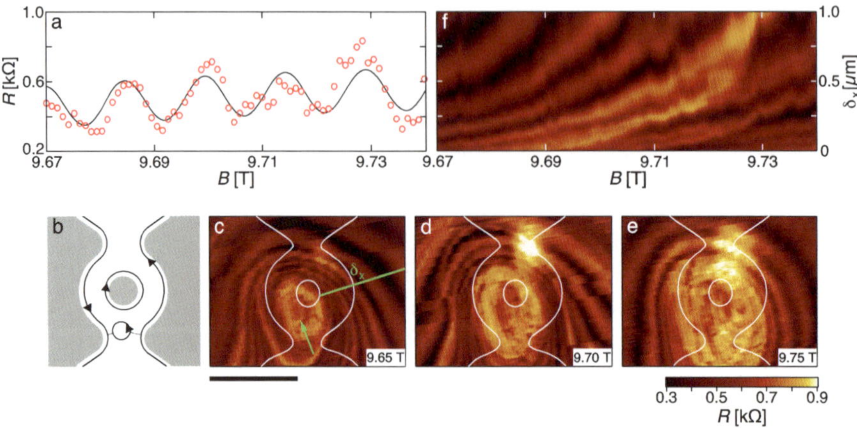

Fig. 5.6 Identification of the center of a Coulomb island inside a QR using SGM. Frame **a** shows R *versus* B in the range $B = [9.67\ \text{T}, 9.74\ \text{T}]$. The periodicity of the fringes is $\Delta B = 17$ mT, which corresponds to a QHI with a surface equivalent to a disc of radius ~ 65 nm located, as sketched in **b**, somewhere in the arms of the 300-nm-wide QR. **c**, **d**, and **e** are SGM resistance maps obtained at $B = 9.65$, 9.70 and 9.75 T, respectively, with $V_{\text{tip}} = -1$ V. The *white lines* correspond to the position of the QR and the scale bar represents 1 μm. This sequence of images reveals the position of the center of the QHI (marked with a *green arrow* in **c**), inside the QR. **f** depicts B-dependence of the $R(x, y)$ profile measured along the *dashed line* in **c**, $V_{\text{tip}} = -1$ V. The fringes share the same periodicity $\Delta B = 17$ mT as in **a**

Three consecutive SGM maps for $B = 9.65$, 9.70 and 9.75 T are shown in Figs. 5.6c to 5.6e, respectively, where the position of the QR is drawn, superimposed on the SGM data. Concentric fringes are observed, centered close to one of the openings of the QR. As B increases the position of resistance fringes evolves. This is clearly illustrated in Fig. 5.6f, where B is swept while scanning with a negatively polarized tip ($V_{\text{tip}} = -1$ V) along the same line, highlighted in green and denoted δ_x in Fig. 5.6c. As B increases the concentric fringes shift away from their center, indicated by the green arrow in Fig. 5.6c. Importantly, we also note that the periodicity $\Delta B = 17$ mT found in Fig. 5.6f is the same as the one extracted from the magnetoresistance curve in Fig. 5.6a. This allows concluding that the QHI, which is at the origin of these "super-period" oscillations, has its center indicated by the green arrow in Fig. 5.6c.

Remarkably, the slope direction in the plane B *versus* δ_x can be used to discriminate between ESs surrounding a potential hill or looping around a potential well. If we assume that a QHI is created around a potential hill, approaching a negatively biased tip will raise the potential and increase the QHI area. On the other hand, in the case of a QHI formed around a potential well, the effect of a negatively biased tip would be to reduce the QHI surface. In Fig. 5.6f, isoresistance lines, that correspond to isoflux states through the QHI, move away from their center as B is raised, which unambiguously indicates that the QHI surrounds a potential hill.

5.5 Revealing an Analog of the Braess Paradox in Branched-Out Rectangular Rings

In the above sections, SGM has been used to image the electron transport in annular shaped QRs. In this section, we demonstrate that SGM can also be used to tune the electron transport by depleting, by gate effect, a conduction channel in a branched-out mesoscopic network, whose primary shape is a rectangular ring. This leads to the discovery of a mesoscopic analog [9] of the Braess paradox [10].

Adding a new road to a congested network can paradoxically lead to a deterioration of the overall traffic situation, i.e. longer trip times for road users. Or, in reverse, blocking certain streets in a complex road network can surprisingly reduce average trip time [51]. This counter-intuitive behavior has been known as the Braess paradox [10]. Later extended to other networks in classical physics, such as mechanical or electrical networks [52, 53], this paradox lies in the fact that adding extra capacity to a congested network can counter-intuitively degrade its overall performance. Initially known for classical networks only, we have extended the concept of the Braess paradox to the quantum world [9]. By combining quantum simulations of a model network and SGM, we have discovered that an analog of the Braess paradox can occur in mesoscopic semiconductor networks, where electron transport is ballistic and coherent.

We consider a simple two-path network in the form of a rectangular ring connected to source and drain via two openings (see Fig. 5.7a for the network topology). In practice, this ring is patterned from a GaInAs heterojunction as for the QRs discussed in the previous sections. The dimensions are chosen to ensure that the embedded 2DEG is in the ballistic and coherent regimes of transport at 4.2 K. The short wires in the ring are chosen to be narrower than the source/drain openings to behave as congested constrictions for propagating electrons. Branching out this ring by patterning a central wire (see Fig. 5.7a) opens a third path to the electrons that bypasses the antidot in the initial rectangular ring. Then, we use SGM to partially block by gate effects the transport through the additional branch. Doing so should intuitively result in a decreased current transmitted through the device, but this is just the opposite behavior that is found in certain conditions, both experimentally and in quantum simulations [9]. Therefore, in a naive picture, electrons in such networks turn out to behave like drivers in congested cities: blocking one path favors "traffic" efficiency.

The above finding is summarized in the simulations of Figs. 5.7a and 5.7d, which show the network geometry and a calculated conductance crosscut as a function of tip position, respectively. Here, the outer width and length of the initial corral are 0.75 and 1.6 µm, respectively, whereas the widths of the lateral, upper/lower, and central (additional) arms are $W = 140$ nm, $L = 180$ nm, and $W3 = 160$ nm, respectively. The width of the source and drain openings are $W0 = 320$ nm. This ensures that electron flow in the lateral arms (in the absence of the central arm) is congested because $2W < W0$. In other words, all injected conduction channels (about 10) into the network cannot be admitted in these arms [9].

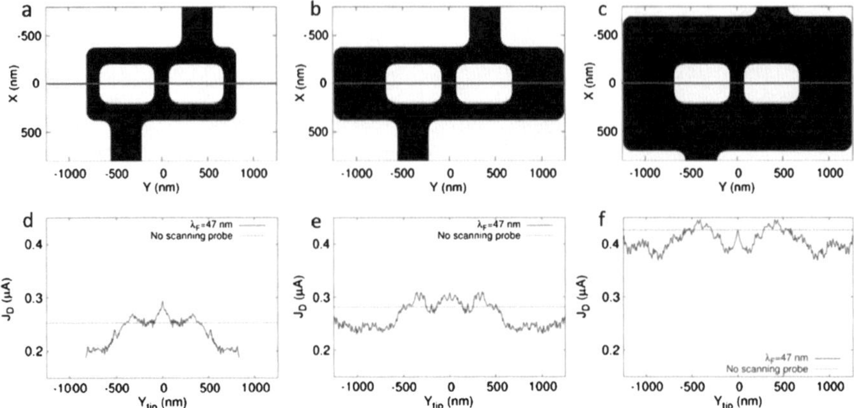

Fig. 5.7 Evidence for a Braess-like paradox in a mesoscopic rectangular network. Frames **a**, **b**, and **c** depict the network geometries with parameters given in the text. Frames **d** to **f** are the corresponding calculated conductance crosscuts in the presence of a depleting SGM tip scanning along the median lines of **a**, **b**, and **c**, respectively. The *horizontal dotted lines* give the unperturbed conductance without tip. The Fermi wavelength is 47 nm. Fluctuations in the conductance profiles are UCFs due to the tip-induced change in the potential felt by electrons propagating through the device

The crosscut in Fig. 5.7d is obtained by computing the network conductance (source-drain voltage $= 1$ mV) as function of the tip position scanned along the median line of Fig. 5.7a. This line crosses the lateral and central arms. The tip potential is mimicked by a point-like potential of -1 V placed at 100 nm above the 2DEG [9], which corresponds to a lateral extension of $\sim$400 nm for the tip-induced potential perturbation at the 2DEG level. This model potential entirely depletes the 2DEG in one arm when the tip passes above it.

It is clear from Fig. 5.7d that depleting the central arm produces a distinctive conductance peak that goes well beyond the unperturbed value. This peak has a much larger amplitude as the universal conductance fluctuations (UCFs) that are seen as small oscillations for any tip position along the median lines, as a consequence of the tip-induced change in the potential felt by electrons propagating through the device [54]. This strong central peak is the signature of the counter-intuitive Braess-like behavior mentioned above. Noteworthy, closing one of the lateral arms reduces the conductance, in agreement with the intuitive expectation: the paradox is seen only when the central branch is closed, not the lateral ones, which stresses the particular role played by this central branch.

Congestion plays a key role in the occurrence of the classical Braess paradox [10, 51, 52]. In order to probe a similar role in the mesoscopic counterpart paradox, we have simulated two additional networks with enlarged lateral arms (Figs. 5.7b and 5.7e: $W = 560$ nm, L unchanged) and with both enlarged lateral and upper-lower arms (Figs. 5.7c and 5.7f: $W = 560$ nm, $L = 500$ nm) [55]. This releases congestion in the lateral arms. It is clear from Figs. 5.7e and 5.7f that releasing congestion smoothens the counter-intuitive conductance peak seen in the congested network when the central arm is blocked. Nevertheless, there is still a slight conductance (current) increase when the tip scans just above the central arms in networks e and f, but it no longer goes beyond the unperturbed conductance for the largest network f. This finding entails the particular roles played by the additional branch and by network congestion in the occurrence of a distinctive Braess-like paradox. Yet, more experimental and theoretical work is needed to put forward a conclusive explanation at the microscopic level for the paradoxical behavior reported here.

5.6 Conclusion

The few examples presented in this chapter show how powerful is SGM in imaging, and possibly tuning, the electronic transport in ring-shaped semiconductor devices and to reveal how electrons behave down there. It gives valuable complementary view on phenomena that are usually considered within a macroscopic experimental scheme. The imaging of AB interferences, the ability of locating precisely compressible Coulomb islands in a quantum Hall interferometer, and the closing of a selected branch in a mesoscopic rectangular ring to induce a Braess-like phenomenon all are illustrative of this claim.

Acknowledgements B.H. and F.M. are associate and postdoctoral researchers respectively with the Belgian FRS-FNRS. This work has been supported by FRFC grant no. 2.4.546.08.F and FNRS grant no 1.5.044.07.F, by the Belgian Science Policy (Interuniversity Attraction Pole Program IAP-6/42) as well as by the PNANO 2007 program of the Agence Nationale de la Recherche, France ("MICATEC" project). V.B. acknowledges the award of a "Chaire d'excellence" by the Nanoscience Foundation in Grenoble.

References

1. B. Hackens, F. Martins, T. Ouisse, H. Sellier, S. Bollaert, X. Wallart, A. Cappy, J. Chevrier, V. Bayot, S. Huant, Nat. Phys. **2**, 826 (2006)
2. F. Martins, B. Hackens, M.G. Pala, T. Ouisse, H. Sellier, X. Wallart, S. Bollaert, A. Cappy, J. Chevrier, V. Bayot, S. Huant, Phys. Rev. Lett. **99**, 136807 (2007)
3. M.G. Pala, B. Hackens, F. Martins, H. Sellier, V. Bayot, S. Huant, T. Ouisse, Phys. Rev. B **77**, 125310 (2008)
4. M.F. Crommie, C.P. Lutz, D.M. Eigler, Science **262**, 218 (1993)
5. M.G. Pala, S. Baltazar, F. Martins, B. Hackens, H. Sellier, T. Ouisse, V. Bayot, S. Huant, Nanotechnology **20**, 264021 (2009)
6. B.J. van Wees, L.P. Kouwenhoven, C.J.P.M. Harmans, J.G. Williamson, C.E. Timmering, M.E.I. Broekaart, C.T. Foxon, J.J. Harris, Phys. Rev. Lett. **62**, 2523 (1989)
7. B. Rosenow, B.I. Halperin, Phys. Rev. Lett. **98**, 106801 (2007)
8. B. Hackens, F. Martins, S. Faniel, C.A. Dutu, H. Sellier, S. Huant, M. Pala, L. Desplanque, X. Wallart, V. Bayot, Nat. Commun. **1**, 39 (2010)
9. M.G. Pala, S. Baltazar, P. Liu, H. Sellier, B. Hackens, F. Martins, V. Bayot, X. Wallart, L. Desplanque, S. Huant, Phys. Rev. Lett. **108**, 076802 (2012)
10. D. Braess, A. Nagurney, T. Wakolbinger, Transp. Sci. **39**, 446 (2005)

11. M.A. Eriksson, R.G. Beck, M. Topinka, J.A. Katine, R.M. Westervelt, K.L. Campman, A.C. Gossard, Appl. Phys. Lett. **69**, 671 (1996)
12. M.A. Topinka, B.J. LeRoy, S.E.J. Shaw, E.J. Heller, R.M. Westervelt, K.D. Maranowski, A.C. Gossard, Science **289**, 2323 (2000)
13. M.A. Topinka, B.J. LeRoy, R.M. Westervelt, S.E.J. Shaw, R. Fleischmann, E.J. Heller, K.D. Maranowski, A.C. Gossard, Nature **410**, 183 (2001)
14. N. Paradiso, S. Heun, S. Roddaro, G. Biasiol, L. Sorba, V. Venturelli, F. Taddei, F. Giovannetti, F. Beltram, Phys. Rev. B **86**, 085326 (2012)
15. C. Obermüller, A. Deisenrieder, G. Abstreiter, K. Karrai, S. Grosse, S. Manus, J. Feldmann, H. Lipsanen, M. Sopanen, J. Ahopelto, Appl. Phys. Lett. **75**, 358 (1999)
16. M. Brun, S. Huant, J.C. Woehl, J.F. Motte, L. Marsal, H. Mariette, J. Microsc. **202**, 202 (2001)
17. K. Karrai, R.D. Grober, Appl. Phys. Lett. **66**, 1842 (1995)
18. K.E. Aidala, R.E. Parrott, T. Kramer, E.J. Heller, R.M. Westervelt, M.P. Hanson, A.C. Gossard, Nat. Phys. **3**, 464 (2007)
19. A. Pioda, S. Kicin, T. Ihn, M. Sigrist, A. Fuhrer, K. Ensslin, A. Weichselbaum, S.E. Ulloa, M. Reinwald, W. Wegscheider, Phys. Rev. Lett. **93**, 216801 (2004)
20. P. Fallahi, A.C. Bleszynski, R.M. Westervelt, J. Huang, J.D. Walls, E.J. Heller, M. Hanson, A.C. Gossard, Nano Lett. **5**, 223 (2005)
21. A.C. Bleszynski, F.A. Zwanenburg, R.M. Westervelt, A.L. Roest, E.P.A.M. Bakkers, L.P. Kouwenhoven, Nano Lett. **7**, 2559 (2007)
22. R. Crook, C.G. Smith, A.C. Graham, I. Farrer, H.E. Beere, D.A. Ritchie, Phys. Rev. Lett. **91**, 246803 (2003)
23. M.R. Connolly, K.L. Chiou, C.G. Smith, D. Anderson, G.A.C. Jones, A. Lombardo, A. Fasoli, A.C. Ferrari, Appl. Phys. Lett. **96**, 113501 (2010)
24. J. Berezovsky, M.F. Borunda, E.J. Heller, R.M. Westervelt, Nanotechnology **21**, 274013 (2010)
25. J. Berezovsky, R.M. Westervelt, Nanotechnology **21**, 274014 (2010)
26. S. Schnez, J. Guttinger, M. Huefner, C. Stampfer, K. Ensslin, T. Ihn, Phys. Rev. B **82**, 165445 (2010)
27. M.R. Connolly, K.L. Chiu, A. Lombardo, A. Fasoli, A.C. Ferrari, D. Anderson, G.A.C. Jones, C.G. Smith, Phys. Rev. B **83**, 115441 (2011)
28. N. Pascher, D. Bischoff, T. Ihn, K. Ensslin, Appl. Phys. Lett. **101**, 063101 (2012)
29. M.R. Connolly, R.K. Puddy, D. Logoteta, P. Marconcini, M. Roy, J.P. Griffiths, G.A.C. Jones, P.A. Maksym, M. Macucci, C.G. Smith, Nano Lett. **12**, 5448 (2012)
30. H. Sellier, B. Hackens, M.G. Pala, F. Martins, S. Baltazar, X. Wallart, L. Desplanque, V. Bayot, S. Huant, Semicond. Sci. Technol. **26**, 064008 (2011)
31. S. Washburn, H. Schmid, D. Kern, R.A. Webb, Phys. Rev. Lett. **59**, 1791 (1987)
32. R.A. Webb, S. Washburn, C.P. Umbach, R.B. Laibowitz, Phys. Rev. Lett. **54**, 2696 (1985)
33. F. Martins, B. Hackens, H. Sellier, P. Liu, M.G. Pala, S. Baltazar, L. Desplanque, X. Wallart, V. Bayot, S. Huant, Acta Phys. Pol. A **119**, 569 (2011)
34. B. Szafran, Phys. Rev. B **84**, 075336 (2011)
35. G. Colas des Francs, C. Girard, J.C. Weeber, C. Chicane, T. David, A. Dereux, D. Peyrade, Phys. Rev. Lett. **86**, 4950 (2001)
36. C. Chicanne, T. David, R. Quidant, J.C. Weeber, Y. Lacroute, E. Bourillot, A. Dereux, G. Colas des Francs, C. Girard, Phys. Rev. Lett. **88**, 097402 (2002)
37. A. Cuche, O. Mollet, A. Drezet, S. Huant, Nano Lett. **10**, 4566 (2010)
38. R.E. Prange, S.M. Girvin, *The Quantum Hall Effect*, 2nd edn. (Springer, New York, 1990)
39. V.J. Goldman, B. Su, Science **267**, 1010 (1995)
40. R.P. Taylor, A.S. Sachrajda, P. Zawadzki, P.T. Coleridge, J.A. Adams, Phys. Rev. Lett. **69**, 1989 (1992)
41. M. Kataoka, C.J.B. Ford, G. Faini, D. Mailly, M.Y. Simmons, D.R. Mace, C.-T. Liang, D.A. Ritchie, Phys. Rev. Lett. **83**, 160 (1999)
42. F.E. Camino, W. Zhou, V.J. Goldman, Phys. Rev. B **72**, 075342 (2005)

43. Y. Zhang, D.T. McClure, E.M. Levenson-Falk, C.M. Marcus, L.N. Pfeiffer, K.W. West, Phys. Rev. B **79**, 241304 (2009)
44. A.J.M. Giesbers, U. Zeitler, M.I. Katsnelson, D. Reuter, A.D. Wieck, G. Biasiol, L. Sorba, J.C. Maan, Nat. Phys. **6**, 173 (2010)
45. C. Zhou, M. Berciu, Phys. Rev. B **72**, 085306 (2005)
46. E. Peled, D. Shahar, Y. Chen, E. Diez, D.L. Sivco, A.Y. Cho, Phys. Rev. Lett. **93**, 236802 (2003)
47. N. Aoki, C.R. da Cunha, R. Akis, D.K. Ferry, Y. Ochiai, Phys. Rev. B **72**, 155327 (2005)
48. A. Baumgartner, T. Ihn, K. Ensslin, K.D. Maranowski, A.C. Gossard, Phys. Rev. B **76**, 085316 (2007)
49. A. Baumgartner, T. Ihn, K. Ensslin, G. Papp, F. Peeters, K.D. Maranowski, A.C. Gossard, Phys. Rev. B **74**, 165426 (2006)
50. N. Paradiso, S. Heun, S. Roddaro, D. Venturelli, F. Taddei, V. Giovannetti, R. Fazio, G. Biasiol, L. Sorba, F. Beltram, Phys. Rev. B **83**, 155305 (2011)
51. H. Youn, M.T. Gastner, H. Jeong, Phys. Rev. Lett. **101**, 128701 (2008)
52. C.M. Penchina, L.J. Penchina, Am. J. Phys. **71**, 479 (2003)
53. D. Witthaut, M. Timme, New J. Phys. **14**, 083036 (2012)
54. R.A. Webb, S. Washburn, C.P. Umbach, R.B. Laibowitz, Phys. Rev. Lett. **54**, 2696 (1985)
55. M. Pala, H. Sellier, B. Hackens, F. Martins, V. Bayot, S. Huant, Nanoscale Res. Lett. **7**, 472 (2012)

Chapter 6
Self-organized Formation and XSTM-Characterization of GaSb/GaAs Quantum Rings

Andrea Lenz and Holger Eisele

Abstract In this chapter we discuss the details of the spatial structure of GaSb/GaAs(001) quantum rings, and how the observed nanostructures were achieved during the self-assembled growth and formation process. In contrast to the In(Ga)As/GaAs material system, for which thin capping of quantum dots followed by a longer growth interruption is necessary to initiate the quantum-dot to quantum-ring transformation, GaSb quantum rings occur already after the material deposition on GaAs(001), sometimes even without capping. Upon capping almost all quantum dots are transformed into quantum rings. The GaSb quantum rings are found to exhibit a clear central opening, which is filled by the pure GaAs capping material. The more direct formation process of the GaSb/GaAs(001) quantum rings can be understood by analyzing the details of the energies acting during the self-assembled growth. Hence, these more distinct quantum-ring structures are even more promising to show corresponding electronic states, as e.g., the *Aharonov-Bohm* effect.

6.1 Introduction

The self-organized growth of pseudomorphic compound semiconductor nanostructures is mainly characterized by the interplay of strain energies and surface free energies [1–7]. Pseudomorphic growth of a material (called film) with a natural lattice parameter $a_{0,f}$ on top of a surface of another material (called matrix) with a different natural lattice parameter $a_{0,m}$ leads to the evolution of strain. As long as no dislocations occur and the film keeps closed in form of a layer-like structure the strain can only be relaxed within the film along growth direction. This is the so-called biaxial strain in the film [8]. Further strain relaxation can occur only by roughening of the interface between the film and the matrix or by roughening of the

A. Lenz (✉) · H. Eisele
Institut für Festkörperphysik, Technische Universität Berlin, Hardenbergstr. 36, 10623 Berlin, Germany
e-mail: andrea.lenz@physik.tu-berlin.de

H. Eisele
e-mail: holger.eisele@physik.tu-berlin.de

V.M. Fomin (ed.), *Physics of Quantum Rings*, NanoScience and Technology,
DOI 10.1007/978-3-642-39197-2_6, © Springer-Verlag Berlin Heidelberg 2014

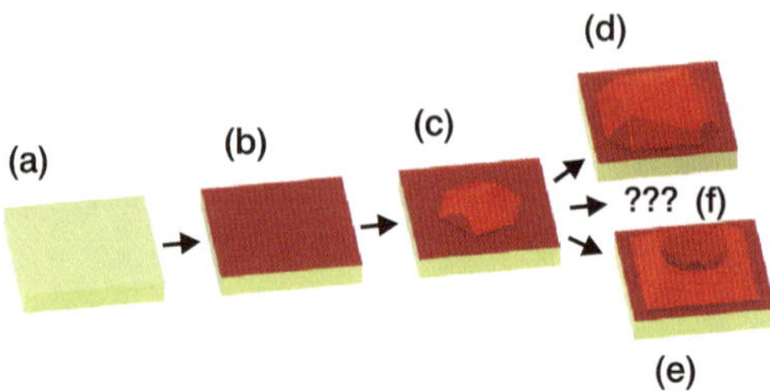

Fig. 6.1 Possibilities of nanostructure evolution upon ongoing material deposition: (**a**) clean matrix material surface, (**b**) wetting layer, (**c**) small quantum dot (*hut* structure). In order to keep the base interface area small either (**d**) large quantum dots (*dome* structure) or (**e**) quantum rings can occur. Maybe there are further, up to now unknown structures (**f**)

film surface. The latter case occurs during the 2-dimensional to 3-dimensional phase transition in the *Stranski-Krastanov* growth mode [9], when the strain energy of the wetting layer reaches a point at which it is energetically more favorable to form a larger surface area on the cost of higher surface energy. This results in the quantum-dot evolution on top of a wetting layer, which occurs typically at 1.4–1.8 ML in the InAs/GaAs material system (7.16 % lattice mismatch) [10] and at about 1.2 ML in the GaSb/GaAs material system (7.83 % lattice mismatch) [11–13].

A quantum dot surrounded by a wetting layer relaxes its strain mainly at the apex, while at the base interface the quantum-dot material is still strained [2, 14]. The larger the quantum dot gets at its base interface to the matrix material upon further material deposition, the more the strain energy at quantum-dot base increases. Especially the strain energy resulting from the shear components increases, since the deposited quantum-dot material is strained at the interface, but relaxes toward the apex. This strain situation can directly be observed in intermixed grown $In_{0.5}Ga_{0.5}As$/GaAs quantum dots, which exhibit a reversed-cone composition profile [15, 16] with the highest InAs content at the upper center and the lowest at the outer base. The strain energy at the base interface, especially at the outer region of the quantum dot, drives the transition from flat quantum dots toward ones with steeper side facets [17] [see Fig. 6.1(c)–(d)], which relaxes the strain energy better, again at the cost of additional surface energy. In order to increase the base interface area without further increasing the strain energy at the base, there is another geometry possible: a ring-like contact between the matrix material and the deposited film, the so-called quantum rings [18] [see Fig. 6.1(e)]. The formation process of such quantum rings and their spatial structure will be presented in the following especially for the GaSb/GaAs(001) material system.

Self-organized quantum-ring formation was at first reported in 1997 by J.M. García et al. for the InAs/GaAs system [18] and in 2003 by T. Raz et al. for the InAs/InP system [19]. In the same year quantum rings were also observed in the Ge/Si system by J. Cui et al. [20] and 2004 Kobayashi et al. primarily demonstrated the self-assembled growth of GaSb/GaAs quantum rings [21]. In unstrained material systems, as e.g., GaAs/AlGaAs, quantum-rings can be formed via droplet epitaxy [22, 23]. Beside many fundamental investigations on the growth process of quantum-ring structures [24–28] and their appearance after the capping process [13, 29–34], also ideas on applications were developed.

Quantum rings in memory devices or in laser devices, for example, exhibit shorter optical emission wavelengths as compared with samples containing quantum dots [35, 36]. Physically even more interesting is the occurrence of the *Aharonov-Bohm* effect and the existence of persistent currents in quantum rings, as introduced in Chap. 1. While the preceding chapters have dealt with the In(Ga)As/GaAs quantum-ring growth as well as its optical and structural characterization [Chaps. 2–5, 7–8], this chapter will focus on the resulting GaSb/GaAs quantum-ring structures and its formation process. The GaSb/GaAs system should be very promising to exhibit a strong Aharonov-Bohm effect since the central opening of the quantum rings is much more pronounced as compared with other material systems [32, 34, 37]. Furthermore, since GaSb/GaAs quantum-ring structures are found to be a type-II quantum system [33], the resulting structural sizes and their stoichiometry might be as well interesting for future theoretical work, as it will be discussed for type-II ZnTe/ZnSe and Zn(Mn)Te/ZnSe nanostructures in Chap. 11.

6.2 Characterization by XSTM

In order to analyze the spatial structure of compound semiconductor nanostructures there are several methods available, providing access to different aspects. None of these methods is universal enough to determine all structural parameters at the same time. Atomic force microscopy (AFM) and especially *in situ* plane-view scanning tunneling microscopy (STM) are very useful to study the structural properties such as the lateral density or the height before any capping by the matrix material occurs [27, 38–40]. High-resolution transmission electron microscopy (HR-TEM) images the lattice displacement and strain profile of the crystal by the interference of columns of about 100s of atoms [41–46]. X-ray diffraction (XRD) is a very versatile method to extract a lot of structural information from such semiconductor nanostructure ensembles [47–50], but not on single species. Atom probe tomography (APT) is a highly interesting method to probe the 3-dimensional composition of nanostructured materials [51], e.g., like metal alloys [52], but unfortunately it does not provide atomic resolution on compound semiconductor materials due to strong group-V material clustering effects [53]. Cross-sectional scanning tunneling microscopy (XSTM) provides atomic-scale resolution on compound semiconductors at a cleavage surface across the nanostructure with the ability of chemical mapping [54–59]. Even if by random cleavage through the crystal only one cross-sectional surface can be studied for each nanostructure, from a statistical analysis over larger traces of these nanostructures at the investigated surface also deeper information can be extracted. In the following we will explain XSTM more in detail and show how one receives detailed structural information. In Sects. 6.3 and 6.4 we will compare these results also with structural information gained from other analysis methods.

Shape, size, and composition of a quantum dot or quantum ring have a strong influence on its electronic properties [60–62]. Theoretical calculations show that a change of the InGaAs composition by only 30 % causes an energy shift of some tens of meV for the electron and/or hole states, while a change of the composition profile affects this shift even stronger by up to 150 meV [62]. For that reason an accurate determination of the quantum-dot or quantum-ring stoichiometry and the composition distribution is a very important input for any calculation and interpretation of their (opto)-electronic properties. Moreover, the resulting composition distribution is important for a detailed understanding of the growth and capping process, and hence for improving the growth for device applications.

6.2.1 Methodology

In STM the image contrast originates from a tunneling current between a conductive surface and a conductive, atomically sharp tip. This tunneling current I is proportional to the integral between the *Fermi* level E_F and the applied bias V_T between tip and sample (times the elementary charge) $E_F + eV$ over the product of both densities of states (the one from the tip and the one from the sample surface) times the tunneling matrix element, describing the probability of electron tunneling from a filled into an empty state. Both densities of states are furthermore folded by the *Fermi-Dirac* distribution, representing the temperature dependence of the experiment. During the scanning process across the sample surface a controller determines the z-value (height) of the tip in a way that the measured current I is kept constant to a predefined set current I_{set}. At areas at the sample surface with higher tunneling probability or an increased density of states the tip needs to be retracted in order to keep the tunneling current constant, resulting in a higher tip-sample separation, higher z-value, or brighter image contrast. At areas with lower tunneling probability or smaller density of states the tip needs to be pushed further to the surface in order to keep the tunneling current constant, resulting in closer tip-sample separation or darker image contrast. In this way a $z(x, y)$ hight map is gained during scanning the tip across the surface along x- and y-axes.

The principle configuration of an XSTM is shown in Fig. 6.2. In (a) the photograph shows a cleaved XSTM sample mounted on a copper plate, facing a tungsten tip mounted on a tip holder. At the cleavage surface some contrast variations can be seen, which belong to steps at the cleavage plane. In order to measure a tunneling current I the sample bias V_T is applied, as it is sketched in Fig. 6.2(b). Thus, XSTM is in principle nothing else than the application of STM at a narrow cleavage surface across some heterostructures.

6.2.2 Image Contrasts

There are several prerequisites for imaging a material contrast with atomic resolution using XSTM: First, the sample needs to be conductive enough to transport

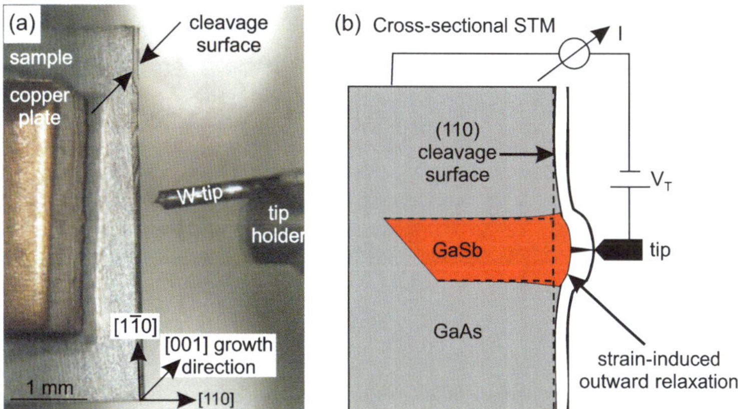

Fig. 6.2 Principle function of an XSTM. (**a**) Photograph of a tungsten tip facing a cleavage sample, and (**b**) sketch of the tip-sample configuration, including the tunneling current I, the sample bias V_T, and the strain-induced outward relaxation of a GaSb/GaAs nanostructure

currents in the order of typically a few pA to nA to a back contact without building up a significant electric potential due to charging. This is fulfilled by metals as well as by doped semiconductors, as they are conductive enough.

Second, in order to achieve atomic resolution the electronic surface states need to be localized in the scale of single atoms. This is fulfilled mostly for dangling covalent bonds and ionic bonds, as they are present at non-polar semiconductors surfaces. Any surface states energetically located within the bulk band gap would lead to disturbances of the atomic contrast. In the case of non-polar cleavage surfaces of compound semiconductors the surface states are typically energetically located within the bulk bands. Independent on the crystal structure—zincblende or wurtzite—they exhibit always a (1×1)-surface unit cell [63–66] with 2 dangling bonds for the zincblende (110) and the wurtzite (10$\bar{1}$0) surfaces and 4 dangling bonds for the wurtzite (11$\bar{2}$0) surface. Half of these dangling bonds are anion-related and half cation-related. Upon cleavage the single electron in every cation-related dangling bond gets transferred to the neighboring anion-related dangling bond, leading to a buckling of the topmost surface atoms. This spatial relaxation of the topmost layer at the surface drives the surface states, which are originally energetically located within the band gap, outside of it into the valence band region for the filled states and into the conduction band region for the empty states [7, 63]. This is the ideal situation, for which the electronic states from the valence band edge and conduction band edge contribute locally to the electronic contrast in XSTM.

The above described electronic contrast by the band edge states is reflected by the so-called atom selective imaging [54] of the compound semiconductor non-polar surfaces: at negative sample biases the electrons tunnel from the localized occupied states of the surface into the tip, while at positive sample biases they tunnel from the tip into the localized empty states of the surface. Different elements from the same chemical group at compound semiconductor surfaces show dangling bonds of

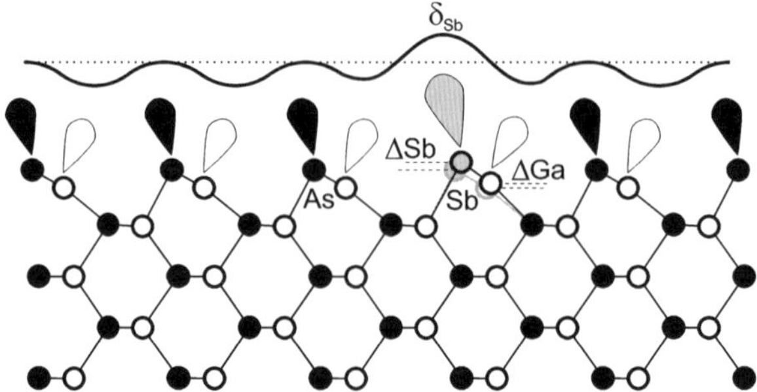

Fig. 6.3 Schematic side view (along [$\bar{1}$10]) on the relaxed GaAs(110) surface, demonstrating electronic and spatial contrast δ_{Sb} of a single uncharged Sb impurity atom. In case of a filled-state XSTM image the occupied dangling bonds of the As (*black*) and Sb (*gray*) atoms are imaged (indicated by the *solid surface corrugation line*), while in case of an empty-state XSTM image the unoccupied dangling bonds of the Ga (*white*) atoms are imaged. At the position of a single Sb atom, both the Sb atom itself and the neighboring Ga atom are relaxed by Δ_{Sb} and Δ_{Ga}

different sizes: typically the higher the chemical period the larger the dangling bond is extended outside the surface toward the vacuum [67].

The side view of an Sb-atom within the relaxed GaAs(110) surface is shown in Fig. 6.3. It illustrates that in case of a filled-state XSTM image the resulting surface corrugation (indicated by the solid line) is determined by the occupied dangling bonds of the As and Sb atoms. The Sb atom exhibits a larger dangling bond (gray) as compared with the As atoms (black). In addition, the longer back-bond shifts the Sb atom outward the surface by Δ_{Sb}. Thus, in order to keep the tunneling current constant, the tip needs to be lifted up above an Sb atom by δ_{Sb}, leading to a brighter image contrast as compared with the surrounding As atoms (darker contrast). Even in case of an empty-state XSTM image, where the unoccupied dangling bonds of the Ga atoms (white) are imaged, an identification of an Sb atom is possible: due to the slight outward shift Δ_{Ga} of the Ga atoms directly beside the Sb atom a weaker, but still slightly brighter contrast appears at XSTM images.

Furthermore, in the presence of localized quantum states, such as quantum-well, quantum-wire, or quantum-dot states, the electronic contrast usually gets enhanced [68, 69]. Such quantum-confined states are typically extended over several atoms, up to several nm. The brighter contrast results from an increased probability for the electrons to tunnel into or out of these states as compared with the matrix material region in their surrounding.

The up to now described electronic contrasts are all bias dependent. The second type of contrast being active is related to the distortion of the lattice, and therewith to the elastic relaxation of atoms out of their natural lattice positions. This relaxation of strain, which was introduced by the self-assembly of compound semiconductor nanostructures during their formation process, occurs due to the additional degree of freedom from the new surface upon cleavage. It occurs in order to achieve the strain energy minimum. This strain-induced outward relaxation is sketched in Fig. 6.2(b) and leads to an additional bright contrast in the XSTM images of nanostructures. The strain relaxation contrast is completely bias independent.

The strain relaxation at cleavage surfaces has two components: the vertical relaxation outward the surface and a lateral component within the topmost surface layers [70]. The vertical component yields information on the total amount of strained material underneath the cleavage surface. Using this, one is able to determine the extension of the part of the quantum ring, which is not cleaved away but remains under the observed surface. The lateral relaxation can be used to measure the local lateral lattice constant, therewith yielding a measure of the local stoichiometric composition [16, 71] in the topmost layers under the cleavage surface. The analysis using this relaxation will be described in the following Sect. 6.2.3.

The strain relaxation contrast and the electronic contrast mechanisms are typically adding each other up in XSTM on compound semiconductor cleavage surfaces. This effect is related to the fact that the covalent bond lengths of all elements is reflected in the dangling bond length as well as in the strain relaxations. Nevertheless, due to the integral dependency of the tunneling current on the density of states, for applied high absolute biases the differences in electronic contrast vanishes. For GaAs-based material systems this is typically fulfilled for absolute biases $|V_T| \geq 2.8$ V [57]. For quaternary InGaAsP the electronic contrast vanishes already at $|V_T| \approx 2.0$ V [72]. Hence it is often possible to extract the strain-related contrast part from images with high absolute biases.

6.2.3 Stoichiometry Determination

The composition of III–V semiconductor nanostructures can be determined from XSTM data based on the local lattice constant, which is modified by strain relaxation due to the lattice mismatch between the active and the matrix material. A correlation between the local lattice parameter and the stoichiometry [11, 15, 59, 70, 73, 74] can be derived from strain relaxation calculations. Thus, the local lattice parameter needs to be evaluated, which is given by the distance between neighboring atomic chains at XSTM images.

Figure 6.4 shows the main steps of the stoichiometry determination procedure [Figs. 6.4(a–c)] as well as the results for the InGaAs wetting layer [Fig. 6.4(d)] and a GaSb quantum dot [Fig. 6.4(e)], both embedded in GaAs matrix material.

The determination of the local lattice parameter starts with taking a height profile along growth direction, for better statistics averaged perpendicular to it across the area indicated by the black box in Fig. 6.4(a). The height profile is plotted in Fig. 6.4(b) (black), showing undulations corresponding to the atomic chains superposed by a broader peak due to the vertical strain relaxation of the InGaAs/GaAs material. The latter is represented by the smoothed vertical background (dark gray). This background is subtracted from the height profile, resulting in the corrugation curve (light gray), which contains the information on the atomic chain positions.

During the next step, a *Gaussian* curve is fit to every maximum of the atomic corrugation, in order to receive the position of the respective atomic chain. Then, the neighboring positions are subtracted from each other, resulting in the atomic chain-to-chain distances [in Fig. 6.4(c)]. The latter are still disturbed by image distortions

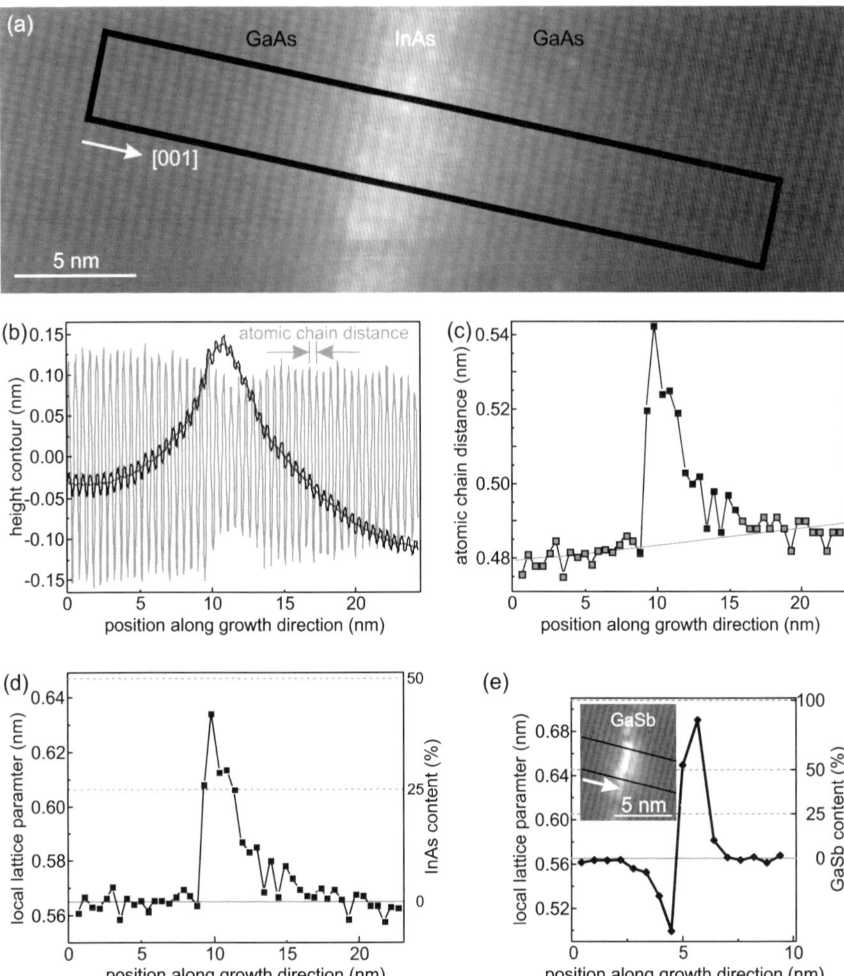

Fig. 6.4 Evaluation of the local stoichiometry of (**a–d**) an InGaAs/GaAs wetting layer and (**e**) a GaSb/GaAs quantum dot [11, 75]. (**a**) XSTM image with the area (*black box*) where the height profile in (**b**) is taken from. (**b**) Height profile along [001] direction averaged along [$\bar{1}$10] direction (*black*), smoothed vertical background (*dark gray*), and the amplified atomic corrugation curve (*light gray*) derived from subtracting the background from the height profile. (**c**) Atomic chain-to-chain distance as measured in (**b**) without calibration (*black*), and linear fit for the matrix material (*gray*). (**d**) Local lattice parameter (*left axis*) of the InGaAs wetting layer (*black*) derived from calibrating (**c**) with the GaAs lattice parameter (*gray, solid line*), and calculated lattice parameters (*gray, dashed lines*) for 25 % and 50 % InAs in GaAs (*right axis*). (**e**) Determination of the local lattice parameter (*left axis*) of the GaSb/GaAs quantum dot (*black*) in the *inset* XSTM image, compared with calculated lattice parameters (*gray, dashed lines*) for 25 %, 50 %, and 100 % GaSb in GaAs (*right axis*). The major difference between a homogeneous, 2-dimensional layer (**d**) and inhomogeneous islands is the local compression of the matrix material underneath the quantum dots visible in (**e**) (see text)

due to non-linearities, creep, and thermal drift of the STM. In order to correct these effects, it is assumed that far away from the layered or island structure the matrix material is pure GaAs, exhibiting its undisturbed lattice parameter. Hence, in a last step, the atomic chain-to-chain distances corresponding to the GaAs matrix (gray-squares) are fitted with a linear or quadratic function (depending on the origin of the image distortions). The such fitted curve (gray line) correspond to the undisturbed lattice parameter of the matrix material, i.e. 0.565 nm for GaAs. The calibrated result of the local lattice parameter is shown Fig. 6.4(d) (black).

The determination of the local lattice parameters as described above is frequently achieved with an interacting software tool developed by E. Lenz [76]. A prerequisite to use this tool is the input of images with high atomic resolution, negligible surface adatoms, and with a considerably extension into the matrix material for accurate calibration.

The resulting data are then compared with reference values from strain relaxation calculations, e.g., local lattice parameters for tow-dimensional layers with 25 % and 50 % InAs in GaAs [70] or for 25 %, 50 %, and 100 % GaSb in GaAs [11]. These reference values are indicated as dashed lines in Figs. 6.4(d, e).

One major difference between a laterally homogeneous 2-dimensional layer and a quantum dot-like island is the occurrence of a compression in the matrix material directly underneath and above the quantum dot. In the case of quantum dot-like islands one should see this compression on both sides, at least if both interfaces are atomically sharp [16, 59]. Typically, the quantum dots or quantum rings form a sharp interface at their bases, but show segregation into the capping layer. Furthermore the quantum dot is usually wider at its base than at its top. Hence, mostly only at the base side of quantum dots and quantum rings the compression of the GaAs matrix material is observed [16, 59], while at the top side the weaker compression is compensated by the segregation. Nevertheless, the described compression of the matrix material leads to a slight underestimation of the actual material concentration in a quantum dot or quantum ring particular at its base. Considering this effect, the actual concentration at the quantum-dot or quantum-ring center is slightly higher than suggested by the plotted data.

6.2.4 Statistical Analysis

Due to the random position of the cleavage surface, in XSTM images only a cut through a semiconductor nanostructure is accessible. Thus, the cross-section of quantum dots or quantum rings occur statistically (see Fig. 6.5). Cutting a quantum ring almost through its center (f, g) it appears at the cleavage surface like two small quantum dots, while a cleavage through the quantum-ring side (c) may appear very similar to a cross-section of a quantum dot. In order to confirm that the investigated sample contains indeed quantum rings instead of close neighboring quantum dots, a statistical analysis has to be performed.

There are two possibilities for such a statistical analysis: on the one hand one can measure the neighboring distance between each pair of quantum dot-like ap-

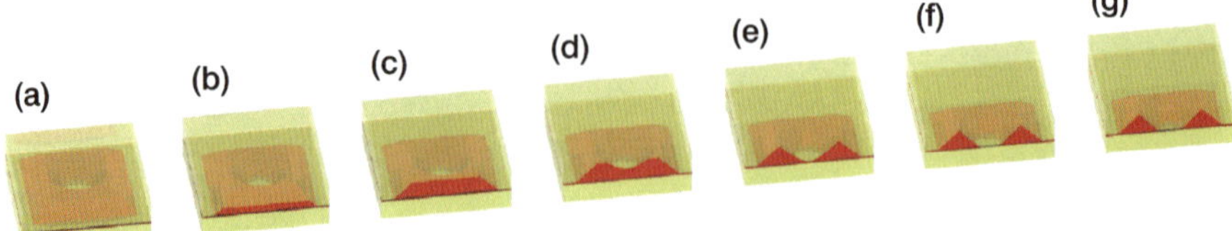

Fig. 6.5 Cross-sections resulting from different cleavage planes through a quantum ring: (**a**) only the wetting layer is visible, (**b**)–(**c**) trapezoidal shape for side cuts, (**d**)–(**e**) kidney shapes from partial cuts through the quantum ring, and (**f**)–(**g**) double triangular or trapezoidal shapes from more central cuts through the quantum ring

pearances along a trace at the cross-sectional surface, and on the other hand one can analyze the appearing diameter of the assumed central opening of a quantum ring. Using the latter method, one can also gain information about the ratio of the central opening versus the outer size [32]. As such an analysis fits with a statistical simulation one can draw conclusions on the amount of quantum dots within the sample, which were transformed into quantum rings, even without top view imaging. For details of the analysis see Refs. [32, 37].

6.3 GaSb/GaAs Quantum-Ring Structure

GaSb/GaAs(001) quantum rings can be grown in a self-assembled way by depositing GaSb on GaAs(001). Nevertheless, the quantum-ring formation does not start from the very first beginning of the deposition, but at least two phase transitions are necessary unless the quantum rings evolve (see. Fig. 6.1).

Depending on the growth conditions, especially the substrate temperature, one starts with a clean GaAs(001) surface, which is either c(4 × 4) or $\beta 2(2 \times 4)$ reconstructed [77]. Upon supplying Sb or GaSb or their precursor materials in molecular beam epitaxy (MBE) or metal organic chemical vapor deposition (MOCVD), first the As-rich surface reconstruction needs to be exchanged into an Sb-rich one, before the GaSb material can grow. This is the so-called Sb-soaking process or Sb-for-As exchange process [32, 37, 78–81]. It should be noted here that after the soaking process already a certain amount of GaSb is present at the growth surface, i.e. the amount of GaSb within the surface reconstruction, which is typically about 0.5–0.75 ML [82, 83]. Such a material exchange at the growth surface is not necessary for switching group-III materials, as e.g., during InAs/GaAs growth, because of the group-V rich surfaces.

Upon further GaSb supply in the next step a wetting layer forms [11, 12, 75]. An XSTM image of the GaSb wetting layer is shown in Fig. 6.6(a). It is clearly visible that this wetting layer is noncontinuous along the [$\bar{1}$10] direction. This might be due to the fact that an incomplete ML was deposited in this case.

At about 1 ML of GaSb deposition, the evolution of very tiny quantum dots could be observed, as it is shown in Fig. 6.6(b) [11, 12]. In the analyzed sample a total material amount of $\sim$ 1.1 ML GaSb was found. Since these quantum dots are very small

Fig. 6.6 XSTM images of GaSb/GaAs(001) layers with increasing amount of deposited GaSb. (**a**) noncontinuous GaSb wetting layer, (**b**, **c**) quantum dots, (**d**, **e**) quantum rings, and (**f**) sketch of the quantum-ring cross-section. (**a**, **b**) are grown by MOCVD at 470 °C without Sb soaking. (**c**, **d**, **e**) are grown by MBE at 490 °C, applying 5 s of Sb soaking before the GaSb deposition. Further parameters are: (**a**) 21 s GaSb deposition ($\sim$1.0 ML), the XSTM image was taken at $V_T = -3.0$ V and $I_T = 80$ pA, (**b**) 25 s GaSb deposition ($\sim$1.1 ML), taken at $V_T = -1.6$ V and $I_T = 80$ pA, (**c**) $\sim$1.5 ML (1 ML + soaking) GaSb, taken at $V_T = -2.5$ V and $I_T = 80$ pA, (**d**) $\sim$3.2 ML (2.7 ML + soaking) GaSb, taken at $V_T = -2.3$ V and $I_T = 90$ pA, (**e**) $\sim$3.6 ML (3.1 ML + soaking) GaSb, taken at $V_T = -3.0$ V and $I_T = 60$ pA

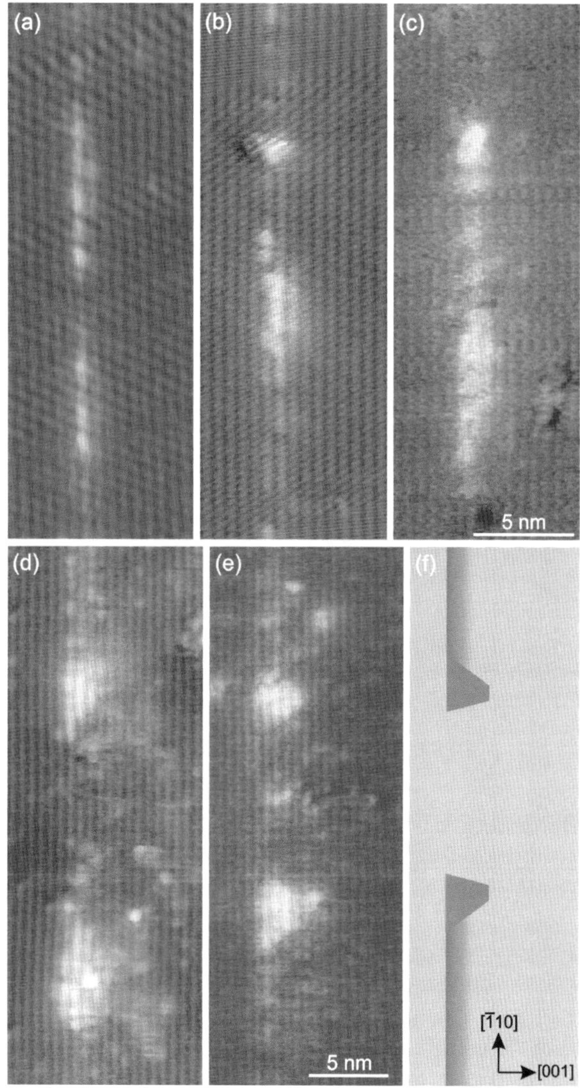

and show a very low areal density, this finding corresponds well with the theoretically calculated critical thickness for quantum-dot formation in GaSb/GaAs(001) of about 1.0–1.2 ML [13]. Starting from about 1.5 ML of GaSb deposition [see Fig. 6.6(c)] larger quantum dots were found with a higher density. At this coverage already large quantum dots may transform into quantum rings upon capping due to the additionally applied strain energy.

For high GaSb coverages above 2.0 ML, as shown in Figs. 6.6(d) and (e), a clear quantum-ring formation could be observed [32, 37]. While at the outside the quantum rings are still connected to the wetting layer [upper and lower parts of

Figs. 6.6(d) and (e)], within the quantum ring the Sb appears to be removed completely: the central opening is mainly filled with GaAs matrix material from the capping [see Fig. 6.6(f)]. Further on, no so-called *second wetting layer* from quantum-dot to quantum-ring transition upon thin capping can be observed [32, 37, 84], as it is typically found in the In(Ga)As/GaAs system [29].

These findings for the quantum-ring formation processes are also confirmed by other investigations on capped GaSb/GaAs depositions using TEM and APT, as well as on uncapped ones by AFM: Upon the deposition of about 3.0 ML of GaSb on GaAs without capping AFM shows sometimes clear quantum-ring structures [21, 85, 86]. The occurrence of the quantum-dot to quantum-ring transition even without capping seems to be dependent on the growth conditions in particular before the GaSb material deposition [86].

For capped samples the transition could be observed already for ~ 2.0 ML of deposited GaSb using TEM [87] and APT [88]. From the APT investigations also a clear central opening is found. Using XSTM the transition from quantum dots to quantum rings was also confirmed by a statistical analysis of the occurrence of the different cross-sections from random cleavage. All cross-sections from quantum rings as described in Fig. 6.5 were found in XSTM images containing the quantum-ring structures [32, 33, 37]. From the statistical analysis it was demonstrated that all observed nanostructures are quantum rings, and the size of the central opening could be quantified to about 42 % of the outer quantum-ring extension [32].

6.3.1 GaSb/GaAs Quantum-Ring Size and Shape

Before the quantum-dot to quantum-ring transition occurs, the resulting GaSb/GaAs quantum dots are quite small as compared with those in the In(Ga)As/GaAs system: Directly after the critical thickness of quantum-dot evolution of about 1.0–1.2 ML very tiny quantum dots appear, measuring only about 4–8 nm of lateral extension and up to max. 2 nm of height [11]. At about 1.5 ML, at which the transformation into quantum rings may start already, the quantum-dot sizes are still comparably small: their lateral extension is typically ~ 13 nm and their heights are limited to about 2.4 nm [12, 37]. While the tiny quantum dots show a triangular cross-section (being related to a pyramidal shape) [11], the larger ones already occur truncated by a typical (001) top facet. This truncation is also observed for the In(Ga)As/GaAs material system and can be related to a material re-arrangement during the capping process [15, 57–59, 71, 89].

Beyond the quantum-dot to quantum-ring transition the resulting quantum rings are slightly larger: while their lateral extension is typically found to be ~ 16 nm, their heights remain at about 2 nm [32]. It should be noted here that the outer extension is found to be quite constant, while their inner opening varies much at the cross-sectional images. This can be related to the random cleavage through a quantum-ring structure, where for off-central cleavages a quantum ring may appear similar to a quantum dot (see Sect. 6.2.4). Furthermore, the quite constant outer extension is a hint for side facets being perpendicular to the (110) cleavage surface, i.e. {111}, {112}, or {113} facets, but no {101} facets. An analysis of the angle between the quantum-ring base and the side facet typically leads to values around 50°, which corresponds well with the nominal 54.7° of {111} facets [32, 37].

The smaller sizes of the quantum dots and quantum rings in the GaSb/GaAs material system as compared with the In(Ga)As/GaAs one can be understood by looking closer to the strain-energy situation: In the GaSb material system the strain energy is about 1/3 higher as compared with the InAs/GaAs system (1.87 eV for GaSb/GaAs vs. 1.40 eV InAs/GaAs per III–V pair). The difference in strain energy originates especially from the higher shear-strain components [13, 32]. Hence the base of the quantum dots and quantum rings (and therewith the contact area with the matrix material) needs to be smaller to compensate the increased strain energy. Consequently, also steeper side facets occur in order to incorporate more material within the nanostructure. The observed steeper {111} facets in the GaSb/GaAs system as compared with the slightly flatter {101} facets typically observed for InAs/GaAs [16, 59, 71] can hence be understood as similarity to the *hut*-to-*dome* structural transition from flat to steep facets for uncapped quantum dots [see also Fig. 6.1(c, d)] [17, 39]. In both cases the amount of agglomerated material within the nanostructure increases without significantly increasing the interface area to the substrate, and hence without increasing the strain energy at the interface too much.

Beside pseudomorphic growth at the interface between GaAs and GaSb also the formation of an interfacial misfit dislocation array (IMF) can occur [90, 91]. The deposited GaSb layer on top of such an IMF can also form quantum dots [90]. Such nanostructures show lateral sizes of about 15–20 nm and heights of up to 6 nm [34, 90]. Thus, they are typically larger as compared with quantum dots grown pseudomorphically. The increased size occurs due to a partial strain relaxation of the material caused by the IMF. Furthermore, the dislocations are not necessarily bound to the IMF, but can also thread along growth direction through the nanostructure or even into the capping layer [34]. In the same sample in areas without IMF formation also quantum rings could be observed, showing similar structural parameters as reported above.

6.3.2 GaSb/GaAs Quantum-Ring Stoichiometry

Typically, the GaSb/GaAs quantum dots and quantum rings are found to consist of high GaSb content: the tiny quantum dots show GaSb contents between 60 % and pure GaSb stoichiometry (see Fig. 6.4) [11, 75], as well as the outer rims of quantum rings (see Fig. 6.7) [32, 37]. At the central opening of the quantum rings almost no GaSb material can be found, see Fig. 6.7, yellow squares. Hence, not only the GaSb from the central part of the quantum ring got removed, but also the base of the quantum ring which originally was part of the wetting layer. This finding is in clear contrast to the In(Ga)As/GaAs material system, for which typically only a reduced In content is found at the upper quantum-ring center, while at the base of the quantum ring the material remains [18, 29, 30].

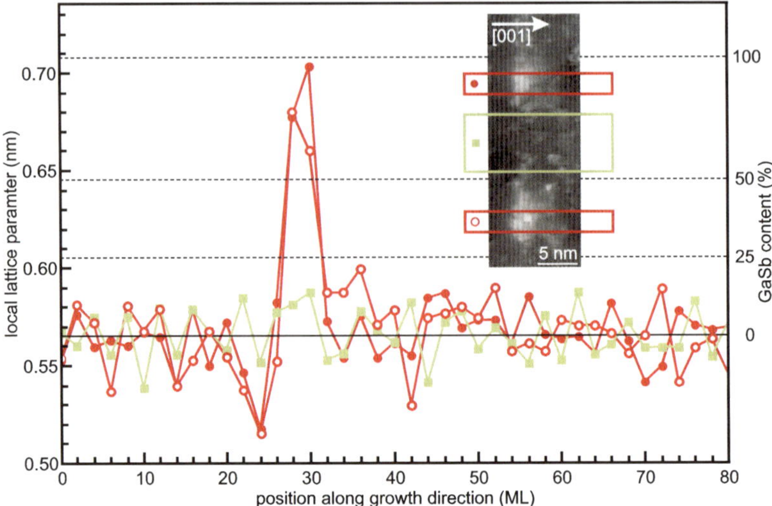

Fig. 6.7 Local lattice parameter (left axis) of the GaSb/GaAs quantum ring shown in the *inset* XSTM image [same as in Fig. 6.6(d)], and the corresponding GaSb content (*right axis*). For the outer rim of the quantum ring a GaSb content of about 80 % (*red circles*) and almost 100 % (*red dots*) is derived, while for the central opening (*yellow squares*) the GaSb content is almost 0 %

For the wetting layer surrounding the quantum dots and quantum rings the stoichiometry analysis leads to a different situation: Typically the wetting layer is found to have a much lower GaSb content as compared with the quantum dots and quantum rings. It is also lower as compared with the InAs content of the wetting layer of similar InAs/GaAs nanostructures. For all capped MBE-grown samples the wetting layer consists of only $\sim$20 % of GaSb at the base interface, followed by an exponential decay along growth direction [32, 37]. Thus, the capping of the wetting layer is accompanied by a strong segregation of the GaSb along growth direction. This is also caused by strain-energy reduction: directly at the growth surface the Sb atoms can relax better as compared with those within GaAs. The observed segregation constant is typically found in the range of 0.80–0.90 [37]. The strong segregation is also confirmed by other investigation methods, such as TEM and APT [45, 88]. It is much higher as compared with typical values for the In(Ga)As/GaAs system of about 0.70–0.78 [89, 92].

The total amount of GaSb found in and (segregated) above the wetting layer is found to be around 1.0 ML. This is in good agreement with the nominal 1.0–1.2 ML being present in the wetting layer, before the capping was initiated. It should be noted here that in MOCVD growth almost no segregation was found [75]. This may be related to the different dissociation process of the precursor materials at the growth surface, which might suppress the atomic exchange processes from the growth surface with the layer underneath. Nevertheless, also for MOCVD grown samples $\sim$1.0 ML of GaSb material was found in the wetting layer regions.

6.3.3 GaSb/GaAs Quantum-Ring Electronic States

Using the scanning tunneling spectroscopy (STS) mode [93] on top of cross-sectional cleaved surfaces (XSTS) further information on the electronic structure of quantum dots [68] and quantum rings [33] can be gained. It should be noted here that in this method only the response from the remaining part of the nanostructure underneath the cleavage surface can be probed, which differs from the electronic properties of the whole nanostructure due to missing material as well as due to partial strain relaxation at the cleavage surface. Nevertheless, XSTS is an interesting approach since the electronic properties can be related with the spatial structure from the XSTM images of the identical nanostructure.

For the surrounding wetting layer as well as for quantum dots the type-II alignment of the GaSb/GaAs could be shown already via bias dependent imaging in XSTM [94]. The emptying of the confined hole states leads to an interfacial band bending of the conduction band in the surrounding matrix material of the nanostructure [84, 94]. The increased tunneling probability due to this band bending can be imaged as apparent broadening of the nanostructures in empty-state XSTM images, while filled-state images show a sharp contrast at the interface [94].

In a similar way also the hole state of GaSb/GaAs quantum rings can be probed directly via XSTS. The XSTS I–V curves as well as the differentiated $(dI/dV)/(\overline{I/V})$–V curves at the position of the GaSb quantum ring show an increased tunneling current below the conduction band edge as compared with the surrounding GaAs matrix material [33]. Due to the intrinsic characteristic of the investigated layer the peak in the $(dI/dV)/(\overline{I/V})$–V curves is caused by tunneling into the emptied hole states upon tip-induced band bending [33]. The GaSb quantum-ring hole-state is found to be energetically located 0.2–0.3 eV above the valence band maximum of the GaAs matrix material. Additionally the conduction band offset could be determined to about 0.1 eV, demonstrating the type-II alignment also for GaSb/GaAs quantum rings.

A comparison of the energetic alignment of the hole state with data generated via photoluminescence spectroscopy (PL) leads to an energetic position of 0.37 eV above the valence band maximum [33]. Taking the above discussed effects of cleaving half of the quantum ring away and of relaxing the cleavage surface into account, this fits well with the result of 0.2–0.3 eV obtained from XSTS. This is a promising result for the observation of *Aharanov-Bohm* oscillation in the future [84].

6.4 Formation Process of GaSb/GaAs Quantum Rings

The formation of GaSb quantum rings in GaAs matrix material occurs via the formation of quantum dots, which turn then into quantum rings due to strain-energy minimization as described above (see Sect. 6.1). In the following, a closer look is presented on the quantum-dot formation and quantum-ring transformation process.

From the statistical evaluation of the quantum dot/ring cross-section (see Sect. 6.3.1) it was found that only for very thin depositions of about 1.0–1.2 ML GaSb on GaAs small quantum dots are found after capping. Exceeding about 1.5 ML of material deposition always quantum rings are present within the samples [32, 37]. This finding is also confirmed by HR-TEM investigations, where almost all capped quantum dots were found to be transformed into quantum rings [87]. The quantum-ring structure itself is also more pronounced in the GaSb/GaAs system as compared with the In(Ga)As/GaAs systems: For the latter, the quantum-ring center is typically found to be not completely free of In. It is only more diluted as compared with the situation at the rim. The GaSb/GaAs quantum rings are typically found to be free of Sb at the center (see Sect. 6.3.2); even more also the wetting layer gets removed within the quantum ring. This situation may be induced by the high strain energy in this system. Furthermore, no additional growth interruption was inserted during the capping, and the capping process was performed quite fast for all samples investigated by XSTM. From such growth conditions one can conclude that either the quantum-dot to quantum-ring transformation process occurs very fast during the initial capping, or they form already before the capping was started. Indeed, even under the limited lateral resolution in AFM, GaSb/GaAs(001) quantum rings were observed prior to any capping process [21, 85, 86]. Up to now, such a behavior was not observed for the In(Ga)As/GaAs material system.

For a more detailed understanding of the GaSb/GaAs quantum-ring formation a closer look to the different energies driving the formation process is useful: On the one hand, the surface free energies—as far as they are known—of the different GaSb surface reconstructions are very similar to each other [82, 95]. Even more, Sb or a thin GaSb layer on GaAs is well know to act as a surfactant—a so-called *surface active agent*—, being also well known for lowering the surface free energy globally. Hence, the free surface energies over all can be estimated to be quite low, most likely in the range of the one of comparable InAs surfaces, or even lower. On the other hand, the lattice mismatch of GaSb on GaAs amounts to 7.83 %. Due to different elastic constants this leads to $\sim$33 % higher strain energy already for the first deposited ML of GaSb [13] as compared with the same amount of InAs on GaAs. Using this information, the critical thickness of quantum-dot formation can be calculated to about 1.2 ML of GaSb on GaAs(001). The increased strain energy for GaSb/GaAs as compared with InAs/GaAs originates especially from the shear-strain components.

The above discussed energetic situation continues for the case of further material deposition. Hence, after the GaSb/GaAs quantum-dot formation the strain energy is much higher as compared with the InAs/GaAs material system. It is further also increased already for small quantum dots, and especially at their base. This increased strain energy makes the quantum-dot to quantum-rings transformation process very likely, even without the additional strain introduced by any capping material.

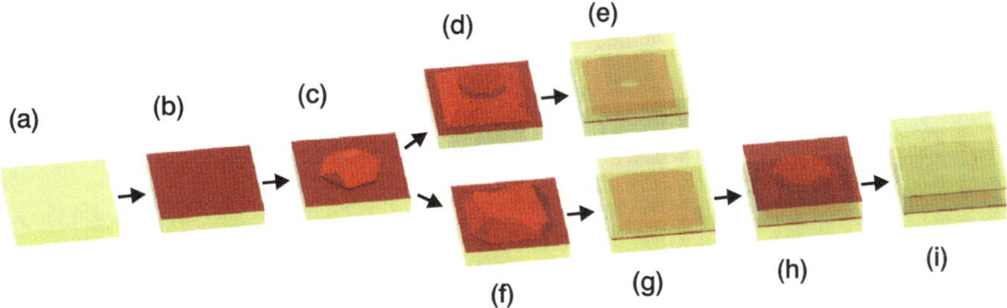

Fig. 6.8 Different paths of the quantum-ring evolution in GaSb/GaAs (**a**–**e**) and In(Ga)As/GaAs (**a**–**c**, **f**–**i**) material systems: After deposition of a wetting layer (**b**) on the clean surface (**a**), small quantum dots occur (**c**) in both cases. In the GaSb/GaAs material system the quantum rings can already evolve from uncapped quantum dots (**c**)–(**d**), and might be further capped (**e**). In the In(Ga)As/GaAs material system first a transition into larger quantum dots occurs (**c**)–(**f**), which need to be capped thinly (**f**)–(**g**). During a growth interruption (**h**) the central opening is formed together with a second wetting layer. This structure might also be further capped (**i**)

6.5 General Conclusions on the Quantum Ring Evolution

In the In(Ga)As/GaAs(001) system at least 2.5 ML of InAs material has typically to be deposited as well as a growth interruption during capping is necessary for the quantum-ring formation process [18, 29, 30] [see Fig. 6.8(a–c, f–i)]. In contrast, in the GaSb/GaAs(001) material system all quantum dots are transformed into quantum rings already for 2.0 ML of GaSb on GaAs(001) [32, 37]. The quantum rings are more pronounced without any Sb at the ring center, while for In(Ga)As/GaAs quantum rings typically a diluted amount of In remains at the central base of the quantum ring. Furthermore, no growth interruption during the capping process is necessary to enhance the quantum-ring formation process. Hence, as a conclusion, it can be stated that the quantum-ring formation process occurs easier in the GaSb/GaAs(001) material system. Even more, for GaSb/GaAs(001) the formation of quantum rings can be observed without capping at all [21, 86], which is a much more direct formation process [see Fig. 6.8(a–e)].

All these finding can be understood by the approximately 1/3 higher strain energy in the GaSb/GaAs system as compared with the InAs/GaAs system, while the surface energies are similar [13]. In the InAs/GaAs system the strain energy before capping is not high enough to initiate the quantum-dot to quantum-ring formation process. Hence, the GaSb/GaAs system with its more direct quantum-ring formation should rather be considered as the model system for quantum-ring formation.

Acknowledgements The authors would like to thank Rainer Timm for XSTM work and analysis, Mario Dähne for fruitful discussion, the Deutsche Forschungsgemeinschaft DFG for the financial support by the collaborative research centers Sfb 296, Sfb 787, by the projects No. Da 408/4, Da 408/8, Da 408/13, and by the SANDiE NoE of the European Commission. One of the authors (H.E.) acknowledge financial support from the Alexander von Humboldt-Foundation.

References

1. E. Pehlke, N. Moll, A. Kley, M. Scheffler, Appl. Phys. A **65**, 525 (1997)
2. N. Moll, M. Scheffler, E. Pehlke, Phys. Rev. B **58**, 4566 (1998)
3. P. Kratzer, E. Penev, M. Scheffler, Appl. Phys. A **75**, 79 (2002)
4. P. Kratzer, E. Penev, M. Scheffler, Appl. Surf. Sci. **216**, 436 (2003)
5. T. Hammerschmidt, P. Kratzer, M. Scheffler, Phys. Rev. B **77**, 235303 (2008)
6. T. Hammerschmidt, P. Kratzer, M. Scheffler, Phys. Rev. B **81**, 159905 (2010)
7. H. Lüth, *Solid Surfaces, Interfaces and Thin Films*, 5th edn. (Springer, Berlin, 2010)
8. P.R.C. Kent, G.L.W. Hart, A. Zunger, Appl. Phys. Lett. **81**, 4377 (2002)
9. I. Stranski, L. Krastanow, Sitz.ber. Wien. Akad. Wiss., Kl. IIb **146**, 797 (1938)
10. C. Prohl, B. Höpfner, J. Grabowski, M. Dähne, H. Eisele, J. Vac. Sci. Technol. B **28**, C5E13 (2010)
11. R. Timm, H. Eisele, A. Lenz, S.K. Becker, J. Grabowski, T.Y. Kim, L. Müller-Kirsch, K. Pötschke, U.W. Pohl, D. Bimberg, M. Dähne, Appl. Phys. Lett. **85**, 5890 (2004)
12. R. Timm, A. Lenz, H. Eisele, L. Ivanova, K. Pötschke, U. Pohl, D. Bimberg, G. Balakrishnan, D.L. Huffaker, M. Dähne, Phys. Status Solidi C **3**, 3971 (2006)
13. H. Eisele, M. Dähne, J. Cryst. Growth **338**, 103 (2012)
14. M. Grundmann, O. Stier, D. Bimberg, Phys. Rev. B **52**, 11969 (1995)
15. N. Liu, J. Tersoff, O. Baklenov, A.L. Holmes Jr., C.K. Shih, Phys. Rev. Lett. **84**, 334 (2000)
16. A. Lenz, R. Timm, H. Eisele, C. Hennig, S.K. Becker, R.L. Sellin, U.W. Pohl, D. Bimberg, M. Dähne, Appl. Phys. Lett. **81**, 5150 (2002)
17. M. Xu, Y. Temko, T. Suzuki, K. Jacobi, J. Appl. Phys. **98**, 083525 (2005)
18. J.M. García, G. Medeiros-Ribeiro, K. Schmidt, T. Ngo, J.L. Feng, A. Lorke, J. Kotthaus, P.M. Petroff, Appl. Phys. Lett. **71**, 2014 (1997)
19. T. Raz, D. Ritter, G. Bahir, Appl. Phys. Lett. **82**, 1706 (2003)
20. J. Cui, Q. He, X.M. Jiang, Y.L. Fan, X.J. Yang, F. Xue, Z.M. Jiang, Appl. Phys. Lett. **83**, 2907 (2003)
21. S. Kobayashi, C. Jiang, T. Kawazu, H. Sakaki, Jpn. J. Appl. Phys. **43**, L662 (2004)
22. N. Koguchi, S. Takahashi, T. Chikyow, J. Cryst. Growth **111**, 688 (1991)
23. V. Mantovani, S. Sanguinetti, M. Guzzi, E. Grilli, M. Gurioli, K. Watanabe, N. Koguchi, J. Appl. Phys. **96**, 4416 (2004)
24. J.S. Lee, H.W. Ren, S. Sugou, Y. Masumoto, J. Appl. Phys. **84**, 6686 (1998)
25. A. Lorke, R. Blossey, J.M. García, M. Bichler, G. Abstreiter, Mater. Sci. Eng. B **88**, 225 (2002)
26. J. Sormunen, J. Riikonen, M. Mattila, J. Tiilikainen, M. Sopanen, H. Lipsanen, Nano Lett. **5**, 1541 (2005)
27. G. Costantini, A. Rastelli, C. Manzano, P. Acosta-Diaz, R. Songmuang, G. Katsaros, O.G. Schmidt, K. Kern, Phys. Rev. Lett. **96**, 226106 (2006)
28. H.S. Ling, C.P. Lee, J. Appl. Phys. **102**, 024314 (2007)
29. A. Lenz, H. Eisele, R. Timm, S.K. Becker, R.L. Sellin, U.W. Pohl, D. Bimberg, M. Dähne, Appl. Phys. Lett. **85**, 3848 (2004)
30. P. Offermans, P.M. Koenraad, J.H. Wolter, D. Granados, J.M. García, V.M. Fomin, V.N. Gladilin, J.T. Devreese, Appl. Phys. Lett. **87**, 131902 (2005)
31. P. Offermans, P.M. Koenraad, J.H. Wolter, D. Granados, J.M. García, V.M. Fomin, V.N. Gladilin, J.T. Devreese, Physica E **32**, 41 (2006)
32. R. Timm, H. Eisele, A. Lenz, L. Ivanova, G. Balakrishnan, D. Huffaker, M. Dähne, Phys. Rev. Lett. **101**, 256101 (2008)
33. R. Timm, H. Eisele, A. Lenz, L. Ivanova, V. Vossebürger, T. Warming, D. Bimberg, I. Farrer, D.A. Ritchie, M. Dähne, Nano Lett. **10**, 3972 (2010)
34. E.P. Smakman, J.K. Garleff, R.J. Young, M. Hayne, P. Rambabu, P.M. Koenraad, Appl. Phys. Lett. **100**, 142116 (2012)
35. R.J. Warburton, C. Schäflein, D. Haft, F. Bickel, A. Lorke, J.M.G.K. Karrai, W. Schoenfeld, P.M. Petroff, Nature **405**, 926 (2000)

36. F. Suárez, D. Granados, M.L. Dotor, J.M. García, Nanotechnology **15**, S126 (2004)
37. R. Timm, A. Lenz, H. Eisele, L. Ivanova, M. Dähne, G. Balakrishnan, D.L. Huffaker, I. Farrer, D.A. Ritchie, J. Vac. Sci. Technol. B **26**, 1492 (2008)
38. J. Márquez, L. Geelhaar, K. Jacobi, Appl. Phys. Lett. **78**, 2309 (2001)
39. A. Rastelli, M. Kummer, H. von Känel, Phys. Rev. Lett. **87**, 256101 (2001)
40. H. Eisele, K. Jacobi, Appl. Phys. Lett. **90**, 129902 (2007)
41. L. Reimer, *Transmission Electron Microscopy*. Springer Series in Optical Sciences, vol. 36 (Springer, Berlin, 1984). ISBN 3-540-11794-6
42. D.B. Williams, C.B. Carter, *Transmission Electron Microscopy—A Textbook for Materials Science* (Plenum Press, New York, 1996). ISBN 0-306-45324-X
43. F. Heinrichsdorff, A. Krost, N. Kirstaedter, M.H. Mao, M. Grundmann, D. Bimberg, A.O. Kosogov, P. Werner, Jpn. J. Appl. Phys. **36**, 4129 (1997)
44. D. Litvinov, D. Gerthsen, A. Rosenauer, M. Schowalter, Phys. Rev. B **74**, 165306 (2006)
45. S.I. Molina, A.M. Beltrán, T. Ben, P.L. Galindo, E. Guerrero, A.G. Taboada, J.M. Ripalda, M.F. Chrisholm, Appl. Phys. Lett. **94**, 043114 (2009)
46. A. Rosenauer, T. Mehrtens, K. Müller, K. Gries, M. Schowalter, P.V. Satyam, S. Bley, C. Tessarek, D. Hommel, K. Sebald, M. Seyfried, J. Gutowski, A. Avramescu, K. Engl, S. Lutgen, Ultramicroscopy **111**, 1316 (2011)
47. M. Sauvage-Simkin, Y. Garreau, P. Pinchaux, M.B. Véron, J.P. Landesman, J. Nagle, Phys. Rev. Lett. **75**, 3485 (1995)
48. A. Krost, F. Heinrichsdorff, D. Bimberg, A. Darhuber, G. Bauer, Appl. Phys. Lett. **68**, 785 (1996)
49. G. Biasiol, S. Heun, Phys. Rep. **500**, 117 (2011)
50. J. Stangl, V. Holý, G. Bauer, Rev. Mod. Phys. **76**, 725 (2004)
51. M.K. Miller, *Atom Probe Tomography: Analysis at the Atomic Level* (Kluwer Academic/ Plenum, Berlin, 2000)
52. B. Gault, M. Müller, A. La Fontaine, M.P. Moody, A. Shariq, A. Cerezo, S.P. Ringer, G.D.W. Smith, J. Appl. Phys. **108**, 044904 (2010)
53. M. Müller, D.W. Saxey, A. Cerezo, G.D.W. Smith, J. Phys. Conf. Ser. **209**, 012026 (2010)
54. R.M. Feenstra, J.A. Stroscio, J. Tersoff, A.P. Fein, Phys. Rev. Lett. **58**, 1192 (1987)
55. M.B. Johnson, M. Pfister, S.F. Alvarado, H.W.M. Salemink, MRS Proc. **332**, 599 (1993)
56. B. Legrand, B. Grandidier, J.P. Nys, D. Stiévenard, J.M. Gérard, V. Thierry-Mieg, Appl. Phys. Lett. **73**, 96 (1998)
57. H. Eisele, O. Flebbe, T. Kalka, M. Dähne-Prietsch, Surf. Interface Anal. **27**, 537 (1999)
58. H. Eisele, O. Flebbe, T. Kalka, C. Preinesberger, F. Heinrichsdorff, A. Krost, D. Bimberg, M. Dähne-Prietsch, Appl. Phys. Lett. **75**, 106 (1999)
59. O. Flebbe, H. Eisele, T. Kalka, F. Heinrichsdorff, A. Krost, D. Bimberg, M. Dähne-Prietsch, J. Vac. Sci. Technol. B **17**, 1639 (1999)
60. P.W. Fry, I.E. Itskevich, D.J. Mowbray, M.S. Skolnick, J.J. Finley, J.A. Barker, E.P. O'Reilly, L.R. Wilson, I.A. Larkin, P.A. Maksym, M. Hopkinson, M. Al-Khafaji, J.P.R. David, A.G. Cullis, G. Hill, J.C. Clark, Phys. Rev. Lett. **84**, 733 (2000)
61. J. Shumway, A.J. Williamson, A. Zunger, A. Passaseo, M. DeGiorgi, R. Cingolani, M. Catalano, P. Crozier, Phys. Rev. B **64**, 125302 (2001)
62. A. Schliwa, M. Winkelnkemper, D. Bimberg, Phys. Rev. B **76**, 205324 (2007)
63. A. Kahn, Surf. Sci. Rep. **3**, 193 (1983)
64. B. Siemens, C. Domke, M. Heinrich, P. Ebert, K. Urban, Phys. Rev. B **59**, 2995 (1999)
65. L. Ivanova, H. Eisele, A. Lenz, R. Timm, M. Dähne, O. Schumann, L. Geelhaar, H. Riechert, Appl. Phys. Lett. **92**, 203101 (2008)
66. H. Eisele, P. Ebert, Phys. Status Solidi RRL **6**, 359 (2012)
67. K.J. Chao, C.K. Shih, D.W. Gotthold, B.G. Streetman, Phys. Rev. Lett. **79**, 4822 (1997)
68. B. Grandidier, Y.M. Niquet, B. Legrand, J.P. Nys, C. Priester, D. Stiévenard, J.M. Gérard, V. Thierry-Mieg, Phys. Rev. Lett. **85**, 1068 (2000)
69. T. Maltezopoulos, A. Bolz, C. Meyer, C. Heyn, W. Hansen, M. Morgenstern, R. Wiesendanger, Phys. Rev. Lett. **91**, 196804 (2003)

70. H. Eisele, *Cross-Sectional Scanning Tunneling Microscopy of InAs/GaAs Quantum Dots* (Wissenschaft & Technik Verlag, Berlin, 2002). ISBN 3-89685-388-0
71. H. Eisele, A. Lenz, R. Heitz, R. Timm, M. Dähne, Y. Temko, T. Suzuki, K. Jacobi, J. Appl. Phys. **104**, 124301 (2008)
72. R.M. Feenstra, Physica B **273**, 796 (1999)
73. R.S. Goldman, R.M. Feenstra, C. Silfvenius, B. Stålnacke, G. Landgren, J. Vac. Sci. Technol. B **15**, 1027 (1997)
74. D.M. Bruls, J.W.A.M. Vugs, P.M. Koenraad, H.W.M. Salemink, J.H. Wolter, M. Hopkinson, M.S. Skolnick, F. Long, S.P.A. Gill, Appl. Phys. Lett. **81**, 1708 (2002)
75. R. Timm, J. Grabowski, H. Eisele, A. Lenz, S.K. Becker, L. Müller-Kirsch, K. Pötschke, U.W. Pohl, D. Bimberg, M. Dähne, Physica E **26**, 231 (2005)
76. E. Lenz, In Berechnung154, version 154, TU Berlin, 2004–2010. AG Dähne
77. A. Ohtake, J. Nakamura, S. Tsukamoto, N. Koguchi, A. Natori, Phys. Rev. Lett. **89**, 206102 (2002)
78. K. Suzuki, R.A. Hogg, Y. Arakawa, J. Appl. Phys. **85**, 8349 (1999)
79. J.P. Silveira, J.M. García, F. Briones, J. Cryst. Growth **227**, 995 (2001)
80. I. Farrer, M.J. Murphy, D.A. Ritchie, A.J. Shields, J. Cryst. Growth **251**, 771 (2003)
81. T. Nakai, S. Iwasaki, K. Yamaguchi, Jpn. J. Appl. Phys. **43**, 2122 (2004)
82. J.E. Bickel, N.A. Modine, C. Pearson, J.M. Millunchick, Phys. Rev. B **77**, 125308 (2008)
83. J. Bickel, C. Pearson, J.M. Millunchick, Surf. Sci. **603**, 14 (2009)
84. R.J. Young, E.P. Smakman, A.M. Sanchez, P. Hodgson, P.M. Koenraad, M. Hayne, Appl. Phys. Lett. **100**, 082104 (2012)
85. C.C. Tseng, S.C. Mai, W.H. Lin, S.Y. Wu, B.Y. Yu, S.H. Chen, S.Y. Lin, J.J. Shyue, M.C. Wu, IEEE J. Quantum Electron. **47**, 335 (2011)
86. A. Martin, T. Saucer, K. Sun, S. Kim, G. Ran, G. Rodriguez, X. Pan, V. Sih, J. Millunchick, J. Vac. Sci. Technol. B **30**, 02B112 (2012)
87. M. Ahmad Kamarudin, M. Hayne, R.J. Young, Q.D. Zhuang, T. Ben, S.I. Molina, Phys. Rev. B **83**, 115311 (2011)
88. A.M. Beltrán, E.A. Marquis, A.G. Taboada, J.M. Ripalda, J.M. García, S.I. Molina, Ultramicroscopy **111**, 1073 (2011)
89. H. Eisele, P. Ebert, N. Liu, A. Holmes, C.K. Shih, Appl. Phys. Lett. **101**, 233107 (2012)
90. G. Balakrishnan, J. Tatebayashi, A. Khoshakhlagh, S.H. Huang, A. Jallipalli, L.R. Dawson, D.L. Huffaker, Appl. Phys. Lett. **89**, 161104 (2006)
91. C.J. Reyner, K.N.J. Wang, A. Lin, M.B. Liang, S. Goorsky, D.L. Huffaker, Appl. Phys. Lett. **99**, 231906 (2011)
92. A. Lenz, H. Eisele, J. Becker, L. Ivanova, E. Lenz, F. Luckert, K. Pötschke, A. Strittmatter, U.W. Pohl, D. Bimberg, M. Dähne, Appl. Phys. Express **3**, 105602 (2010)
93. R.M. Feenstra, Semicond. Sci. Technol. **9**, 2157 (1994)
94. R. Timm, R. Feenstra, H. Eisele, A. Lenz, L. Ivanova, E. Lenz, M. Dähne, J. Appl. Phys. **105**, 093718 (2009)
95. O. Romanyuk, F. Grosse, W. Braun, Phys. Rev. B **79**, 235330 (2009)

Chapter 7
Fabrication of Ordered Quantum Rings by Molecular Beam Epitaxy

Jiang Wu and Zhiming M. Wang

Abstract Quantum rings have attracted a lot of attention due to their unique properties and have been under extensive theoretical and experimental investigations. For example, Aharonov-Bohm effect has been observed in quantum rings which shows potential to realize quantum computational devices. In addition, quantum rings have found application in optoelectronics. Due to the ring-shaped morphology altered from dots, the vertical confinement in nanorings is stronger than in quantum dots. Laser and infrared photodetectors have recently been demonstrated by using quantum rings. To meet the urgent demands for quantum rings, various effects have been devoted to quantum ring fabrication techniques. There are two of the most used bottom-up fabrication methods of self-assembled rings using molecular beam epitaxy (MBE). Semiconductor quantum rings can be created by conventional molecular beam epitaxy and Droplet Epitaxy technique. Despite great efforts devoted to quantum ring fabrication using these techniques, alignment of quantum rings is not well documented. Fabrication of ordered quantum ring is of high priority for theoretical as well as practical investigations, such as persistent current and photodetectors. Recently, both vertically and laterally ordered quantum rings have been demonstrated. In this chapter, the growth mechanisms and fabrication techniques for aligned quantum rings grown are reviewed.

7.1 Introduction

Unique properties of nanomaterials have attracted considerable interest over the last two decades in various fields in physics, engineering, chemistry, and biology, which present the promise of realizing next generation electronic and optoelectronic devices.

J. Wu (✉) · Z.M. Wang
State Key Laboratory of Electronic Thin Film and Integrated Devices, University of Electronic Science and Technology of China, Chengdu 610054, P.R. China
e-mail: jiangwu@uestc.edu.cn

Z.M. Wang
e-mail: zhmwang@uestc.edu.cn

V.M. Fomin (ed.), *Physics of Quantum Rings*, NanoScience and Technology,
DOI 10.1007/978-3-642-39197-2_7, © Springer-Verlag Berlin Heidelberg 2014

Among various nanomaterials, quantum rings show interesting electronic, magnetic and optical properties and attracted considerable attention [1–8]. For example, quantum phase coherence effects, such as the Aharonov-Bohm and Aharonov-Casher effects, have been observed in quantum rings [4]. The optical Aharonov-Bohm effect has also been predicted and demonstrated, which can be potentially used for applications in quantum information processing systems [9–11]. In addition, research efforts on quantum rings have also led to various practical applications in the last few years. Quantum ring infrared photodetectors have been reported in the mid-infrared and THz spectral range [12, 13]. Nanorings have shown promise in high density magnetic memory applications [14]. Quantum ring lasers have also been reported [15].

The increasing interest in quantum rings gives rise to a variety of fabrication techniques [16–19]. Currently, quantum rings are mainly fabricated by two major fabrication methods, conventional molecular beam epitaxy and Droplet Epitaxy technique. Generally, nanostructures fabricated by these two methods are randomly distributed. However, for many applications, it is advantageous to achieve ordered nanostructures in order to provide optimum performance. For example, ordered quantum dot arrays can result in a higher absorption and higher responsivity for quantum dot infrared photodetectors [20]. Three-dimensionally aligned quantum dot array can provide a strong electron wave function overlap, which is considered to be beneficial for intermediate band solar cells [21]. A lot of ordered nanostructures, such as nanowires, quantum dots and nanoparticles are fabricated by different methods [22–27], including template growth and drying mediated self-assembly [28–30]. However, there are only a few reports on ordered quantum rings.

In this chapter, we present different growth techniques to fabricate ordered quantum rings. We first review fabrication of quantum ring chains on GaAs (100) surface. After that, in Sect. 7.2.2, fabrication of two-dimensionally aligned quantum rings grown on a high index surface is presented. In Sect. 7.3, fabrication of GeSi quantum rings on pre-patterned substrate is overviewed. Finally, future prospects of the ordered quantum rings are presented in Sect. 7.4.

7.2 Fabrication of Laterally Ordered Quantum Rings on Quantum Dot Superlattice Template

7.2.1 Fabrication of Ordered Quantum Ring Chains on GaAs (100) Surface

Generally, quantum rings can be fabricated from S-K quantum dots and by Droplet Epitaxy technique. By using S-K growth mode, strained quantum dots, which are partially capped by a thin layer of a substrate material, undergo a morphological transformation from quantum dots to ring-shaped nanostructures. Alternatively, ring-shaped nanostructure can be formed by Droplet Epitaxy through control over the crystallization process of nano-droplets. The morphology of crystallized nanostructures can be well controlled from quantum dots to quantum rings. However, both processes in general result in randomly distributed quantum rings.

Nevertheless, growth of ordered quantum dots has been demonstrated [31]. One approach to obtain ordered quantum dots is based on multiple quantum dot stacking. The strain field introduced into the system by quantum dots results in increasing size and spacing uniformity in successive quantum dot layers because the nucleation rate is strongly dependent on the strain field. For example, Wang et al. have produced laterally ordered quantum dot chains on GaAs(100) by stacking multiple quantum dot layers. Similarly, multiple quantum dot layers can be adopted for quantum ring growth and used to form a template for laterally ordered quantum rings. Currently, the formation of laterally ordered quantum dots by strain field engineering has been well studied [32]. In this section, lateral ordering of self-assembled quantum rings is presented.

A molecular Beam Epitaxy (MBE) system has been used for fabricating ordered quantum ring chain samples on semi-insulating GaAs (100) substrates. First, the native oxide is removed at 610 °C under an As_4 flux for ten minutes. The deoxidized GaAs substrate temperature is then changed to 590 °C. A GaAs buffer layer (200 nm) is grown, after which the substrate temperature is changed again to 540 °C. The formation of ordered quantum ring chains is divided into two major steps, (i) formation of ordered quantum dot chain template and (ii) quantum ring conversion. Formation of quantum dot chains has been investigated by growth of multilayers of InGaAs/GaAs quantum dots. Quantum dot chains with length over five microns have been demonstrated. The growth procedure for quantum ring chains consists of formation of quantum dot chains as a template. For the quantum ring chain template growth, $In_{0.4}Ga_{0.6}As$ (8.4 ML) is deposited first and then growth interruption (10 s) is introduced. The $In_{0.4}Ga_{0.6}As$ quantum dots are sequentially capped by 20 ML GaAs. Another 60 ML thick GaAs spacer is deposited after raising the substrate temperature back to 580 °C. The $In_{0.4}Ga_{0.6}As$ (8.4 ML)/GaAs (80 ML) quantum dot structure is repeated eight times in order to improve the vertical correlation and to obtain uniform quantum dot chains. Deposition of multi-layer InGaAs quantum dots creates an uniform strain field. The quantum dot superlattice strain field in turn, can lead to the ordered arrangement of the last layer of quantum rings [33]. After completion of the eight periods of multiple $In_{0.4}Ga_{0.6}As$/GaAs quantum dots, the substrate temperature is reduced to 540 °C again. At the same time, the As cracker is operated at a temperature about 900 °C to produce As_2. When the growth conditions are ready, the RHEED screen shows the c(4 × 4) surface reconstruction [34]. In order to create quantum rings, InAs quantum dots are grown using the S-K mode by depositing 2.1 ML InAs. After formation of InAs quantum dots, the growth is interrupted and the sample is annealed for a few tens of seconds. The growth interruption has been found to be helpful in alignment of quantum dots along the [01-1] direction. In addition, the interruption also promotes nucleation of quantum dots more uniformly. After formation of well-aligned quantum dot chains, those chains are immediately covered by 4 nm GaAs. The as-grown InAs quantum dots are about 6–8 nm in height. The deposition of 4 nm GaAs only

Fig. 7.1 AFM images of quantum dot chains and quantum ring chains. After Ref. [36]

partially covers the quantum dots, which results in unbalanced surface forces. These forces convert nanodots into ring-shaped nanostructures [35].

Figure 7.1(a) shows an example of the Atomic Force Microscopy (AFM) image of InAs quantum dot and quantum rings chains. The growth of multiple quantum dot layers increases the uniformity of the quantum dot arrangement. Stacking of multiple quantum dot layers promotes the vertical correlation of quantum dots due to the nucleation of quantum dots at the strain minima [31]. The resulted ordering pattern is related to the morphology of the quantum dots and the elastic medium in which the quantum dots are included. The anisotropy of GaAs matrix and InAs quantum dots introduces a strain field along the [01-1] direction. Therefore, the multiple stacking of the quantum dots led to formation of quantum dot chains along the [01-1] direction.

According to the "dewetting model" proposed by Blossey and Lorke [35], it is the instability of quantum dots introduced by the unbalanced surface forces that causes the conversion of quantum dots into quantum rings during capping. The process of quantum ring formation based on the "dewetting model" is briefly reviewed in Chaps. 2 and 3. An AFM image of the ordered quantum rings converted from partially capped quantum dots is shown in Fig. 7.2. The lateral ordering of quantum rings is well preserved. Figure 7.2 shows that the quantum ring chain is formed along the [01-1] direction and the average chain length is about 1.0 μm. The quantum ring chains share similar surface morphology to the quantum dot chains because the rings are directly converted from the quantum dots. The length and regularity of the quantum ring chain may be further improved by improving the regularity of the quantum dot chain template. In addition, the process of dot-to-ring transformation can be also improved by using different growth conditions, such as thickness of the GaAs capping layer.

To obtain a better characterization of the quantum ring chains, Fig. 7.3 compares two typical AFM images of quantum ring chains and quantum dot chains as well as AFM line profiles of a typical quantum ring and quantum dot. The partial capping truncates the quantum dots and, thus, quantum rings become lower than quantum

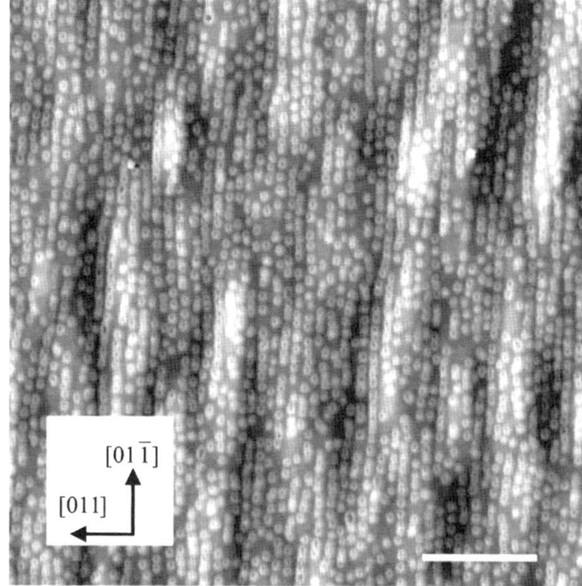

Fig. 7.2 Tapping mode 5.0×5.0 µm AFM image of quantum ring chains grown on an $In_{0.4}Ga_{0.6}As/GaAs(100)$ multi-layered quantum-dot superlattice. The scale bar is 1 µm. After Ref. [36]

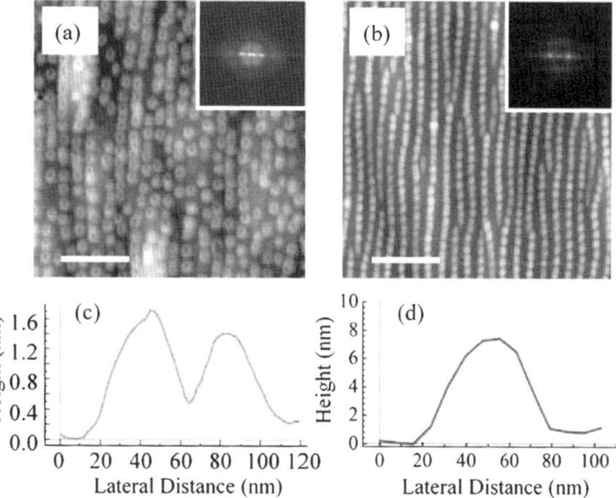

Fig. 7.3 (**a**) Tapping mode AFM image of quantum ring chains. (**b**) Tapping mode AFM image of quantum dot chains. The *insets* are Fourier transforms from the AFM image of quantum-ring chains and quantum-dot chains. (**c**) Cross-sectional line profile of a quantum ring. (**d**) Cross-sectional line profile of a quantum dot. The scale bar is 500 nm. After Ref. [36]

dots. However, the outward reconfiguration of the nanostructure makes rings slightly larger than the quantum dots. Quantum rings also show an oval morphology along the [01-1] direction, which is likely because of an isotropic redistribution of materials during the ring-to-dot transformation process [37]. The insets in Figs. 7.3(a) and (b) show corresponding Fourier transforms of quantum ring and quantum dot chains. The Fourier transforms confirm the long-range one-dimensional ordering of quantum rings and quantum dots.

The photoluminescence measurement of an capped quantum ring sample (50 nm GaAs cap layer) suggests a strong emission from ordered quantum ring chains, as

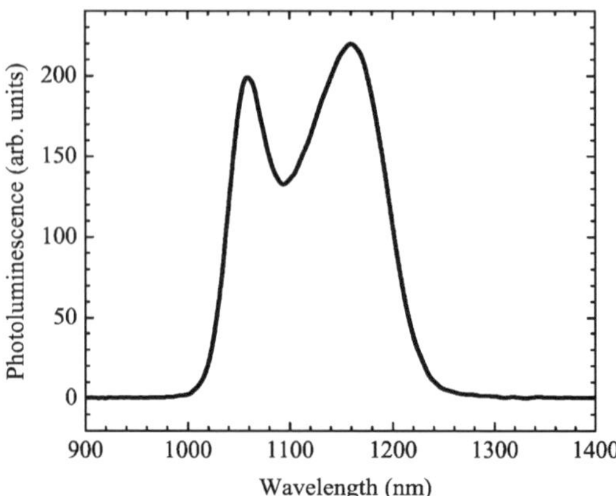

Fig. 7.4 PL spectrum of the ordered quantum ring chains on the quantum dot superlattice template measured at 77 K. After Ref. [36]

shown in Fig. 7.4. Two photoluminescence peaks, at 1060 nm and 1163 nm, of Gaussian profiles are observed. The photoluminescence peak at 1060 nm is due to the multiple $In_{0.4}Ga_{0.6}As$ quantum dot layers. The peak at 1163 nm is assigned to the quantum rings based on a narrower bandgap of the InAs. The difference in the strain and quantum confinement between the multiple quantum dots and quantum rings are also accounted for the shift of photoluminescence emission peak. The emission from the quantum rings also shows a broader line-width than the emission spectrum from $In_{0.4}Ga_{0.6}As$ quantum dots. This can be explained by the reduced quantum ring uniformity. As shown in the AFM images, after the partial capping process, there are imperfect rings formed which broadens the photoluminescence spectrum. In order to improve the quantum ring chain quality, a number of factors require a further improvement. First, improvement of ordering and uniformity of the initial quantum dot chains would be helpful to obtain better quantum ring chains.

Second, optimization the capping layer thickness and post-growth annealing also plays a critical role in improving quantum ring chain quality.

7.2.2 Fabrication of Laterally Ordered Quantum Ring Arrays on GaAs High Index Surfaces

To investigate control over the lateral quantum ring ordering by the "self-organized anisotropic strain engineering" technique [38], fabrication of quantum rings on high index surfaces is performed. Compared with one-dimensional quantum ring chains observed on GaAs (100) surface, a periodic two dimensional quantum ring array has been demonstrated on high index surfaces. Likewise, the semi-insulating GaAs substrates are used for sample preparation by molecular beam epitaxy [39]. The growth conditions are the same as those for quantum ring chain growth. In this study, GaAs (311)B and (511)B high index surfaces are used in addition to GaAs (100) surface. Figure 7.5 shows the AFM images of the laterally alignment of quantum rings formed on different GaAs surfaces. Interestingly, quantum ring chains are formed

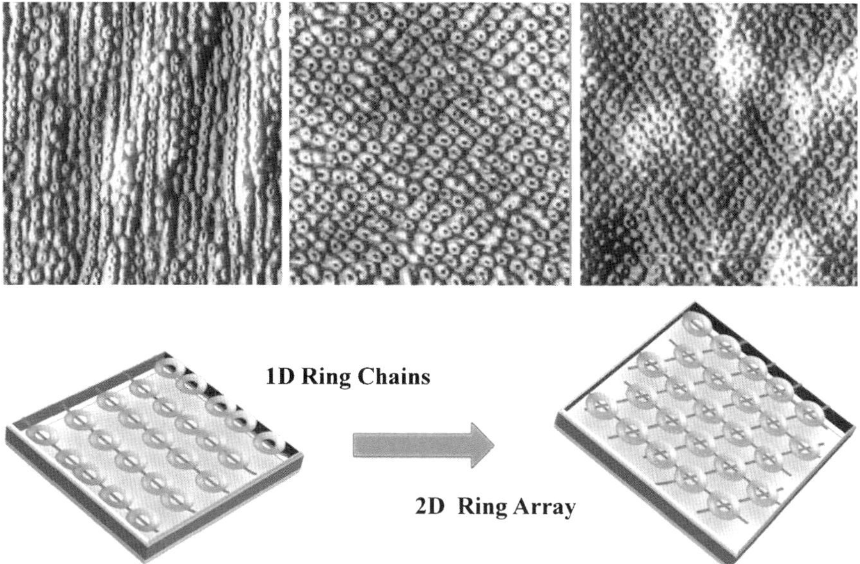

Fig. 7.5   AFM images of quantum rings grown on GaAs (100), (311)B, and (511)B surfaces. The *bottom illustration* shows a transition from a one-dimensional ordering on the (100) surface to a two-dimensional array on high index surfaces. After Ref. [40]

on the (100) surface while different patterns are formed on the high index surfaces, (311)B and (511)B.

Even though the quantum rings are grown under the same conditions, the multiple quantum dots and then, the over-grown quantum rings show quite different surface morphology on different surfaces. The distinct morphological differences embrace the shape of quantum rings and lateral alignment pattern of quantum rings. Typical AFM images and cross-sectional pro files of quantum rings grown on different surfaces are shown in Fig. 7.6. The quantum rings grown on (100), (311)B and (511)B surfaces show similarity in size and density as listed in Table 7.1. However, slightly elongated quantum rings are observed on GaAs (100) surfaces. On the other hand, quantum rings formed on high index surfaces appear more spherical but with a more asymmetrical height distribution. This difference results mainly from two factors. First, the surface diffusion of In adatoms, surface energy, and step bunches vary on surfaces with different Miller indices, which results in different shape and size of InAs quantum dots [41–43]. On high index surfaces, the diffusion along the [01-1] direction can be tuned to match with diffusion along the [011] direction [44]. Second, as mentioned in the previous Section, the surface energy is orientation-dependent and facet-dependent. The net force acting upon the quantum dot during the ring transformation differs on different surfaces [45]. As shown in Figs. 7.6(b) and (c), the AFM line profiles of quantum rings show different heights along the [23-3] and [25-5] directions. The height distribution of quantum rings on high index surface is attributed to the process of dot-to-ring conversion because the

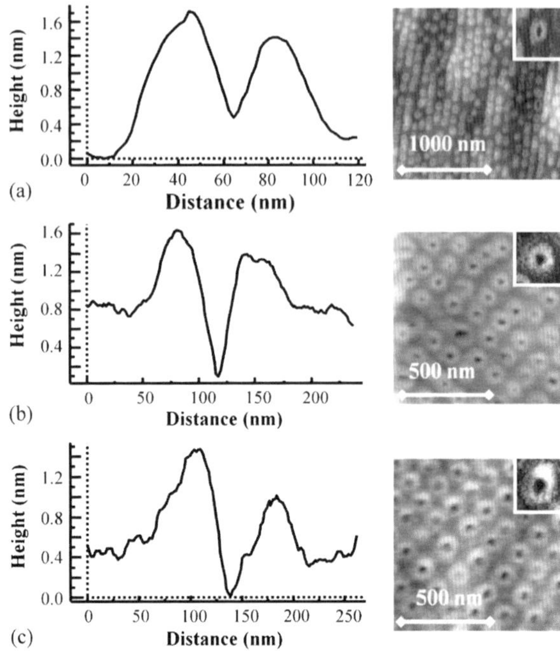

Fig. 7.6 Cross-sectional line-profiles of quantum rings grown on (**a**) GaAs (100) surface; (**b**) GaAs (311)B surface; (**c**) GaAs (511)B surface. The line-profiles are taken along the [01-1], [23-3], and [25-5] for the (100), (311)B, and (511)B, respectively. The right side of the line-profiles are the AFM images of quantum rings grown on the (100), (311)B, and (511)B surfaces. The insets are AFM images of a single quantum ring grown on each surface. After Ref. [40]

Table 7.1 Average radii and density of quantum rings on GaAs (100), (311)B, and (511)B surfaces

Surface	(100)	(311)B	(511)B
Average radius (nm)	26.3 ± 8.2	37.6 ± 10.7	35.2 ± 7.5
Density (cm^{-2})	7.9×10^9	5.0×10^9	5.6×10^9

initial shape of quantum dots on high index surfaces is rather spherical. Since the initial quantum dots are expected to be a round dome shape, the anisotropic rings may be attributable to the process of dot-to-ring transformation. As mentioned earlier, the anisotropic ring morphology can be affected by different facets, orientation-dependent free energy, and strain effects, which may be the main factors leading to direction-dependent forces and thus, uneven material distribution of quantum rings on high index surfaces.

More importantly, quantum rings formed on the (100) surface align in a chain-shaped pattern, while a two-dimensional periodical alignment of rings is observed on high index surfaces when a strain quantum dot superlattice template is present. Generally, it is considered that the elastic interactions of multiple layers of quantum dots result in the lateral ordering of quantum dots or quantum rings [31, 46]. On GaAs (100) surface, the adatoms mainly diffuse along the [01-1] direction and quantum dots are more relaxed along the [01-1] direction. Therefore, the strain field transferred to the subsequent quantum dot layer is elliptical. The elliptical field tends to align the subsequent quantum dots along the [01-1] direction. Similarly, there are surface steps with variable separations perpendicular to the $[2n - n]$ directions on

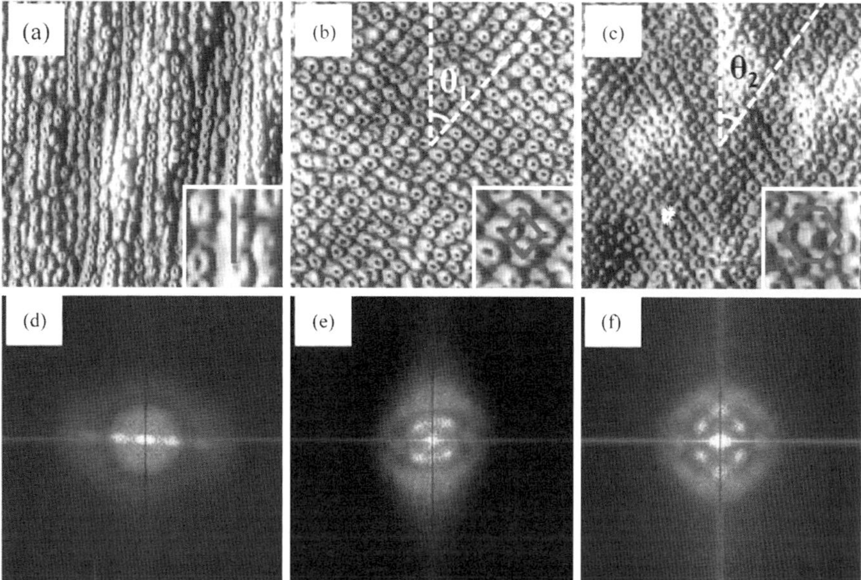

Fig. 7.7 AFM images of ordered quantum rings on the top of multilayered $In_{0.4}Ga_{0.6}As$ quantum dot templates grown on (**a**) GaAs (100), (**b**) (311)B, and (**c**) (511)B surfaces. The *second row* images are fast Fourier transforms taken from the corresponding AFM images: (**d**) GaAs (100), (**e**) (311)B, and (**f**) (511)B surfaces. The *insets* are magnified AFM images showing different quantum ring lattice structures. After Ref. [40]

high index $(n11)$B surfaces. The nominal step separation for a GaAs $(n11)$ surface consisting of the (001) terraces, denoted as S, is equal to $0.2 \times n$ nm. Therefore, the nominal step separation can be adjusted from 0.6 nm [(311)B surface] to 1.8 nm [(911)B surface]. The change in surface steps affects the adatom surface diffusion as well as strain field correlation between adjacent quantum dot layers. On high index surfaces, the diffusion along the [01-1] and [011] directions can be adjusted to closely match [44]. Consequently, control over S creates the opportunity to manipulate the surface adatom migration pattern, which, therefore, is promising for tuning the nanostructure ordering pattern.

Figure 7.7 shows AFM images (2.5 μm × 2.5 μm) and the corresponding fast Fourier transforms of quantum ring arrays grown on the (100), (311)B, and (511)B surfaces. The Fourier transform from quantum rings on the (100) surface confirms long range one-dimensional ordering while the Fourier transforms from quantum rings on the (311)B and (511)B surfaces confirm two-dimensional ordering due to more isotropic strain field of quantum dot superlattices. The Fourier transforms of AFM images show quantum ring array patterns with two symmetry axes for (311)B surface and with four symmetry axes for (511)B surface. These Fourier transform patterns and the magnified AFM images indicate that the quantum rings on the (311)B and (511)B surfaces align in rhombic and hexagonal symmetry, respectively. By adjusting the nominal surface step separation, the strain is most relaxed along an in-plane direction which is deflected by about 40 degrees to the [23-3] (or [25-5]) direction. From the AFM images, the measured angles between the strain relaxation direction and the [2n − n] direction are 41.6 [(311)B surface] and 36.2 [(511)B surface] degrees. To sum up, lateral alignment of quantum ring arrays can be achieved by strain engineering and quantum ring conversion through partial capping. The order pattern can also be tuned by choosing substrates with different Miller Indices.

7.3 Fabrication of Quantum Rings on Pre-patterned Substrates

7.3.1 Simulations of Formation of Ordered Quantum Dots and Quantum Rings Through Pre-patterning

Alternative approach to self-assembly of aligned nanostructures is to create ordered artificial template by pre-patterning. A few methods have been proposed to fabricate ordered nanostructures on pre-patterned substrates, which consist of ordered pits, or ordered humps, or regularly distributed strain energy profiles [47, 48]. The idea is to form nanostructures on the patterned pit or hump with one-to-one correlation in the subsequent growth. However, it faces great challenges to achieve well-aligned nanostructure arrays through surface pre-patterning. For example, it is challenging to obtain one-to-one correlation between pattern and nanostructures, Phase diagrams for heteroepitaxy of quantum structures on patterns are theoretically simulated by a model [49]. The phase diagrams for fabrication of ordered quantum dots and rings provide insight in obtaining the one-to-one correlation between patterns and the subsequently grown quantum structures. The model assumes a fixed pre-patterned substrate surface and a thin transition layer with a linearly varied mismatch strain atop. The surface morphology evolves with deposition and the surface diffusion, which is governed by surface chemical potential. A small random noise to the deposition is added to the simulation of growth with random fluctuation. Cosine-shaped pits or humps are considered on the pre-patterned substrate. The simulation also assumes that the radius of the pits (or humps) is comparable with the surface roughness wavelength.

The simulated phase diagrams for pit and hump pre-patterning with the variation of the normalized deposition rate R^* and diagonal edge-to-edge distance P^* are shown in Fig. 7.8. The phase diagrams for pit and hump pre-patterning show three and five phase zones, respectively. On the phase diagram for pit pre-patterning, as shown in Fig. 7.8(a), the one-to-one relation between the pits and quantum dots is maintained in zone 1. The formed dots are stable against growth noise during the growth process. However, no one-to-one correlation is observed between pits and formed dots. In zone 2, additional dots or structures with different morphologies are formed between the patterned pits. However, one-to-one correlation can be still obtained because the additional dots disappear with further growth. In the zone 1 of the phase diagram for hump pre-patterning, one-to-one correlation is also achieved but with shape transition during the growth process. First, quantum rings are formed on

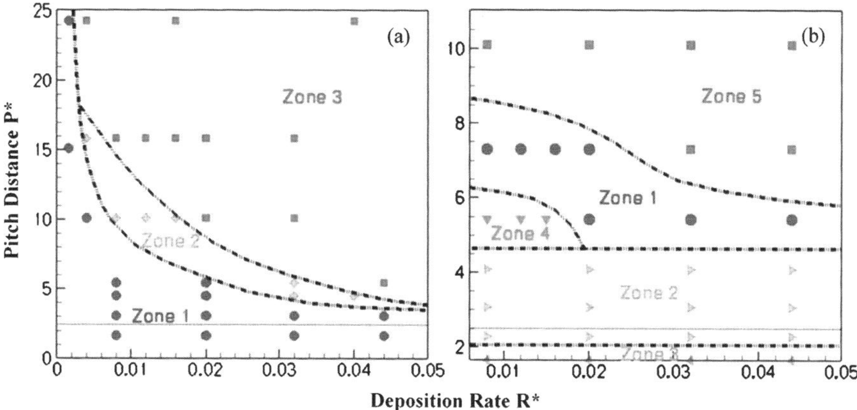

Fig. 7.8 Calculated phase diagram for (**a**) pit and (**b**) hump pre-patterning. *Points on the diagrams denote the calculated cases*

the humps. Sequentially, the rings gradually shrink and eventually change into dots when the growth proceeds. In zones 2 and 3, only dots form between humps during growth. However, the nucleation positions vary for zones 2 and 3. Incomplete rings form at the early stage and further break up into dots without order arrangements for zone 4. Both dots and rings form in zone 5. There is no one-to-one correlation for zones 4 and 5.

Two examples of nanostructures formed on pits and humps are shown in Fig. 7.9. Figures 7.9(a1) and (a2) show the cases in zone 1 and zone 3 of the phase diagram for pit pre-patterning, respectively. Quantum dot arrays are formed with one-to-one correlation in zone 1. However, surface morphology conversion between dots to ripples is observed for zone 3 without one-to-one correlation. Figure 7.9(b) shows an example of nanostructures formed on humps. Clearly, shape transitions have occurred during the growth. The surface chemical potential profiles play an important role in the formation of the one-to-one correlation between patterns and nanostructures. The simulated phase diagrams can serve as guideline for optimized fabrication of ordered quantum dot or quantum ring arrays.

7.3.2 Fabrication of GeSi Nanorings on Patterned Si (100) Substrate

Laterally aligned nanostructures can be obtained on a pre-patterned substrate with ordered nanopatterns. A number of methods have been developed to create nanopatterns, including nanoimprint lithography, holographic lithography, anodic oxidation nanolithography, and nanosphere lithography. Nanosphere lithography is a cost-effective and high throughput nanofabrication technique. Nanosphere lithography

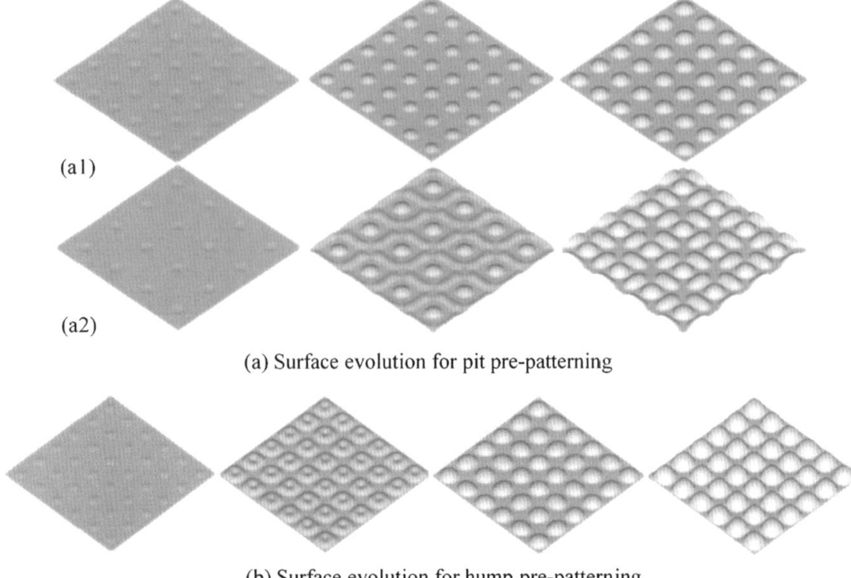

(a1)

(a2)

(a) Surface evolution for pit pre-patterning

(b) Surface evolution for hump pre-patterning

Fig. 7.9 (**a**) Snapshots of surface evolution for pit pre-patterning: (**a1**) a case in zone 1 and (**a2**) a case in zone 3. (**b**) Snapshots of surface evolution for hump pre-patterning: a case in zone 1. After Ref. [49]

has been used to create ordered GeSi QDs. The GeSi nanorings are formed by capping the GeSi QDs with a thin Si capping layer [50].

Figure 7.10 shows a schematic illustration for nanopatterning using nanosphere lithography. Nanopatterning consists of four major steps.

1. The polystyrene sphere (diameter of 430 nm) suspension is mixed with methanol (1:1). Weekes' method is used to assemble polystyrene nanospheres on the surface of deionized water in a close-packed monolayer, which is then transferred onto a clean p-type Si (001) substrate with hydrogen-terminated surface by draining the de-ionized water.
2. The diameter of polystyrene spheres is shrunk down to about 80 nm by the reactive ion etching.
3. Au-Si alloy and SiO_2 mask are formed via Au-catalyzed oxidation after deposition of a thin Au film (1 nm) on a surface covered with polystyrene spheres. The polystyrene spheres are removed from the substrate by immersing in tetrahydrofuran under ultrasonic treatment.
4. After formation of the mask, ordered inverted pyramid-like pits with {111} facets are created by etching the substrate in KOH solution. The substrate is immersed in $KI:I_2:H_2O$ (4:1:40) solution for 10 h to remove the Au and Au-Si alloy mask and then cleaned and passivated.

The sample structure of ordered quantum rings is illustrated in Fig. 7.11. The growth recipe for GeSi quantum rings is as follows. First, the substrate is thermally

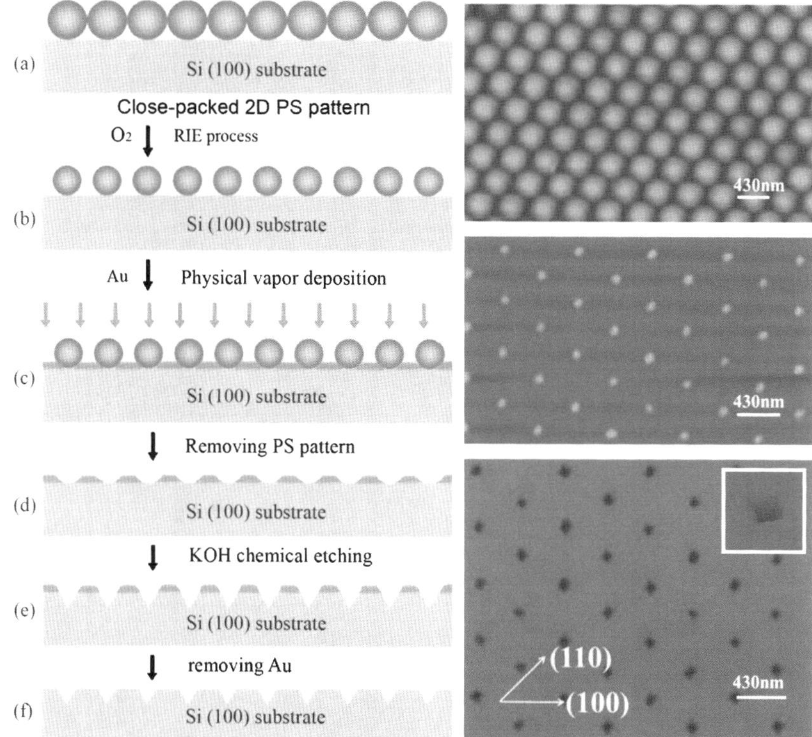

Fig. 7.10 Schematic illustration for the fabrication of ordered pit-pattern. (**a**) Closed-packed PS single ML pattern. (**b**) PS pattern after O_2 RIE. (**c**) Au film deposition. (**d**) Removing PS pattern in THF. (**e**) KOH selective etching. (**f**) Inverted pyramid-like pits pattern with {111} facets after Au has been removed. The *panels at the right side* show the AFM images at corresponding stages. After Ref. [50]

treated at 860 °C for 3 min. Buffer layer of 130 nm thick Si is deposited while the substrate temperature is increasing from 400 to 500 °C. A quantum dot layer is grown by a two-step deposition of Ge; the first 5 ML Ge are deposited when growth temperature increases from 500 to 640 °C while the subsequent 7 ML Ge are deposited at 640 °C. After the formation of the first quantum dot layer, the substrate temperature is ramped down to 500 °C again and then the substrate temperature is immediately ramped back to 640 °C while a 20 nm thick Si spacer is grown. A second layer of GeSi quantum dots is formed by depositing 8 ML Ge at 640 °C. Sequentially, a thin Si capping layer is deposited over the GeSi quantum dots to transform the ordered GeSi quantum dots into ordered GeSi nanorings. Figure 7.11 also shows the quantum dots formed in the first layer and quantum rings formed in the second layer.

As discussed in the previous Section, the distance between adjacent nanopatterns plays a critical role in fabrication of an ordered nanostructure array.

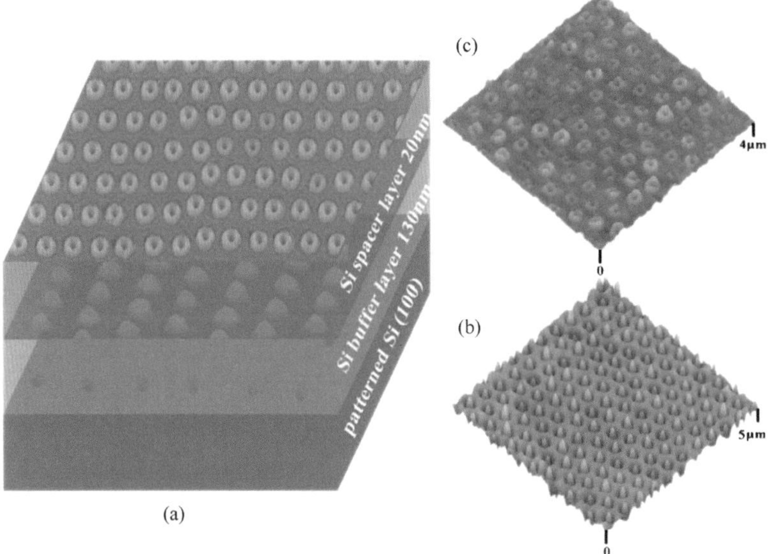

Fig. 7.11 (**a**) Schematic sample structure. AFM images of (**b**) ordered GeSi QDs and (**c**) ordered GeSi nanorings. After Ref. [50]

When the distance is small, e.g. 200 nm, the Ge adatoms also accumulate between adjacent pits in addition to assembling in the pits. In such a case, dot-to-ring transformation is hardly attainable. Using polystyrene spheres as large as 430 nm, the quantum dots are mainly formed in the pits. However, reactive ion etching is required to shrink the size of the nanospheres in order to reduce the size of the nanopits. If the size of the pits is much larger than that of the quantum dots, multiple quantum dots tend to nucleate in a single pit.

The AFM images of the GeSi nanostructure samples capped with Si thin films of different thickness are shown in Fig. 7.11. A thin Si capping layer (<2 nm) only results in dots with shallow dips in the center or a small portion of converted rings. When the Si capping layer thickness is over 3 nm, most quantum dots can be converted into quantum rings.

The surface morphology of nanostructures after the capping process also varies at different capping temperature. When a thin Si capping layer is deposited at 610 °C, the mean lateral size of transformed nanorings is 175 nm. The value decreases to 165 nm when the capping layer is deposited at 640 °C. The standard deviation of lateral size also decreases from 24.3 (capping at 610 °C) to 21.8 nm (capping at 640 °C). After annealing the sample with 2 nm Si capping layer at 610 °C for 30 min, it has been shown that a part of the quantum dots are converted into quantum rings. However, other quantum dots disappear due to mass migration at long-term annealing. Under a further increase of the annealing time to 60 min, the effects involving mass migration and intermixing cause formation of superdomes and no quantum rings remain visible anymore. High growth temperature and long-term annealing can significantly modify the surface morphology.

7.4 Perspectives and Future Work

For optoelectronic and electronic devices, both the well control in the quantum ring shape and uniformity and the precise positing of quantum rings play a critical role in optimizing the device performance. In addition, investigation of the physical properties have been focused on a single quantum ring. New collective behavior may be revealed from coupled quantum ring arrays. Therefore, fabrication of laterally ordered quantum ring arrays opens the opportunities for investigations of new physics and device applications. The major research efforts on quantum rings can be directed towards developing large scale well-ordered quantum ring arrays and their applications.

First, additional fabrication techniques, such as "interference lithography", can be explored for fabricating of ordered quantum ring arrays in addition to self-assembly and pre-patterning overgrowth. For example, wafer-scale metallic nanoring array has been obtained by using "interference lithography" [51, 52]. Ordered arrays of Au, Ni, and Si nanorings have also demonstrated by using a porous alumina mask. Improvements in the array periodicity, the control over quantum ring size and shape, and the control over array pattern will be the key factors in future efforts for fabrication of ordered arrays of quantum rings.

Second, it is interesting to study the collective effects of ordered quantum ring arrays. The ordered nanostructure array is considered to show an interesting "atomic states" analogy to atoms in a periodic lattice. Moreover, quantum rings show distinct properties, such as persistent currents, which cannot be found in other systems. The experimental investigation in ordered quantum ring arrays may reveal further new phenomena.

Finally, ordered quantum ring arrays also open wide possibilities for functional devices, such as quantum computing and optoelectronic devices. Quantum rings possess unique optical properties which can be employed for novel photodetectors and lasers. For example, the shallow bound-state energy levels of the quantum rings can be employed to detect photons in the terahertz regime [53]. Polarization sensitive photoconductivity could be realized in quantum ring based photodetectors as a result of the unique morphology of quantum rings. Moreover, Aharonov-Bohm and unique magnetic-optical effects have been found in nano-rings [54, 55]. Quantum rings also have high stability of spin states and spin-dependent transport properties [56, 57]. One-qubit spintronic quantum gates has been fabricated using quantum rings, and thus, being assisted by resonant tunneling transport and spin-orbit interaction, quantum information processing can be achieved [58]. The development of ordered quantum ring arrays may assist in realizing semiconductor-based quantum computing devices.

References

1. R. Leturcq, L. Schmid, K. Ensslin, Y. Meir, D.C. Driscoll, A.C. Gossard, Phys. Rev. Lett. **95**, 126603 (2005)
2. A. Fuhrer, S. Luscher, T. Ihn, T. Heinzel, K. Ensslin, W. Wegscheider, M. Bichler, Nature **413**, 822 (2001)
3. V.M. Fomin, V.N. Gladilin, J.T. Devreese, N.A.J.M. Kleemans, P.M. Koenraad, Phys. Rev. B **77**, 205326 (2008)
4. M. Zarenia, J.M. Pereira, F.M. Peeters, G.A. Farias, Nano Lett. **9**, 4088 (2009)
5. A.J.M. Giesbers, U. Zeitler, M.I. Katsnelson, D. Reuter, A.D. Wieck, G. Biasiol, L. Sorba, J.C. Maan, Nat. Phys. **6**, 173 (2010)
6. W. Chang, C. Lin, Y. Fu, T. Lin, H. Lin, S. Cheng, S. Lin, C. Lee, Nanoscale Res. Lett. **5**, 680 (2010)
7. A. Lorke, R. Johannes Luyken, A.O. Govorov, J.P. Kotthaus, J.M. Garcia, P.M. Petroff, Phys. Rev. Lett. **84**, 2223 (2000)
8. N.A.J.M. Kleemans, I. Bominaar-Silkens, V.M. Fomin et al., Phys. Rev. Lett. **99**, 146808 (2007)
9. A.O. Govorov, S.E. Ulloa, K. Karrai, R.J. Warburton, Phys. Rev. B **66**, 081309 (2002)
10. I.R. Sellers, V.R. Whiteside, I.L. Kuskovsky, A.O. Govorov, B.D. McCombe, Phys. Rev. Lett. **100**, 136405 (2008)
11. M.D. Teodoro, V.L. Campo, V. Lopez-Richard et al., Phys. Rev. Lett. **104**, 086401 (2010)
12. S. Bhowmick, G. Huang, W. Guo, C.S. Lee, P. Bhattacharya, G. Ariyawansa, A.G.U. Perera, Appl. Phys. Lett. **96**, 231103 (2010)
13. J. Wu, Z. Li, D. Shao, M.O. Manasreh, V.P. Kunets, Z.M. Wang, G.J. Salamo, B.D. Weaver, Appl. Phys. Lett. **94**, 171102 (2009)
14. Z.C. Wen, H.X. Wei, X.F. Han, Appl. Phys. Lett. **91**, 122511 (2007)
15. T. Mano, T. Kuroda, K. Mitsuishi, M. Yamagiwa, X. Guo, K. Furuya, K. Sakoda, N. Koguchi, J. Cryst. Growth **301–302**, 740 (2007)
16. Y.S. Jung, W. Jung, C.A. Ross, Nano Lett. **8**, 2975 (2008)
17. C. Somaschini, S. Bietti, N. Koguchi, S. Sanguinetti, Nano Lett. **9**, 3419 (2009)
18. D. Granados, J.M. García, Appl. Phys. Lett. **82**, 2401 (2003)
19. A.Z. Li, Z.M. Wang, J. Wu, G.J. Salamo, Nano Res. **3**, 490 (2010)
20. P. Martyniuk, A. Rogalski, Proc. SPIE **6940**, 694004 (2008)
21. Q. Shao, A.A. Balandin, A.I. Fedoseyev, M. Turowski, Appl. Phys. Lett. **91**, 163503 (2007)
22. Y. Chang, S. Jian, J. Juang, Nanoscale Res. Lett. **5**, 1456 (2010)
23. Z.M. Wang, S. Seydmohamadi, J.H. Lee, G.J. Salamo, Appl. Phys. Lett. **85**, 5031 (2004)
24. Q. Wei, J. Lian, W. Lu, L. Wang, Phys. Rev. Lett. **100**, 076103 (2008)
25. C.M. Müller, F.C.F. Mornaghini, R. Spolenak, Nanotechnology **19**, 485306 (2008)
26. J.L. Baker, A. Widmer-Cooper, M.F. Toney, P.L. Geissler, A.P. Alivisatos, Nano Lett. **10**, 195 (2010)
27. Z. Huang, T. Shimizu, S. Senz, Z. Zhang, X. Zhang, W. Lee, N. Geyer, U. Gösele, Nano Lett. **9**, 2519 (2009)
28. S. Lee, C. Mao, C.E. Flynn, A.M. Belcher, Science **296**, 892 (2002)
29. Z. Fan, J.C. Ho, Z.A. Jacobson, R. Yerushalmi, R.L. Alley, H. Razavi, A. Javey, Nano Lett. **8**, 20 (2008)
30. M. Junkin, J. Watson, J.P.V. Geest, P.K. Wong, Adv. Mater. **21**, 1247 (2009)
31. J. Tersoff, C. Teichert, M.G. Lagally, Phys. Rev. Lett. **76**, 1675 (1996)
32. H. Wen, Z.M. Wang, G.J. Salamo, Appl. Phys. Lett. **84**, 1756 (2004)
33. Z.M. Wang, K. Holmes, Y.I. Mazur, G.J. Salamo, Appl. Phys. Lett. **84**, 1931 (2004)
34. Z.M. Wang, G.J. Salamo, Phys. Rev. B **67**, 125324 (2003)
35. R. Blossey, A. Lorke, Phys. Rev. E **65**, 021603 (2002)
36. J. Wu, Z. Wang, K. Holmes, E. Marega Jr., Y. Mazur, G. Salamo, J. Nanopart. Res. **14**, 919 (2012)
37. Z.M. Wang, H. Churchill, C.E. George, G.J. Salamo, J. Appl. Phys. **96**, 6908 (2004)

38. H. Lan, Y. Ding, Nano Today (2012)
39. L. Hrivnák, Czechoslov. J. Phys. **34**, 436 (1984)
40. J. Wu, Z.M. Wang, K. Holmes, E. Marega Jr., Z. Zhou, H. Li, Y.I. Mazur, G.J. Salamo, Appl. Phys. Lett. **100**, 203117 (2012)
41. B.L. Liang, Z.M. Wang, K.A. Sablon, Y.I. Mazur, G.J. Salamo, Nanoscale Res. Lett. **2**, 609–613 (2007)
42. Z.M. Wang, V.R. Yazdanpanah, C.L. Workman, W.Q. Ma, J.L. Shultz, G.J. Salamo, Phys. Rev. B **66**, 193313 (2002)
43. Z. Li, J. Wu, Z. Wang, D. Fan, A. Guo, S. Li, S. Yu, O. Manasreh, G. Salamo, Nanoscale Res. Lett. **5**, 1079 (2010)
44. M. Schmidbauer, S. Seydmohamadi, D. Grigoriev, Z.M. Wang, Y.I. Mazur, P. Schäfer, M. Hanke, R. Köhler, G.J. Salamo, Phys. Rev. Lett. **96**, 066108 (2006)
45. R. Blossey, A. Lorke, Phys. Rev. E **65**, 021603 (2002)
46. G. Springholz, M. Pinczolits, V. Holy, S. Zerlauth, I. Vavra, G. Bauer, Physica E, Low-Dimens. Syst. Nanostruct. **9**, 149 (2001)
47. H. Brune, M. Giovannini, K. Bromann, K. Kern, Nature **394**, 451 (1998)
48. R. Li, P. Dapkus, M.E. Thompson, W. Jeong, C. Harrison, P. Chaikin, R. Register, D. Adamson, Appl. Phys. Lett. **76**, 1689 (2000)
49. P. Liu, Y.W. Zhang, C. Lu, Appl. Phys. Lett. **90**, 071905 (2007)
50. Y. Ma, J. Cui, Y. Fan, Z. Zhong, Z. Jiang, Nanoscale Res. Lett. **6**, 205 (2011)
51. R. Ji, W. Lee, R. Scholz, U. Gösele, K. Nielsch, Adv. Mater. **18**, 2593 (2006)
52. K.L. Hobbs, P.R. Larson, G.D. Lian, J.C. Keay, M.B. Johnson, Nano Lett. **4**, 167 (2004)
53. H. Ling, S. Wang, C. Lee, M. Lo, J. Appl. Phys. **105**, 034504 (2009)
54. I.R. Sellers, A.O. Govorov, B.D. McCombe, J. Nanoelectron. Optoelectron. **6**, 4–19 (2011)
55. A.O. Govorov, A.V. Kalameitsev, R. Warburton, K. Karrai, S.E. Ulloa, Physica E, Low-Dimens. Syst. Nanostruct. **13**, 297 (2002)
56. E. Zipper, M. Kurpas, J. Sadowski, M.M. Maska, J. Phys. Condens. Matter **23**, 115302 (2011)
57. R. Citro, F. Romeo, Phys. Rev. B **75**, 073306 (2007)
58. S. Bellucci, P. Onorato, Phys. Rev. B **78**, 235312 (2008)

Chapter 8
Self-assembled Semiconductor Quantum Ring Complexes by Droplet Epitaxy: Growth and Physical Properties

Stefano Sanguinetti, Takaaki Mano, and Takashi Kuroda

Abstract Extremely complex semiconductor quantum ring structures, as single ring, multiple concentric quantum rings and coupled ring/disk structures, can be easily designed and grown by Droplet Epitaxy. In this paper, the fabrication and the characterization of such quantum ring complexes is reviewed. Electronic structure, single photon emission, carrier dynamics and magnetic properties in ring structures will be discussed.

8.1 Introduction

Ring geometries have longly fascinated the physics community, as electron confinement in nanometric rings induces a topological quantum mechanical coherence, the Ahronov-Bohm effect [1]. Since the exciton has a non zero orbital magnetic moment, exciton AB effect has been predicted as well [2]. It is also possible to fabricate semiconductor quantum ring molecules, made by concentric quantum ring structures: these structures could permit to explore the magneto-optical excitations due to the Rashba spin orbit interaction [3]. Magnetic field level dispersion in quantum ring is different from quantum dots and useful: the ground state total angular momentum changes from zero to non zero applying a magnetic field [4, 5]. This also produces a different energy dispersion of the excitons depending on ring radius. Moreover, charge tunneling between states of different angular momentum is strongly suppressed in concentric quantum ring by selection rules. Therefore, con-

S. Sanguinetti (✉)
LNESS and Dipartimento di Scienza dei Materiali, Universitá di Milano Bicocca, Via Cozzi 53, 20125 Milano, Italy
e-mail: stefano.sanguinetti@unimib.it

T. Mano · T. Kuroda
National Institute for Materials Science, 1-2-1 Sengen, Tsukuba 305-0047, Japan

T. Mano
e-mail: Mano.Takaaki@nims.go.jp

T. Kuroda
e-mail: Kuroda.Takashi@nims.go.jp

V.M. Fomin (ed.), *Physics of Quantum Rings*, NanoScience and Technology,
DOI 10.1007/978-3-642-39197-2_8, © Springer-Verlag Berlin Heidelberg 2014

centric quantum ring are very relevant in the research on semiconductor-based quantum computational devices because they offer the control of effective coupling of direct-indirect excitons [6]:this could be a promising way to fabricate multiple two level states devices with switchable interaction. The possibility to fill the ring with few electrons has recently allowed to detect the Ahronov-Bohm effect through the magnetic oscillation in the persistent current carried by the single electron states [7].

Two different Molecular Beam Epitaxy (MBE) growth methods can be used to self-assemble compound semiconductor quantum rings. The first one is based on the Stranki-Krastanov (SK) growth of InAs/GaAs, InAs/InP or GaSb/GaAs(001) nano-islands; a partial GaAs capping and a subsequent high temperature annealing produce a significant mass transfer between the center and the edge of the island [8–12] thus producing a ring shaped nanostructure. Due to the high complexity of the phenomena, this method offers only a limited degree of freedom for the design of the nanostructure. This rules out a precise on demand control of size and shape of the quantum ring, and thus of their electronic properties. Ring geometries in In(Ga)As materials have been also fabricated by *in-situ* etching of InGaAs SK nano-islands [13]. In both cases the achievement coupled ring geometries remains impracticable. An alternative fabrication method is based on Droplet Epitaxy (DE) [14]. This is a flexible growth technique which allows for the fabrication of quantum dots with governable size and density in strain free materials [15–17]. Furthermore, a designable wetting layer [18] avoids well know issues in SK dots [19–24], and the method has been demonstrated on many growth surface, including (311)A [25] and (111)A [26]. In addition, DE allows to obtain a large variety of three dimensional nanostructures with different geometries, ranging from rings to complex dot configurations [27–33]. This intrinsic design flexibility is due mainly to the splitting, in time, of the III-column and V-column element supply, with an independent choice, for each of the two elements, of the best growth conditions. DE allows the fabrication of rings from strained [34–40] or unstrained heterostructure, with single and multiple concentric quantum ring geometries [27, 28, 41] as well as more exotic dot/ring structures [42].

In this review we will summarize the main results about growth design of DE ring shaped nanostructures, theoretical study of their electronic properties and experimental characterization of DE-quantum rings.

The review is organized as follows. Section 8.2 describes the fabrication of quantum rings with different shapes by DE, Sect. 8.3 is dedicated to the electronic structure of the DE-quantum rings. The photoluminescence properties of the DE rings are shown in Sect. 8.4 while in Sect. 8.5 carrier dynamics experimental results in DE-quantum rings are reported.

8.2 The Droplet Epitaxy

Droplet Epitaxy (DE), an MBE based growth technique for the fabrication of three-dimensional quantum nanostructures [14], has demonstrated an unmatched ability

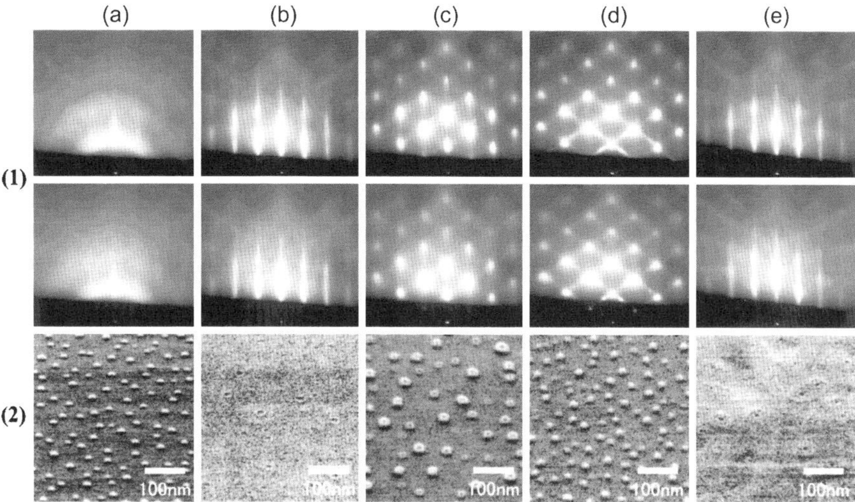

Fig. 8.1 (**1**) and (**2**) are the RHEED patterns and surface morphologies (Scanning Electron Microscopy) of the samples at each stage of the growth process, respectively. In (**1**), *upper column*: the electron beam along [110]; *lower column*: the electron beam along [1$\bar{1}$0]. (**a**) is after the Ga deposition at 200 °C. (**b**), (**c**), (**d**) and (**e**) are after subsequent As$_4$ molecular-beam irradiation with 4×10^{-7} Torr at 200 °C, 4×10^{-5} Torr at 200 °C, 4×10^{-5} Torr at 150 °C and 4×10^{-7} Torr at 150 °C, respectively. Reprinted from Ref. [15], Copyright (2000), with permission from Japanese Applied Physical Society)

to assemble quantum ring semiconductor nanostructures with complex geometries. Nanostructures morphologies range from single rings [28], concentric double rings [27], concentric triple rings [41] and coupled ring/disks [43]. DE exploits the sequential supply of group-III and group-V elements, unlike standard MBE growth, where the elements are simultaneously supplied to the substrate. In DE growth of GaAs, first Ga irradiation in absence of As leads to the formation of nanometric droplets with homogeneous size then an As flux is supplied for the crystallization of the metallic Ga droplets into GaAs quantum nanostructures. A proper choice of the growth conditions allows to obtain different morphologies of the GaAs quantum nanostructures. Figure 8.1 clarifies, in a qualitative way, the effect of the growth conditions on the obtained nanostructures and shows how DE gives a large freedom in the shape design.

The authors present a systematic study on the formation of GaAs/AlGaAs quantum nanostructures: starting from an identical set of Ga droplets, crystallization is performed at different substrate temperatures and As fluxes. As clearly shown, the conditions for the crystallization are crucial for the shape control of the GaAs quantum nanostructures. Indeed, in the case of lower As flux or higher temperatures, it was not possible to obtain 3D island growth, because of the annihilation of droplets. The position occupied by the droplets before As crystallization is marked by a shallow holes surrounded by a tiny ring. The presence of avoided growth below the Ga droplet has been attributed to the preferential As crystallization in the droplet at the triple point [41]. Additional effects related to Ga etching may be possible [44], although strongly quenched by the low growth temperature. By irradiation with a higher As flux, 3D dimensional growth occurs, as a result of incorporation of arsenic atoms inside the Ga droplet, thus giving rise to GaAs nanocrystals with dome (Fig. 8.1d) and ring (Fig. 8.1c) shapes.

DE therefore enables the fabrication of 3D nanostructures with ring shapes. Mano et al. [27] reported a wide sampling of the growth conditions space and demonstrated, for the first time, the self-assembly of GaAs/AlGaAs double rings. No other techniques are available for the fabrication of such complex semiconductor ring structures. Recent advances in DE permit an even larger degree of freedom in the design of the nanostructures, allowing the assembly of multi-ring and coupled ring-disk structures [41, 43].

A conventional MBE apparatus is used for the fabrication by DE of all the GaAs/AlGaAs ring structures. After the growth an $Al_{0.3}Ga_{0.7}As$ barrier layer by conventional MBE, the substrate temperature is reduced to a range between 200 and 350 °C, depending on requirements of the size and density of the nanostructures [15, 17], As valve is closed and the As pressure in the growth chamber depleted. When pressure is low enough, only cation (Ga) atoms are supplied. In the experiment here reported, As_4 was used. In absence of As, Ga atoms are incorporated into the As terminated surface, resulting in the transition to a Ga stabilized reconstruction [45]. Subsequent Ga deposition gives rise to the formation of tiny Ga droplets on the surface. Density and size of the Ga droplets depends on substrate temperature and total amount of Ga supplied. Since Ga diffusion on the Ga terminate substrate is a temperature activated process, the higher the temperature the smaller droplets density. This step sets the initial condition from which the DE-rings will be fabricated. During As supply, metallic Ga droplets act as nanometer scale local reservoirs of group III material.

8.2.1 Fabrication of Ring Structures

In order to get single ring and double ring structures, after the deposition of the droplets, the substrate temperature was set around 200 °C. Rings formation can be obtain by the irradiation of an As_4 molecular beam with a fluxes of ≈ 0.8–1×10^{-5} Torr beam equivalent pressure (BEP) for single ring [46] and ≈ 0.5–2×10^{-6} Torr BEP for double ring [27]. In this growth step, Ga contained in the droplets fully crystallizes in form of rings. Real-time process observation is possible through reflection high-energy electron diffraction (RHEED): the complete change of the pattern from halo to spots corresponds to the crystallization of the Ga contained in the droplets. After crystallization, an annealing of 10 min at 350 °C at constant As flux was performed. This last step ensured the complete crystallization of the metallic Ga reservoirs on the surface.

Figure 8.2 shows AFM images of a single nanostructure of the uncapped single ring and double ring samples. Well defined, topologically non-connected, rings are clearly visible. The latter characteristic is peculiar only of DE-quantum rings.

Fig. 8.2 *Left panel*: AFM image of a single ring. *Right panel*: AFM image of a single double ring. (From Kuroda et al. [28])

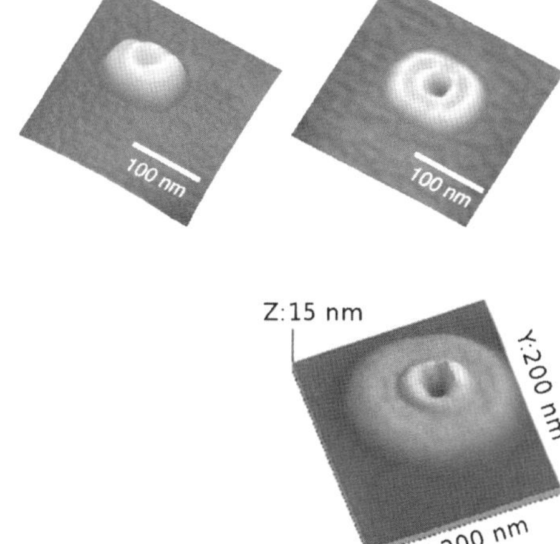

Fig. 8.3 AFM image of GaAs ring/disk crystallized under an As flux of 8×10^{-6} Torr. (Reprinted from [43], Copyright (2010), with permission from Institute of Physics))

Coupled structures, as a nanometer-high holed flat disk with a diameter of hundreds of nanometres on which a ring structure lays down, can also be grown by DE. These coupled ring/disk nanostructures show how it is possible to obtain localized states with different dimensionality and tunable coupling in a designable structure [43, 47].

Ring/disks growth was performed in close resemblance of the single ring and double ring procedure. An atomically smooth AlGaAs surface was prepared by MEE [48] at 350 °C. After supplying 10 MLs of Ga on the substrate, the formation of nearly hemispherical Ga droplets occurred. Straight afterwards, an As flux in the range 8×10^{-7}–8×10^{-6} Torr was irradiated on the substrate at the constant temperature of 350 °C for 20 minutes.

An AFM image of an uncapped sample is shown in Fig. 8.3. A well defined ring/disk is visible, made up of a disk with height of ≈ 6 nm and diameter of around 300 nm and an inner ring marked by a 10 nm high ridge and a diameter of around 80 nm. A decrease of the As flux leads to the expansion of the disk and to its thinning, whereas the central hole diameter remains unaltered [43].

DE method can be extendend to produce triple rings and multiple quantum rings. The key process for realize these complex nanostructures is the use of multiple short time As pulses at different substrate temperatures in order to partially crystallize the metallic Ga in the droplet [41]. The process to obtain GaAs/AlGaAs triple ring is as follows. After the growth of the AlGaAs barrier, substrate temperature was lowered to 350 °C and As valve was closed. At this temperature, the surface had an As-rich $c(4 \times 4)\beta$ surface reconstruction [49]. A subsequent three-step growth procedure was performed. Step 1 requires the supply of 10 MLs of Gallium at 350 °C in absence of As, step 2 corresponds to an As flux supply equal to 8×10^{-7} Torr at 250 °C for 20 seconds and step 3 consists of the supply of an As flux of the same intensity at 300 °C for 20 minutes (thus on until complete crystallization of the deposited Ga).

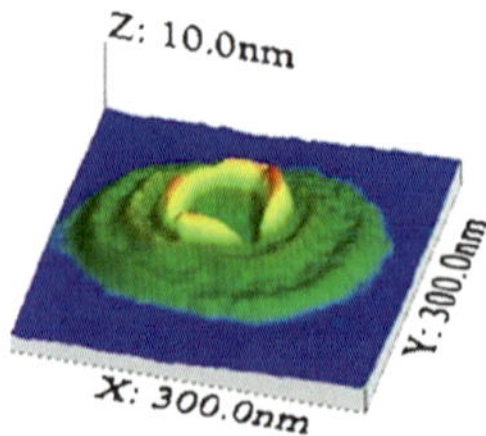

Fig. 8.4 AFM images of a typical triple ring. The crystallographic axis are parallel to the image border: [1$\bar{1}$0] (horizontal on image) and [110] (vertical on image) (Reprinted from [41], Copyright (2009), with permission from American Chemical Society)

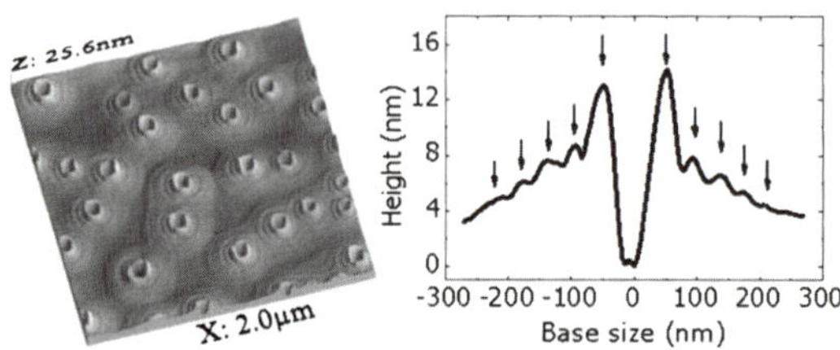

Fig. 8.5 *Left panel*: AFM images of a typical five-ring structures. *Right panel*: Line profile along the [0$\bar{1}$1] direction of a GaAs five-ring structure. Starting from the inner and moving to the outermost ring, the ring radii of the structures are around 50, 90, 130, 170, and 210 nm, while the heights are around 13, 8, 7, 5.5, and 4.5 nm, respectively. (Reprinted from [41], Copyright (2009), with permission from American Chemical Society)

Figure 8.4 is the AFM image of the nanostructure fabricated by such three-steps growth DE procedure. Ga droplets evolved in well-defined GaAs triple rings structures with good rotational symmetry. Inner, middle and outer ring have diameters of around 80 nm, 140 nm and 210 nm respectively and heights around 7 nm for the inner rings, 4 nm for middle rings and 3 nm for the outer rings. A slight elongation of $\approx$11 % along the [0$\bar{1}$1] direction can be observed, probably due to the anisotropic surface migration of Ga on the (100) GaAs surface [50]. As in the ring/disks, the inner ring diameter is nearly equal to that of the original Ga droplet. Triple rings density matches that of the original droplets, confirming that also in this case all Ga droplets transforms into GaAs triple rings. More complex ring structures, obtained through a four step DE procedure, are reported in Fig. 8.5 [41].

8.2.2 Growth Model

Growth dynamics of triple rings was investigated by stopping the process at different steps and morphological characterizing the samples via ex-situ AFM measurements. The quenching was just after: (i) droplet formation, (ii) the short time As supply; (iii) the full crystallization of the droplet. Selective wet etching was used

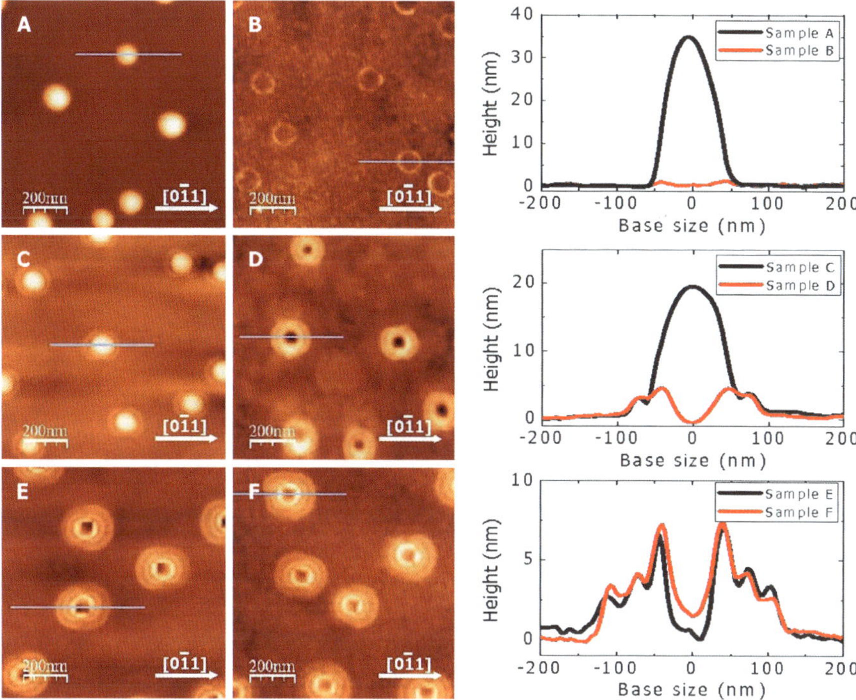

Fig. 8.6 AFM images of as-grown samples S1, S2, and S3 (*left panels*), etched samples S1–E, S2–E and S3–E (*center panels*) and corresponding line profiles taken along [0-11] direction (*right panels*), after 10 MLs Ga supply at 350 °C (*top panels*), after 8×10^{-7} Torr As supply at 250 °C for 20 seconds (*middle panels*) and after 8×10^{-7} Torr As supply at 300 °C for 20 minutes (*bottom panels*). (Reprinted from [41], Copyright (2009), with permission from American Chemical Society)

to remove metallic Ga on the surface [41]. AFM images and typical line profiles of the six samples are reported in Fig. 8.6. 10 MLs Ga were supplied at 350 °C (sample S1, Fig. 8.6A) and numerous nearly hemispherical Gallium droplets were formed. Average diameter was around 80 nm, height around 35 nm and density around 8×10^8 cm^{-2}.

At first step, Ga etching allows to observe the presence of a GaAs ring structure just under the original droplet, due to the crystallization at the droplet's edge (Fig. 8.6B). The partial crystallization step, when the initial Ga droplets are irradiated with an Arsenic flux of 250 °C for 20 seconds, led to a structure formed by a central dome of unreacted Ga, with the same radius of the initial Ga droplet, surrounded by a shallow ring of ≈ 140 nm diameter (Fig. 8.6C). Ga etching shows the formation of a GaAs double ring structure, whose inner ring is lying just under the edge of the metallic Ga droplet (Fig. 8.6D). The third step, the final As supply at 300 °C for 20 minutes, completely crystallizes Ga atoms, forming the outermost third ring structure with a diameter of around 210 nm. Therefore, a complete GaAs triple ring structure was obtained (Fig. 8.6E–F).

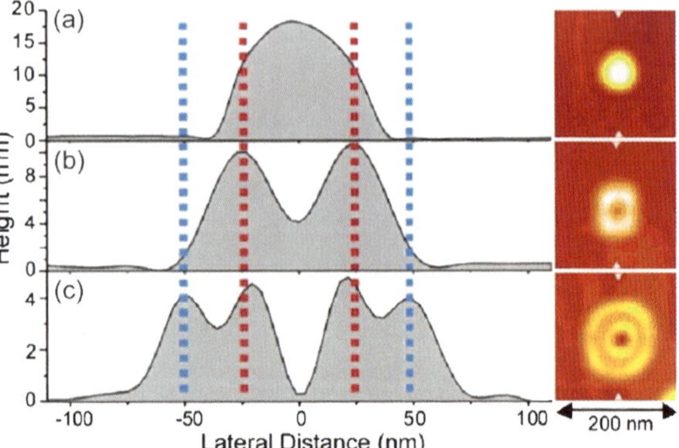

Fig. 8.7 AFM images of the as-grown droplet (*upper panel*), single ring (*center panel*) and double ring (*bottom panel*) nanostructures obtained from the same initial droplet configuration. (Reprinted from [27], Copyright (2005), with permission from American Chemical Society)

A discussion of the phenomenology is now presented. The semiconductor quantum ring which were characterized exhibit three concentric rings. However, the rings do not share the same origin. The two crystallization steps in presence of an As flux form the two external rings, whereas the inner ring is formed during the Ga droplet formation, when no intentional As flux is supplied to the sample. The ring lies underneath the Ga droplet, at its edge, and is not altered by the subsequent steps of the fabrication process. The formation of the inner ring can be accounted for by an internal transport of As atoms incorporated at the droplet edge. Therefore, the inner ring might be due to the low solubility of As in the metallic Ga and of its accumulation on the GaAs tiny ring, found just after the droplet deposition, because of internal convection flux. The same diameter is thus preserved for all the growth conditions. It is worth noting that the inner ring shares the same origin of the inner ring in double rings, because its formation does not depend on the specific conditions used during the arsenization step [27]. Mano et al. [27] already showed clearly this behaviour and pointed out that radius of single ring, inner ring in double ring and Ga metallic droplet are identical (see Fig. 8.7)

To the outer rings, which are formed during the arsenization steps, can be associated a different formation mechanism. In DE a reservoir of the III-column species (the droplet) resides permanently on the surface and the V-column species is supplied in form of distributed flux. Therefore the balance between the Ga migration from the droplet and As flux determines the growth of the GaAs nanostructure. Then, following the changes in average surface reconstruction during the growth process is the utmost importance in order to understand such mechanism. These changes have been determined by [41], through the analysis of the RHEED pattern and of the RHEED specular beam intensity. During the deposition step, after 1.7 MLs supply of Ga molecular-beam Ga-rich (4×6) surface reconstruction [51] appeared, while during As supply the surface changed from the Ga-rich limit (4×6) to As-stabilized (2×4) and finally to As-rich $c(4 \times 4)$.

The interplay between the As adsorption on the Ga-rich (4×6) surface and the Ga migration on the As-stabilized (2×4) is supposed to be the key point for the formation of quantum ring structures. Figure 8.8 shows a schematic diagram of the

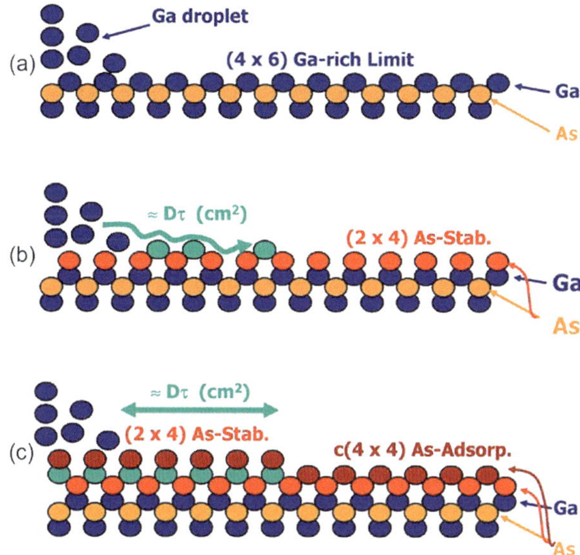

Fig. 8.8 Schematic explanation of the proposed growth mechanism for the formation of outer rings structures. Ga droplets are formed on a Ga-rich (4 × 6) surface reconstruction (**a**). During As supply a (2 × 4) surface reconstruction appears all over the substrate on the top of which the Ga atoms, coming from the droplets, can migrate covering a mean displacement area of $\approx D\tau$ (**b**). Far away from the droplet the surface turns to the As-rich c(4 × 4). The border of this area act as a pinning site for the migration of the Ga atoms (**c**). The detailed atomic arrangements for the different surface reconstructions are ignored for simplicity. (Reprinted from [41], Copyright (2009), with permission from American Chemical Society)

proposed mechanism. 10 MLs Ga are supplied: just after 1.7 MLs the c(4 × 4) reconstruction changes to a Ga-rich (4 × 6) surface and subsequently droplets are formed (Fig. 8.8a). As soon as Arsenic molecular beam is supplied, the substrate starts to change to an As-stabilized (2 × 4) surface reconstruction and nearly simultaneously some Ga atoms migrate and form a monolayer of GaAs around the droplet [52] (Fig. 8.8b). Because of the cylindrical symmetry of the diffusion dynamics, Ga atoms can cover a mean displacement area $\approx D\tau$ where D is the surface diffusion coefficient of Ga atoms and τ the average time interval between arrival and adsorption of As atoms at a specific site. At the same time, far from the Ga droplets, the Arsenic adsorption on the surface not affected by Ga diffusion makes the reconstruction to change to c(4 × 4). The border of the As-rich c(4 × 4) region act as a pinning site for the Ga atoms diffusing from the droplet on the As-stabilized (2 × 4) surface, as here Ga atoms can easily find excess As atoms (Fig. 8.8c): this causes the formation of the outer ring structure. It's worth noting that step 2 and step 3 starts from similar configurations, characterized by a reservoir of metallic Ga in the position of the droplet and a Ga-rich (4 × 6) surface reconstruction. The change in surface reconstruction from As-rich to Ga-rich between step 2 and 3 might be due to a diffusion of Ga from the droplets on the As-rich surface, forming a 2D GaAs thin layer, in absence of an intentional As flux.

Although the formation kinetics of the inner and outer rings is different, the composition of the rings is the same. Also the formation dynamics of the double rings and ring/disks can be easily explained [27, 43] by this model. The double ring and ring/disk are realized via a two-step growth process: first a deposition of Ga into droplets in absence of As flux and then a step where As with moderate flux ($\approx 1 \times 10^{-7}$ Torr BEP) is irradiated on the sample at 250 °C until full crystallization is achieved. The inner ring has, like in our triple ring structure, the same radius of the initial droplet and therefore should be formed just after the droplet deposition [27]. The subsequent arsenization step is responsible of the formation of the outer ring.

It is important to point out that there is a significant difference, in the case of multi-step growth, between the amount of Ga atoms initially supplied (typically 10 ML in the case of triple rings) and the equivalent amount of Ga contained inside the final structure. This difference suggests that only a fraction of the initially supplied Ga atoms effectively concur to the formation of the 3D nanostructure, while the other part, estimated to be around 6–7 MLs, might be consumed in another process. The reason of this discrepancy might be found considering the fabrication procedure for the formation of triple rings. As mentioned before, the three main steps of the growth are performed at different temperatures: 350 °C for the droplets formation, 250 °C for the first As supply and 300 °C for the second As supply. The change in substrate temperature requires growth interruption times of about one hour for each change. During this waiting time a portion of the Ga atoms stored in the droplets might be consumed to form a 2D GaAs thin layer all over the substrate. We believe this phenomenon to be caused by a slow 2D crystallization of Ga atoms diffusing from the droplets, even in absence of an intentional As supply. Indeed, an As background pressure of around 1×10^{-9} Torr is present during the whole procedure, thus providing an unintentional As pressure which promotes the partial crystallization of Ga atoms contained in the droplet during the growth interruptions. A slow GaAs crystallization all over the substrate might take place also in case of very low As pressure [43]. In these conditions of very low As flux, the surface mobility of Ga atoms is so large that an uniform layer of GaAs might be formed all over the substrate surface. In a capped sample, embedded in an $Al_{0.3}Ga_{0.7}As$ barrier, this layer can act as a quantum well, confining carriers and eventually being optically active. Photoluminescence investigation on capped triple ring substrates confirm this assumption (see Fig. 8.9) showing that in the region where the emission from quantum confined GaAs structures is expected, two peaks, corresponding to emission from the triple ring ensemble ($E_{\text{triple ring}} = 1.55$ eV) and from the expected 6–7 MLs-thick GaAs QW ($E_{\text{QW}} = 1.76$ eV), originating from the GaAs layer formed during growth interruptions, are present.

8.2.3 Ring Anisotropy

At high growth temperatures, where a (2×4) As rich surface reconstruction is expected to dominate [45], the well known anisotropy in the diffusion coefficient of

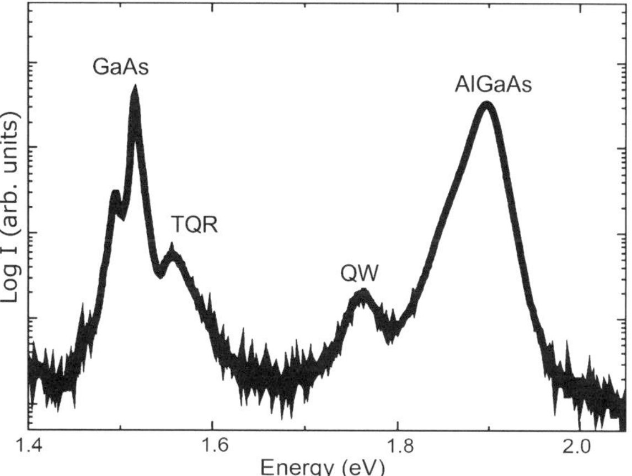

Fig. 8.9 Photoluminescence spectra of triple rings (TQR) recorded at 15 K. GaAs and $Al_{0.3}Ga_{0.7}As$ peaks appeared at 1.52 eV and 1.90 eV, respectively. Between them additional two features, attributed to triple rings (1.55 eV) and to a 7MLs QW (1.76 eV). Data from Ref. [53]

Ga between the two [110] and [1$\bar{1}$0] directions [50] leads to GaAs nanostructures with a shape anisotropy. In order to quantify this effect on the droplet ring shape, Somaschini et al. [54] realized a five-ring sample with As crystallization temperatures between 380 and 320 °C. The AFM image of the five-ring is shown in Fig. 8.10. As expected the anisotropy is clearly visible in the image. The Ga atoms migration from the original droplet perimeter during As adsorption as a function of temperature was characterized via the diffusion equivalent area $D\tau$. The results are shown in Fig. 8.10 for both [110] and [1$\bar{1}$0] directions. The exponential dependence of $D\tau$ on the temperature is confirmed in both cases. By the Arrenhius plot the values of 1.28 eV and 0.85 eV for the activation energy for the Ga migration along [110] and [1$\bar{1}$0] respectively were obtained. These values should not be considered as universal values since the Ga migration happened on a non perfectly flat GaAs (001) surface. In fact the surface of the outer ring shows a sub-nanometre surface roughness. Although the detailed mechanism is not yet clear, the roughness of the

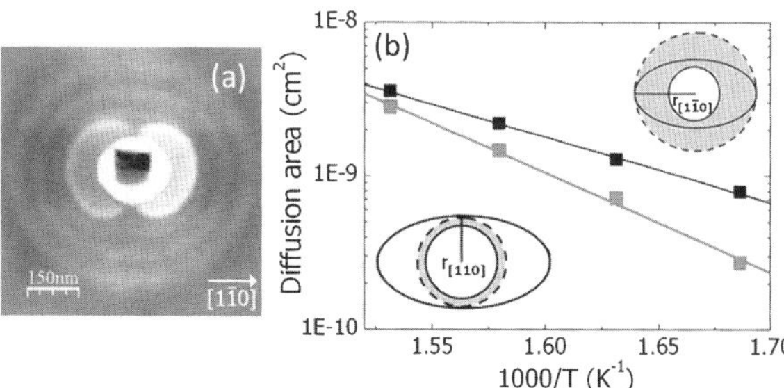

Fig. 8.10 (**a**) AFM image of a single GaAs five-ring structure fabricated in Ref. [53]. (**b**) Arrenhius plot of the diffusion area covered by Ga atoms during their migration from the Ga droplets as a function of the reverse temperature. In the *insets*: scheme of the procedure to calculate the diffusion area (*blue zone* delimitated by the *dashed line*) for both [1$\bar{1}$0] (*top panel*) and [110] (*bottom panel*). *Solid lines* represent the inner ring and an outer ring with a marked anisotropy. Data from Ref. [54]

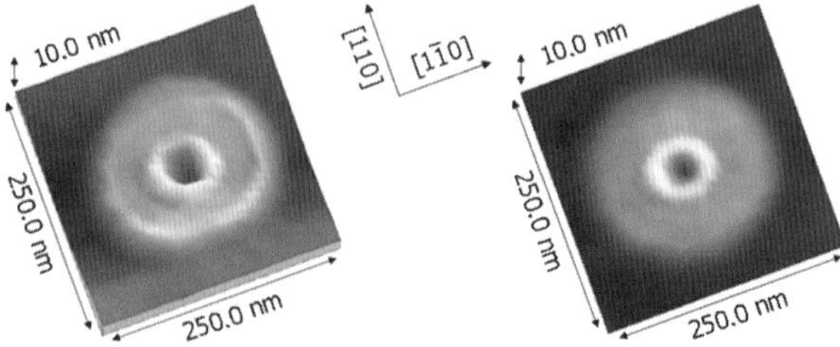

Fig. 8.11 Magnified AFM images of double ring (*left*) and ring/disk (*right*). The crystallization conditions were: 4×10^{-7} Torr at 300 °C for the double ring and 4×10^{-7} Torr at 350 °C for the ring/disk. Data from Ref. [57]

actual surface over which gallium atoms can migrate might in general affect their diffusion length.

8.2.4 Ring vs. Disk Control

As shown in the previous sections, by a suitable selection of the growth conditions it is possible to carefully determine the final shape of the nanocrystals. In particular, ring shaped nanostructures like double ring or ring/disk, two structures characterized by the presence of an inner ring encircled by an outer region with cylindrical symmetry, are of particular interest as high efficiency optoelectronic devices have been produced based on these nanostructures [55, 56]. Even though these two systems are sharing the main morphological features, double ring and ring/disk can be clearly distinguished, since in the first case, a clear peak-and-valley line profile is present, while in the second case, a flat disk appears around the inner ring. In Ref. [57], scanning the growth parameter area where double ring and ring/disk fabrication has been reported [27, 43, 58, 59] and, via detailed in-situ and ex-situ morphological characterizations, the mechanism at the basis of the difference between the two structures was identified. The formation of a clear peak-and-valley height profile, typical of the double ring structure, is related to the dynamics of the surface reconstruction around the Ga droplet during the arsenization step. In particular, the establishment and the subsequent ordering speed of the As-rich c(4×4) surface reconstruction around the Ga droplets during the crystallization process is the crucial ingredient for the differentiation between double ring and ring/disk.

The double ring and ring/disk show a good cylindrical symmetry (see Fig. 8.11), with an outer part which depends on the exact growth conditions and an inner ring, at the centre of the nanostructure, which displays the same radius irrespective of the sample preparation and which marks the original droplet position. As previously demonstrated, the inner ring is the result of the crystallization of Ga at the droplets edge, initiated by a partial dissolution of As inside liquid Ga, even in absence of intentional As supply [41]. On the contrary, the morphology of the outer circular regions, developed around the initially formed droplets, strongly depends on the crystallization conditions. A clear transition from the ring-like to the disk-like geometry is observed as the growth temperature is increased. It's worth noting that a deep change in nanostructure morphology takes place in a rather narrow tempera-

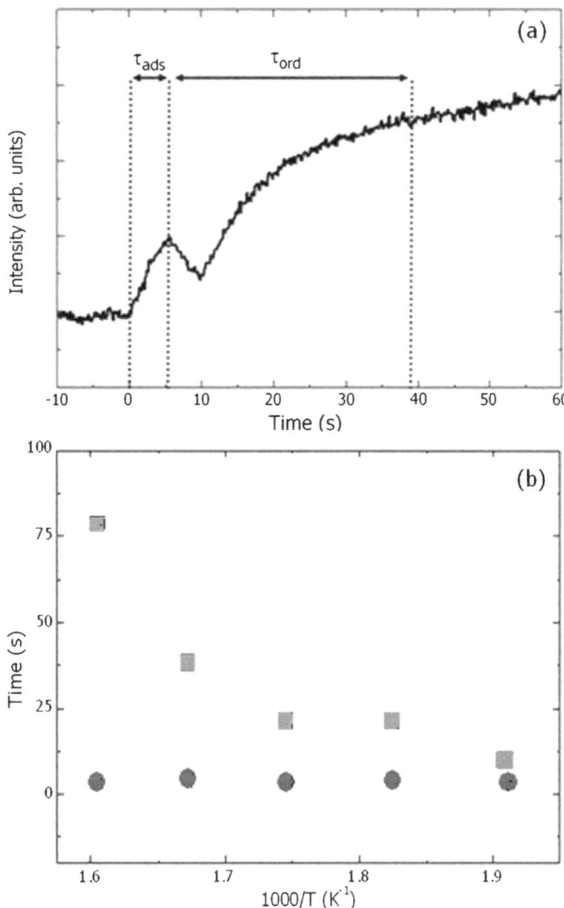

Fig. 8.12 Typical specular beam RHEED intensity change during As (4×10^{-7} Torr) adsorption on a droplets-free (4×6) reconstruction (**a**). τ_{ads} and τ_{ord} are indicated on the graph. Temperature dependence of τ_{ads} (*circles*) and τ_{ord} (*squares*) (**b**). Data from Ref. [57]

ture window, between 300 °C and 350 °C. We are therefore in presence of a growth process rapidly changing with the temperature.

What marks the difference is the structural configuration of the substrate surface as it is seen by each Ga droplet during the crystallization process. The changes in the surface structure and the specular beam intensity as a function of time, during As adsorption on a droplets-free (4×6) reconstruction can be understood following the typical specular beam intensity change during an As irradiation (Fig. 8.12). Just after opening the As valve the specular beam increases, showing a maximum corresponding to the establishment of a (2×4) reconstruction, decreases and increases again after the initiation of a c(4×4) reconstruction which progressively orders until the specular beam intensity shows the saturation [60]. τ_{ads} is the time interval between the As cell opening and the formation of a (2×4) surface reconstruction and, in a similar way, τ_{ord} is the time interval between the establishment of the (2×4) and the ordering of the c(4×4). τ_{ord} corresponds to $1 - 1/e^2$ times the saturation value of the specular beam intensity. In Fig. 8.12 the temperature dependence of

τ_{ads} (circles) and τ_{ord} (squares) is reported. While τ_{ads} does not show any dependence on the temperature, τ_{ord} increases with increasing the substrate temperature. As matter of fact, the structural changes that are needed to transform the (2×4) into the $c(4 \times 4)$ show a temperature dependence. The appearance of $c(4 \times 4)$ regions is faster for lower substrate temperatures, where the ring morphology is predominant.

The variation in the growth dynamics between the case of double ring and ring/disk can therefore arise from the different configuration of the substrate surface around the Ga droplets. Initially Ga droplets are sitting on the (4×6) surface reconstruction and, after the As supply, As adsorption promotes Ga atoms diffusion from the droplets. The size of the area covered by this diffusion is set by the average Ga migration distance which is determined by the diffusion coefficient and by the As adsorption time τ_{ads}, which is almost constant in our growth temperature range. At distances much larger than the Ga migration length, no Ga originated from the droplets can be found. In these areas the substrate surface will complete the transition to (2×4) and finally to the $c(4 \times 4)$ reconstruction, being the speed of this phenomenon, described by τ_{ord}, strongly dependent on the substrate temperature, as shown before. While Ga can easily diffuse on (2×4) surface, $c(4 \times 4)$ regions can act as preferential nucleation sites due to the large amount of As present on this As-rich configuration. Therefore the boundary of the $c(4 \times 4)$ region constitute a pinning point for Ga atoms diffusion. At lower temperatures, Ga atoms migrating from the droplet and reaching the border to the quickly formed $c(4 \times 4)$ region, will preferentially nucleate there and this phenomenon gives rise to the accumulation of GaAs at the distance from the droplets which marks the As rich region. This is what gives rise to the peak-and-valley line profile and it is what we normally call "outer ring". At higher temperatures, the establishment of the $c(4 \times 4)$ regions around the Ga droplets is a slow process and the crystallization of the available Ga atoms might finish before the formation of the As-rich regions. Ga atoms migrating from the droplets will not find any preferential site for the nucleation and thus will give rise to the disk-like feature. Therefore the speed of the transition to the As-rich $c(4 \times 4)$ surface reconstruction plays a key role in determining the final shape of the nanostructures, meaning that the surface around the Ga droplets cannot be considered as inert during the crystallization process.

8.3 Electronic Properties

8.3.1 Theoretical Predictions

The energy levels of the ring can be evaluated in the framework of a single-band effective-mass envelope model approximation [61, 62]. Although being a simple method, compared to more complex pseudopotential and multiband $k \cdot p$ methods [63, 64], the model catches the relevant electronic characteristics of the droplet epitaxy quantum nanostructures. The actual shape measured by AFM is adopted as the potential of quantum confinement; for simplicity, the ring is assumed to hold a

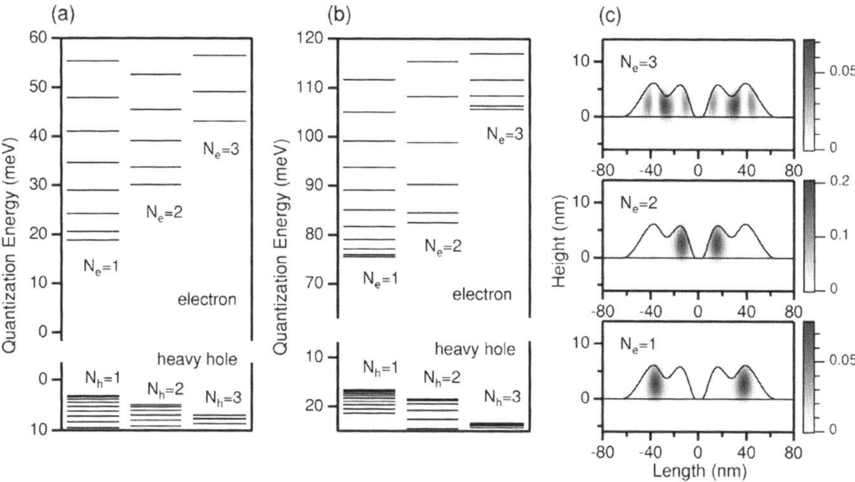

Fig. 8.13 Single-carrier energy levels in (**a**) single ring and (**b**) double ring. Quantization energies for an electron (a heavy hole) with the three lowest z-radial quantum numbers, $N_{e(h)}$ and various angular momenta (up to 10) are presented. (**c**) Cross-sectional imaging of electronic probability density in double ring for $N_e = 1$, 2, and 3 with $L = 0$. The *line* represents the potential of confinement used for calculation. (Reprinted from [28])

cylindrical symmetry, although, in most of the cases, a slight (within 15 %) asymmetry is present. In DE *lattice-matched* GaAs/(Al, Ga)As rings, strain effects are negligible. Thus the simple effective-mass approach is expected to provide accurate energy levels. In the case of SK grown dots, instead, the electronic structure is strongly modified by complex strain effects [65]. The reliability of the present method is seen in Refs. [17, 66], where good agreement is shown between the asymmetric photoluminescence (PL) lineshape in a GaAs/(Al, Ga)As quantum dot ensemble and the calculation, which takes into account the morphologic distribution of dots. Since DE rings are sufficiently small, confinement effects are dominant and Coulomb interaction can be treated as a constant shift in the transition energies, independent of the choice of an electron state and the hole state.

The expected eigenfunctions, taking into account the rotational symmetry of the Hamiltonian, are of form Φ_L, where $L(= 0, \pm 1, \ldots)$ is the azimuthal quantum number, is expanded in terms of a complete set of the base functions, $\xi_{i,j}^L$, formed by products of Bessel functions of integer order L and sine functions of z,

$$\Phi_L(z, r, \theta) = \sum_{i,j > 0} A_{i,j}^L \xi_{i,j}^L(z, r, \theta), \tag{8.1}$$

$$\xi_{i,j}^L(z, r, \theta) = \beta_i^L J_L(k_i^L r) e^{iL\theta} \sin(K_j z), \tag{8.2}$$

where $k_i^L R_c$ is the ith zero of the Bessel function of integer order $J_L(x)$, $K_j = \pi j / Z_c$, and β_i^L is appropriate normalization factors [61],

Figure 8.13(a) shows a series of single carrier levels of single ring. Because the system has cylindrical symmetry, each carrier level is specified by the z-radial quantum number, $N(=1, 2, \ldots)$, and an azimuthal quantum number L, corresponding to the angular momentum. The L-dependent sequence of quantized levels shows a typical signature of ring-type confinement. For an ideal ring with infinitesimal width, being treated as a one-dimensional system with translational periodicity, the level series is expressed as

$$E_L = \frac{\hbar^2}{2m^*R^2}L^2,\tag{8.3}$$

where R and m^* respectively represent the radius of the ring and the carrier mass. The line sequence in Fig. 8.13(a) reflects clearly the bilinear dependence of the level series, shown in (8.3). In addition, the L-dependent sequence is less crowded for larger N. According to (8.3), the situation corresponds to an increase in the effective value of R, implying a stretched orbital trajectory caused by in-plane centrifugal force.

Figure 8.13(b) shows the energy levels in double ring. Due to the smaller height, the quantization energies are larger than those of quantum ring. The level sequence of $N = 1$ is more densely populated than that of $N = 2$, suggesting a large difference in carrier trajectories between the two levels. This difference is confirmed by the wavefunctions shown in Fig. 8.13(c), which illustrates the envelope wavefunction of an electron with various values of N. They are of zero angular momentum. The electron of $N = 1$ is confined mainly in the outer ring. That of $N = 2$ is in the inner ring, and that of $N = 3$ is situated in both rings. That differential confinement engenders remarkable changes in their trajectory. The amount of penetration for the electron of $N = 1$ to the inner ring is found to be $\sim$0.1, whereas that of $N = 2$ to the outer ring is $\sim$0.05.

In the case of triple ring, predicted ground state wavefunction is completely localized within the inner ring, while for more excited states wavefunctions localized within the two external rings can be found [41]. The expected electronic properties of a ring/disk are more complex. These structures constitute the good example of nanostructures with coupled localized-extended states with cylindrical symmetry (the protrusion at the inner ring edge acts, in fact, as three-dimensional electronic carrier confinement potential, thus being like a ring laid down on top of quantum disk). Ring/disks thus offer additional degrees of freedom for the control of effective coupling between excitons entrapped in quantum nanostructures [6]. The calculated ring/disk ground electron and hole states appear to be confined in the ring structure at the edge of the inner ring/disk hole. The ring/disk excited state is, on the other side, a quantum well like state extended along the disk [67].

8.3.2 Beyond Effective Mass Approximation

In the EMA approach mentioned above, Coulomb interactions between carriers has been ignored. Coulomb effects can be included using the configuration mixing (CM) method, where the interaction Hamiltonian is expanded by the single-carrier wavefunction configurations, then diagonalized [65]. With the CM method it is possible to correctly describe the binding energies of excitons in strongly-confined quantum dots, such as the (In, Ga)As/GaAs systems, where the Coulomb terms is treated as weak perturbation on the single-carrier states. For weakly-confined quantum dot systems, on the other hand, the CM method often fails, thus it is hard to reach sufficient convergence even with large CM matrices available by practical computational capacity. The quantum ring systems would belong to the latter category, since the single-particle levels are densely packed in spectra compared to the typical energy range of Coulomb binding (a few tens of meV).

An alternative method to overcome this conversion problem is the quantum Monte-Carlo (QMC) method, which enables to determine the exact many-body ground-state energies under the condition of EMA. The details of this method are given by [68]. In this approach single-particle energies were calculated as well as multi-particle energies on an equal footing, including all correlation effects. The multiple exciton spectra of DE grown GaAs quantum dots calculated by means of QMC show a size dependence which quantitatively agrees with the experimental results [69].

8.4 Photoluminescence Emission

8.4.1 Broad Area Photoluminescence

In Fig. 8.14 typical broad area PL spectra of the ring samples capped with an $Al_{0.3}Ga_{0.7}As$ barrier are reported. It is worth mentioning that the capping procedure does not induce strong changes in the droplet epitaxy made nanostructure shape, as demonstrated in Ref. [70]. Thermal annealing processes, required to recover the crystalline quality of the barrier, induces some limited intermixing at the barrier with less than 0.5 nm diffusion length even at the highest annealing temperatures [71].

In all samples PL is characterized by a broad emission band located between the GaAs and the $Al_{0.3}Ga_{0.3}As$ energy gaps.

The experiment reported by [72] shows samples ranging from quantum dots to double ring obtained from an identical droplet configuration. In this way it is possible to observe the effect of structure evolution at constant volume, determined by the Ga content of the initial droplets.

The nanostructure emission blue-shifts as the structure evolves from quantum dot to double ring. The structural evolution [72] is accompanied both by an increase of the nanostructure radius, by a strong reduction of the nanostructure height and by the presence of an additional lateral confinement in ring structures. The anti-correlation of E_{emi} and height in the DE-nanostructures (see Fig. 8.15) shows that height reduction is the main factor affecting the emission energy, whereas additional lateral confinement given by the ring width in quantum ring structures plays a minor role. A correlation between the nanostructure shape and its PL full width at half maximum (FWHM) can be observed. Being the initial droplet configuration the same, such PL broadening cannot be related to shape dependent nanostructure

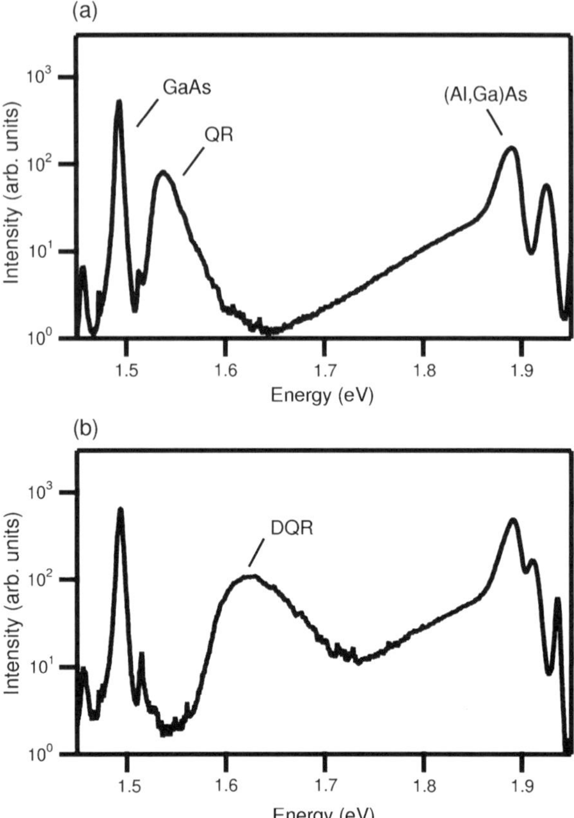

Fig. 8.14 Far-field emission spectra of the sample with (**a**) single rings and (**b**) double rings at 5 K plotted on a logarithmic scale. The excitation density is 50 mW/cm^2. (Reprinted from [28])

volume fluctuations. Bietti et al. [72] thus attributes such dependence of PL FWHM to shape fluctuation. In single ring and double ring fabrication, the presence of a slower crystallization step reduces shape disorder.

Figure 8.16a shows the ensemble optical emission of triple ring structures embedded. A clear emission peak is visible at $E_A = 1.56$ eV (band A) with a full width at half maximum 30 meV, above the excitonic GaAs signature at 1.519 eV. As excitation power density (P_{exc}) is increased (Fig. 8.16a) a second band (band B) appears on the high energy side of the fundamental band ($E_B = 1.58$ eV). The intensity of this band increases superlinearly with the laser excitation power. The theoretically predicted triple ring ground state transition energy is $E_{GS} = 1.58$ eV. The ground state wavefunction is completely localized in the inner ring. The first radial excited state is located 20 meV above the ground state ($E_{ES} = 1.60$). Also in this case, the wavefunction is localized within the inner ring volume. The predicted transition energy E_{GS} lies well within the A line bandwidth. Band A thus belongs to the ensemble emission from the triple ring ground states. Moreover, the energy difference between A and B bands matches the energy difference $E_{ES} - E_{GS} = 20$ meV. In addition, the P_{exc} behavior of band A and B is very similar to that shown by quantum dot ensembles where the additional band appearing at high P_{exc} is attributed to

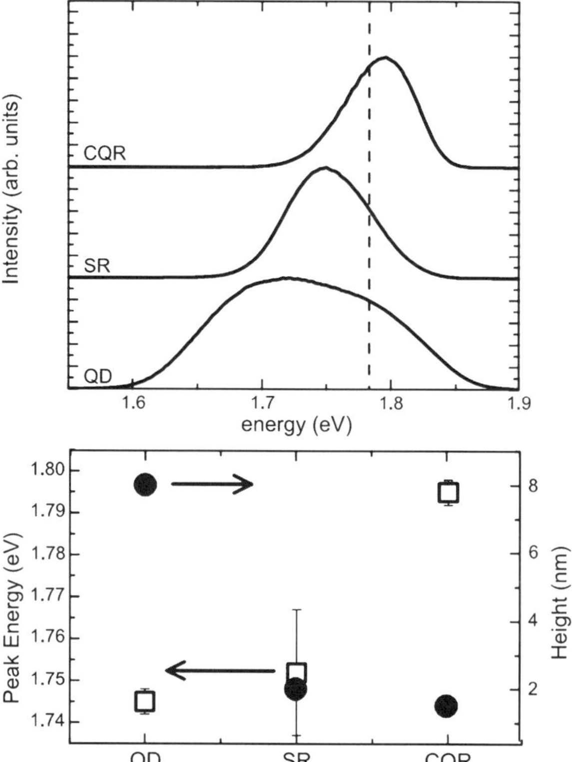

Fig. 8.15 *Top panel*: PL of quantum dot, single ring and double ring samples. The *vertical line* indicates the laser excitation energy for RPL measurements. *Bottom panels*: PL peak energy (*open squares*) and height (*black circles*) of the three nanostructure. (Reprinted from [72], Copyright Wiley-VCH Verlag GmbH and Co. KGaA (2009), reproduced with permission)

excited states emission. The excited state population in quantum dots is linked to the ground state by a waterfall-like chain, thus being visible only when the ground state of the quantum dot is occupied. On this basis, we attribute band B to first excited transition. It is worth noting that the linked dynamics between the ground state and the first excited state arises from the fact that the two are localized within the same ring, therefore showing an agreement with what has been found in single ring structures [28].

Ring/disks are nanostructures with coupled localized-extended states with cylindrical symmetry. In Fig. 8.17 the PL spectra at $T = 14$ K of the sample is reported. An intense and broad band is clearly visible at 1.55 eV, with a full width at half maximum of ≈ 30 meV. The band shows a shoulder at 1.60 eV. The observed PL peak value is in good agreement with the calculated emission energy ($E_{\mathrm{GS}}^{\mathrm{th}} = 1.56$ eV). The theoretically calculated ring/disk ground electronic and hole states appear to be confined in the ring structure which is formed at the edge of the inner ring/disk hole. The ring/disk excited state is, on the other side, a quantum well like state extended along the disk ($E_{\mathrm{EX}}^{\mathrm{th}} = 1.59$ eV).

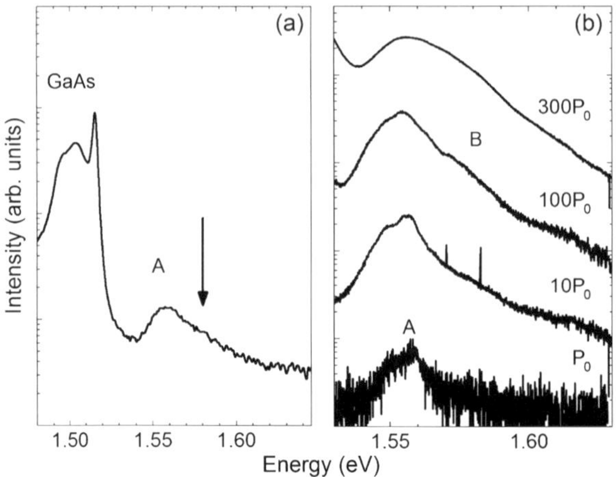

Fig. 8.16 (**a**) PL spectra of the triple ring sample measured at $T = 15$ K and $P_{exc} = 10$ W/cm^2. The arrow indicates the theoretical prediction based on a typical triple ring AFM image. (**b**) PL spectra of the triple ring sample measured at $T = 15$ K as a function of P_{exc} in the range 5–1500 W/cm^2. Here $P_0 = 5$ W/cm^2. A and B labels indicate triple ring ground state and excited state emission, respectively. (Reprinted from [73], Copyright (2009), with permission from Institute of Physics)

Fig. 8.17 PL spectrum of the ring/disk sample at low temperature ($T = 14$ K). *Upper right corner*: AFM image of a single ring/disk. The emission at 1.55 eV is attributed to carriers confined in the ring protrusion of the ring/disk, while the shoulder at 160 eV to states belonging to the disk. (Reprinted from [67])

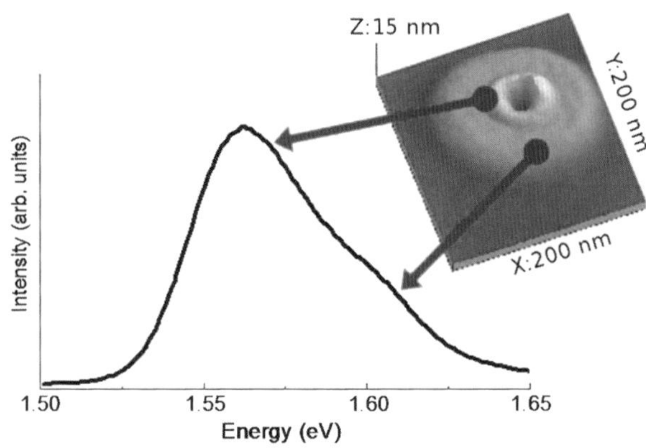

8.4.1.1 Resonant Photoluminescence

In Fig. 8.18 the resonant PL spectra (RPL) of the single ring and double ring samples are illustrated in comparison with that of a quantum dot. All the spectra are characterized by large, unstructured bands located at decreasing $\Delta E \equiv E_{exc} - E_{emi}$ energies. The PL emission from the samples shows its maximum intensity at $\Delta E = 84$ meV, $\Delta E = 38$ meV and $\Delta E = 29$ meV for quantum dot, single ring and double ring, respectively. From the RPL spectra some informations on the carrier dynamics in the DE-nanostructure can be deduced. The ΔE values strongly depends on the actual DE-nanostructure. Such dependence allows us to safely at-

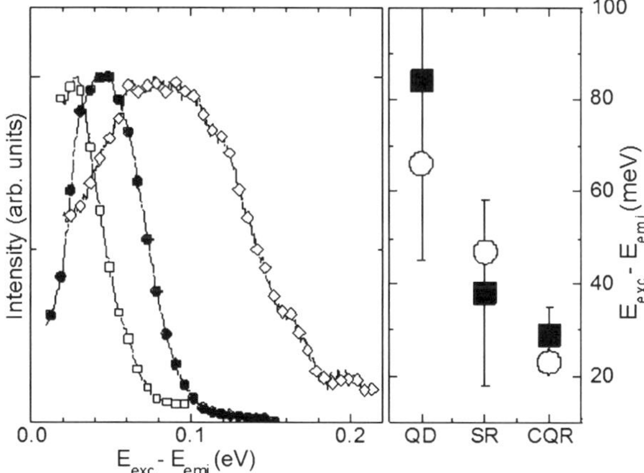

Fig. 8.18 *Left panel*: RPL spectra of quantum dot (*open diamonds*), single ring (*full circles*) and double ring (*open squares*) samples. Experimental conditions were: $E_{\mathrm{exc}} = 1.78\,\mathrm{eV}$ and $T = 10$ K. All the spectra are normalized to their maximum. *Right panel*: Experimental RPL peak energy (*full squares*) and theoretical calculations of the energy difference between the ground and the excited state of the DE-nanostructure (*circles*). The *error bars* indicate the experimental band FWHM. (Reprinted from [72], Copyright Wiley-VCH Verlag GmbH and Co. KGaA (2009), reproduced with permission)

tribute the RPL peak, according to [74], to the emission of the ground state of DE-nanostructure, whose excited state energy corresponds to the laser excitation energy E_{exc}. Moreover, the absence, in the spectra, of any modulation with a spacing corresponding the GaAs LO-phonon energy (36 meV) implies that carrier relaxation in DE-nanostructure is not affected by polaronic [75, 76] or phonon-bottleneck effects [77, 78]. The reduction of the FWHM of the RPL spectra as the shape evolves from quantum dot to double ring should be related to the narrower density of states, owing to the smaller size dispersion, of the latter DE-nanostructure, as shown by the PL spectra.

In Fig. 8.18 the calculated energy separation between the DE-nanostructure ground and excited states is reported, together with the experimental ΔE values. There is a rather good agreement between the theoretical predictions and the experimental results. The maxima of the RPL spectra always correspond to the ΔE values determined by absorption of photons on the first excited state. The energy separation between the ground and the excited state decreases as the nanostructure evolves from quantum dot to double ring (see Fig. 8.18). Theoretical calculations state that ΔE is determined by the radius of the DE-nanostructure. In fact, in all the three DE-nanostructure, the first excited state corresponds to a change in the radial quantum number. In quantum dot the ground and excited state corresponds to a s- and p-type modulations of both electronic and hole radial wavefunction 3D confined in the dot. In single ring structures, the wavefunction is naturally confined in the ring. The first excited state shows a p-type modulation of the radial part of the wavefunction [28]. In the double ring the ground state is localized in the outer ring, while the first excited state lies in the inner ring, around 4 meV above the outer ring state [27]. However, the RPL resonance is located at 29 ± 6 meV, thus much higher than the inner ring resonance. Such energy corresponds, for both rings, to the energy difference between the ground state and the first excited state whose wavefunction lies in the same ring. In double ring, this excited state corresponds, in analogy to the single ring case, to a p-type modulation of the radial part of the ring wavefunction. This means a suppression, in double ring, of the two ring coupling.

8.4.2 Single Nanostructure Photoluminescence

8.4.2.1 Spectroscopy of a Single Quantum Ring

In Fig. 8.19a, typical single ring PL spectra (and their dependence on excitation intensity) are reported. Quantum ring density is of the order of 6×10^8/cm^2 (DE allows to control density about four orders of magnitude) and is thus possible to measure single nanostructure emission with standard micro-PL apparatus [28]. In single ring at low excitation, we find a single emission line appearing at 1.569 eV, due to recombination of an electron and a hole both occupying the ground state of the ring. Increasing excitation intensity, a new emission line, indicated by an arrow, emerges at 1.582 eV. Further increase in excitation density causes saturation in the intensity of the original line along with a nonlinear increase in the new line. Superlinear dependence of the emission intensity suggests that the satellite line comes from the electron-hole recombination from an excited level of the ring. Thus, the energy difference between the ground and the excited state in the single ring is 13 meV.

In addition to the state-filling feature associated with photoinjection, the ground-state emission is shifted to lower energies. This is a signature of multi-carrier effects. If multiple carriers are present inside a ring, their energy levels are modified by the Coulomb interaction among them. Because the energies are renormalized according to exchange corrections for *parallel-spin* carriers, this results in spectral red shift of the emission, depending on the number of carriers. Similar features have been reported in GaAs quantum dots [16] and InAs quantum rings [79]. At high excitation, we also find spectral broadening, which is attributed to carrier collision processes.

Single ring emission is considerably large compared with those observed in quantum dots [16]. Line broadening can be ascribed to spectral diffusion, *i.e.*, an effect of the local environment that surrounds each quantum ring. Our samples are expected to contain a relatively large density of imperfections and excess dopants, associated with low-temperature growth. This causes local-field fluctuation, leading to efficient broadening of the PL spectra. Detailed discussion on the origin of line broadening in quantum nanostructure is presented in Ref. [80].

Figure 8.19b shows PL spectrum of a single double ring. As in the case of single ring, the spectra consist of discrete lines, *i.e.*, a main peak and a satellite one, which is on the high energy side of the main peak. The former is associated with

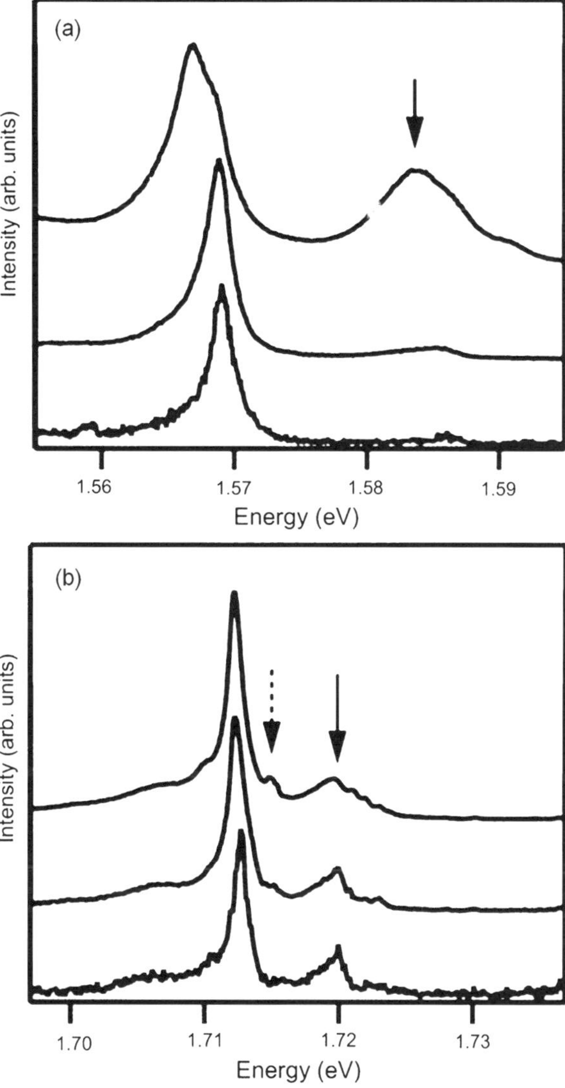

Fig. 8.19 Emission spectra for a single GaAs single ring (**a**) and double ring (**b**). Their respective excitation densities were, from bottom to top, 1, 10, and 30 W/cm^2. Spectra are normalized to their maxima and offset for clarity. (From [28])

recombination of carriers in the ground state, whereas the latter comes from the excited states. The energy difference between ground-state and excited-state lines is 7.2 meV. In contrast to the single ring case, we observe the satellite peak even at the lowest excitation, where the estimated carrier population inside a ring should be less than 0.1, according to [28]. Observation of the excited state emission suggests a reduction of carrier relaxation from the excited level to the ground level. This feature will be discussed later. At high excitation, we find several additional lines superimposed on the spectra, as shown by the broken arrows. These contributions imply the presence of fine energy structures in double ring.

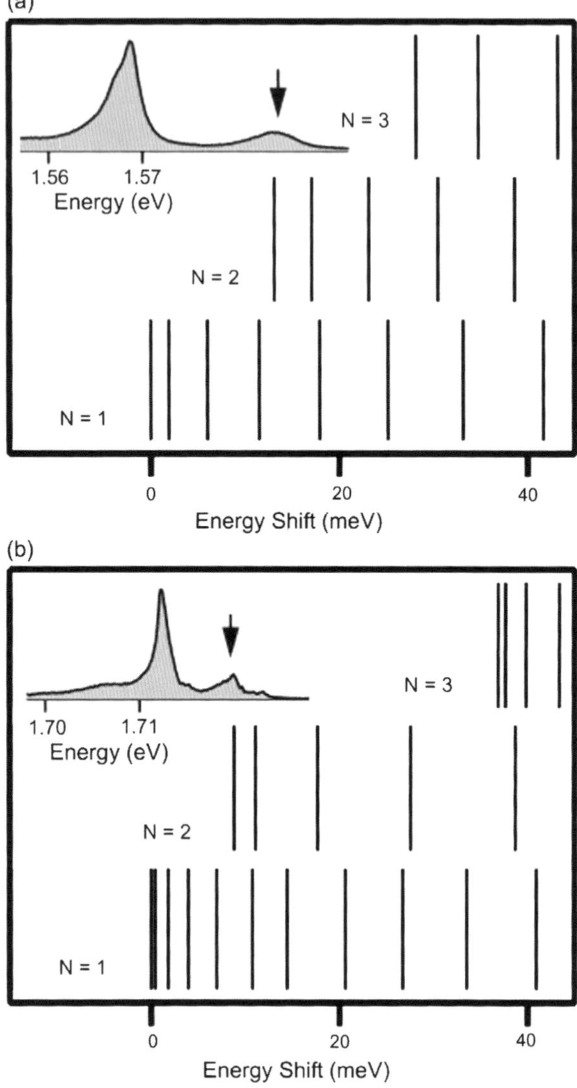

Fig. 8.20 (**a**) A series of optical transition energies in quantum ring, obtained by the calculation. The PL spectrum of a quantum ring at 15 W/cm^2 is shown in the *inset*. (**b**) The energies of optical transitions in double ring, together with the PL spectrum of a double ring at 10 W/cm^2 for comparison. (Reprinted from [28])

Figure 8.20(a) shows a series of transition energies for single ring. For comparison, the emission spectra of quantum ring are plotted. The main peak and the high-energy satellite in the observed spectra are assigned respectively to the recombination of the *e-h* pair in the lowest state, $(N, L) = (1, 0)$, and to that of the first excited *z-radial* state, $(2, 0)$. The split between the two transitions is estimated in 13.1 meV, in agreement with the energy shift obtained by experiments. The emissions associated with high angular momenta are not present, which suggests rapid relaxation of angular momentum, whose process is quite faster than transition between *z*-radial quantization levels or recombination between an electron and a hole.

A possible origin for fast angular momentum relaxation is structural asymmetry of the ring, due to elongation, impurity and surface roughness. In this case, angular momentum is not a good quantum number, and scattering between different L levels efficiently occurs.

Comparison between the experimental spectra of double ring and calculation is reported in Fig. 8.20(b). As in the case of quantum ring, the main PL peak and the satellite one are attributed respectively to the transition of $(N, L) = (1, 0)$ and that of $(2, 0)$. The energy split deduced from calculation is 8.8 meV, which agrees with the experimental value. It is worth noting that, in double ring, the wavefunction of $N = 1$ is localized mainly in the outer ring, while that of $N = 2$ is localized in the inner ring. Thus, the two peaks in the observed spectra come from the two rings which form the nanostructure. The excited-state emission from the inner ring appears even when the carrier population is less than one. This constitutes direct evidence for the carrier confinement into the two rings. Tunneling probability between inner and outer ring is not very large, engendering the observation of the excited-state emission.

8.5 Carrier Dynamics in Ring Structures

Double ring emission consists of a doublet emission, attributed to the inner ring (IR), the higher energy line, and to the outer ring (OR), the lower energy line. The relative intensity, at low temperature, of the two lines does not depend on laser excitation power density P_{exc}. This behavior is the fingerprint of a decoupled dynamics of the two rings, although a strong phonon bottleneck effect between the ground and excited state of the double ring could give rise to similar CW spectra. Fundamental insights on possible coupling mechanisms between double ring excited states may be obtained by time resolved state filling experiments on single double rings. Reference [81] reports the time resolved emission, at low P_{exc}, of the two lines of different double ring doublets. The rise-time of the IR and OR lines is $\tau_R = 120 \pm 40$ ps, thus being about four times larger than the commonly measured τ_R for quantum dot structures, even in the case of GaAs/AlGaAs quantum dots [23]. Despite small differences (within 30 %) between IR and OR decay times (τ_D), the two lifetimes are quite similar for each pairs of rings, although the overall lifetime varies by more than a factor four in different double rings (200 ps $< \tau_D <$ 900 ps), possibly due to different defectivity of the environment [80]. Time resolved PL measurements as a function of P_{exc}, reported in Fig. 8.21, can give more information on the nature of the mutiplets origin. Increasing power density, strong increases of the rise PL times of the fundamental optical transition are found, as due to state filling condition. At high P_{exc}, correlated dynamics between two lines is observed, which originates from the multiexcitonic states in the OR (lines L1 and L2 in Fig. 8.21), in close resemblance to the quantum dot case [16, 82, 83]. Risetime of the line L1 corresponds to decaytime of line L2, thus demonstrating a link between the carrier population in the two states. Such correlation between decay and rise times arises from states which

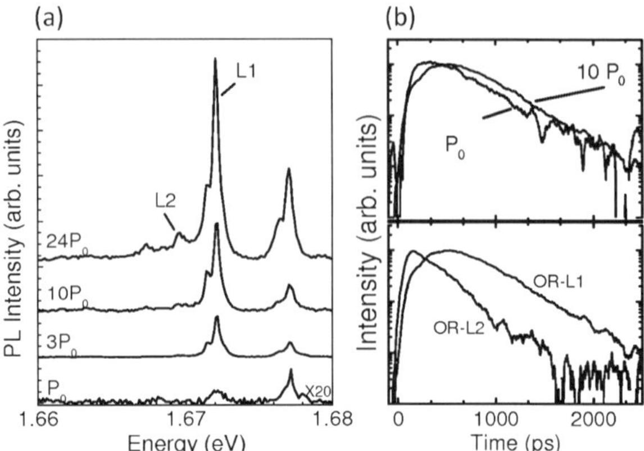

Fig. 8.21 (**a**) Time integrated spectra of a double ring at different excitation power densities. $T = 10$ K and $P_0 = 3$ W/cm^2. (**b**) *Top panel*: time resolved PL of the OR line of a double ring at $T = 10$ K and at different P_{exc}. (From [81])

are bound in a cascade-like behaviour. As far as the carrier dynamics is concerned, IR fundamental line shows a similar behavior, indicating an independent state filling mechanism with respect to the OR case, although the cascade-like dynamics in the IR multiplet is partially hidden by OR recombination present on the low energy side of the IR line.

The overall phenomenology clearly excludes a carrier transfer between IR and OR. Recombination kinetics of carriers in the inner and outer rings are then decoupled. In fact, if the carrier dynamics in the two structures were in some way correlated, a hierarchical order in the PL decay, with the shorter decay time belonging to the higher state, should be observed, denoting a cascade mechanism associated to the carrier relaxation paths. On the contrary, the IR recombination lifetimes is always larger than the risetime of the OR time resolved emission. Moreover, almost the same decay times are observed in IR and OR recombination for each double ring, even if large variations are observed for different double rings. The analysis of carrier dynamics under large optical injection suggests similar conclusions. For both IR and OR emission we observe the increase of rise times when increasing P_{exc}, a clear evidence of saturation effects associated to state filling conditions. This dynamics is illustrated by the lines L1 and L2 present at high P_{exc} in the emission of the OR of double ring. In this case, the τ_D of L2 corresponds to the rise time of L1 implying that the two emissions come from energy states connected in a cascade type dynamics, where the higher states in the ladder act as feeders of the ground state. As reported before, the L1 and L2 lines of the OR multiplet are ascribed to single and multi exciton recombination and the difference in their emission energy stems from a different occupation of the ring, which changes the number of spectator excitons from several (line L2) to zero (line L1), thus making the dynamics of two lines strictly correlated. At the same time, comparing the time dependence of the IR and OR at $P_{\text{exc}} = 70$ W/cm^2 lines, we cannot find any correlation between rise and decay times of the emission, thus demonstrating that the carrier dynamics in the two rings is decoupled.

The decoupling of the carrier kinetics happens despite the proximity of the two rings in a double ring structure. Supposing that the exciton are free to move over the whole ring, the lack of coupling could be traced back to the lack of resonance conditions between states having the same angular momentum value inside the IR and OR.

8.5.1 Ring Shape Disorder Effects

As far as the energy relaxation efficiency of the electron-hole pairs photogenerated in the quantum rings is concerned, it is worth noting that the low P_{exc} measurements show relative large values of τ_R, compared to the quantum dot case. This can be explained as a less effective relaxation channel in the double ring, compared to the quantum dot case, despite the much closer spacing of the ring states which should prevent a phonon-bottleneck effect. These are quite controversial considerations, linked with the puzzling large broadening of 1 meV of the PL lines. It should be noted that quantum rings possess an electronic structure which is a crossover between the dot and the wire cases due to their rather peculiar annular shape [28], since the linear extension of a ring of 80 nm diameter is already quite large (250 nm). Therefore quantum rings can be considered as a warped analogous of quantum wires (QWi). Confinement energy fluctuations, whose magnitude is significantly lower than the exciton binding energy [84], can be observed in QWi, due to their size disorder. Such disorder principally affects the exciton center of mass (COM) part of the exciton wavefunction, giving rise to states with a spatially localized COM motion. The presence of such states has strong effects on the optical properties of QWi: the more evident is, naturally, the inhomogeneous broadening of the emission lines, which is related to the exciton energy disorder. Moreover, the exciton COM localization in QWi induces a serious reduction of the phonon scattering rates [84]. At the same time, as far as the kinetics of excitons in localized states induced by disorder is concerned, it is well known that an increase of the emission risetime associated to the exciton motion toward the state of minimum energy is observed. There are two main effects of the exciton COM localization on the coupling of OR and IR. On one side, excitonic COM localization makes possible the coupling between states with different angular momentum of the two rings because it partially relaxes the angular momentum conservation in exciton transitions. But, on the other side, it increases the average spatial separation between the exciton states, which are now localized in small regions of the rings, thus making the coupling between states belonging to OR and IR even more difficult.

8.5.2 Magneto-Photoluminescence

Magneto PL measurement were performed on single double rings [85]. A series of polarized PL spectra with varying magnetic field in Faraday geometry is shown in Fig. 8.22. The left and right series corresponds to σ_- and σ_+ polarized signals, respectively. Both polarizations have identical spectra at 0 T. They consist of a sharp

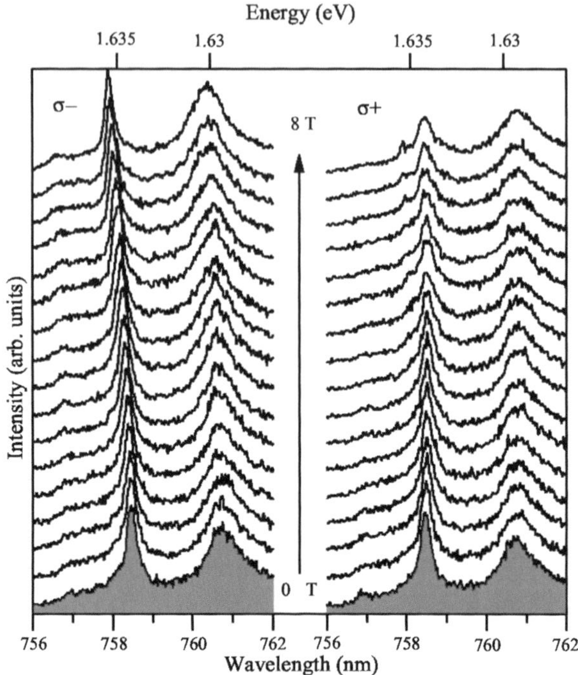

Fig. 8.22 Circularly polarized PL spectra of a double ring as a function of magnetic field from 0 T to 8 T (from bottom to top) with 0.5 T step. A *left panel* shows the magnetic dependence for the σ_- component, and a *right panel* shows that for the σ_+ component. (Reprinted from [85], Copyright Wiley-VCH Verlag GmbH and Co. KGaA (2009), reproduced with permission)

line at 758.6 nm and a broad line at 760.8 nm, which are assigned to the recombination of carriers inside the inner ring and the outer ring of a double ring.

Increasing the magnetic field, the σ_- spectra shift to shorter wavelengths, while the σ_+ spectra slightly move to longer wavelengths, as expected for the Zeeman effect of quantum ring excitons. According to a fit to a quadratic dependence on magnetic field, we estimated an exciton g factor $g_X = 2.4$ (± 0.3), and a diamagnetic coefficient 9.5 (± 1) $\mu eV/T^2$. The value of g_X is almost the same reported by previous studies on GaAs lens-shaped dots, while the diamagnetic coefficient is around two times larger for the present quantum ring than the dots [86]. Difference in the diamagnetic shift is probably due to lateral expansion of carrier confinement in quantum rings.

A striking feature is that the intensity of the σ_+ signals is significantly quenched by around 20 %, compared to that of the σ_- signals, at fields more than 6 T. Note that the σ_+ signal, which shows the PL quench, arises from the lower Zeeman level. We can therefore rule out the possibility of carrier relaxation between the Zeeman levels.

Two processes can be accountable for the PL quench at high magnetic field were proposed. The first is the exciton AB effect, which comes from level crossing between different orbital states in quantum ring. According to the theoretical proposal of Ref. [87], the transition strength decreases to zero when relative orbital angular momentum is changed from 0 to 1, being realized at a particular field. The field of level crossing is given by a equation, $\Delta\Phi/\Phi_0 = 1/2$, where $\Delta\Phi$ is the number of magnetic flux which inserts relative trajectory for electron and hole motion, and Φ_0 is a flux quantum. As a rough estimation which ignores Coulomb interactions and assuming a ring shape studied in [28], the transition field is expected to be ~ 10 T, in reasonable agreement with the present result.

The second process is related to the level crossing between bright and dark trions. Because of unintentional doping, our sample can possibly be p type, leading to the formation of positive trions. In this case, the ground-state trion is bright at zero field, and a dark trion with triplet holes should be present at a higher energy. The g factor for the relevant emission is given by $|g_c|$ for the bright cold trion, and $|g_c - 2g_v|$ for the dark hot trion, the latter becoming larger than the former. Thus, the bright-to-dark transition can appear at sufficiently high field, in contrast to the case of neutral excitons. The value of the field of this transition is related to the energy split from a cold trion to a hot trion, which is essentially given by the s-p orbital splitting for a hole. This splitting is on the order of 10 meV for conventional quantum dots, requiring an extremely high field for the transition. In the double rings, on the other hand, the orbital split is as small as ≤ 1 meV, because of their expanded shape. Therefore, the magnetically-induced transition can be realized at a field available in this experiment.

8.5.3 Single-Photon Emission

In order to investigate the peculiar phenomena typical of quantum rings [1, 3–5], they are usually considered as an ideal quantum system. However, in real semiconductor quantum devices disorder cannot be neglected. Effects related to disorder are largely discussed in quantum wells and quantum wire literature, leading to the concept of exciton localization [88]. On the other hand, carrier confinement in quantum dots occurs over a spatial region much smaller that the exciton Bohr radius. For this reason disorder does not play a role and the single quantum dot electronic properties are well described as a two level system. This is clearly demonstrated by antibunching measurements [89]. In quantum rings, the length of the circumference is usually larger than the exciton Bohr radius. This means that quantum rings electronic structure is a crossover between the dot and the wire cases, due to quantum rings rather peculiar annular shape which. Intensity time correlation experiments based on Hanbury, Brown and Twiss interferometer can address this issue. The second order correlation function $g^{(2)}$ well characterizes the quantum nature of the emitter by the presence or not of photon antibunching [89–91].

The second-order autocorrelation function for both the OR and IR emissions of an individual double ring is reported in [92] (see Figs. 8.23a and 8.23b). The reduced intensity of the peak at zero time delay indicates that there is a small probability of finding two or more photons inside each emitted pulse. This is the signature of a single-photon emitter under pulsed excitation. While $g^{(2)}(0) = 0$ is expected for a ideal two levels system, the condition of $g^{(2)}(0) < 0.5$ is required for a single-photon emitter and the condition $g^{(2)}(0) < 1$ still denotes the quantum nature of the

Fig. 8.23 Panel (**a**) and (**b**) are pulsed excitation measurement of $g^{(2)}(\tau)$ for OR and IR, respectively. The error bar associated to each peak is the coincidence count square root normalized to the average value of the peaks intensity. σ is the standard deviation of the average value of the peak intensity (except for the zero-delay one). Panel (**c**): Summary of all the measured $g^{(2)}(0)$ for IR (*squares*) and OR (*circles*); the *error bars* are the measured standard deviation σ. (Reprinted from [92])

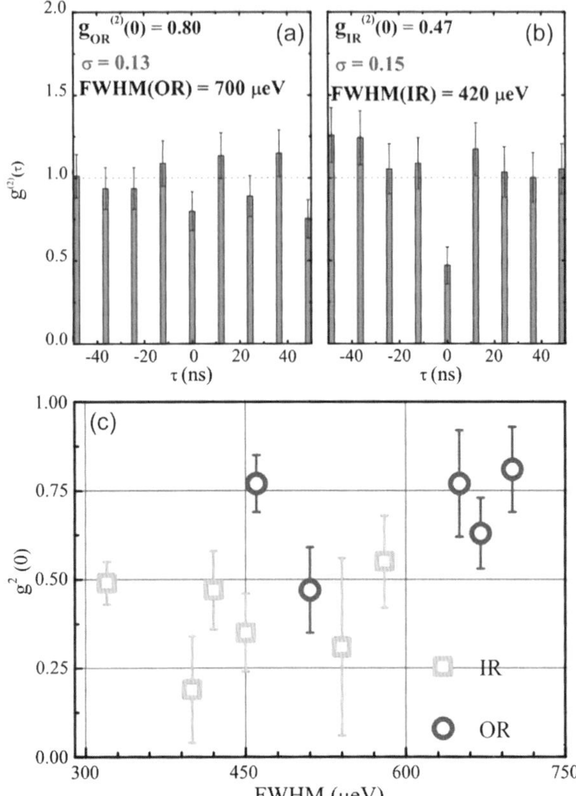

emitter [89, 93]. In our case, the correlation function at zero time delay, $g^{(2)}(0)$, is estimated to be 0.47 (± 0.15) for the IR recombination and 0.80 (± 0.13) for the OR recombination.

The same analysis was performed on several double rings (Fig. 8.23c), showing an increase of $g^{(2)}(0)$ with the increasing linewidth. The condition $g^{(2)}(0) < 0.5$ is usually fulfilled by the IR but not by the OR. Finally, cross correlation intensity measurements between IR and OR give $g_{IR\text{-}OR}^{(2)}(0) = 1$ within the experimental error. These results are not unexpected because of the independent dynamic already reported on these nanostructures [81].

On the basis of the previous results, Abbarchi et al. [92] concluded that IR is usually small enough to contain well-separated quantum states and gives rise to an optical transition with antibunching features. On the contrary the OR is sufficiently large to be influenced by structural disorder, resulting in an inhomogeneously broadening of the emission band. Therefore, even if disorder turns out to be an important issue for large quantum rings, for sufficiently small ring diameter, DE quantum rings can be considered as an almost ideal quantum system.

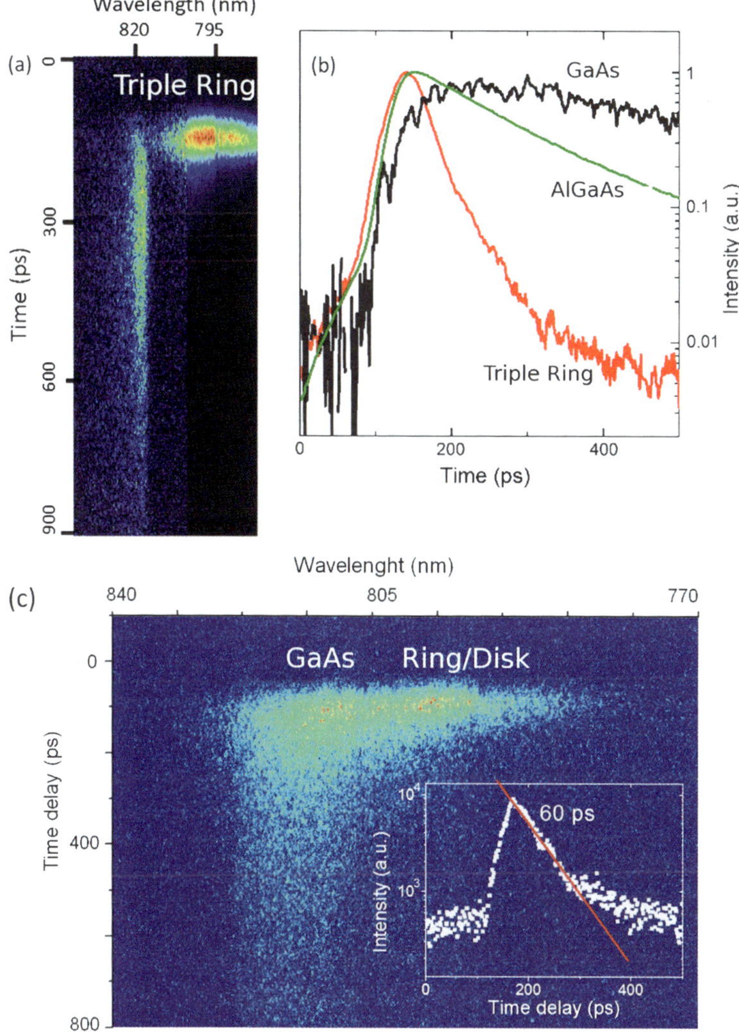

Fig. 8.24 (**a**): Temporally and spectrally resolved images of the triple ring sample. GaAs and triple ring emissions are labelled. Data from Ref. [94]. (**b**): Time resolved PL traces of the GaAs, AlGaAs and triple ring emissions. (**c**) Temporally and spectrally resolved image of the ring/disk emission. GaAs and ring/disk emissions are labelled. *Inset*: Time resolved PL traces of the ring/disk emission. The exponential fit, with a $\tau_D = 60$ ps is also shown (*red line*). Data from Ref. [96]

8.5.4 Fast Exciton Dynamics in Complex Nanostructures

As reported in Ref. [94] (see Fig. 8.24), triple rings show an extremely fast decay time $\tau_D = 40$ ps, to be compared with the usual DE quantum dot and quantum ring values, which range between 300 and 500 ps [16, 81, 95]. The short decay time is a general feature of all the triple ring structures measured in the sample. The fast optical response cannot be likely attributed to non radiative processed arising from the defected triple ring barrier, because of the much longer decay times of the GaAs ($\tau_{GaAs} = 300$ ps) and of the $Al_{0.3}Ga_{0.7}As$ barrier ($\tau_{AlGaAs} = 200$ ps), measured in the same sample at the same conditions.

Similar results are observed in the time resolved PL of ring/disk materials grown on Si/Ge substrates [96] (see Fig. 8.24). As in the case of triple ring, ring/disk show an extremely fast decay time of $\tau_D = 60$ ps and, also here, the observed fast optical response cannot be attributed to non radiative processed arising from the defected ring/disk barrier ($\tau_{AlGaAs} = 200$ ps, ($\tau_{GaAs} = 300$ ps)).

The puzzling short τ_D was attributed in both structures to the outcome of an intrinsic decay mechanism in the triple ring and ring/disk such as the strong electron-hole overlap and large transition dipole matrix element in such extended, complex nanostructures. Ring based nanostructures are then promising self-assembled materials for ultrafast optical switches for high-bit-rate operations.

8.6 Conclusions

The DE-quantum rings are an intriguing example of designable rings. Very complex ring structures can be devised and fabricated using the highly flexible DE process. Some examples of this large design flexibility in semiconductor nanostructure fabrication by DE are concentric multiple ring structures and coupled ring disks.

The growth mechanism is based on the control of Ga surface diffusion and As incorporation in tiny (nanometer size) Ga droplets formed in the first step of the DE process. Due to this multi-step kinetic controlled fabrication, DE-quantum rings are designable in size, shape and density.

Electronic structures of the DE-quantum rings have been identified using an optical approach. In the small quantum rings, carriers are quantized along two orthogonal degrees of freedom: radial motion and rotational motion. Optical transition takes place on recombination of an electron and a heavy hole, being in the ground state of the ring and in the excited radial state. In concentric double quantum rings, emission originating from the outer and from the inner ring are observed distinctly. Results of effective-mass calculations well reproduce the emission spectra.

As far as carrier dynamics is concerned, despite the small spatial separation between the two rings of the double rings, the carrier relaxation dynamics and the exciton kinetics in the IR and OR are decoupled. Moreover, a significant increase of the FWHM and rise time of the emission curve, with respect to the quantum dot case, has been observed. This phenomenology is attributed to the exciton center of mass localization induced by structural disorder along the ring.

PL emission from single coaxial GaAs rings as a function of magnetic field exhibited polarized signals which showed exciton AB effects and level crossing between bright and dark hot trions.

Since well separated quantum states are present in DE-quantum rings, optical transitions exhibiting antibunching features has been. Thus DE-quantum rings can be considered as an almost ideal quantum system. Circular nanostructures show a large variety of fascinating phenomena based on the quantum carrier confinement, of the utmost relevance for the possibility of their exploitation in the research of quantum computational devices.

DE technique therefore enables to produce singly-addressable, self-assembled, complex quantum ring structures that could pave the way to multiple two level states devices with switchable interaction.

Acknowledgements The authors gratefully acknowledge collaboration with K. Sakoda, M. Abbarchi, S. Adorno, S. Bietti, M. Gurioli, M. Guzzi, E. Grilli, G. Kido, X. Marie, F. Minami, C. Somaschini, B. Urbaszek, A. Vinattieri and K. Watanabe. This work was supported in Italy by the CARIPLO Foundation (prj. SOQQUADRO—no. 2011-0362) and in Japan by Grant-in-Aid from Ministry of Education, Sports, Science and Technology of Japan.

References

1. Y. Aharonov, D. Bohm, Phys. Rev. **115**(3), 485 (1959)
2. M. Grochol, F. Grosse, R. Zimmermann, Phys. Rev. B **74**(11), 115416 (2006)
3. W.H. Kuan, C.S. Tang, C.H. Chang, Phys. Rev. B **75**(15), 155326 (2007)
4. A. Lorke, R.J. Luyken, A.O. Govorov, J.P. Kotthaus, J.M. Garcia, P.M. Petroff, Phys. Rev. Lett. **84**(10), 2223 (2000)
5. A. Fuhrer, S. Lüscher, T. Ihn, T. Heinzel, K. Ensslin, W. Wegscheider, M. Bichler, Nature **413**(6858), 822 (2001)
6. L. Dias da Silva, J. Villas-Bôas, S. Ulloa, Phys. Rev. B **76**(15), 155306 (2007)
7. N. Kleemans, I. Bominaar-Silkens, V. Fomin, V. Gladilin, D. Granados, A. Taboada, J. García, P. Offermans, U. Zeitler, P. Christianen, J. Maan, J. Devreese, P. Koenraad, Phys. Rev. Lett. **99**(14), 146808 (2007)
8. J.M. Garcia, G. Medeiros-Ribeiro, K. Schmidt, T. Ngo, J.L. Feng, A. Lorke, J. Kotthaus, P.M. Petroff, Appl. Phys. Lett. **71**(14), 2014 (1997)
9. D. Granados, J.M. Garcia, Appl. Phys. Lett. **82**(15), 2401 (2003)
10. T. Raz, D. Ritter, G. Bahir, Appl. Phys. Lett. **82**(11), 1706 (2003)
11. S. Kobayashi, C. Jiang, T. Kawazu, H. Sakaki, Jpn. J. Appl. Phys. **43**(5B), L662 (2004)
12. R. Timm, H. Eisele, A. Lenz, L. Ivanova, G. Balakrishnan, D. Huffaker, M. Dähne, Phys. Rev. Lett. **101**(25), 256101 (2008)
13. F. Ding, L. Wang, S. Kiravittaya, E. Muller, A. Rastelli, O.G. Schmidt, Appl. Phys. Lett. **90**(17), 173104 (2007)
14. N. Koguchi, T. Satoshi, T. Chikyow, J. Cryst. Growth **111**, 688 (1991)
15. K. Watanabe, N. Koguchi, Y. Gotoh, Jpn. J. Appl. Phys. **39**(2), L79 (2000)
16. T. Kuroda, S. Sanguinetti, M. Gurioli, K. Watanabe, F. Minami, N. Koguchi, Phys. Rev. B **66**(12), 121302(R) (2002)
17. V. Mantovani, S. Sanguinetti, M. Guzzi, E. Grilli, M. Gurioli, K. Watanabe, N. Koguchi, J. Appl. Phys. **96**(8), 4416 (2004)
18. S. Sanguinetti, K. Watanabe, T. Tateno, M. Gurioli, P. Werner, M. Wakaki, N. Koguchi, J. Cryst. Growth **253**(1–4), 71 (2003)
19. S. Sanguinetti, M. Padovani, M. Gurioli, E. Grilli, M. Guzzi, A. Vinattieri, M. Colocci, P. Frigeri, S. Franchi, Appl. Phys. Lett. **77**(9), 1307 (2000)
20. D. Colombo, S. Sanguinetti, E. Grilli, M. Guzzi, L. Martinelli, M. Gurioli, P. Frigeri, G. Trevisi, S. Franchi, J. Appl. Phys. **94**(10), 6513 (2003)
21. S. Sanguinetti, M. Gurioli, E. Grilli, M. Guzzi, M. Henini, Appl. Phys. Lett. **77**(13), 1982 (2000)

22. S. Sanguinetti, T. Mano, M. Oshima, T. Tateno, M. Wakaki, N. Koguchi, Appl. Phys. Lett. **81**(16), 3067 (2002)
23. S. Sanguinetti, K. Watanabe, T. Tateno, M. Wakaki, N. Koguchi, T. Kuroda, F. Minami, M. Gurioli, Appl. Phys. Lett. **81**(4), 613 (2002)
24. M. Gurioli, S. Sanguinetti, M. Henini, Appl. Phys. Lett. **78**(7), 931 (2001)
25. T. Mano, T. Kuroda, K. Mitsuishi, Y. Nakayama, T. Noda, K. Sakoda, Appl. Phys. Lett. **93**, 203110 (2008)
26. T. Mano, M. Abbarchi, T. Kuroda, B. McSkimming, A. Ohtake, K. Mitsuishi, Y. Nakayama, T. Noda, K. Sakoda, Appl. Phys. Express **3**, 065203 (2010)
27. T. Mano, T. Kuroda, S. Sanguinetti, T. Ochiai, T. Tateno, J. Kim, T. Noda, M. Kawabe, K. Sakoda, G. Kido, N. Koguchi, Nano Lett. **5**(3), 425 (2005)
28. T. Kuroda, T. Mano, T. Ochiai, S. Sanguinetti, K. Sakoda, G. Kido, N. Koguchi, Phys. Rev. B **72**(20), 205301 (2005)
29. M. Yamagiwa, T. Mano, T. Kuroda, T. Tateno, K. Sakoda, G. Kido, N. Koguchi, F. Minami, Appl. Phys. Lett. **89**(11), 113115 (2006)
30. K.A. Sablon, J.H. Lee, Z.M. Wang, J.H. Shultz, G.J. Salamo, Appl. Phys. Lett. **92**(20), 203106 (2008)
31. Z.M. Wang, B. Liang, K.A. Sablon, J. Lee, Y.I. Mazur, N.W. Strom, G.J. Salamo, Small **3**(2), 235 (2007)
32. T. Mano, T. Kuroda, K. Mitsuishi, M. Yamagiwa, X.J. Guo, K. Furuya, K. Sakoda, N. Koguchi, J. Cryst. Growth **301–302**, 740 (2007)
33. T. Mano, T. Kuroda, K. Kuroda, K. Sakoda, J. Nanophotonics **3**, 031605 (2009)
34. N. Pankaow, S. Thainoi, S. Panyakeow, S. Ratanathammaphan, J. Cryst. Growth **323**(1), 282 (2011)
35. W. Jevasuwan, P. Boonpeng, S. Panyakeow, S. Ratanathammaphan, Microelectron. Eng. **87**(5–8), 1416 (2010)
36. N. Pankaow, S. Panyakeow, S. Ratanathammaphan, J. Cryst. Growth **311**(7), 1832 (2009)
37. C. Zhao, Y.H. Chen, B. Xu, C.G. Tang, Z.G. Wang, F. Ding, Appl. Phys. Lett. **92**(6), 063122 (2008)
38. T. Noda, T. Mano, M. Jo, T. Kawazu, H. Sakaki, J. Appl. Phys. **112**(6), 063510 (2012)
39. P. Alonso-Gonzalez, L. Gonzalez, D. Fuster, Y. Gonzalez, A.G. Taboada, J.M. Ripalda, A.M. Beltran, D.L. Sales, T. Ben, S.I. Molina, Cryst. Growth Des. **9**, 1216 (2009)
40. T. Noda, T. Mano, Appl. Surf. Sci. **254**(23), 7777 (2008)
41. C. Somaschini, S. Bietti, N. Koguchi, S. Sanguinetti, Nano Lett. **9**(10), 3419 (2009)
42. C. Somaschini, S. Bietti, N. Koguchi, S. Sanguinetti, Nanotechnology **22**(18), 185602 (2011)
43. C. Somaschini, S. Bietti, S. Sanguinetti, N. Koguchi, A. Fedorov, Nanotechnology **21**(12), 125601 (2010)
44. Z.M. Wang, B.L. Liang, K.A. Sablon, G.J. Salamo, Appl. Phys. Lett. **90**(11), 113120 (2007)
45. A. Ohtake, Surf. Sci. Rep. **63**(7), 295 (2008)
46. T. Mano, N. Koguchi, J. Cryst. Growth **278**(1–4), 108 (2005)
47. S. Bietti, C. Somaschini, S. Sanguinetti, N. Koguchi, G. Isella, D. Chrastina, Appl. Phys. Lett. **95**(24), 241102 (2009)
48. Y. Horikoshi, M. Kawashima, H. Yamaguchi, Jpn. J. Appl. Phys. **27**(2), 169 (1988)
49. A. Ohtake, N. Koguchi, Appl. Phys. Lett. **83**(25), 5193 (2003)
50. K. Ohta, T. Kojima, T. Nakagawa, J. Cryst. Growth **95**, 71 (1989)
51. A. Ohtake, P. Kocán, K. Seino, W. Schmidt, N. Koguchi, Phys. Rev. Lett. **93**(26), 266101 (2004)
52. K. Kanisawa, J. Osaka, S. Hirono, N. Inoue, Appl. Phys. Lett. **58**, 2363 (1991)
53. C. Somaschini, S. Bietti, A. Fedorov, N. Koguchi, S. Sanguinetti, Nanoscale Res. Lett. **5**(12), 1897 (2010)
54. C. Somaschini, S. Bietti, A. Fedorov, N. Koguchi, S. Sanguinetti, Nanoscale Res. Lett. **5**, 1865 (2010)
55. J. Wu, Z. Li, D. Shao, M.O. Manasreh, V.P. Kunets, Z.M. Wang, G.J. Salamo, B.D. Weaver, Appl. Phys. Lett. **94**(17), 171102 (2009)

56. J. Wu, D. Shao, Z. Li, M.O. Manasreh, V.P. Kunets, Z.M. Wang, G.J. Salamo, B.D. Weaver, Appl. Phys. Lett. **95**(7), 071908 (2009)
57. C. Somaschini, S. Bietti, N. Koguchi, S. Sanguinetti, Appl. Phys. Lett. **97**, 203109 (2010)
58. S. Huang, Z. Niu, Z. Fang, H. Ni, Z. Gong, J. Xia, Appl. Phys. Lett. **89**(3), 031921 (2006)
59. J.H. Lee, Z.M. Wang, Z.Y. Abuwaar, N.W. Strom, G.J. Salamo, Nanotechnology **17**(15), 3973 (2006)
60. C. Deparis, J. Massies, J. Cryst. Growth **108**, 157 (1991)
61. J.Y. Marzin, G. Bastard, Solid State Commun. **92**(5), 437 (1994)
62. M. Califano, P. Harrison, Phys. Rev. B **61**(16), 10959 (2000)
63. J.W. Luo, G. Bester, A. Zunger, Phys. Rev. B **79**(12), 125329 (2009)
64. R. Heitz, H. Born, F. Guffarth, O. Stier, A. Schliwa, A. Hoffmann, D. Bimberg, Phys. Status Solidi (a) **190**(2), 499 (2002)
65. O. Stier, M. Grundmann, D. Bimberg, Phys. Rev. B **59**(8), 5688 (1999)
66. S. Sanguinetti, K. Watanabe, T. Kuroda, F. Minami, Y. Gotoh, N. Koguchi, J. Cryst. Growth **242**(3–4), 321 (2002)
67. S. Bietti, C. Somaschini, E. Sarti, N. Koguchi, S. Sanguinetti, G. Isella, D. Chrastina, A. Fedorov, Nanoscale Res. Lett. 1650–1653 (2010)
68. T. Tsuchiya, S. Katayama, Solid-State Electron. **42**, 1523 (1998)
69. M. Abbarchi, T. Kuroda, T. Mano, K. Sakoda, C.A. Mastrandrea, A. Vinattieri, M. Gurioli, T. Tsuchiya, Phys. Rev. B **82**(20), 201301(R) (2010)
70. J.G. Keizer, J. Bocquel, P.M. Koenraad, T. Mano, T. Noda, K. Sakoda, Appl. Phys. Lett. **96**(6), 062101 (2010)
71. S. Sanguinetti, T. Mano, A. Gerosa, C. Somaschini, S. Bietti, N. Koguchi, E. Grilli, M. Guzzi, M. Gurioli, M. Abbarchi, J. Appl. Phys. **104**(11), 113519 (2008)
72. S. Bietti, C. Somaschini, M. Abbarchi, N. Koguchi, S. Sanguinetti, E. Poliani, M. Bonfanti, M. Gurioli, A. Vinattieri, T. Kuroda, T. Mano, K. Sakoda, Phys. Status Solidi (c) **6**(4), 928 (2009)
73. C. Somaschini, S. Bietti, S. Sanguinetti, N. Koguchi, A. Fedorov, M. Abbarchi, M. Gurioli, IOP Conf. Ser., Mater. Sci. Eng. **6**, 012008 (2009)
74. M. Bissiri, G. Baldassarri Höger Von Högersthal, M. Capizzi, P. Frigeri, S. Franchi, Phys. Rev. B **64**(24), 245337 (2001)
75. A.S. Bhatti, M. Capizzi, A. Frova, M. Bissiri, G. Baldassarri Höger Von Högersthal, P. Frigeri, S. Franchi, Phys. Rev. B **62**(7), 4642 (2000)
76. R. Heitz, I. Mukhametzhanov, O. Stier, A. Madhukar, D. Bimberg, Phys. Rev. Lett. **83**, 4654 (1999)
77. R. Heitz, M. Grundmann, N.N. Ledentsov, L. Eckey, M. Veit, D. Bimberg, Appl. Phys. Lett. **68**(3), 361 (1996)
78. S. Sanguinetti, D. Colombo, M. Guzzi, E. Grilli, M. Gurioli, L. Seravalli, P. Frigeri, S. Franchi, Phys. Rev. B **74**(20), 1 (2006)
79. R.J. Warburton, C. Schaflein, D. Haft, F. Bickel, A. Lorke, K. Karrai, J. Garcia, W. Schoenfeld, P. Petroff, Nature **405**(6789), 926 (2000)
80. M. Abbarchi, F. Troiani, C. Mastrandrea, G. Goldoni, T. Kuroda, T. Mano, K. Sakoda, N. Koguchi, S. Sanguinetti, A. Vinattieri, M. Gurioli, Appl. Phys. Lett. **93**(16), 162101 (2008)
81. S. Sanguinetti, M. Abbarchi, A. Vinattieri, M. Zamfirescu, M. Gurioli, T. Mano, T. Kuroda, N. Koguchi, Phys. Rev. B **77**(12), 125404 (2008)
82. E. Dekel, D. Gershoni, E. Ehrenfreund, J.M. Garcia, P.M. Petroff, Phys. Rev. B **61**(16), 11009 (2000)
83. E. Dekel, D.V. Regelman, D. Gershoni, E. Ehrenfreund, Phys. Rev. B **62**(16), 11038 (2000)
84. A. Feltrin, J.L. Staehli, B. Deveaud, V. Savona, Phys. Rev. B **69**, 233309 (2004)
85. T. Kuroda, T. Mano, T. Belhadj, B. Urbaszek, T. Amand, X. Marie, S. Sanguinetti, K. Sakoda, N. Koguchi, Phys. Status Solidi (b) **246**(4), 861 (2009)
86. M. Abbarchi, T. Kuroda, T. Mano, K. Sakoda, M. Gurioli, Phys. Rev. B **81**(3), 035334 (2010)
87. A.O. Govorov, S.E. Ulloa, K. Karrai, R.J. Warburton, Phys. Rev. B **66**(8), 081309 (2002)
88. V. Savona, J. Phys. Condens. Matter **19**(29), 295208 (2007)

89. C. Becher, A. Kiraz, P. Michler, A. Imamoglu, W. Schoenfeld, P. Petroff, L. Zhang, E. Hu, Phys. Rev. B **63**(12), 1 (2001)
90. T. Kuroda, M. Abbarchi, T. Mano, K. Watanabe, M. Yamagiwa, K. Kuroda, K. Sakoda, G. Kido, N. Koguchi, C. Mastrandrea, L. Cavigli, M. Gurioli, Y. Ogawa, F. Minami, Appl. Phys. Express **1**, 042001 (2008)
91. T. Kuroda, T. Belhadj, M. Abbarchi, C. Mastrandrea, M. Gurioli, T. Mano, N. Ikeda, Y. Sugimoto, K. Asakawa, N. Koguchi, K. Sakoda, B. Urbaszek, T. Amand, X. Marie, Phys. Rev. B **79**(3), 035330 (2009)
92. M. Abbarchi, C. Mastrandrea, A. Vinattieri, S. Sanguinetti, T. Mano, T. Kuroda, N. Koguchi, K. Sakoda, M. Gurioli, Phys. Rev. B **79**(8), 085308 (2009)
93. R. Loudon, *The Quantum Theory of Light* (Oxford University Press, New York, 1983)
94. M. Abbarchi, L. Cavigli, C. Somaschini, S. Bietti, M. Gurioli, A. Vinattieri, S. Sanguinetti, Nanoscale Res. Lett. **6**(1), 569 (2011)
95. M. Abbarchi, M. Gurioli, S. Sanguinetti, M. Zamfirescu, A. Vinattieri, N. Koguchi, Phys. Status Solidi (c) **3**(11), 3860 (2006)
96. L. Cavigli, S. Bietti, M. Abbarchi, C. Somaschini, A. Vinattieri, M. Gurioli, A. Fedorov, G. Isella, E. Grilli, S. Sanguinetti, J. Phys. Condens. Matter **24**(10), 104017 (2012)

Part II
Aharonov-Bohm Effect for Excitons

Chapter 9
New Versions of the Aharonov-Bohm Effect in Quantum Rings

A.V. Chaplik and V.M. Kovalev

Abstract The Aharonov-Bohm effect in semiconductor quantum rings is considered with accounting for electron-electron interaction, spin-orbit coupling, influence of acoustic and electro-magnetic external fields. We discuss also oscillatory dependence of characteristic parameters of composite particles and collective excitations on external magnetic field.

9.1 Introduction

A 'canonical' manifestation of the Aharonov-Bohm (AB) effect occurs in an experiment with an electron beam embracing infinitely thin solenoid that bears the magnetic flux Φ (see Fig. 9.1). The scattering cross-section of the electrons periodically depends on Φ with the period $\Phi_0 = hc/e$—the magnetic flux quantum (h—the Plank constant, c—speed of light, e—the electron charge). Just this mental experiment has been analyzed by Aharonov and Bohm in their classical work [1]. Since then a great variety of papers has been published demonstrating various aspects of this remarkable quantum phenomena (for review see [2]). However a really new breathing the AB effect acquired when it was established in solid state physics for nanostructures—quantum dots, quantum rings (QR) and nanotubes. As to quantum ring, the most direct and the simplest manifestation of the AB effect is oscillations of magnetization of QRs [3] and periodic dependence of the current flowing through the QR placed in a perpendicular magnetic field **H** as depicted in Fig. 9.2 [4].

The present paper is devoted to the AB-type effects in more complicate situations. We will consider system of many interacting electrons in QR, the influence

A.V. Chaplik (✉) · V.M. Kovalev
Institute of Semiconductor Physics, 630090 Novosibirsk, Russia
e-mail: chaplik@isp.nsc.ru

V.M. Kovalev
e-mail: vadimkovalev@isp.nsc.ru

A.V. Chaplik · V.M. Kovalev
Novosibirsk State University, 630090 Novosibirsk, Russia

V.M. Fomin (ed.), *Physics of Quantum Rings*, NanoScience and Technology,
DOI 10.1007/978-3-642-39197-2_9, © Springer-Verlag Berlin Heidelberg 2014

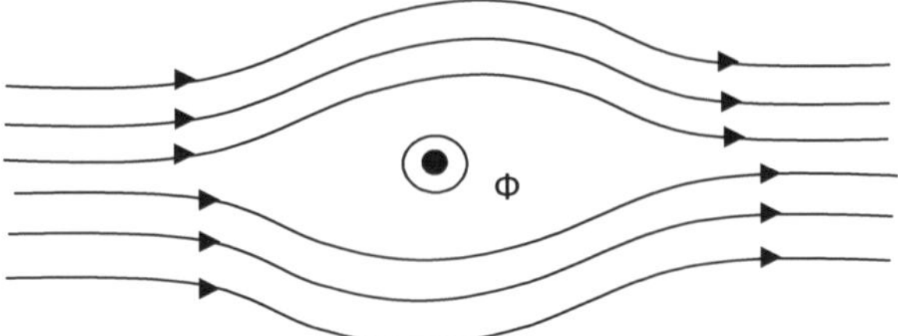

Fig. 9.1 An Aharonov-Bohm effect

Fig. 9.2 Schematic sketch of
the typical ring lateral
transport experiment setup

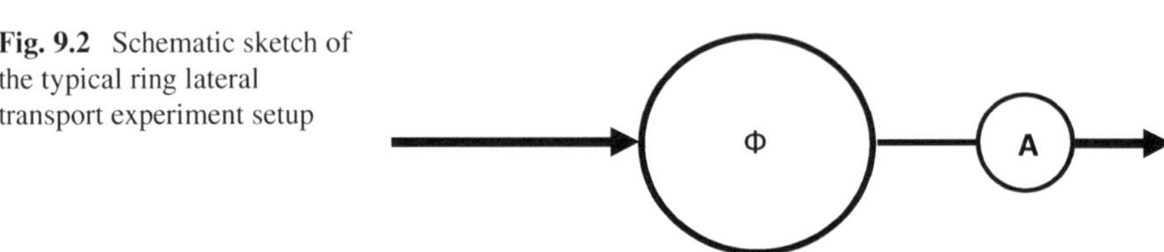

of the spin-orbit interaction (SOI) on the AB oscillations, the AB effect for composite particles (like excitons and trions) and for collective excitations—plasmons. Excitons in a QR under external electromagnetic radiation, electronic ring in the field of the surface acoustic wave will be also analysed. Moreover we will discuss the 'vertical' magnetotransport (as opposite to the one of Fig. 9.2)—the AB effect for tunnel current flowing perpendicularly to the plane of the ring residing in a dielectric barrier. We intentively do not discuss a very interesting and experimentally thoroughly investigated area: AB effect for radially polarized excitons in QR. This subject is considered in the present book—see Chaps. 10, 11, and 12, as well as the papers [5, 6] and [7].

9.2 Electron-Electron Interaction and Persistent Current in a Quantum Ring [8]

In the seminal work by Büttiker, Imry and Landauer the idea of the persistent current in a QR has been formulated for the first time [9]. Büttiker et al. considered a single electron picture and the experiment [3] showed (unexpectedly) that the results observed are described well by the theory of free, i.e., noninteracting, electrons. Significantly, the mean free path of the electrons in the original material was four or five times the diameter of the rings; i.e., the experiments were carried out in the ballistic regime.

These experiments stimulated several theoretical papers, in which authors demonstrated, with various degrees of generality, that the persistent current was independent of an electron-electron interaction. For example, Müller-Groeling et al. [10, 11] asserted that an electron-electron interaction W does not affect persistent currents if the commutation relation $[L_z, W] = 0$ holds, where L_z is the projection of the resultant orbital angular momentum of the system onto the magnetic field. The field was directed perpendicular to the plane of the ring. It is easy to see that this assertion is too strong. The condition offered is satisfied by any interaction which depends on only pairs of differences between the coordinates of the electrons.

Let us assume that there are only two electrons, which interact with each other not by a simple Coulomb mechanism but by some 'molecular' mechanism (e.g., a Lennard-Jones mechanism). The electrons then form a pair, and the equilibrium distance between the electrons is determined by the parameters of the potential. This distance may be much smaller than the radius of the ring. In this case, the oscillations of the persistent current as a function of the magnetic flux will clearly occur with a period $\Phi_0/2$ which is by no means the same as in the case of noninteracting electrons, in which the period is Φ_0.

Krive et al. [12] studied a 1D ring Wigner crystal, using as a model Hamiltonian a continuous string, which is valid in the long-wave approximation. They derived correct expressions for the ballistic persistent current, but their associated comments give the impression that there is agreement with the free-electron theory in the high-temperature limit, also. However, a simple comparison with that theory [13] reveals a substantial difference (the argument of an exponential function!) between these two expressions at $T \geq NB$, where N is the number of electrons in the ring, and $B = \hbar^2/2ma^2$ is the rotational quantum (m is the effective mass and a is the radius of the QR).

In the present paragraph we consider a discrete model: an N-electron molecule that forms, quite naturally, at a sufficiently low density of particles, at which the correlation energy is much larger than the Fermi energy. A corresponding estimate is found by comparing the kinetic energy $N^2\hbar^2/(ma^2)$ with the potential energy $N\tilde{e}^2/a$, where $\tilde{e}$ is the effective charge, i.e., the charge incorporating the dielectric constant. We see that Coulomb effects are predominant under the condition $a > Na_B$, where a_B is the effective Bohr radius.

A ring molecule has a single rotational degree of freedom and $N - 1$ vibrational degrees of freedom (the ballistic regime, without pinning). It is easy to show that the latter degrees of freedom do not affect the total persistent current. To show this, we start from the original Hamiltonian of the model in a magnetic field H:

$$\hat{H} = \sum_{k=1}^{N} \frac{1}{2m}\left(\hat{P}_k - \frac{e}{c}A_k\right)^2 + W \tag{9.1}$$

In this Hamiltonian we transform to Jacobi variables

$$\varphi_0 = \frac{1}{\sqrt{N}}\sum_{k=1}^{N}\varphi_k, \qquad \theta_1 = \frac{\varphi_1 - \varphi_2}{\sqrt{1\cdot 2}}, \qquad \theta_2 = \frac{\varphi_2 + \varphi_3 - 2\varphi_3}{\sqrt{2\cdot 3}},\dots \tag{9.2}$$

Here φ_k are azimuthal angles which specify the positions of the particles on a circle; we have $\hat{P}_k = -i(\hbar/a)\partial_{\varphi_k}$ and we also have $A_k = Ha/2$ (A_k is independent of k). In these new variables, and for the new unknown function

$$\chi = \exp\left(-i\varphi_0\lambda\sqrt{N}\right)\psi(\varphi_0, \theta_1, \ldots, \theta_{N-1}), \quad \lambda = \Phi/\Phi_0, \tag{9.3}$$

we find an equation which does not contain the magnetic field

$$-B\left(\frac{\partial^2}{\partial\varphi_0^2} + \sum_{k=1}^{N-1}\frac{\partial^2}{\partial\theta_k^2}\right)\chi + W\chi = E\chi \tag{9.4}$$

The obvious solution is $\chi = \exp(i\kappa\varphi_0)f(\theta_k)$, and the energy is $E = E_{\text{int}} + B\kappa^2$, where E_{int} is the purely vibrational energy corresponding to the internal degrees of freedom θ_k. This vibrational energy does not depend on the magnetic flux Φ. The condition for periodicity under rotation of the entire system as a whole through an angle 2π ($\varphi_0 \to 2\pi\sqrt{N}, \theta_k \to \theta_k$) yields an expression from which we can determine κ

$$2\pi(\lambda\sqrt{N} + \kappa) = 2\pi J/\sqrt{N}, \tag{9.5}$$

where J is an integer. Hence the rotational energy is

$$E_{rot} = E - E_{\text{int}} = \frac{B}{N}(J - N\lambda)^2. \tag{9.6}$$

Expression (9.6) corresponds to the energy of a plane rotator with a radius a, a mass Nm, and a charge Ne in a magnetic field H. In general, this expression yields a persistent current which oscillates with a period Φ_0/N as a function of H. However, the symmetry of the Wigner molecule and the Pauli principle impose restrictions on the values of the rotational quantum number J. For a two-electron 'molecule', we thus have the familiar problem of ortho and para states: $J = 0, \pm2, \pm4, \ldots$, if the total spin is $S = 0$; or $J = \pm1, \pm3, \ldots$, for $S = 0$. In the case of three electrons, the situation is analogous to the NH_3 molecule [14]. In the superortho state ($S = 3/2$), the symmetry of the term is A_2, and the possible values are $J = 0, \pm3, \pm6, \ldots$. In the ortho-para state ($S = 1/2$), the symmetry is E, and we have $J = \pm1, \pm2, \pm4, \pm5, \ldots$.

We consider an N-electron ring molecule in the superortho state $S = N/2$. By virtue of the symmetry of the equilibrium positions of the particles, a rotation through an angle $2\pi/N$, which contributes a factor $\exp(i2\pi J/N)$ to the wave function, is equivalent to a cyclic permutation of N identical fermions. Such a permutation introduces a factor $(-1)^{N-1}$. We can thus find the possible values of J

$$J = nN + N(N+1)/2 \tag{9.7}$$

where n is an arbitrary integer. For odd N we find $J = N\times$(an integer); for even N we should have $J = N\times$(a half-integer). Other values of J are possible only if $S < N/2$.

We would also point out the substantial dependence of the properties of a 1D system of fermions on the parity of the number of particles—a dependence which is well known from other examples.

To find the current it is sufficient to calculate the rotational part of the partition function of the system. Since we are dealing with a single molecule (a single ring) at thermal equilibrium with its surroundings (a heterostructure plus a substrate, etc.), we should use a Gibbs distribution for the subsystem. For odd N we have

$$Z_N = \sum_{k=-\infty}^{+\infty} \exp\left[-\beta B N (k - \lambda)^2\right], \quad \beta = 1/T, \tag{9.8}$$

and for even N we have

$$Z_N = \sum_{k=-\infty}^{+\infty} \exp\left[-\beta B N (k - \lambda - 1/2)^2\right]. \tag{9.9}$$

At low temperatures $(T \ll N B)$, only a single term is important in series (9.8) and (9.9); for the persistent current we find the expression

$$I = cT \frac{\partial \ln Z}{\partial \Phi} = (-1)^N \frac{2eBN}{h} \left[\frac{\Phi}{\Phi_0} - \mathrm{Int}\left(\frac{\Phi}{\Phi_0} + \frac{1}{2}\right)\right], \tag{9.10}$$

which is indeed the same as the result for noninteracting electrons. Under the condition $T \geq N B$, however, we find the following from (9.8) and (9.9)

$$I = (-1)^N \frac{4\pi cT}{\Phi_0} \exp\left(-\frac{\pi^2 T}{N B}\right) \sin\left(2\pi \frac{\Phi}{\Phi_0}\right). \tag{9.11}$$

Here we have written only the leading term of the Fourier series of I in Φ/Φ_0, which follows from (9.8) and (9.9) after we apply the Poisson summation formula. Expression (9.11) is the same as that derived by Krive et al. [12], who examined a system of 'spinless' fermions. In other words, they actually considered the case of a total spin polarization $(S = N/2)$. However, that expression differs from the high-temperature limit for free electrons (see Ref. [13]; in the expression in that paper, we need to make the obvious transformation from a thin cylinder to a 1D ring)

$$I = -\frac{4\pi cT}{\Phi_0} N \exp\left(-\frac{\pi^2 T}{B}\right) \sin\left(2\pi \frac{\Phi}{\Phi_0}\right) \tag{9.12}$$

in both the T dependence and the N dependence. This situation is completely natural, since expression (9.11) corresponds to high temperatures, but still to a molecule which has not dissociated into free particles.

We see from the discussion above that the results for highly correlated electrons are by no means always the same as the results for free electrons. They are the same only at a sufficiently low temperature and this coincidence is caused by the high-symmetry equilibrium configuration of the particles, that itself is a consequence of the strong repulsion of the particles. Furthermore, as can easily be verified in the cases of two- and three-electron systems, the persistent current depends on the total spin in the case $T = 0$, changing sign as we go from $S = 1$ to $S = 0$ or from $S = 3/2$ to $S = 1/2$. These points must be kept in mind in a discussion of the problem in the model of spinless fermions, which is frequently used.

9.3 Electronic Absorption of the Surface Acoustic Wave by QR in a Magnetic Field [15]

We have shown above that PC as a function of T and N behaves differently for the Wigner molecule and for free electrons. This is the D.C. (zero frequency) phenomenon. In this paragraph we consider yet another possibility to distinguish the two regimes: electron transitions in QRs caused by a high frequency spatially nonuniform field, namely the piezoelectric field accompanying a surface acoustic wave (SAW). The crucial point is different mechanism of the electronic absorption: for free electrons the dipole matrix element has nonzero value while in the Wigner molecule the absorption is allowed due to only quadruple and higher multipolarity transitions.

To be specific consider the following structure. On the GaAs substrate an array of the InAs QR is fabricated. The SAW propagates over the surface and a constant uniform magnetic field is applied perpendicular to the surface. For the sake of simplicity we will consider the model of 1D ring and we will ignore spin-dependent effects: Zeemann energy and SO interaction. These effects result in small splitting of the single-particle energy levels while our goal is to demonstrate the importance of the Coulomb interaction and qualitative difference in the SAW absorption by independent particles and by a strongly correlated system—the ring Wigner molecule.

9.3.1 Noninteracting Electrons

In this section, we consider briefly the situation where only a single electron is present in a quantum ring. Let the substrate occupy the half-space $z < 0$, and let a homogeneous magnetic field $H = (0, 0, H)$ be applied perpendicularly to the surface. The magnetic flux penetrating through the quantum ring is defined as $\Phi = H\pi a^2$. The SAW propagates in the (x, y) plane and has a wave vector of $\mathbf{q} = (\omega/s, 0)$. Here, ω and s are the SAW frequency and velocity, respectively. We can then represent the perturbation, which acts on an electron, as

$$V(x) = V_0 \exp(iqx - i\omega t) + c.c. \tag{9.13}$$

where $V_0 = e\varphi_{\text{SAW}}$ for piezoelectric interaction; here, φ_{SAW} is the amplitude of the SAW electric field potential. Henceforth, we will drop multipliers of the form $\exp(i\omega t)$.

The Schrödinger equation has the following form in the model of a 1D quantum ring

$$-\frac{\hbar^2}{2ma^2}\left(\frac{d}{d\varphi} - i\frac{\Phi}{\Phi_0}\right)^2 \Psi_j(\varphi) = \varepsilon_j \Psi_j(\varphi). \tag{9.14}$$

The wave functions and the spectrum are given by

$$\Psi_j(\varphi) = (2\pi)^{-1/2} \exp(ij\varphi),$$

$$\varepsilon_j = \frac{\hbar^2}{2ma^2}\left(j + \frac{\Phi}{\Phi_0}\right)^2, \quad j = 0, \pm 1, \pm 2, \ldots, \tag{9.15}$$

where $\Phi_0 = hc/e$ is the magnetic flux quantum. In order to calculate the SAW absorption coefficient, we use the expression

$$\Gamma = \frac{2\pi\omega}{I_0} \sum_{jj'} |V_{j'j}|^2 \delta(\varepsilon_{j'} - \varepsilon_j - \omega). \tag{9.16}$$

Here, I_0 is the SAW intensity and $V_{j'j}$ is the matrix element of perturbation (9.13) over states (9.15). In expression (9.16), the magnetic flux appears only in combination $j + \Phi/\Phi_0$ in terms of the energy levels ε_j. It follows immediately that Γ depends periodically on Φ with a period of Φ_0. The results of calculating the absorption coefficient using formula (9.16) are discussed later.

9.3.2 Two-Electron Wigner Molecule

The Hamiltonian of a system consisting of two interacting electrons in a quantum ring is given by

$$H = \sum_{\alpha=1,2} \frac{1}{2m}\left(p_\alpha + \frac{e}{c}A_\alpha\right)^2 + V(x_1, x_2) + U\big(|x_1 - x_2|\big), \tag{9.17}$$

where $V(x_1, x_2) = V_0(\exp(iqx_1 - i\omega t) + \exp(iqx_2 - i\omega t)) + c.c.$ is the potential of electron interaction with a SAW, $U(|x_1 - x_2|)$ is the potential of Coulomb interaction between electrons, $p_\alpha = -i\hbar\nabla_\alpha$ is the momentum operator of the αth electron, and $A_\alpha = Ha/2$ is the angular component of the vector potential at the ring. On separating the variables in a polar coordinate system, we obtain the equations

$$-\frac{\hbar^2}{4ma^2}\left(\frac{d}{d\varphi_c} - 2i\frac{\Phi}{\Phi_0}\right)^2 F(\varphi_c) = E_1 F(\varphi_c),$$

$$-\frac{\hbar^2}{ma^2}\frac{d^2}{d\theta^2}G(\theta) + U(\theta)G(\theta) = E_2 G(\theta) \tag{9.18}$$

for the center-of-mass motion and relative motion of electrons, respectively. In (9.18)

$$U(\theta) = \frac{e^2}{\varepsilon a\sqrt{2(1 - \cos\theta)}}, \tag{9.19}$$

$$E = E_1 + E_2$$

is the interaction energy and the total energy of the system, $\varphi_c = (\varphi_1 + \varphi_2)/2$ is the center-of-mass coordinate, and $\theta = \varphi_1 - \varphi_2$ is the coordinate of relative motion of electrons.

The wave function of the system is expressed as

$$\psi(\varphi_c, \theta) = (2\pi)^{-1/2} F(\varphi_c) G(\theta), \quad F(\varphi_c) = \exp(i J \varphi_c) \tag{9.20}$$

where $G(\theta)$ is the solution to the second equation in (9.18) with a periodic potential and, as such, should have the Bloch form

$$G(\theta) = e^{ip\theta} u(\theta), \quad u(\theta) = u(\theta + 2\pi).$$

In order to solve the second equation in (9.18), we consider the limiting case where the Coulomb interaction potential is much larger than the kinetic energy of the electrons; i.e., we require that the following condition be satisfied

$$\frac{e^2}{\varepsilon a} \gg \frac{\hbar^2}{ma^2}.$$

It is easy to understand that the validity of this condition makes it possible to use the tight-binding approximation

$$G(\theta) = \frac{1}{N} \sum_{m=-N/2}^{N/2} e^{ip\theta_m} \chi(\theta - \theta_m), \quad N \to +\infty, \tag{9.21}$$

where $\theta_m = \pi(2m + 1)$ and $-1/2 < p \le 1/2$.

In the same approximation, we can represent the potential $U(\theta)$ as

$$U(\theta) = \frac{e^2}{\varepsilon a \sqrt{2(1 - \cos\theta)}}$$

$$\approx \frac{e^2}{2\varepsilon a} + \sum_{m=-\infty}^{+\infty} \frac{m\Omega^2 a^2}{2} (\theta - \theta_m)^2 \Theta(\pi - |\theta - \theta_m|), \tag{9.22}$$

where $\Omega^2 = e^2/8\pi\varepsilon_0\varepsilon m^* a^3$ and $\Theta(x)$ is the unit-step function. As a result, the Schrödinger equation for relative motion becomes oscillatory and we obtain

$$E_2 = \hbar\Omega(n + 1/2) + \Delta(p), \quad n = 0, 1, 2\ldots,$$

$$\chi_n(\theta - \theta_m) = C_m \exp\left(-\frac{(\theta - \theta_m)^2}{2x_0^2}\right) H_n\left(\frac{\theta - \theta_m}{x_0}\right),$$

where $x_0 = [\hbar/(m\Omega a^2)]^{1/2}$ is the oscillation length; $H_n(x)$ is the Hermitian polynomial; $C_n = 1/(2^n n! \pi^{1/2} x_0)^{1/2}$ is the normalizing constant; and $\Delta(p)$ is a small (in the tight-binding approximation) addition to the relative motion energy and accounts for tunneling of electrons through the potential barrier toward each other.

Substituting expression (9.20) into the first of (9.18), we obtain the energy of the center-of-mass motion as

$$E_1 = \hbar\omega_0(J + 2\Phi/\Phi_0)^2, \quad \omega_0 = \hbar/2ma^2.$$

The parameters J and p should be independently determined from the conditions for periodicity of the total wave function with respect to φ_1 and φ_2. These conditions are written as

$$\pi J + 2\pi p = 2\pi N_1;$$
$$\pi J - 2\pi p = 2\pi N_2, \tag{9.23}$$

where N_1 and N_2 are integers. As a result, we find that J is an integer and p is either an integer or a half-integer; i.e., within the first 'Brillouin zone', we have either $p = 0$ if J is even-valued and $p = 1/2$ if the total-momentum number is odd-valued.

In order to determine the ground state and the wave function of the system, we have to take into account the Pauli exclusion principle, which gives rise to a correlation between the quantum numbers of orbital motion and the total spin S of the system; i.e.

$$J = 2\ell + 1, \quad \ell = 0, \pm 1, \pm 2, \dots \qquad p = 1/2 \quad \text{at } S = 1,$$
$$J = 2\ell, \quad \ell = 0, \pm 1, \pm 2, \dots \qquad p = 0 \quad \text{at } S = 0. \tag{9.24}$$

Thus, if external perturbation applied to the system is independent of electron spins ($\Delta S = 0$), only the transitions with changes of J by an even number are possible: $\Delta J = 2\ell$, where $\ell = 0, \pm 1, \pm 2, \dots$.

Since we do not restrict our consideration only to the dipole approximation, we express the external perturbation in terms of expansion in the Bessel functions $J_n(x)$ (here and everywhere further all special functions are determined in accord with [16]) when calculating the SAW absorption

$$V(\varphi_1, \varphi_2) = V_0 \sum_k i^k J_k(qa) \left(e^{ik\varphi_1} + e^{ik\varphi_2} \right) + c.c. \tag{9.25}$$

where subscripts 1 and 2 correspond to the first and second electrons, respectively. Passing to the center-of mass system, we obtain

$$V(\varphi_c, \theta) = 2V_0 \sum_k i^k J_k(qa) e^{ik\varphi_c} \cos\left(\frac{k\theta}{2} \right) \tag{9.26}$$

In order to calculate the SAW absorption coefficient in the case of two electrons in a quantum ring, it is not sufficient just to replace the matrix element $V_{j'j}$ in expression (9.16) with the perturbation matrix element (9.26). The point is that the total spin S of the system's ground state changes periodically as the magnetic flux through the quantum ring varies. Figure 9.3 shows the dependence of energy levels of a Wigner molecule in a quantum ring on the magnetic flux; it can be seen from Fig. 9.3 that the role of the ground state with the total spin $S = 0$ is replaced by the state with $S = 1$. This phenomenon (the so-called singlet-triplet oscillations) was predicted by Wagner et al. [17] for the first time. Taking into account these oscillations, we can express the SAW absorption as

$$\frac{\Gamma}{\Gamma_0} = g_{\text{singlet}} \sum_{\eta\eta'} \frac{A_{\eta\eta'} A_{\eta\eta'}^*}{1 - (\omega_{\eta'\eta} - \omega)^2 \tau^2} + g_{\text{triplet}} \sum_{\eta\eta'} \frac{A_{\eta\eta'} A_{\eta\eta'}^*}{1 - (\omega_{\eta'\eta} - \omega)^2 \tau^2}. \tag{9.27}$$

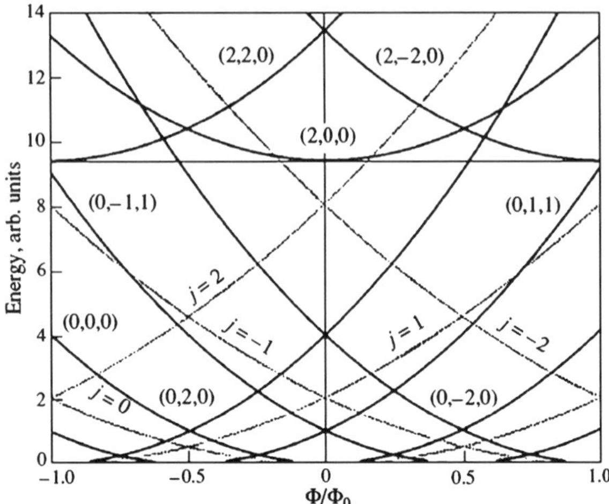

Fig. 9.3 Energy levels of a free electron (*dashed lines*) and two interacting electrons (*solid lines*) in relation to the magnetic flux. Quantum numbers are given in the form (n, J, S)

The following designations are used here

$$A_{\eta\eta'} = \frac{V_{\eta\eta'}}{2V_0}, \qquad V_{\eta\eta'} = \langle \eta' | V(\varphi_c, \theta) | \eta \rangle,$$

$$\omega_{\eta'\eta} = (E_{\eta'} - E_\eta)/\hbar, \qquad \Gamma_0 = 64\pi N_R \frac{e^2 \tau}{\hbar(\varepsilon + 1)} K_{\text{eff}}^2,$$

η is the totality of quantum numbers, i.e., the total angular momentum quantum number J and the oscillation quantum number n; τ is the relaxation time; K_{eff} is the electromechanical-coupling coefficient; N_R is the surface density of quantum rings; ε is the relative permittivity of the substrate; and $g_{\text{singlet}} = 1$ and $g_{\text{triplet}} = 3$ are the weighting factors of the singlet and triplet states, respectively. It is noteworthy that the summation in expression (9.27) is performed over even values of J in the first sum and over odd values of J in the second sum according to the rules of (9.24). Referring to hybrid structures of the type described by Rotter et al. [18, 19], we assume that $N_R = 10^6$ cm^{-2}, $\varepsilon = 50$, $\tau = 5 \times 10^{-11}$ s, and $K_{\text{eff}} = 0.237$; as a result, we obtain $\Gamma_0 = 2.4 \times 10^3$ cm^{-1}.

When deriving expression (9.27), we took into account that the initial state $|\eta\rangle = |n, J\rangle$ is always occupied, whereas the final state $|\eta'\rangle = |n', J'\rangle$ is always unoccupied. Taking into account (9.20) and (9.21), we obtain the following expression for the matrix element $V_{\eta'\eta}$ (see calculations in [15]):

$$\langle n', J' | V | n, J \rangle = \frac{V_0}{\sqrt{2^{n'} n'!}} \sum_k (-i)^k J_k(qa) \left(\frac{ikx_0}{2} \right)^{n'}$$

$$\times \exp\left(-\frac{k^2 x_0^2}{16} \right) [1 + (-1)^{n'}] \delta_{J', J'+k} \delta_{S'S} \tag{9.28}$$

Since the transitions involving J are possible only for an even value of ΔJ ($\Delta J = 2\ell$), it follows from (9.28) that $k = 2\ell$. It is worth noting that, in the model

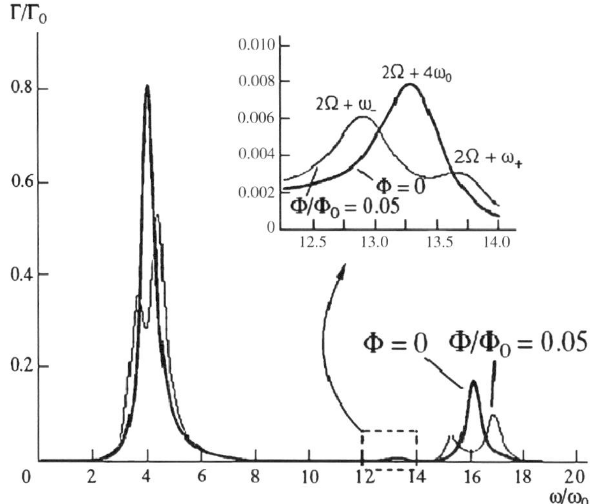

Fig. 9.4 Dependence of the absorption coefficient on the frequency of a SAW for two values of the magnetic flux and for the system's total spin $\Phi = 0$. *Thick solid lines* correspond to absorption at $\Phi = 0$ and *thin solid lines* correspond to absorption at $\Phi = 0.05\Phi_0$. Dependence of the absorptance by vibrational degrees of freedom of a Wigner molecule on the SAW frequency is shown in the *insert*

under consideration, the matrix element (9.28) is independent of the magnetic flux, which threads through the quantum ring; consequently, we may posit that the heights of the absorption peaks are not sensitive to variations in the magnetic field, at least in the approximation of a 1D quantum ring. The results of numerical calculation of the absorption coefficient on the basis of formulas (9.16) and (9.27) are shown in Figs. 9.4, 9.5(a) and 9.5(b) and are discussed below. When performing calculations, we assumed that $a = 800A$ and $m = 0.04 \cdot m_0$.

9.3.3 Discussion

The energy spectrum of a system consisting of two electrons is shown in Fig. 9.3 in relation to the magnetic flux; the states of a Wigner molecule are described using the quantum numbers n, J, and S. Here, $n = 0, 1, 2, \ldots$ is the oscillation quantum number; $J = 0, +2, +4$: is the azimuthal quantum number (the total angular moment of the system); and $S = 0$ (or 1) is the total spin. The values of the energy are given in units of the rotational quantum $\hbar\omega_0$. The angular electron momentum j is indicated for the states of free electron. The energy levels of the type $(1, J, S)$ are not shown in Fig. 9.3; the reason is that (as can be seen from expression (9.28)) these levels are not involved in absorption owing to the selection rules $n' = 0, 2, 4, \ldots$. The frequency dependence of the coefficient of SAW absorption $\Gamma(\omega)$ by a Wigner molecule in a quantum ring at $S = 0$ and two values of magnetic flux is illustrated in Fig. 9.4.

It can be seen that, aside from the high-intensity peaks corresponding to the transitions $(0, 0, 0) \to (0, \pm 2, 0)$, there are additional peaks (phonon replicas) related to absorption of SAWs by the vibrational degree of freedom of a Wigner molecule (see the inset in Fig. 9.4). The first (low-frequency) group of peaks corresponds to a single transition for $\Phi = 0$ (the resonance frequency is $4\omega_0$) and two transitions at $\Phi = 0.05\Phi_0$ since the degeneracy with respect to momentum is removed

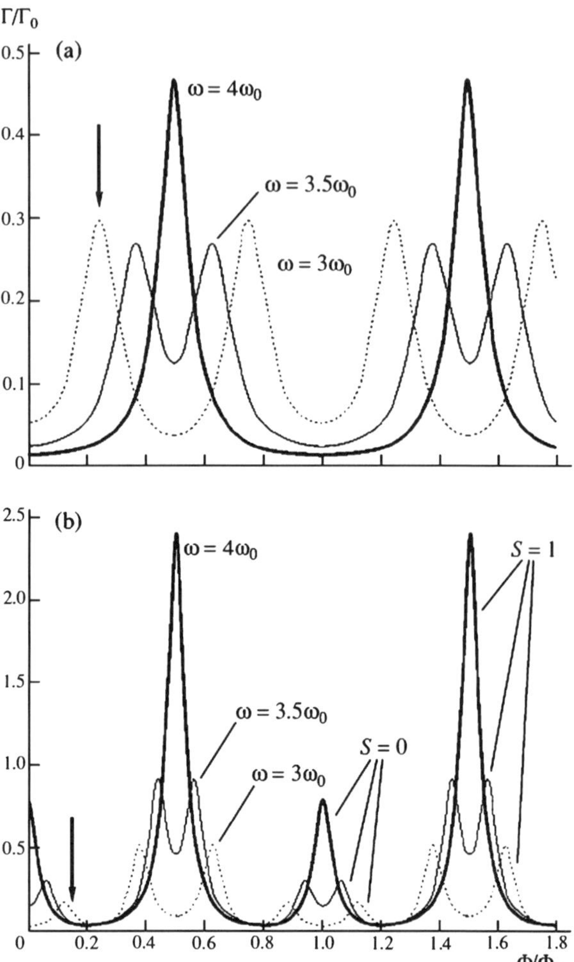

Fig. 9.5 Dependences of the absorption coefficient on the magnetic flux at frequencies $\omega = 4\omega_0$, $\omega = 3.5\omega_0$, and $\omega = 3\omega_0$: (**a**) for a single electron and (**b**) for two interacting electrons

in this case; as a result, two peaks at the frequencies designated as ω_+ and ω_- in Fig. 9.4 are observed in the absorption spectrum. The phonon replicas are shifted by the frequency of allowed vibrational transition 2Ω and also include three peaks at $\omega = 2\Omega + \omega_-$, $\omega = 2\Omega + 4\omega_0$, and $\omega = 2\Omega + \omega_+$. A group of high-frequency peaks in the vicinity of $\omega = 16\omega_0$ corresponds to transitions $(0, 0, 0) \rightarrow (0, \pm 4, 0)$. It is noteworthy that the parameter $\omega a/s$ is much larger than unity ($4\omega_0 a/s = 2.22$) for all peaks shown in Fig. 9.4; this means that the dipole approximation is certainly inapplicable. The most interesting (and simplest for experimental observation) is the dependence of the SAW absorption spectra on the magnetic flux through a quantum ring. These spectra are shown in Figs. 9.5(a) and 9.5(b) for the cases of a single electron and two strongly interacting electrons in a quantum ring. It can be seen that, when the second electron appears in the quantum ring (i.e., at a certain voltage), positions of the peaks change jumpwise (see the thick arrows in Figs. 9.5(a) and 9.5(b)) since the effective rotational quanta differ from each other by a factor of 2. The height of the peaks is determined by the quantity $J_k(\omega a/s)$. If the parameter $\omega a/s$ is not small compared to unity (as in our calculations), the peak heights vary rapidly as the values of ω or a vary; a reverse ratio (with respect to that shown in Fig. 9.5) between the heights of absorption peaks for one electron or two interacting electrons can also be observed.

9.4 Effect of Spin-Orbit Interaction on Persistent Currents in QR [20]

The main question usually discussed when speaking on persistent currents (PC) problem is the periodicity of the function $J(\Phi)$—magnetic flux dependence of PC. In the diffusion regime (electron free path is small as compared with the ring radius) the period is $\Phi_0/2$, while in the opposite limiting case (ballistic regime) one expects the full quantum periodicity Φ_0. This result can be derived both with and without accounting for electron-electron interaction, but in both cases the spin-orbit (SO) interaction is usually ignored. Now we show that a special kind of the SO-interaction, which is actual for two-dimensional (2D) systems, may essentially change the Φ-dependence of PC, namely, may suppress the first Fourier harmonic and simulate the $\Phi_0/2$—periodicity. The above mentioned special kind of the SO-interaction is a very well known Rashba term and can be written as

$$V_{\mathrm{so}} = \frac{\alpha}{\hbar}[\boldsymbol{\sigma}, \mathbf{p}]\mathbf{n}, \tag{9.29}$$

where α is the phenomenological constant (dependent on the properties of the interface), $\mathbf{n}$ is the normal to the 2D electron system, $\boldsymbol{\sigma}$ and $\mathbf{p}$ are Pauli matrices and momentum operator correspondingly; the characteristic parameter $\alpha \neq 0$ only for asymmetric in z-direction (i.e. along the normal) 2D system (e.g. heterojunction GaAs/AlGaAs).

Consider now a 1D quantum ring placed in the perpendicular magnetic field H (z-direction). The total single electron Hamiltonian reads

$$H = H_0 + V_{\mathrm{so}} + H_z, \tag{9.30}$$

where $H_0 = \pi^2/2m + u(r)$, $\boldsymbol{\pi} = \mathbf{p} + e\mathbf{A}/c$, $\mathbf{A}$ is the vector potential of the external magnetic field $\mathbf{H}$, $u(r)$ is the lateral potential, m is the effective mass, $H_z = g\mu_B\sigma_z H/2$ (g is the effective Lande factor); V_{so} is given by (9.29) with the change $\mathbf{p} \to \boldsymbol{\pi}$. In polar coordinates (r, φ) we have

$$H_0 = H_r + H_\varphi,$$

$$H_r = -\frac{\hbar^2}{2m}\frac{1}{r}\frac{\partial}{\partial r}\left(r\frac{\partial}{\partial r}\right) + u(r),$$

$$H_\varphi = -\frac{\hbar^2}{2m}\frac{1}{r^2}\frac{\partial^2}{\partial\varphi^2} - i\frac{\hbar\omega_c}{2}\frac{\partial}{\partial\varphi} + \frac{m^2\omega^2 r^2}{8}, \tag{9.31}$$

and

$$V_{so} = \frac{\alpha}{\sqrt{2}}\left\{(\sigma_+ e^{-i\varphi} + \sigma_- e^{i\varphi}), \left(\frac{r}{2l^2} - \frac{i}{r}\frac{\partial}{\partial\varphi}\right)\right\}$$
$$+ \frac{\alpha}{\sqrt{2}}(\sigma_+ e^{-i\varphi} - \sigma_- e^{i\varphi})\left(\frac{1}{2r} + \frac{\partial}{\partial r}\right).$$

Here $\sigma_\pm = (\sigma_x \pm i\sigma_y)/\sqrt{2}$, $\omega_c = eH/mc$ is the cyclotron frequency, $l = \sqrt{c\hbar/eH}$ is the magnetic length, figure brackets in (9.31) stand for the operation of symmetrization $\{A, B\} = (AB + BA)/2$ which is necessary to conserve the hermitiancy of the Hamiltonian; we have chosen the gauge $A_\varphi = Hr/2$, $A_r = A_z = 0$.

Let us consider the lateral potential of the form $u(r) = 0$ for $a - \delta \leq r \leq a + \delta$ ($\delta \ll a$) and $u(r) = \infty$ for $|r - a| > \delta$. This potential corresponds to a narrow (1D) ring with the radius a. By averaging the third and the fourth equations in (9.31) over the ground state of the Hamiltonian H_r we come to the effective Hamiltonian of our problem

$$H = B\left(-i\frac{\partial}{\partial\varphi} + \frac{\Phi}{\Phi_0}\right)^2 + \Lambda\left\{\begin{pmatrix} 0 & e^{-i\varphi} \\ e^{i\varphi} & 0 \end{pmatrix}, \left(-i\frac{\partial}{\partial\varphi} + \frac{\Phi}{\Phi_0}\right)\right\} + H_z$$
$$(9.32)$$

Here $B = \hbar^2/2ma^2$, $\Lambda = \alpha/a$, and $\Phi = \pi a^2 H$ is the magnetic flux through the ring. Now we have to solve the Schrödinger equation $H\Psi = E\Psi$, where Ψ is the two-component spinor. This equation allows a solution in the form

$$\Psi = \begin{pmatrix} \psi_1 e^{i(m-1/2)\varphi} \\ \psi_2 e^{i(m+1/2)\varphi} \end{pmatrix} \tag{9.33}$$

with $m = $ half-integer to provide the periodicity of the wave function with respect to the motion along the ring: $\varphi \to \varphi + 2\pi$. The coefficients $\psi_{1,2}$ are determined by the system of equations

$$B(m \mp 1/2 + \bar{\Phi})\psi_{1,2} + \Lambda(m + \bar{\Phi})\psi_{2,1} = (E \pm \varepsilon_s/2)\psi_{1,2} \tag{9.34}$$

($\bar{\Phi} = \Phi/\Phi_0$). The electron energy spectrum follows from (9.34)

$$E_m^\pm = B\left(\lambda_m^2 + 1/4\right) \pm \sqrt{(B\lambda_m + \varepsilon_s/2)^2 + \Lambda^2\lambda_m^2}, \tag{9.35}$$

where $\lambda_m = m + \bar{\Phi}$. Thus, we have two sets of the energy levels due to spin splitting caused by SO-interaction and by Zeemann term H_z. PC, as well as any thermodynamic property of the system, is expressed by the sum over all the energy states, i.e. it contains contributions from both E_m^+ and E_m^-. At $\Lambda = g = 0$ (i.e. in the absence of SO-interaction and Zeemann splitting) we have from (9.35) $E_m^+ = B(|\lambda_m| \pm 1/2)^2$ which is equivalent to the well known result $E_n = B(n + \bar{\Phi})^2$ with $n = 0, \pm 1, \pm 2, \ldots$.

In what follows we neglect the Zeeman splitting: the value ε_s is rather small in GaAs for magnetic fields of interest [3]. Then the expression (9.35) can be simplified

$$E_m^\pm = B\left(|\lambda_m| \pm \gamma \pm 1/2\right)^2 - B\bar{\Lambda}^2/4, \tag{9.36}$$

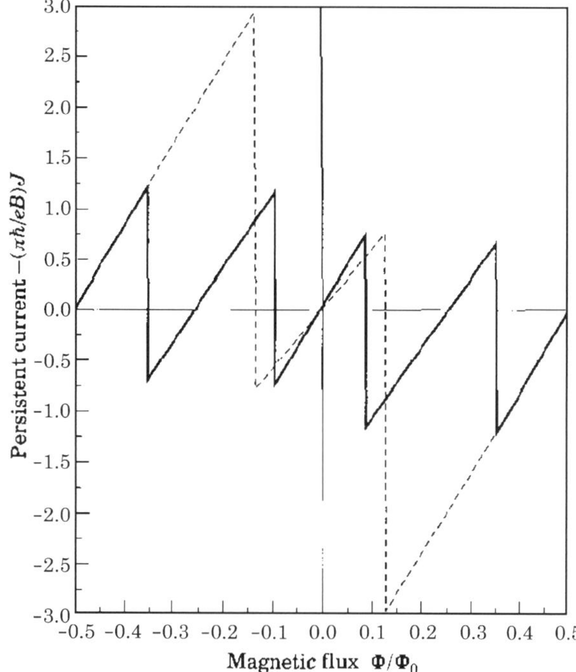

Fig. 9.6 Persistent current for the GaAs ring with $a = 1$ mkm, $\varepsilon_F = 3.42$ B, $\gamma = 1.78$ at $T = 0$ K (*solid line*); the *dotted line* corresponds to the case of no SO-interaction ($\gamma = 0$)

where $\bar{\Lambda} = \Lambda/B$, $\gamma = (\sqrt{1 + \bar{\Lambda}^2} - 1)/2$. It is easy to see that the energy levels given by (9.36) can be renumbered in the following way

$$\tilde{E}_m^\pm = B(n + \bar{\Phi} \pm \gamma)^2 - B\,\bar{\Lambda}^2/4, \qquad (9.37)$$

where now n is integer. Hence, the persistent current, as well as any thermodynamically equilibrium value, with SO-interaction allowed for, can be written in the form

$$J(\Phi/\Phi_0) = \left[J_0(\Phi/\Phi_0 + \gamma) + J_0(\Phi/\Phi_0 - \gamma)\right]/2, \qquad (9.38)$$

where $J_0(\Phi/\Phi_0)$ is the PC in the system without SO-interaction. This result formally coincides with one derived by Meir et al. [21] in the framework of the tight-binding model with SO-scattering.

In the Boltzmann case, where only a few Fourier harmonics of $J(\Phi/\Phi_0)$ are essential, (9.38) gives an additional factor $\cos(2\pi\gamma n)$ for the nth harmonics. In the GaAs ring with $a = 1$ mkm the first harmonics is essentially suppressed ($\cos(2\pi\gamma) \approx 0.19$; we have used for the parameter α the value 2.5×10^{-10} eV cm^{-1}, then $\gamma = 1.78$). The second harmonics changes its sign ($\cos(4\pi\gamma) \approx -0.92$). In the Fermi case the shape of the curve $J(\Phi/\Phi_0)$ drastically differs from $J_0(\Phi/\Phi_0)$ (see Fig. 9.6).

All the above discussed results correspond to an ideal (without electron scattering) quantum ring. Consider now the situation with spin-independent scattering potential $V(\varphi)$. We look for the solution of the problem with disorder as linear combination $\Psi = \sum_{n,\pm} A_n^\pm \Psi_n^\pm$ of the spinors for an ideal ring with the energy spectrum given by (9.37)

$$\Psi_n^+ = \frac{1}{t}\begin{pmatrix} \frac{2\gamma}{\bar\Lambda}e^{i(n-1/2)\varphi} \\ e^{in\varphi} \end{pmatrix}, \qquad \Psi_n^- = \frac{1}{t}\begin{pmatrix} e^{in\varphi} \\ -\frac{2\gamma}{\bar\Lambda}e^{i(n+1/2)\varphi} \end{pmatrix} \qquad (9.39)$$

$(t = \sqrt{1 + 4\gamma^2/\bar\Lambda^2})$. By making use of (9.39) it is easy to obtain the Schrödinger equation in the matrix form

$$\left[B(n + \bar\Phi \pm \gamma)^2 - B\bar\Lambda^2/4 - \varepsilon\right]A_n^\pm + \sum_{n'} A_{n'}^\pm V_{n-n'} = 0, \qquad (9.40)$$

where $V_l = (1/2\pi)\int_0^{2\pi} d\varphi\, V(\varphi)\exp(il\varphi)$. It is evident from (9.40) that if the energy spectrum $\varepsilon_n^{(0)}(\bar\Phi)$ without SO-coupling is known, the eigenvalues of (9.40) can easily be derived as

$$\varepsilon_n^\pm(\bar\Phi) = \varepsilon_n^{(0)}(\bar\Phi \pm \gamma). \qquad (9.41)$$

Therefore, the PC is described by (9.38) even with the disorder allowed for.

In Ref. [21] a 1D ring-shaped lattice has been considered in the nearest neighbors tight-binding approximation. Each site of the chain is characterized by a certain spin-flip amplitude (SO-scattering) and the Zeemann term is not included (Aharonov-Bohm geometry). It was shown that there are two sets of the energy levels phase shifted in the flux dependence from those of the problem without SO-scattering (just like it follows from (9.41)) but the phase shift γ has not been considered explicitly. Our consideration based on the exact solution of the Schrödinger equation (continual representation) shows that this result has a more general character, but simultaneously one can see limitations for the validity of the statements made: the exact energy levels given by (9.36) with accounting for the Zeemann term ε_s can not be renumbered in such a way in order to form two groups with phase shifted dependence on the magnetic flux. Correspondingly, PC and other thermodynamic characteristics can not be written in the form of (9.38). Moreover, our consideration gives a nontrivial dependence of the phase shift γ on the ring radius; this is essential for interpretation of experimental data.

Thus the influence of spin-orbit interaction on PC in quantum rings is rather important. The magnetic flux dependence of PC can be essentially changed as compared with the case without SO-coupling problem. The resultant dependence is rather sensitive to the strength of SO-interaction and the value of the ring radius.

9.5 AB Effect for Composite Particles

Till now we have been considering charged particles (electrons) as the AB phase shift in the wave function is proportional to the charge e. A principal question arises whether at all and how can the AB effect be manifested for neutral but composite entities consisting of charged particles. The simplest example is an exciton in a QR [22] placed in a perpendicular magnetic field.

9.5.1 Excitons in 1D Ring [22]

Let us start with an idealized model: an electron and a hole in a 1D ring placed in a magnetic field. The Hamiltonian of the problem can be written in the form

$$H = \frac{\hbar^2}{2m_e a^2}\left(-i\frac{\partial}{\partial\varphi_e} + \lambda\right)^2 + \frac{\hbar^2}{2m_h a^2}\left(-i\frac{\partial}{\partial\varphi_h} - \lambda\right)^2 - \frac{e^2}{2\varepsilon a|\sin(\frac{\varphi_e-\varphi_h}{2})|} \tag{9.42}$$

where $\lambda = \Phi/\Phi_0$, m_e, m_h are the effective masses of the electron and hole, respectively; φ_e, φ_h are the azimuthal coordinates of the particles, and ε is the dielectric constant, a is a ring radius. To separate the internal motion in the exciton from that of the center of mass, one needs to go over to the variables

$$\varphi_c = \frac{m_e\varphi_e + m_h\varphi_h}{M}, \qquad \theta = \varphi_e - \varphi_h, \quad M = m_e + m_h. \tag{9.43}$$

Then the Hamiltonian reads

$$H = -B\frac{\partial^2}{\partial\varphi_c^2} + \beta\left(i\frac{\partial}{\partial\theta} - \lambda\right)^2 + U_c(\theta), \tag{9.44}$$

where $B = \hbar^2/2Ma^2$, $\beta = \hbar^2/2\mu a^2$, $\mu = (m_e + m_h)/M$, and $U_c(\theta)$ is the Coulomb energy. The total wave function may be written as $\Psi = \exp(iJ\varphi_c + i\lambda\theta)\chi(\theta)$, where J is a real number and $\chi(\theta)$ obeys the 1D Schrödinger equation

$$-\beta\frac{\partial^2\chi}{\partial\theta^2} + U_c(\theta)\chi = \left(E - BJ^2\right)\chi = w\chi. \tag{9.45}$$

Here E is the total energy of the exciton and w is its internal energy.

To determine J and the allowed solutions of (9.45) we have to make the total wave function independently periodic in φ_e and φ_h with the period 2π. On the other hand, (9.45) has, formally, the Bloch-type solutions

$$\chi = e^{ip\theta}v(\theta), \quad -1/2 < p \le 1/2, \tag{9.46}$$

where v is a periodic function of θ with the same period 2π, because this is the period of the potential: $U_c \sim |\sin\theta/2|^{-1}$. By adding 2π to φ_e and φ_h independently we come to the relations

$$J\frac{m_e}{M} + \lambda + p = N_e, \qquad J\frac{m_h}{M} - \lambda - p = N_h, \tag{9.47}$$

where N_e, N_h are arbitrary integers. It follows from (9.47) that $J = N_e + N_h$, and thus J is an integer—the rotational quantum number of the exciton as a whole.

Now consider (9.45). In a small vicinity of the point $\theta = 0$, as well as at $\theta = \pm 2\pi, \pm 4\pi, \ldots$, one can expand the Coulomb energy: $U_c \approx e^2/\varepsilon a\theta$, so we have a 1D hydrogen "atom" with charge e^2/ε and effective mass μ. It is known that the ground state of such a system is logarithmically deep:

$$E_0 = -\frac{2\mu e^4}{\varepsilon^2\hbar^2}\ln^2(a_B/a_0), \tag{9.48}$$

where a_0 is a "cutoff" radius. For a 3D exciton in a strong magnetic field this radius a_0 equals the magnetic length a_H. In our case a_0 is obviously related to the size (width or thickness) of the ring.

Besides the ground state, a 1D exciton has the hydrogen series of energy levels $E_n = -\mu e^4/2\varepsilon^2 \hbar^2 n^2$, $n = 1, 2, 3, \ldots$. All these levels form (approximately!) the set of allowed values for the eigenvalue w in (9.45) provided that the effective Bohr radius $a_B = \varepsilon \hbar^2/e^2 \mu$ is much smaller than the distance $2\pi a$ to the next well in the Coulomb energy U_c. For the QRs reported up till now the condition $a_B \ll a$ is very well satisfied, and we may solve (9.45) in the tight-binding approximation.

Any solution of the type (9.46) corresponds to the energy

$$w = E_n - \Delta_n \cos(2\pi p), \quad n = 0, 1, 2, \ldots, \quad \Delta_n > 0, \tag{9.49}$$

where the "quasimomentum" should be determined by (9.47). Then for the total energy of the exciton one obtains

$$\begin{aligned} E(J, n) &= BJ^2 + E_n - \Delta_n \cos 2\pi (\Phi/\Phi_0 + Jm_e/M) \\ &= BJ^2 + E_n - \Delta_n \cos 2\pi (\Phi/\Phi_0 - Jm_h/M); \quad n = 0, 1, 2, \ldots \end{aligned} \tag{9.50}$$

The quantity Δ_n in (9.50) is an amplitude corresponding to the tunneling of the electron to the hole (or vice versa) around the QR. In order of magnitude it is $\Delta_n \sim |E_n| \exp(-2\pi a/na_B)$ for $n \geq 1$ and $\Delta_0 \sim |E_0| \exp[(-2\pi a/a_B)\ln(a_B/a_0)]$ for the ground state. The possibility of such tunneling shifts the energy levels from their unperturbed positions E_n and results in a periodic dependence of the binding energy of the ring exciton on the magnetic flux, with the period Φ_0. The latter statement has a general character and is not connected with the tight-binding approximation used. The eigenvalues w of (9.45) are always periodic functions of p from (9.46), and via (9.47) they are periodic functions of the magnetic flux.

In the same tight-binding approximation the dependence of the exciton creation probability $\Im$ on the magnetic flux can be found. $\Im$ is determined by the integral over φ of wave function at coincides arguments $\varphi_e = \varphi_h = \varphi$ and the nonzero value of $\Im$ is at $J = 0$

$$\Im \sim |v(0)|^2 \delta_{J,0},$$

$$|v(0)|^2 = \frac{1 - \delta^2}{1 - 2\delta \cos(2\pi\lambda) + \delta^2}; \quad \delta = e^{-2\pi R/a}. \tag{9.51}$$

Thus, the intensity of the exciton line oscillates with magnetic flux with the period Φ_0. The relative modulation rate of the intensity equals

$$\frac{\Im_{max}}{\Im_{min}} = \left(\frac{1 + \delta}{1 - \delta}\right)^2. \tag{9.52}$$

9.5.2 Accounting for the Finite Width [23]

Consider now a more realistic model and take into account the radial motion of the electrons and holes confined to move within the same QR. In order to take into account the radial degree of freedom of electrons and holes, we will use the parabolic model of a ring proposed by Chakraborty and Pietilahen [24]. The potential energy of a particle has the form

$$U_i(\rho) = \frac{m_i \omega_i^2 (\rho - R_0)^2}{2}, \quad i = e, h, \tag{9.53}$$

where the subscripts e and h mark an electron or a hole, and R_0 is the electron (and hole) radius of the ring. Thus, the position of the minimum is assumed to be the same for electrons and holes. In the present work, we will be interested in orbital (diamagnetic) effects, leaving aside the contribution from the spin degree of freedom. The latter depends on the specific form of the system through the g-factors of electrons and holes, and a transition, say, from a quantum point to the ring is not critical for this contribution. On the contrary, the orbital motion is sensitive to a change in the object topology (the Aharonov-Bohm effect), and we will be interested here in such an effect of a magnetic field on the system behavior.

According to estimates of the above parameters of a quantum ring, the radial motion is characterized by a much smaller amplitude as compared to the azimuthal motion, for which the corresponding size is just $2\pi a$. In this sense, the ring is narrow, and we will assume that the condition $\hbar \omega_i \gg W_i$, is satisfied, where $W_i = \hbar^2/2m_i a^2$ is the rotational motion quantum (an ideal one-dimensional ring corresponds to the limit $W_i/\hbar \omega_i \to 0$). In this case, the problem can be solved in the approximation which is well known from molecular theory: first, the rotational energy levels are determined for fixed nuclei (in our case, for fixed radial coordinates ρ_e and ρ_h of an electron and a hole), after which the small vibrations of nuclei are taken into account. In this case, effects of the type of a nonrigid rotator and the interaction between vibration and rotation are observed. Similar corrections to energy levels also appear in the problem on excitons in a finite-width quantum ring.

We begin our analysis with a system containing N electrons in a ring. This is the final state of the initial system $(Ne + h)$ after recombination, and hence the magnetic-field dependence of its energy makes a significant contribution to the diamagnetic shift of the exciton luminescence line. In an ideal one-dimensional ring, the position of electrons is defined by azimuthal angles $(\varphi_j, \ j = 1, \ldots, N)$. It is convenient to introduce the Jacobi variables in accordance with the formulas (9.2).

In new variables, the Hamiltonian of the system taking into account the magnetic field B normal to the ring assumes the form

$$H_{1D} = \frac{W_{e0}}{N} \left[-i \frac{\partial}{\partial \varphi_0} + N \bar{\Phi}(a) \right]^2 - W_{e0} \left(\frac{\partial^2}{\partial \psi_1^2} + \cdots + \frac{\partial^2}{\partial \psi_{N-1}^2} \right)$$
$$+ V_{ee}(\psi_1, \ldots, \psi_{N-1}), \tag{9.54}$$

where $W_{e0} = \hbar^2/2m_e a^2$ is the rotational quantum, $\bar{\Phi} = \pi H a^2/\Phi_0$ (Φ_0 is the magnetic flux quantum), and V_{ee} is the electron-electron interaction energy, which is naturally independent of φ_0. Thus, the magnetic-field dependent component of the system energy is defined by the first term in (9.54)

$$E = \frac{W_e}{N} \left(J + N \bar{\Phi}(a) \right)^2, \tag{9.55}$$

where $J = 0, \pm 1, \pm 2, \ldots$ is the total angular momentum of the system.

It is important that the value of E is independent of the electron-electron interaction. For this reason, the diamagnetic shift, for instance, for an interacting system with $J = 0$ (the ground state in a weak field) is equal just to the n-fold shift of a single particle:

$$E_{J=0} = N W_{e0} \bar{\Phi}^2 = N \frac{(\hbar \Omega_e)^2}{16 W_{e0}}, \tag{9.56}$$

where Ω_e is the cyclotron frequency of an electron.

Let us now take into account the radial degrees of freedom and write the Hamiltonian in the polar coordinates φ_j and ρ_j

$$H_{2D} = \sum_j \left\{ T(\rho_j) + W_{ej} \left[-i \frac{\partial}{\partial \varphi_j} + \bar{\Phi}(\rho_j) \right]^2 + U_e(\rho_j) \right\} + V_{ee}, \tag{9.57}$$

where $T(\rho_j)$ is the kinetic energy operator for the radial motion, $W_{ej} = \hbar^2/2m_e \rho_j^2$, $\bar{\Phi}(\rho_j) = \pi H \rho_j^2 / \Phi_0$, $U(\rho_j)$ is defined by formula (9.53), and V_{ee} is now a function of all φ_j and ρ_j. The second term in (9.57) containing the magnetic field can be transformed as follows

$$\sum_j W_{ej} \left[-i \frac{\partial}{\partial \varphi_j} + \bar{\Phi}(\rho_j) \right]^2$$

$$= W_{e0} \sum_j \left[-i \frac{\partial}{\partial \varphi_j} + \bar{\Phi}(R_0) \right]^2 + \sum_j (W_{e0} - W_{ej}) \frac{\partial^2}{\partial \varphi_j^2}$$

$$+ W_{e0} \bar{\Phi}(R_0) \sum_j \delta \bar{\Phi}_j, \tag{9.58}$$

where $\delta \bar{\Phi}_j = \pi B(\rho_j^2 - a^2)/\Phi_0$. We now apply transformation (9.2) to the right-hand side of (9.58). Since φ_0 does not appear explicitly in the Hamiltonian (V_{ee} depends only on pairwise difference $\varphi_j - \varphi_k$ in which φ_c does not appear), we can single out the "magnetic" component of the Hamiltonian, i.e.,

$$\Delta H = \frac{W_{e0}}{N} \left[-i \frac{\partial}{\partial \varphi_0} + N \bar{\Phi}(R_0) \right]^2 + W_{e0} \bar{\Phi}(R_0) \sum_j \delta \bar{\Phi}_j, \tag{9.59}$$

which completely determines the diamagnetic shift of energy levels, the effects of the finite ring width appearing only in the second term on the right-hand side of (9.59). Introducing the vibrational coordinates $\xi_j = \rho_j - R_0$, we obtain the following expression for perturbation:

$$\Delta H_{\text{pert}} = \pi W_{e0} \bar{\Phi}(a) H \sum_j \left(2a\xi_j + \xi_j^2 \right). \tag{9.60}$$

It was mentioned in the beginning of this subsection that for $\hbar \omega_e \gg W_{e0}$, the corrections to energy associated with ΔH_{pert} can be calculated in perturbation theory. Strictly speaking, the unperturbed wave function of radial motion must include the interaction V_{ee} between particles. However, the Coulomb interaction strongly perturbs the azimuthal and weakly radial motion of particles in the approximation adopted by us. Indeed, expanding the paired interaction of particles separated by a distance of the order of R_0 from one another into a power series in their radial coordinates ξ_i, we can easily obtain the estimates

$$\Delta\xi \sim l_0 \frac{W_{e0}}{\hbar\omega_e}\left(\frac{R_y^*}{\hbar\omega_e}\right)^{1/2},\tag{9.61}$$

where $l_0 = \sqrt{\hbar/m\omega_e}$ is the amplitudes of vibrations in potential (9.53) and R_y^* is the effective Rydberg energy, for the shift of the equilibrium position, and

$$\Delta\omega^2 \sim \omega_e^2\left(\frac{R_y^*}{\hbar\omega_e}\right)^{1/2}\left(\frac{W_{e0}}{\hbar\omega_e}\right)^{3/2}\tag{9.62}$$

for the correction to the square of frequency. For the ring parameters indicated at the beginning of the article, $R_y^* \sim \hbar\omega_e$, and hence both corrections are small in the parameter $W_{e0}/\hbar\omega_e$, and the vibrational component of diamagnetic shift can be calculated using the unperturbed vibrational functions. This gives

$$\Delta E_{\text{vibr}}(B) = N\left[\frac{(\hbar\Omega_e)^2}{8\hbar\omega_e}\left(v_e + \frac{1}{2}\right) - \frac{(\hbar\Omega_e)^4}{64W_{e0}(\hbar\omega_e)^2}\right],\tag{9.63}$$

where v_e is the vibrational quantum number.

In this formula, we have also taken into account the second approximation in the linear term appearing in (9.60) (the contribution proportional to H^4). It can be seen that the vibrational component of diamagnetic shift is small as compared to the rotational component [formula (9.56)] in the same parameter $W_{e0}/\hbar\omega_e$.

9.5.2.1 Neutral Exciton X^0

The rotational component of diamagnetic shift in the energy of a neutral formation X^0 connected with the motion of an exciton as a whole obviously vanishes. For an ideal one-dimensional ring, there exists a contribution to the exciton binding energy ε, which is a periodic function of the magnetic flux and emerges as a result of the tunneling of an electron and a hole towards each other along the ring. In the strong coupling approximation ($a_B \ll 2\pi a$, where a_B is the effective Bohr radius) and for zero total angular momentum $J = 0$, the binding energy is given by

$$\varepsilon = \varepsilon_0 - \Delta\cos(2\pi\Phi/\Phi_0),\tag{9.64}$$

where $\varepsilon_0 < 0$ is the energy level of a one-dimensional exciton and $\Delta > 0$ is the tunneling amplitude (see above). For weak fields (fluxes), (9.64) gives a positive (diamagnetic) energy shift. However, this shift can be considerably smaller than the vibrational diamagnetic shift which does not contain an exponential smallness. Besides, radial vibrations of particles lead to the attenuation of the oscillating energy component due to the destructive interference of trajectories embracing various magnetic fluxes.

In order to take these effects into consideration, we introduce the coordinate φ_c of the exciton "center of gravity" and the angular separation φ between an electron and a hole given by (9.43) as well as the radial coordinates $\xi = \rho_e - a$, $\eta = \rho_h - a$.

To avoid cumbersome expressions, we confine our analysis to the case of zero orbital angular momentum of an exciton as a whole (for an optical transition, the intensity of the exciton line is determined by the integral $\int d\varphi_e \Psi \ (\varphi_e = \varphi_h)$ which differs from zero only for $J = 0$). The magnetic-field dependent component of the Hamiltonian has the form

$$H = -(W_e + W_h)\left(\frac{\partial}{\partial\varphi} + i\,\frac{W_e\bar{\Phi}_e + W_h\bar{\Phi}_h}{W_e + W_h}\right)^2 + \frac{W_e W_h(\bar{\Phi}_e - \bar{\Phi}_h)^2}{W_e + W_h} \qquad (9.65)$$

where $\bar{\Phi}_{e,h} = \pi H \rho_{e,h}^2/\Phi_0$, $W_{e,h} = \hbar^2/2m_{e,h}\rho_{e,h}^2$.

The phase transformation of the wave function

$$\psi = \chi e^{-i\lambda\varphi}, \quad \lambda(\xi,\eta) = \frac{W_e\bar{\Phi}_e + W_h\bar{\Phi}_h}{W_e + W_h}, \qquad (9.66)$$

makes it possible to eliminate the second term in the brackets in (9.65) and to obtain the Schrodinger equation for the relative motion of an electron and a hole in the adiabatic approximation, i.e., for fixed radial coordinates ξ and η. Following the same way as above (the conditions for the wave function periodic in φ_e and φ_h), we obtain the oscillating contribution to energy of the form $\Delta(\xi,\eta)\cos[2\pi\lambda(\xi,\eta)]$. Averaging this contribution with the vibrational functions of radial motion, we obtain the following expression for attenuation of the Aharonov-Bohm oscillations of the exciton binding energy

$$\varepsilon = \varepsilon_0 - \bar{\Delta}e^{-H^2/H_0^2}\cos(2\pi\Phi/\Phi_0), \qquad (9.67)$$

where we have used the following expression for the damping decrement of the vibrational ground state

$$\frac{1}{H_0^2} = \left(\frac{2\pi^2 a}{M\Phi_0}\right)^2 \hbar\left(\frac{m_h^2}{m_e\omega_e} + \frac{m_e^2}{m_h\omega_h}\right), \qquad (9.68)$$

and $\bar{\Delta}$ is the tunneling amplitude averaged over radial vibrations. Thus, in the case under investigation, oscillations attenuate according to the Gaussian law, and the damping decrement $1/H_0$ increases linearly with the ring radius.

The vibrational diamagnetic shift for an exciton is determined by the last term in Hamiltonian (9.65). After averaging over vibrations, we obtain

$$\Delta E_{\mathrm{vibr}}(X^0) = \frac{\hbar}{2M}\left[\frac{m_e\Omega_e^2}{\omega_e}\left(v_e + \frac{1}{2}\right) + \frac{m_h\Omega_h^2}{\omega_h}\left(v_h + \frac{1}{2}\right)\right], \qquad (9.69)$$

where v_e and v_h are the quantum numbers for the vibrational states of electrons and holes, and Ω_e and Ω_h are their cyclotron frequencies.

9.5.3 Trion X^- and Multiply Charged Excitons

The complex e–e–h (trion) is known to form a bound state of three particles, i.e., is more advantageous from the energy point of view than an individual electron and an exciton. Since a trion has a charge, its motion as a whole along a ring in a magnetic field must lead to a rotational diamagnetic shift of energy levels. Besides, as in the case of an exciton, we can expect an oscillating magnetic-flux dependence of the trion internal energy in a one-dimensional ring. Indeed, introducing the variables

$$\varphi = \frac{m_e(\varphi_1 + \varphi_2) + m_h\varphi_h}{M_{tr}}, \qquad \alpha = \frac{\varphi_1 + \varphi_2}{2} - \varphi_h, \qquad \beta = \varphi_1 - \varphi_2, \quad (9.70)$$

where $M_{tr} = 2m_e + m_h$ is the trion mass and $\varphi_{1,2}$ are the coordinates of the electrons, we can easily derive the following expression for the total energy of the trion

$$E_{tr} = W_{tr}(J - \bar{\Phi})^2 + \varepsilon_{int}. \qquad (9.71)$$

Here, $W_{tr} = \hbar^2/2M_{tr}a^2$, ε_{int} is the internal energy, which is the eigenvalue of the equation

$$-\left(\frac{1}{2}W_{e0} + W_{h0}\right)\frac{\partial^2\chi}{\partial\alpha^2} - 2W_e\frac{\partial^2\chi}{\partial\beta^2} +$$
$$+ \left[V(\beta) - V(\alpha + \beta/2) - V(\alpha - \beta/2)\right]\chi = \varepsilon_{int}\chi, \qquad (9.72)$$
$$V(\gamma) = \frac{e^2}{2R_0|\sin(\gamma/2)|}.$$

The wave function ψ of the trion is connected with the solution $\chi(\alpha, \beta)$ of (9.72) through the relation

$$\Psi = \chi \exp\left(i(J\varphi_c + \lambda\alpha)\right), \quad \lambda = \bar{\Phi}\frac{2(W_{e0} + W_{h0})}{W_{e0} + 2W_{h0}}. \qquad (9.73)$$

The potential energy in (9.72) is a periodic function on the plane (α, β). The basis vectors of the corresponding lattice are given by

$$\bar{\mathbf{e}}_1 = \boldsymbol{\alpha}_0\pi + \boldsymbol{\beta}_02\pi, \qquad \bar{\mathbf{e}}_2 = \boldsymbol{\alpha}_02\pi + \boldsymbol{\beta}_0 \qquad (9.74)$$

where α_0 and β_0 are the unit vectors of the Cartesian system of coordinates in the plane (α, β). Equations (9.74) describe a rectangular body-centered lattice (in the two-dimensional sense). The reciprocal lattice is of the same type with the basis vectors

$$\mathbf{g}_1 = 0\cdot\boldsymbol{\alpha}_0 + 1\cdot\boldsymbol{\beta}_0, \qquad \mathbf{g}_2 = 1\cdot\boldsymbol{\alpha}_0 + \left(-\frac{1}{2}\right)\boldsymbol{\beta}_0. \qquad (9.75)$$

The general solution of (9.72) is characterized by the quasimomentum vector $p\boldsymbol{\alpha}_0 + q\boldsymbol{\beta}_0$, and energy ε_{int} is a function of its components p and q. Subjecting the total function φ to the conditions of periodicity in φ_1, φ_2, and φ_h, we can find the allowed values of p and q, and thus determine the energy levels of a trion in a quantum ring. The periodicity conditions have the form

$$\frac{J m_e}{M_{\mathrm{tr}}} + \frac{\lambda + p}{2} + q = n_1, \qquad \frac{J m_e}{M_{\mathrm{tr}}} + \frac{\lambda + p}{2} - q = n_2,$$

$$\frac{J m_h}{M_{\mathrm{tr}}} - \lambda - p = n_3, \tag{9.76}$$

were n_1, n_2, and n_3 are integers. It follows hence that J and $2q$ are integers, while $p = J m_h / M_{\mathrm{tr}} - \lambda + \text{integer}$. The energy $\varepsilon_{\mathrm{int}}$ is periodic on the reciprocal lattice, and hence it depends periodically on Φ through λ. By way of an example, we write the result in the strong-coupling approximation:

$$\varepsilon_{\mathrm{int}} = \varepsilon_0 - 2\Delta(2\pi) \cos 2\pi (2 m_e J / M_{\mathrm{tr}} + \lambda)$$
$$- 4\Delta\left(\pi \sqrt{5}\right) \cos \pi (2 m_e J / M_{\mathrm{tr}} + \lambda) - 2\Delta(4\pi). \tag{9.77}$$

The arguments 2π, $\sqrt{5}\pi$, and 4π of the tunneling amplitudes indicate the distances from the initial site to their nearest neighbors in the lattice which is defined by (9.74), and ε_0 is the binding energy of a trion in a rectilinear quantum wire (i.e., the limit $R_0 \to \infty$). All the tunneling amplitudes in (9.77) are positive. Thus, the binding energy of the trion oscillates with the magnetic flux (two harmonics are present in the model under investigation). The period of oscillations $\Delta\Phi$ is determined by the value of λ from (9.73), i.e., is a function of the ratio of effective masses; for the fundamental harmonic, we have

$$\Delta\Phi = \Phi_0 \frac{2 m_e + m_h}{M_{\mathrm{tr}} + m_h}. \tag{9.78}$$

To our knowledge this is so far the only example of non-universal period of the AB oscillations: $\Delta\Phi$ depends on the ratio of the effective masses.

In the opposite limit of weak coupling, the Coulomb interaction between particles can be neglected outside the region of forbidden bands in the spectrum. In the first Brillouin zone, we must put $n_1 = n_2 = n_3 = 0$ in formulas (9.76), whence we obtain $J = 0, q = 0$, and $p = -\lambda$. It follows from (9.72) that

$$\varepsilon_{\mathrm{int}} = \lambda^2 (W_{h0} + W_{e0}/2), \tag{9.79}$$

while for the total energy we obtain the expression

$$E_{\mathrm{tr}} = W_{\mathrm{tr}} \bar{\Phi}^2 + \lambda^2 (W_{h0} + W_{e0}/2) = (2 W_2 + W_h) \bar{\Phi}^2 \tag{9.80}$$

as should be expected in the case of three free particles. A comparison of this result with equation from (9.71) for $J = 0$ shows that the Coulomb interaction reduces considerably the rotational component of diamagnetic shift (to zero for X^0).

The vibrational diamagnetic shift is of the order of $\hbar \Omega_e^2 / \omega_e$ as in the case of an exciton, but the main contribution for a trion comes from rotations defined by the first term on the right-hand side of (9.71). The observable quantity is the shift of the exciton luminescence line, which is equal to the difference between energies (9.71) and (9.56) in the principal order in $W_e / \hbar \omega_e$ for $N = 1$

$$\Delta \nu_{\mathrm{tr}} = (W_{\mathrm{tr}} - W_{e0}) \bar{\Phi}^2, \qquad J = 0. \tag{9.81}$$

Since $M_{\mathrm{tr}} > m_e$, the obtained diamagnetic shift turns out to be negative. Naturally, the total shift contains a contribution from spin splitting; the obtained value of Δv_{tr} describes the behavior of the center of gravity of the spin multiplet.

Finally, the same computations (separation of the rotation of the system as a whole and the gradient transformation of the wave function) can be used to obtain the following magnetic-flux dependent contribution to the total energy for a multiply charged complexes (hole $+ N$ electrons)

$$E_N = W_M\big[J + (N-1)\bar{\Phi}\big]^2, \quad M = Nm_e + m_h, \quad W_M = \hbar^2/2MR_0^2. \quad (9.82)$$

The corresponding shift of a line upon the transition $(h + Ne) \to (N-1)e$ is given by

$$\Delta v_N = (N-1)\big[W_M(N-1) - W_e\big]\bar{\Phi}^2 = -\frac{(N-1)\hbar^2}{2R_0^2}\left(\frac{m_e + m_h}{m_e M}\right)\bar{\Phi}^2 \quad (9.83)$$

In formulas (9.82) and (9.83), we have omitted small corrections to $E_N(B)$ and Δv_N associated with radial vibrations and the magnetic-field dependent component of the internal energy of the complex. Numerical estimates for $m_e = 0.07m_0$, $m_h = 0.35m_0$, $\hbar\omega_e = 12$ meV (i.e., close to R_y^*), and $a = 14$ nm give $\Delta v(\mathrm{X}^0) = 7 \times 10^{-3}H^2$ meV for the frequency shift for a neutral exciton (the ground state in radial vibrations) and $\Delta v(\mathrm{X}^-) = -5 \times 10^{-2}H^2$ meV for a trion, where H is measured in teslas. The values of ω_e and a are borrowed from [25].

Let us summarize the obtained results. The diamagnetic shift of the energy of a system comprising electrons only in a quantum ring of a small but finite width in a transverse magnetic field does not depend on the electron-electron interaction. Conversely, the Coulomb interaction in the X^{n-} complexes reduces the diamagnetic shift appreciably. In the case of a neutral exciton, we are left only with the vibrational component of the shift and with the exponential small contribution associated with the Aharonov-Bohm effect. The diamagnetic shift of the exciton luminescence line is positive for X^0 (and small in the parameter $W_e/\hbar\omega_e$) and negative for all charged excitons. The period of oscillations of the binding energy of charged complexes in magnetic flux differs from Φ_0 and depends on the ratio of effective masses of electrons and holes.

9.6 AB Effect for Plasmons [26]

Yet another example of the AB effect for neutral formations relates to magneto-plasmon oscillations. In other words we speak now about collective excitations of the Bose-type bearing no electric charge and we will show that in such a situation the oscillating dependence of the eigenfrequency on the magnetic field is possible either. AB effect for plasmons in nanotubes has considered in [27]. In nanotubes the plasmon spectrum is continues due to the motion of electrons along the tube axis. The spectrum is characterized by the continuous quantum number—the longitudinal momentum of the plasmon, and by the discrete azimuth momentum $\ell = 0, \pm 1, \pm 2, \ldots$.

Now we deal with QR of finite width where the plasmon spectrum is totally discrete and depends on two quantum numbers ℓ and r, where r labels radial modes of the plasma oscillations. The finite width of the ring means that the magnetic flux through the ring is not specified. It follows from the general considerations that the destructive interference of the contributions made by different electron trajectories to the wave function phase should distorts the strict periodicity of the Aharonov-Bohm oscillations. This distortion truly takes place; however, a particular implementation of this scenario is of interest. As will be shown below, the period of the Aharonov-Bohm oscillations in the finite-width ring specifically depends on the magnetic field, and the amplitude of the Aharonov-Bohm oscillations increases slowly in the low magnetic field region (estimated below).

Since the motion of the particles in the ring is finite in all directions and the single-particle spectrum is discrete, we consider the collective oscillations of the intersubband plasmon type. The spectrum of these excitations in the ring is also discrete, and the role of the longitudinal momentum of a plasmon in a rather narrow ring (the ring width is much shorter than the average radius) is played by its angular momentum ℓ. When an electromagnetic plane wave is incident normally to the ring plane, only plasmons with $\ell = \pm 1$ are obviously excited (this is an analog of the selection rules for dipole transitions in a uniform field), and higher harmonics with $|\ell| > 1$ can be excited if the electric field of the wave is modulated in the ring plane. According to the expansion of the plane wave in cylindrical harmonics, the plasmon absorption intensity with the momentum ℓ is proportional to $J_\ell^2(ka)$, a is the average radius of the ring, and k is the momentum of the exciting field in the ring plane.

To calculate the plasma frequencies, we use the model [28] allowing for an analytical solution of the single-particle Schrodinger equation. The potential energy of the electrons in the ring is written as

$$U(\rho) = \frac{A}{\rho^2} + B\rho^2 \tag{9.84}$$

Introducing the average radius of the ring a (the value of ρ at which potential energy (9.84) has a minimum) and the frequency of the low-amplitude radial oscillations ω_0, we rewrite (9.84) as

$$U(\rho) = \frac{m_e \omega_0^2 a^2}{8} \left(\frac{\rho^2}{a^2} + \frac{a^2}{\rho^2} \right) \tag{9.85}$$

Then, the condition of the smallness of the ring width, which will be used below, is represented in the form

$$\beta \equiv \frac{m_e \omega_0 a^2}{2\hbar} \gg 1 \tag{9.86}$$

Expression (9.86) means that the oscillatory length for the frequency ω_0 is much shorter than the circle radius at which $U(\rho)$ is minimal. The single particle spectrum and the wave functions are defined as

$$\psi(\mathbf{r}) = C_M \rho^M e^{-\rho^2/4\Delta^2} e^{im\varphi} F\left(-r; M+1; \rho^2/2\Delta^2\right),$$

$$E_{r,m} = \alpha\omega_0\left(r + \frac{1+M}{2}\right) + \frac{m\omega_c}{2},$$

$$M = \sqrt{m^2 + \beta^2}, \ m = 0, \pm 1, \pm 2, \ldots \tag{9.87}$$

where r and m are the radial and azimuthal quantum numbers, respectively; $\alpha = \sqrt{1 + \omega_c^2/\omega_0^2}$, ω_c is the cyclotron frequency; and $\Delta = 1/\sqrt{m_e\omega_0}$.

The perturbation of the electron density in the linear response theory has the form

$$\delta n(\mathbf{r}) = \sum_{ij} \frac{V_{ij}\psi_i^0(\mathbf{r})\psi_j^0(\mathbf{r})(f_i^0 - f_j^0)}{E_i - E_j + \omega} = \sum_\ell \delta n_\ell(\rho)e^{i\ell\varphi}, \tag{9.88}$$

where $\psi_i^0(\mathbf{r})$ are the unperturbed wave functions of the single particle problem; $\mathbf{r} = (\rho, \varphi)$, V_{ij} are the matrix elements of monochromatic excitation; $i = (r', m')$, $j = (r, m)$, f^0 are the Fermi occupation numbers. Substituting $\delta n(\mathbf{r})$ into the Poisson equation $\Delta\varphi = -4\pi e\delta n(\mathbf{r})\delta(z)$, we write its formal solution (at $z = 0$) in terms of the point source function

$$V^{\text{ind}}(\mathbf{r}) = e \int \frac{\delta n(\mathbf{r}')\,d\mathbf{r}'}{|\mathbf{r} - \mathbf{r}'|} = \sum_\ell V_\ell^{\text{ind}}(\rho)e^{i\ell\varphi} \tag{9.89}$$

Here, $V^{\text{ind}}(\mathbf{r})$ is the density excitation induced part of the electrostatic potential. For the ℓth component, we have

$$V_\ell^{\text{ind}}(\rho) = \int_0^\infty G_\ell(\rho, \rho')\delta n_\ell(\rho')\rho'd\rho', \tag{9.90}$$

Here,

$$G_\ell(\rho, \rho') = \int_0^\infty J_\ell(k\rho)J_\ell(k\rho')\,dk = \frac{1}{\pi\sqrt{\rho\rho'}}Q_{\ell-1/2}\left(\frac{\rho^2 + \rho'^2}{2\rho\rho'}\right). \tag{9.91}$$

where $Q_n(x)$ is a spherical function of the second kind. To close the system of equations, we find the matrix elements of the induced potential $V^{\text{ind}}(\mathbf{r})$ and obtain

$$V_{s,n}^{s',n+\ell} = e^2 \sum_{r,r',m} \frac{V_{r,m}^{r',m+\ell}(f_{r,m} - f_{r',m+\ell})F_{r,m;r',m+\ell}^{s,n;s',n+\ell}}{E_{r,m} - E_{r',m+\ell} - \omega}, \tag{9.92}$$

where

$$F_{r,m;r',m+\ell}^{s,n;s',n+\ell}$$
$$= \int_0^\infty \rho\,d\rho \int_0^\infty \rho'\,d\rho'\, R_{s,n}(\rho)R_{s',n+\ell}(\rho)G_\ell(\rho, \rho')R_{r,m}(\rho')R_{r',m+\ell}(\rho').$$
$$\tag{9.93}$$

Here, $R(\rho)$ are the radial wave functions in potential (9.85). The nontriviality condition of the solution of the system of equations (9.92) yields the plasmon spectrum $\omega_{r,\ell}$ where r is the radial mode number.

Now, we use the ring narrowness condition $\beta \gg 1$. Then, the energy intervals between the levels with different radial numbers in the single particle spectrum are much larger than the intervals between the levels with different ℓ values. In what follows, we consider the plasmons of the lowermost radial subband, i.e., consider the system of equations (9.92) with $r = r' = 0$. In the same approximation, $G_\ell(\rho, \rho')$ are significant only at ρ, ρ' close to a: $|\rho - a| \sim |\rho' - a| \leq \Delta = 1/\sqrt{m_e \omega_0}$, and the radial wave functions for $r = 0$ can be approximated as follows:

$$R_{0,m}(\rho) \sim \rho^\beta e^{-\frac{\rho^2}{\Delta^2}} \sim e^{\beta \ln \rho - \rho^2/\Delta^2} \sim e^{-(\rho-a)^2/2\Delta^2}. \tag{9.94}$$

Then, the integrals in (9.93) are readily calculated, and the form factors $F^{0,n;0,n+\ell}_{0,m;0,m+\ell}$ appear to depend only on ℓ ($F^{0,n;0,n+\ell}_{0,m;0,m+\ell} \equiv F_\ell$),

$$F_\ell = \frac{1}{\pi a}\left(\ln \frac{2\sqrt{2}a}{e^{C/2}\Delta} - \psi(\ell + 1/2)\right), \tag{9.95}$$

where $\psi(x)$ is the Euler function. We immediately note that the problem for $\ell \gg 1$ is semiclassical in the azimuthal degree of freedom, the quantity ℓ/a is transformed to the momentum of the one-dimensional plasmon k, and the known logarithmic singularity of the dispersion law of the 1D plasmon arises in the long-wavelength limit $\omega \gg k v_F$ because $\psi(\ell \gg 1) \sim \ln \ell$. We recall also that the long-wavelength asymptotic form of the 1D plasmon dispersion $\omega^2 \sim k^2 \ln k$ is the same both in the RPA method we use and in the Luttinger liquid model [29].

After all of the simplifications made, the dispersion equation becomes

$$1 = e^2 F_\ell \sum_m \frac{f(E_m) - f(E_{m+\ell})}{E_m - E_{m+\ell} - \omega} \tag{9.96}$$

In the same narrow ring approximation ($\beta \gg 1$), the energy levels are given by the expression

$$E_m \approx \alpha B \left(m + \frac{\Phi}{\alpha \Phi_0}\right)^2 - B\frac{\Phi^2}{\alpha \Phi_0^2} + \frac{\alpha \omega_0}{2}\left(1 + \frac{\beta}{2}\right), \tag{9.97}$$

then $E_{m+\ell} - E_m = \alpha B \ell^2 + 2\alpha B \ell (m + \frac{\Phi}{\alpha \Phi_0})$. Now, the condition $\omega \gg |E_{m+\ell} - E_m|$ corresponds to the long-wavelength period in the plasmon spectrum. In the second order in this parameter, the frequency ω_ℓ is represented in the form

$$\omega_\ell^2 + e^2 F_\ell \sum_m \left(f(E_m) - f(E_{m+\ell})\right)\left(E_m - E_{m+\ell} + \frac{(E_m - E_{m+\ell})^2}{\omega_\ell}\right) = 0. \tag{9.98}$$

In this expression, the chemical potential in the distribution function is determined by the equality

$$\sum_m f(E_m) = N, \tag{9.99}$$

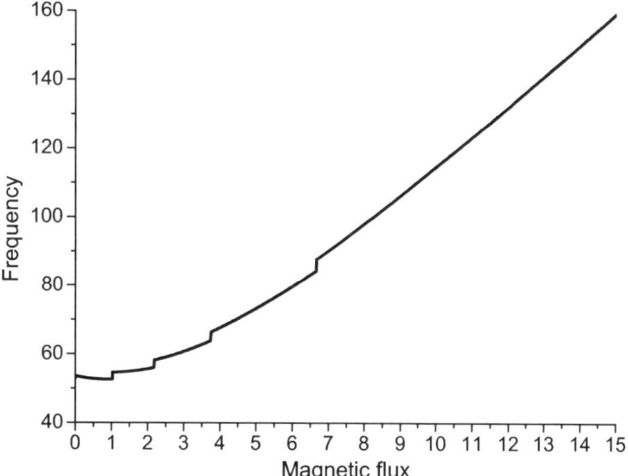

Fig. 9.7 Plasma frequency at $\ell = 1$ versus the magnetic flux through the middle circle of the ring. The magnetic flux is specified in the units of quantum Φ_0, the plasma frequency, in the units of rotating quantum B. The ring parameters are $a = 10a_B$, where a_B is the effective Bohr radius, and $\Delta = a_B$. The number of electrons in the ring is $N = 50$

where N is the total number of electrons in the ring. To calculate the sum in equality (9.99), we use the Poisson summation formula. The equation for the chemical potential becomes

$$\frac{1}{\pi} \sum_k \frac{\sin(2\pi k \sqrt{(\mu - \mu_0)/\alpha B})}{k} \cos\left(2\pi k \frac{\Phi}{\alpha \Phi_0}\right) = N, \qquad (9.100)$$

This equation implies that the main contribution to the left-hand side of (9.100) at $N \gg 1$ comes from the term with $k = 0$, which determines the smooth part of the chemical potential $\mu - \mu_0 = \alpha B N^2/4$. The remaining terms give small oscillating addends.

Applying again the Poisson summation formula and solving (9.98) by iterations, we obtain

$$\omega_\ell^2 = 2e^2 F_\ell \alpha B N \ell^2 + \frac{32}{\sqrt{2}} \sqrt{\frac{e^2 F_\ell}{N}} \alpha^{3/2} B^{3/2} \ell^2 \sum_{k=1}^{\infty} A_k \sin\left(2\pi k \frac{\Phi}{\alpha \Phi_0}\right), \quad (9.101)$$

where

$$A_k = \frac{\sin(\pi k N) - \pi k N \cos(\pi k N)}{4\pi^2 k^2}. \qquad (9.102)$$

The first term in (9.101) presents the monotonic part of plasmon dispersion and corresponds to the long-wavelength asymptotic expression $k^2 \ln k$. Both the period and amplitude of the oscillating part increase with the magnetic field, but the amplitude increases only in the region corresponding to the long-wavelength limit, while the period increases throughout the magnetic field range, as is seen in the Fig. 9.7, which presents the result of the numerical solution of dispersion equation (9.96).

Therefore, the magnetic field dependence of the plasmon frequency in the finite-width ring has both the monotonic part and the Aharonov-Bohm oscillations with the period and amplitude varying with the magnetic field. The characteristic magnetic field is determined by comparing the magnetic length and the ring width.

9.7 Polaronic Effect in QRs [30]

The electron-phonon interaction in reduced-dimension systems has been discussed in the literature since the very beginning of the physics of low-dimensional systems as an independent branch of solid-state physics. The effect of acoustic phonons on the mobility of electrons in an inversion channel was considered as early as 1982 by Ando et al. [31] by example of a two-dimensional electron system in silicon. In low-dimensional A_3B_5 systems (just as in bulk samples), polaron phenomena due to optical phonons attract considerable interest. These phenomena have been discussed in [32–34] (quantum wires) and in [32–35] (quantum dots) with regard to the effect of a strong (in the sense that the magnetic length is of the same order of magnitude or less than the size of the domain where the charge carriers move) magnetic field. By example of a quantum dot, it was demonstrated in [35] that polaron phenomena become more prominent as the size of a quantum dot is reduced: a shift in the electron energy due to the coupling to polar optical phonons is inversely proportional to the radius of a quantum dot.

Quantum rings occupy a special position among nanoobjects. The main topological feature of these rings is that the domain where an electron moves is not simply connected. This fact leads to AB oscillations in a magnetic field. It is well known that AB oscillations arise even if an electron is not subject to a Lorentz force (a thin solenoid inside a ring). However, from the experimental point of view, a typical situation is when a uniform magnetic field is applied to the system and the ring has a finite width. In this case, the magnetic field may significantly affect the radial motion of particles.

Polaron phenomena in quantum rings of finite width in a magnetic field must be characterized by distinctive features. The main feature is the nonmonochromaticity of the Aharonov-Bohm oscillations of a polaron shift, which is attributed to the difference in the magnetic fluxes that are enclosed by different electron trajectories (see preceding paragraph). Moreover, when an exciton is generated, the contributions of the electron and the hole to the polarization of the medium have opposite signs, and it is important that the finiteness of the ring should be taken into account when calculating the net effect determined by the wave functions of the particles. The present paragraph is devoted to the theoretical study of the formation of magnetopolarons in a quantum ring with regard to the radial motion of particles and the effect of polaron phenomena on optical interband transitions.

9.7.1 Electron and Hole Polarons

We use the same model (9.84) for the radial confinement, and, in additional to electrons we have now the wave functions and energy levels of holes

$$\psi_N = C_N r^N \exp\left(-\frac{r^2}{4a_h^2}\right)e^{il_h\varphi}, \qquad E_h = \hbar\Omega_h\left(\frac{N+1}{2}\right) + l_h\frac{\hbar\omega_{Bh}}{2}$$

$$C_N = \frac{1}{a_h^{N+1}\sqrt{2\pi}}\sqrt{\frac{1}{2^N\Gamma(N+1)}}, \quad N^2 = l_h^2 + 2m_h A/\hbar^2. \tag{9.103}$$

Here l_h is the eigenvalue of the projection operator of the angular momentum of a hole (here we use $l_{h,e}$ notation for angular momenta of electron and hole while m_e and m_h remain for effective masses), $\Omega_h^2 = \omega_0^2 + \omega_{Bh}^2$ is the combined frequency, ω_{Bh} is the cyclotron frequency of a hole, $a_h = \sqrt{\hbar/m_h\Omega_h}$ is the oscillatory length of an electron at the combined frequency Ω_h. Electron wave functions have the same form like (9.103). In further calculations, we assume, for simplicity, that the parameters A, B in (9.84) are the same for electrons and holes (unless otherwise stated). This assumption has a small effect on the results.

As pointed out above, the strong localization of particles in quantum rings may enhance their interaction with longitudinal optical phonons and thereby significantly change the quantization energies of the particles (a polaron shift). To calculate the coupling energy of a polaron, one should add the Hamiltonian of free phonons and electron-phonon interaction to the Hamiltonian of electron localized in QR. Free phonons and electron-phonon coupling are described by

$$H = T_e + \sum_{\mathbf{q}} \hbar\omega_{\mathbf{q}} b_{\mathbf{q}}^+ b_{\mathbf{q}} + \sum_{\mathbf{q}} F_{\mathbf{q}}(\mathbf{r})\left(b_{-\mathbf{q}}^+ + b_{\mathbf{q}}\right),$$

$$F_{\mathbf{q}}(\mathbf{r}) = D_{\mathbf{q}}(\mathbf{r})e^{i\mathbf{q}\mathbf{r}} = \frac{e}{q}\sqrt{\frac{2\pi\hbar\omega_{\mathbf{q}}}{\varepsilon^*}}e^{i\mathbf{q}\mathbf{r}} \tag{9.104}$$

Here, T_e is a free electron Hamiltonian in a QR, $b_{\mathbf{q}}^+ (b_{\mathbf{q}})$ are the creation (annihilation) operators of phonons with the wavevector $\mathbf{q}$, e is the electron charge, $\varepsilon^{*-1} = \varepsilon_\infty^{-1} - \varepsilon_0^{-1}$ is the effective optical permittivity, and $\omega_{\mathbf{q}}$ is the phonon frequency. In the case of a weak-coupling polaron, the polaron correction can be calculated by perturbation theory:

$$\Delta E_l = -\sum_{q,l'} \frac{|P_{ll'}(q)|^2}{E_{l'} - E_l + \hbar\omega_q - \Delta},$$

$$\Delta = \Delta E_l - \Delta E_G, \tag{9.105}$$

where $P_{ll'}(q)$ is the matrix element of the electron-phonon interaction. The quantity Δ depends on the type of perturbation theory. It is well known that the Rayleigh-Schrodinger perturbation theory ($\Delta = 0$) well describes only a correction to the ground state of the system. For excited states, one should apply the Wigner-Brillouin perturbation theory with $\Delta = \Delta E_l - \Delta E_G$. Here, ΔE_G is a polaron correction to the ground state calculated for $\Delta = 0$. In the case of dispersion-free optical phonons, ($\omega_q = \omega_{\mathrm{opt}} = \mathrm{const}$), the polaron correction to the electron states is given by (see calculation in [36])

$$\Delta E_{l_e} = -\frac{e^2\hbar\omega_{\mathrm{opt}}}{\varepsilon^* a_e} \sum_{l'_e} \frac{A_{l_e,l'_e}}{E_{l'_e} - E_{l_e} + \hbar\omega_q - \Delta},$$

$$A_{l_e,l'_e} = \frac{\Gamma^2(\Delta l_e/2 + M'/2 + M/2 + 1)}{\Gamma^2(\Delta l_e + 1)\Gamma(M + 1)\Gamma(M' + 1)}$$

$$\times \int_0^\infty \frac{dt}{2\pi} t^{\Delta l_e} F\left(\Delta l_e/2 + M'/2 + M/2 + 1; \Delta l_e + 1; -t^2\right), \tag{9.106}$$

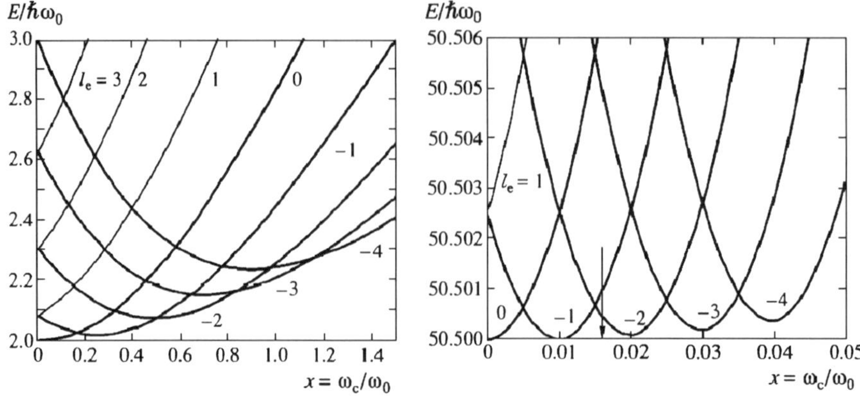

Fig. 9.8 Energy spectrum of an electron in a wide (*left panel*) and narrow (*right panel*) QR

where F is a degenerate hypergeometric function and Γ is the gamma function. The result for a hole polaron has the same form, except the change $a_e \to a_h$. Figure 9.8 shows the energy spectrum of an electron in a quantum ring for $\omega_0 = 12\hbar/2m_e a^2$. One can see that, when the finiteness of the ring width is taken into account, the spectrum significantly differs from the spectrum of the one-dimensional model,

$$E_l = \frac{\hbar^2}{2m_e a^2}(l + \Phi/\Phi_0)^2, \tag{9.107}$$

where $\Phi_0 = hc/e$ is the quantum of a magnetic flux. Naturally, one should expect that, as the ratio of the radial quantum to the rotational one increases, the spectrum of the system ever more closely approaches the spectrum E_l of the one-dimensional ring. This fact is illustrated in Fig. 9.8, where $\omega_0 = 400\hbar/2m_e a^2$. The arrow in this figure indicates the value of the magnetic field equal to 1 T. The values of the quantum number l_e are shown in the figure. In Fig. 9.8, energy is normalized by the radial quantum $\hbar\omega_0$. The polaron corrections to the states with different values of l_e calculated by formula (9.106) are shown in Fig. 9.9. One can see that, in addition to the oscillatory component, there exists a smooth envelope that comes from the magnetic field dependence of the radial wave functions of an electron. Figure 9.10(a) represents the magnetic field dependence of the ground state of an electron in a quantum ring, and Fig. 9.10(b) shows the polaron correction (ΔE_G) to this dependence calculated by formula (9.106). One can see that this dependence is also oscillatory. Similar to Fig. 9.9, the graph has an envelope; however, this envelope much more weakly depends on magnetic field than that for excited states. All the calculations shown in Figs. 9.9 and 9.10 are performed for the ratio of parameters given by $\omega_0 = 400\hbar/2m^* r_0^2$.

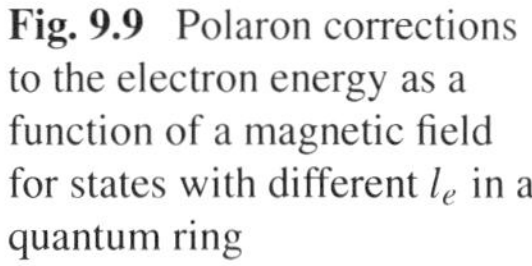
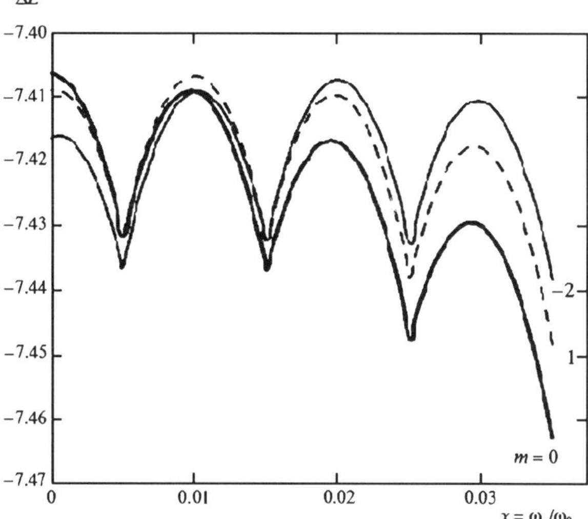

Fig. 9.9 Polaron corrections to the electron energy as a function of a magnetic field for states with different l_e in a quantum ring

9.7.2 *Excitonic Polaron and Interband Optical Transitions*

The absorption of light quantum with energy greater than the bandgap energy of a material gives rise to an electron-hole pair in a quantum ring. Each particle polarizes the medium, the polarization having opposite signs for the electron and the hole. As a result, these polarization wells partially compensate each other. As is shown in [35], in spherical quantum dots, the compensation is not exact when only the degeneracy of the valence band of the material is taken into account. In the case under consideration, due to the strong quantization in the vertical direction, the degeneracy is removed; however, the polarizations are not fully compensated because of the strong difference between the effective masses of an electron and a heavy hole.

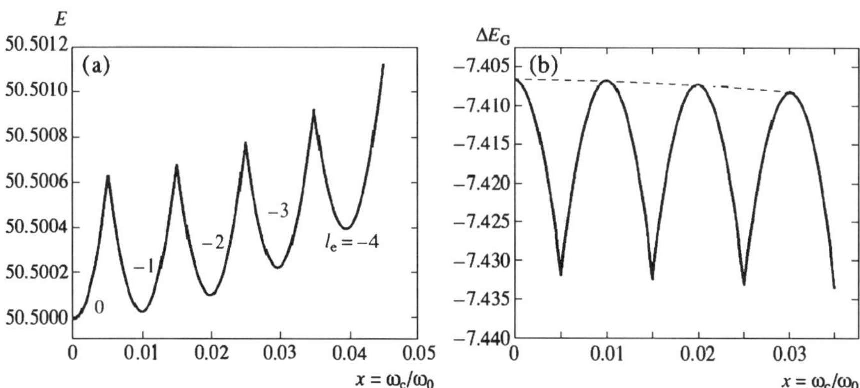

Fig. 9.10 (**a**) The ground state of an electron in a quantum ring and (**b**) a polaron correction to it as functions of a magnetic field

Now, we consider interband transitions. Before the absorption of a photon, a quantum ring has no charged particles and, hence, the medium is nonpolarized. After the absorption (in the final state of the system), there is an electron in the conduction band and a hole in the valence band; this leads to the polarization of the material. This means that the wavefunctions of the oscillators of the medium before $|\Phi^i\rangle$ and after $|\Phi^f\rangle$ the absorption are not orthogonal: $\langle\Phi^i|\Phi^f\rangle \neq 0$, because the functions $|\Phi^f\rangle$ have a displaced equilibrium state due to the polarization [35]. Taking into account this circumstance, we can represent the absorption probability of a quantum $\hbar\omega$ of light in the dipole approximation as

$$W_{\text{abs}} = \frac{2\pi}{\hbar}|p_{cv}|^2|I_{l_e,l_h}|^2 \sum_K W_K \delta(E_{eh} - E_K - \hbar\omega), \qquad (9.108)$$

where E_{eh} is the transition energy, $E_K = K\hbar\omega_{\text{opt}}$ (K is an integer) is the total energy of phonons that are involved in the interband transition, W_K is the emission probability of K phonons during the transition, I_{l_e,l_h} is the overlap integral of the envelope functions of the electron and hole, and p_{cv} is the Bloch amplitudes matrix element of the interband transition. The emission probability W_K of phonons is determined by the scalar product of the oscillation functions of the medium, $\langle\Phi^i|\Phi^f\rangle \neq 0$. In the case of dispersion-free optical phonons $\hbar\omega_q = \hbar\omega_{\text{opt}}$, this probability is determined by (see [35] for the details of calculation)

$$W_K = \frac{S_0^K}{2^K K!}e^{-S_0^2/2}, \quad S_0 = \delta E_{\text{ex}}/\hbar\omega_{\text{opt}}, \qquad (9.109)$$

where δE_{ex} is the total polaron shift of the exciton, which is defined below. The expression for the emission probability of a light quantum

$$W_{\text{em}} = \frac{2\pi}{\hbar}|p_{cv}|^2|I_{l_e,l_h}|^2 \sum_K W_K \delta(E_{eh} - E_K - \Delta E_s - \hbar\omega), \qquad (9.110)$$

differs from the absorption probability (9.108) by the Stokes shift ΔE_s, which is given by (see [35])

$$\Delta E_s = 2\delta E_{\text{ex}}. \qquad (9.111)$$

Formulas (9.109) and (9.111) show that both the intensity of phonon replicas and the Stokes shift of the spectrum are determined by the quantity ΔE_{ex}, which represents the renormalization of the exciton spectrum due to the formation of a polaron state. Below, we will show that this quantity is an oscillating function of a magnetic field (similar to the electron spectrum considered above); this leads to the oscillations of the Stokes shift and of the intensities of phonon replicas. Note that the oscillations of the intensity of exciton luminescence in quantum rings have been predicted in [37] (see also (9.51) of Sect. 9.5.1 of this review).

To determine ΔE_{ex}, we write out an expression for the Hamiltonian of two particles, an electron and a hole, in a quantum ring that interact with longitudinal phonons

$$H = T_e + T_h + V(|\mathbf{r}_e - \mathbf{r}_h|) + \sum_q \hbar\omega_q b_q^+ b_q + \sum_q [F_q(\mathbf{r}_e) - F_q(\mathbf{r}_h)](b_{-q}^+ + b_q),$$

$$(9.112)$$

here $T_{e(h)}$ is a free electron (hole) Hamiltonian in QR.

Since the radius a may vary within rather wide limits during the formation of quantum rings, one should consider two possible situations. For sufficiently small radii ($a \ll a_B$ where B is the effective Bohr radius of an exciton in the ring material), the Coulomb interaction can be neglected. In this case, the dynamics of the electron and the hole are independent, and the Schrödinger equation is decomposed into two equations, for the electron and the hole. The polaron corrections to the spectrum of the electron-hole pair are calculated by a formula similar to (9.105), in which the matrix element $P_{ll'}(q)$ should be replaced by a sum of the matrix elements of the electron-phonon and hole-phonon interactions (the third term in (9.112)). The calculations yield

$$\delta E_{\mathrm{ex}} = \Delta E_e + \Delta E_h + \Delta E_{\mathrm{mix}}. \tag{9.113}$$

Here, ΔE_e and ΔE_h are the polarization shifts of an electron and a hole, respectively, and ΔE_{mix} is a combined shift. The one-particle energy shifts ΔE_e and ΔE_h are calculated by formula (9.106); for ΔE_{mix}, we obtain

$$\Delta E_{\mathrm{mix}} = -\frac{e^2 \hbar \omega_{\mathrm{opt}}}{\varepsilon^*} \sum_{\alpha'} \frac{B_{\alpha\alpha'}}{E_{\alpha'} - E_\alpha + \hbar\omega_q - \Delta},$$

$$B_{\alpha\alpha'} = \frac{a_e^{\Delta l_e} a_h^{\Delta l_h}}{2^{\Delta l_e + \Delta l_h}}$$

$$\times \frac{\Gamma(1 + \frac{\Delta l_e + M' + M}{2})\Gamma(1 + \frac{\Delta l_h + N' + N}{2})}{\Gamma(\Delta l_e + 1)\Gamma(\Delta l_h + 1)\sqrt{\Gamma(M+1)\Gamma(M'+1)\Gamma(N+1)\Gamma(N'+1)}}$$

$$\times \int_0^\infty \frac{dq_x}{2\pi} q_x^{\Delta l_e + \Delta l_h} F\left(\Delta l_e/2 + M'/2 + M/2 + 1; \Delta l_e + 1; -q_x^2 a_e^2/2\right)$$

$$\times F\left(\Delta l_h/2 + N'/2 + N/2 + 1; \Delta l_h + 1; -q_x^2 a_h^2/2\right) \tag{9.114}$$

where α is the set of quantum numbers (l_e, l_h). Thus, since the particles move independently for $a \ll a_B$ and each term in (9.113) is an oscillating function of a magnetic field, we should expect that the Stokes shift ΔE_s and the intensity W_K will oscillate according to (9.109) and (9.111).

In the opposite case, when $a \gg a_B$, we should take into account the effect of the Coulomb potential on the dynamics of particles; therefore, we fail to obtain an exact analytic result. Let us carry out a qualitative analysis. By analogy with Sect. 9.4 of this review, we assume that the motion is adiabatic in the radial direction. Following the calculation procedure described in that paragraph, we obtain the following relations for the exciton spectrum:

$$E_J = E_{\mathrm{rad}} + \langle B \rangle J^2 + \langle E_{\mathrm{rel}}(\Phi) \rangle,$$

$$B = \frac{\hbar^2}{2(m_e r_e^2 + m_h r_h^2)}. \tag{9.115}$$

Here, $J = l_e + l_h$ is the total momentum of the exciton. The first term in the first equation is the energy of the radial motion of particles, the second term is the energy of motion of an exciton as a whole, and the last term represents the energy of the bound state of the particles. The angular brackets denote averaging with respect to the radial coordinates r_e and r_h of the particles. Note that $\langle B \rangle$ depends on the magnetic field only via the effective oscillation length $a_e = \sqrt{\hbar/m_e \sqrt{\omega_0^2 + \omega_{Be}^2}}$. However, Fig. 9.8 shows that, for sufficiently narrow rings, the variation limits of the magnetic field are such that $\omega_{Be}/\omega_0 \ll 1$, and the quantity $\langle B \rangle$ very weakly depends on the magnetic field. As regards the last term in (9.115), according to (9.49) and (9.50), the flux-dependent correction to the electron-hole interaction energy is proportional to

$$\langle E_{\mathrm{rel}}(\Phi) \rangle \sim e^{-2\pi a/a_B} \tag{9.116}$$

In the limit of $a \gg a_B$, this expression is exponentially small, and the effect of a magnetic field on the polaron corrections to the spectrum (9.115) is negligible. It was implicitly assumed in (9.115) that an electron and a hole in the ring move along circles of identical radii. According to [38], if we take into account that the radius of a circle for an electron, a^e, is different from the radius of a circle for a hole a^h, we obtain

$$E_J = E_{\mathrm{rad}} + \langle B(J + \Delta\Phi/\Phi_0)^2 \rangle + \langle E_{\mathrm{rel}}(\Phi) \rangle, \tag{9.117}$$

instead of (9.115); here, $\Delta\Phi$ is proportional to $-e(a^h - a^e)B$, i.e., to the product of the dipole moment of an exciton multiplied by the magnetic field. This correction leads to a considerable reconstruction of the exciton spectrum [38].

Thus, in both situations, the spectra of an electron- hole complex (for $a \ll a_B$ and of an exciton (9.117) for $a^e, a^h \gg a_B$, considered as functions of a magnetic field, have intersecting branches. In particular, the ground state oscillates with a magnetic field [38]. This leads to oscillations of ΔE_G in expression (9.105) for the polaron correction and, hence, to the oscillations of the polaron correction ΔE_{ex} to the optically active state $J = l_e + l_h = 0$ of the exciton. As pointed out above, the Stokes shift in the absorption and emission spectra of light, as well as the intensity of phonon replicas, are determined by the quantity ΔE_{ex} and, hence, oscillate with a magnetic field. However, when $a^e, a^h \gg a_B$, the polaron shift may exhibit appreciable oscillatory phenomena only for a radially polarized exciton $a^e \neq a^h$.

9.8 Vertical Transport Through QR (Longitudinal Magnetoresistance) [39]

As we mentioned in the beginning of this chapter the situation of the Fig. 9.2 is the most typical for the experiments on AB effect in d.c. measurements. This means that one measures the *transversal* magnetoresistance (current is orthogonal to the magnetic field). In the present paragraph of this chapter we will consider the *longitudinal* magnetoresistance that can be realized if a QR is burred in a dielectric barrier. We study the tunnel (vertical) transport and the magnetic field is normal to the ring, i.e. parallel to the current. We suppose (Fig. 9.11) that QR is fabricated of

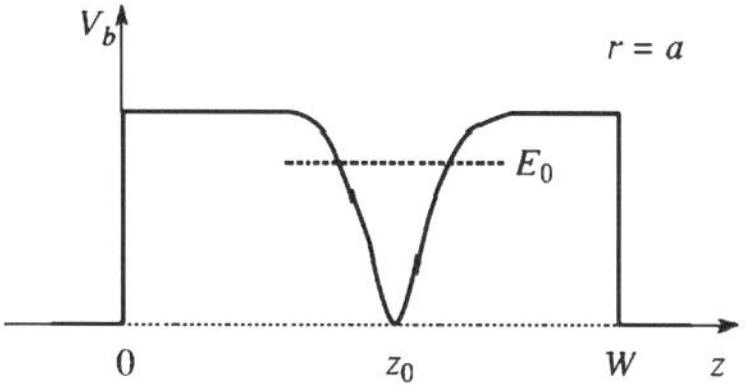

Fig. 9.11 The energy diagram of the structure with QR

a narrow band material (as compared with that of the barrier), hence, the ring plays a role of attracting impurity in the barrier with quasistationary bound states, possibility of the resonance tunneling and all such things. As we will show the resonance tunneling in the structure described is accompanied by the AB oscillations.

9.8.1 Spectrum and Wave Functions of a QR

We assume that a ring consists of a narrow band material (for example, InAs) and is immersed into a barrier, which is a wideband semiconductor (GaAs). The energy diagram of the system, corresponding to the distance from the ring center equal to its radius, is shown in Fig. 9.11; the coordinate z is measured along a normal to the heterostructure (the tunnel current flows in this direction). Thus, the interior and exterior regions of the ring are occupied with the barrier material, and the resonant tunneling corresponds to the particle trajectories that pass through the points on the ring.

Obviously, the model of one-dimensional ring $|\Psi|^2 \sim \delta(r-a)\delta(z-z_0)$ is insufficient for solving the problem posed because the resonant tunneling is determined by the overlap of the wave function of a particle impinging on the barrier and that of a particle bound in the ring. We will represent the potential of the ring in cylindrical coordinates by the expression

$$U(\mathbf{R}) = -U_0\delta(r-a)u(z-z_0), \tag{9.118}$$

where $U_0 > 0$, a is QR radius, z_0 is the position of the ring in the barrier plane, and $u(z)$ is a dimensionless function that is everywhere positive and has a sharp maximum at $z = z_0$ (see Fig. 9.11). Assume that the potential well is sufficiently narrow along z so that the wave function of the bound state cannot appreciably change over a distance on the order of the width of the function $u(z)$. Let $u(q)$ be the Fourier transform of $u(z)$. Then, after separating the variable φ, the Schrödinger equation in the q representation over z is expressed as

$$\nabla_r^2 \psi(r,q) - \left(k^2 + q^2 + \frac{m^2}{r^2}\right)\psi(r,q) + \frac{2m_eU_0}{\hbar^2}\delta(r-a)\psi(r,z_0) = 0. \tag{9.119}$$

Here, $E = -\hbar^2k^2/2m_e$; m_e is the electron mass in the barrier, where the electron mainly resides according to the model (9.118); $m = 0, \pm1, \pm2$: is the azimuthal quantum number; and

$$\psi(r, z_0) = \int \frac{dq}{2\pi} \psi(r, q) e^{iqz_0}. \tag{9.120}$$

Applying a conventional method to solving the radial equation (9.119) with the potential in the form of a δ-function and determining $\psi(r, z_0)$ in a self-consistent manner, we obtain from (9.120) the following equation for the energy eigenvalues in the ring:

$$1 = \lambda a \int_{-\infty}^{+\infty} dq\, K_{|m|}\left(a\sqrt{k^2 + q^2}\right) I_{|m|}\left(a\sqrt{k^2 + q^2}\right) u(q), \tag{9.121}$$

where $I_{|m|}$ and $K_{|m|}$ are Bessel functions of imaginary argument of the first and third kind, respectively. As is clear from (9.121), one cannot replace $u(z)$ by a δ-function because all the energy levels E_m in the ring will tend to minus infinity in this case. However, as will be clear from the analysis below, the main, resonant, part of the rotational spectrum and the tunneling conductivity depends on the form of the potential through a single parameter E_0. Hence, the ring model (9.118) provides a fairly reasonable approximation. For moderately small radii of the ring, the rotational quantum $\hbar^2/2m_e a^2$ is much less than the depth of the ground state $|E_0| = \hbar^2 k_0^2/2m_e$; i.e., $ka \gg 1$. In this limit, one can easily derive the following asymptotic expression from (9.121) (the argument of the Bessel functions is greater than their index):

$$E_m = E_0 + \frac{\hbar^2 m^2}{2m_e a^2} + \frac{\hbar^2 m^2 (m^2 + 5/2)}{2m_e a^2 (k_0 a)^2}. \tag{9.122}$$

Here, the second term corresponds to the spectrum of a one-dimensional ring (a plane rotator), while the third term represents a correction due to the finite width of the electron wave function spread out near the circle $r = a$.

In what follows, we need the wave functions of the bound states corresponding to energy E_m. These states are solutions to (9.119) that are finite at zero and decrease as $r \to \infty$; they are represented as follows

$$\psi_m(r < a, z) = c_m \int_{-\infty}^{\infty} dq\, K_{|m|}\left(a\sqrt{k_m^2 + q^2}\right) I_{|m|}\left(r\sqrt{k_m^2 + q^2}\right) e^{iq(z-z_0)},$$

$$\psi_m(r > a, z) = c_m \int_{-\infty}^{\infty} dq\, I_{|m|}\left(a\sqrt{k_m^2 + q^2}\right) K_{|m|}\left(r\sqrt{k_m^2 + q^2}\right) e^{iq(z-z_0)}, \tag{9.123}$$

where c_m is a normalizing factor.

One can see that the wave functions depend on the form of $u(z)$ only through k_m, i.e., ultimately through the parameter E_0.

9.8.2 Tunnel Current in the Model of a δ-Shaped Solenoid

Suppose that an infinitely thin solenoid with magnetic flux Φ passes through a quantum ring situated inside the barrier of a tunnel structure. The flow of tunneling electrons is parallel to the solenoid. The energy levels in the one-dimensional ring are classified according to the momentum $m = 0, \pm 1, \pm 2, \ldots$. Accordingly, a plane wave incident to the system is expanded in terms of cylindrical harmonics:

$$\Psi(\mathbf{r}) = \frac{1}{\sqrt{2\pi\,\Omega}} \sum J_\nu(k_{\|}r)\exp(im\varphi), \tag{9.124}$$

where Ω is a normalizing volume, J_ν is a Bessel function, $\nu = |m + \Phi/\Phi_0|$, r and φ are cylindrical coordinates in the plane of the structure, and $k_{\|}$ is the projection of the wave vector of the incident wave onto this plane; the number m is preserved during tunneling.

The wave functions of the bound states in the ring and the spectrum of appropriate energies are given by formulas (9.121)–(9.123) in which the index m of the Bessel functions should be replaced by ν.

To determine the tunnel current, we apply the Bardeen method and calculate the tunnel width Γ of the bound state and the transition amplitudes T_p and $T_{p'}$ from the state in the ring to the left and right contacts, respectively. As a result, we obtain $\Gamma = \Gamma_l + \Gamma_r$, where

$$\Gamma_{l(r)} = \pi \sum_{p(p')} \left| T_{p(p')}(E_{p(p')}) \right|^2 \delta(E_{p(p')} - E_m), \tag{9.125}$$

where E_p and $E_{p'}$ are, respectively, the energies of the electrons impinging on and transmitted through the barrier and $T_{p(p')} = E_m \langle \Psi_{p(p')} | \Psi_m \rangle$ are the overlapping integrals of the functions of the bound and free states, respectively. The tunneling probability from state p (on the left) to state p' (on the right of the barrier) through the level E_m is given by

$$\omega_{pp'} = \frac{2\pi}{\hbar} \frac{|T_p|^2 |T_{p'}|^2}{(E_p - E_m)^2 + \Gamma^2} \delta(E_p - E_m). \tag{9.126}$$

For the tunnel current, we have

$$I = e \sum_{p,p'} \omega_{pp'} \left[f(E_p - \mu) - f(E_{p'} - \mu + eV) \right], \tag{9.127}$$

where f is the Fermi-Dirac distribution function, μ is the chemical potential, and V is the voltage applied to the barrier.

For the conductivity of the system at $T = 0$, we have

$$g(\Phi) = g_0 \sum_m E_m^2 \frac{I_{|m+\Phi/\Phi_0|}(k_m^2 a^2 \theta_1^2) I_{|m+\Phi/\Phi_0|}(k_m^2 a^2 \theta_2^2)}{(1 + \frac{\pi}{6} \frac{(m+\Phi/\Phi_0)^2 - 1/4}{k_m a})[(\mu - E_m)^2 + \Gamma^2]}, \tag{9.128}$$

where

$$\theta_1^2 = \frac{V_b \sqrt{1 - E_m/V_b}}{k_b z_0 E_m}, \qquad \theta_2^2 = \frac{V_b \sqrt{1 - E_m/V_b}}{k_b(W - z_0) E_m},$$

V_b is barrier height, W is its width, and $k_b = \sqrt{2m_e V_b/\hbar^2}$.

Recall that, here, the energies E_m are functions of the combination $|m + \Phi/\Phi_0|$; therefore, the tunneling conductivity $g(\Phi)$, being a sum over m from $-\infty$ to $+\infty$, is

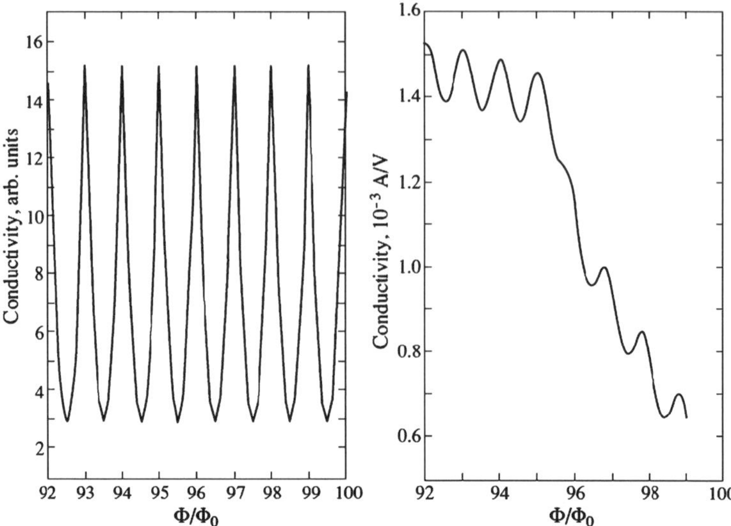

Fig. 9.12 Tunneling conductivity vs magnetic flux: thing solenoid (*left panel*) and uniform magnetic field (*right panel*)

a periodic function of the magnetic flux with period Φ_0. The numerically calculated values of $g(\Phi)$ are shown in Fig. 9.12. Because the model with a δ-shaped solenoid has a purely illustrative character, we performed the calculation for the case when electrons inside and outside the barrier have identical effective masses.

9.8.3 Uniform Magnetic Field

We consider the tunneling from a 3D emitter to a 3D collector. Outside the barrier, tunneling electrons are described by the Fock-Darwin wave functions (when the moment with respect to the normal to the system and the momentum along the field are specified). The general case involves very laborious calculations; therefore, we restrict the analysis to the limit case $H < \hbar\omega_c$, where $\hbar\omega_c$ is the Landau quantum. Obviously, the above inequality is equivalent to the condition $4\Phi \gg \Phi_0$ (the number of flux quanta passing through the ring is large). Suppose that the Fermi level lies between the lowest and the first Landau levels of the impinging electron; i.e., let us restrict ourselves to the range of $\delta\Phi$ that contains several quanta Φ_0 but is significantly less than the value of Φ such that $\hbar\omega_c \sim \mu$. Then, the functions $\Psi_{0,m}(\mathbf{r})$, which correspond to the lowest Landau level and $m \in (-\infty, 0)$, serve as the wave functions of the initial state (on the left of the barrier). Hence, one can see that the tunneling conductivity is not a periodic function of the magnetic flux since it is expressed in terms of a sum over a semi-infinite interval of m.

The tunneling widths are determined by the expressions

$$\Gamma_l = \frac{4}{3}\left(\frac{m_{e1}}{2\hbar^2}\right)^{1/2}\frac{k_m^2 a^2}{a}\frac{E_m^2 G^2(\nu)}{1+\frac{\pi}{6}\frac{\nu^2-1/4}{ka}}$$

$$\times \frac{\sqrt{E_m - \hbar\omega_c/2}}{(1-\frac{m_{e1}}{m_{e2}})(E_m - \hbar\omega_c/2) + V_b\frac{m_{e1}}{m_{e2}}}$$

$$\times \exp\left(-2k_b z_0\sqrt{1 - \frac{E_m - \hbar\omega_c/2}{V_b}}\right) \tag{9.129}$$

For Γ_r we have to replace $z_0 \rightarrow W - z_0$; m_{e1}, m_{e2} are the electron effective masses outside and inside the barrier, respectively.

In (9.129),

$$G(\nu) = \int_0^\infty \frac{J_\nu(sa)}{s^2 + k^2} R_\nu(s) s\, ds,$$

$$R_\nu(s) = \frac{2^{1+|m|/2}}{\sqrt{|m|!}} a_H^{1+\nu} \frac{\Gamma(1+\nu/2+|m|/2)}{\Gamma(1+\nu)} s^\nu$$

$$\times F\left(1+\nu/2+|m|/2; 1+\nu; -a_H^2 s^2\right) \tag{9.130}$$

is the Fourier-Bessel transform of the wave functions of the state outside the barrier for the lowest Landau level with a moment $m < 0$.

The conductivity ($T = 0$) is given by

$$g(\Phi) = \frac{16\pi}{9}\frac{e^2}{\hbar}\sum_{m=-\infty}^{m=0}\frac{m_{e1}k_m^4 a^4}{\hbar^2 a^2}\frac{E_m - \hbar\omega_c/2}{[(1-\frac{m_{e1}}{m_{e2}})(E_m - \hbar\omega_c/2) + V_b\frac{m_{e1}}{m_{e2}}]^2}$$

$$\times \frac{E_m^4 G^4(\nu)}{(1+\frac{\pi}{6}\frac{(m+\Phi/\Phi_0)^2-1/4}{k_m a})[(\mu - E_m)^2 + \Gamma^2]}$$

$$\times \exp\left(-2k_b W\sqrt{1 - \frac{\mu - \hbar\omega_c/2}{V_b}}\right) \tag{9.131}$$

The numerical calculation was performed for the following parameters: $\mu = 1.03|E_0|$, $m_{e1} = 0.025m_0$ (InAs), $m_{e1} = 0.07m_0$ (GaAs), $a = 300$ Å, $V_b = 0.55$ eV, $W = 100$ Å, and $z_0 = W/2$.

The results of calculations (Fig. 9.12) show that the oscillating character of $g(\Phi)$ is preserved. The characteristic interval between the spikes is equal to Φ_0, and $\Gamma \sim 0.002$–0.01 eV for a magnetic field strength of $H \sim 4.04$–4.4 T.

A decrease in the conductivity as the magnetic field increases is attributed to the fact that, for a fixed total energy of a tunneling electron, its "longitudinal" energy (the continuous part of the spectrum) decreases with the field. Therefore, the interval between this energy and the barrier ridge increases, and the decay length of the wave function along z decreases. As a result, the overlapping integral T_p decreases.

Thus, we have shown that a tunnel current through a heterostructure with quantum rings exhibits a specific Aharonov-Bohm effect. The tunnel current as a function of magnetic field for a given voltage across the structure has the form of modulated oscillations with a characteristic period Φ_0 in the flux. In the magnetic field scale, this interval is much less than the scale related to the Landau quantization for rings with radii on the order of 10–100 nm in fields of the order of 1–10 T.

9.9 Ring Excitons Under External Electromagnetic Radiation [40]

Elementary excitations in nanostructures can be controlled and tuned by means of external fields. This possibility makes the corresponding domain of low-dimensional physics interesting both from fundamental and applied point of view. In the last paragraph of this chapter we consider effect that experienced by excitons in the QR without static uniform magnetic field but nevertheless occurs for exciton moving in ring-shape potential. We will consider excitons in a quantum ring in the presence of low frequency electromagnetic (EM) radiation. The electric field of the EM wave is supposed to be uniform because the wave length is much greater then the ring radius. The uniform field cannot change the momentum **P** of an electrically neutral system but in the quantum ring we deal with the three-body problem: electron, hole and the ring itself. Due to the curvature the center-of-mass degree of freedom is characterized by the angular momentum J rather then the momentum **P** and, as we will show, the uniform electric field affect the free rotation of the exciton as a whole. Moreover this motion turns out coupled with the internal motion in excitons and this coupling results in a specific fine structure of the luminescence line.

9.9.1 Model and Hamiltonian

To quantitatively analyze the problem we consider an exciton in 1D quantum ring perturbed by the uniform circularly-polarized electromagnetic field $\mathbf{F}(t) = F_0(\cos(\omega t), \sin(\omega t))$. Interaction of the exciton with the field $\mathbf{F}(t)$ is given by

$$V(\varphi_e, \varphi_h, t) = eF_0 a\big[\cos(\varphi_e - \omega t) - \cos(\varphi_h - \omega t)\big], \qquad (9.132)$$

where φ_e and φ_h define the positions of the electron and hole in the ring ($0 \le \varphi_{e,h} \le 2\pi$). After some simple algebra the interaction term in the Hamiltonian can be written as $V(t) = 2eF_0 a \sin(\varphi_C + \gamma\theta - \omega t)\sin(\theta/2)$. The Hamiltonian of the problem without irradiation reads:

$$H_0 = -B\frac{\partial^2}{\partial\varphi_C^2} - \beta\frac{\partial^2}{\partial\theta^2} - \frac{e^2}{2\varepsilon a|\sin(\theta/2)|}, \qquad (9.133)$$

where the center-of-mass rotational quantum $B = \hbar^2/2Ma^2$ and that of the internal motion $\beta = \hbar^2/2\mu a^2$ are introduced and $\mu^{-1} = m_e^{-1} + m_h^{-1}$ is a reduced exciton mass. The last term in (9.133) stands for Coulomb attraction between electron and (heavy) hole with dielectric constant ε. Two methods for solving the

Schrödinger equation $i\hbar\partial_t\Psi = (H_0 + V(t))\Psi$ are used: the perturbation theory and the adiabatic approximation. The appropriate condition for perturbation approach is $eF_0a \ll \min(B, \mu e^4/\varepsilon^2\hbar^2)$. For the adiabatic method we accept that the internal motion is faster in compare with the center-of-mass one: $B \ll \mu e^4/\varepsilon^2\hbar^2$ and, at the same time, the ratio between B and $2eF_0a$ is assumed arbitrary.

9.9.2 Perturbation Theory Calculations

The solution of the Schrodinger equation in this case can be found by usual procedure known from quantum mechanics textbooks [14]. Without interaction term (9.132) the exciton dynamics is described by the total exciton angular momentum J and internal quantum number ℓ. The total unperturbed wave function can be presented as $\Psi_{J,\ell}^{(0)} = \exp(iJ\varphi_C)\psi_\ell(\theta)\exp(-iE_{J,\ell}t)$ where $E_{J,\ell} = E_g + BJ^2 + \varepsilon_\ell$, E_g is a band gap of the quantum ring material, and ε_ℓ is an internal motion energy. If the effective Bohr radius a_B is much smaller then the ring circumference $a_B \ll 2\pi a$ then only small θ are essential ($\theta \ll 1$) and we come to the 1D hydrogen atom problem. The exact wave function $\psi_\ell(\theta)$ and energy ε_ℓ of 1D hydrogen atom can be found, for example, in [41]. In the first order in respect of the interaction $V(t)$ the exciton wave function $\Psi_{J,\ell}^{(1)}$ for the arbitrary state (J, ℓ) is $\Psi_{J,\ell}^{(1)} = \Psi_{J,\ell}^{(0)} + \sum_{J',\ell'} A_{J',\ell'}^{J,\ell}\Psi_{J',\ell'}^{(0)}$. We consider here only the low-frequency external field $\hbar\omega \ll \mu e^4/\varepsilon^2\hbar^2$, which cannot excite the internal motion of the exciton. In this case the expansion coefficients $A_{J',\ell'}^{J,\ell}$ for $\ell' = \ell = 0$ are given by

$$A_{J',0}^{J,0} = \frac{2eF_0a}{\hbar}\left(C_{J'J}^{(1)}\frac{e^{i(\omega_{J'J}-\omega)t}}{i(\omega_{J'J}-\omega)}\right) + \left(C_{J'J}^{(2)}\frac{e^{i(\omega_{J'J}+\omega)t}}{i(\omega_{J'J}+\omega)}\right),$$

$$C_{J'J}^{(1)} = \frac{4(-1)^{J-J'+1}}{i(J-J'+1)}\int_0^{2\pi}d\theta\left|\psi_0(\theta)\right|^2\sin\left(\frac{\theta}{2}\right)\sin\left(\frac{\gamma\theta}{2}\right)\sin\left((J-J'+1)\frac{\theta}{2}\right),$$

$$C_{J'J}^{(2)} = \frac{4(-1)^{J-J'-1}}{i(J-J'-1)}\int_0^{2\pi}d\theta\left|\psi_0(\theta)\right|^2\sin\left(\frac{\theta}{2}\right)\sin\left(\frac{\gamma\theta}{2}\right)\sin\left((J-J'-1)\frac{\theta}{2}\right).$$

$$(9.134)$$

To calculate the probability w_J per unit time of the exciton interband transition with emission of the photon with wave vector $\mathbf{q}$ and a frequency $\Omega_{\mathbf{q}}$ from the initial state of the system $|J, \ell; N_{\mathbf{q}} = 0\rangle$ to the final state $|\text{vac}; N_{\mathbf{q}} = 1\rangle$ we use the relation [42]

$$w_J = \left(\frac{eA_0p_{cv}}{\hbar m_0c}\right)^2$$

$$\times \lim_{t\to\infty}\frac{1}{t}\left|\int_0^t d\tau e^{i\Omega_{\mathbf{q}}\tau}\int_0^{2\pi}\int_0^{2\pi}d\varphi_e d\varphi_h\delta(\varphi_e - \varphi_h)\Psi_{J,\ell}^{(1)}(\varphi_C,\theta,\tau)\right|^2$$

$$(9.135)$$

Here A_0 is the amplitude of the vector potential of the emitted photon and m_0 is the bare electron mass. After integration we obtain

$$w_J = (2\pi)^2 \left(\frac{eA_0 p_{cv}}{\hbar m_0 c}\right)^2 |\psi_{\ell=0}(0)|^2 \Big(\delta_{J,0}\delta(\Omega_{\mathbf{q}} - E_{0,0})$$

$$+ \frac{(2eF_0 a)^2 |C_{0,J}^{(1)}|^2}{\hbar^2(\omega_{0,J} - \omega)^2}\delta(\Omega_{\mathbf{q}} - E_{J,0} - \omega)$$

$$+ \frac{(2eF_0 a)^2 |C_{0,J}^{(2)}|^2}{\hbar^2(\omega_{0,J} + \omega)^2}\delta(\Omega_{\mathbf{q}} - E_{J,0} + \omega)\Big) \tag{9.136}$$

We see from this equation that luminescence spectrum acquires a "fine structure". The intensities of the satellites are proportional to the intensity of the external EM field F_0^2. It should be noted that (9.136) is only applicable far from the resonances $|\omega_{0,J} \pm \omega| \gg eF_0 a$.

9.9.3 Resonance Frequency of the External Field

Let us now consider the resonance case $\omega = \omega_{J',J}$. The wave function is a linear combination of the wave functions of the resonantly coupled states J' and J [14]:

$$\Psi(t) = \cos(\Omega_{JJ'}t)\Psi_{J'}^{(0)}(t) - \frac{i\Omega_{JJ'}^*}{\Omega_{JJ'}}\sin(\Omega_{JJ'}t)\Psi_J^{(0)}(t), \tag{9.137}$$

where $\Omega_{JJ'} = -ieF_0 a C_{J',J}^{(1)}$ is a Rabi frequency. In the limit $\theta \ll 1$ the coefficient $C_{J',J}^{(1)}$ can be easily calculated from (9.134), and for Rabi frequency we have $\Omega_{JJ'}^* = \Omega_{JJ'} = 3(-1)^{J-J'}eF_0 a\gamma/4\kappa^4\hbar$, where $\kappa = a/a_B \gg 1$. In (9.137) we assume that initial ($t = 0$) state of the exciton is $\Psi_{J'}^{(0)}$. With this wave function we compute the transition probability:

$$w_{J,J'} = \pi^2 \left(\frac{eA_0 p_{cv}}{\hbar m_0 c}\right)^2 |\psi_{\ell=0}(0)|^2 \Big(\delta_{J',0}\delta(\Omega_{\mathbf{q}} + \Omega_{J0} - E_{0,0})$$

$$+ \delta_{J',0}\delta(\Omega_{\mathbf{q}} - \Omega_{J0} - E_{0,0})$$

$$+ \delta_{J,0}\delta(\Omega_{\mathbf{q}} + \Omega_{0J'} - E_{0,0}) + \delta_{J,0}\delta(\Omega_{\mathbf{q}} - \Omega_{0J'} - E_{0,0})\Big). \tag{9.138}$$

As we can see from this equation there is splitting of the exciton line typical for any Rabi situation. It should be noted that at least one of the states J', J must be a ground state $J' = 0$ or $J = 0$ in order to get a nonzero probability of the transition.

9.9.4 Adiabatic Approximation

Taking into account the relation $B \ll \mu e^4/\varepsilon^2\hbar^2$ we can simplify the Schrödinger equation by means of averaging it over the ground state of the internal motion. To do this we first make a transform to the rotating coordinate system applying the unitary transformation to the exciton wave function $\bar{\Psi} = \hat{S}\Psi$, where $\hat{S} = \exp(i\omega t \hat{L}_z)$, and $\hat{L}_z = -i\partial/\partial\varphi_C$. This transformation procedure only affects the perturbation operator $\bar{V}(t) = \hat{S}^{-1}V(t)\hat{S} = 2eF_0a\sin(\varphi_C + \gamma\theta)\sin(\theta/2)$ and leaves unchanged the operator H_0: $\hat{S}^{-1}H_0\hat{S} = H_0$. Keeping in mind that $\hat{S}\psi_0(\theta) \equiv \psi_0(\theta)$ the total wave function of the exciton in rotating system is $\bar{\Psi}(\varphi_C,\theta,t) = \chi(\varphi_C,t)\psi_0(\theta)$. By averaging over the ground state of the internal motion we obtain the equation for center-of-mass wave function $\chi(\varphi_C,t)$, which holds for arbitrary ratio B and $2eF_0a$. For $\theta \ll 1$ we have:

$$i\hbar\frac{\partial}{\partial t}\chi = \left(-B\frac{\partial^2}{\partial\varphi_C^2} + i\hbar\omega\frac{\partial}{\partial\varphi_C} + eF_0a\gamma\cos(\varphi_C)\langle\theta^2\rangle\right)\chi,$$

$$\langle\theta^2\rangle = \int_{-\infty}^{+\infty}\theta^2|\psi_0(\theta)|^2d\theta = \frac{1}{2\kappa^2}. \tag{9.139}$$

Here we used $\psi_0(\theta) = \kappa^{1/2}\exp(-\kappa|\theta|)$. To solve this equation we restrict ourselves to the strong intensity limit $2\kappa^2 B \ll eF_0a\gamma$. Expanding $\cos(\varphi_C)$ in the vicinity of the minimum $\varphi_C \sim -\pi$ up to the harmonic oscillator potential, the solution of (9.139) can be written as

$$\chi(\varphi_C,t) = e^{-i\mathcal{E}_n t}\sqrt{\sqrt{\frac{1}{\pi L^2}}\frac{1}{2^n n!}}\,e^{-\frac{1}{2}(\frac{\varphi_C+\pi}{L})^2}e^{-i\frac{\omega}{\omega_0}\varphi_C}H_n\left(\frac{\varphi_C+\pi}{L}\right), \tag{9.140}$$

where $H_n(x)$ is the Hermite polynomial, $L = (\frac{4\kappa^2 B}{eF_0a\gamma})^{1/4} \ll 1$ is a dimensionless oscillatory length, $\mathcal{E}_n = E_g + \frac{B}{2L^2}(n+1/2) - B(\frac{\omega}{\omega_0})^2 - \frac{eF_0a\gamma}{2\kappa^2} + \varepsilon_0$, and $\omega_0 = 2B/\hbar$. With this wave function and the relation $\Psi = \hat{S}^{-1}\chi\psi_0$ we have for the probabilities of transitions from nth exciton state:

$$w_n = \left(\frac{eA_0 p_{cv}}{\hbar mc}\right)^2\sqrt{\frac{1}{\pi L^2}}\frac{4\pi^2}{2^n n!}|\psi_0(0)|^2 e^{-(\frac{\omega}{\omega_0}L)^2}H_n^2\left(\frac{\omega}{\omega_0}L\right)\delta(\Omega_{\mathbf{q}} - \mathcal{E}_n). \tag{9.141}$$

We see that in this case there is also a set of resonances in the luminescence spectrum of the exciton. For $\omega > \omega_0/L$ intensities of the resonances exhibit an oscillatory dependence on the frequency ω.

Thus we have the probabilities of the intersubband transitions (exciton luminescence) in the situation when the initial state wave function is modulated by the low-frequency electromagnetic radiation. This modulation results in the "fine structure" of the luminescence line. Of course the predicted fine structure can be wiped off by inevitable disorder and phonons. We suppose that the scattering processes do not destroy totally, at least, single-electron quantization in the ring and the spectrum $(\hbar^2/2m_e a^2)m^2$ with integer m "survives". At sufficiently low temperature only emission of the phonons contributes to the broadening of the exciton lines. Then it is easy to show that the ratio of the phonon emission rate of an electron $1/\tau_e$ to that of an exciton $1/\tau_{ex}$ reads:

$$\frac{\tau_{ex}}{\tau_e} \sim \left(\frac{C_e}{C_e + C_h}\right)^2\left(\frac{m_e + m_h}{m_e}\right)^2 \tag{9.142}$$

for the deformation potential interaction with acoustic phonons; C_e and C_h are the deformation potential constants. For GaAs quantum ring the numerical estimate gives $\tau_{\mathrm{ex}}/\tau_e \sim 60$. This result relates the case of low intensity of the external field when the center-of-mass commits nearly free rotation. Strongly external field results in localization of the center-off-mass wave function in the rotating frame of reference. We have considered a simple model: 1D harmonic oscillator interacting with 3D acoustic phonons. The ratio of the phonon emission rate to the oscillator frequency decreases monotonically when this frequency increases (localization becomes stronger). Thus, the external low frequency radiation only weakens the destructive effect of phonons.

Thus it was shown that exciton spectrum of the quantum ring under external electromagnetic radiation has a specific fine structure. Intensities of the satellites are calculated as functions of the intensity and frequency of the external field. In the case of strong external field the satellites' intensities oscillate versus external field frequency.

In conclusion: the AB effect has appeared in physics as something exotic, strange and very difficult for experimental realization. However its solid state versions— electron transport through QRs, excitons, trions and plasmons in QRs and nanotubes made stand this effect in a row of quite "normal" physical phenomena. Their experimental investigation looks rather perspective and promises interesting physical results and technical applications.

References

1. Y. Aharonov, D. Bohm, Phys. Rev. **115**, 485 (1959)
2. H. Batelaan, A. Tonomura, Phys. Today **38** (2009)
3. D. Mailly, C. Chapelier, A. Benoit, Phys. Rev. Lett. **70**, 2020 (1993)
4. A.G. Aronov, Y.B. Lyanda-Geller, Phys. Rev. Lett. **70**, 343 (1993)
5. I.R. Sellers, A.O. Govorov, B.D. McCombe, J. Nanoelectron. Optoelectron. **6**, 4 (2011)
6. F. Ding, B. Li, N. Akopian et al., J. Nanoelectron. Optoelectron. **6**, 51 (2011)
7. M.D. Teodoro, V.L. Campo Jr., V. Lopez-Richard et al., Phys. Rev. Lett. **104**, 086401 (2010)
8. A.O. Govorov, A.V. Chaplik, L. Wendler, V.M. Fomin, JETP Lett. **60**, 643 (1994)
9. M. Büttiker, Y. Imry, R. Landauer, Phys. Lett. A **96**, 365 (1983)
10. D. Müller-Groeling et al., Europhys. Lett. **22**, 193 (1993)
11. A. Müller-Groeling, H.A. Weidemüller, Phys. Rev. **49**, 4752 (1994)
12. I.V. Krive, R.I. Shekhter, S.M. Girvin et al., Phys. Scr. **54**, 123 (1994)
13. I.O. Kulik, JETP Lett. **11**, 275 (1970)
14. L.D. Landau, E.M. Lifshitz, *Quantum Mechanics: Non-relativistic Theory* (Pergamon, New York, 1977)
15. V.M. Kovalev, A.V. Chaplik, Semiconductors **37**, 1195 (2003)
16. I.S. Gradshtein, I.M. Ryshik, *Tables of Integrals, Series and Products* (Academic, New York, 1965)
17. M. Wagner, U. Merkt, A.V. Chaplik, Phys. Rev. B **45**, 1951 (1992)
18. M. Rotter, A.V. Kalameitsev, A.O. Govorov et al., Phys. Rev. Lett. **82**, 2171 (1999)
19. A.O. Govorov, A.V. Kalameitsev, V.M. Kovalev et al., Phys. Rev. Lett. **87**, 226803 (2001)
20. A.V. Chaplik, L.I. Magarill, Superlattices Microstruct. **18**, 321 (1995)
21. Y. Meir, Y. Gefen, O. Entin-Wohlman, Phys. Rev. Lett. **63**, 798 (1989)

22. A.V. Chaplik, JETP Lett. **62**, 900 (1995)
23. A.V. Chaplik, J. Exp. Theor. Phys. **92**, 169 (2001)
24. T. Chakraborty, L. Pietilainen, Phys. Rev. **50**, 8460 (1994)
25. A. Lorke, R.J. Luyken, A.O. Govorov et al., Phys. Rev. Lett. **84**, 2223 (2000)
26. V.M. Kovalev, A.V. Chaplik, JETP Lett. **90**, 679 (2009)
27. A.I. Vedernikov, A.O. Govorov, A.V. Chaplik, J. Exp. Theor. Phys. **93**, 853 (2001)
28. W.C. Tan, J.C. Inkson, Phys. Rev. B **53**, 6947 (1996)
29. H.J. Schultz, in *Field Theories for Low-Dimensional Systems*, ed. by G. Marandy et al. (Springer, Berlin, 2000)
30. V.M. Kovalev, A.V. Chaplik, J. Exp. Theor. Phys. **101**, 686 (2005)
31. T. Ando, A.B. Fowler, F. Stern, Rev. Mod. Phys. **54**, 437 (1982)
32. L. Wendler, A.V. Chaplik, R. Haupt, O. Hipolito, J. Phys. Condens. Matter **5**, 4817 (1993)
33. I.P. Ipatova, A.Yu. Maslov, O.V. Proshina, Phys. Solid State **37**, 991 (1995)
34. I.P. Ipatova, A.Yu. Maslov, O.V. Proshina, Phys. Low-Dimens. Semicond. Struct., Nos. 4–5 (1996)
35. I.P. Ipatova, A.Y. Maslov, O.V. Proshina, Semiconductors **33**, 765 (1999)
36. V.M. Kovalev, A.V. Chaplik, J. Exp. Theor. Phys. **101**, 686 (2005)
37. A.V. Chaplik, JETP Lett. **75**, 292 (2002)
38. A.O. Govorov, S.E. Ulloa, K. Karrai, R.J. Warburton, Phys. Rev. B **66**, 081309(R) (2002)
39. V.M. Kovalev, A.V. Chaplik, J. Exp. Theor. Phys. **95**, 912 (2002)
40. V.M. Kovalev, A.V. Chaplik, Europhys. Lett. **77**, 47003 (2007)
41. H.N. Nunez Yepez, C.A. Vargas, A.L. Salas Brito, Eur. J. Phys. **8**, 189 (1987)
42. L.V. Keldysh, Sov. Phys. JETP **20**, 1307 (1965)

Chapter 10
Aharonov-Bohm Effect for Neutral Excitons in Quantum Rings

M.D. Teodoro, V.L. Campo Jr., V. Lopez-Richard, E. Marega Jr.,
G.E. Marques, and G.J. Salamo

Abstract Quantum interference patterns predicted by theory due to the finite structure of neutral excitons in InAs quantum rings are corroborated experimentally in the magneto-photoluminescence spectra of these nanostructures. The effects associated to built in electric fields and to the temperature on these Aharonov-Bohm-like oscillations are described and confirmed by complementary experimental procedures.

10.1 Introduction

Attempts to verify fundamental quantum mechanical phenomena experimentally are some times hampered by serious scale limitations of a system. Nonetheless, the astonishing progress observed in the synthesis and growth of nanoscopic systems has opened opportunities for raising anew old questions while developing the technical ways for finding answers to them. This has certainly been the case of the search for optical implications associated to Aharonov-Bohm (AB) (Refs. [1, 2]) effects in nanoscale ring structures, or quantum rings (QRs), that gained a significant impetus in the last decades [3–6]. Their peculiar rotational symmetry has encouraged the search for a myriad of effects related to the quantization of the angular momentum and to the unique spatial distribution of the wave function [7–10].

There is however, at this moment, an important shortcoming in the experimental emulation of the original conditions for AB-effect in nanoscopic QRs. Unlike the

M.D. Teodoro · V.L. Campo Jr. · V. Lopez-Richard · G.E. Marques
Departamento de Física, Universidade Federal de São Carlos, 13565-905 São Carlos, São Paulo, Brazil

E. Marega Jr. (✉)
Instituto de Física de São Carlos, Universidade de São Paulo, 13566-590 São Carlos, São Paulo, Brazil
e-mail: euclydes@ifsc.usp.br

G.J. Salamo
Arkansas Institute for Nanoscale Materials Science and Engineering, University of Arkansas, Fayetteville, AR 72701, USA

V.M. Fomin (ed.), *Physics of Quantum Rings*, NanoScience and Technology,
DOI 10.1007/978-3-642-39197-2_10, © Springer-Verlag Berlin Heidelberg 2014

famous picture of an AB scattering situation, for the available experimental configurations, the carriers are confined within QR regions with finite values of magnetic field. Although experimentally we shall renounce, for now, to test the seeming "paradox" of carriers being affected by a static vector potential in a region with zero field, we still consider an observed effect as of the AB-type if it can be explained assuming that the magnetic field is ideally concentrated in the middle of the QRs. Thus, we shall demonstrate that the experimental effects described bellow arise essentially from potential vector-mediated quantum interference.

As highlighted in previous chapters, in stationary systems, the "interference" patterns that will be assigned to an AB-nature are beyond the angular momentum quantization that affects the modulation with the external magnetic field of a single carrier selfenergy and we will focus on the excitonic states that can be characterized as proposed theoretically in Refs. [11–14]. Instead of looking only at the oscillatory dependence on the magnetic flux of the electron-hole recombination energy during photo-luminescence (PL) [15], we also consider the excitonic oscillator strength whose oscillatory behavior reflects directly the changes in the exciton wave function as the magnetic flux increases [16]. A similar experimental work was reported in Ref. [4] for type-II QRs, however, here we are considering both electron and hole moving inside the ring so that the correlation between them becomes crucial to the oscillatory behavior found in the PL integrated intensity.

We will report the observation of a very unusual and difficult to be detected effect, since neutral excitons display a weak sensitivity to magnetic fields due to their neutrality [17]. Their coupling to the vector potential (or magnetic field) will be revealed when their finite structure is considered. Thus, interesting properties of coupled electrons and holes will rise during light absorption or emission of these QRs that can, in turn, be affected by structural factors such as piezoelectricity [18] and strain fields [19], valence mass anisotropy [19, 20], or temperature. We will contrast the experimental observations with the theoretical predictions. We were able to detect, analyze, and discuss systematically the different patterns observed in the oscillation of the integrated PL as function of the magnetic flux (not just the magnetic field) given that the ring size turns into a critical scaling factor for the experimental observation of this phenomenon, as reported in Ref. [21].

10.2 Experiment

The samples were grown by molecular beam epitaxy on a semi-insulating GaAs (100) substrate. After the growth of 0.5 μm GaAs buffer layer at 580 °C, cycles of 0.14 ML of InAs plus 2 s interruption under As_2 flux were deposited until the total formation of 2.2 ML of InAs QDs at 520 °C. Then, QDs were annealed during 30 s to improve the size distribution. In order to obtain the QRs structures, the dots were partially capped with 4 nm of GaAs cap layer grown at 520 °C. The growth rate was measured by RHEED at 1 and 0.065 ML/s for GaAs and InAs respectively. Two different samples were grown with the same conditions. The first one contain uncapped quantum rings for morphological studies and the second one, a 50 nm GaAs cap layer was deposited to investigate the optical emission. Structural and morphological analysis of both samples were carried out by atomic force microscopy (AFM) and transmission electron microscopy (TEM). The magneto-photoluminescence experiments were performed at 2 K, with magnetic field up to 15 T, using a laser (532 nm) with 10 mW as excitation source. The luminescence was detected by a liquid-nitrogen-cooled InGaAs charged coupled device camera.

10.3 Results and Discussion

Figure 10.1(a) shows an atomic force microscopy image of the uncapped InAs ring sample. A 3D image of a single quantum ring is also displayed in Fig. 10.1(b), which represents the statistical average of ring sizes in the sample: 0.9 nm in height, 15 nm inner and 60 nm outer diameters. In Fig. 10.1(c) is shown a plan-view TEM image from the sample with 50 nm GaAs cap layer that shows clearly the ring-like shape. A reduction in the quantum ring dimensions, if compared with the sample without cap layer, can be clearly observed but the origin for causing this change is not yet well understood. As will be discussed later, the ring radius of the capped samples is compatible with the optical experimental results. A cross-section TEM image, taken closer to $\langle 110 \rangle$ direction, is shown in Fig. 10.1(d), where a 30 nm distance between the rims of the torus-like structure can be measured [22].

The PL spectra of the QRs, taken at 2 K and for $B = 15$ T (gray line), is shown in Fig. 10.2(a). Only two radiative recombination channels due to a bi-modal size of ring distributions are observed: a shoulder at the low energy side and a main peak at higher energy side, labeled QR1 and QR2, respectively. The typical qualitative aspects for the shapes of these rings are shown on the left side of the figure. Both spectra were fit with two Gaussian lines to evaluate the evolution of the peaks energy and the area below the curves (integrated intensities) as a function of the applied magnetic field.

Under an external magnetic field, applied along the growth direction, we observed the expected blue shift of both emission bands, shown in Fig. 10.2(b). The experimental points for the magnetic shift were obtained from the peak position at given B-value and subtracted the energy value at zero field. As the magnetic field increases, the ground-state changes from angular momentum from $l = 0$ to $l = 1$, from $l = 1$ to $l = 2$, and etc., and the energy being an slightly oscillatory function of magnetic field [10]. The small oscillations of the peak position with increasing magnetic field were detected clearly for both ring emissions. However, being the amplitude of the oscillations of the QR1 structure $\sim$0.5 meV, or too small when compared to the linewidth of the PL band, $\sim$60 meV, the uncertainty of the energy oscillation becomes large for high fields. Yet, unambiguous oscillations of the integrated intensities are much easier to be detected.

The oscillations of the intensity, as the magnetic field increases, are depicted in Fig. 10.3 [23]. One must note that the period and the sequences of minima and

Fig. 10.1 (**a**) Atomic force microscope image from the InAs uncapped quantum rings sample. (**b**) 3D AFM image from a single quantum ring. (**c**) TEM plan-view image, taken slightly off the ⟨110⟩ direction. (**d**) A cross-section TEM image from the buried quantum ring samples. *Yellow dashed-line* is just a guide to the eyes

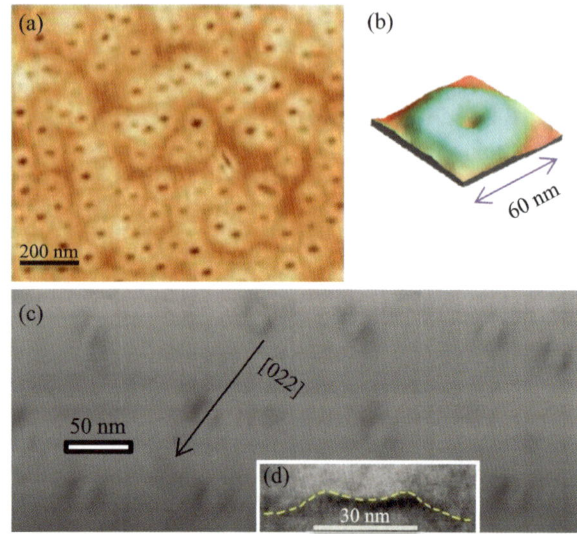

Fig. 10.2
(**a**) Photoluminescence spectra of InAs quantum ring sample, taken at 2 K and for $B = 15$ T (*gray line*). There are two radiative channels (*triangles*—peak 1 and *circles*—peak 2) identified as the recombination from two different size of quantum rings, and with morphologies represented by the *pictures on the left side*. (**b**) The magnetic shift taken from the PL spectra at increasing magnetic field. One angular momentum transition is clearly observed at $B = 5.5$ T. Anther occurs near $B = 13$ T but only becomes clear in the integrated intensity

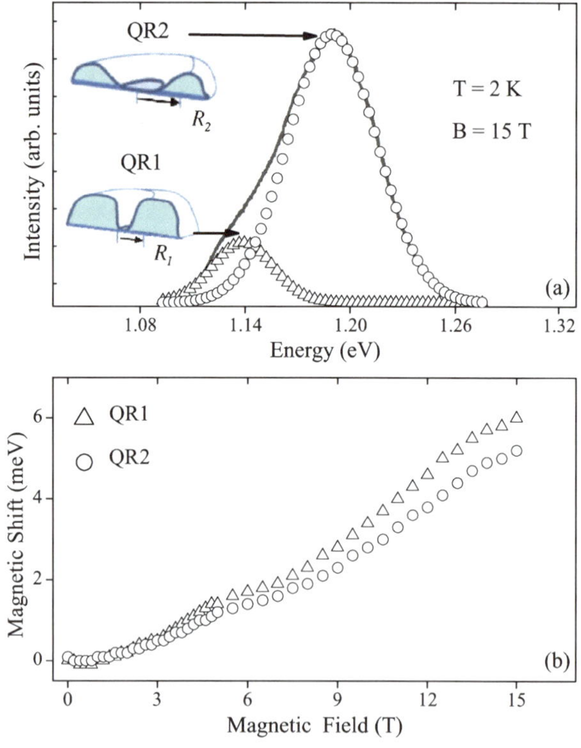

Fig. 10.3 Integrated intensities of the two emission lines as a function of the magnetic field. (**a**) In-phase AB-oscillations of the lower energy emission line of QR1 rings show a period corresponding to an effective radius $R = 11.6$ nm. (**b**) Counterphase oscillations of the higher energy emission line of QR2 rings show a period corresponding to an effective radius $R = 19$ nm

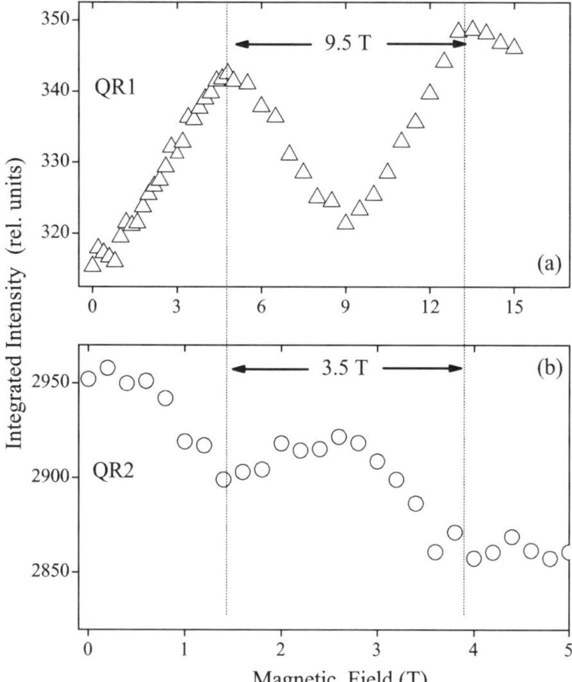

maxima differ in these PL intensities. To understand their nature we will adapt the theoretical model of Römmer and Raikh [12–14] to the structural conditions and parameters of our samples. According to this model, the electron and the hole move in a ring of radius R and zero cross-section width threaded by a magnetic flux $\Phi = B\pi R^2$ and they interact by means of a contact potential. The corresponding two-particle Hamiltonian is given by

$$\widehat{H}_0 = \frac{(\widehat{P}_n + \frac{e\Phi}{2\pi Rc})^2}{2m_n} + \frac{(\widehat{P}_p - \frac{e\Phi}{2\pi Rc})^2}{2m_p} - 2\pi V \delta_P(\varphi_n - \varphi_p), \qquad (10.1)$$

where $\varphi_{n(p)}$ is the angular position of electron (hole) with orbital angular momentum $\widehat{P}_{n(p)} = \frac{\hbar}{iR}\frac{\partial}{\partial \varphi_{n(p)}}$.

The last term, representing an attractive short-range interaction, the function $\delta_P(\varphi_n - \varphi_p)$ is a 2π-periodic Dirac's δ-function. The strength of the interaction parameter V was chosen so that the exciton ground-state energy obtained by this expression fits the reported value of the exciton binding energy, and we have used for the exciton binding energy, the value 4.35 meV as reported in Ref. [24]. As the strength of V increases, a more tightly bound exciton is produced, becoming thus less sensitive to magnetic and electric fields. The theoretical study of the model above will follow the work in Ref. [11], which Chaplik and Kovalev, in Chap. 9 of the present book, extend in several ways. While we consider a contact interaction between electron and hole, Chaplik and Kovalev treat the more realistic Coulomb interaction, providing analytical results for the strong-coupling regime.

If we change to center-of-mass and relative coordinates,

$$\varphi_c = (m_n\varphi_n + m_p\varphi_p)/M, \tag{10.2}$$

$$\theta = \varphi_n - \varphi_p, \tag{10.3}$$

with $M = m_n + m_p$ being the total mass, and θ being the angular separation between electron and hole positions on the ring. The Hamiltonian becomes,

$$\widehat{H}_0 = \widehat{H}_0^{\varphi_c} + \widehat{H}_0^{\theta}, \tag{10.4}$$

where

$$\widehat{H}_0^{\varphi_c} = -\frac{\hbar^2}{2MR^2}\frac{\partial^2}{\partial\varphi_c^2}, \tag{10.5}$$

$$\widehat{H}_0^{\theta} = \frac{\hbar^2}{2\mu R^2}\left(\frac{1}{i}\frac{\partial}{\partial\theta} + \frac{\Phi}{\Phi_0}\right)^2 - 2\pi V\delta_P(\theta), \tag{10.6}$$

and μ is the exciton reduced mass defined by $1/\mu = 1/m_n + 1/m_p$.

The solution for the center-of-mass motion equation is given by

$$\psi_c(\varphi_c) = \frac{e^{iJ\varphi_c}}{\sqrt{2\pi}} \tag{10.7}$$

with the corresponding eigenenergies

$$E_c(J) = \frac{\hbar^2 J^2}{2MR^2}. \tag{10.8}$$

As already stated, the main effects will be connected to the relative motion, i.e., to the internal structure of the exciton. Firstly, we can make a gauge transformation from an eigenfunction $\phi_{\text{int}}(\theta)$ of the Hamiltonian $\widehat{H}_0^{\theta}$ to a new function $\chi(\theta)$, as

$$\phi_{\text{int}}(\theta) = e^{-i\frac{\Phi}{\Phi_0}\theta}\chi(\theta). \tag{10.9}$$

One should note that both $|\phi_{\text{int}}(\theta)|^2$ or $|\chi(\theta)|^2$ give the probability density of finding electron and hole angular positions differing by θ. In terms of χ, the eigenequation for the relative motion reads

$$-\frac{\hbar^2}{2\mu R^2}\chi''(\theta) - 2\pi V\delta_P(\theta)\chi(\theta) = w\chi(\theta) \tag{10.10}$$

which is an eigenequation for a particle moving in a 2π-periodic potential. So, following Chaplik [11], the function χ satisfies Bloch's theorem and can be written as

$$\chi(\theta) = e^{ip\theta}v(\theta), \tag{10.11}$$

where v is a 2π-periodic function. The Bloch function $\chi(\theta)$ satisfies the relation

$$\chi(\theta + 2\pi) = e^{i2p\pi}\chi(\theta) \tag{10.12}$$

and p can be restricted to the first Brillouin zone $(-1/2, 1/2]$. The total exciton wave function will be

$$\Psi_J(\Lambda, \theta) = \frac{e^{iJ\Lambda}}{\sqrt{2\pi}}\, e^{i(-\frac{\Phi}{\Phi_0}+p)(\theta)}\, v(\theta). \tag{10.13}$$

Given (10.12), we just need to determine the function $\chi(\theta)$ in the interval $[-\pi, \pi]$ using the torsional boundary conditions

$$\chi(\pi) = e^{i2p\pi} \chi(-\pi), \tag{10.14}$$

$$\chi'(\pi) = e^{i2p\pi} \chi'(-\pi). \tag{10.15}$$

One could understand the torsional boundary condition in terms of the AB effect: imagine the hole being at the origin. When the electron leaves the origin, it can arrive at the diametrically opposite point in the ring moving clockwise in one half or counterclockwise in the other half of the ring. Due to the magnetic field, its wave function will acquire different phase factors according to these paths, so at the opposite point, there will be interferences. For zero or for $p = 1/2$ phase shifts along the half cycle paths, the interference will be constructive or destructive, a situation similar to the original AB effect detected in transport.

In order to determine the values for angular momentum J in the center of mass eigenfunction and p parameter phase value in the function χ, we need to go back to the coordinates φ_n and φ_p and impose that the total wave function be periodic if the angular positions φ_n or φ_p change by 2π. The periodicity of $\Psi(\varphi_c, \theta)$ in φ_n and φ_p yields

$$Jm_n/M + \left(p - \frac{\Phi}{\Phi_0}\right) = N_n \in \mathbf{Z}, \tag{10.16}$$

$$Jm_p/M - \left(p - \frac{\Phi}{\Phi_0}\right) = N_p \in \mathbf{Z}. \tag{10.17}$$

The addition of these last equations yields

$$J = N_n + N_p \in \mathbf{Z}, \tag{10.18}$$

which means that the center-of-mass angular momentum, measured in units of $\hbar$, assumes integer values $J = 0, \pm 1, \pm 2, \ldots$, what is totally reasonable because it is exactly what would happen for a single particle of mass M moving in the ring.

By subtracting (10.16) from (10.17) we are lead to

$$\gamma J + 2\left(\frac{\Phi}{\Phi_0} - p\right) = N_p - N_n \in \mathbf{Z}, \tag{10.19}$$

where we have introduced the constant

$$\gamma = \frac{m_p - m_n}{M}. \tag{10.20}$$

Given an integer value for J, the parameter p can be determined uniquely in the interval $(-1/2, 1/2]$ because the integer $\gamma J + 2(\frac{\Phi}{\Phi_0} - p) = N_p - N_n$ has the same parity as J. Once we have determined p, we can solve the eigenequation (10.10) for

χ under the boundary conditions (10.14) and (10.15) to find several internal eigenfunctions. Analogously to band structure calculations, for each Bloch wavevector p we have several eigenfunctions, one for each band. The fact that the parameter p depends on the effective masses becomes a relevant feature when dealing with excitons that involve either heavy or light holes. We will demonstrate that the hole in-plane effective mass can affect the way the AB oscillations are affected by external factors such as temperature or applied electric field.

At this point it is easy to understand that several physical properties of this system will be invariant when the magnetic flux through the ring changes by a multiple of the quantum flux. The ratio $\frac{\Phi}{\Phi_0}$ enters in (10.19) and determines p. If $\frac{\Phi}{\Phi_0}$ changes by an integer n, we can consider (10.18) and (10.19) with N_h increased by n and N_e decreased by n, so that the eigenvalues for J and the value for p remain unchanged. Therefore, the function χ and the internal energy w will be unchanged, while the total wave function will change by a local phase factor $e^{-in\theta}$. Any property which is not affected by this local phase factor will oscillate periodically as a function of the magnetic field.

To study the function χ giving the internal structure of the exciton in more detail, it is convenient to rewrite (10.10) in terms of dimensionless quantities,

$$-\chi'' - V_0 \delta(\theta)\chi = \epsilon_{\text{int}}\chi, \tag{10.21}$$

where $\epsilon_{\text{int}} = w/\epsilon_0$ and $V_0 = 2\pi V/\epsilon_0$, with $\epsilon_0 = \hbar^2/2\mu R^2$. If we write negative energies as $\epsilon_{\text{int}} = -\kappa^2$ and positive energies as $\epsilon_{\text{int}} = k^2$, we get the following transcendental equations to determine the energies

$$\cosh(2\pi\kappa) - \cos(2\pi p) = \pi V_0 \frac{\sinh(2\pi\kappa)}{2\pi\kappa}, \tag{10.22}$$

for negative energies and

$$\cos(2\pi k) - \cos(2\pi p) = \pi V_0 \frac{\sin(2\pi k)}{2\pi k}, \tag{10.23}$$

for positive energies.

The corresponding eigenfunctions are

$$\chi_\kappa(\theta) = \begin{cases} N_{\kappa,p}[\sinh(\kappa\theta) - e^{-i2p\pi}\sinh(\kappa(\theta - 2\pi))], & 0 \le \theta \le \pi \\ N_{\kappa,p}e^{-i2p\pi}[\sinh(\kappa(\theta + 2\pi)) - e^{-i2p\pi}\sinh(\kappa\theta)], & -\pi \le \theta \le 0, \end{cases} \tag{10.24}$$

and

$$\chi_k(\theta) = \begin{cases} N_{k,p}[\sin(k\theta) - e^{-i2p\pi}\sin(k(\theta - 2\pi))], & 0 \le \theta \le \pi \\ N_{k,p}e^{-i2p\pi}[\sin(k(\theta + 2\pi)) - e^{-i2p\pi}\sin(k\theta)], & -\pi \le \theta \le 0 \end{cases} \tag{10.25}$$

where $N_{\kappa,p}$ and $N_{k,p}$ are normalization factors.

One needs to be careful when considering the cases $p = 0$ and $p = 1/2$. From (10.24) and (10.25), the corresponding functions χ will display even parity, i.e., $\chi(-\theta) = \chi(\theta)$. However, there are also special odd functions. When $p = 0$, we have special solutions of (10.23) which correspond to $k = 1, 2, 3, 4, \ldots$. Substituting these values of k in (10.24), one apparently gets functions identically equal to zero. However, a careful account of the dependence of the normalization factor on k implies that at these special values of k, the wave functions are, in fact, given by $\chi(\theta) = \sin(k\theta)/\sqrt{2\pi}$, which display odd parity. Analogously, when $p = 1/2$, we have special solutions of (10.23) corresponding to $k = 1/2, 3/2, 5/2, \ldots$, whose wave functions are also given by $\chi(\theta) = \sin(k\theta)/\sqrt{2\pi}$. It is easy to understand these odd solutions directly from (10.21). Being equal to zero at the origin, implies that the solutions will not feel the delta potential. Therefore, the equation is simply $-\chi'' = \epsilon_{exc}\chi$, which allows solutions of form $\sin(k\theta)$. To satisfy the boundary conditions, in (10.14) and (10.15), k must be integer when $p = 0$ and half-integer when $p = 1/2$.

The oscillator strength of the ground-state with wave function $\Psi_0(\Lambda, \theta)$ (see (10.13)), will be given by

$$\mathcal{O}_0 = \left| \int_{-\pi}^{\pi} \Psi_0(\varphi_c, 0) d\varphi_c \right|^2, \tag{10.26}$$

and, therefore, is a periodic function on the magnetic flux through the ring since upon increasing the magnetic flux by Φ_0 the excitonic wave function changes by the phase-factor $e^{i\theta}$, which is equal to one for $\theta = 0$. This periodic behavior is displayed in Fig. 10.4(a). In Chap. 9, Chaplik and Kovalev found oscillations of the intensity of the exciton line along with those of the biding energy in the strong-coupling case.

According to the contact interaction model just presented, the oscillations in Fig. 10.4(a) can be understood as follows. The ground-state has a center-of-mass angular momentum $J = 0$ and it is an even function. When the magnetic flux is zero, the χ function is 2π-periodic, on the other hand, for magnetic flux equal to $\frac{\Phi}{\Phi_0} = 1/2$, the χ function is anti-periodic. This fact combined with the fact that the ground-state wave function should not have any node, implies that $\chi(-\pi) = \chi(\pi) = 0$ and therefore the function χ for $\Phi = \Phi_0/2$ must be more concentrated around $\theta = 0$ than the χ function corresponding to the ground-state in the absence of magnetic flux. See this effect depicted in Fig. 10.4(b). So, the oscillator strength should have a maximum at $\frac{\Phi}{\Phi_0} = 1/2$ according to the model presented above. It worths noticing that the oscillation pattern just described is qualitatively similar for both heavy and light-hole excitons. However, quantitatively, the way the pattern is affected by external factors may depend drastically on the character of the valence band state involved. The oscillations of the oscillator strength of the excitonic recombination are a direct consequence of the correlation of electrons and holes due to the Coulomb attraction, leading to the peculiar AB interference obtained experimentally. In the absence of Coulomb attraction, electron and hole would be independent particles and the corresponding oscillator strength would be independent of the magnetic flux through the ring. Yet, the sequence of minima and maxima obtained after this brief theoretical discussion would correspond only to the picture displayed in Fig. 10.3(a) of QR1 with an effective radius $R = 11.6$ nm. The oscillations corresponding to the QR2 emission appear in counterphase. Thus, additional ingredients must be added to the model as well as complementary experimental facts must be

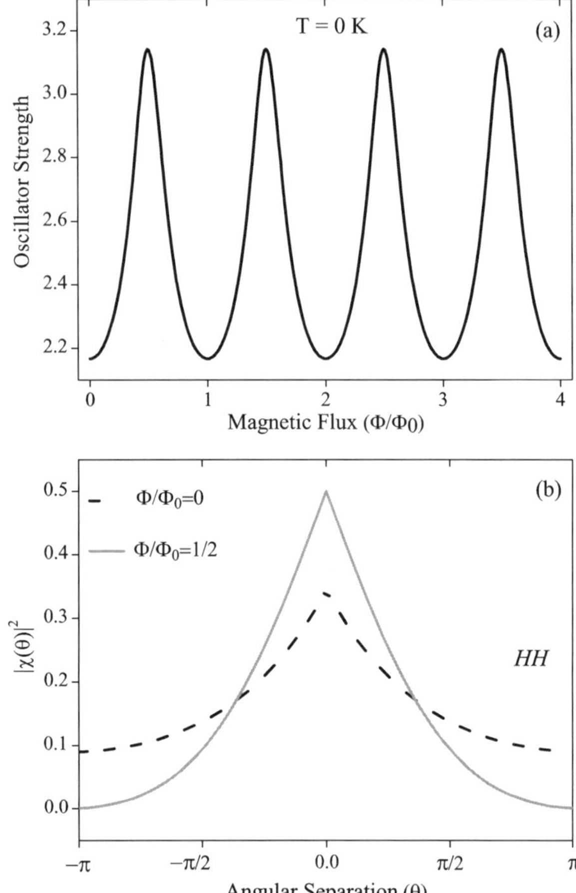

Fig. 10.4 Oscillator strength as a function of magnetic flux for an heavy-hole exciton binding energy, $E_b = 4.35$ meV. (**b**) Probability density of finding the electron and the hole angular positions differing by θ for heavy-hole. *Dotted black line* correspond to magnetic flux equal to zero and gray curve corresponds to magnetic flux equal to $\Phi_0/2$. In this case, the value of the magnetic flux requires that the ground-state wave function be equal to zero at $\theta = \pm\pi$, and to keep the normalization, a more pronounced peak appears at $\theta = 0$ with direct impact on the oscillator strength

sought to understand such an anomalous behavior. If their period was also to be defined by the flux quantum, their effective QR radius would be $R = 19$ nm. The fact that the emission band QR2 has higher transition energy and larger average radius than QR1 suggests that QR2 may present smaller ring width and height than those for QR1. Also, note in Fig. 10.3(b), that a monotonic decrease of the center of the oscillations takes place. Such an effect cannot be understood within our one-dimensional ring model, where the rings have zero width. The combination of finite ring width and certainly potential localization effects cannot be discarded as causes for the background shifts of the PL oscillations.

10.4 Inquiring for Reasons of AB-Oscillations in Counterphase

We turn next to a discussion of an effect which may potentially invert the sequence of minima and maxima of the oscillator-strength oscillations: the heating. In order to

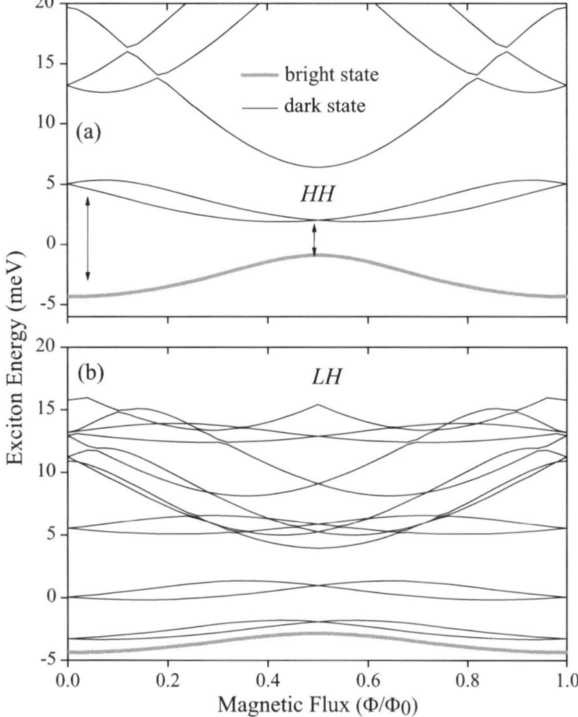

Fig. 10.5 Magnetic flux dependence of the first states of the exciton in the case of (**a**) heavy hole and (**b**) light hole. Exciton binding energy is $E_b = 4.35$ meV. The *gray curves* denote the bright excitons (which have center-of-mass angular momentum $J = 0$), while *black curves* correspond to dark ones

take into account the effect of temperature, one should generalize (10.26) including all excitonic states,

$$\mathcal{O}(T) = \frac{\sum_n \mathcal{O}_n e^{-E_n/k_B T}}{\sum_n e^{-E_n/k_B T}}, \tag{10.27}$$

where $\mathcal{O}_n$ is the oscillator strength for the nth state (whose wave function is Ψ_n),

$$\mathcal{O}_n = \left| \int_{-\pi}^{\pi} \Psi_n(\varphi_c, 0) d\varphi_c \right|^2. \tag{10.28}$$

The electronic structure engineering in 0-dimensional structures appears as an effective tool for tuning the magnetic properties of these systems [25–27]. In particular, strain fields and confinement configurations can be adjusted so that the character of the valence band ground-state can be changed from heavy- to light-hole character. As demonstrated below, this will have implications in the way the exciton recombination responds to the magnetic field and temperature. In Fig. 10.5, the exciton energy levels are displayed for HH and LH excitons, where the only parameter changed has been the in-plane hole mass. We must note that their in-plane effective masses (in the parabolic approximation) are related to the Luttinger parameters in the following way: $m_{HH} = 1/(\gamma_1 + \gamma_2)$ and $m_{LH} = 1/(\gamma_1 - \gamma_2)$ [25]. The character of the valence band involved in the exciton recombination will, thus, affect the picture of the exciton energy dispersion and modify the shape of these

Fig. 10.6 Average oscillator strength as a function of magnetic flux for an exciton binding energy equal to $E_b = 4.35$ meV and for various temperatures. Panel (**a**) corresponds to the heavy-hole case and panel (**b**) corresponds to the light hole case. In case (**a**), for magnetic flux equal to $\Phi_0/2$, the maximum oscillator strength gets converted into a minimum. In case (**b**), since the total mass $(m_h + m_e)$ is large, the center-of-mass states are almost degenerated and the thermal average mix so many states that the oscillator strength decreases and becomes flat

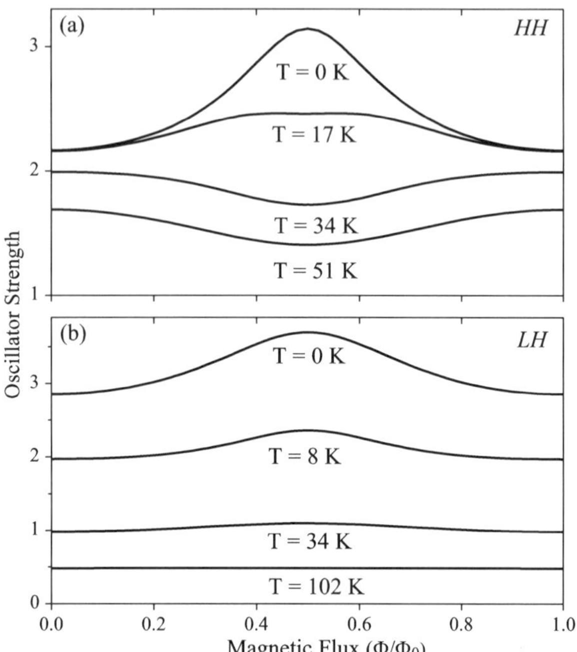

levels as the magnetic field is varied. In the calculations, the InAs band parameters used are $m_e = 0.026$, $\gamma_1 = 20.4$, and $\gamma_2 = 8.3$ [24].

As the temperature rises, the occupation of excited levels of the exciton becomes more probable. Note that the first excited levels correspond all to dark excitons. As the dark exciton levels approach the bright one the net occupation of the ground-state changes and reduces the thermalized oscillator strength. At relatively high temperatures, this effect transforms the maximum of the oscillator strength at $\frac{\Phi}{\Phi_0} = 1/2$ into a minimum for the HH exciton as displayed in Fig. 10.6(a). As the temperatures rises, the net occupation of the ground-state decreases as these levels become closer, what takes place at $\frac{\Phi}{\Phi_0} = 1/2$ for the HH-exciton. This, in turn, reduces the thermalized oscillator strength and its amplitude. Yet, for the LH exciton, such a reversion of the maximum is not observed for temperatures up to 34 K (see Fig. 10.6(b)). This is due to the peculiar electronic structure of the first LH-exciton levels displayed in Fig. 10.5(b). According to our model, the temperature inversion of the maximum appears practically undetectable since the oscillator strength becomes essentially flat before inversion. Therefore the heating may invert the sequence of maxima and minima of the AB-oscillations, at least in the case of the heavy-hole excitons. However, this would take place at temperatures well above the 2 K and most of our magneto-optical measurements were performed below these temperatures. Hence, an alternative hypothesis must be searched in order to explain the experimental observation. Before considering another mechanism for the inversion in the sequences of maxima and minima, we should point out that Ref. [4] reports measurements of oscillatory behavior with the magnetic field at temperatures as high as 180 K for type II QDs. It is interesting that the data in Ref. [4] in fact show the inversion in the sequence of maxima and minima when the temperature is raised from 4 K to 60 K (and, apparently, another inversion when the temperature is raised to 180 K) despite the difference between the system there and our system.

Another effect that can lead to the reversion of the maximum of the oscillator strength of the exciton ground-state, even at low temperatures, is the appearance of an in-plane electric field. Such an electric field can be caused by uniaxial strains and by eccentricity of the QR. We can explain this straightforwardly. The effect of an in-plane electric field F along x-axis can be added to the Hamiltonian used in the previous explanations as

$$\widehat{H} = \widehat{H}_0 + eFR\big(\cos(\varphi_n) - \cos(\varphi_p)\big). \tag{10.29}$$

In order to compute the eigenfunctions in the presence of the in-plane electric field, we write down the Hamiltonian matrix in the basis of the eigenfunctions of $\widehat{H}_0$ and proceed with a numerical diagonalization. Since we are interested in low temperatures, we can truncate the matrix dimension after including all basis states whose energies are lower than some high enough cut-off value. It is easy to show that the electric field can only couple states whose center-of-mass angular momenta differ by one, an this simplifies significantly the task of building the Hamiltonian matrix.

With respect to the dependence of the matrix elements on the magnetic flux, we see immediately that the diagonal elements are periodic, since they correspond to the energies in the absence of the electric field, which remain unchanged when the magnetic flux changes by a multiple of the flux quantum Φ_0, as discussed in the previous section. The off-diagonal elements $\langle \Psi_m | eFR(\cos(\varphi_n) - \cos(\varphi_p)) | \Psi_{m'} \rangle$ are also periodic. This is due to the fact that the eigenfunctions Ψ_m and $\Psi_{m'}$, in the absence of electric field, change by the same phase factor $e^{in(\varphi_n - \varphi_p)}$ when the magnetic flux is changed by $n\Phi_0$ (n integer). This common phase-factor is canceled out when computing the off-diagonal matrix element above. Therefore the exciton energies and the oscillator strength in the presence of an in-plane electric field will be periodic functions of the magnetic flux with period equal to Φ_0. In general, we expect that the oscillator strength will decrease when we increase the electric field, since the angular separation between the electron and the hole becomes larger.

For the particular cases of magnetic flux equal to zero or equal to $\Phi_0/2$, we can get a more detailed understanding exploiting a symmetry of the Hamiltonian. The Hamiltonian in (10.29) is invariant under a reflection relative to the x-axis combined with an inversion of the magnetic field. For zero magnetic flux, this implies that the eigenfunctions will be even or odd. For magnetic flux equal to $\Phi_0/2$, the symmetry under simultaneous reflexion and inversion of the magnetic field leads to a relation between the eigenfunctions for $\Phi = \Phi_0/2$ and for $\Phi = -\Phi_0/2$,

$$\Psi(-\varphi_c, -\theta, \Phi = -\Phi_0/2) = \pm \Psi(\varphi_c, \theta, \Phi = \Phi_0/2). \tag{10.30}$$

On the other hand we know, from the previous discussions, that increasing the magnetic flux by Φ_0 adds a phase-factor to the eigenfunction,

$$\Psi(\varphi_c, \theta, \Phi = \Phi_0/2) = e^{-i\theta}\, \Psi(\varphi_c, \theta, \Phi = -\Phi_0/2). \tag{10.31}$$

With (10.30) and (10.31), we get

$$\Psi(\varphi_c, \theta, \Phi = \Phi_0/2) = \pm e^{-i\theta}\Psi(-\varphi_c, -\theta, \Phi = \Phi_0/2), \qquad (10.32)$$

and after making the same gauge transformation as before, defining

$$\tilde{\Psi}(\varphi_c, \theta, \Phi = \Phi_0/2) = e^{i\frac{\Phi}{\Phi_0}\theta}\Psi(\varphi_c, \theta, \Phi = \Phi_0/2), \qquad (10.33)$$

we find that these functions $\tilde{\Psi}$ will have defined parity when $\Phi = \Phi_0/2$,

$$\tilde{\Psi}(\varphi_c, \theta, \Phi = \Phi_0/2) = \pm\tilde{\Psi}(-\varphi_c, -\theta, \Phi = \Phi_0/2). \qquad (10.34)$$

Naturally, the same is true for any half-integer value of Φ/Φ_0. For such special values of magnetic flux, an even ground-state will exist as we increase the electric field until some critical electric field, F_C, is achieved. It is natural that at low electric fields the ground-state be even, since at zero electric field the ground-state has center-of-mass angular momentum $J = 0$ and an even function χ. At F_C, there is a level-crossing between an even and an odd state and, above F_C, the ground-state becomes odd, which is also natural because being odd it will have significant projection on zero electric field states whose corresponding χ functions are odd, and only these odd functions give rise to electron and hole well separated as we must have at high electric fields.

At $\Phi = \Phi_0/2$, by (10.33) and (10.34), the odd ground-state wave functions satisfies $\Psi(\varphi_c, 0) = -\Psi(-\varphi_c, 0)$, which implies zero oscillator strength when we consider (10.26). Therefore, if for zero electric field, we had a maximum in the oscillator strength at $\Phi = \Phi_0/2$, for electric field above the critical value, we have a minimum in the oscillator strength at $\Phi = \Phi/2$ as illustrated in Fig. 10.7. The inversion of the maximum in the curves of OS against magnetic flux at $T = 0$ K occurs as the result of increasing the in-plane electric field above certain critical value. This effect was first discussed in Ref. [14], where the electric field was supposed applied in a controllable fashion. Here, however, the electric field is supposed to be present in the sample due to uniaxial strains and eccentricity of the QR.

According to the description above, the stronger the electric field becomes the higher is the projection of the ground-state over states with electron and hole more separated, decreasing the oscillator strength as well as its sensitivity to the magnetic field. This is particularly relevant at $\Phi/\Phi_0 = 1/2$ beyond the critical field, when the ground-state becomes a pure odd function with null oscillator strength [14]. For QR with finite width this reduction to zero is not expected.

Since the AB-oscillations in counterphase appear only for the QR2 emission, a set of complementary measurements can be put in place to elucidate whether or not in-plane electric fields are present in one quantum ring set and not in the other.

Throughout the course of the synthesis of self-assembled quantum dots and rings, strain fields are formed in the (001) plane. This has been used to control the alignment and elongation of these 0-dimensional structures along certain crystallographic directions. Given the spatial differences of In migration velocities [28], the $[1\bar{1}0]$ appears as the preferential direction for QR elongation and a built-in electric field may appear due to the piezoelectric effect [29]. The potential appearance of built-in

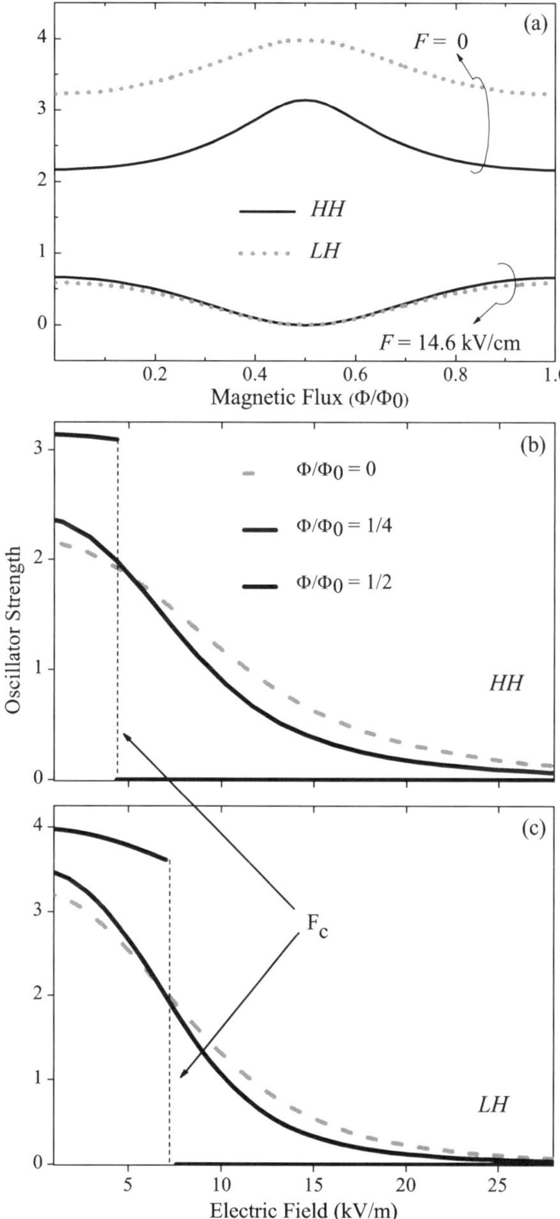

Fig. 10.7 (**a**) Average oscillator strength for the ground-state excitons as a function of the magnetic flux. Top curves calculated for $F = 0$ and bottom curves for $F = 14.6$ kV/cm. For electric fields above some critical value, F_c, the oscillator strengths become equal to zero at the magnetic flux equal to $\Phi_0/2$. Panels (**b**) and (**c**): Oscillator strength for heavy- and light-hole exciton ground-states as a function of the electric field for some values of magnetic flux. For $\frac{\Phi}{\Phi_0} = 1/2$, there is an abrupt change of the oscillator strength due to level crossing happening at $F \simeq 4.8$ kV/cm: for lower electric fields, the ground-state displays even parity, for higher electric fields, the ground-state changes to odd parity. Since the symmetry is exact at this particular magnetic flux, even and odd states do not couple and, thus, the occurrence of level crossing with consequent abrupt changes in the oscillator strengths. In this figure, we considered a ring of radius 10 nm for both heavy and light holes

electric fields in such structures has been a topic discussed previously [18]. Since the piezoelectric polarization leads to the red-shift of the PL emission line (Stark effect), increasing the free-carrier density one may screen the built-in electric field provoking a subsequent blue-shift of the transition energy. This can be achieved by varying the excitation intensity. In Fig. 10.8 we have displayed the PL spectra vs.

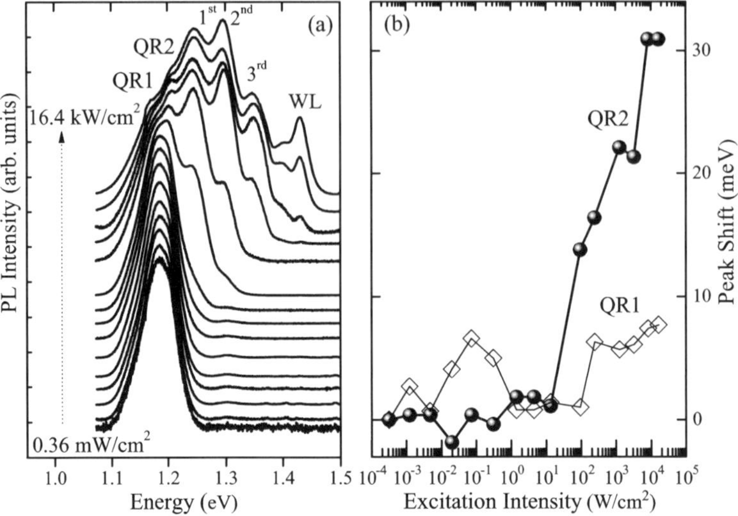

Fig. 10.8 (**a**) Series of the PL emission for increasing excitation powers from 0.36 mW/cm² (*bottom*) to 16.4 kW/cm² (*top*). (**b**) Position of emission peaks vs. excitation power

excitation intensity. Since exciton excited states decay non-radiatively much faster than by photon emission, the hypothesis of bimodal subsets of QRs prevails. In the PL spectra displayed in Fig. 10.8, we see that only for very high power excitation it is possible to see emission from excited states. In Fig. 10.2, the power excitation was very low, reinforcing the interpretation of bimodal QRs. The excitation power increase has provoked the appearance of additional emission bands corresponding to the excited states of QR2 band.

In Fig. 10.8(b), we plotted the energy shift of the QR1 and QR2 versus excitation intensity. Note the blue shift of the emission band QR2 at the high excitation regime while the emission band QR1 remains practically constant. This would accord with the idea that a built-in electric field has been screened in QR2 set and not in QR1 which is consistent with the contrasting AB-interference patterns. We are however faced with the question: why the built-in electric fields related to the band QR1 are negligible if the effect of the elongation should affect all rings? The potential answer to that question may reside in different ring widths corresponding to each emission band: for QR1 it is larger than that for QR2, as mentioned above, leading to a more significant strain field relief for the former than for the latter. This would ultimately result in a much lower piezoelectric field in the QR1 set. Yet, one may still argue that this cannot be considered a definitive proof.

To construct a more convincing argument for the presence of the built-in electric field in QR2 while expecting negligible values in QR1, additional optical experiments were performed: PL measurements experiments vs. temperature are displayed in Fig. 10.9(a). The variation of the integrated intensity for each emission band, as a function of temperature, is plotted in Fig. 10.9(b). It is reasonable to assume that a decrease of the PL intensity with temperature should take place when dark exci-

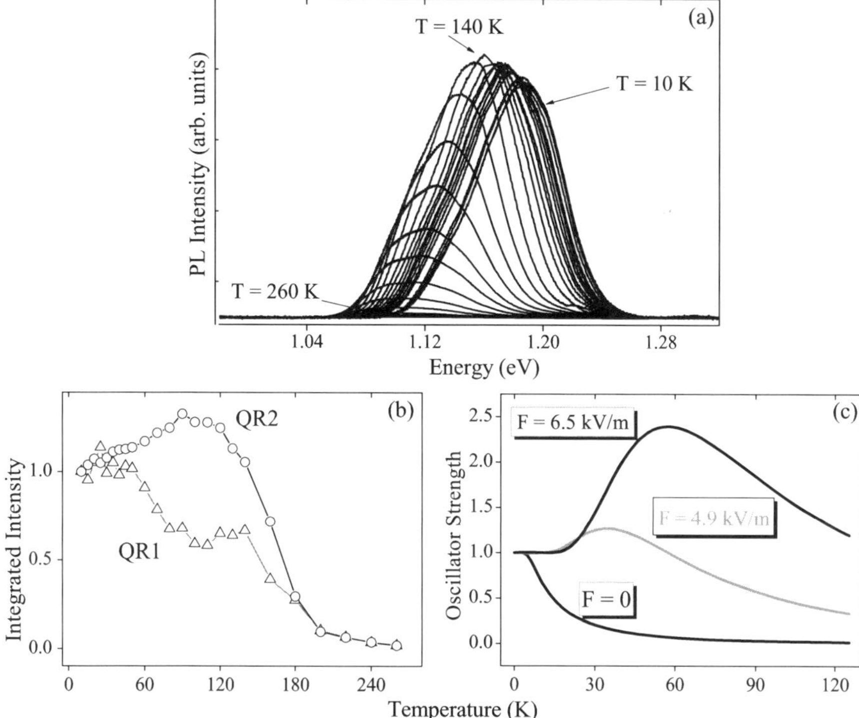

Fig. 10.9 (**a**) PL spectra obtained for several temperatures. At low temperature, the PL spectra are dominated by the emission band QR2, whereas above 160 K, it becomes dominated by QR1, which is attributed to the transference of carriers between QRs, via wetting layer (WL), favoring the lower energy states, as also observed in QD systems. (**b**) and (**c**) PL integrated intensity normalized to the value at $T = 10$ K, for the emission bands QR1 and QR2, respectively. (**c**) Calculated OS (normalized to the value at $T = 0$) for three values of the in-plane electric field F considering a ring of radius 19 nm. The increase of PL intensity at low temperatures can be understood as an electric field effect

tonic states become thermally occupied. The QR1 emission follows this expected pattern. Yet, an intensity increase takes place for QR2 at $T < 100$ K and we may prove that this is another signature of the presence of an in-plane electric field in this set of ring samples. By using our previously described model we are able to calculate the oscillator strength as a function of temperature for finite values of the in-plane field and the results are shown in Fig. 10.9(c). According to this model, the oscillation strength increases by the activation of more efficient channels for optical recombination at excited states, where the electron-hole angular separation is smaller if compared to the ground-state. It is interesting that this complementary experiment confirms that the counterphase AB-oscillations for the band QR2 (Fig. 10.3(b)) cannot be ascribed to occupation effects induced by temperature for HH-excitons, since only for $T > 150$ K an effective reduction of the ground-state population takes place, which is far above the temperature of 2 K where the AB-oscillations were detected. It is common, in systems that display bimodality, to ascribe the increase in intensity of the lower energy emission with respect to the one at higher temperature to thermal redistribution of carriers to the lower energy ensemble with increasing temperature. However, in the results displayed in Fig. 10.9, it is the QR2 emission, with higher energy, that suffers the relative increase for temperatures between 10 and 140 K. This reinforces the conviction of the presence of an in plane electric field in the QR2 set that opposes the mechanism of thermal activation transference of carriers from QR2 to QR1.

Finally, other attempts were made to find such oscillations in neutral excitons using similar samples like the one described here [30]. According to Ref. [21], the main condition for observation AB effect in type-I excitons in quantum rings, is the presence of an effective confinement region close to 7 nm, given by the difference between the outer and inner radii. This is approximately the value reported in Fig. 10.1(d).

References

1. Y. Aharonov, D. Bohm, Phys. Rev. **115**, 485 (1959)
2. Y. Aharonov, D. Bohm, Phys. Rev. **123**, 1511 (1961)
3. M. Grochol, F. Grosse, R. Zimmermann, Phys. Rev. B **74**, 115416 (2006)
4. I.R. Sellers, V.R. Whiteside, I.L. Kuskovsky, A.O. Govorov, B.D. McCombe, Phys. Rev. Lett. **100**, 136405 (2008)
5. I.R. Sellers, A.O. Govorov, B.D. McCombe, J. Nanoelectron. Optoelectron. **6**, 4 (2011)
6. P.A. Orellana, M. Pacheco, Phys. Rev. B **71**, 235330 (2005)
7. J.M. García, G. Medeiros-Ribeiro, K. Schmidt, T. Ngo, J.L. Feng, A. Lorke, J. Kotthaus, P.M. Petroff, Appl. Phys. Lett. **71**, 2014 (1997)
8. A. Lorke, R.J. Luyken, A.O. Govorov, J.P. Kotthaus, J.M. García, P.M. Petroff, Phys. Rev. Lett. **84**, 2223 (2008)
9. F.M. Alves, C. Trallero-Giner, V. Lopez-Richard, G.E. Marques, Phys. Rev. B **77**, 035434 (2008)
10. A. Lorke, R.J. Luyken, A.O. Govorov, J.P. Kotthaus, J.M. García, P.M. Petroff, Phys. Rev. Lett. **84**, 2223 (2000)
11. A. Chaplik, Pis'ma Zh. Eksp. Teor. Fiz. **62**, 885 (1995) [JETP Lett. **62**, 900 (1995)]
12. R.A. Römer, M.E. Raikh, Phys. Rev. B **62**, 7045 (2000)
13. R.A. Römer, M.E. Raikh, Phys. Status Solidi (b) **221**, 535 (2000)
14. A.M. Fischer, V.L. Campo Jr., M.E. Portnoi, R.A. Röomer, Phys. Rev. Lett. **102**, 096405 (2009)
15. E. Ribeiro, A.O. Govorov, W. Carvalho Jr., G. Medeiros-Ribeiro, Phys. Rev. Lett. **92**, 126402 (2004)
16. C. Gonzalez-Santander, F. Dominguez-Adame, R.A. Romer, Phys. Rev. B **84**, 235103 (2011)
17. A.O. Govorov, S.E. Ulloa, K. Karrai, R.J. Warburton, Phys. Rev. B **66**, 081309(R) (2002)
18. J.A. Barker, R.J. Warburton, E.P. O'Reilly, Phys. Rev. B **69**, 035327 (2004)
19. J.I. Climente, J. Planelles, F. Rajadell, J. Phys. Condens. Matter **17**, 1573 (2005)
20. J.I. Climente, J. Planelles, W. Jaskólski, Phys. Rev. B **68**, 075307 (2003)
21. M. Tadić, N. Cukarić, V. Arsoski, F.M. Peeters, Phys. Rev. B **84**, 125307 (2011)
22. M.D. Teodoro, A. Malachias, V. Lopes-Oliveria, D.F. Cesar, V. Lopez-Richard, G.E. Marques, E. Marega Jr., M. Benamara, Yu.I. Mazur, G.J. Salamo, J. Appl. Phys. **112**, 014319 (2012)
23. M.D. Teodoro, V.L. Campo Jr., V. Lopez-Richard, E. Marega Jr., G.E. Marques, Y. Galvao Gobato, F. Iikawa, M.J.S.P. Brasil, Z.Y. AbuWaar, V.G. Dorogan, Yu.I. Mazur, M. Benamara, G.J. Salamo, Phys. Rev. Lett. **104**, 086401 (2010)

24. O. Madelung, W. Martienssen (eds.), *Landölt-Börnstein Comprehensive Index* (Springer, Berlin, 1996)
25. E. Margapoti, F.M. Alves, S. Mahapatra, T. Schmidt, V. Lopez-Richard, C. Destefani, E. Menendez-Proupin, F. Qu, C. Bougerol, K. Brunner, A. Forchel, G.E. Marques, L. Worschech, Phys. Rev. B **82**, 2053181 (2010)
26. E. Margapoti, L. Worschech, S. Mahapatra, K. Brunner, A. Forchel, F.M. Alves, V. Lopez-Richard, G.E. Marques, C. Bougerol, Phys. Rev. B **77**, 073308 (2008)
27. E. Margapoti, F.M. Alves, S. Mahapatra, V. Lopez-Richard, L. Worschech, K. Brunner, F. Qu, C. Destefani, E. Menendez-Proupin, C. Bougerol, A. Forchel, G.E. Marques, New J. Phys. **14**, 043038 (2012)
28. V. Baranwal, G. Biasiol, S. Heun, A. Locatelli, T.O. Mentes, M.N. Orti, Phys. Rev. B **80**, 155328 (2009)
29. M. Hanke, Yu.I. Mazur, E. Marega Jr., Z.Y. AbuWaar, G.J. Salamo, P. Schäfer, M. Schmidbauer, Appl. Phys. Lett. **91**, 043103 (2007)
30. N.A.J.M. Kleemans, J.H. Blokland, A.G. Taboada, H.C.M. van Genuchten, M. Bozkurt, V.M. Fomin, V.N. Gladilin, D. Granados, J.M. García, P.C.M. Christianen, J.C. Maan, J.T. Devreese, P.M. Koenraad, Phys. Rev. B **80**, 155318 (2009)

Chapter 11
Optical Aharonov-Bohm Effect in Type-II Quantum Dots

I.R. Sellers, I.L. Kuskovsky, A.O. Govorov, and B.D. McCombe

Abstract The chapter describes the principles of and reviews recent progress on the Optical Aharonov-Bohm effect in quantum dots focusing on type-II quantum dot structures. Type-II quantum dots have been predicted to display the Optical Aharonov-Bohm effect (OABE) due to the strong polarization of an exciton that results from the spatial separation of the electrons and holes in such systems. The Aharonov-Bohm effect for type-II II–VI Zn(Mn)Te/ZnSe quantum dots is the principal topic of this chapter, with additional discussion of InGaAs/GaAs, InP/GaAs and GeSi type-II quantum dots, in which these effects have also been observed.

11.1 Introduction

11.1.1 Background

The optical Aharonov-Bohm effect (OABE) is an interesting physical phenomenon that occurs in *neutral* excitonic systems in which the charge carriers are confined and

I.R. Sellers (✉)
Department of Physics & Astronomy, University of Oklahoma, 440W Brooks Street, Norman, OK 73017, USA
e-mail: sellers@ou.edu

I.L. Kuskovsky
Department of Physics, Queens College, City University of New York, 65-30 Kissena Blvd, Flushing, NY 11376, USA
e-mail: igor.kuskovsky@qc.cuny.edu

A.O. Govorov
Department of Physics & Astronomy, Clippinger Research Laboratory, Ohio University, Athens, OH 45701, USA
e-mail: Govorov@helios.phy.ohiou.edu

B.D. McCombe
Department of Physics, Fronczak Hall, University at Buffalo, The State University of New York, Buffalo, NY 14260, USA
e-mail: mccombe@buffalo.edu

V.M. Fomin (ed.), *Physics of Quantum Rings*, NanoScience and Technology,
DOI 10.1007/978-3-642-39197-2_11, © Springer-Verlag Berlin Heidelberg 2014

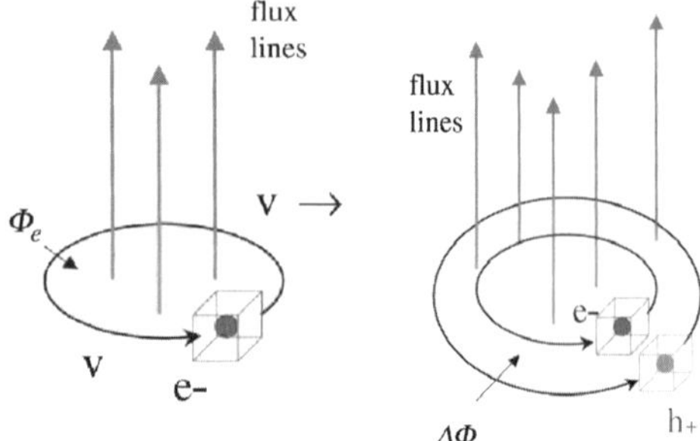

Fig. 11.1 (**a**) The AB effect for a charged particle depends on the magnetic flux through the area orbited by the particle. (**b**) The very different situation for a neutral quasi-particle (exciton) is illustrated. The origin of the excitonic Aharonov-Bohm effect in which the constituent particles orbit the magnetic flux in a closed cylindrical geometry depends on the difference in their relative rotation, and the flux between them. This produces the effective difference in phase of the carriers and an observable oscillation in transition energy and/or intensity. © 2002, reproduced courtesy of the American Physical Society [1]

polarized in a cylindrical geometry [1, 2] (Figs. 11.1(a) and (b)). In such systems a magnetic field penetrating the area enclosed by the paths of the constituent carriers results in a phase shift of their wavefunctions proportional to the flux associated with this field and also makes a contribution to the energy of the system. Although the *conventional* Aharonov-Bohm effect [3] is a property of charged systems, it has been predicted that neutral excitons in semiconductor nanostructures (such as certain types of quantum dots (QDs) or quantum rings (QRs)) are sufficiently polarized (along the radial direction of a circular orbit) that an applied field penetrating the area enclosed by their individual trajectories will act to strongly modulate the emission of the system as a result of optical selection rules and the relative phase of the constituent particles [1].

In Type-II systems the strong localization of one carrier within the QD, coupled by the Coulomb attraction to the otherwise free carrier of opposite charge in the matrix defines the ring-like nature of the system. Specifically, in a magnetic field the carrier within the barrier/matrix orbits the periphery of the QD with a phase proportional to the number of flux quanta (hc/e) threading the area separating the carrier in the matrix from the strongly localized carrier in the QD center.

Due to the cylindrical symmetry of the confinement, the exciton ground state has zero orbital angular momentum projection ($L = 0$) at zero magnetic field, but with increasing magnetic flux the ground state changes to states of higher orbital angular momentum ($L = -1, -2, -3$). As a result, the ground state energy oscillates, and the intensity changes from strong (bright excitonic transition with $L = 0$) to weak (dark excitonic transitions for states with $L \neq 0$) with increasing magnetic field [1].

Recent progress in epitaxial growth and device fabrication techniques has permitted the creation of structures in which the Aharonov-Bohm effect has been demonstrated (both in metallic and semiconductor ring geometries) in magneto-transport measurements [4–7] and for self-assembled quantum rings by far infrared

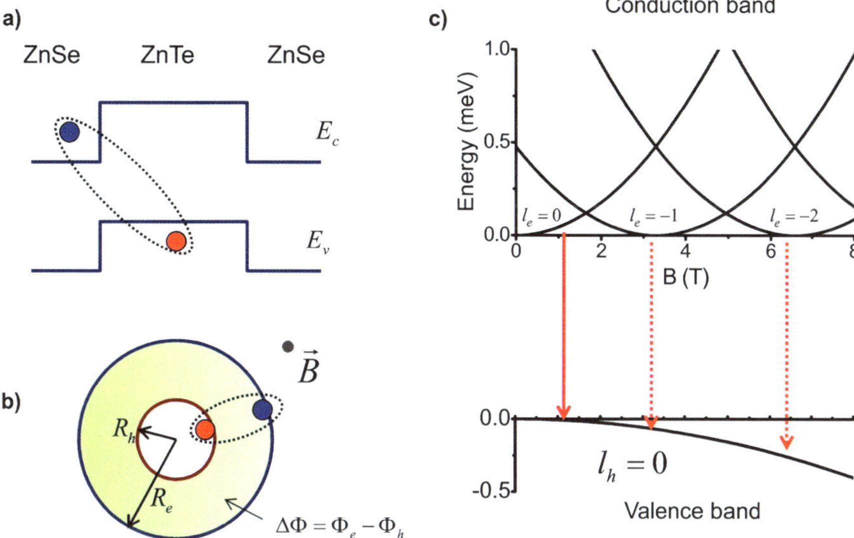

Fig. 11.2 (**a**) Schematic band structure of a type-II QD. (**b**) Theoretical model for an electron–hole pair in a type-II quantum dot or quantum ring. Electron and hole are moving on concentric ring trajectories in the presence of a perpendicular magnetic field. The magnetic flux $\Delta\Phi$ through the area (in *yellow*) between the two trajectories creates a topological Aharonov-Bohm phase in the exciton wavefunction. (**c**) Calculated spectra of electron and hole in this system for $R_h = 0$ and $R_e = 20$ nm for the parameters of the Zn(TeSe) system. *Solid* and *dotted vertical arrows* depict optically allowed and forbidden transitions

and capacitance methods [8], and magnetization spectroscopy [9]. In this chapter we describe the behavior of *type-II semiconductor QD systems*, whose unique properties have resulted in the highest temperature demonstration of the optical Aharonov-Bohm effect (OABE). Type-II QDs are particularly suitable for observing the OABE, since the constituent particles of the exciton, the electron and hole, are strongly polarized due to the spatial separation of the carriers in such systems. The OABE has been confirmed experimentally in a number of Type-II QD systems recently, including InP/GaAs QDs [10] and Ge/Si QDs [11], as well as the II–VI QDs [12–14].

11.1.2 Basic Theory

One simple model for a neutral exciton, a bound state of a Coulomb-correlated electron-hole pair, in type-II QDs and QRs is a system with two circular orbits for the carriers (Figs. 11.2(a) and (b)). If we assume that the electron and hole are essentially uncoupled (very weak Coulomb attraction) and move on two rings with different radii, R_e and R_h, in the presence of an external magnetic field, the energy of the electron-hole pair takes the form [1, 2]:

$$E_{\text{exc}} = E_g + \frac{\hbar^2}{2m_e R_e^2}\left(l_e + \frac{\Phi_e}{\Phi_0}\right)^2 + \frac{\hbar^2}{2m_h R_h^2}\left(l_h - \frac{\Phi_h}{\Phi_0}\right)^2, \tag{11.1}$$

where E_g is a term that includes the band gap energy and any confinement energies, and l_e and l_h represent the angular momenta of the electron and hole, respectively. Φ_0 is the flux quantum (h/e). The magnetic field B enters (11.1) through the magnetic fluxes $\Phi_e = \pi R_e^2 B$ and $\Phi_h = \pi R_h^2 B$. These fluxes describe the quantum phases accumulated by the wave functions of particles as they travel along the ring(s). Equation (11.1) assumes non-interacting electron and hole and, therefore, an independent motion of the carriers. If a strong Coulomb attraction between the particles is included, they will rotate together as shown in Fig. 11.1(b); the exciton spectrum becomes [1, 2]

$$E_{\text{exc}} = \tilde{E}_g + \frac{\hbar^2}{2M R_0^2}\left(L + \frac{\Delta\Phi}{\Phi_0}\right)^2, \tag{11.2}$$

where $L = l_e + l_h$ is the total momentum of an exciton, $R_0 = (R_h + R_e)/2$, and $M = (m_e R_e^2 + m_h R_h^2)/R_0^2$. Importantly, the magnetic field enters this equation now via the magnetic flux, $\Delta\Phi$, through the area *between* the particle trajectories (blue area in Fig. 11.1(b)):

$$\Delta\Phi = \Phi_e - \Phi_h = \pi B\left(R_e^2 - R_h^2\right). \tag{11.3}$$

In the case of a type-II QD with a hole strongly confined inside the QD, we can assume $R_h \to 0$ and obtain: $M R_0^2 = m_e R_e^2$ and $\Delta\Phi = \Phi_e$. Quantum dots in the ZnTe/ZnSe system have such properties; a hole is strongly confined inside the QD potential, whereas the electron, due to the type-II structure, is not confined by the QD potential, and orbits around the QD, attracted to the hole in the QD by the Coulomb force (Figs. 11.1(a) and 11.1(b)). An example of the energy spectrum of such an exciton in a type-II QD with a strongly confined hole is shown in Fig. 11.1(c). The main feature of this system is that the electron in the presence of a normal magnetic field changes its ground state with B, whereas the hole always has the ground state with $l_h = 0$. This unequal behavior may dramatically change the optical emission since an emission process in a direct-gap semiconductor requires the total momentum $L = l_e + l_h = 0$. Since the lowest hole state always has $l_h = 0$, the only optically active states of the exciton should be those with $l_e = 0$. Thus all ground states of the electron in the conduction band with $l_e = -1, -2, -3, \ldots$ are dark states (electric-dipole transitions to the hole ground state are forbidden.). Therefore, one can predict that the exciton emission from a type-II QD with cylindrical symmetry in a magnetic field will be suppressed for $B > B_{\text{AB}}$, where B_{AB} corresponds to the field at which the energy of the electron ground state crosses from $l_e = 0 \to l_e = -1$. The field B_{AB} is given by the condition $\Phi_e = \Phi_0/2$. In real systems, the suppression of emission may not take place, or may be only partially effective, because QDs always have some defects (or shapes) that destroy the cylindrical symmetry and the conservation of angular momentum selection rule. We expect rather that the energy and intensity of exciton lines of type-II QDs will oscillate with increasing magnetic field [1, 15]. This behavior has been observed in several experiments [10–14].

11.1.3 Influences of Type-II Quantum Dots and Quantum Rings

It is interesting to compare the physics of magneto-excitons in type-I and type-II quantum dots. In the type-I system, both electron and hole are localized in a 3D quantum box at the same position and they cannot therefore easily form the necessary rotation to develop the OABE. Consequently, an applied magnetic field cannot force the confined carriers to change their angular momenta in the ground states. In other words, both carriers in the ground state will always satisfy the condition $l_e = l_h = 0$, and the spectrum in the magnetic field will behave in the same manner as that of the confined holes in Fig. 11.2(c), thus the OABE is not expected for the type-I systems. In this scenario there is a diamagnetic shift of the exciton energy, and a stronger localization of the carriers in high B. As such, it is clear that the magneto-optical properties of type-I and type-II QDs are very different. Type-II quantum dots and rings are also different than the type-I quantum rings [16–18] in which the electron and hole move on the same circular path and, therefore, the exciton is not polarized in the radial direction. In such structures, another type of OABE that results from carriers tunneling along a circular path in a 1-D ring has been proposed [16, 17]. In this case the OABE oscillations may be exponentially small in amplitude, since an electron in this effect should tunnel along a circular path, which is typically a relatively long distance.

11.2 Experimental Evidence of the Optical Aharonov-Bohm Effect in Type-II Quantum Dots

11.2.1 Zn(TeSe)/ QDs in ZnTe/ZnSe Superlattices

In this section we describe experimental evidence of AB oscillations in II–VI Zn(TeSe) QDs. In addition, the hole is strongly confined in the dot, while the electron wavefunction is distributed in the ZnSe matrix. In this system it has been shown that through the creation of a columnar QDs, the OABE can be made remarkably robust [13]. Prior to the presentation of the magneto-optical properties of the Zn(TeSe) QDs the growth is described, which is crucial to understanding the formation and structure of QDs in this system.

The samples discussed are Molecular Beam Epitaxy (MBE)-grown ZnTe/ZnSe superlattices, consisting of 120–240 layers. The growth followed a procedure called migration-enhanced-epitaxy (MEE), in which the alloys produced during growth result from in the deposition of a sequence of the constituent sources with short period between to promote preferential migration of the surface atoms to their most energetically favorable positions on the growth surface (Fig. 11.3(a)). In the case

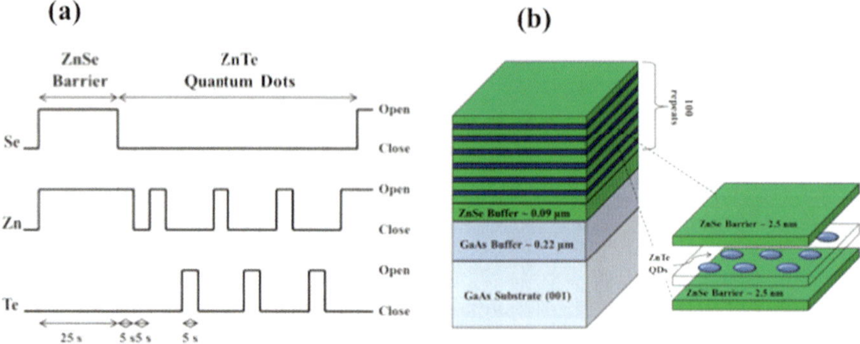

Fig. 11.3 (**a**) MBE Shutter sequence and (**b**) resulting structure for migration-enhanced-epitaxy of ZnTe/ZnSe QD-containing superlattice structures

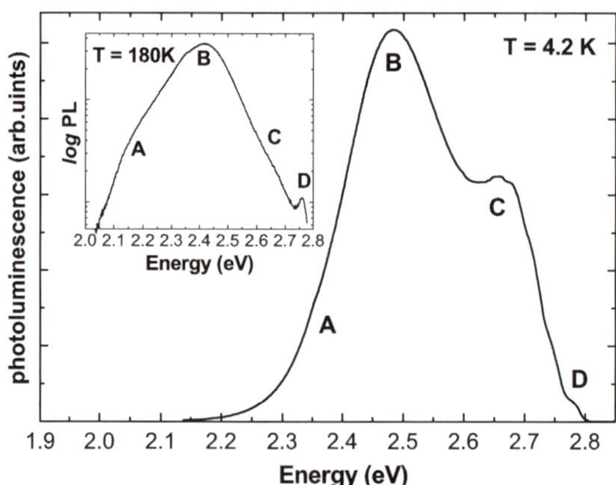

Fig. 11.4 Photoluminescence of the ZnTe/ZnSe superlattice structure at 4.2 K. The *inset* shows the PL at 180 K. The relative contribution of the Zn(TeSe) QDs (A), Te-clusters (B) and pairs of Te-based isoelectronic centers (C) are indicated. Also observed is the emission from free excitons in the Zn(TeSe) matrix (D), which increases at elevated temperatures due to the ionization of bound excitons (see *inset*). © 2008, reproduced courtesy of the American Physical Society [13]

of ZnTe, these dynamics result in the clustering of Te-atoms, and eventually to the formation of Te-rich Zn(TeSe) QDs [19, 20]. The deposited layer containing the QD structures is approximately 0.7 nm thick. This region is then capped with a 2.4 to 3 nm ZnSe layer. A schematic of the growth process and structure created is shown in Fig. 11.3(b). Further details of the growth technique can be found in [21].

To fully understand the origin of QD formation in this multilayer superlattice structure the relationship of certain characteristic features of the photoluminescence (PL) spectra to structures in the superlattice must be explained. The PL at 4.2 K, presented in Fig. 11.4, displays the well known two-band behavior of the Zn(TeSe) system, which has been studied extensively [19–24]. The shoulder around 2.60–2.75 eV (C) is generally accepted to be the result of recombination from excitons

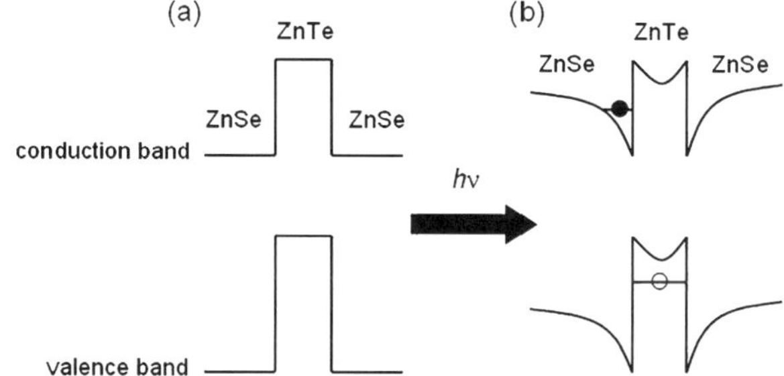

Fig. 11.5 Schematic representation of the type-II band alignment for Zn(TeSe) quantum dots (**a**) before and (**b**) after generation of a high density of carriers

bound to pairs of Te-atoms [19, 24], which substitute isoelectronically for Se [25]. The feature at $\sim$2.79 eV (D) is related to free excitons in the alloy Zn(TeSe) alloy matrix [21], while the dominant feature around 2.50 eV (B) has been historically attributed to the contribution of clusters of larger numbers of these isoelectronic centers. Recently it has been shown that these Te-clusters evolve into type-II Zn(SeTe) QDs in ZnTe/ZnSe superlattice structures if the Te deposition exceeds $\sim$3 monolayers (MLs) [19–21, 26].

Since QDs are formed as a result of clustering of Te-atoms, they are physically larger than the isoelectronic Te_n complexes, which results in lower energy emission relative to that of isoelectronic bound excitons (IBE). The low energy shoulder, labeled (A) in Fig. 11.4, along with some contribution from the low energy tail of (B), is attributed to QD emission. This can be observed more clearly in the inset to Fig. 11.4, which shows the PL at 180 K, on a semi-log scale. The increased contribution of the QDs is more evident at elevated temperatures because a redistribution of carriers to lower energy *larger QDs* occurs at elevated temperatures in confined systems [27]. Further evidence of the higher thermal energy in the system is demonstrated by the increased contribution of free excitons from the ZnSe matrix (D) due to the ionization of carriers bound to isoelectronic centers at higher temperature. Furthermore, it has also been shown that the QDs, which evolve from the clustering of Te-atoms in the ZnTe/ZnSe multilayer structure described here, also align vertically and therefore form columnar QD-like nanostructures [12].

To further confirm the existence of QDs in ZnTe/ZnSe superlattice structures power dependent cw and time-resolved PL (TRPL) measurements were performed. Power dependent PL was been used previously to demonstrate the formation of type-II transitions due to the unique properties of these systems [28]. Specifically, upon increasing excitation the PL of a type-II system exhibits a large blue shift in energy due to the formation of a dipole layer at the interface between the bound and confined carriers [21, 28]. The accumulation of carriers at the interface results in the formation of a triangular potential well for electrons on the barrier material, the depth of which increases with increasing photo-excitation. This is shown schematically in Fig. 11.5.

This increase (blue-shift) in the transition energy of the system is visible in photoluminescence measurements as the power of excitation, and therefore carrier concentration is increased. Evidence of such behavior in ZnTe/ZnSe superlattice is pre-

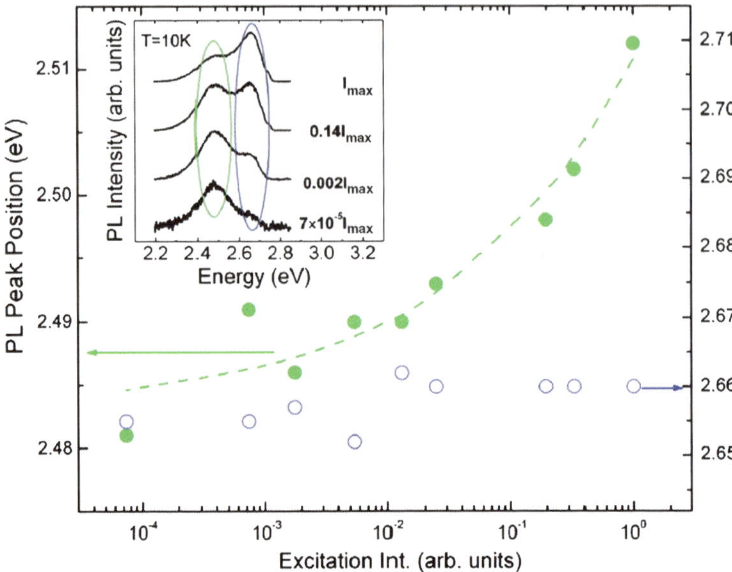

Fig. 11.6 Peak position of emission of QDs (*closed green circles*) and excitonic complexes (*open blue circles*) as a function of increasing photo-excitation. The *inset* shows the relative contribution of the two emission channels at 10 K over five orders of magnitude of excitation intensity. © 2005, reproduced courtesy of the American Physical Society [19]

sented in Fig. 11.6, which shows the effect of power on the different complexes observed in the PL over five orders of magnitude (see inset).

With increasing power, the high-energy excitonic peak (open circles) displays negligible shift, as expected. However, the lower energy QD peak (closed circles) displays a $P^{1/3}$ dependency and blue shift of more than 30 meV, consistent with the formation of a dipole layer in a type-II system [28]. To confirm further the contribution of the shift of the lower energy peak to type-II QD formation, time-resolved photoluminescence measurements were also performed. Since in type-II systems their exists a spatial separation of the carriers, the radiative lifetime of the excitons should be evident via extended PL lifetimes in time-resolved PL measurements. The PL lifetime was studied at 14 K under excitation by a pulsed laser system emitting linearly polarized light at 400 nm (repetition rate ~250 kHz, pulse duration ~300 fs). The initial lifetime of the PL traces was extracted in 5 nm intervals in the wavelength region between 545 nm and 450 nm to determine the lifetime across the integrated PL spectrum, which consists of emission from both QD and excitonic complexes. The PL lifetime traces taken at four positions across the PL spectrum are shown in Fig. 11.7. The green band, dominated by emission from QDs, and the blue band, from excitonic complexes, are represented by the 2.53 and 2.67 eV traces, respectively. The high energy trace at 2.73 eV represents the position at which the contribution of QDs is negligible, while the trace at 2.38 eV is dominated by large type-II QDs. Clearly, as the contribution of the QDs increases, the lifetime is also extended.

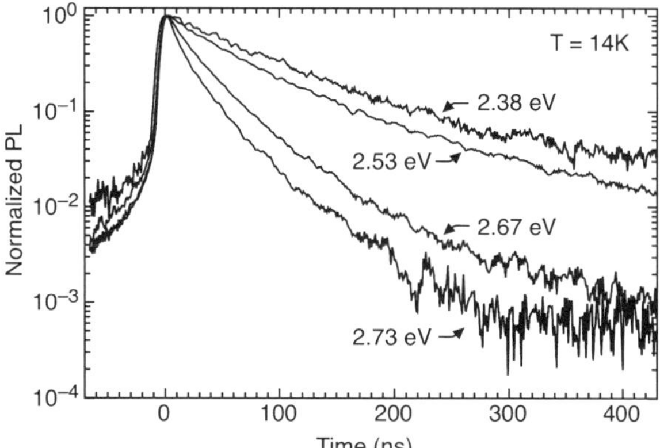

Fig. 11.7 PL decay near the tail of the "*green band*" (2.38 eV), the peak of the "*green band*" (2.53 eV), the peak of the "*blue band*" (2.67 eV), and the side of the "*blue band*" (2.73 eV)

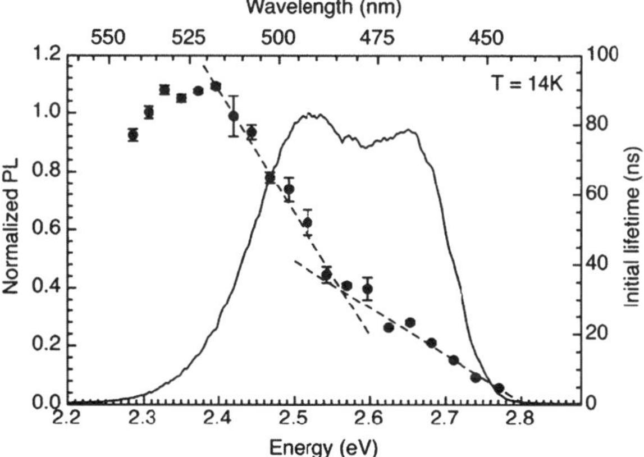

Fig. 11.8 Initial PL lifetime at 14 K at 5 nm intervals (*closed circles*) spanning the background of the time integrated normalized PL (*solid line*). The *dashed lines* display the lifetime per unit energy of the *green* (1) and *blue* (2) band, respectively. © 2008, reproduced courtesy of the American Physical Society [26]

In Fig. 11.8 the initial lifetime of the PL is shown at a series of photon energies at which the lifetime was evaluated, which span the emission spectrum shown. At high energies, where emission due to excitons bound to pairs and clusters of Te-atoms dominates, the lifetime is shorter ($<$1–5 ns at 2.8 eV). However, as the contribution of the low energy states attributed to QDs increases, the PL lifetime increases significantly, such that a peak lifetime $>$80 ns is evident at $\sim$2.35 eV. Interestingly, this is exactly the position of the emission attributed to QDs in the temperature dependent PL (see Fig. 11.4). The extremely long lifetime of the PL at low energies is consistent with a type-II band alignment, for which the lifetime is enhanced due to the spatial separation of the carriers; coupled with the blue shift observed in power dependent PL measurements, this confirms the formation of Zn(TeSe) QDs in the ZnTe/ZnSe superlattice structures.

In addition to the work that demonstrates that QDs form in the ZnTe/ZnSe superlattices, structural studies have shown that these QDs align vertically and therefore form a columnar structure [12]. Recently, similar effects were also observed in ZnMgTe/ZnSe QD superlattices, for which the presence of Mg resulted in higher scattering intensity in the x-ray diffraction patterns, permitting the observation of

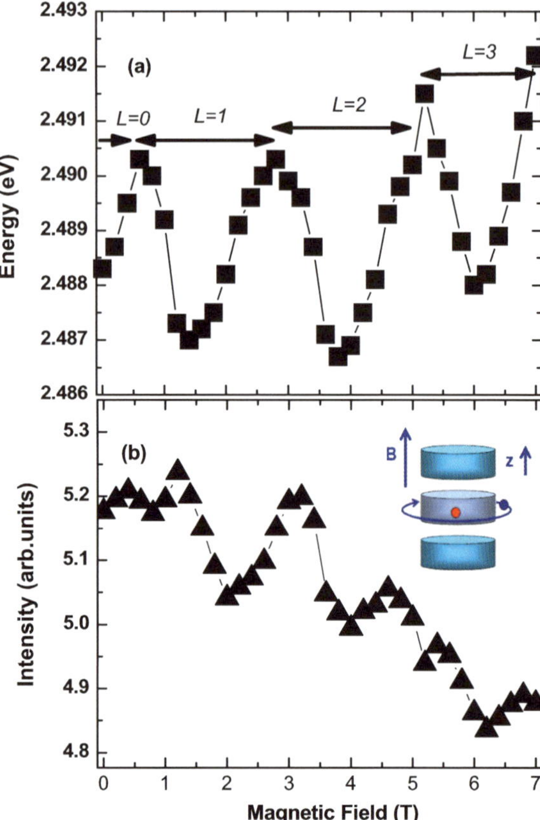

Fig. 11.9 (**a**) Magneto-PL energy (**a**) and intensity (**b**) at 4.2 K for columnar QDs. $L > 0$ in (**a**) represent the total orbital momentum characterizing the ground states as a function of magnetic field, and the *arrows* indicate transitions from one ground state to the next. The inset to Fig. 11.12(b) shows a schematic of the columnar QD geometry. © 2008, reproduced courtesy of the American Physical Society [13]

higher order superlattice peaks in both the (004) and (002) reflections; this allowed for more detailed studies of vertical coupling of QDs [29]. This vertical alignment of QDs is not unexpected since the narrow spacer layers (2.4–3.0 nm) between the QD-containing layers promote strain interaction and therefore alignment (stacking) of QDs in successive layers [29–31], which is a well known phenomenon in multi-layer QD systems [32]. It is shown below that this geometry, coupled with the type-II band alignment, is particularly suitable for facilitation of strong AB interference.

In type-II Zn(SeTe) QDs, the hole is strongly confined by the nanostructure potential. It has been shown theoretically [12] that the electron within an exciton in a single layer type-II QD would be localized either above or below the QD due to the pancake-like geometry of the typical QDs and the Coulomb attraction to the hole. In this configuration, no AB effect would be expected. Thus, again, the columnar geometry of the present system is of particular importance, since it dictates energetically the ring-like geometry of the correlated charge carrier motion (shown schematically in Fig. 11.9(b) because of the large cost in kinetic energy for locating the electron above or below the dots. In addition, numerical calculations have also shown that the tunneling between QDs in the column is negligible [12] and therefore the exciton cannot move along the growth direction.

Based on the preceding discussion, we assume that in our stacked Zn(TeSe) QDs an individual exciton is localized in one of the dots. As such, the confined exciton forms a radial dipole with the constituent particles orbiting over different geometries described in the simplest case by (11.2); the hole is strongly localized with $R_h \approx 0$. According to (11.2) the rotating dipole created by the orbit of the bound electron results in transitions from one ground state (characterized by a particular orbital angular momentum projection and energy) to another (with a different angular moment projection and energy) as a function of magnetic field; the angular momenta at zero field $L = 0$ progress to $L = -1$, $L = -2$, etc. with increasing field. Therefore, the energy of the lowest excitonic state oscillates as a function of magnetic field. Correspondingly, the radius of the orbit of the electron can be inferred from the period of oscillation as a function of magnetic field, allowing for comparison with structural measurements.

To determine the effects of magnetic field upon the PL spectrum, the sample was mounted in the Faraday geometry in an optical access superconducting magnet system. The PL spectra were recorded in 0.2 T steps as a function of temperature between 4.2 and 180 K. The energy and intensity of the two dominant features (A and B in Fig. 11.4) were evaluated at each field through careful Gaussian fitting of the individual peaks. Since isoelectronic centers do not possess type-II geometry, we concentrate on feature (A) here, where the dominant contribution is emission from type-II QDs. The effect of the magnetic field on the PL energy at 4.2 K is shown in Fig. 11.9(a). Here, three, well-resolved oscillations are evident; these oscillations are undamped within the field range studied. Since the period of oscillation should be approximately given by $\delta B = \Phi_0 / \pi R_e^2$, a radius of $R_e = 23.5$ nm is obtained from the data, a reasonable value compared with the results of structural measurements, which yield a lateral QD diameter of $\sim$20 nm. The effect upon the intensity of the PL under the same conditions is presented in Fig. 11.9(b). Again, clear oscillations are apparent. However, in this case the oscillations are relatively weaker, and show 'damping', an overall decrease in the PL intensity with increasing magnetic field. This reduction in PL intensity has recently been attributed to the increased occupation of higher order states in the QDs, and an increased contribution of the dark excitons related to heavy-hole transitions [32].

Oscillations in the PL intensity of quantum rings due to the OABE, were first predicted by Govorov et al. [1]. In that work, Govorov and co-workers predicted that for a strongly correlated electron and hole the *independent* rotation of the constituent particles over different trajectories results in a periodic switching of the ground state with $L = 0$ to $L \neq 0$, in specific windows of magnetic field. However, since the hole is strongly localized in the QD core in the present case, rotation of the hole and therefore transitions in its angular momentum are unlikely. Indeed, from this perspective the Zn(TeSe) QDs are rather reminiscent of the conventional AB effect, in which the phase of the wavefunction of an individual charged particle is changed by the applied field. Here, however, the AB effect is still observed via the recombination of *both* carriers, and therefore the exciton, through the emission of photons.

Based on the above description, it is unlikely that the intensity oscillations shown in Fig. 11.9(b) can be due to the uncorrelated rotation of the constituent particles

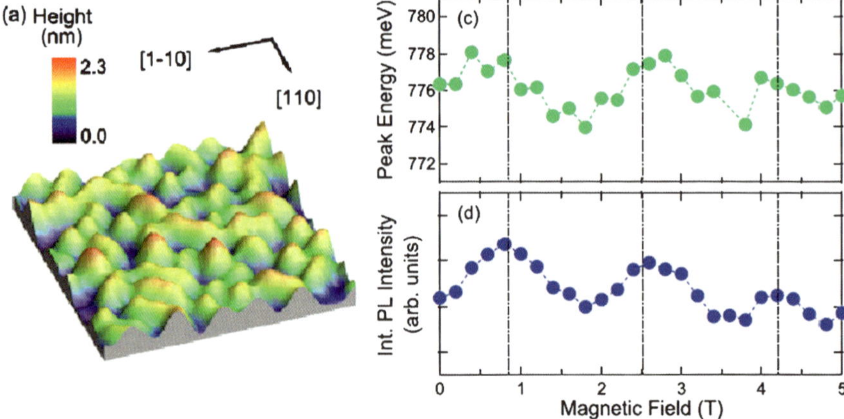

Fig. 11.10 (**a**) Atomic Force image of a single GeSi QD layer. Aharonov-Bohm oscillations in (**b**) photoluminescence energy and (**c**) intensity at 2 K from a multilayer GeSi QD structure. © 2010, reproduced courtesy of the American Physical Society [11]

about the QD; the PL intensity would be expected to quench at $L \neq 0$, approximately 0.6 T in Fig. 11.9(a). However, as mentioned above, the selection rules for transitions in total angular momentum are strictly valid only for the situation of perfect rotational symmetry. In *real* QD structures this is highly improbable since the growth processes result in elongated QD structures, which has been confirmed in TEM for the Zn(TeSe) QDs. Therefore, the selection rules are likely to be relaxed in the QDs studied in this work resulting in the observation of $L \neq 0$ higher order states. In addition, impurity scattering has also been predicted to relax the selection rules relevant to the optical ABE in nano-rings resulting the mixing of higher order angular momentum states and the appearance of higher order states [15]. Since material fluctuations and defect formation are probable in the present system such effects also appear likely.

Recently, the observations Aharonov-Bohm oscillations in both energy and intensity have also been observed in type-II Ge/Si QDs grown by MBE [11] in this group-IV system where, similar to Zn(TeSe) dots, the hole is strongly confined within the (Ge-rich) QDs, while the electron preferentially rests in the (silicon) spacer layers. Since these Group-IV systems have indirect bandgaps, multiple layers of uncoupled QDs were grown to increase the optical absorption. Evidence of energy and intensity oscillations due to AB coherence is shown for the GeSi QDs in Fig. 11.10. This result *further* demonstrates the transferability (and suitability) of type-II QDs from various material systems for the observation of the OABE. In the case of the GeSi type-II dots, the relaxation of the selection rules is attributed to the non-uniformity of the system evidenced in the atomic force image shown in Fig. 11.10(a).

The robust nature of the OABE in these Zn(TeSe) QDs is further demonstrated by remarkably persistent behavior of the effect observed at higher temperature. The effect of temperature on the PL energy and intensity at 60 K is shown in Figs. 11.11(a) and (b), respectively. With increasing temperature the oscillations decrease in mag-

Fig. 11.11 (**a**) Magneto-PL energy (**a**) and intensity (**b**) at 60 K for the Zn(SeTe) QDs. The *inset* to (**a**) shows the energy dependence at 180 K, while the *inset* to Fig. (**b**) shows the temperature dependence of the integrated PL intensity from the dots. © 2008, reproduced courtesy of the American Physical Society [13]

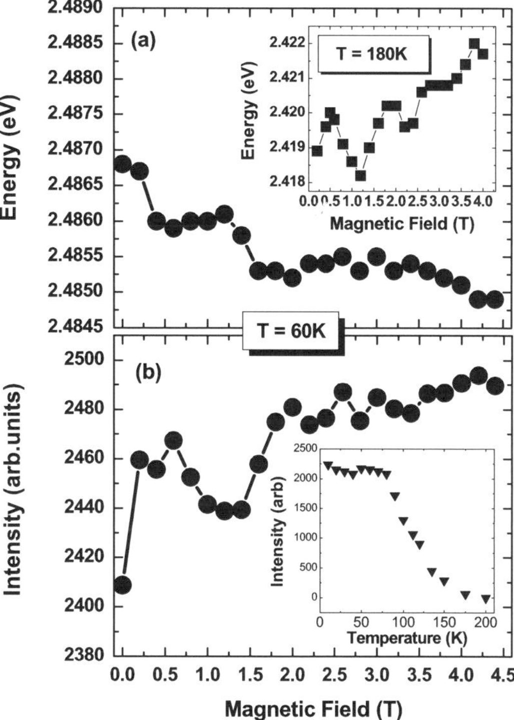

nitude and damp at higher magnetic fields due to decoherence and fluctuations in the system. In the GeSi [11] system the damping of the oscillations for higher order orbital momenta transitions has been attributed to the non-uniformity in size of the QDs in the system studied, which theory indicates introduces decoherence of the ensemble [11]. However, in the case of GeSi QDs, the effects were observed at low temperature ($T < 4$ K). In the case of the Zn(TeSe) QDs (see Fig. 11.9) it is clear that little evidence of damping is evident at low temperature. This indicates in the Zn(TeSe) systems, at least, the source of decoherence is mostly thermal. Specifically, these behaviors appear to be driven *predominately* by ionization of the type-II excitons at elevated temperatures. Indeed, the temperature dependent PL shown as an inset to Fig. 11.11(b) clearly illustrates that above $\sim$70 K the overall PL intensity decreases rapidly. This also is supported by detailed temperature dependent studies reported in Ref. [19].

This is consistent with the thermal ionization of the type-II QD excitons, which have a binding energy of 7.3 meV [19, 21]. However, as is shown as the inset to Fig. 11.11(a), despite the loss of overall PL intensity evidence of the OABE is observed up to 180 K. This remarkably persistent behavior suggests that where type-II excitons exist in these columnar QD structures, AB coherence will occur. We believe this robust behavior can be attributed to the unique columnar system studied, coupled with the fact that the orbiting particle in our system is an electron, rather than a hole (the case in other systems studied [10, 11, 32, 33]. This is supported by

Fig. 11.12 Schematic illustrations of (**a**) conduction and valence band profiles, indicating the spatial separation of electrons and holes. (**b**) Top view of the QD plane, indicating the holes confined to the ring around the QD due to coulomb interaction with the electron trapped in the dot. © 2004, reproduced courtesy of the American Physical Society [10]

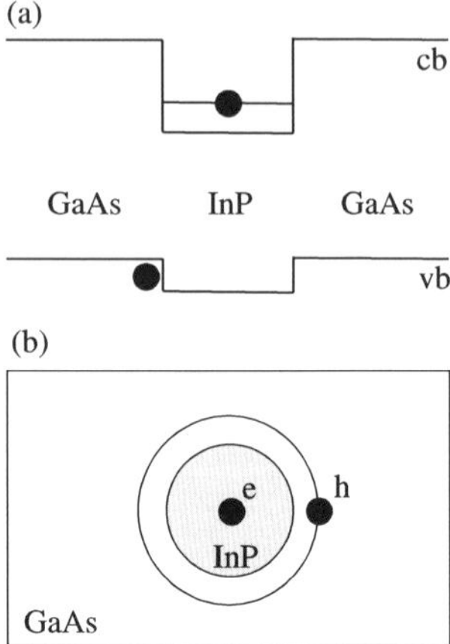

the fact that in the GeSi system [11], in which an uncoupled multilayer structure was studied, the effect was restricted to cryogenic temperatures (2 K). In addition, in the InP/GaAs system, the presence of a rotating hole has also been indicated to decrease the coherence [11].

11.2.2 InP/GaAs Quantum Dots

The first type-II system to display evidence of the OABE predicted by Govorov et al. was InP/GaAs quantum dots [10]. The energy band alignment of this heterostructure is shown in Fig. 11.12(a).

Here, the electron is strongly confined in the QD, while the hole resides in the GaAs matrix. Upon the application of a magnetic field, and at temperature below the electron-hole ionization energy, the hole orbits the QD periphery producing an observable OABE. The sample under investigation was grown by metal-organic-vapor-phase epitaxy (MOVPE). A single InP QD layer was deposited via the well-known Stranski-Krastanov process with PH_3 as a precursor, on epi-ready GaAs substrates at 600 °C. The QD layer was followed by a 50 nm GaAs cap deposited at the same temperature. The effect upon the photoluminescence energy is shown in Fig. 11.13. Figure 11.13(a) shows the behavior of the PL peaks extracted from the raw data spectra, where an oscillatory behavior is evident. The oscillations are more evident in Fig. 11.14(b), where the rising background due to diamagnetic behavior of the "free" electron has been subtracted from the data of (a).

The period of oscillation for the rotation of the electron as a function of increasing orbital momenta obtained from (11.2) [1] is also shown in Fig. 11.13(b). In this simplest case, the higher order ($L > 2$) orbital momenta observed experimentally are not well reproduced by this equation. The damping evident is attributed to the inhomogeneous broadening associated with the distribution of the InP/GaAs QD sizes. To account for this size dispersion quantitatively, the radius (R) in (11.2),

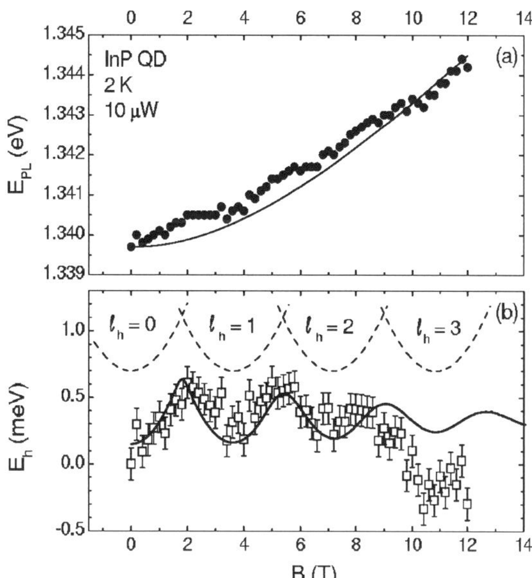

Fig. 11.13 (a) The PL peak position as a function of the applied magnetic field (*solid circles*). (**b**) Hole energy dependence (with background removed) on the applied magnetic field (*open squares*), showing the Aharonov-Bohm oscillations. The *error bars* indicate the uncertainties in the peak position after Gaussian fits. The *dashed curves* show the parabolas according to (11.2). The *solid line* shows the amended fit through the incorporation of a Gaussian distribution of the QD size. © 2004, reproduced courtesy of the American Physical Society [10]

which is related to dominant QD radius in the distribution, is replaced by $2\Delta R$ to account for a Gaussian distribution of the QDs of $\sim$1.6 nm. Upon this correction a clear damping in the AB coherence is evident, consistent with the data. Since this data was measured at cryogenic temperatures, dot inhomogeneity would appear to be the dominant source of decoherence in these structures. Indeed, the fact that energy and intensity oscillation have been observed in the Zn(TeSe) and GeSi/Si systems, in which the electron orbits the strongly confined hole, while only energy oscillations are evident in the InP/GaAs systems, with the opposite geometry (confined electron, rotating hole), suggests the lighter mass of the electron aids the magnitude of the AB coherence in type-II QDs, presumably due to the enhanced mobility and reduced scattering cross section of electrons, which reduces decoherence.

11.2.3 The InGaAs/GaAs Fluctuation Dots

Very recently AB-like oscillations have also been reported for in InGaAs quantum wells (QWs) due to the presence of indium-rich InGaAs *fluctuation quantum dots* [34]. Simultaneously with the magneto-PL spectra, long lifetimes were reported for the excitons in the PL peak reported responsible for the OABE [34]. This is suggested to support the scenario of a type-II quantum dot formed in the plane of the QW with indium-rich InGaAs islands (Fig. 11.14). According to the model proposed in Ref. [34], the localization of the electrons in the 2D circular ring around an In-rich island comes from to joint action of the Coulomb attraction and the strain-induced potential (Fig. 11.14(c)). The resulting potential creates a type-II confinement for

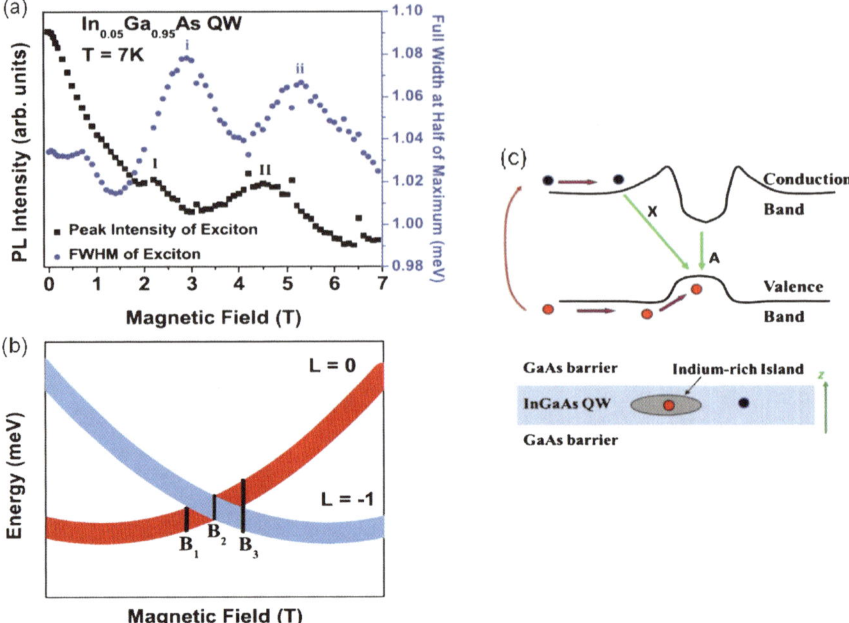

Fig. 11.14 (**a**) Plot of the exciton PL intensity (*squares*) and the exciton FWHM (*circles*) as a function of magnetic field applied along the z-axis. (**b**) Schematic diagram of the reported energies of the $L = 0$ and $L = -1$ exciton states in the vicinity of their crossing. (**c**) Schematic of the conduction and valence band edges in the vicinity of an indium-rich island and schematic of the spatially indirect exciton associated with indium-rich islands. The hole is confined at the center of the island, while the electron is held outside by the electron hole Coulomb attraction. © 2012, reproduced courtesy of the American Physical Society [34]

an exciton: the hole is localized, whereas the electron orbits around the hole in the plane of the QW. This physical situation is again proposed to provide the necessary conditions for the OABE in PL emission.

11.3 Effects of Inhomogeneity in Type-II Quantum Dots

Before moving on to the effect of the OABE in magnetic systems, we briefly return to the question of the high level of uniformity evident in the low temperature magneto-PL energy-spectra in the Zn(TeSe) columnar QDs (Fig. 11.9), and, specifically, the question of QD inhomogeneity. As discussed previously, if the damping observed in the InP/GaAs and GeSi QDs at low temperature is the result of QD size distribution—with a contribution of increase decoherence of the rotating holes in the InP/GaAs system—then the results at low temperature suggest a low inhomogeneity for the Zn(TeSe). This conclusion is actually supported be recent experimental work on these systems in which the spectral dependence of the AB coherence was investigated [35, 36].

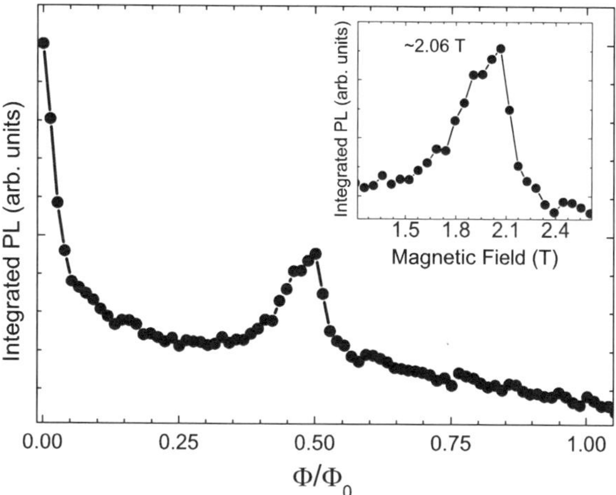

Fig. 11.15 Normalized integrated PL intensity as a function of normalized magnetic flux. *Insets* show enlarged peak region as a function of the magnetic field. © 2012, reproduced courtesy of the American Physical Society [36]

11.3.1 Spectral Analysis of Type-II Columnar Zn(TeSe) Quantum Dots

11.3.1.1 Magneto-Photoluminescence—Integrated Intensity

In Fig. 11.15 we present the normalized integrated PL intensity as a function of the normalized magnetic flux, with an inset that shows an enlargement of the peak at $B_{AB} \sim 2.06$ T. Using (11.1) (Sect. 11.1.2), we find the characteristic radius of the electronic orbit to be 17.8 nm for these QDs. Considering that the stacking of QDs results in averaging out variations in individual QD size in the column, and that at low temperature $L = 0$ (i.e. ground state) dominates, the magnetic flux was calculated for the corresponding area enclosed within the electron trajectory with the goal of comparing our experimental observations with the theoretical calculations for quantum rings. For the ABE the oscillation peak occurs at $\Phi/\Phi_0 = 1/2$. In Fig. 11.15 the magnitude of the AB oscillation (relative to the background) in integrated PL is $\sim$4 %, and the full-width-at-half-maximum (FWHM) of the oscillation in the normalized magnetic flux is $\sim$0.06. This magnitude of the AB oscillation peak is larger [14, 37] and the peaks are significantly narrower [32, 39–41] than the theoretical values previously reported for both QDs [14, 37], and quantum rings [38].

We compare the widths of the Aharonov-Bohm 'peaks' with theoretical predictions available for quantum rings[1] even though there are critical differences between the two system, since such calculations currently do not exist for type-II systems. Without the inclusion of an electric field, a FWHM of $\sim$0.3 (in units of normalized magnetic flux) has been predicted for bright excitons in these systems. This is substantially wider than that observed in Fig. 11.14; however, if one assumes a presence of a strong in-plane electric field ($u_0 = 0.3$ in Ref. [38] notation) then the observed FWHM can easily be explained and understood, since the application of an in plane electric field has been shown to enable control of the excitonic oscillator strength, to the extent that whether the system is optically active (or not) can be controlled [38]. The application of an electric field results in a strong narrowing of the peak associated with the transition between orbital momenta states, L. Physically this occurs due to an increased contribution of higher order dark state that become increasingly occupied with the increase in an in-plane electric field (or temperature) [38]. The electron and hole must be within the same angular co-ordinate ($L = 0$) for the transition to be allowed, and therefore optically active. The application of the electric field polarizes the exciton and narrows the range of magnetic flux values when $L \neq 0$ [38, 39].

Such electric field control was observed recently to enhance the AB effect in InAs nano-rings [18, 32]. Despite the obvious difference between type-I and type-II systems, such effects may also explain the strong peak narrowing and increase in OABE strength observed in the Zn(TeSe) system described here (Fig. 11.15). For instance, in the presence of the built-in electric field, a relatively large ($\sim$9 %) Aharonov-Bohm oscillation (in intensity) has been observed for the InAs (type-I) quantum rings discussed, and described in Ref. [33]. We note that in Refs. [18, 32], and the references therein, the oscillator strength of the Aharonov-Bohm oscillation is observed experimentally within the ground state of the exciton, with total angular momentum $L = 0$; although we observe an oscillation in PL intensity mainly due to transition from the $L = 0$ to non-zero states. In our case, this is attributed to the presence of defects, and possibility of mixed states, due to the non-ideality of real systems. The built-in electric field in a zinc-blende heterogeneous system grown along [001] can be of piezoelectric origin as a result of the anisotropic strain in the crystal [40, 41]. This is known to result in elongation of QDs in the plane in III–V systems [27]. In the Zn(TeSe) epitaxial system, in which such properties exist, additional strain is induced by the lattice mismatch between the ZnTe epilayer (forming the dots) and ZnSe matrix. The anisotropic nature of the strain in the system considered is supported by the following observations: (i) the QDs are elongated preferentially along the [110] and [1 10] planes [12], which leads to anisotropic strain, similar to that reported for InGaAs/GaAs quantum rings [18]; (ii) a vertical correlation among the QDs is evident, driven by strain communication between the stacked layers of QD superlattices [30, 42]. Vertical correlation of stacked QDs in a similar system of type-II ZnMgTe/ZnSe superlattice was reported recently [43]. Further support can be inferred from recent work on different types of ZnTe/ZnSe QDs grown and studied for purpose of analyzing a built-in strain [43]; a number of conclusions in this work directly imply strong strain anisotropy in Zn(TeSe) QD systems [44]. To date the behavior of the oscillator strength in these II–VI systems is not well understood. Indeed, since theoretical investigations specifically pertaining to type-II *quantum dots* are currently limited, the effects due to the presence or absence of an electric field, and/or due to thermal processes remains unknown. Such theoretical investigations are now the focus of various investigations in these type-II QD systems.

Fig. 11.16 PL Intensity versus magnetic field at three spectral positions (*blue*—2.540 eV; *red*—2.598 eV; *green*—2.664 eV) for the spectrum of (**a**). The *inset* shows an enlarged picture of the "double AB peak" as seen at 2.52 eV (red spectrum—main panel). © 2012, reproduced courtesy of the American Physical Society [36]

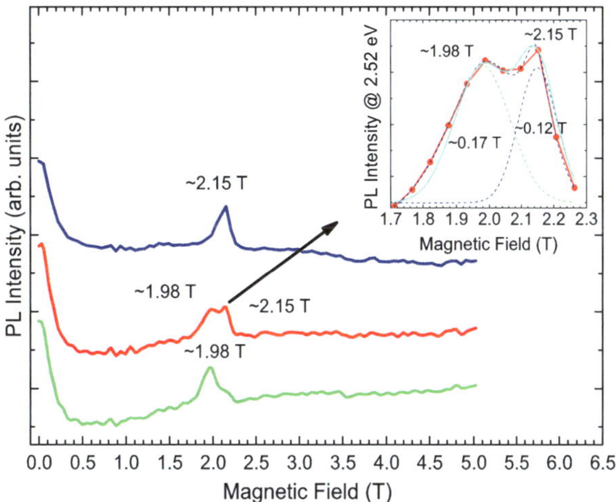

11.3.1.2 Magneto-Photoluminescence—Spectral Dependence of the Optical Aharonov-Bohm Effect

To gain further insight into the material structure of the investigated Zn(TeSe) QDs, a detailed investigation of the spectral dependence of the OABE coherence was also performed. This was achieved through careful tracking of the OABE at various energies across the PL spectrum.

In Fig. 11.16 the AB oscillations in PL intensity, at three different spectral energies, corresponding to low-and high-energy positions across the PL spectrum, are shown. Also shown inset is an expansion of the peak of the Aharonov-Bohm feature in PL at 2.52 eV. This analysis yields some interesting information regarding the AB coherence in these QDs. Specifically, the AB oscillation investigated shows a distinctive change in values of B_{AB} (the critical field for the OABE). Moreover, careful examination of the transition region shows that there is a "double Aharonov-Bohm peak," as seen clearly for the 2.52 eV feature in the inset to Fig. 11.16.

Since B_{AB} is determined by the radius of the electronic orbit, we infer that the electronic orbit changes across the PL spectrum, which indicates the presence of more than one type of columns formed by the QDs, with different effective radii for the orbiting electrons. As such, the size of the type-II exciton can be extracted from (11.1) with sub-nanometer precision [35]. The FWHM of each of the AB peaks ($\sim$0.12 to 0.17 T, shown in inset of Fig. 11.16) is comparable to the FWHM of the AB peak ($\sim$0.14 T) in a sample where only one AB peak is observed [36]. This suggests that all Zn(TeSe) QD samples grown by MEE display a similar built-in electric field, which is consistent with the fact that such fields originate from the lattice mismatch between ZnTe and ZnSe. This is supported by strain communication of QDs grown on preceding layers, that align due to the strain fields in the vertical direction, and the elongation of the QDs along the inplane [110] and [110] directions. The strong evidence for anisotropic strain distribution in ZnTe/ZnSe QDs was shown recently in Ref. [44], and can be clearly inferred from general discussion in

Ref. [44] where it was shown that ZnTe and ZnSe have particularly large elastic anisotropy.

To complete the picture, in Fig. 11.17(a–c) the behavior of B_{AB}, and the magnitude of AB oscillations, as a function of the PL spectral energy. We observe that B_{AB} (Fig. 11.17(b) changes from a lower value of $\sim$1.98 T, at low energy (below $\sim$2.47 eV), to a higher value of $\sim$2.15 T, at higher energy (above $\sim$2.55 eV). The error in determination of B_{AB} is less than the observed change (also displayed in the plot). An average of the B_{AB} is determined from the two peaks (as observed in the 'double AB peak' (Fig. 11.16(b)), to indicate the transition in B_{AB}. The 'double peak' B_{AB} values are also shown in Fig. 11.15(b) (blue and green markers), for the transition spectral region. The transition of B_{AB} from a lower value to higher value is observed within a relatively narrow spectral range—between 2.49 and 2.54 eV. To relate such an observation to the PL spectra, three Gaussian peaks were used to decompose the PL; the 'green' band is formed by two Gaussians with comparable weights, while the third Gaussian represents a "combined" 'blue' band for the higher energies dominated by the isoelectronic centers originating from the barriers [36]. Under such conditions, the best fitting resulted in a "reduced" overlap of the lower energy Gaussians in close proximity of the spectral region of interest, indicating that the change in transition field is observable due to the subdued emission from the two overlapping PL bands. The contribution from two different type-II excitons also explains the asymmetry of the AB peaks in the integrated intensity (Fig. 11.15), as the emission of one of the excitons dominates, in a specific spectral region (the almost constant size of each stack is clearly inferred from almost constant value of the 'critical field' across a given spectral range). Typically, such large PL linewidths (as observed in Figs. 11.4, 11.8, and 11.17(b)), evident for these Zn(TeSe) structures described, are attributed to inhomogeneity in the size, strain, and composition of the material. However, the evidence here is that the QD distribution is actually rather homogeneous (or low) in the columnar QDs investigated. If this hypothesis is true then the observed broadening must have an alternative origin. Recently, the broadening of the PL has been shown to be strongly correlated with thermal processes, via phonon interaction, which may at least partially describe the large PL linewidths observed. Although this conclusion is still under investigation, it has been proposed previously for these systems [21] and other similar systems [29] as well as well understood for bulk ZnSeTe [21].

The difference in the PL peak energy between these two stacks, estimated from the peaks of the two Gaussians used to form the green band, is $\sim$150 meV. Such a large difference in energy is unrealistic considering only the change due to a difference in QD radius [35, 36]; thus other factors, such as size, shape and chemical composition of the QDs, must also be play a role in determining these effects upon the PL characteristics for such systems. Thus, we suggest that the two QD columns are distinguishable not only in terms of radius of electronic orbit, or the lateral sizes of QDs, but also due to the change in QD thickness and/or change in the band offsets because of the variations in Te composition.

The spectral variation in the magnitude of the Aharonov-Bohm peak is shown Fig. 11.17(c). The magnitude of an AB oscillation is directly related to changes in

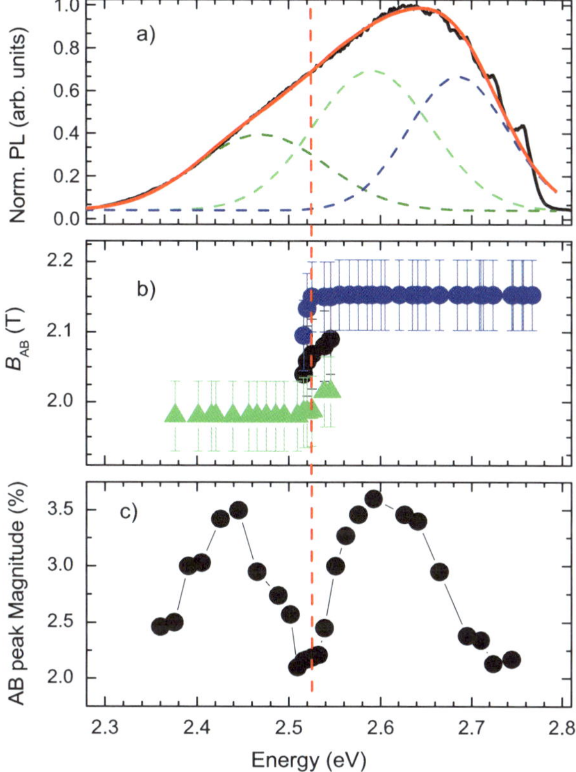

Fig. 11.17 Analysis of Aharonov-Bohm oscillations as a function of PL emission energy: (**a**) PL spectrum decomposed into three Gaussians; (**b**) transition magnetic field, B_{AB}; (**c**) magnitude of AB oscillations with respect to the background. © 2012, reproduced courtesy of the American Physical Society [36]

the oscillator strength (overlap of the wavefunction of the rotating particles) of the exciton during the Aharonov-Bohm transition. The relative change in the oscillator strength is predicted to be smaller with increasing oscillator strength in quantum rings, due to a more favorable Aharonov-Bohm configuration: to date (to the best of our knowledge) no such theoretical predictions are available *specifically* for type-II QDs. Our observations, however, contradict the conclusions derived for more conventional quantum-rings (at least partially), as it would then be expected that for 'smaller' dots, the stronger oscillator strength [39, 45] would give rise to a smaller relative Aharonov-Bohm amplitude. This is explained in terms of the contribution of the background signal, related to the overall interaction, which is larger in smaller rings, and tends to overwhelm the additional (relatively small) signal related to the Aharonov-Bohm interaction. Although we observe trends in the magnitude of oscillations in some samples similar to that predicted, for other samples we observe a variation in the magnitude of the Aharonov-Bohm oscillations that is comparable for both lower and higher energy sides of the peak of the QD spectrum, with a decrease in the Aharonov-Bohm amplitude at the transition spectral region, indicating the physics in these type-II *quantum dots* to be more subtle than standard quantum-ring configurations. As such, further work is needed to improve our understanding of type-II QDs.

11.4 Optical Aharonov-Bohm Effect in Magnetic (ZnMn)Te/ZnSe Type-II Quantum Dots

11.4.1 Magnetic Properties

In the previous section, we have presented clear experimental evidence of optical Aharonov-Bohm effects in non-magnetic type-II *columnar QDs* up to remarkably high temperatures. In the this section we further elucidate the suitability of *stacked QD* structures for Aharonov-Bohm coherence in type-II systems; in addition, we show that in magnetic systems these effects can be controlled, enhanced or destroyed, by the magnetic order of the system [14].

The samples studied were again grown by MBE on (001) GaAs substrates by migration-enhanced techniques, somewhat similar to those used for the non-magnetic samples presented previously. However, in the case of the magnetic samples studied here, the formation of the QDs resulted from a self-assembly process driven by the 7 % mismatch in lattice constant between the ZnSe matrix and the (ZnMn)Te QD layers, rather than by Te-clustering. Again, details of the growth process are described elsewhere [46]; we provide a brief summary here for clarity. After growth of a planarizing GaAs buffer layer at 580 °C, the sample temperature was reduced to ~300 °C to deposit the ZnSe buffer layer. The (ZnMn)Te QDs were then formed via the deposition of 2.6 MLs of (ZnMn)Te. Each of the QD layers was then capped with 5 nm of ZnSe. In the sample discussed below, five layers of QDs were deposited separated by 5 nm spacer layers, resulting in the formation of stacks of five QDs (columnar QDs), as confirmed from TEM measurements [46]. The nominal Mn-composition in the dot containing layers was estimated to be ~5 %, as determined from X-ray diffraction lattice constant measurements. The optimization of the magnetic QD growth and material characterization is described fully in Ref. [46].

Diluted Magnetic Semiconductors (DMS) in the II–VI System have been studied extensively for more than 20 years, driven by their interesting semiconducting and magnetic properties, as well as by potential applications [47]. The substitutional incorporation of Mn ions for the cations in II–VI materials results in a strong exchange interaction between the Mn ions and carrier spins. At low Mn concentrations the alloys are paramagnetic, while at larger concentrations spin glass and antiferromagnetic (AF) phases are formed due to the AF Mn-Mn exchange interaction. In the paramagnetic phase the Mn-carrier exchange interaction results in extremely strong magneto-optical properties including giant Zeeman shifts, and large degrees of optical polarization [47], which can be described at low fields in terms of a giant g-factor of the carriers.

Despite the extensive work performed on bulk and quantum wells based on (ZnMn)Te and (ZnMn)Se, type-II (ZnMn)Te/ZnSe QDs are, *interestingly*, not well developed. Indeed, DMS QDs in general have not been as comprehensively studied as their bulk or QW counterparts [47]. This is mainly due to the inferior optical properties of such systems in comparison to higher dimensional DMS systems (poor photoluminescence), which occurs due to the intrinsic effect of the dominance of the

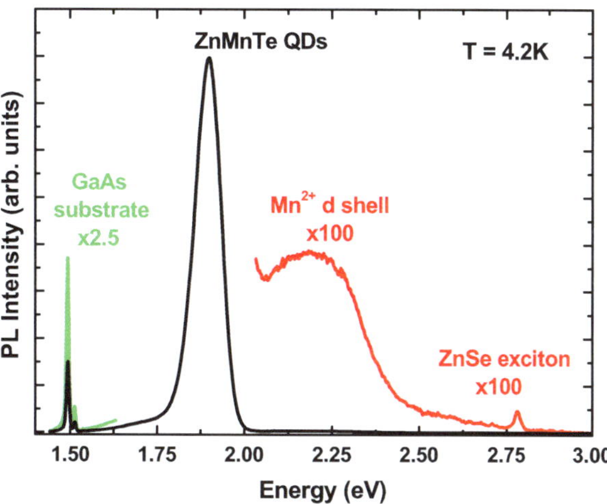

Fig. 11.18 Photoluminescence of (ZnMn)Te/ZnSe QDs at 4.2 K. The PL is dominated by the QDs because of their type-II nature. The emission from the Mn-electronic transitions is evident at a higher energy ($\sim$2.2 eV) than that of the QDs and as such has negligible PL intensity. Also evident are features associated with the GaAs substrate at $\sim$1.42 eV and the ZnSe matrix at $\sim$2.8 eV [49]

Mn-electronic transitions upon Mn-incorporation in the II–VI QD matrix. Specifically, the Mn-electronic transitions reside within the II–VI band gap and their properties are dominated by an orange-band (typically 2.1 eV) associated with the Mn-states. In QDs, in particular, it has been observed that a rapid transfer of carriers from the excitonic QD transitions to the Mn-electronic states occurs with deleterious effects on the band-edge optical properties of the QDs. The exact nature of this effect remains controversial, although it appears to be related to an Auger-like process, with a strong dependence on spin degeneracy of the system [48].

An interesting property of type-II (ZnMn)Te/ZnSe QDs is that, although the Mn-electronic states occur at energies below the direct ZnTe and ZnSe energy gaps, the indirect (in real space) nature of the lowest energy transition in ZnTe/ZnSe QDs results in emission from the QDs that is lower in energy than the Mn-states inhibiting transfer of carriers non-radiatively from the QDs to the Mn states. This is illustrated in Fig. 11.18, which shows the zero field photoluminescence of the (ZnMn)Te/ZnSe QDs at 4.2 K excited using a HeCd laser (325 nm) [49]. The PL of the QDs occurs at $\sim$1.9 eV, well below the ZnTe energy gap, and results from a combination of the type-II offset and quantum confinement effects. A schematic of the energy diagram for these QDs is present in Fig. 11.19(d).

In Fig. 11.18 the emission due to the Mn^{2+} states is observed at $\sim$2.2 eV, and is considerably weaker than that of the QDs. This occurs because of the type-II nature of the QDs and their concomitantly lowered recombination energy, and clearly demonstrates the decoupling of the non-radiative transfer of carriers to the Mn-states in these QDs. Confirmation of the nature of these transitions has been made using magneto-PL which show negligible Zeeman shifts in the Mn-related transitions due to their low g-factor ($g = 2$), while in contrast (as will be discussed below) the properties of the QD are affected significantly by the application of a magnetic field. Also evident in Fig. 11.18 are emission features from the ZnSe matrix at 2.78 eV and the GaAs substrate at $\sim$1.42 eV, the appearance of which perhaps demonstrates the low QD coverage of the samples studied.

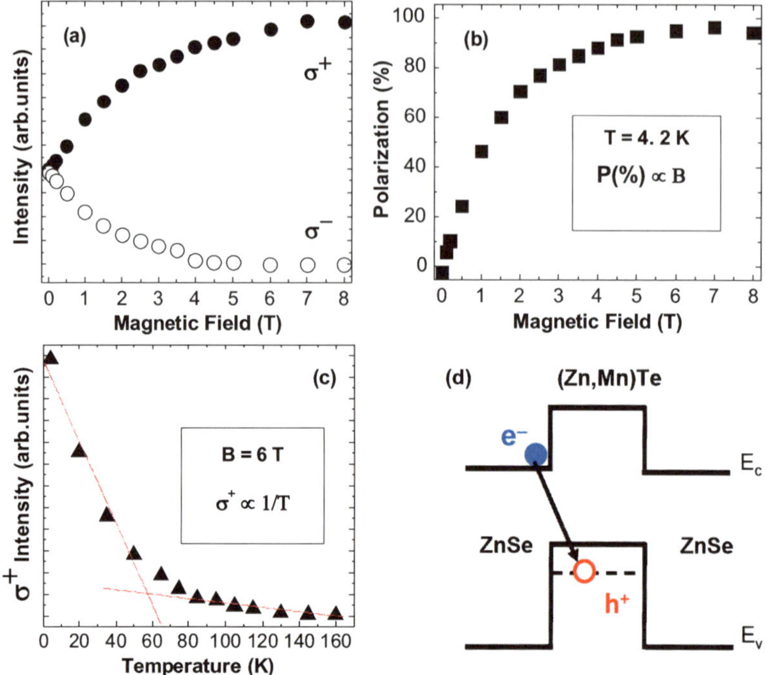

Fig. 11.19 (**a**) Intensity of the σ^+ (*closed circles*) and σ^- (*open circles*) photoluminescence components versus magnetic field at 4.2 K. (**b**) The degree of optical polarization versus magnetic field at 4.2 K. (**c**) Photoluminescence intensity of the dominant σ^+ component at 6 T as a function of temperature. (**d**) Schematic representation of the energy band offsets for (ZnMn)Te/ZnSe QDs

As is alluded to above, the magnetic nature of the QDs is confirmed directly through the study of the magneto-optical properties of their QD emission. Figure 11.19(a) shows the integrated photoluminescence intensity of the σ^+ and σ^- components in the Faraday geometry. As the magnetic field increases, the electron ($m_s = \pm 1/2$) and hole ($m_j = \pm 3/2$) states split due to the combined effect of re-alignment of the Mn- spins in the direction of the applied field and their strong exchange interaction with the individual carrier spin. This process results in the preferential relaxation of carriers to the lowest energy spin state ($m_s = -1/2$, $m_j = 3/2$), and an emitted polarization that represents the total carrier spin orientation due to the selection rules of the transition. Here, since the dominant transition in (ZnMn)Te QDs is associated with an exciton formed between (spin-down) electrons ($m_s = -1/2$) and (spin up) holes ($m_j = 3/2$), the total angular momentum projection of the transition is equal to $+1$, and as such the emission is dominated by left circularly polarized light. Figure 11.19(b) shows the degree of optical polarization, $P = (I_+ - I_-)/(I_+ + I_-)$, where $I_+(I_-)$ represent the intensity of the photons dominated by $\sigma^+(\sigma^-)$ luminescence at 4.2 K.

Since II–VI DMS systems are paramagnetic, their Zeeman energy is characterized by the well-known B/T dependence of such materials. The effect of the magnetic field, B upon P (%), and therefore the magnetic order, can be clearly observed in Fig. 11.19(b), while the effect of temperature is evident in Fig. 11.19(c), which shows the effect of increasing the temperature on the intensity of the dominant σ^+ PL component at a fixed applied field of 6 T.

Despite the applied field, increasing the temperature reduces the intensity of the luminescence such that zero *effective* polarization is observed at $T > 55$ K. This behavior results since there is a randomization of the Mn-spin bath at low B, or high T, in paramagnetic systems, reducing the net magnetic order to zero.

This behavior can be described qualitatively by considering the average magnetic moment of a single Mn impurity and the strength of the spin fluctuations. The average moment can be written as

$$\bar{M}_z = \langle M_z \rangle = S \cdot g_{Mn} \cdot \mu_B \cdot B_{5/2}(x) \quad \text{and}$$

$$\sqrt{\Delta M^2} = \sqrt{\langle (M_z - \bar{M}_z)^2 \rangle} = \sqrt{k_B T \frac{\partial \bar{M}_z}{\partial B}}, \tag{11.4}$$

where $\langle M_z \rangle$ is the thermal average over the states of Mn spin with $S = 5/2$; $B_{5/2}(x)$ is the Brillouin function, $x = g_{Mn} \cdot S \cdot \mu_B \cdot B / k_B T$, and B is the magnetic field. We take $T = 4.2$ K and $g_{Mn} = 2$.

The results of these calculations are shown in Fig. 11.20(b) and clearly illustrate the improvement of magnetic order and the suppression of spin fluctuations above 4 T, and also demonstrate the B/T dependence of Mn-based systems. This is further demonstrated by the inset to Fig. 11.20(a). At 40 K, which is below the temperature at which the optical polarization is quenched, weak OABE is still evident. At temperatures above which the optical polarization is removed, evidence of the OABE is also gone.

11.4.2 Effect of Magnetic Disorder on the Optical Aharonov-Bohm Effect in (ZnMn)Te/ZnSe Quantum Dots

Since the (ZnMn)Te QDs have columnar type-II geometry, we expected to observe Aharonov-Bohm coherence in this system. We have found the magnetic order in the system to be crucial to the behavior of the OABE; indeed it can be used to control the coherence.

Figure 11.21 shows the raw PL data from the (ZnMn)Te/ZnSe columnar QD sample discussed above at a number of magnetic fields; note the clear increase of peak PL intensity with B. The corresponding integrated intensity of the PL at 4.2 K is plotted vs. magnetic field in Fig. 11.22(a). With increasing magnetic field the overall PL intensity increases significantly. In a simple picture this increase with field can be attributed to a shrinking of the exciton envelope function with magnetic field in the plane normal to the field. This tends to force the electron envelope function in the ZnSe matrix closer to the QD interface because the hole is strongly confined in the (ZnMn)Te QD (the behavior is similar to that of a shallow hydrogenic donor

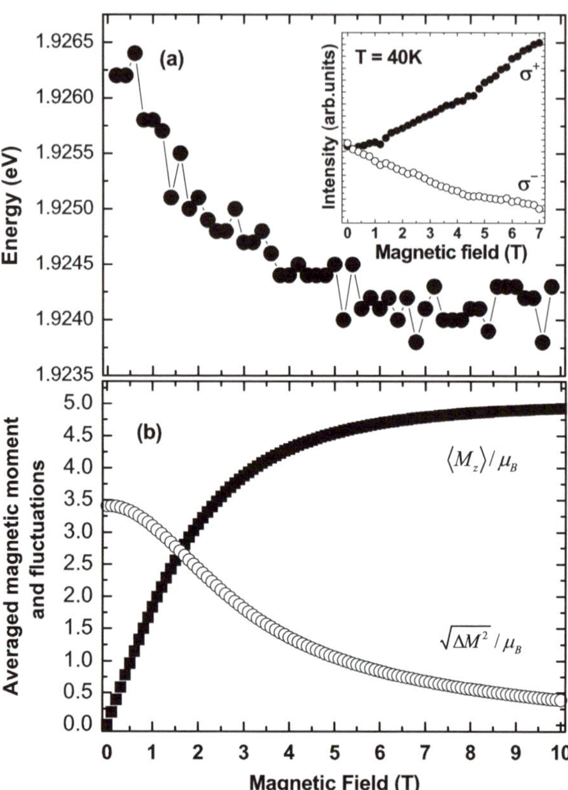

Fig. 11.20 (**a**) Photoluminescence energy versus magnetic field 4.2 K. The shift represents $\frac{1}{2}$ the total Zeeman energy. The *inset* shows the intensity of the σ^+ and σ^- components of the photoluminescence at 40 K. (**b**) Average magnetic moment $\langle Mz \rangle$ (*closed squares*) and fluctuations ΔM^2, as a function of magnetic field. © 2008, reproduced courtesy of the American Physical Society [14]

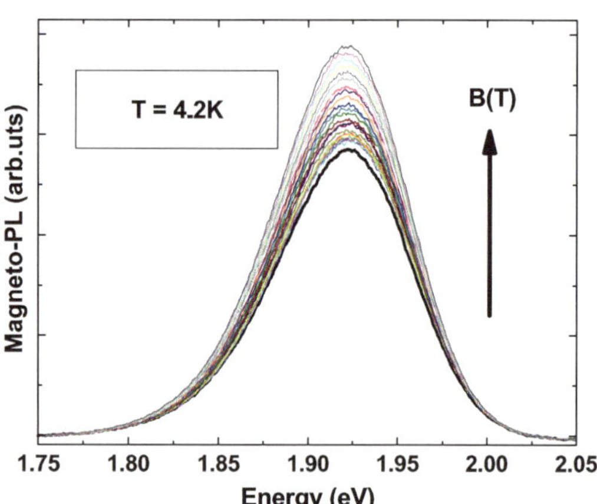

Fig. 11.21 PL spectra at various values of magnetic field at 4.2 K for (ZnMn)Te QDs. These data were used to determine the points plotted in the main panel of Fig. 11.22(a). © 2008, reproduced courtesy of the American Physical Society [14]

impurity in semiconductors). This leads to increased penetration of the electron envelope function into the QD, which increases the electron-hole envelope function overlap, thereby increasing the radiative recombination rate.

Fig. 11.22 (**a**) Integrated PL intensity versus magnetic field for (ZnMn)Te QDs at 4.2 K. *Inset* shows a schematic of the correlated electron and hole. (**b**) Peak intensity extracted from data in (**a**) by subtracting the monotonically increasing background. © 2008, reproduced courtesy of the American Physical Society [14]

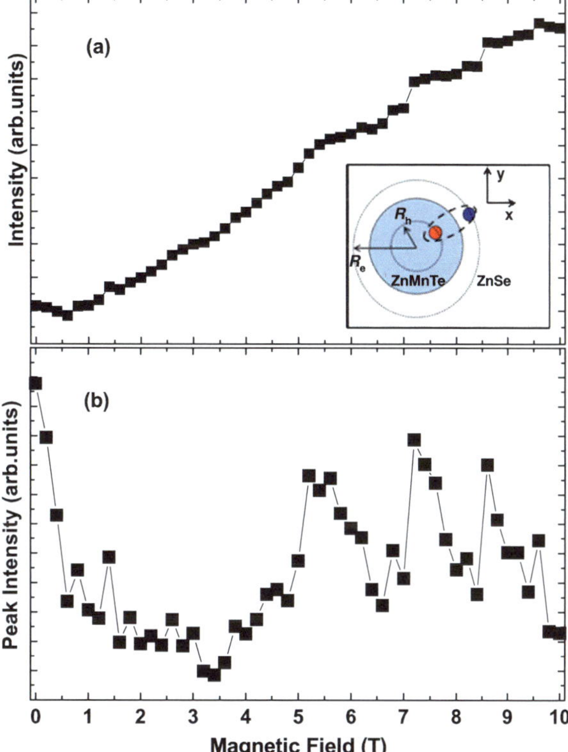

Another possible mechanism should also be considered. A systematic increase in the PL intensity with magnetic field is a well known property of DMS systems that has been attributed to the suppression of carrier transfer to the low energy Mn d-band electronic states lying in the forbidden gap, as discussed in the previous section. It has been shown by Falk et al. [48] that at high magnetic fields the non-radiative transfer of carriers from the excitonic state to Mn d-band states is suppressed, so that the excitonic PL of DMS systems is considerably enhanced. However, as stated above, the availability of low energy indirect excitons has been shown to circumvent the transfer of carriers to the Mn d-band states quite effectively, and the excitonic transitions dominate the PL spectra in the present case (see Fig. 11.18 and discussion). While the mechanism discussed by Falk et al. cannot be discounted entirely (transfer of carriers must indeed occur to produce the Mn d-band PL), it is extremely weak and its suppression cannot lead to the large changes in excitonic PL intensity observed with increasing magnetic field.

More important for the present consideration are the periodic oscillations in intensity that are visible, superimposed upon the rising intensity background of the PL in Fig. 11.22(a). These oscillations are seen more clearly in Fig. 11.22(b), where the monotonically increasing background has been subtracted. Once again, these oscillations have their origin in the ring-like geometry of the columnar QDs, which creates a rotating dipole (shown schematically in the inset to Fig. 11.22(a)) and an OABE. This is consistent with the non-magnetic columnar QDs discussed previously [12, 13]. Since the hole is strongly localized, only the electron experiences changes in total orbital angular momentum projection with increasing magnetic field. Therefore in the simplest case this can be qualitatively described by Fig. 11.2. As discussed above, with increasing magnetic field the L value characterizing the ground state orbital angular momentum projection changes from $L = 0$ to $L \neq 0$, and in structures with perfect rotational symmetry the PL should be strongly suppressed. However, the columnar QD structures investigated here are far from perfect, with elongation in the plane and surface defect incorporation likely, both of which relax the angular momentum selection rules. For these DMS QDs interaction with magnetic impurities and clusters, and spin-orbit coupling, must also be considered. Magnetic impurities play a strong role in breaking rotational symmetry, as discussed in more detail below, however, spin-orbit coupling is rather weak for electrons, particularly in ZnSe, and we consider it to play a negligible role in the present structures.

Importantly, in contrast to the non-magnetic QDs, the (ZnMn)Te QDs exhibit no visible oscillations in the PL peak energy. This is the result of their relatively large diameter ($\sim$28 nm), which results in a very small predicted amplitude of energy oscillation, only $\sim$0.04 meV from (11.3). Such weak energy oscillations are not observable, they are obscured by the large inhomogeneous broadening of the excitonic peak ($\sim$80 meV). The effective radius of the electrons determined from the period of the AB oscillations in Fig. 11.22 is estimated to be $\sim$28 nm, consistent with TEM images taken from the structures [46].

Returning to the oscillations in emission intensity, it can be seen from Fig. 11.22(a) that the strength of the oscillations appears to increase substantially at the highest magnetic fields. This can be seen more clearly in Fig. 11.22(b), in which the monotonic PL background has been subtracted. In this figure, three well resolved oscillations are evident with peaks at $\sim$5.4, 7.4, and 9 T. This enhancement of the OABE at higher fields appears strongly related to the spin polarization of the carriers (see Fig. 11.19) and therefore to the magnetization of the dots; at large magnetic fields the randomness of the magnetic moments contributed by individual dots in the columns is decreased so the effective field seen by an individual dot in the ensemble becomes more uniform throughout the ensemble.

In the paramagnetic situation at low magnetic fields the Mn-spins are randomly oriented and fluctuating due to the finite temperature of the system ($\sim$4 K), and the low blocking temperature of (ZnMn)Te [50]. As the magnetic field is increased and the spin disorder is reduced, the Aharonov-Bohm oscillations increase in concert with the increase in the average magnetic moment along the columns, and the fluctuations decrease, as described qualitatively for a PM system in (11.3) and (11.4) and shown in Fig. 11.20(b). Since the hole in the exciton and the Mn spins are coupled through an exchange interaction, fluctuations in the Mn-spins will create fluctuations in the exchange potential for the orbiting electron, even if the ZnSe matrix is (as expected) Mn-free. In addition, the finite penetration of the electron wavefunction into the QD will further enhance these effects. Such fluctuations apparently have a destructive effect on the visibility of the OABE in the ensemble Aharonov-Bohm phase, considerably reducing its strength. The sensitivity of the OABE to the magnetic properties of the QDs is further evidenced through studies as a function of temperature, as shown in the inset to Fig. 11.20(a). With increasing temperature the Aharonov-Bohm oscillations decrease in amplitude, but, although weak, remain evident at 40 K (Fig. 11.20(a)). Above 50 K, which is the temperature at which the degree of optical polarization is completely randomized (see Fig. 11.19(c)), the Aharonov-Bohm oscillations are also lost.

11.5 Summary, Conclusions and Outlook

In this chapter we have summarized theoretical and experimental work on the optical Aharonov-Bohm effect in type-II QDs. This effect has been observed in several systems, including III–V InP/GaAs, group-IV GeSi, and II–VI Zn(Mn)Te QDs. Recently, it has also been reported in fluctuation quantum dots in the InGaAs/GaAs system. Here, particular focus has been given to the II–VI Zn(Mn)Te/ZnSe systems, with reference also made to the major contributions of the InP/GaAs and GeSi systems. Evidence of optical Aharonov-Bohm behavior was first observed experimentally in InP/GaAs QDs in the PL energy. This was quickly followed by the observation in both the PL energy *and* intensity in Zn(Mn)Te/ZnSe QDs, and more recently in GeSi dots. In non-magnetic ZnTe/ZnSe superlattice structures, which contain Zn(TeSe) QDs in a predominantly ZnSe matrix, the AB coherence is remarkably robust due to the natural formation of columnar QDs, in which the ring-like geometry is defined by the lateral position of the rotating electron relative to the QD. In these systems Aharonov-Bohm oscillations have been observed up to 180 K, and they appear to be limited only by thermal ionization of the excitons. The OABE in Type-II (ZnMn)Te/ZnSe QDs results in an interesting interplay between the strength of the Aharonov-Bohm oscillations and the magnetic properties of the (ZnMn)Te QDs. In particular, the magnetic properties of these QDs can either quench (magnetically disordered) or enhance (magnetically ordered) the ensemble OABE depending on the applied field and/or temperature; consistent with the B/T dependence of the magnetization. Evidence of coherence in both the PL energy and intensity has been observed for Zn(Mn)Te/ZnSe and GeSi QDs. The absence of evidence of AB oscillations in the intensity for InP/GaAs QDs, is not fully understood, but may indicate a reduced coherence of the orbiting holes relative to the electrons in the other systems (which have a higher mobilities), distinguishing this system from the others.

Acknowledgements   B.D. McCombe acknowledges partial support from the National Science Foundation Grant: DMR 1008138, and the Office of the Provost, University at Buffalo. I.L. Kuskovsky acknowledges support from the National Science Foundation: DMR-1006050, and the Department of Energy, Office of Basic Energy Science (Division of Materials Science and Engineering), Award No. SC003739.

References

1. A.O. Govorov, S.E. Ulloa, K. Karrai, R.J. Warburton, Phys. Rev. B **66**, 081309(R) (2002)
2. A.V. Kalameitsev, A.O. Govorov, V.M. Kovalev, JETP Lett. **68**, 669 (1998)
3. Y. Aharonov, D. Bohm, Phys. Rev. **115**, 485 (1959)
4. R.A. Webb, S. Washburn, C.P. Umbach, R.B. Laibowitz, Phys. Rev. Lett. **54**, 2696 (1985)
5. G. Timp, A.M. Chang, J.E. Cunningham, T.Y. Chang, P. Mankiewich, R. Behringer, R.E. Howard, Phys. Rev. Lett. **58**, 2814 (1987)
6. C.J.B. Ford, T.J. Thornton, R. Newbury, M. Pepper, H. Ahmed, D.D. Peacock, R.A. Ritchie, J.E.F. Frost, Appl. Phys. Lett. **54**, 21 (1989)
7. A. Fuhrer, S. Luscher, T. Ihn, T. Heinzel, K. Ensslin, W. Wegscheider, M. Bichler, Nature (London) **413**, 822 (2001)
8. A. Lorke, R.J. Luyken, A.O. Govorov, J.P. Kotthaus, J.M. Garcia, P.M. Petroff, Phys. Rev. Lett. **84**, 2223 (2000)
9. N.A.J.M. Kleemans, I.M.A. Bominaar-Silkens, V.M. Fomin, V.N. Gladilin, D. Granados, A.G. Taboada, J.M. Garcia, P. Offermans, U. Zeitler, P.C.M. Christianen, J.C. Mann, J.T. Devreese, P. Koenraad, Phys. Rev. Lett. **99**, 146808 (2007)
10. E. Ribeiro, A.O. Govorov, W. Carvalho Jr., G. Medeiros-Ribeiro, Phys. Rev. Lett. **92**, 126402 (2004)
11. S. Miyamoto, O. Mountanabbir, T. Ishikawa, M. Eto, E.E. Haller, K. Sawano, Y. Shiraki, K.M. Itoh, Phys. Rev. B **82**, 073306 (2010)
12. I.L. Kuskovsky, W. MacDonald, A.O. Govorov, L. Moroukh, X. Xei, M.C. Tamargo, M. Tadic, F.M. Peeters, Phys. Rev. B **76**, 035342 (2007)
13. I.R. Sellers, V.R. Whiteside, I.L. Kuskovsky, A.O. Govorov, B.D. McCombe, Phys. Rev. Lett. **100**, 136405 (2008)
14. I.R. Sellers, V.R. Whiteside, A.O. Govorov, W.-C. Fan, W.-C. Chou, I. Khan, A. Petrou, B.D. McCombe, Phys. Rev. B **77**, 241302(R) (2008)
15. L.G.G.V. Dias da Silva, S.E. Ulloa, A.O. Govorov, Phys. Rev. B **70**, 155318 (2004)
16. A. Chaplik, JETP Lett. **62**, 900 (1995)
17. R.A. Romer, M.E. Raikh, Phys. Rev. B **67**, 121304 (2003)
18. A.V. Maslov, D.S. Citrin, Phys. Rev. B **67**, 121304 (2003)
19. Y. Gu, I.L. Kuskovsky, M. Van der Voort, G.F. Neumark, X. Zhou, M.C. Tamargo, Phys. Rev. B **71**, 045340 (2005)
20. M. Jo, M. Endo, H. Kumano, I. Suemune, J. Cryst. Growth **301–302**, 277 (2007)
21. Y. Gu, I.L. Kuskovsky, G.F. Neumark, *Wide Gap Light Emitting Materials and Devices*, 1st edn. (Wiley-VCH, New York, 2007), pp. 147–176
22. V. Akimova, A.M. Akhekyan, V.I. Kozlovsky, Y.V. Korostelin, P.V. Shapin, Sov. Phys., Solid State **27**, 1041 (1985)
23. Q. Fu, D. Lee, A.V. Nurmikko, L.A. Kolodziejski, R.L. Gunshor, Phys. Rev. B **39**, 3173 (1989)
24. A. Muller, P. Bianucci, C. Piermarocchi, M. Fornari, I.C. Robin, R. André, C.K. Shih, Phys. Rev. B **73**, 081306(R) (2006)
25. M.D. Pashley, K.W. Haberern, W. Friday, J.M. Woodall, P.D. Kirchner, Phys. Rev. Lett. **60**, 2176 (1988)
26. M.C.-K. Cheung, A.N. Cartwright, I.R. Sellers, B.D. McCombe, I.L. Kuskovsky, Appl. Phys. Lett. **92**, 032106 (2008)
27. I.R. Sellers, H.Y. Liu, T.J. Badcock, K.M. Groom, D.J. Mowbray, M. Guttiérrez, M. Hopkinson, M.S. Skolnick, Physica E **26**, 382 (2005)
28. F. Hatami, M. Grundmann, N.N. Ledenstov, F. Heinrichsdorff, R. Heitz, J. Böhrer, D. Bimberg, S.S. Ruvimov, P. Werner, V.M. Ustinov, P.S. Kop'ev, Zh.I. Alferov, Phys. Rev. B **57**, 4635 (1998)
29. U. Manna, I.C. Noyan, Q. Zhang, I.F. Salakhutdinov, K.A. Dunn, S.W. Novak, R. Moug, M.C. Tamargo, G.F. Neumark, I.L. Kuskovsky, J. Appl. Phys. **111**, 033516 (2012)

30. H.Y. Liu, I.R. Sellers, R.J. Airey, M.J. Steer, P.A. Houston, D.J. Mowbray, J. Cockburn, M.S. Skolnick, B. Xu, Z.G. Wang, Appl. Phys. Lett. **80**, 3769 (2002)
31. M. Sugawara, K. Mukai, Y. Nakata, K. Otsubo, H. Ishilkawa, IEEE J. Sel. Top. Quantum Electron. **6**, 462 (2000)
32. M.D. Teodoro, V.L. Campo, V. Lopez-Richard, E. Marega, G.E. Marques, Y. Galvao Gobato, F. Iikawa, M.J.S.P. Brasil, Z.Y. Abu Waar, V.G. Dorogan, Y.I. Mazur, M. Benamara, G.J. Salamo, Phys. Rev. Lett. **104**, 086401 (2010)
33. M. Bayer, M. Korkusinski, P. Hawrylak, T. Gutbrod, M. Michel, A. Forchel, Phys. Rev. Lett. **90**, 186801 (2003)
34. L. Schweidenback, T. Ali, A.H. Russ, J.R. Murphy, A.N. Cartwright, A. Petrou, C.H. Li, M.K. Yates, G. Kioseoglou, B.T. Jonker, A.O. Govorov, Phys. Rev. B **85**, 245310 (2012)
35. B. Roy, H. Ji, S. Dhomkar, F.J. Cadieu, L. Peng, R. Moug, M.C. Tamargo, I.L. Kuskovsky, Appl. Phys. Lett. **100**, 213114 (2012)
36. B. Roy, H. Ji, S. Dhomkar, F.J. Cadieu, L. Peng, R. Moug, M.C. Tamargo, Y. Kim, D. Smirnov, I.L. Kuskovsky, Phys. Rev. B **86**, 165310 (2012)
37. M.H. Degani, M.Z. Maialle, G. Medeiros-Ribeiro, E. Ribeiro, Phys. Rev. B **78**, 075322 (2008)
38. A.M. Fisher, V.L. Campo Jr., M.E. Portnoi, R.A. Romer, Phys. Rev. Lett. **102**, 096405 (2009)
39. B. Li, F.M. Peeters, Phys. Rev. B **83**, 115448 (2011)
40. A.G. Kontos, N. Chrysanthakopoulos, M. Calamiotou, T. Kehagias, P. Komninou, U.W. Pohl, J. Appl. Phys. **90**, 3301 (2001)
41. R.E. Balderas-Navarro, K. Hingerl, A. Bonanni, D. Stifter, J. Vac. Sci. Technol. B **19**, 1650 (2001)
42. V. Holy, G. Springholz, M. Pinczolits, G. Bauer, Phys. Rev. Lett. **83**, 356 (1999)
43. U. Manna, I.C. Noyan, Q. Zhang, I.F. Salakhutdinov, K.A. Dunn, S.W. Novak, R. Moug, M.C. Tamargo, G.F. Neumark, I.L. Kuskovsky, J. Appl. Phys. **111**, 033516 (2012)
44. S.J. Kim, B.C. Juang, W. Wang, J.R. Jokisaari, C.Y. Chen, J.D. Phillips, X.Q. Pan, J. Appl. Phys. **111**, 093524-8 (2012)
45. J.M. Rorison, Phys. Rev. B **48**, 4643 (1993)
46. M.C. Kuo, J.S. Hsu, J.L. Shen, K.C. Chiu, W.-C. Fan, Y.C. Lin, C.H. Chia, W.-C. Chou, M. Yasar, R. Mallory, A. Petrou, H. Luo, Appl. Phys. Lett. **89**, 263111 (2006)
47. J.K. Furdyna, J. Appl. Phys. **64**, R29 (1988)
48. H. Falk, J. Hübner, P.J. Klar, W. Heimbrodt, Phys. Rev. B **68**, 165203 (2003)
49. V.R. Whiteside, Optical Aharonov-Bohm effect in type-II ZnTe/ZnTe quantum dots. Ph.D. Thesis, Department of Physics & Astronomy, University at Buffalo—SUNY, 2011
50. A. Tardowski, C.J.M. Denissen, W.J.M. Dejonge, A.T.A.M. Dewaele, M. Demianiuk, R. Triboulet, Solid State Commun. **59**, 199 (1986)

Chapter 12
Excitons Confined in Single Semiconductor Quantum Rings: Observation and Manipulation of Aharonov-Bohm-Type Oscillations

F. Ding, B. Li, F.M. Peeters, A. Rastelli, V. Zwiller, and O.G. Schmidt

Abstract We report on a magneto-photoluminescence study of single neutral excitons confined in single self-assembled semiconductor quantum rings. Oscillations in the exciton radiative recombination energy and in the emission intensity are observed under an applied magnetic field. Special emphasis is placed on the manipulation of this Aharonov-Bohm-type oscillations with a vertical electric field. We observe that both the exciton oscillator strength and the periodicity of the oscillation can be tuned. This tunability is explained by calculating the single particle wave function in both unstrained and strained semiconductor quantum rings in the presence of external electrical fields.

F. Ding (✉) · O.G. Schmidt
Institute for Integrative Nanosciences, IFW Dresden, Helmholtzstr. 20, 01069 Dresden, Germany
e-mail: f.ding@ifw-dresden.de

O.G. Schmidt
e-mail: o.schmidt@ifw-dresden.de

B. Li · F.M. Peeters
Departement Fysica, Universiteit Antwerpen, Groenenborgerlaan 171, 2020 Antwerpen, Belgium

B. Li
e-mail: bin.li@ua.ac.be

F.M. Peeters
e-mail: francois.peeters@ua.ac.be

A. Rastelli
Institute of Semiconductor and Solid State Physics, Johannes Kepler University Linz,
Altenbergerstr. 69, 4040 Linz, Austria
e-mail: armando.rastelli@jku.at

V. Zwiller
Kavli Institute of Nanoscience, Delft University of Technology, P.O. Box 5046, 2600 GA Delft,
The Netherlands
e-mail: v.zwiller@tudelft.nl

V.M. Fomin (ed.), *Physics of Quantum Rings*, NanoScience and Technology,
DOI 10.1007/978-3-642-39197-2_12, © Springer-Verlag Berlin Heidelberg 2014

12.1 Introduction

The Aharonov-Bohm (AB) effect [1] demonstrates the phase change of a charged particle traveling around a region enclosing a magnetic field. This effect has been observed independently for electrons and holes in micro and nanoscale structures [2–4]. Theoretical investigations have predicted that it is also possible to observe the AB effect when considering a single neutral exciton (an electron-hole pair) confined in a quantum ring (QR) structure. The phases accumulated by the electron and the hole will be different after one revolution, leading to modulations between different quantum states [5]. This so called neutral exciton AB effect can be probed from the photoluminescence (PL) emission in semiconductor QRs, since the change in the phase of the exciton wave function is accompanied by a change in the exciton total angular momentum, making the PL emission magnetic field dependent [6–10]. Recently it was also proposed that this effect can reduce the oscillator strength of an exciton to zero, which may leads to the realization of an optical quantum memory [11].

Self-assembled semiconductor nanostructures are ideal test beds for the experimental observation of this neutral exciton AB effect. The composite nature of excitons enables the electron and the hole making up the exciton to acquire different phases as they propagate inside the ring-like structure, leading to an observable modification of the exciton recombination energy and lifetime. Experimental verification of this effect has been reported for radially polarized neutral excitons confined in a type-II quantum dots (QDs) structure [12–14]. However, in a type-I QD or QR structure the electron-hole separation is relatively small, making the experimental observation for such an effect a difficult task. Recently there have been reports on optical AB-type oscillations in type-I InAs QR ensemble [15] and in a single InGaAs QR [16] under increasing magnetic field. In both cases the radial polarization of the exciton, either by build-in electric field or due to asymmetric effective confinement for the electrons and holes, plays an important role.

In this chapter we study single self-assembled In(Ga)As/GaAs QRs fabricated by molecular beam epitaxy combined with *in situ* AsBr$_3$ nanohole drilling [17]. A detailed investigation of the morphology of buried In(Ga)As/GaAs nanostructures unveils a progressive morphological evolution from QDs to volcano-shaped QRs, as the AsBr$_3$ etching depth is increased. Due to the asymmetry in the effective confinement for electrons and holes, we expect a radial polarization of the neutral exciton. These QRs show magnetic field dependent oscillations in the PL energy and intensity of a *single* neutral exciton, in agreement with previous theoretical predictions. With a vertical electric field the radial polarization and the oscillator strength of the exciton can be modified, leading to a tunable AB-type effect [16]. In the end of this chapter we consider a theoretical interpretation of the tunable AB effect by calculating single particle wavefunctions in both unstrained and strained semiconductor QRs in the presence of vertical electrical fields.

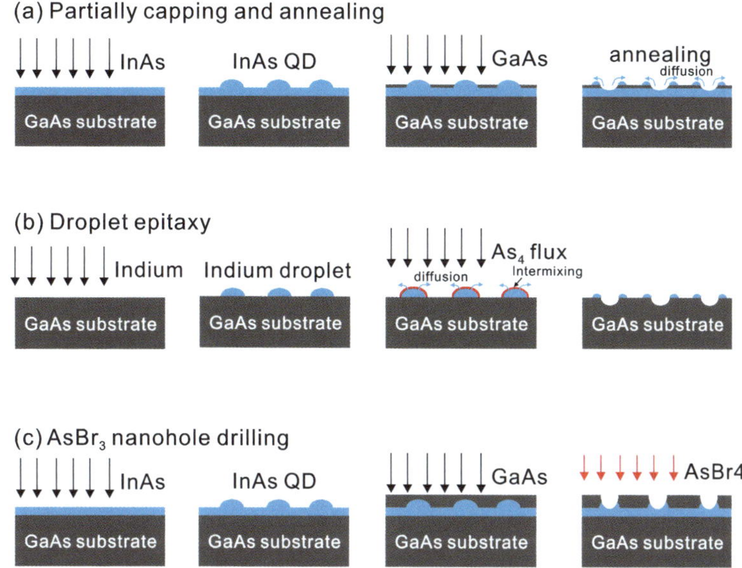

Fig. 12.1 Schematic representations of the fabrication techniques of In(Ga)As/GaAs QRs with molecular beam epitaxy

12.2 Fabrication Techniques of Self-assembled Quantum Rings

12.2.1 AsBr₃ in situ Nanohole Drilling

In the previous chapters the formation mechanism of self-assembled QRs, especially by the droplet epitaxy method, is discussed. In fact there are currently several important techniques available for the fabrication of semiconductor QRs with molecular beam epitaxy, see Fig. 12.1. Our employed technique bases on AsBr$_3$ *in situ* nanohole drilling [Fig. 12.1(c)], and it is different from the commonly used partially capping and annealing technique [18] [Fig. 12.1(a)] and the droplet epitaxy technique [19–21] [Fig. 12.1(b)], which have different formation mechanism and morphology.

The QR samples examined in this chapter are grown on GaAs (001) substrates in a solid-source molecular-beam epitaxy system equipped with an AsBr$_3$ gas source. After oxide desorption, a 300-nm GaAs buffer layer is grown at 570 °C. A 1.8-monolayer (ML) InAs layer is then deposited at 500 °C, using an In growth rate of 0.01 ML/s. After 30 s growth interruption, the substrate temperature is lowered to 470 °C and a 10-nm GaAs cap is deposited at a rate of 0.6 ML/s while the temperature is ramped back to 500 °C. After GaAs deposition, the AsBr$_3$ etching gas is subsequently supplied to drill the nanoholes. The etching rate of GaAs was 0.24 ML/s, which was calibrated by reflection high-energy electron diffraction intensity oscillations.

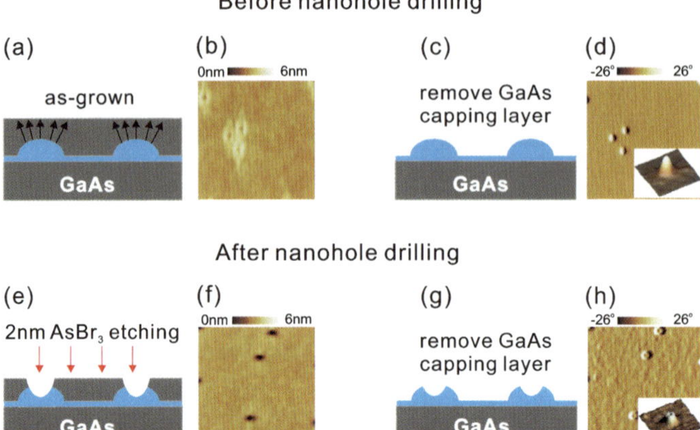

Fig. 12.2 Morphology studies of the buried QDs and QRs with AFM. The schematic representations and the surface morphology (500×500 nm^2) of the as-grown QDs are shown in (**a**) and (**b**) respectively. After removing the GaAs capping layer by selective chemical etching, see the schematic in (**c**), the surface morphology is shown in (**d**). Note that (**b**) and (**d**) are taken from the same area on the sample. The nanohole drilling process is shown in the second row. The *insets* in (**d**) and (**h**) show the three dimensional images of a single QD and a single volcano-shaped QR

12.2.2 *Structural Characterization*

During the deposition of 10 nm GaAs capping layer on the as-grown QDs, the surface morphology develops into a rhombus-shaped structure with a tiny hole in the middle [Figs. 12.2(a) and (b)]. After supplying the AsBr$_3$ etching gas, there is a preferential removal of the central part of the *buried* QDs due to the local strain field induced by the buried QDs, see Figs. 12.2(e) and (f) [17]. In order to characterized the underlying structures, a selective chemical etchant [ammonium hydroxide (28 % NH$_4$OH), hydrogen peroxide (31 % H$_2$O$_2$) and deionized water (1:1:25)] is used to remove the GaAs capping layer. NH$_4$OH:H$_2$O$_2$ solutions are known to etch selectively GaAs over InGaAs alloys [22–25]. The mixture used here etches GaAs at a rate of $\sim$10 nm/s [23]. Figures 12.2(d) and (h) show the morphology after the removal of GaAs capping layer and suggest that the selective etching process significantly slows down at the In(Ga)As wetting layer (WL), see also Fig. 12.3(a). Prolonged etching results in the removal of the WL and the undercut of the QDs. We verified that different etching experiments with the same nominal duration (1 second) performed on the same sample yield compatible results. Slower etching rates of GaAs are achievable with diluted solutions [26].

The selective removal of capping layer combined with AFM enables a statistical structural analysis of the uncapped In(Ga)As nanostructures as a function of the nominal AsBr$_3$ etching depth, see Fig. 12.3(b). A cross sectional transmission electron microscopy (TEM) image of a single QR from a similar sample is also presented [Fig. 12.3(a)], it provides comparable information to that from the AFM studies. Therefore we argue that the selective chemical etching combined with AFM

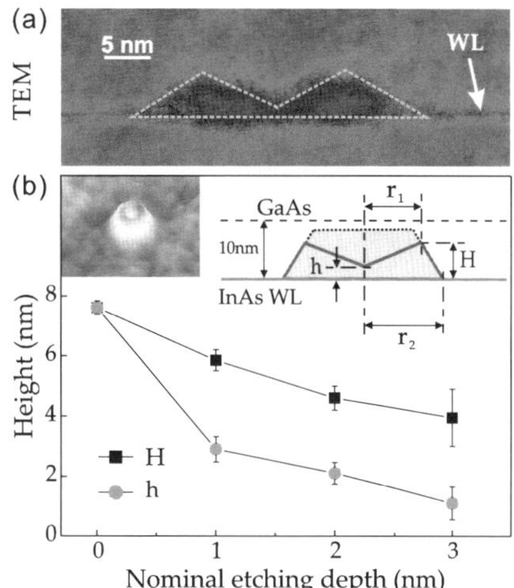

Fig. 12.3 (**a**) Cross-sectional TEM image shows the QRs with nominal AsBr$_3$ etching depth of 3 nm, it provides structural information comparable to that from the AFM image shown in the *inset* of (**b**). The statistical structural analysis of the uncapped In(Ga)As nanostructures as a function of the nominal etching depth is shown in (**b**). H is the height of the embedded nanostructures, and h is the distance between the InAs WL and the bottom of the dip at the nanostructure apex

is a reliable, fast and convenient way to characterize buried In(Ga)As/GaAs nanostructures, it provides information on the three-dimensional morphology and does not require any special sample preparation.

We focus now on the QRs with a nominal etching depth of 3 nm. Both AFM and TEM confirm that the average inner radius r_1 of the QR is about 8.5 nm, while the average outer radius r_2 is about 19 nm, and the height H is about 4 nm. The inner height h is less than 1 nm. Such a ring structure could produce asymmetric confinement potentials for the electron and the hole, thus the neutral exciton AB can be expected.

12.3 Magneto-Photoluminescence

The photoluminescence under magnetic fields can be used to probe the energy changes and intensity changes that related to the nature of the excitonic ground state. The samples are placed in a cryostat tube which is evacuated and then refilled with a small amount of helium exchange gas. The tube is then inserted into a 4.2 K helium bath dewar equipped with a superconducting magnet capable of providing fields up to 9 tesla (T). The sample is excited by a 532 nm CW laser beam and the PL is collected by an objective with 0.85 numerical aperture. The PL signal is then dispersed by a spectrometer with 0.75 m focal length equipped with a 1800 g/mm grating, and captured by a liquid nitrogen cooled Si charge-coupled device (CCD). Figures 12.4(a) and (b) show the PL spectra for a single QR and a single QD under magnetic fields from $B = 0$ T to 9 T in Faraday configuration. The system has been carefully optimized and tested in order to minimize the drifts associated with magnetic field ramping. An optimum resolution of 2.5 μeV can be achieved by fitting the PL peaks with Lorentzian functions [16]. With this system we can detect a fine structure splitting of a neutral exciton state of 13 μeV with an error of ±2.5 μeV,

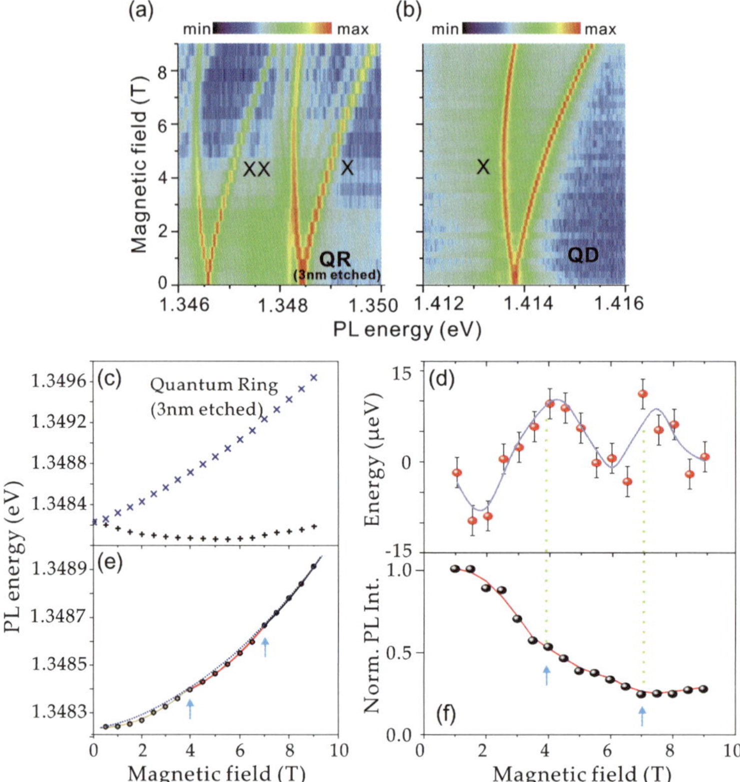

Fig. 12.4 Magneto-PL from a single QR (nominal etching depth of 3 nm) (**a**) and a single QD (**b**) under different magnetic fields. The fitted PL peak position versus B field before (**c**) and after (**e**) averaging the Zeeman splitting for the neutral exciton emission from the single QR. The *blue dashed line* in (**e**) indicates a parabolic fit. Oscillations in PL energy and corresponding normalized PL intensity as a function of B field are shown in (**d**) and (**f**), respectively. The *solid lines* represent the smoothed data using a Savitzky-Golay filter. The *arrows* in (**d**) and (**f**) indicate changes in the ground state transitions

which is near the resolution limit of our PL system and would be undetectable without the Lorentzian fitting process.

In Figs. 12.4(a) and (b) we can observe the characteristic Zeeman splitting as well as the diamagnetic shift of the PL emission energy for both single QD and QR. The *neutral* exciton emission energies of the QR before and after averaging the emission energy values of the Zeeman split lines are plotted in Figs. 12.4(c) and (e). It is observed that the emission energy does not scale quadratically with increasing B field. Figure 12.4(e) looks quite similar to the one shown in a very recent report [15], in which the AB oscillations in *ensemble* InAs/GaAs QRs were observed. However, the diamagnetic shift for a single QR studied in our work is ∼6 times smaller than the value for the *ensemble* QRs and the oscillation amplitude is much smaller. Also, no Zeeman splitting was observed by Teodoro *et al.* because of the broad ensemble emission. In order to visualize the magnetic field induced oscillations, we subtract a single parabola from Fig. 12.4(e) and plot the results in Fig. 12.4(d). There are two maxima at ~4 T and ~7 T (the solid line here, which represents the smoothed data using a Savitzky-Golay filter, is guide to the eye), suggesting changes in the ground state transitions.

Another significant feature observed is the change in PL intensity with increasing magnetic field, see Fig. 12.4(a). The change in PL intensity is one of the signatures expected for a QR and is attributed to oscillations in the ground state transitions [6, 12, 13, 15]. Two knees can be observed at ~4 T and ~7 T [Fig. 12.4(f)], which correspond well with the peak energy oscillation maxima in Fig. 12.4(d) (indicated by arrows).

All the observations here are well above the system resolution, and they strongly contrast with the experimental results on a conventional QD [see Fig. 12.4(b)], where no peak energy oscillation and PL intensity quenching are observed [16]. Especially, the PL intensity of this QR quenches by more than 70 % with increasing magnetic field, which is confirmed by repeating the measurements and similar observations in other QRs. A recent study on single GaAs QR obtained by droplet epitaxy also revealed a significant reduction in the PL intensity by more than 20 % at $B > 6$ T [27], the exciton AB effect may account for the observation.

12.4 Gate-Controlled AB-Type Oscillations

There are great fundamental and application interests to have a control knob for this optical AB effect. Fischer et al. proposed that a combination of magnetic fields and electrical fields can be used to control the exciton oscillator strength and therefore switch a single exciton between optical bright and dark states [11]. In our experiments some of the QRs are embedded in an n-i-Schottky structure [see Fig. 12.5(a)], consisting of 20 nm n^+ GaAs layer followed by a 20 nm thick spacer layer under the QRs, 30 nm i-GaAs and a 116 nm thick AlAs/GaAs short period superlattice. With a 5 nm thick semi-transparent Ti top gate, an electric field can be applied to modify the energy band of the ring [28, 29]. When a forward bias V_g is scanned from 0.3 to 0.32 V (with steps of 0.5 mV), the neutral exciton emission energy from a single QR is tuned by about 30 μeV [see Fig. 12.5(b)]. It is worth mentioning that both raw peak positions and Lorentzian fitted peak positions are plotted in Fig. 12.5(b), in order to illustrate the peak determination procedure mentioned above.

This gated-QR structure can therefore be used to demonstrate the possibility of controlling the magnetic field dependent PL with an external electric field. The oscillations for a single QR (with a nominal AsBr$_3$ etching depth of 3 nm) under a forward bias of 2.8 V are clearly seen in Fig. 12.6(a), with the two transition points at 2.6 T and 8 T. When we decrease the bias from 2.8 V to 0.8 V, it is clear that the transition points shift smoothly to higher magnetic fields [Fig. 12.6(b)], indicating that the effective radii of the electron and the hole are modified by the external gate potential. The PL intensity is also strongly modified at different electric fields and

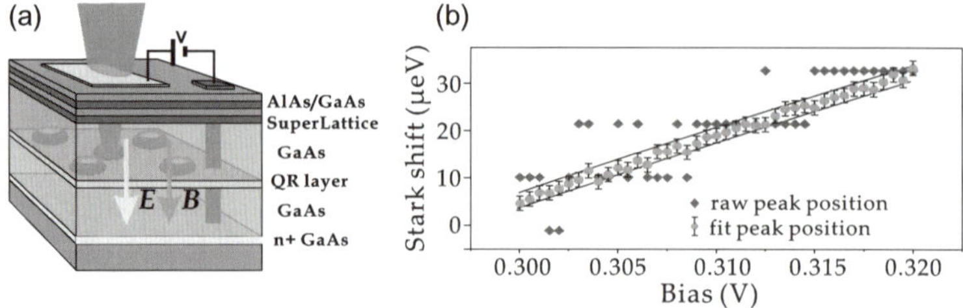

Fig. 12.5 (**a**) Schematic of the n-i-Schottky structure used to control the QR emission under magnetic fields. (**b**) When the bias changes from 0.3 V to 0.32 V, the PL emission shifts by $\sim$30 μeV. Both the raw and the Lorentzian fitted peak positions are plotted

Fig. 12.6 (**a**) Under a forward bias of 2.8 V the PL energy shows clear transitions at $\sim$2.6 T and $\sim$8 T. (**b**) When the bias changes from 2.8 V to 0.8 V, the transitions shift smoothly to higher magnetic fields. (**c**) The PL intensity shows modulation under increasing magnetic fields

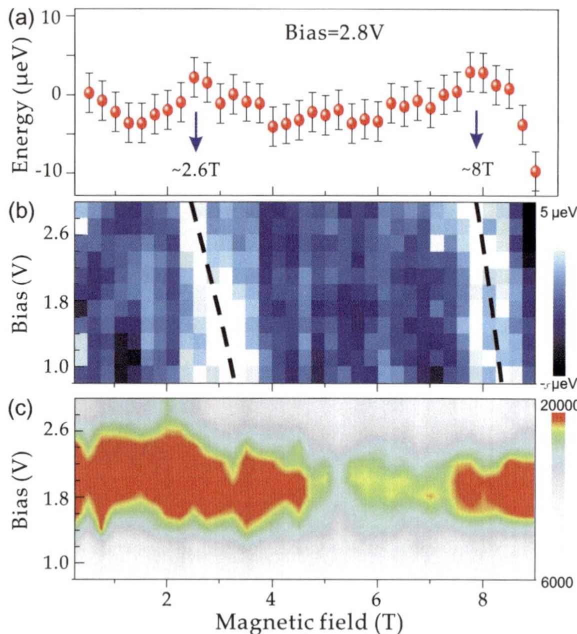

magnetic fields, see Fig. 12.6(c). At $\sim$1.8 V the PL intensity shows clear modulation by the magnetic field, i.e., first quenches by more than 25 % at $\sim$5.2 T and then recovers fully to its original value with increasing magnetic field. This observation is quite similar to the long-existed theoretical prediction for a weakly bound neutral exciton confined in a QR, again, indicating a consequence of the optical AB effect [6].

12.5 The Physical Origin of Tunable Optical AB Effect

Theoretically it is important to distinguish between excitons composed of strongly and weakly bound electron-hole pairs [6]. In the strongly bound regime where the

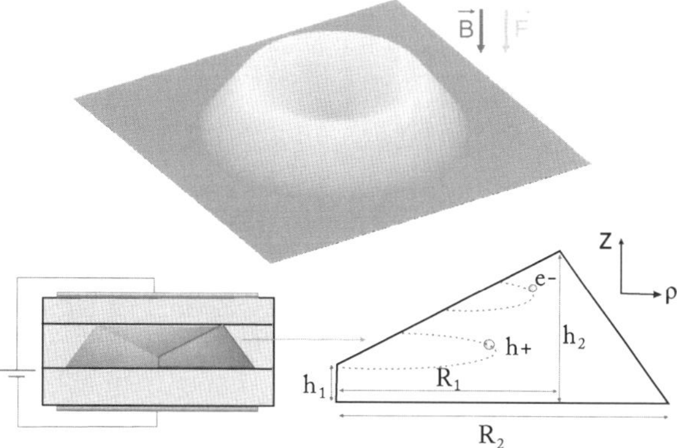

Fig. 12.7 Schematics of the volcano-like quantum ring. A perpendicular electric and magnetic field are applied in the vertical direction. *Bottom*: the structure of the quantum ring in the ρ-z plane

electron-hole Coulomb interaction is dominant, the exciton moves along the ring as one tightly bound particle and AB effect manifests itself in a bright to dark transition in the ground state with increasing magnetic field. While in the opposite limit of a weakly bound electron-hole pair, there is strong radial polarization of the exciton and therefore, an interesting modulation between different bright and dark states will appear. We believe that the electron-hole pairs in above-mentioned QRs are close to the weakly bound regime, this is because: (i) in the ring geometry the ratio of Coulomb to kinetic energies is proportional to the ring radius, while for small rings ($R < a^*$, the effective Bohr radius) the quantization due to kinetic motion is strong [6, 30] and the Coulomb interaction becomes weaker; (ii) in our structure the n-doped layer which locates 20 nm under the QRs (also, the metal contact in the gated structure) can sufficiently screen the effectiveness of the Coulomb interaction [7]; (iii) in experiment we also observe that the changes in the binding energies $\Delta E_B(\mathrm{XX}) \approx \Delta E_B(\mathrm{X}^+)$ when E changes (not shown here), indicating that the strongly confined few-particle picture is valid in our QRs and, the effective radii of electron and hole are very different [31–33].

In the following sections we use the finite-element method to calculate the electron and the hole wave functions and energies in III–V semiconductor quantum rings. Both lattice-matched rings (GaAs/AlGaAs) and lattice-mismatched rings (InAs/GaAs) are considered in our calculations, in order to study the effect of strain on the optical AB effect. The total energy of the exciton is then calculated in the presence of Coulomb interaction by diagonalizing the total Hamiltonian in the space spanned by the product of the single particle states, and study the AB effect in such quantum rings. The tunability of the AB effect for both the single particles and the exciton energy by applying a perpendicular external electric field is also shown.

12.5.1 Model

Figure 12.7 shows the investigated geometry of the three dimensional ring, A volcano shaped ring is embedded in a barrier material which is different from the ring material. In a three dimensional semiconductor quantum ring, the full Hamiltonian of the exciton within the effective mass approximation is given by

$$H_{\text{tot}} = \sum_{j=e,h} (\mathbf{P}_j - q_j \mathbf{A}_j) \frac{1}{2m_j} (\mathbf{P}_j - q_j \mathbf{A}_j) + V_c(\mathbf{r}_e - \mathbf{r}_h)$$

$$+ \sum_{j=e,h} \delta E_j(\mathbf{r}_j) + \sum_{j=e,h} V_j(\mathbf{r}_j) - eFz_e + eFz_h, \qquad (12.1)$$

where $V_j(\mathbf{r}_j)$ is the confinement potential of the electron (hole) due to the band offset of the two materials which will be different inside and outside the ring. $V_c(\mathbf{r}_e - \mathbf{r}_h) = e^2/4\pi\varepsilon|(\mathbf{r}_e - \mathbf{r}_h)|$ is the Coulomb potential between the electron and the hole, and $\delta E_j(\mathbf{r}_j)$ is the strain-induced shift of the energy which depends on the strain tensor ϵ_{ij}. We did not take the piezoelectric potential into account since it is negligible compared to the other terms in our case. The last two terms of (12.1) are the potential energy in the presence of the perpendicular top to bottom directed electric field F.

12.5.2 GaAs/AlGaAs Quantum Ring

12.5.2.1 Single Particle Wave Functions

We consider first a volcano shaped GaAs ring surrounded by AlGaAs [34] where strain is negligible. We assume that only the lowest electronic subband and the highest hole band (heavy hole) is occupied. In the GaAs, the electron and the hole have effective mass $m_e/m_0 = 0.063$ and $m_h/m_0 = 0.51$, respectively. The static dielectric constant is $\varepsilon = 12.5\varepsilon_0$ and the band gap is $E_g = 1.42$ eV at helium temperature. While for AlGaAs, we have $m_e/m_0 = 0.082$, $m_h/m_0 = 0.568$, $\varepsilon = 12.5\varepsilon_0$, and a band gap of $E_g = 1.78$ eV. This results in a band gap difference of $\Delta E_g = 360$ meV between GaAs and AlGaAs, which leads to a conduction band offset of about $\Delta E_c = 250$ meV and a valence band offset of about $\Delta E_v = 110$ meV, so we can take $V(\mathbf{r}_{e(h)}) = 0$ inside the ring and $\Delta E_{c(h)}$ outside the ring. The parameters used were taken from Ref. [35]. As the dielectric constant is practically the same everywhere and the difference of the lattice constant for GaAs and AlGaAs is very small, we may ignore the dielectric mismatch effect, and the strain induced term in (12.1).

Because of cylindrical symmetry we rewrite the Hamiltonian in cylindrical coordinates. Assume the wave function of the single particles to be $\Psi(\rho, z, \theta) = \psi_{e(h)}(\rho, z)e^{-il_{e(h)}\theta}$, after averaging out the angular part of the wave function, we obtain the 2D single particle Schrödinger equation:

$$\left(-\frac{\hbar^2}{2m_{e(h)}} \left(\frac{\partial^2}{\partial z^2} + \frac{\partial^2}{\partial \rho^2} + \frac{1}{\rho}\frac{\partial}{\partial \rho} - \frac{l^2_{e(h)}}{\rho^2} - \frac{q^2 B^2 \rho^2}{4m_{e(h)}} \right) - \frac{q B l_{e(h)} \hbar}{2m_{e(h)}} \right.$$

$$\left. + V(\mathbf{r}_{e(h)}) + q F z_{e(h)} \right) \psi_{e(h)}(\rho, z) = E_{\rho,z} \psi_{e(h)}(\rho, z), \qquad (12.2)$$

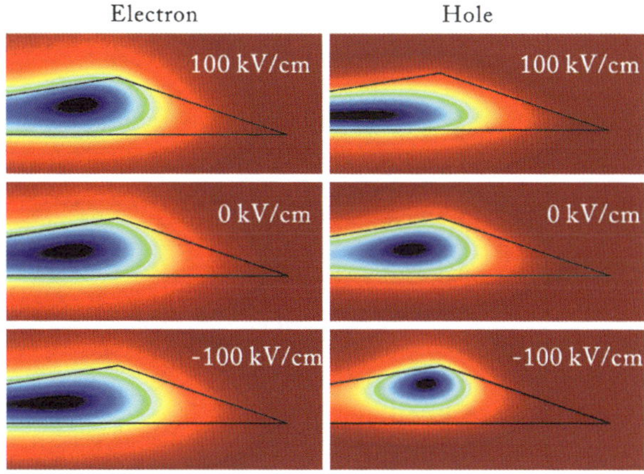

Fig. 12.8 Contour plot of the ground state wave function of the electron and the hole for different values of the perpendicular electric field F in the (ρ, z) plane. Here $B = 0$, $l_{e(h)} = 0$ and the electric field F from top to bottom are 100 kV/cm, 0, -100 kV/cm. The contour of the ring is shown by the *black curve*

here q is $-e$ for the electron and e for the hole. Taking advantage of cylindrical symmetry we only need to solve the problem in a half section of the ring as shown at the bottom inset of Fig. 12.7. We use the finite-element methods to obtain the ground state energy for different angular moment $l_{e(h)}$ and electric field F as a function of magnetic field B. For a better understanding and further consideration of the transition, we first write down the single particle Schrödinger equation in dimensionless form (assume $B = 0$ and $l_{e(h)=0}$):

$$\left(-\frac{m_{e0(h0)}}{m_{e(h)}} \left(\frac{\partial^2}{\partial z^2} + \frac{\partial^2}{\partial \rho^2} + \frac{1}{\rho}\frac{\partial}{\partial \rho} \right) + V'(\mathbf{r}_{e(h)}) + A_F F' z_{e(h)} \right) \psi_{e(h)}(\rho, z)$$
$$= E'_{\rho,z} \psi_{e(h)}(\rho, z), \tag{12.3}$$

where $E'_{\rho,z} = E_{\rho,z}/E_0$, and $E_0 = \hbar^2/2m_{e0(h0)}R_0^2$ is the energy unit. We take $R_0 = 1$ nm as the length unit, and $m_{e0(h0)}$ is the effective mass of the electron (hole) in the barrier material GaAs. The effective band offset $V'(\mathbf{r}_{e(h)}) = V(\mathbf{r}_{e(h)})/E_0$ is 0.425 for the electron and 1.472 for the heavy hole. We take the unit of the electric field to be $F_l = 1$ kV/cm and the coefficient $A_F = F_l R_0 e/E_0 = 0.00263 m_{e0(h0)}$.

Figure 12.8 shows the contour plot of the electron and the hole wave function in the ρ-z plane. Here the size of the ring is chosen to be $h_1 = 4$ nm, $h_2 = 6$ nm, $R_1 = 12$ nm and $R_2 = 30$ nm. When the electric field changes from 100 kV/cm to -100 kV/cm the wave function of the hole changes from a QD-like to a ring-like wave function, while the wave function of the electron changes from ring-like to QD-like. The hole wave function is more sensitive to the electric field, which can be understood from (12.3). The term related to the electric field is $0.00263 m_{e0(h0)} F'z$, which is proportional to the effective mass. Since the heavy hole has a larger effective mass, the effect of the electric field will be larger.

12.5.2.2 Constraints for Dot-Ring Wave Function Transition

From (12.3) and Fig. 12.8 we can determine the constraints in order to have an electric field tunable AB effect: (1) The electric field term $A_F F' dz$ must be com-

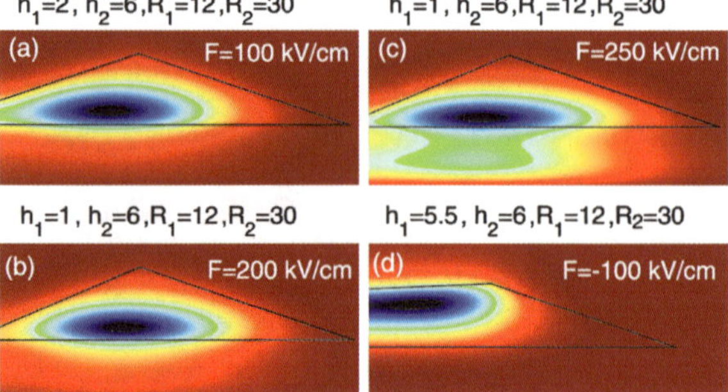

Fig. 12.9 Contour plot of the ground state wave function of the heavy hole for different sizes of the ring. The size (in unit of nm) and the corresponding electric field are specified above each figure

parable to the difference of the confinement energy of the QD-like and ring-like wave function, which is proportional to $1/h_1^2 - 1/h_2^2$. When passing from the QD-like to the ring-like wave function, the difference of the z coordinate is proportional to $(h_2 - h_1)/2$, so the first constraint condition is $A_F F'(h_2 - h_1)/2 \sim 1/h_1^2 - 1/h_2^2$. For example in our case, $h_1 = 4$ nm, $h_2 = 6$ nm and for the heavy hole with an effective mass $m_h = 0.51m_0$ the electric field should be at least $2(1/h_1^2 - 1/h_2^2)/A_F/(h_2 - h_1) * F_l = 2(h_1 + h_2)/(h_1^2 h_2^2 A_F) * F_l$, which is several tens of kV/cm. If we decrease h_1, then the electric field should be larger. We see from Fig. 12.9(a) that if $h_1 = 2$ nm, the wave function of the hole is still ring-like when $F = 100$ kV/cm, and we need to increase the electric field in order to have a QD-like wave function, which leads to the second constraint condition: (2) the confinement energy of the electron (hole) when it is QD-like (i.e. which is $\sim 1/h_1^2$) should be smaller than the band offset $V'(\mathbf{r}_{e(h)})$ of the quantum ring, otherwise the electron (hole) will move out of the ring to the barrier for large electric field values, as can been seen from Figs. 12.9(b) and (c). The wave function is still ring like when $F = 200$ kV/cm, but when we increase the electric field to $F = 250$ kV/cm, the wave function starts to penetrate in the barrier region. The AB effect is still tunable, but the effect of the electric field will be very small in case of a small ring height h_2. (3) The confinement energy difference of the QD-like and ring-like states, which are proportional to $1/h_1^2 - 1/h_2^2$ should not be very small, otherwise the wave function of the electron (hole) will always be QD-like and shows no observable AB effect (as we can see from Fig. 12.9(d), when $h_1 = 5.5$ nm and $h_2 = 6$ nm the wave function is still QD-like when $F = -100$ kV/cm). The above three constraints were obtained for holes, but they are also valid for electrons.

12.5.2.3 Exciton Energy Including Coulomb Interaction

The total exciton energy is calculated using the configuration interaction (CI) method. We first construct the total exciton wave function as a product of linear combinations of single electron and hole wave functions, and then calculate the matrix elements of the exciton Hamiltonian. By diagonalizing the obtained matrix we obtain the exciton energies and their corresponding wave functions.

The Hamiltonian of the exciton is

$$H_{\text{tot}} = H_e + H_h + U_c, \tag{12.4}$$

here H_e and H_h are the single particle Hamiltonians,

$$U_c = U_0 * \frac{1}{\sqrt{\rho_e'^2 + \rho_h'^2 - 2\rho_e'\rho_h' \cos(\varphi_e - \varphi_h) + (z_e' - z_h')^2}}, \tag{12.5}$$

is the Coulomb energy between the electron and the hole, where $U_0 = -e^2/4\pi\varepsilon R_0$.

As the total angular momentum is a good quantum number because of cylindrical symmetry, we assume the wave function of the exciton to be:

$$\Psi_L(\mathbf{r}_e, \mathbf{r}_h) = \sum_k C_k \Phi_k(\mathbf{r}_e, \mathbf{r}_h), \tag{12.6}$$

for given total angular moment L. Here, $\Phi_k(\mathbf{r}_e, \mathbf{r}_h) = \psi_{n_e} e^{-il_e\varphi_e} \psi_{n_h} e^{-il_h\varphi_h}$, and the exciton wavefunction is constructed out of single particle eigenstates, and k stands for the indices (n_e, n_h, l_e, l_h). $l_e + l_h = L$ should always be satisfied for fixed L, $n_e(n_h)$ is the quantum number of the single particle radial wave function. As the energy of the single particle eigenstates with large angular moment quantum number l_e (l_h) and quantum number n_e (n_h) is much larger as compared to the ones with smaller angular moment quantum number, we can limit ourselves to several tens of single particle eigenstates. With this wave function, we can construct the matrix of the total Hamiltonian and by diagonalizing the obtained matrix, we can get the eigenvalues and eigenvectors.

For a fixed total angular moment, we have

$$U_c(k, j) = C_k C_j \langle \Phi_k(\mathbf{r}_e, \mathbf{r}_h) | U_c | \Phi_j(\mathbf{r}_e, \mathbf{r}_h) \rangle$$

$$= C_k C_j \int \int \int \int \psi_{n_e} \psi_{n_h} \psi_{m_e} \psi_{m_h}$$

$$\times A_{l_{ej}, l_{hj}, l_{ek}, l_{hk}} \rho_e' d\rho_e' \rho_h' d\rho_h' dz_e' dz_h', \tag{12.7}$$

and where the angular part of the integral is

$$A_{l_{ej}, l_{hj}, l_{ek}, l_{hk}} = U_0 \int_0^{2\pi} \int_0^{2\pi} d\varphi_e d\varphi_h$$

$$\times \frac{e^{-i(l_{ej} + l_{hj} - l_{ek} - l_{hk})\varphi_e} e^{i(l_{hj} - l_{hk})(\varphi_e - \varphi_h)}}{\sqrt{\rho_e'^2 + \rho_h'^2 - 2\rho_e'\rho_h' \cos(\varphi_e - \varphi_h) + (z_e' - z_h')^2}}. \tag{12.8}$$

We know from (12.8) that, as the Coulomb interaction between the particles does not break the cylindrical symmetry, $l_{ej} + l_{hj} - l_{ek} - l_{hk}$ should be zero, which also makes the integral nonzero. If we take $\varphi_e - \varphi_h$ as a new variable in (12.8), it can be changed into an elliptic integral, which is easy to calculate numerically. The remaining integral is easy to calculate numerically. The effective exciton Bohr radius is $a_B = 4\pi\varepsilon\hbar^2/(\mu e^2) = 11.7673$ nm ($\mu = 1/(1/m_e + 1/m_h)$), and we take a small volcano shaped ring ($R_1 = 10$ nm, $R_2 = 16$ nm, $h_1 = 2$ nm and $h_2 = 4$ nm), whose size is comparable to the exciton Bohr radius.

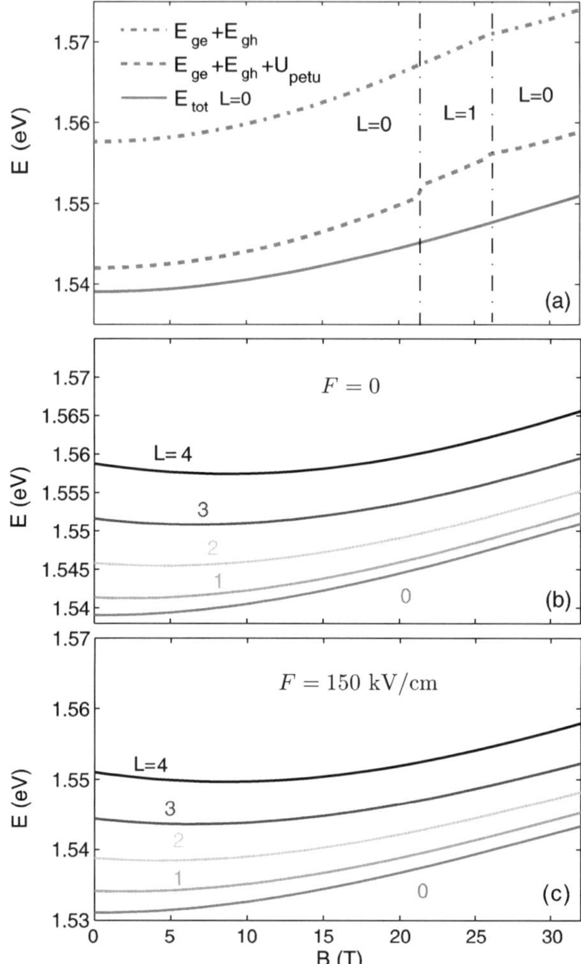

Fig. 12.10 (**a**) Exciton energy for total angular momentum $L = 0$ as obtained from the CI method (*solid line*) and compared to the results of the ground exciton energy without Coulomb interaction (*dot-dashed line*), and the ground state energy by taking the Coulomb energy as a perturbation (*dashed line*). Here $F = 0$. (**b**) Ground state energy and the four lowest excited exciton energies as a function of the magnetic field for $F = 0$ and (**c**) for $F = 150$ kV/cm. Here $R_1 = 10$ nm, $R_2 = 16$ nm, $h_1 = 2$ nm and $h_2 = 4$ nm

The results of the total exciton energy for $F = 0$ and are shown in Fig. 12.10, and as a comparison, we also show, in Fig. 12.10(a), the results of the ground exciton energy without Coulomb interaction and the total ground state energy by taking the Coulomb energy as a perturbation (we took only the lowest electron and hole state into account, the result is much closer to the real one for a smaller ring when the Coulomb energy is small as compared to the kinetic energy). From Fig. 12.10(a) we know that the Coulomb interaction has a large effect on the total exciton energy, and unlike the perturbation result, the exciton energy from the CI method shows a monotonous increase with magnetic field. There are no oscillations as function of the magnetic field. Figures 12.10(b) and (c) show the CI results, for the exciton energies with different values of the total angular momentum L, when $F = 0$ and $F = 150$ kV/cm. It shows that the state with total angular moment $L = 0$ is always the ground state, which is different from the perturbation theory result shown in

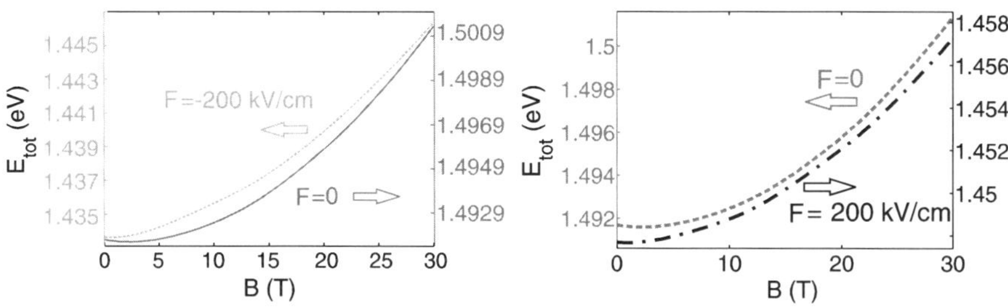

Fig. 12.11 Exciton ground state energy as a function of the magnetic fields at different electric fields. (*Left*) $F = -200$ compared with $F = 0$ kV/cm, and (*Right*) $F = 200$ kV/cm compared with $F = 0$ kV/cm

Fig. 12.10 where the total angular moment of the ground state switches between $L = 0$ and $L = 1$. We notice that the energy of the exciton states show no oscillation with increasing magnetic field, both with or without perpendicular electric field. The effect of the electric field on the exciton leads to a decrease of the exciton energy and a larger difference between the energy of the lowest exciton states.

Till now, we did not observe any optical AB effect in the exciton ground state energy and any large influence of the perpendicular electric field on the oscillation. In order to see the effect of the electric field more clearly, we calculate the energy for a quantum ring with a larger height and a much larger electric field. Since the period of the oscillation for the previous case is large, we increase the radius of the ring in order to decrease the period. Figure 12.11 shows the exciton ground state energy with respect to the magnetic field in the case of $F = 0$ kV/cm, $F = -200$ kV/cm and $F = 200$ kV/cm. Here $R_1 = 14$ nm, $R_2 = 18$ nm, $h_1 = 1$ nm and $h_2 = 10$ nm (the size is still comparable to the effective Bohr radius). As the state with total angular moment $L = 0$ always has the lowest energy, the ground state is the same as the $L = 0$ state here. It is clear that although the $F = 0$ state shows no oscillation at all, the ground state energy in the presence of an electric field shows a weak but observable oscillation, especially for the case of $F = -200$ kV/cm. And the ground state energy is smaller in the presence of a top to bottom directed electric field.

For a better understanding of the existence of the ground state AB oscillations and the difference of the oscillation period for different values of the electric field, we calculate the average value $\langle \rho \rangle$ in the radial direction and $\langle z \rangle$ in the z direction of the electron and the hole in the exciton ground state, which we show in Fig. 12.12. From Fig. 12.12(a) we see that $\langle \rho \rangle$ and $\langle z \rangle$ for both electron and hole have no clear step-like behavior, but they rather change smoothly. Furthermore, the difference of the average values of the electron and the hole in both radial and z direction is rather small, especially in the z direction. This makes then the Coulomb interaction energy very large and the polarity of the exciton extremely small, which greatly weakens the oscillations in the exciton ground state energy. That is the reason why we can not observe any oscillation of the exciton energy in the absence of the electric field. But in the presence of a perpendicular electric field, we see from Figs. 12.12(b) and 12.12(c) that the difference of the average values between the electron and the

Fig. 12.12 Average distance $\langle\rho\rangle$ (*left y scale* in same color) and $\langle z\rangle$ (*right y scale*) in unit of nm of the electron (hole) as a function of magnetic field for (**a**) $F = 0$, (**b**) $F = -200$ kV/cm and (**c**) $F = 200$ kV/cm. In (**b**) and (**c**) the curves of the electron and the hole are separated in different sub figures

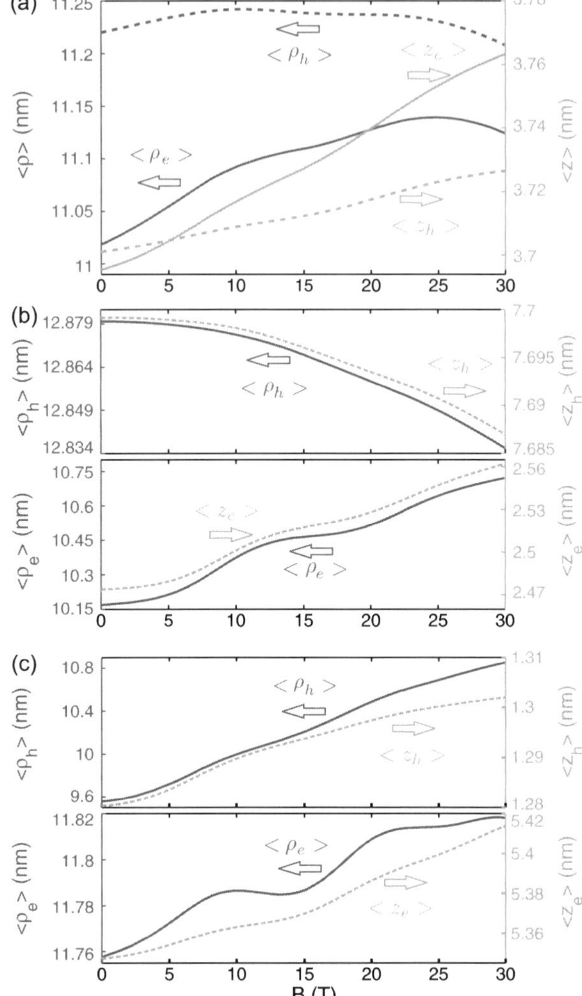

hole becomes much larger. And the average values $\langle\rho\rangle$ and $\langle z\rangle$ show pronounced increasing (or decreasing) step-like behavior with increasing magnetic field, especially for the electron. In both cases of $F = 200$ kV/cm and $F = -200$ kV/cm, $\langle\rho_e\rangle$ and $\langle z_e\rangle$ show obviously oscillations with increasing magnetic field, and their oscillation period is more or less the same as the period of the oscillation in the exciton energy. That is because the coupling of the electron and hole states with different angular momentum becomes much smaller as a result of the weak Coulomb interaction. If we look at the normalization constant C_k in (12.6) we know that the step-like behavior in $\langle\rho_e\rangle$ and $\langle z_e\rangle$ in Figs. 12.12(b,c) originate from the angular momentum transition of the main contributing basis function in the total exciton wave function. With increasing magnetic field, the angular momentum pair (l_e, l_h) in the state

$\Phi_k(\mathbf{r}_e, \mathbf{r}_h)$ which has the largest contribution to the total wave function changes from $(0, 0)$ to $(-1, 1)$ or to even larger values, and this transition becomes more notable in the presence of a perpendicular electric field as the coupling between the different contributing basis functions is much weaker.

From Fig. 12.12 we also see that there are some differences between the cases $F = -200$ kV/cm and $F = 200$ kV/cm. In the case of $F = -200$ kV/cm, the electron is in the center area of the ring while the wave function of the hole has a ring shape. The Coulomb interaction makes the electron (hole) move to the top (center) area of the ring towards the hole (electron). When the magnetic field is increased, the electron (hole) wave functions change towards the area with larger (smaller) values of ρ in which the hole (electron) is located, but as a result of the strong electric field and confinement energy, only $\langle \rho_e \rangle$ is mainly changed. Moreover, compared to the electron, the averages $\langle \rho \rangle$ and $\langle z \rangle$ for the hole change more smoothly with increasing magnetic field. The behavior for $F = 200$ kV/cm is different. The electron now is in the top area of the ring, ρ_e and z_e change over a rather small range as a result of the strong confinement in the top area of the ring. Moreover, with increasing magnetic field, $\langle \rho \rangle$ and $\langle z \rangle$ for both the electron and the hole increase towards the area with larger value of ρ, but the change for the electron is very small. This is different from the case of $F = -200$ kV/cm.

In order to study the exciton recombination, we calculate the oscillator strength (the dimensionless quantity that expresses the strength of the transition) for the state with total angular momentum $L = 0$. The oscillator strength for the exciton ground state is defined as [36, 37]

$$f_g = \frac{2}{m} \frac{|\langle \mathrm{ex}| \sum_i p_{xi} |0\rangle|^2}{E_{\mathrm{ex}} - E_0}, \tag{12.9}$$

where m is the free electron mass and $|\mathrm{ex}\rangle$ ($|0\rangle$) is the exciton state (single electron and hole pair). By using the envelope-function approximation, we can derive [38]

$$f_g = \frac{2P^2}{m(E_{\mathrm{ex}} - E_0)} \left| \int \Psi_g(\mathbf{r}_e, \mathbf{r}_e) d\mathbf{r}_e \right|^2, \tag{12.10}$$

here P includes all intra matrix-element effects, $\Psi_g(\mathbf{r}_h, \mathbf{r}_e)$ is the exciton ground state wave function. For simplicity, here we will focus on the main variable part of the oscillator strength which is $O_f = |\int \Psi_g(\mathbf{r}_e, \mathbf{r}_e) d\mathbf{r}_e|^2$, named the overlap integral. The result of the overlap integral for the exciton ground state is shown in Fig. 12.13, for $F = -200$ kV/cm, $F = 0$ kV/cm and $F = 200$ kV/cm. In the presence of the electric field, O_f exhibits more structure, and we can clearly see steps in the overlap integral. These have the same period as the AB oscillation, which also originates from the angular momentum transition. In fact, there is also an angular momentum transition for the exciton ground state in case of $E = 0$, but neither on oscillation in the ground state energy nor steps in the Overlap integral squared are found. The insets of Fig. 12.13 give the 3D contour plot of the electron, hole and exciton probability distribution in the ρ-z plane for $B = 13$ T in the presence (absence) of an external electric field. These figures explain the basic mechanism for the existence of the AB oscillation and oscillator strength steps when $E = -200$ kV/cm.

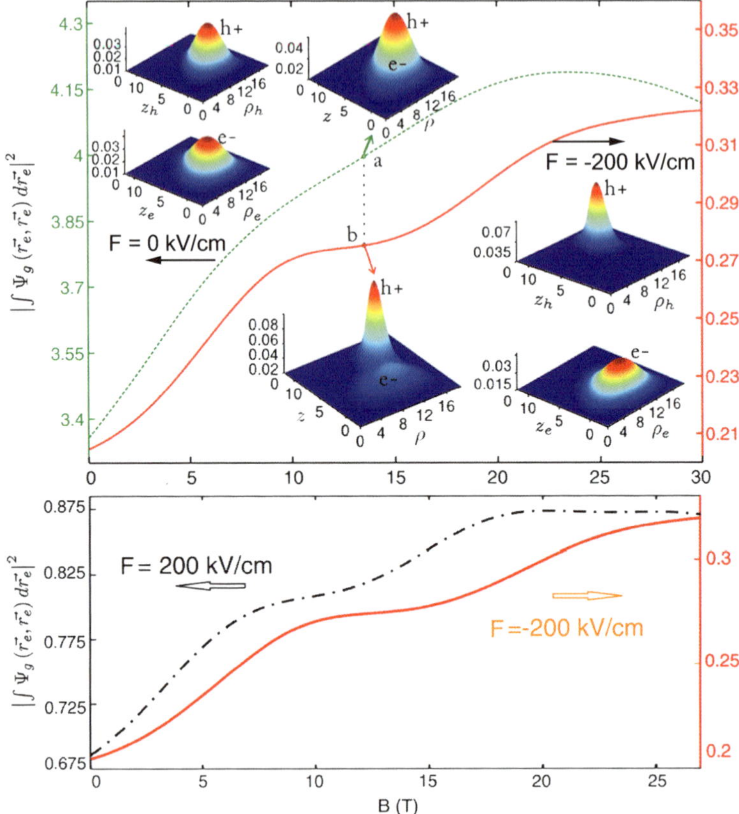

Fig. 12.13 Overlap integral $O_f = |\int \Psi_g(\mathbf{r}_e, \mathbf{r}_e)d\mathbf{r}_e|^2$ of the exciton ground state energy as a function of the perpendicular magnetic field. *Green dashed, red solid,* and *black dash-dotted (lower figure) curves* are for $E = 0$, $E = -200$, and $E = 200$ kV/cm, respectively. The *insets* of the *upper figure* show the 3D contour plot of the probability distribution for the electron and the hole when $B = 13$ T, the *upper (bottom) three insets* are for the electric field $E = 0$ ($E = -200$) kV/cm. Note the different scales for the three curves

Notice that in the absence of electric field, the electron and the hole probability distributions strongly overlap, they only differ slightly in the localization (as a result of larger effective mass, the hole has a larger localization). The average position of the hole in the radial direction is $\langle r_h \rangle = 11.24$ nm which is a little larger than the electron $\langle r_e \rangle = 11.1$ nm (the difference in the perpendicular direction $\langle z_h \rangle = 3.711$ and $\langle z_e \rangle = 3.723$ nm is even smaller). The Coulomb interaction between the electron and the hole is large, which greatly weakens the AB oscillation in the exciton ground state energy. This is very different from the case of $E = -200$ kV/cm as we can see from the bottom insets of Fig. 12.13 for the electron and the hole probability distributions. The strong external bottom to top directed electric field pushes the electron and the hole wave function in opposite z-direction, increasing the polarity of the exciton. Compared to the electron, the hole has a much larger localization and

Fig. 12.14 Second derivative of the exciton ground state energy with respect to the magnetic field for different values of the perpendicular field $F = 0$ (*black solid line*), $F = -100$ kV/cm (*red solid line*), $F = 100$ kV/cm (*red dash-dotted line*), $F = -200$ kV/cm (*light blue solid line*), $F = 200$ kV/cm (*light blue dashed line*)

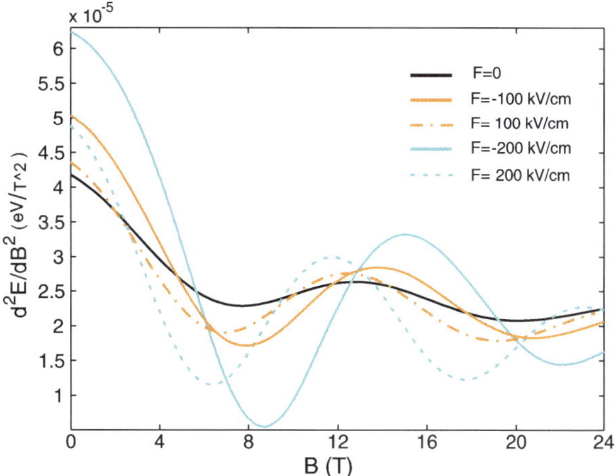

stays at the top area of the ring with a large radius and height. The average values $\langle r_h \rangle = 12.87$ nm and $\langle z_h \rangle = 7.697$ nm for the hole are much larger than the values for the electron ($\langle r_e \rangle = 10.45$ nm and $\langle z_e \rangle = 2.513$ nm). Furthermore, the Coulomb energy between the electron and the hole is reduced and the exciton is not bound and the electron and hole are only weakly correlated. The areas encircled by the electron and the hole are very different and the electron and hole pick up a very different phase resulting in a non-vanishing AB effect.

Figure 12.13 also shows that the overlap integral for the case of $F = 0$ is much larger, which is a consequence of the extremely small polarity of the exciton. O_f in case of $F = -200$ kV/cm shows the smallest value as compared to the other two cases, which is due to the fact that the exciton has the largest polarity and is most stable, and the AB effect is the strongest. We can also find that for all three cases, the overlap integral increases with increasing magnetic field, which means the exciton has a smaller polarity and also a weaker AB oscillation. This is quite different from the case of a single particle in a one dimensional ring where the AB oscillation has almost the same strength. The reason is that by increasing the magnetic field, the Zeeman term makes the angular momenta (absolute value) of both the electron and the hole to increase, and consequently the electron and the hole will have a larger effective radius (note that in the one dimensional ring the radius is fixed). However, because the confinement potential is different inside the 3D ring it prevents the electron and the hole from moving to the outer part of the ring which has a radius larger than the top area of the ring. This leads, as shown in Fig. 12.12, to a smaller difference in the electron and hole position inside the ring with increasing magnetic field, and consequently, a larger value of the overlap integral and weaker AB oscillation.

The external perpendicular electric field does not only enhance the AB oscillation of the exciton ground state energy, but also changes the period of the AB oscillation, as we can clearly see from the second derivative of the exciton ground state energy with respect to magnetic field. Figure 12.14 shows the results for different values of the external perpendicular electric field. We already know from Fig. 12.13 that the hole has a slightly larger radius than the electron in the absence of an external electric field, the applied bottom to top directed electric field makes the hole (electron) have an even larger (smaller) radius and height, this will largely decrease the Coulomb energy and strongly enhance the AB oscillation of the exciton when the electric field is very strong. However from Fig. 12.14 we know the bottom to top directed electric field will increase the period of the oscillation which requires a larger magnetic field to observe it.

When the electric field is top to bottom directed, i.e., the ones with positive values, as the hole (electron) is pushed to the bottom (top) area of the ring, the polarity of the exciton decreases when a very small electric field is applied (the results of such a small electric field are not shown here), which possibly weakens the AB oscillation. But when the top to bottom directed electric field is strong, the exciton can have a large polarity, and the AB oscillation could be enhanced just like the case of bottom to top directed electric field (the oscillation of course is weaker than the bottom to top directed electric field with the same absolute value). Moreover, in the presence of the top to bottom electric field, the period of the AB oscillation could be largely decreased, e.g. we can see from Fig. 12.14 that when $F = 250$ kV/cm, the first oscillation happens at $B \approx 6$ T, which is much smaller than the one without an applied electric field (which happens at $B \approx 8$ T). The period change is a consequence of the change of average radius of the electron and/or hole. However, it is hard to obtain this result only from the magnetic field dependence of $\langle \rho_e \rangle$ and $\langle \rho_h \rangle$. We notice from Figs. 12.12(b) and 12.12(c) that $\langle \rho_e \rangle$ and $\langle z_e \rangle$ is smaller in the case of $F = -200$ kV/cm as compared to $F = 0$ kV/cm, but $\langle \rho_h \rangle$ and $\langle z_h \rangle$ is larger, which probably decreases the AB oscillation period of the total exciton energy. Moreover, $\langle \rho_e \rangle$ and $\langle z_e \rangle$ is larger in the case of $F = 200$ kV/cm, but $\langle \rho_h \rangle$ and $\langle z_h \rangle$ is smaller. The point is that as the hole has a much larger effective mass, the effective radius of the exciton is mainly determined by the electron. The period of the oscillation is proportional to $m_h/\langle \rho_e \rangle^2 + m_e/\langle \rho_h \rangle^2$, and is mainly determined by the effective radius of the electron (this is valid for the case when the oscillation origins from the $(0, 0)$ to $(-1, 1)$ and even larger angular momentum pair $(l, -l)$ transition).

12.5.3 InGaAs/GaAs Quantum Ring

Now let us turn to a different system where strain is important: $In_{1-x}Ga_xAs$ ring surrounded by GaAs, the concentration of Ga is proportional to the coordinate z inside the ring [17], which is $x = 0.4 - 0.05z$. We take $R_1 = 15$ nm, $R_2 = 22$ nm, $h_1 = 0.5$ nm and $h_2 = 4$ nm for the ring.

For $In_{1-x}Ga_xAs$ [35], we have the effective masses $m_e/m_0 = 0.023 + 0.037x + 0.003x^2$, $m_h/m_0 = 0.41 + 0.1x$, dielectric constant $\varepsilon = (15.1 - 2.87x + 0.67x^2)\varepsilon_0$, and a band gap of $E_g = 0.36 + 0.63x + 0.43x^2$ eV. This results in a band gap difference of $\Delta E_g = 1.06 - 0.63x - 0.43x^2$ eV between GaAs and $In_{1-x}Ga_xAs$, we take 25 % of ΔE_g be the valence band offset and 75 % be the conduction band offset.

Since $x = 0.4 - 0.05z$ inside the ring, the band gap difference will be the largest at the top of the ring (which is $(1.06 - 0.63 \cdot (0.2) - 0.43(0.2)^2) = 0.9168$ eV), we assume $V(\mathbf{r}_h) = 0$ when $z = 4$ inside the ring, then we find that $V(\mathbf{r}_h) = 0.25 \cdot 0.9168 = 0.275$ eV outside the ring and $V(\mathbf{r}_h) = 0.275 - 0.25(1.06 - 0.63x - 0.43x^2) = 0.05324 - 0.01461z + 0.00032z^2$ eV inside the ring, while for the conduction band offset we have $V(\mathbf{r}_e) = 0.5032 + 0.9168 \cdot 0.75 = 1.145$ eV outside the ring and $V(\mathbf{r}_e) = 1.145 - 0.75(1.06 - 0.63x - 0.43x^2) = 0.62756 - 0.03409z + 0.00075z^2$ eV inside the ring. For convenient of our calculation, we take the value of the hole confinement potential $V(\mathbf{r}_h)$, as in case of the electron, which make the hole also be confined in a potential well, larger value of $V(\mathbf{r}_h)$ stands for higher energy. As a simplification, we do not take the dielectric mismatch effect into account because it is nearly the same in the ring and in the barrier material, but just assume $\varepsilon = 12.5\varepsilon_0$ inside and outside the quantum ring structure.

As in Sect. 12.5.2, we will solve the single particle Schrödinger equation first. As the lattice constant [35] is $a_1 = 0.56533$ nm for GaAs and $a_2 = 0.60583 - 0.0405x$ nm for $\text{In}_{1-x}\text{Ga}_x\text{As}$ (x from 0.2 to 0.4), there is a lattice mismatch $(a_2 - a_1)/a_1$ of 6 %, which results in a large strain. The difference from the previous case is that in a strained quantum ring the total confinement potential terms $V(\mathbf{r}_{e(h)})$ in (12.2) now comes from the band offset energy due to the band gap difference, and the strain induced term. We calculated the strain by adapting a method developed by John Davies which is based on Eshelby's theory of inclusions [39, 40], where the elastic properties are assumed to be isotropic and homogeneous. In our model the lattice mismatch between the two materials is $\varepsilon_0 = (0.243 + 0.02025z)/5.6533$ inside the ring and 0 outside. By using the finite-element method, it is easy to obtain the eigenstrain. The strain in our case will change the potential of the electron (hole) thus modify the band structure including the conduction and the valence edge energies which are among the most important parameters characterizing them.

We assume that the conduction band is decoupled from the valence band. The edge of the conduction band responds only to the hydrostatic strain, the total confinement potential for the electron now becomes [41]:

$$V(\mathbf{r}_e) = V_{e,\text{off}} + U_c$$
$$= V_{e,\text{off}} + a_c\varepsilon_{\text{hyd}}, \tag{12.11}$$

where $a_c = -7.17$ eV [42–45] for GaAs and $-2.09x - 5.08$ eV for $\text{In}_{1-x}\text{Ga}_x\text{As}$ is the hydrostatic deformation potential for the conduction band, and $\varepsilon_{\text{hyd}} = \varepsilon_{xx} + \varepsilon_{yy} + \varepsilon_{zz}$ denotes the hydrostatic strain ε_{hyd}.

The total confinement potential for the heavy hole becomes [41]:

$$V(\mathbf{r}_h) = V_{h,\text{off}} + U_v$$
$$= V_{h,\text{off}} + P + \text{sgn}(Q)\sqrt{Q^2 + RR^\dagger + SS^\dagger},$$
$$P = a_v(\epsilon_{xx} + \epsilon_{yy} + \epsilon_{zz}),$$
$$Q = \frac{b}{2}(\epsilon_{xx} + \epsilon_{yy} - 2\epsilon_{zz}),$$

Fig. 12.15 The *left* (*right*) *column* of figures are the electron confinement potential in the ρ (z) direction for different values of z (ρ). *Solid curves* are the band offset (confinement potential without strain) of the electron, while the *dashed curves* are those for the case when strain is included. Here the confinement potentials are in units of eV

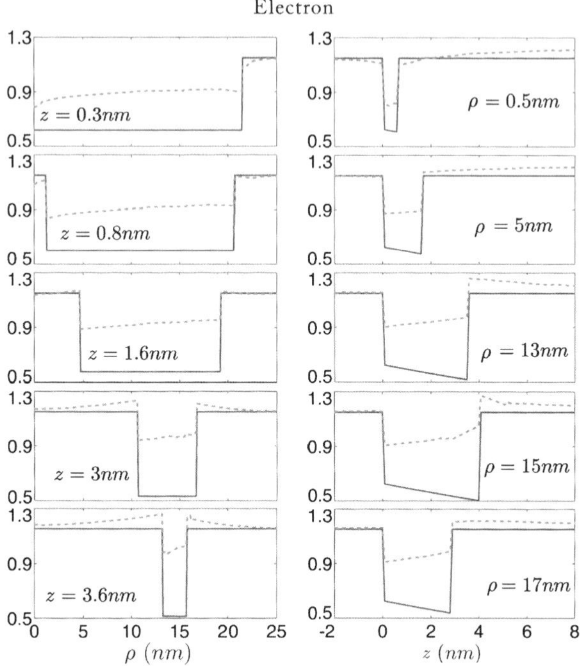

$$R = \frac{\sqrt{3}b}{2}(\epsilon_{xx} - \epsilon_{yy}) + id\epsilon_{xy},$$

$$S = -d(\epsilon_{xz} - i\epsilon_{yz}), \tag{12.12}$$

here $a_v = 1 + 0.16x$ eV, and $b = -1.8 - 0.2x$ eV for $In_{1-x}Ga_x As$ are the deformation potentials of the valence band and $d = -3.6 - 1.2x$ eV [42–45].

Figure 12.15 shows the confinement potential of the electron in the $In_{1-x}Ga_x As$ quantum ring with (red dashed curves) and without (blue solid curves) strain induced energy shift. The figures in the left column show the confinement potential in the ρ direction for a fixed value of z, while the right figures are for the confinement potential in the z direction but with fixed value of ρ. we found, in the absence of strain, that the electron confinement potential is a constant in the ρ direction, as the concentration of Ga depends only on the z coordinate. Moreover, the electron always has the lowest confinement potential in the place with the largest value of z, manifesting that the electron is well confined in the top area of the ring. In the presence of strain, the confinement potential in the ρ is never a constant both in and outside the quantum ring. The confinement potential with fixed value of z has a clear decreasing tendency with decreasing ρ, and reaches its minimum value at the central region of the ring. In the z direction, we have the opposite result from the case without strain. The electron always has the lowest confinement potential at the bottom of the volcano-shaped quantum ring as a result of the lowest strain induced shift there. Thus, in the presence of strain, the electron may move from the

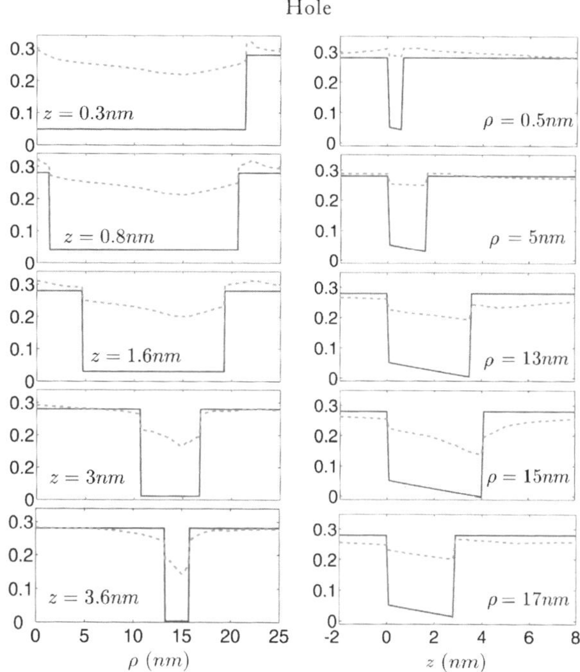

Fig. 12.16 The *left* (*right*) *column* of figures are the hole confinement potential in the ρ (z) direction for different values of z (ρ). *Solid curves* are the band offset (confinement potential without strain) of the hole, while the *dashed curves* are those for the case when strain is included. Here the confinement potentials are in units of eV

top area of the ring towards the bottom of the ring, and towards the central region of the volcano-shaped quantum ring.

The strain included total confinement potential for the hole is completely different. As we can see from the left figures of Fig. 12.16, the strain induced energy shift makes the hole have the lowest confinement potential at the place with $\rho \simeq 15$ nm inside the ring. While in the z direction, the hole has its minimum confinement potential inside the ring for the largest value of z. As a result, different from the electron, the hole has the lowest confinement potential in the top area of the volcano-shaped quantum ring. Furthermore, we can find from Fig. 12.16 that the strain decreases the band offset of the hole. The strong confinement makes the electron wave function difficult to extend in the top area of the ring with large ρ; while the result for the hole is opposite. Therefore, the electron and the hole have completely different confinement potential and localization, and the exciton may exhibit a large polarity that will give rise to an observable AB oscillation as function of an extend perpendicular magnetic field.

With the obtained strain induced confinement potential, we solved the single particle Hamiltonian in the three dimensional ring. The results for the electron and the hole energy are shown in Fig. 12.17. We can clearly see that the electron and hole spectra show similar patterns in the case without taking strain into account [Figs. 12.17(a) and (b)], but they are quite different when strain is present. The first two transitions in Figs. 12.17(a) and (b) do not occur at the same B-value, but the difference is very small, especially by comparing them to the bottom two figures.

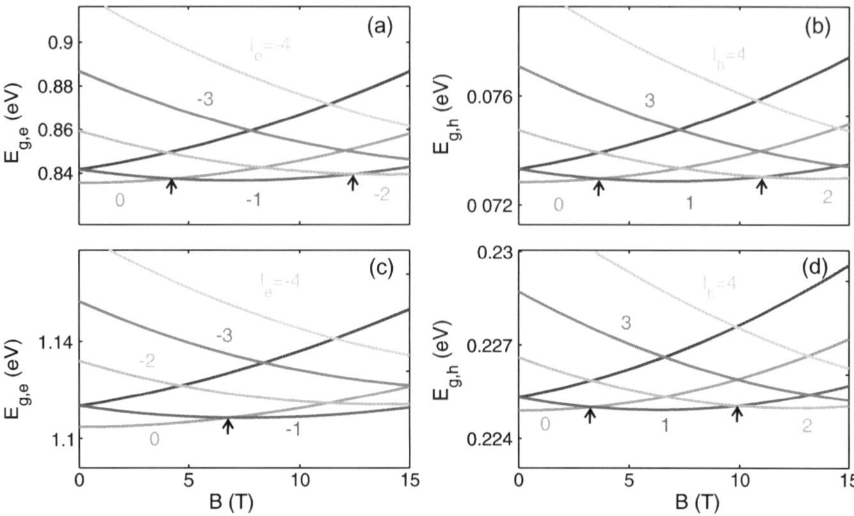

Fig. 12.17 Single particle ground state energy for different values of angular momentum as a function of magnetic field. (**a**) and (**b**) are the six lowest electron and hole energy levels without strain, while in (**c**) and (**d**) strain was taken into account. The *arrows* indicate angular momentum transitions in the ground state

As shown in Figs. 12.17(c) and (d) the first transition for the electron takes place for B around 7 T, while for the hole it takes place for B around 3 T. Moreover, we only have one transition for the electron below $B = 15$ T, the electron is in the region with very small radius while the hole prefers the top of the ring, as we forecasted. Thus, as a result of the large strain distribution difference, the wave function distributions for the electron and the hole are much more different when strain is present.

Since the ring is small and we have large confinement potentials, the Coulomb interaction energy is smaller than the kinetic energy, and the exciton will be more polarized as is verified in Fig. 12.18, where the total exciton ground state energy with and without strain are shown. The ground state energy is parabolic like with increasing magnetic field in the absence of strain, and the amplitude of the AB oscillation is small. When strain is included, the exciton ground state energy is no longer quadratic in B. The AB oscillation is more pronounced as seen from the right figure of Fig. 12.18, where the second derivative of the exciton ground state energy with respect to the magnetic field is shown. By taking the strain into account, the AB oscillation is obviously enhanced, but the period of the oscillation becomes larger, as the strain induced potential confines the electron more towards the center of the ring.

12.5.3.1 Gate Tunable Optical AB Oscillation

In the presence of top to bottom directed electric field, the electron (hole) is strongly attracted towards (repelled from) the top area of the ring, as a result, the polarity

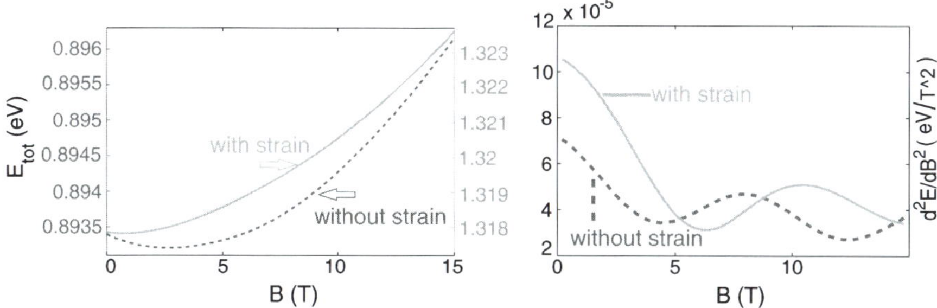

Fig. 12.18 (*Left*) Exciton ground state energy as a function of the magnetic field without strain (*dashed line*, with *y*-axis labeling on the left) and with strain (*solid line*, with *y*-axis labeling on the right). (*Right*) the second derivative of the exciton ground state energy with respect to the magnetic field

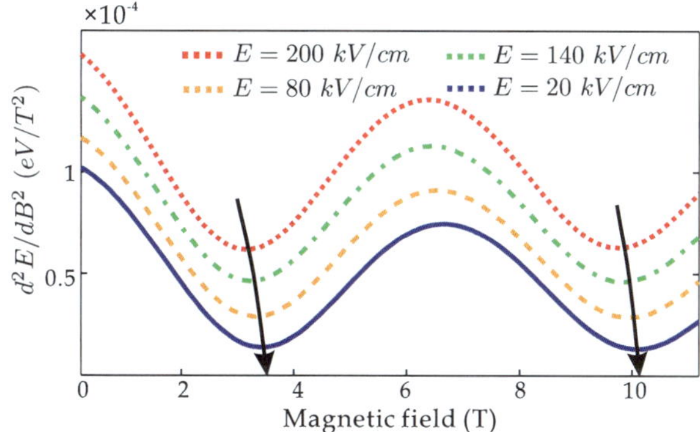

Fig. 12.19 Second derivative of the exciton energy of the state with $L = 0$. Note that the curves are shifted vertically to observe the tendency clearly

of the exciton can be reduced if the electric field is not very strong. Figure 12.19 shows the second derivative of the exciton ground state energy with the total angular momentum $L = 0$ for different values of electric field (here, like in GaAs quantum ring, we found the $L = 0$ state is always the ground state for the exciton, thus the states with other values of L are not shown). It is clearly shown that by decreasing the top to bottom directed electric field from $E = 200$ kV/cm to $E = 20$ kV/cm, the amplitude of the oscillation is slightly decreased, and the transition shifts smoothly to the region with large magnetic field. These numerical results well reproduce the shift tendency of the experimental results in Fig. 12.6(b).

In Fig. 12.20(a) we show the effective radius of the electron (hole) $\langle \rho_e \rangle$ ($\langle \rho_h \rangle$) as defined by $\int \Psi_L^*(\mathbf{r}_e, \mathbf{r}_h)\rho_{e(h)}\Psi_L(\mathbf{r}_e, \mathbf{r}_h)d\mathbf{r}_e d\mathbf{r}_h$ (which represents the electron and hole positions inside the ring). Here red solid (20 kV/cm) and red dashed curves (200 kV/cm) are for the electron, while the blue curves correspond to the hole. From Fig. 12.20(a) we know that with decreasing the vertical electric field from 200 to 20 kV/cm, the electron is attracted to the bottom area of the ring, decreasing

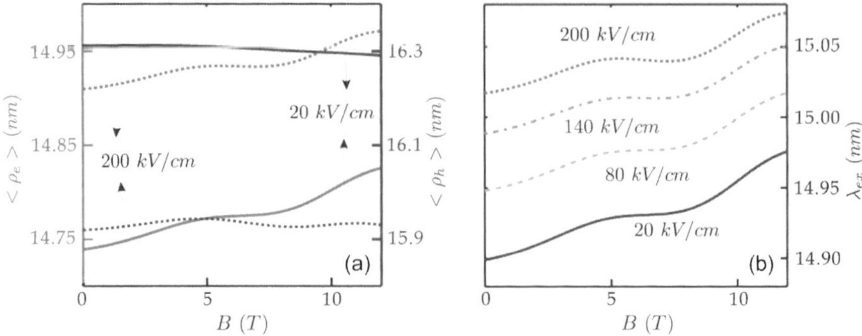

Fig. 12.20 (a) The effective radius of the electron (hole) $\langle \rho_e \rangle$ ($\langle \rho_h \rangle$) as defined by $\int \Psi_L^*(\mathbf{r}_e, \mathbf{r}_h)\rho_{e(h)}\Psi_L(\mathbf{r}_e, \mathbf{r}_h)d\mathbf{r}_e d\mathbf{r}_h$. (b) The effective radius $\lambda_{ex} = [(m_e + m_h)/(m_h/\langle \rho_e \rangle^2 + m_e/\langle \rho_h \rangle^2)]^{0.5}$, which corresponds to the exciton radius definition for a 3D system

its effective radius, while the hole is pushed to the top of the ring and its effective radius increases. From the change in $\langle \rho_e \rangle$ and $\langle \rho_h \rangle$ alone we cannot conclude that the period of the AB oscillation decreases. Theoretical study reveals that the above two transitions, within the $B = 10$ T range, of the exciton ground state come from the angular momentum transition of the main contributing single-particle basis function in the total exciton wave function, but not from the single electron or hole angular momentum. With increasing magnetic field from $B = 0$ to $B = 10$ T, the angular momentum pair (l_e, l_h) in the state $\Phi_k(\mathbf{r}_e, \mathbf{r}_h)$ which has the largest contribution to the total wave function $\Psi_L(\mathbf{r}_e, \mathbf{r}_h)$ changes from $(0, 0)$, to $(-1, 1)$, and then from $(-1, 1)$ to $(-2, 2)$. The period of the oscillation is not only related to the effective radius of the electron and the hole but also to their effective masses.

Figure 12.20(b) shows the exciton radius which is defined, by $\lambda_{ex} = [(m_e + m_h)/(m_h/\langle \rho_e \rangle^2 + m_e/\langle \rho_h \rangle^2)]^{0.5}$, for different values of electric field. A close investigation shows that the magnetic field at which the first transition takes place, as shown in Fig. 12.19, is proportional to $\hbar/e\lambda_{ex}^2$. Because the effective mass of the hole is much larger than that of the electron (also because the electron and the hole radii change within the same order), λ_{ex} should have a similar behavior as the electron effective radius $\langle \rho_e \rangle$. This is clearly observed in Fig. 12.20 and λ_{ex} decreases monotonously with decreasing electric field. As a result, the first transition takes place at a larger magnetic field when we decrease the electric field.

The oscillator strength for the neutral exciton in the current system is also calculated, in order to study the exciton recombination and the strength of the transition. The results for the oscillator strength of the ground state for different values of electric field are shown in Fig. 12.21(a). (As in the previous GaAs ring, here the overlap integral is used instead of the oscillator strength.) We find that the oscillator strength always has a step-like increase tendency, and its value does not change monotonously with increasing electric field. When the electric field increases from $E = 20$ kV/cm to $E = 80$ kV/cm, the overlap integral increases, manifesting an increasing wave function overlap and a decreasing polarity of the neutral exciton. If

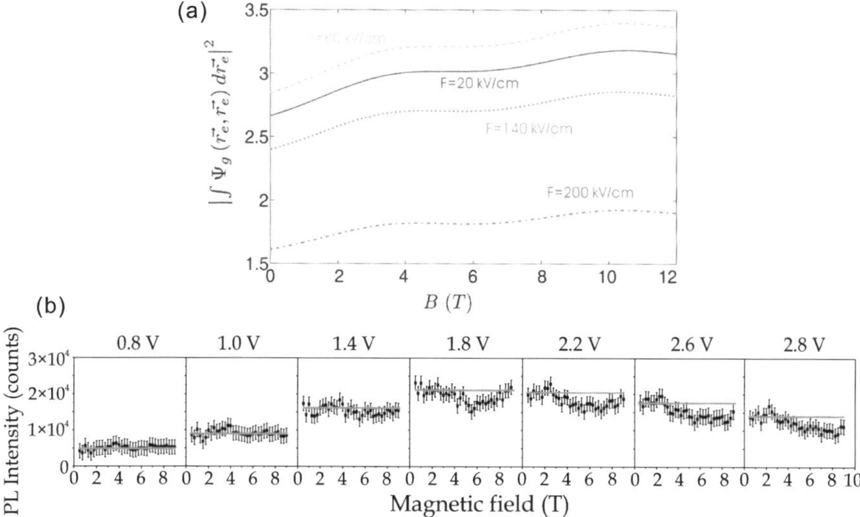

Fig. 12.21 (**a**) Overlap integral of the neutral exciton in the presence of four different values of electric field. (**b**) The experimental result for the PL intensity as a function of magnetic field and for several different electric field. Here *solid lines* mark the general shift of the PL intensity with increasing electric field [16]

the electric field continues to increase, the overlap integral starts to have a decreasing tendency instead. The reason is that in the absence of electric field, the hole has a larger effective radius than the electron. The electron and the hole move towards each other, switch their position and then move apart from each other (in the z direction) by increasing the electric field. These results can be verified from the experimental data for the PL intensity, as we show in Fig. 12.21(b). The PL intensity of the neutral exciton increases at first by increasing the electric field, as a result of a larger wave function overlap; after the PL intensity reaches its maximum (when the electric field is around 2 V), it starts to have a decreasing tendency as the electron and the hole already switches their position and move in the opposite direction with further increase of the electric field.

For a strained $In_{1-x}Ga_xAs$ quantum ring we have a similar tunable AB effect as previously found for GaAs quantum rings. The period of the AB oscillation as a function of applied electric field is shown in Fig. 12.22, the dependence of the AB amplitude (we define it by the difference of the value of the second derivative of the total energy at $B = 0$ from the value at the magnetic field where the first transition takes place) on the electric field is also plotted, but with the y-axis labeling on the right. The period of the AB oscillation, as we found in the case of a GaAs/AlGaAs quantum ring, decreases when we increase the top to bottom directed electric field. The electron, which determines the period of the exciton AB oscillation, prefers to stay close to the center of the ring because the strain included total potential is small there. When we apply a bottom to top directed electric field (negative value of electric field here), the electron is pushed much closer to the center with a smaller

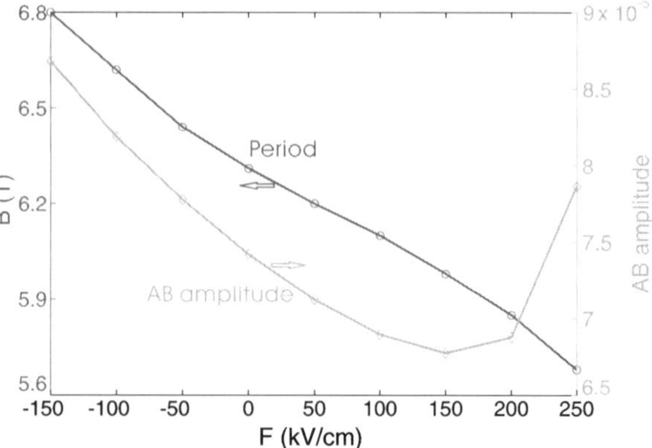

Fig. 12.22 *Solid line marked with circle symbols* is the period of the AB oscillation as a function of applied electric field, while *solid line marked with prism* corresponds to the AB amplitude (*y*-axis labeling on the right)

effective radius, thus the period increases; and as the electric field pushes the hole in the opposite direction, the enlarged polarity of the exciton should increase the amplitude of the AB oscillation, which can be seen from Fig. 12.22. If we change the direction of the electric field, the period will decrease and the AB amplitude will weaken, since the electron and the hole are pushed towards each other. However, the AB amplitude increases when the electric field is so strong (here, when larger than 150 kV/cm) that the electron and the hole switch their position: the electron attains a larger effective radius than the hole. The bottom to top directed electric field has a larger effect on the AB oscillation, because the strain induced potential counteracts (enhance) the influence of the top (bottom) to bottom (top) directed electric field. When comparing with Fig. 12.14, the effect of the electric field here is much smaller than in previous unstrained quantum ring as the height of the ring is smaller.

Figure 12.22 shows similar results as Fig. 12.5 in Ref. [16]. By decreasing the top to bottom directed electric field, the period of the AB oscillation increases. We notice that the second period of the AB oscillation increases more slowly, and does not take place for a magnetic field value three times larger than the first one, which is very different from the case of an ideal one dimensional ring [6].

12.6 Conclusion

In conclusion, we studied the optical properties of the single neutral excitons confined in a self-assembled In(Ga)As/GaAs QR structure. These novel QRs are fabricated by molecular beam epitaxy combined with *in situ* AsBr$_3$ etching. The morphology of In(Ga)As nanostructures embedded in GaAs matrix is unveiled by selective wet chemical etching combined with AFM. The photoluminescence of single QRs reveal oscillations in both PL energy and intensity under an applied magnetic field. We also show that the oscillations can be tuned by applying a vertical electric field which modifies the electron and hole effective radii. In order to explain the experimental results we calculate the single particle energy of a semiconductor quantum ring and find that the AB oscillations of the single particle energies can be tuned by an applied perpendicular electric field when the ring dimensions satisfy some constraint conditions. In addition, the strain inside the self-assembled quantum ring changes the confinement potential of the electron and the hole, and as a result, the polarity of the exciton is increased and the AB effect is enhanced.

Acknowledgement   We acknowledge N. Akopian, U. Perinetti, A. Govorov, C.C. Bof Bufon, C. Deneke, V. Fomin, A. Govorov, and S. Kiravittaya for their help and fruitful discussions.

References

 1. Y. Aharonov, D. Bohm, Phys. Rev. **115**, 485 (1959)
 2. A.G. Aronov, Y. Sharvin, Rev. Mod. Phys. **59**, 755 (1987)
 3. S. Zaric et al., Science **304**, 1129 (2004)
 4. A. Fuhrer et al., Nature **413**, 822 (2001)
 5. A. Chaplik, JETP Lett. **62**, 900 (1995)
 6. A.O. Govorov, S.E. Ulloa, K. Karrai, R.J. Warburton, Phys. Rev. B **66**, 081309 (2002)
 7. L.G.G.V. Dias da Silva, S.E. Ulloa, T.V. Shahbazyan, Phys. Rev. B **72**, 125327 (2005)
 8. R.A. Römer, M.E. Raikh, Phys. Rev. B **62**, 7045 (2000)
 9. M. Grochol, F. Grosse, R. Zimmermann, Phys. Rev. B **74**, 115416 (2006)
10. A.V. Kalameitsev, A.O. Govorov, V.M. Kovalev, JETP Lett. **68**, 669 (1998)
11. A.M. Fischer, J. Vivaldo, L. Campo, M.E. Portnoi, R.A. Römer, Phys. Rev. Lett. **102**, 096405 (2009)
12. E. Ribeiro, A.O. Govorov, W. Carvalho, G. Medeiros-Ribeiro, Phys. Rev. Lett. **92**, 126402 (2004)
13. I.R. Sellers et al., Phys. Rev. Lett. **100**, 136405 (2008)
14. I.R. Sellers, J. Nanoelectron. Optoelectron. **6**, 4 (2010)
15. M.D. Teodoro et al., Phys. Rev. Lett. **104**, 086401 (2010)
16. F. Ding et al., Phys. Rev. B **82**, 075309 (2010)
17. F. Ding et al., Appl. Phys. Lett. **90**, 173104 (2007)
18. A. Lorke et al., Mater. Sci. Eng. B **88**, 225 (2002)
19. C. Zhao et al., Appl. Phys. Lett. **92**, 063122 (2008)
20. S. Sanguinetti, J. Nanoelectron. Optoelectron. **6**, 34 (2011)
21. C. Heyn, J. Nanoelectron. Optoelectron. **6**, 62 (2010)
22. D.G. Hill, K.L. Lear, J.S. Harris Jr., J. Electrochem. Soc. **137**, 2912 (1990)
23. S.-J. Paik et al., Jpn. J. Appl. Phys. **42**, 326 (2003)
24. Z.M. Wang, L. Zhang, K. Holmes, G.J. Salamo, Appl. Phys. Lett. **86**, 143106 (2005)
25. B.N. Zvonkov et al., Nanotechnology **11**, 221 (2000)
26. L. Wang et al., New J. Phys. **10**, 045010 (2008)
27. T. Kuroda et al., Phys. Status Solidi (b) **246**, 861 (2009)
28. R.J. Warburton et al., Nature **405**, 926 (2000)
29. B.D. Gerardot et al., Appl. Phys. Lett. **90**, 041101 (2007)
30. M. Korkusiński, P. Hawrylak, M. Bayer, Phys. Status Solidi (b) **234**, 273 (2002)
31. F. Ding et al., Phys. Rev. Lett. **104**, 067405 (2010)
32. M.-F. Tsai et al., Phys. Rev. Lett. **101**, 267402 (2008)
33. F. Ding et al., J. Nanoelectron. Optoelectron. **6**, 51 (2010)
34. B. Li, F.M. Peeters, Phys. Rev. B **83**, 115448 (2011)
35. M. Levinshtein, S. Rumyantsev, M. Shur, *Handbook Series on Semiconductor Parameters* (World Scientific, Singapore, 1999). Chap. 3
36. G. Bryant, Phys. Rev. B **37**, 8763 (1988)

37. M. Tadić, F.M. Peeters, J. Phys. Condens. Matter **16**, 8633 (2004)
38. C.H. Henry, K. Nassau, Phys. Rev. B **1**, 1628 (1970)
39. W. Dreyer, W.H. Müller, J. Olschewski, Acta Mech. **136**, 171 (1999)
40. K.L. Janssens, B. Partoens, F.M. Peeters, Phys. Rev. B **66**, 075314 (2002)
41. C. Pryor, Phys. Rev. B **57**, 7190 (1998)
42. L. Börnstein, *Numerical Data and Functional Relationships in Science and Technology*, Group III, vol. 17a–b (Springer, New York, 1982)
43. B. Jusserand, M. Cardona, *Light Scattering in Solid V: Superlattices and Other Microstructures*. Topics in Applied Physics, vol. 66 (Springer, Berlin, 1989)
44. C.G. Van de Walle, Phys. Rev. B **39**, 1871 (1989)
45. L.W. Wang, J. Kim, A. Zunger, Phys. Rev. B **59**, 5678 (1999)

Part III
Theory

Chapter 13
Strained Quantum Rings

Pilkyung Moon, Euijoon Yoon, Won Jun Choi, JaeDong Lee,
and Jean-Pierre Leburton

Abstract Electronic structures of quantum rings strongly depend on the strain profiles caused by the material environment around the ring. We will investigate the
strain distributions and electronic structures of quantum rings capped by a support
material of which lattice constant is smaller than those of the substrate and active
material, and compare the results with a conventional quantum ring. The support
material considerably weakens the longitudinal strains and biaxial strain of quantum rings as well as the hole confinement potentials. The unique band alignment of
the structure enables the coexistence of type-I and type-II band alignments.

13.1 Introduction

In semiconductor heterostructures charge carriers exhibit electronic structures and
optical properties different from those of bulk materials because of the reduction

P. Moon (✉)
Department of Physics, Tohoku University, Sendai, Japan
e-mail: pilkyung.moon@gmail.com

E. Yoon
Department of Materials Science and Engineering, Seoul National University, Seoul, Korea
e-mail: eyoon@snu.ac.kr

W.J. Choi
Center for OptoElectronic Convergence System, Korea Institute of Science and Technology,
Seoul, Korea
e-mail: wjchoi@kist.re.kr

J.D. Lee
Department of Emerging Materials Science, Daegu Gyeongbuk Institute of Science and
Technology, Daegu, Korea
e-mail: jdlee@dgist.ac.kr

J.-P. Leburton
Department of Electrical and Computer Engineering, University of Illinois at Urbana-Champaign,
Urbana, IL, USA
e-mail: jleburto@illinois.edu

V.M. Fomin (ed.), *Physics of Quantum Rings*, NanoScience and Technology,
DOI 10.1007/978-3-642-39197-2_13, © Springer-Verlag Berlin Heidelberg 2014

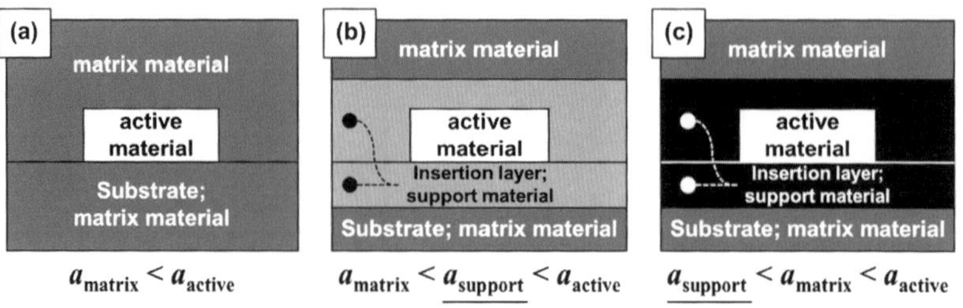

$$a_{\mathrm{matrix}} < a_{\mathrm{active}} \qquad a_{\mathrm{matrix}} < \underline{a_{\mathrm{support}}} < a_{\mathrm{active}} \qquad \underline{a_{\mathrm{support}}} < a_{\mathrm{matrix}} < a_{\mathrm{active}}$$

Fig. 13.1 Schematic geometry of (**a**) the conventional heterostructure composed of two different materials (active material and matrix material) and (**b**), (**c**) the structures with insertion layers above and/or below the active material. a_{matrix}, a_{active}, and a_{support} denote the lattice constants of the matrix material, active material, and support material, respectively

of the spatial dimensions of the active region to a few nanometers, which is well suited for optoelectronic applications. In these nanostructures spatial confinement of charge carriers (electrons and holes) is achieved by pseudomorphically growing two or more materials with different band gaps on top of one another. In this context, recent advances in the growth of self-assembled systems has enabled the fabrication of non-flat nanostructures, such as zero-dimensional quantum dots (QDs) and quantum rings (QRs), in which a dot or ring of one material is coherently strained in a matrix of host material [1].

Figure 13.1(a) shows a conventional semiconductor nanostructure composing of two materials with different band gaps. The material with a smaller band gap (active material) is embedded in the material with a larger band gap (matrix, barrier material) and forms a potential well for charge carriers. The energy levels of charge carriers, which determine the optical emission spectra of nanostructures, are controlled by the band edge energies E_c and E_v of the conduction and valence bands that vary with material composition and lattice distortion (strain), but also by the quantum confinement energies, which are determined by the nanostructure dimensionality. However, in most nanostructures grown by self-assembling techniques, the degree of lattice distortion, but also the sizes of the nanostructures are to some extent intrinsic functions of the materials. Thus, although varying the composition, size, and shape of the nanostructures are widely utilized to obtain specific physical properties, there are a limited number of material combinations available when dealing with only two materials [Fig. 13.1(a)]. And it is for instance, still a challenge to access to optical fiber communication wavelengths of 1.55 μm, since even InAs/InP QDs, which give the closest wavelength, exhibit emission wavelengths above 1.6 μm at room temperature [2].

Therefore, nanostructures like Fig. 13.1(b), which modify the environment in the vicinity of the active region by introducing a third material (support material), have received great attention during the past decade. Hence, several groups [3–8] have reported that a wide range of optical wavelengths can be obtained by growing a layer of support material right above and/or below the nanostructure containing the active region. Moreover, such structure can also influence the optical emission efficiency by increasing the carrier capture rate, thereby enhancing electron-hole recombination [9].

In order to obtain stable and controllable nanostructures during the growth process, a support material made of layered structures is desirable, especially when the active material region is three-dimensional. The layer of support material is usually called the 'insertion layer', since it is often placed between the active material and the matrix material. In order to minimize the strain energy of the structure by reducing the difference between the lattice constants, research on the insertion layer has focused on support materials with a lattice constant (a_{support}) that lies between that of the active material (a_{active}) and that of the matrix material (a_{matrix}); $a_{\text{matrix}} < a_{\text{support}} < a_{\text{active}}$. In this particular case, the strains of both the active and support materials have the same sign. And, in most cases, charge carriers prefer to reside in the active region rather than in the insertion layer, since the band offset of the support material lies between those of the active material and matrix material.

Recently, semiconductor structures for which the lattice constant of the reference matrix is comprised between the one of a support material and the one of the active material i.e. $a_{\text{support}} < a_{\text{matrix}} < a_{\text{active}}$ [Fig. 13.1(c)], have been proposed [2, 10, 11]. Model calculations have shown that the support material can effectively confine one type of charge carriers (i.e. electrons or holes), as well as acting as a second active material for the same type of carriers [12, 13]. Moreover, the structural configuration also allows the control of the ground state among the different electronic bands in both the active and the support materials, by either changing the thickness of the insertion layer, or by applying an external electric field across the structure [14]. Since energy bands of different characters e.g. light (LH) and heavy hole (HH) bands, even in the same material, exhibit different electronic properties, nanostructures similar as Fig. 13.1(c) are well suited to investigate combinations of band offsets and various types of energy bands for applications in opto-electronic devices. This kind of nanostructures can be fabricated with two different designs according to the lattice constant of the active material, which can be either larger, or smaller than that of the matrix material, if resonance conditions between the states which reside in the active and the support materials are satisfied.

In this book chapter, we describe the electronic properties of quantum rings (QR) obtained by the two growth approaches i.e. with the active (support) materials larger (smaller) lattice constant than the matrix materials. Specifically, we will show how the quantum confinement energies of charge carriers, as well as the characters (LH or HH) of the energy states and their spatial distribution in QRs can be tailored by strained-layer engineering i.e. by using a tensile strained insertion layer on compressively strained layers. This technique is made possible by the widely available semiconductor material combinations and the general tendency of III–V materials with a larger lattice constant to have a smaller band gap, so that they can be used as active materials.

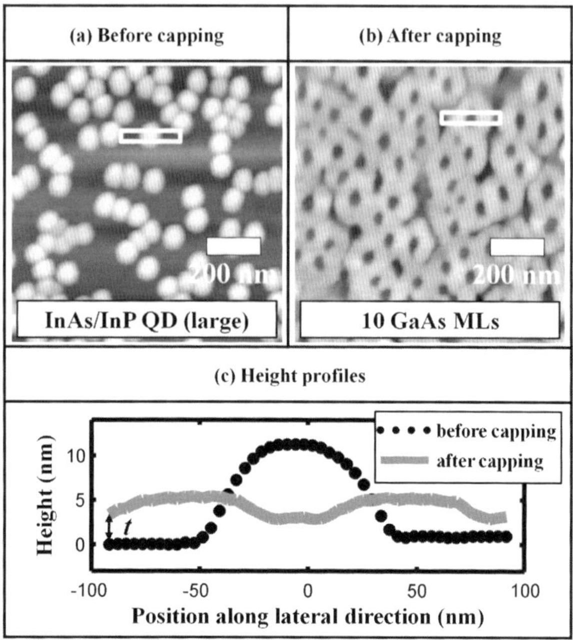

Fig. 13.2 Morphologies of InAs/InP islands (**a**) before and (**b**) after the capping of 10 GaAs MLs obtained by atomic force microscopy. (**c**) Height profiles along the lines in Figs. 13.3(a) and 13.3(b). '*t*' indicates the offset determined by the amount of GaAs deposited. Reprinted figures (**a**) and (**b**) with permission from [12]. Copyright 2009, WILEY-VCH Verlag GmbH & Co. KGaA, Weinheim

13.2 Geometry and Method of Calculation

13.2.1 Model Structures

A QR is a nanometer size island with a geometric hole at its center. It is mainly fabricated by capping a QD with a layer thinner than the dot height, and subsequent growth interruption. There have been many researches on QRs with various combinations of materials, such as InAs/GaAs [15–17], InAs/InP [18], GaAs/AlAs [19] and SiGe/Si [20], however, most of the experimental and theoretical studies have been concentrated on QRs capped by the same material as the substrate [Fig. 13.1(a)]. Recently, Moon et al. [13] reported that QR can also be fabricated by capping a QD with a material different from the substrate [Fig. 13.1(c)]. Figures 13.2(a) and 13.2(b) show the morphological changes of InAs/InP islands before and after the deposition of the thin (10 MLs) GaAs layer, respectively. The geometry of the island before the deposition is a lens-shaped dot with an average width of 88 nm and height of 11.3 nm. After the deposition of the GaAs layer, each dot evolves to a ring-shaped island with a geometric hole at its center. The mean depth of the holes measured from the top of the surrounding rings is 2.36 nm, and the areal density of the holes coincides with that of the dots in Fig. 13.2(a). Figure 13.2(c) shows the height profiles along the lines shown in Figs. 13.2(a) and 13.2(b), where the offset (*t*) reflects the thickness of the GaAs layer deposited. The AFM data indicate the QR evolution that a large portion of the InAs at the center of the dot is removed or redistributed outward by the deposition of the GaAs layer.

Fig. 13.3 Morphologies of (**a**) InAs/InP QD and (**b**) InAs/In$_{0.53}$Ga$_{0.47}$As QD before the growth of an insertion layer, (**c**), (**d**), and (**e**): (**c**) and (**d**) QR grown by capping (**a**) with an insertion layer of 10 GaAs MLs and 10 In$_{0.25}$Ga$_{0.75}$P MLs, respectively, (**e**) QR grown by capping (**b**) with an insertion layer of 10 GaAs MLs. Reprinted figures (**a**) and (**c**) with permission from [11]. Copyright 2009, WILEY-VCH Verlag GmbH & Co. KGaA, Weinheim. Reprinted figures (**b**) and (**e**) with permission from [12]. Copyright 2005, American Institute of Physics

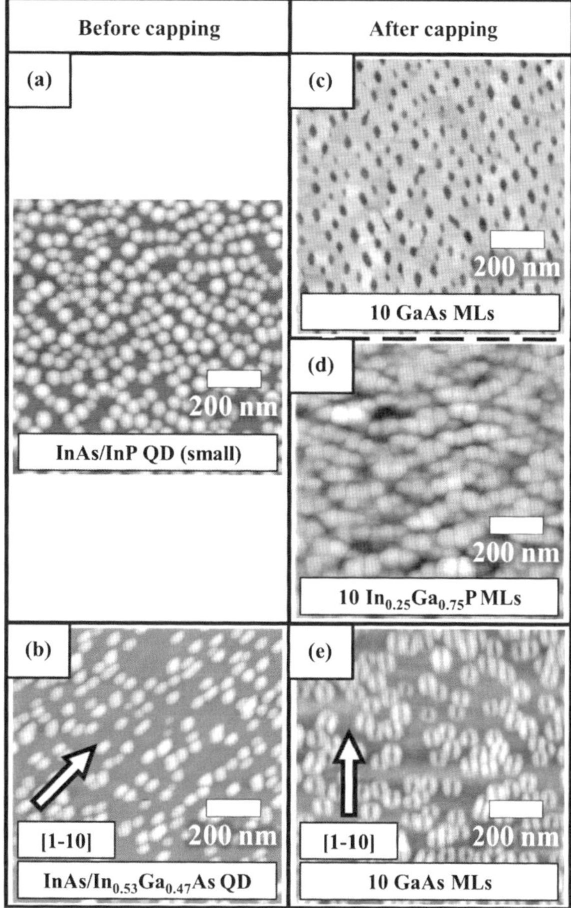

Figure 13.3 shows the morphological changes of InAs islands by the deposition of various materials different from the substrate. Figures 13.3(a) and 13.3(b) show the morphologies (AFM) of InAs QDs grown on InP and In$_{0.53}$Ga$_{0.47}$As substrates, respectively. The size of the dots in Fig. 13.3(a) is different from that in Fig. 13.2(a), since the growth conditions are changed. Figures 13.3(c), 13.3(d), and 13.3(e) show the dramatic morphological changes occurred by capping the dots with a thin (10 MLs) GaAs or In$_{0.25}$Ga$_{0.75}$P layer. Each dot evolves to a ring-shaped island with a hole [Fig. 13.3(c)] or a trench [Fig. 13.3(e)] at its center, or a paired dot-shaped island [Fig. 13.3(d)]. The areal density of the holes and the trenches coincides with that of the dots in Figs. 13.3(a) and 13.3(b).

In this chapter, we will investigate the electronic structures and band characters of an InAs/InP QR by varying the thickness of a GaAs insertion layer. Figure 13.4 shows the (110) and (1-10) planes of the model structure. The lines with and without circles represent the height profiles of the InAs ring and GaAs insertion layer, respectively, where the height is measured from the basal plane of the ring. We used

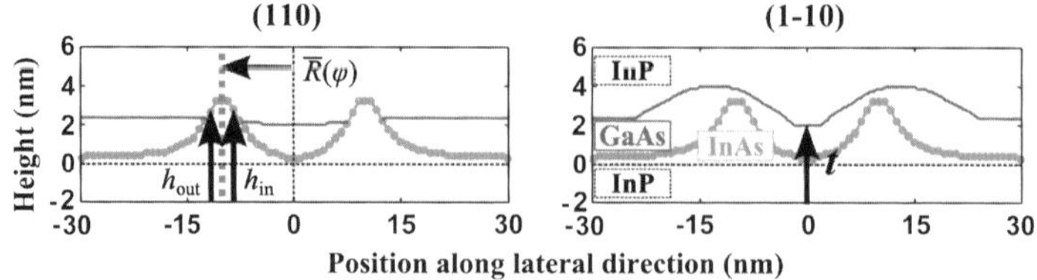

Fig. 13.4 (110) and (1-10) cross sections of the model structure investigated in this chapter. *Line with and without circles* represent the height profiles of the InAs ring and GaAs insertion layer, respectively, where the height is measured from the basal plane of the ring. See text for the notations

the height profile proposed by Offermans et al. [21] and Fomin et al. [22] (and also see Chap. 4) with a little modification.

$$h_{\text{in}}(\rho, \varphi) = h_0 + \frac{[\overline{h}_M(\varphi) - h_0]\{1 - [\rho/\overline{R}(\varphi) - 1]^2\}}{\{[\rho - \overline{R}(\varphi)]/\zeta(\varphi)\}^2 + 1}, \quad \rho \leq \overline{R}(\varphi) \quad (13.1)$$

$$h_{\text{out}}(\rho, \varphi) = h_\infty + \frac{\overline{h}_M(\varphi) - h_\infty}{\{[\rho - \overline{R}(\varphi)]/\overline{\gamma}_{\text{out}}(\varphi)\}^2 + 1}, \quad \rho > \overline{R}(\varphi) \quad (13.2)$$

$$\overline{h}_M(\varphi) = h_M(1 + \xi_h \cos 2\varphi) \quad (13.3)$$

$$\overline{R}(\varphi) = R \times (1 + \xi_R \cos 2\varphi) \quad (13.4)$$

$$\zeta(\varphi) = \sqrt{\frac{\overline{R}(\varphi)^2 \overline{\gamma}_{\text{in}}(\varphi)^2}{\overline{R}(\varphi)^2 - 2\overline{\gamma}_{\text{in}}(\varphi)^2}} \quad (13.5)$$

$$\overline{\gamma}_{\text{in}}(\varphi) = \gamma_{\text{in}} \times (1 + \xi_{\text{in}} \cos 2\varphi) \quad (13.6)$$

$$\overline{\gamma}_{\text{out}}(\varphi) = \gamma_{\text{out}} \times (1 + \xi_{\text{out}} \cos 2\varphi) \quad (13.7)$$

Here, h_{in} and h_{out} correspond to the inner and outer heights of the InAs rim, ρ to the radial coordinate, $\bar{R}$ to the radial coordinate of the InAs ring at the peak, and R to the average of $\bar{R}$, φ to the angular coordinate measured in the [100] direction, $h_0(h_\infty)$ to the thickness of InAs layer at the center of (far away from) the ring, $\bar{h}_M(\varphi)$ to the height of the rim, and h_M to the average of $\bar{h}_M(\varphi)$. The parameters $\overline{\gamma}_{\text{in}}(\varphi)$ and $\overline{\gamma}_{\text{out}}(\varphi)$ determine the points of the half maximum of the rim height:

$$h_{\text{in}}(\rho, \varphi) = \frac{h_0 + \overline{h}_M(\varphi)}{2}, \quad \text{at } \rho = \bar{R}(\varphi) - \bar{\gamma}_{\text{in}}(\varphi), \quad (13.8)$$

$$h_{\text{out}}(\rho, \varphi) = \frac{h_\infty + \overline{h}_M(\varphi)}{2}, \quad \text{at } \rho = \bar{R}(\varphi) + \bar{\gamma}_{\text{out}}(\varphi), \quad (13.9)$$

and γ_{in} and γ_{out} are the average of $\overline{\gamma}_{\text{in}}(\varphi)$ and $\overline{\gamma}_{\text{out}}(\varphi)$, respectively. ξ_h, ξ_R, ξ_{in}, and ξ_{out} describe the anisotropy of the ring. We used $h_0 = h_\infty = 0.29$ nm, $h_M = 3.48$ nm, $\gamma_{\text{in}} = \gamma_{\text{out}} = 3.00$ nm, and assumed an isotropic ring by setting ξ_h, ξ_R, ξ_{in}, and ξ_{out} to zero. To investigate the effects of the insertion layer thickness t on the electronic structures of QR, we performed numerical simulations on the systems with varying t from 0 nm to 4.69 nm, while keeping the geometry of the InAs ring. Although the height profiles of the GaAs insertion layer in an

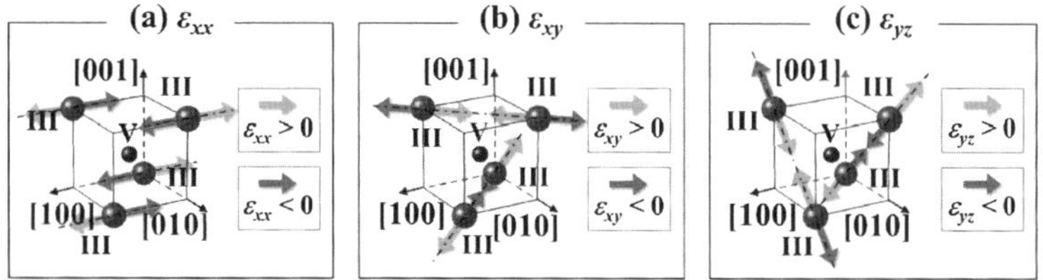

Fig. 13.5 Schematic view of the distortions of atomic coordinates of (**a**) the longitudinal strain (ε_{xx}) and (**b**) the in-plane (ε_{xy}) and (**c**) vertical-plane (ε_{yz}) shear strains. 'III' and 'V' represent group III (cation) and V (anion) atoms, respectively. Arrows with light and dark colors describe the distortions of atomic coordinates for positive and negative strains, respectively

InAs/InP system does not exhibit a geometric anisotropy [Figs. 13.2(b) and 13.3(c)], we used an anisotropic height profile, as that of the InAs/In$_{0.53}$Ga$_{0.47}$As system [Fig. 13.3(e), [13]], to investigate the effects of the anisotropy on the strain and electronic structures of the ring: the GaAs layer exhibits mounds on both sides of the [110] corners [*i.e.*, (1-10) cross section], as can be seen from Fig. 13.3(e), of which height is proportional to t.

13.2.2 Equilibrium Atomic Positions

Longitudinal strain ($\varepsilon_{xx}, \varepsilon_{yy}, \varepsilon_{zz}$) and shear strain ($\varepsilon_{xy}, \varepsilon_{yz}, \varepsilon_{zx}$) is the measure of lattice distortion along $\langle 100 \rangle$ and $\langle 110 \rangle$ directions, respectively. Here, x and y represent the [100] and [010] axes, respectively, and we chose the z axis ([001]) as the growth direction of the nanostructures. Figure 13.5 illustrates the longitudinal strain (ε_{xx}) and the in-plane (ε_{xy}) and vertical-plane (ε_{yz}) shear strains. Atoms labeled 'III' and 'V' represent group III (cation) and group V (anion) atoms, respectively. Orange and blue arrows describe the distortions of atomic coordinates for positive and negative strains, respectively. Positive longitudinal strain reflects the dilation of the atomic spacing along the $\langle 100 \rangle$ direction, while positive in-plane (vertical-plane) shear distortion represents an increase in the bond angle along the [110] ([011]) direction, and/or the decrease in the bond angle along the [1-10] ([01-1]) direction. The opposite is true for negative strains.

Accurate strain profiles are prerequisites for the electronic structure calculations of nanostructures, since the electronic structures strongly depend on the lattice distortions of constituent materials. In highly strained semiconductor materials, the amount of the energy shift by lattice distortion is even comparable to the quantum confinement energies and the band gap of the materials. In this work, we calculated the strain profiles employing a valence force field method with Keating's potential [23, 24] and Martin's parameters [24]. The benefits of the method are that it considers both the anharmonic effects and actual atomic level symmetry (C_{2v}) of the zincblende structure.

We imposed a fixed boundary condition on the basal plane of the supercell, a free-standing boundary condition on the top, and periodic boundary conditions on the four sides. Owing to the long-range behavior of the lattice distortion [25], we used a sufficiently large supercell, which is composed of $100 \times 100 \times 180$ unit cells, to relax the atomic positions and avoid undesirable interactions with periodic images in the lateral directions. We calculate the equilibrium atomic positions by minimizing the elastic energy with a conjugate gradient method and obtain the local strain tensor at each atomic site from the coordinates of its four nearest-neighbor atoms [26].

13.2.3 Piezoelectric Potential

The piezoelectric potential is induced by the charge separation in response to a crystal deformation. Only the shear component of strains contributes to the piezoelectricity. We calculate the potential V_{Piezo} by solving Poisson's equation,

$$\nabla^2 V_{\text{Piezo}}(r) = -\frac{1}{\varepsilon(r)} \rho_P(r). \tag{13.10}$$

Here, $\varepsilon(r)$ is the dielectric constant of the material at position r and $\rho_p(r)$ is the piezoelectric charge density,

$$\rho_P(r) = \text{div}\left(\sum_{j,k} e_{ijk}(r) \varepsilon_{jk}(r) \right) \quad (i, j, k = x, y, z), \tag{13.11}$$

induced by the shear deformation $\varepsilon_{jk}(r)$, and $e_{ijk}(r)$ is the first-order (to the shear strain tensors) piezoelectric constant.

13.2.4 Electronic Structures

In semiconductor materials, the band gap separate the conduction bands derived from s-type atomic orbitals and the valence bands derived from p-type atomic orbitals. An electron excited across the band gap leaves an empty state in otherwise full valence bands. The empty state is often treated as if it is a charged particle ('hole') whose charge is the same magnitude and opposite sign as the electron charge. A spin-orbit interaction splits the sixfold degeneracy of the p-type orbitals ($l = 1$ with a spin $s = 1/2$) at the zone center into fourfold degenerate heavy-hole (HH) plus light-hole (LH) bands with total angular momentum $J = 3/2$, and doubly degenerate split-off (SO) bands with $J = 1/2$. Together with the doubly degenerate electron bands (conduction band; CB), the six hole bands comprise the eight basis bands near the band gap of semiconductor materials. We calculated the electronic structures of QR by solving a k · p Hamiltonian with the eight basis bands [27],

$$H = H_{\text{k}\cdot\text{p}} + H_{\text{strain}} + H_{\text{spin-orbit}} - eV_{\text{piezo}}, \tag{13.12}$$

in the framework of Burt and Foreman's exact envelope-function model [28–30]. The eight-band k · p method is widely used to describe the electronic and optical properties of zincblende and wurtzite materials, especially for structures that are large enough to be described by the wave functions at the Γ point ($k = 0$) [31–34]. The effects of strain and spin-orbit coupling are taken into account by following Bahder's model [35]. All the parameters used in the calculation are taken from Vurgaftman et al. [36], except the Luttinger parameters. To avoid the spurious solutions, we use the parameters by Saïdi et al. [37].

13.3 Results and Discussion

13.3.1 Strain Modified Bulk Band

Most of the three-dimensional heterostructures grown by epitaxial methods have very small aspect ratio (height/base length). Thus, the investigation of the effects of lattice distortion in a flat two-dimensional layer often provides useful information on the study of the effects of strains in three-dimensional structures.

Hydrostatic (isotropic) strain,

$$H = \varepsilon_{xx} + \varepsilon_{yy} + \varepsilon_{zz}, \tag{13.13}$$

in semiconductor materials gives rise to a change in the electronic structure in particular modification of the CB edge E_c and the average of valence bands (VBs) E_v by,

$$\Delta E_c = a_c H, \tag{13.14}$$

$$\Delta E_v = -a_v H, \tag{13.15}$$

whereas biaxial (anisotropic) strain,

$$B = \varepsilon_{zz} - 0.5 \times (\varepsilon_{xx} + \varepsilon_{yy}), \tag{13.16}$$

lifts the degeneracy between HH and LH by [38]

$$\Delta E_{HH\text{-}LH} = -\frac{1}{2}\left(3bB - \Delta + \sqrt{\Delta^2 + 2\Delta bB + 9b^2 B^2}\right). \tag{13.17}$$

Here, a_c and a_v indicate the CB and VB deformation potentials, respectively, b the uniaxial deformation potential for (001) strain, and Δ the spin-orbit splitting energy [26]. Note that the definition and sign of the biaxial strain (13.16) vary with papers. This simplified model, in which both CB and HH are decoupled from other bands, is not used in our actual calculation of the electronic structures, but introduced only to provide an illustrative view of the strain effects on the band edge potentials. For sufficiently small Δ, HH and LH band edge potentials shift by

$$\Delta E_{HH} = -bB, \tag{13.18}$$

$$\Delta E_{LH} = bB, \tag{13.19}$$

Fig. 13.6 Strain modified
bulk bands of InAs and GaAs
pseudomorphically grown on
the substrate with a lattice
constant ranging from that of
GaAs to that of InAs. *Lines*
correspond to the eight-bands
near the fundamental gap:
(CB: electron, HH: heavy
hole, LH: light hole, and SO:
spin-orbit split-off band).
VBM of unstrained InAs is
taken as a reference energy.
Horizontal (*dotted*) *line*
represents the VBM of
unstrained InP

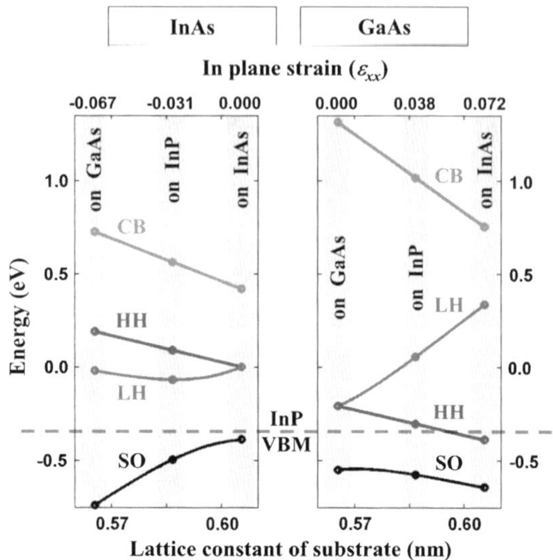

respectively. As a result, either band becomes the valence band maximum (VBM)
depending on the sign of B, and the energy difference between CB and VBM is
reduced by $|bB|$.

Figures 13.6(a) and 13.6(b) show the strain-modified band structures of InAs and
GaAs pseudomorphically grown on the substrate whose lattice constant ranges from
that of bulk GaAs (0.565 nm) to that of InAs (0.606 nm). The vertical yellow bars
indicate the points where the lattice constant of the substrate corresponds to that of
unstrained GaAs, InP, and InAs. Each band is doubly degenerate by spin, and VBM
of unstrained InAs is taken as a reference energy.

A two-dimensional layer, and also a three-dimensional structure with a small
aspect ratio, pseudomorphically grown on a substrate is strained in plane (ε_{xx} and
ε_{yy}) to adjust its own lattice constant to that of the substrate, and also deformed
along the growth direction (ε_{zz}) with a strain of opposite sign (Poisson effect).

As the lattice constant of the substrate increases, the hydrostatic strain of the
pseudomorphically grown material increases from compression ($H < 0$) to tension
($H > 0$), and the biaxial strain decreases from positive to negative values. Thus,
with increasing lattice constant of the substrate, CB shifts downward [$a_c < 0$ and
(13.14)] and the average of VBs moves upward [$a_v < 0$ and (13.15)]. However, HH
decreases monotonically, since the shift by the biaxial strain (13.18) overcomes the
shift of the average of VBs; usually, $|H| < |B|$ and $|a_v| < |b|$. LH exhibits a more
complex behavior. The LH of InAs exhibits a minimum when the lattice constant
of the substrate is 0.587 nm, whereas that of GaAs monotonically increases within
a given range of the lattice constant. HH becomes VBM when the material is in
lateral compression, since LH is pushed down in energy through biaxial strain. The
opposite is true for the material in lateral tension.

13.3.2 Longitudinal Strain

Figure 13.7 shows the (110) and (1-10) cross sections of the strain profiles of the structures in Fig. 13.4 with varying GaAs layer thickness $t = 0, 1.17, 2.35, 3.52, 4.69$ nm. And Fig. 13.8 shows the line profiles of the strains along the lines L1 and L2 (Fig. 13.7), which pass the thickest position of the ring in (110) and (1-10) cross sections, respectively.

The first row in each figure represents the hydrostatic strain, the second row the biaxial strain, and the third and fourth rows show the longitudinal strains in plane (ε_{xx}) and growth (ε_{zz}) directions, respectively. The longitudinal strain along y direction (ε_{yy}) is omitted, since its profile resembles ε_{xx}.

The dotted lines in Fig. 13.7 represent the outlines of the InAs ring and the upper interface of the GaAs layer.

As the first row shows, the InAs ring is compressively strained since the lattice constant of InAs is larger than that of the InP matrix, whereas the GaAs layer is tensile strained due to the small lattice constant of GaAs.

(i) Although the hydrostatic strain inside the ring is slightly enhanced with increasing t, the magnitude of the enhancement is very small considering the large difference between the lattice constants of InAs and GaAs (more than 6 %). Above all else, both the longitudinal strains are weakened as t increases. This peculiar feature, i.e., the relaxation of InAs strains with increasing GaAs layer thickness, is connected to the three-dimensional geometry of the ring and the relation between the lattice constants of the matrix (a_{InP}), ring (a_{InAs}), and insertion layer (a_{GaAs}). At $t = 0$ nm, the ring is most strongly compressed in plane and dilated in growth direction, than the other structures with finite values of t. As t increases, GaAs fills not only the top of the ring but also the lateral sides of the ring, since the GaAs layer is grown on a three-dimensional ring structure. In the presence of GaAs at lateral sides of the InAs ring, the layer with the combination of InAs and GaAs is pseudomorphically strained on the InP substrate. Thus, unlike the fully strained InAs in $t = 0$ nm, the in-plane dilation of GaAs partially compensate the compression of InAs, and reduces the biaxial strain. In a schematic description, the tensile strained GaAs provides space for the atomic relaxation of otherwise fully strained InAs.

(ii) The biaxial strain inside the ring considerably decreases as the thickness of the GaAs layer increases: the strain in the structure with $t = 2.35$ nm becomes $1/3$ of the value in the absence of the insertion layer ($t = 0$ nm). Such a large decrease in biaxial strain with increasing t reflects the reduction of longitudinal strain components ε_{xx}, ε_{yy}, and ε_{zz}, since biaxial strain is the quantitative measure of the difference between the degrees of lateral compression and vertical tension. Note that, the biaxial strain even exhibits a negative value along L1 at $t \geq 3.52$ nm.

(iii) When t exceeds 2.35 nm, the strains at the basal plane of the ring no longer changes with t. This is because, when the GaAs layer is sufficiently thick, the environment near the base of the ring is almost unchanged with increasing t.

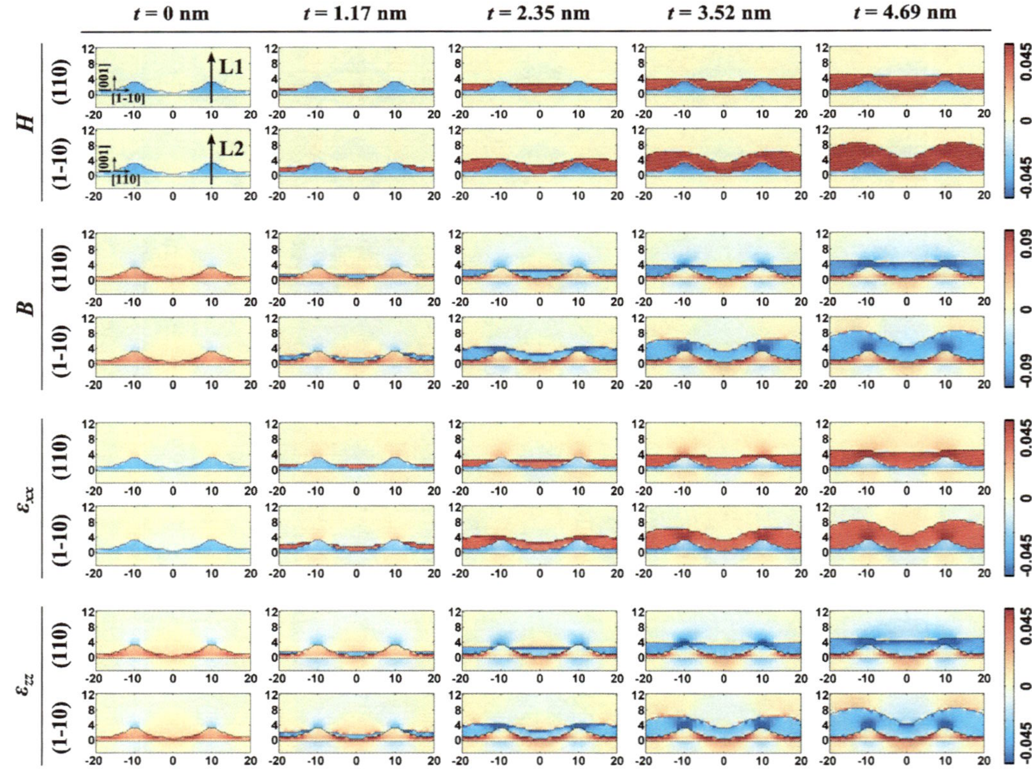

Fig. 13.7 (110) and (1-10) cross sections of the strain profiles of the structures in Fig. 13.4 with varying GaAs layer thickness $t = 0$, 1.17, 2.35, 3.52, 4.69 nm. The *first*, *second*, *third*, and *fourth rows* represent the hydrostatic strain, biaxial strain, longitudinal strains in plane (ε_{xx}) and growth (ε_{zz}) directions, respectively. All lengths are in units of nanometers

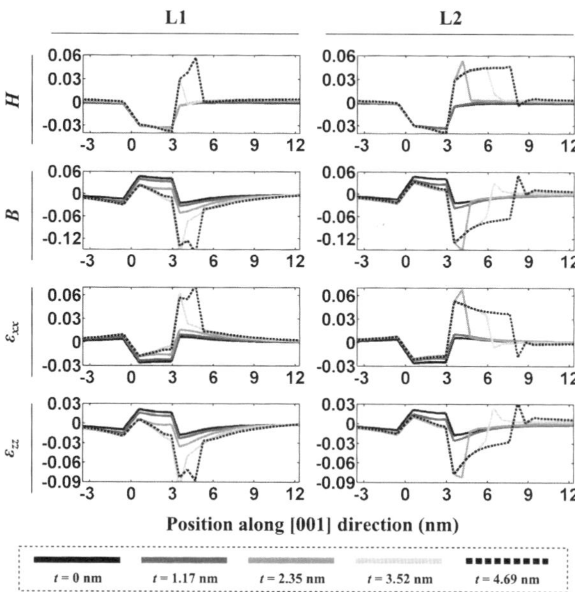

Fig. 13.8 Line profiles of the strains along the lines L1 and L2 in Fig. 13.7. The *first, second, third,* and *fourth row* represent the hydrostatic strain, biaxial strain, longitudinal strains in plane (ε_{xx}) and growth (ε_{zz}) directions, respectively. *Horizontal axis* are in units of nanometers

Moreover, the region, of which strains independent on t, expands to upper side of the ring as t increases; e.g., structures with $t = 3.52$ nm and 4.69 nm show similar strain profiles for the most part of the ring.

(iv) The GaAs layer is under a negative biaxial strain, but the layer is not uniformly strained (Fig. 13.7). The strongest biaxial strain, of which magnitude is two times larger than that of the ring, is found at right above the top of the ring. In addition, the GaAs layer along L1 is more strained than that along L2, whereas the InAs ring along L1 is more relaxed (thus, exhibits weaker strain) than that along L2 (Fig. 13.8). Moreover, as the distances from the ring and insertion layer increase, the hydrostatic strain of the matrix decays fast, while the biaxial strain decays slowly. Thus, the InP matrix right above the insertion layer and right below the ring are biaxially strained over a large volume.

13.3.3 Strain-Modified Band Edge Potentials of Nanostructure

Figure 13.9 shows the strain-modified band edge potentials of the structures with $t = 0, 1.17, 2.35, 3.52, 4.69$ nm, along L1 and L2 in Fig. 13.7. The lines above 0.4 eV represent the CB potentials, in which electrons are confined, and the lines with and without circle marker below 0.4 eV show the LH and HH potentials, respectively. SO potentials are omitted since they are below -0.4 eV. HH band resembles the biaxial strain with maxima in the ring and minima in the insertion layer [(13.18), $b < 0$], while the opposite is true for LH band. The hole band character of VBM is HH in the compressively strained ring (with the exception of $t \geq 3.52$ nm) and

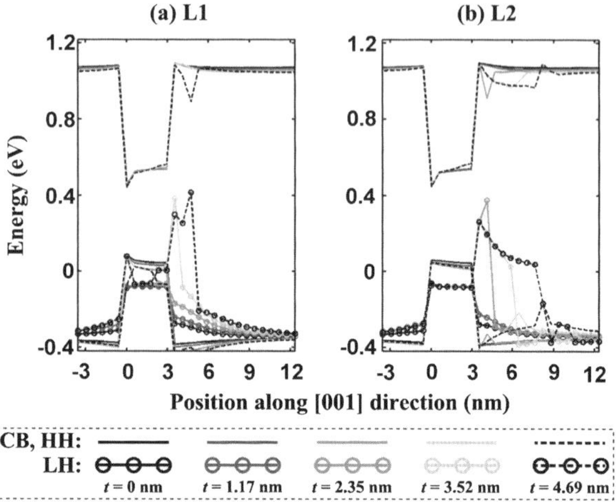

Fig. 13.9 Strain modified band edge potentials in (110) and (1-10) planes of the structures in Fig. 13.4

LH in the tensile strained insertion layer. CB moves upward by the enhancement of H with increasing t, while VBM moves downward by the enhancement of H and the reduction of HH-LH splitting due to the decrease of B ((13.15) and (13.18)). Although, H changes less than B with increasing t, the amount of the shift of CB is comparable to that of VB, since $|a_c|$ is more than 5 times larger than $|a_v|$.

If LH potential of the tensile strained GaAs layer is energetically higher (i.e., favorable) than the HH potential of the compressively strained InAs ring and has sufficient extent in real space, then holes will prefer to be localized in the GaAs layer, and exhibit a LH character. As the strain profiles in the cross sections (Fig. 13.7) show, the GaAs right above the top of the ring is most strongly strained, thus, it will become the most favorable site for LHs (Fig. 13.9). However, this is the case for a relatively thick GaAs layer, since the top of the ring is not covered by GaAs for the structures with a thin GaAs layer ($t \leq 1.17$ nm). The LH potential profiles in Fig. 13.9 show the "hole pockets" in GaAs, which cover a part of the ring ([110] corners of the ring, L2) for 1.17 nm $\leq t < 3.52$ nm, and the entire ring for $t \geq 3.52$ nm.

One of the most distinctive features of the structures in Fig. 13.1(c) is that the structure exhibits both type-I and type-II band alignments due to the coexistence of a compressively strained material and a tensile strained material. As can be seen from Fig. 13.9, both electrons and HHs are confined in the ring, while LHs prefer to be localized in the GaAs layer. Thus, the band alignment of CB and HH is type-I, while that of CB and LH is type-II.

13.3.4 Energy Levels and Band Probability in the Absence of Piezoelectric Field

Figure 13.10 shows the electron and hole energy levels in the absence of piezoelectric potentials for the structures with t ranges from 0 to 4.69 nm. Each state is doubly

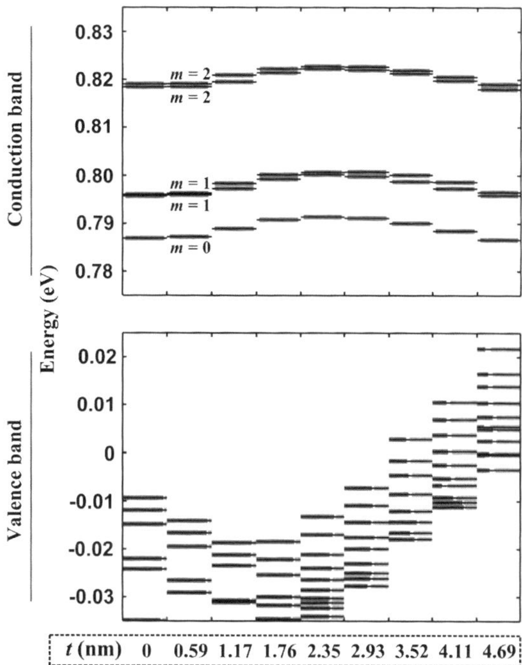

Fig. 13.10 Electron (conduction band) and hole (valence band) energy levels in the absence of piezoelectric potentials for the structures with t ranges from 0 to 4.69 nm. Electron states are labeled by their axial angular momentum (m). The lengths of the bars in the electron states represent the CB characters of each electron state, whereas those of the *dark* and *light bars* in the hole states give the HH and LH percentages of each hole state, respectively

degenerated by spin. We labeled the states with their axial angular momentum (m), since the wave functions of QR have ring-like distributions. Energetically, the states with the same values of m form groups of non-degenerate levels (shell structures), since they have similar probability distributions; the first and second excited states ($m = 1$), as well as the third and fourth states ($m = 2$), are grouped together, whereas the electron ground state ($m = 0$) is isolated from the other states.

The length of the red bar in each electron state represents the CB character of that state, whereas those of the blue and orange bars in each hole state give the HH and LH percentages of that state, respectively. SO fraction (not shown) in each state is negligible, since the SO band is far below the energies.

We plot the HH and LH band percentages in the ring (InAs) and insertion layer (GaAs) in Fig. 13.11(a) to show the probability distributions of each band with varying t. Also we plot the average height of the HH and LH probability densities measured from the basal plane of QR to show the spatial shift of the carriers with increasing t.

The electron state energies strongly depend on the potential profiles in and around the InAs ring, since electron is mainly confined in the ring. The electron energies increase with t, up to $t = 2.35$ nm, due to the enhancement of the compressive strain in the ring. However, the energies decrease as t exceeds 2.35 nm, at which the InAs ring becomes fully capped by the GaAs layer. This is because the CB of the tensile strained GaAs layer, right above the ring, is lower than that of unstrained InP matrix [Fig. 13.9]. Thus, as the extent in real space grows with increasing t, the

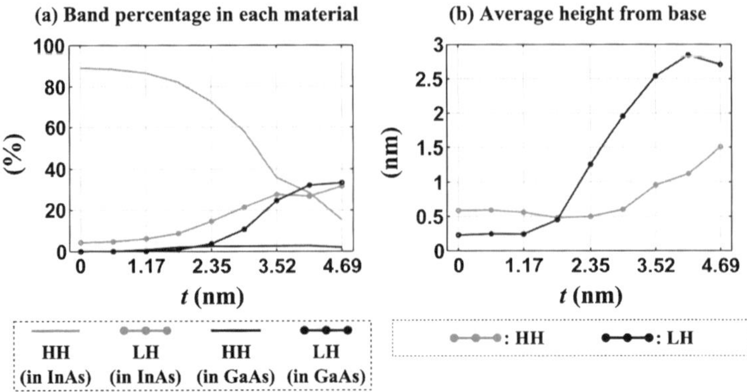

Fig. 13.11 (a) HH and LH band percentages in the ring (InAs) and insertion layer (GaAs) as a function of t. (b) Average height of the HH and LH probability densities measured from the basal plane of QR

insertion layer weakens the confinement of electron, and reduces the energies of the CB states.

Each electron state is mainly composed of CB, and the fractions of CB barely change with the energies, since only one doubly degenerated CB is implemented in an eight-band $k \cdot p$ method (this is justified by the fact that the other CBs are separated very far from the CB used in the method), and the other bands (i.e., VBs) are energetically distant from the electron states.

The energies and spatial distributions of the hole state depend on the potential profiles of both the InAs ring and GaAs layer, since the VBM of the ring (HH) and the insertion layer (LH) are comparable in energies.

For $t \leq 1.76$ nm, each hole state is mainly composed of HH, and mostly confined in the ring [Fig. 13.11(a)]. In such structures, it is often a good approximation to treat the ground and lower excited hole states as if they have only one type of hole character with a pseudospin $1/2$, and the hole energies are sensitive to the HH band edge potential of the ring. The hole energies move downward as t increases (Fig. 13.10), since VBM of the ring decreases by the enhancement of H and weakening of B (Fig. 13.9). The higher excited states of the hole (i.e., the states far from VBM) exhibit slightly enhanced LH probabilities as the hole energies approach the LH band edge of InAs.

As t exceeds 2.35 nm, LH pockets appear at the GaAs layer in the vicinity of the top of the ring (Fig. 13.9), thus the hole energies move upward as t increases. In $t = 2.35$ nm, the LH probability of the hole ground state is 4 times larger than that in $t = 0$ nm, and the increase in the average height of the probability distribution indicates that the LH component moves toward the top of the ring. Although more than 5 % of the hole probability is found at the GaAs layer, the major part of the hole is still localized in the ring due to the small dimension of the hole pocket. Thus, the GaAs layer affects the hole energies mainly by weakening the confinement potential.

In structures with $t \geq 3.52$ nm, the ground hole states have dominantly LH character (>53.5 %). More than 27 % of the hole probability is found at the GaAs layer, and the average distance of the LH probability distribution measured from the base of the ring (2.5 nm) approaches the top of the ring. The simplification of a pseudospin $1/2$ cannot be applied to these structures, and full consideration of the hole bands is necessary. In this regime, the GaAs insertion layer has sufficient space to confine the carrier. Thus, the GaAs layer attracts holes and raises the hole energies by its high band edge potentials. In this regime, HH character increases for higher excited hole states.

13.3.5 Shear Strain and Piezoelectric Potential

Figure 13.12 shows the (110) and (1-10) cross sections of the shear strain profiles of the structures with varying GaAs layer thickness $t = 0, 1.17, 2.35, 3.52, 4.69$ nm. The first and second rows represent the shear in plane (ε_{xy}) and vertical plane (ε_{yz}). The shear strain in another vertical plane (ε_{zx}) is omitted, since its profile resembles ε_{yz}. The dotted lines in Fig. 13.12 represent the outlines of the InAs ring and the upper interface of the GaAs layer.

The in-plane shear strain (ε_{xy}) exhibits extrema inside the ring, which reflect the stretching of the ring in radial directions. As the thickness of the GaAs layer increases, the shear strain is also enhanced since the tensile strained GaAs both at the center and out corner of the ring provides more space for atomic relaxation than the InP matrix. The shear strain in (011) plane (ε_{yz}) exhibits extrema at the interface of the ring and insertion layer, and their magnitude increase as the thickness of the insertion layer increases. Note that, such a relaxation of InAs in radial directions results in the enhancement for the shear strain but the weakening for the longitudinal strain (Figs. 13.7 and 13.8), due to the difference between the definition of the shear and longitudinal strains.

We plot the piezoelectric potentials of the structures with $t = 0, 1.17, 2.35, 3.52, 4.69$ nm in Fig. 13.13. The piezoelectric potential profiles are closely related to the shear strain distributions; e.g., positive e_{xy} induces negative and positive potentials above and below, respectively, and negative e_{xy} induces positive and negative potentials above and below, respectively.

Note that, when it compares the samples with the same thickness of the GaAs layer, the piezoelectric potentials of the InAs/InP ring considered in this chapter is much weaker than those of the InAs/In$_{0.53}$Ga$_{0.47}$As ring in [13]. This is because the size of the ring considered in this chapter is smaller than that in [13], and the piezoelectric constant of InP (-0.035 C/cm^2) is much smaller than that of In$_{0.53}$Ga$_{0.47}$As (-0.099 C/cm^2), since the piezoelectric potential scales with the size of the nanostructure and the piezoelectric constants of constituent materials.

QRs with a thin insertion layer ($t \leq 1.17$ nm) has 16 piezoelectric potential extrema (eight for [110] corners, and eight for [1-10] corners of the ring), as reported for a square profile ring [39] and volcano-like ring [40]. While the square profile ring exhibits potential extrema outside the inner and outer rims, the structures in Fig. 13.13 show potential extrema right above and below the ring owing to the high e_{xy} value inside the ring.

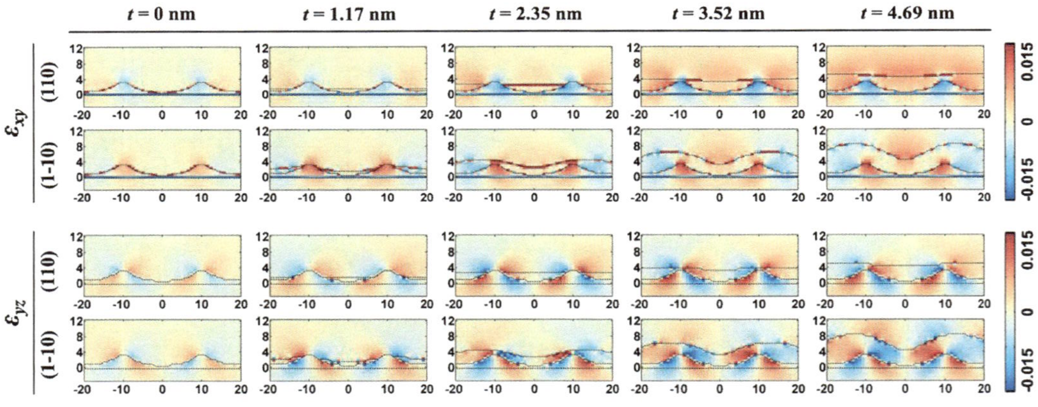

Fig. 13.12 (110) and (1-10) cross sections of the shear strain profiles of the structures with $t = 0$, 1.17, 2.35, 3.52, 4.69 nm. The *first* and *second rows* represent the shear strains in plane (ε_{xy}) and vertical plane (ε_{yz}), respectively. All lengths are in units of nanometers

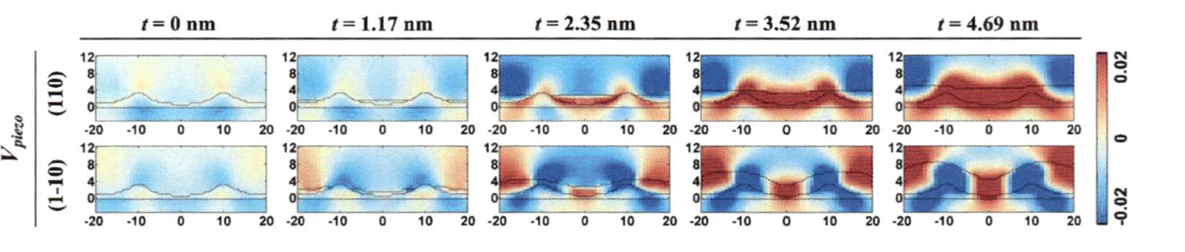

Fig. 13.13 (110) and (1-10) cross sections of the piezoelectric potentials of the structures with $t = 0$, 1.17, 2.35, 3.52, 4.69 nm. All lengths are in units of nanometers, and energy units are in electron volts

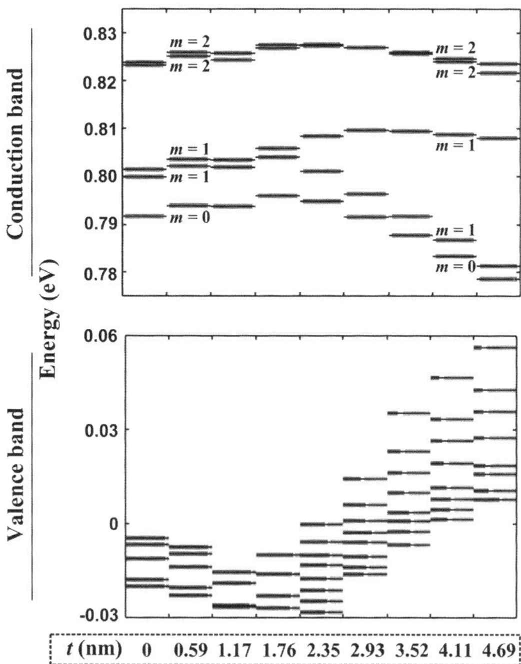

Fig. 13.14 Electron (conduction band) and hole (valence band) energy levels in the presence of piezoelectric potentials for the structures with t ranges from 0 to 4.69 nm. Electron states are labeled by their axial angular momentum (m). The lengths of the bars in the electron states represent the CB characters of each electron state, whereas those of the *dark* and *light bars* in the hole states give the HH and LH percentages of each hole state, respectively

As t increases, the piezoelectric potential is considerably enhanced, since GaAs has the largest piezoelectric constant in this material combination. Moreover, the piezoelectric potential changes its spatial distribution, since the GaAs layer exists only at one side of the ring, i.e., the upper side of the ring. Only the eight piezoelectric potential extrema at the upper side of the ring within the 16 extrema are enhanced, and the GaAs at the center of the ring exhibits a strong positive potential. As t increases, the piezoelectric potential inside the ring changes its sign; at small t, the [1-10] sides [i.e., (110) cross section] and [110] sides of the ring are strongly affected by negative and positive piezoelectric potentials, respectively, while at large t, the sign of the potentials are reversed. The GaAs layer above the top of the ring is affected by the piezoelectric potential with the same sign to that of the underlying ring.

13.3.6 Energy Levels and Band Probability in the Presence of Piezoelectric Field

Figure 13.14 shows the electron and hole energy levels in the presence of piezoelectric potentials for the structures with t ranges from 0 to 4.69 nm. As in Fig. 13.10, each state is doubly degenerated by spin, and labeled with its axial angular momentum (m). The length of the red bar in each electron state represents the CB character of that state, whereas those of the blue and orange bars in each hole state give the

HH and LH percentages of that state, respectively. SO fraction (not shown) in each state is negligible, since the SO band is far below the energies.

In ordinary QD, which does not have a geometric hole in its center, electron and hole states respond in different ways to piezoelectric potentials [41]. The energies and spatial distributions of electron states are less sensitive to the potential, while those of hole states follow even a small potential perturbation very closely to minimize their potential energy. This is because the two different types of charge carriers have clearly distinct effective masses. Electrons have smaller effective mass than holes, thus, the kinetic energies,

$$E_{\mathrm{kinetic}} = -\frac{\hbar^2}{2m^*}\nabla^2, \tag{13.20}$$

of electrons rise much faster than those of holes by the rapid changes of wave functions in a real space. Thus, electrons prefer to keep their wave functions as smooth as possible, while holes are eager to follow the detailed profiles of the potentials.

However, the situation changes to some degrees in the QR considered in this chapter; the piezoelectric potential affects the shell structures of CB. At $t \leq 1.76$ nm, the electron states form non-degenerate levels which are similar to those in the absence of piezoelectric potentials; the first and second excited states ($m = 1$), as well as the third and fourth states ($m = 2$), are grouped together, whereas the ground state ($m = 0$) is isolated from the other states. As t exceeds 2.35 nm, the $m = 1$ shell gets untied, and one of the $m = 1$ states approaches the energy of the ground state.

Such an energy shift originates from the isotropic geometries of the InAs ring, and the sign inversion of the piezoelectric potentials between [110] and [1-10] directions.

At small t, the electron ground state has an almost isotropic, ring-like probability distribution due to the weak piezoelectric potentials. For the same reason, the first and second excited states are close in energies, while they exhibit p-orbital like distributions oriented along [110] and [1-10] directions, respectively.

As t increases, the piezoelectric potential enhances, thus electron prefers to be localized at the corners where the piezoelectric potential lowers its energy. The electron ground state becomes partially localized at [1-10] corners, thus its spatial distribution looks similar to that of the $m = 1$ state oriented along [1-10] direction. Since the strong piezoelectric potential penetrating the ring considerably lowers the energies of the two states, the ground state and one of the $m = 1$ states oriented along [1-10] direction form a new shell structure of non-degenerate states. Another $m = 1$ state, which is oriented along [110] direction, is energetically apart from the two states, since the negative piezoelectric potential at [1-10] corners keeps its energy high.

The piezoelectric potential also affects the hole states. At weak piezoelectric potentials, the hole state energies are similar to those in the absence of the piezoelectric potentials. As t increases, holes prefer to be localized at the [110] corners of the ring, where the negative potential raises the hole energies. Moreover, since the hole energies become closer to VBM of GaAs (LH) than that of InAs (HH), the LH characters are considerably enhanced at large t.

13.4 Conclusion

In this chapter, we investigated the strain and electronic structures of QRs capped with an insertion layer with a lattice constant smaller than those of the matrix and the active materials so that $a_{\text{support}} < a_{\text{matrix}} < a_{\text{active}}$. In particular, we showed the weakening of longitudinal strains and biaxial strain of QR with increasing GaAs insertion layer thickness. As the thickness of GaAs layer increases, the band characters of hole states vary from HH in compressively strained InAs to LH in tensile strained GaAs, and the weighted average of LH occupation probabilities move toward the GaAs layer. Due to the coexistence of a compressively strained nanostructure and a tensile strained layer, the QR structures analyzed in this chapter exhibit a coexistence of type-I (for electrons and HHs) and type-II (for electrons and LHs) band alignments. Since the band characters (and corresponding effective masses) of nanostructures determine their optoelectronic as well as their magnetic properties (e.g., diamagnetic shift) [42], the use of insertion layers with the smallest lattice constant compared to that of the active material with reference to that of the matrix (i.e. $a_{\text{support}} < a_{\text{matrix}} < a_{\text{active}}$) will provide wide flexibility and versatility for the design of new opto-electronic devices. Finally, the hole band characters are very sensitive to the effective potential changes caused by material combinations as well as the geometry of the ring. Those who are interested in the effects of the geometric changes to hole mixing effects are particularly encouraged to refer to the Chap. 17.

Acknowledgements   This work was partially supported by Korea Institute for Advanced Study (P.M.) grant funded by the Korea government, and also by Internal Program (2E23910) of Korea Institute of Science and Technology (W.J.C.).

References

1. V.M. Fomin, J. Nanoelectron. Optoelectron. **6**(1), 1–3 (2011)
2. Q. Gong, R. Nötzel, P. Van Veldhoven, T. Eijkemans, J. Wolter, Appl. Phys. Lett. **84**, 275 (2004)
3. K. Nishi, H. Saito, S. Sugou, J.S. Lee, Appl. Phys. Lett. **74**, 1111 (1999)
4. V.M. Ustinov, N.A. Maleev, A.E. Zhukov, A.R. Kovsh, A.Yu. Egorov, A.V. Lunev, B.V. Volovik, I.L. Krestnikov, Yu.G. Musikhin, N.A. Bert, P.S. Kop'ev, Zh.I. Alferov, N.N. Ledentsov, D. Bimberg, Appl. Phys. Lett. **74**, 2815 (1999)
5. N.T. Yeh, T.E. Nee, J.I. Chyi, T. Hsu, C. Huang, Appl. Phys. Lett. **76**, 1567 (2000)
6. E.T. Kim, Z. Chen, A. Madhukar, Appl. Phys. Lett. **81**, 3473 (2002)
7. R. Jia, D. Jiang, H. Liu, Y. Wei, B. Xu, Z. Wang, J. Cryst. Growth **234**(2), 354–358 (2002)
8. Z.Y. Zhang, B. Xu, P. Jin, X.Q. Meng, Ch.M. Li, X.L. Ye, Z.G. Wang, J. Appl. Phys. **92**, 511 (2002)
9. P.S. Wong, B.L. Liang, V.G. Dorogan, A.R. Albrecht, J. Tatebayashi, X. He, N. Nuntawong, Y.I. Mazur, G.J. Salamo, S.R.J. Brueck, D.L. Huffaker, Nanotechnology **19**, 435710 (2008)
10. Q. Gong, R. Nötzel, P. Van Veldhoven, T. Eijkemans, J. Wolter, Appl. Phys. Lett. **85**, 1404 (2004)
11. K. Park, P. Moon, E. Ahn, S. Hong, E. Yoon, J.W. Yoon, H. Cheong, J.-P. Leburton, Appl. Phys. Lett. **86**, 223110 (2005)
12. P. Moon, K. Park, E. Yoon, J.P. Leburton, Phys. Status Solidi RRL **3**(2–3), 76–78 (2009)

13. P. Moon, W.J. Choi, K. Park, E. Yoon, J.D. Lee, J. Appl. Phys. **109**, 103701 (2011)
14. P. Moon, W.J. Choi, J. Lee, Phys. Rev. B **83**(16), 165450 (2011)
15. J.M. García, G. Medeiros-Ribeiro, K. Schmidt, T. Ngo, J.L. Feng, A. Lorke, J. Kotthaus, P.M. Petroff, Appl. Phys. Lett. **71**, 2014 (1997)
16. S. Suraprapapich, S. Panyakeow, C. Tu, Appl. Phys. Lett. **90**, 183112 (2007)
17. F. Ding, L. Wang, S. Kiravittaya, E. Müller, A. Rastelli, O.G. Schmidt, Appl. Phys. Lett. **90**, 173104 (2007)
18. T. Raz, D. Ritter, G. Bahir, Appl. Phys. Lett. **82**, 1706 (2003)
19. B.C. Lee, C.P. Lee, Nanotechnology **15**, 848 (2004)
20. F.H. Li, Z.S. Tao, J. Qin, Y.Q. Wu, J. Zou, F. Lu, Y.L. Fan, X.J. Yang, Z.M. Jiang, Nanotechnology **18**, 115708-1 (2011)
21. P. Offermans, P.M. Koenraad, J.H. Wolter, D. Granados, J.M. García, V.M. Fomin, V.N. Gladilin, J.T. Devreese, Appl. Phys. Lett. **87**, 131902 (2005)
22. V. Fomin, V. Gladilin, S. Klimin, J. Devreese, N. Kleemans, P. Koenraad, Phys. Rev. B **76**(23), 235320 (2007)
23. P. Keating, Phys. Rev. **145**(2), 637 (1966)
24. R.M. Martin, Phys. Rev. B **1**(10), 4005 (1970)
25. S. Lee, F. Oyafuso, P. Von Allmen, G. Klimeck, Phys. Rev. B **69**(4), 045316 (2004)
26. C. Pryor, J. Kim, L. Wang, A. Williamson, A. Zunger, J. Appl. Phys. **83**, 2548 (1998)
27. O. Stier, M. Grundmann, D. Bimberg, Phys. Rev. B **59**(8), 5688–5701 (1999)
28. M. Burt, J. Phys. Condens. Matter **4**, 6651 (1992)
29. B.A. Foreman, Phys. Rev. B **56**(20), 12748–12751 (1997)
30. H.B. Wu, S. Xu, J. Wang, Phys. Rev. B **74**(20), 205329 (2006)
31. P. Moon, E. Yoon, W. Sheng, J.P. Leburton, Phys. Rev. B **79**(12), 125325 (2009)
32. W. Sheng, J.P. Leburton, Phys. Rev. Lett. **88**(16), 167401 (2002)
33. W. Sheng, J.P. Leburton, Phys. Rev. B **63**(16), 161301 (2001)
34. W. Sheng, J.P. Leburton, Appl. Phys. Lett. **80**(15), 2755–2757 (2002)
35. T.B. Bahder, Phys. Rev. B **41**(17), 11992 (1990)
36. I. Vurgaftman, J. Meyer, L. Ram-Mohan, J. Appl. Phys. **89**, 5815 (2001)
37. I. Saïdi, S. Ben Radhia, K. Boujdaria, J. Appl. Phys. **104**(2), 023706 (2008)
38. C.Y.P. Chao, S.L. Chuang, Phys. Rev. B **46**(7), 4110 (1992)
39. J. Barker, R. Warburton, E. O'Reilly, Phys. Rev. B **69**(3), 035327 (2004)
40. S. Tomić, A.G. Sunderland, I.J. Bush, J. Mater. Chem. **16**(20), 1963–1972 (2006)
41. A. Schliwa, M. Winkelnkemper, D. Bimberg, Phys. Rev. B **76**(20), 205324 (2007)
42. J. Climente, J. Planelles, J. Nanoelectron. Optoelectron. **6**(1), 81–86 (2011)

Chapter 14
Theoretical Modelling of Electronic and Optical Properties of Semiconductor Quantum Rings

Oliver Marquardt

Abstract A variety of the most common theoretical models that have been applied to investigate electronic properties of quantum rings is presented in this chapter. The advantages and disadvantages of these approaches that cover atomistic as well as continuum modelling are discussed and example simulations on different quantum ring systems are reviewed.

14.1 Modelling and Designing of Quantum Rings

The large variety of quantum rings of different shapes, sizes and materials creates a strong demand for theoretical models, that allow not only for providing a detailed understanding of the different properties of these fascinating systems, but also for tailoring quantum rings suited to specific applications, following theoretical predictions. The vast amount of different experimentally observed quantum ring systems induces a correspondingly large set of methods to approach their properties from theory (see e.g. Refs. [1] and [2] and references therein). Apart from analytical models that can describe many of these properties in ideal ring-shaped systems [3–6], numerical approaches of different levels of sophistication exist for the theoretical description of elastic, electronic, magnetic and optical properties.

These formalisms can be classified as atomistic approaches and non-atomistic, continuum models. Whereas an atomistic description can cause a huge computational effort for large systems, most continuum models lack a proper description of interface effects and underlying crystal lattice, what can lead to severe inaccuracies in theoretical studies of quantum rings.

It is therefore essential to understand the advantages and disadvantages of the different numerical approaches available for the description of optoelectronic properties of quantum rings and to identify a suitable computational approach for experimentally observed systems and their accessible modifications.

In this chapter, the most common numerical models available for the description of electronic and optical properties of quantum rings will be briefly introduced. The

O. Marquardt (✉)
Tyndall National Institute, Cork, Ireland
e-mail: marquardt@pdi-berlin.de

V.M. Fomin (ed.), *Physics of Quantum Rings*, NanoScience and Technology,
DOI 10.1007/978-3-642-39197-2_14, © Springer-Verlag Berlin Heidelberg 2014

capabilities and shortcomings of these formalisms will be analysed as well as their applicability and limitations for modelling various different quantum ring systems. In this context, a selection of studies will be reviewed and discussed.

14.2 Atomistic Empirical Pseudopotential Models

The empirical pseudopotential method (EPM) [7, 8] is a sophisticated, atomistic approach, which has been successfully applied to describe and predict the electronic properties of a wide variety of semiconductor quantum dots [9–11], that realise a three-dimensional carrier confinement. For quantum rings, however, only few studies based on this highly accurate method are available [12, 13]. The empirical pseudopotential method allows for a description of the system under consideration within an atomistic picture, which means that the underlying crystal structure as well as interface effects and lattice defects can be properly analysed. On the other hand, it is clear that atomistic models can become quite computationally expensive and cumbersome for larger structures.

Within the EPM, a nanostructured system is expressed in terms of a total screened potential $V(\mathbf{r})$ by a superposition of atomic pseudopotentials $V_a(|\mathbf{r}|)$ [9]:

$$V(\mathbf{r}) = \sum_{\mathbf{R}_a} V_a\big(|\mathbf{r} - \mathbf{R}_a|\big) \tag{14.1}$$

where the $\mathbf{R}_a$'s are the atomic positions. The atomic pseudopotentials are commonly adjusted to fit *ab-initio* calculated or experimental reference data. The electronic wave functions in the system are then represented in a plane wave basis as:

$$\Psi_j(\mathbf{r}) = \sum_{\mathbf{G}} B_j(\mathbf{G}) e^{i\mathbf{G}\cdot\mathbf{r}} \tag{14.2}$$

with $\mathbf{G}$ being a vector in reciprocal space and $B_j(\mathbf{G})$ the corresponding expansion coefficients, that are commonly determined variationally. Using the total potential from (14.1), the Schrödinger equation is solved with the Hamiltonian:

$$\hat{H}\Psi_j(\mathbf{r}) = \varepsilon\Psi_j(\mathbf{r}) \quad \text{with } \hat{H} = -\frac{\hbar^2 \mathbf{k}^2}{2m_0} + V(\mathbf{r}). \tag{14.3}$$

As this approach is computationally extremely expensive, reasonable simplifications can be made such as the limitation to a narrow energy window around the band gap of the system, such that only the electronic states energetically highest within the valence band and those energetically lowest in the conduction band are computed, without the need to compute all the remaining electronic states inside the valence band [9], and the wave function basis set can be reduced to consider only the physically important bands and wave vectors, which allows for a dramatic reduction of computational effort (Linear Combination of Bloch Bands, LCBB) [7].

Elastic properties that commonly modify the electronic structure of the system can be calculated within an atomistic picture using the valence force field method (VFF) to obtain the strained atom positions [14, 15], and then enter the EPM calculation of the electronic properties [7] via the deformation potentials.

Fig. 14.1 Biaxial strain in the (001) plane for (**a**) $R_{in} = R_{out}$ (lens-shaped QD), (**b**) $R_{in} = 3$ nm, (**c**) $R_{in} = 0$ nm, and (**d**) $R_{in} = 3$ nm. Reprinted with permission from [12]. Copyright 2008, European Physical Society

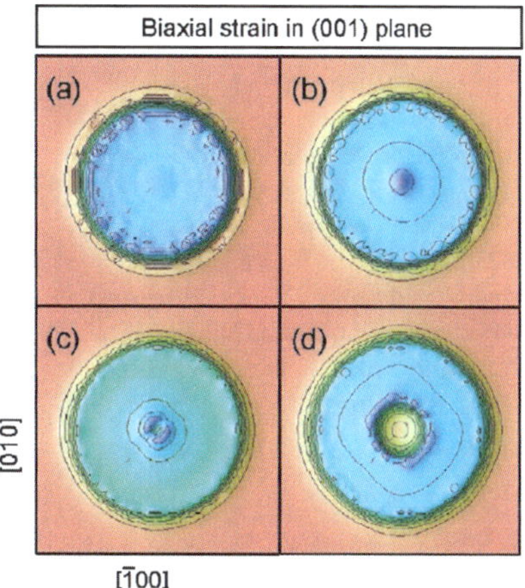

14.2.1 Electronic Structure of Realistic Self-assembled InAs/GaAs Quantum Rings

Few studies on self-assembled InAs/GaAs quantum rings have been carried out using the EPM [12, 13]. In Ref. [12], the electronic and optical properties of InAs quantum rings in a GaAs matrix have been systematically investigated for a set of quantum rings of different inner diameters. Throughout the study, the height of the quantum ring has been fixed to 2.5 nm, and the outer radius to 12.5 nm. The inner radius of the quantum ring is then varied from -12.5 nm to 6 nm. In this notation, $R_{in} > 0$ identifies those systems where the quantum ring vanishes completely inside the inner radius, and $R_{in} < 0$ correspond to quantum rings that have a nonzero height in the centre of the ring, with $R_{in} = -12.5$ nm being a lens-shaped quantum dot. Within all systems under consideration, a wetting layer of one monolayer has been taken into account.

14.2.1.1 Elastic Properties

Using the VFF, the elastic properties of the systems have been computed. The biaxial strain, $B = \sqrt{(\varepsilon_{xx} - \varepsilon_{yy})^2 + (\varepsilon_{zz} - \varepsilon_{xx})^2 + (\varepsilon_{yy} - \varepsilon_{zz})^2}$ is shown in Fig. 14.1 for a lens-shaped quantum dot (Fig. 14.1a) and quantum rings of different inner radii (Fig. 14.1b–d). Whereas the lens-shaped quantum dot in Fig. 14.1(a) yields an almost isotropic biaxial strain, it is clearly visible that already the elastic properties of the quantum ring system become visibly anisotropic along the [110] and the [1$\bar{1}$0] directions, when increasing the inner radius. This behaviour results from the atomistic nature of the InAs/GaAs interfaces within the zinc-blende lattice, as the ratio between the quantum ring's surface area and its volume increases with a larger inner radius, i.e. less confining space inside the ring. As a result of this strain anisotropy, a corresponding anisotropic behaviour of the charge carrier confinement (in particular the hole state confinement) can be expected.

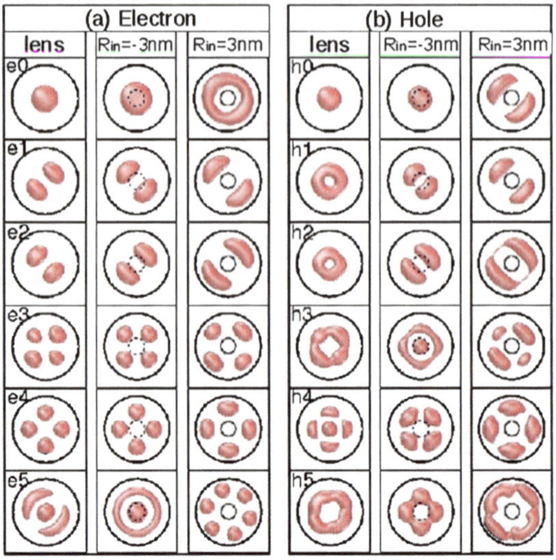

Fig. 14.2 Top view of the squared wave functions of the confined (**a**) electron and (**b**) hole states in the quantum rings. The wave functions at $R_{in} = 12.5, -3, 3$ nm are shown. The iso-surface is chosen to enclose 50 % of the density of the state. Reprinted with permission from [12]. Copyright 2008, European Physical Society

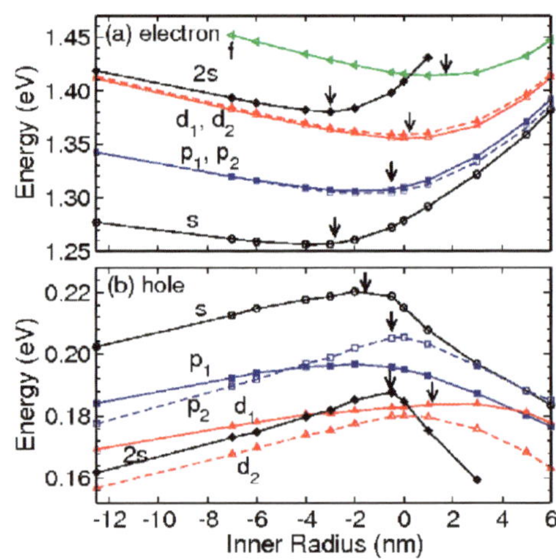

Fig. 14.3 The single-particle energy levels of confined (**a**) electron and (**b**) hole states. The reference energy is chosen to be the valence band edge of GaAs. Reprinted with permission from [12]. Copyright 2008, European Physical Society

14.2.1.2 Single-Particle Electronic Properties

The single-particle states and eigenenergies in Ref. [12] have been computed using the LCBB version of the EPM [7]. The electron and hole charge densities of the first six states closest to the band gap are shown in Fig. 14.2 for a lens-shaped quantum dot and two quantum rings of different inner radii. The eigenenergies of these states are shown in Fig. 14.3 as a function of the inner radius.

The electron eigenenergies of all states considered first decrease to a minimum energy at a certain inner radius and then increase above this radius again. A similar behaviour is observed for the hole states, where the energies increase to a maximum at a certain inner radius and then decrease again. The decreasing of the electron and increasing of the hole energy levels is explained by the influence of strain inducing a stronger confinement when increasing the inner radius of the quantum ring [12]. Above a turning point of -3 nm to $+2$ nm (this point is not the same for different energy levels), the electron eigenenergies increase again, whereas the hole energy levels decrease, as the confinement region becomes more and more narrow. This effect is much more pronounced for those states that have a charge density maximum in the center of the quantum ring, namely the s-like electron and hole ground states and the higher excited 2s-like states. The lack of confinement potential in the center of the dot for positive inner radii forces these charge density maxima into the outer regions, thus inducing a stronger influence of the inner radius on these states than on the p- and d-like states that permanently have their charge density maxima in the outer regions of the quantum ring.

Furthermore, the asymmetry of the strain in the quantum ring leads to an energy splitting of the p- and d-like electron states p_1, p_2 and d_1, d_2, that are almost degenerate in a lens-shaped quantum dot. It is clearly visible in Fig. 14.3, that this splitting increases with larger inner radii, as the atomistic nature of the surface regions becomes important, whereas in the larger confinement area of the lens-shaped quantum dot and quantum rings with negative inner radii the bulk properties of InAs dominate the electronic structure. The strong influence of the strain anisotropy can also be seen in Fig. 14.2, where the p-like electron and hole charge densities switch their orientation from the $[110]$ to the $[1\bar{1}0]$ direction (p_1), and vice versa (p_2) between a lens-shaped quantum dot and a quantum ring.

14.2.1.3 Optical Properties

Based on a configuration-interaction scheme (CI) [16], single-exciton absorption spectra have been calculated from quantum rings with different inner radii. These spectra are shown for light polarisations along $[110]$ (black) and $[1\bar{1}0]$ (red) in Fig. 14.4. It is clearly visible, that strong differences occur between the p-p and the d-d transitions when increasing the inner radius from a lens-shaped quantum dot to a quantum ring. Moreover, a strong polarisation anisotropy can be observed with increasing inner radius. For an inner radius of 6 nm, the transition in the $[1\bar{1}0]$-direction is approximately three times larger than that in the $[110]$-direction [12]. This large anisotropy is another consequence of the strain anisotropy, and it is much larger than in quantum dots of similar size and materials composition.

14.3 The Tight Binding Method

The empirical tight binding method (ETBM) [17, 18] is another atomistic approach to the electronic properties of semiconductor nanostructures. Like the EPM, it has been employed to study the electronic and optical properties of a wide variety of semiconductor quantum wells, quantum wires and quantum dots [19–22]. Within

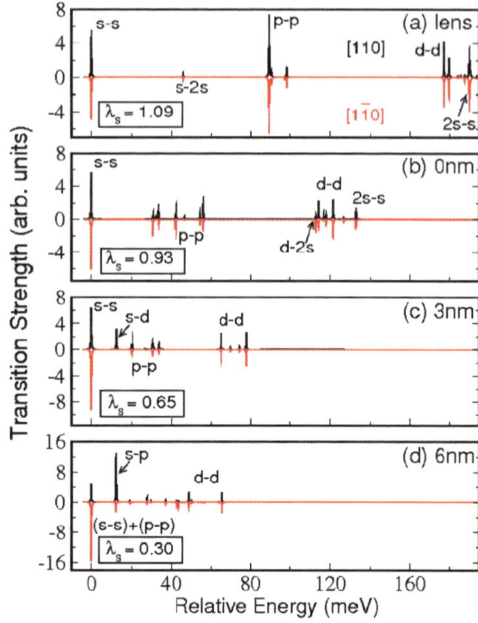

Fig. 14.4 The single exciton absorption spectrum for quantum rings with different inner radius. The primary exciton energy at each panel is shifted to zero. The spectra are shown for four different systems: a lens-shaped quantum dot (**a**), and three quantum rings with inner radii of 0 nm (**b**), 3 nm (**c**) and 6 nm (**d**). Reprinted with permission from [12]

the ETBM, the Schrödinger equation (14.3) is solved using a tight-binding Hamiltonian that is expressed in terms of on-site energies ε_i at the atom positions and hopping elements t_{ij} between the atoms [23]:

$$H = \sum_i \varepsilon_i c_i^\dagger c_i + \sum_{\langle ij \rangle} t_{ij} c_i^\dagger c_j + h.c. \qquad (14.4)$$

and the basis is constructed as a linear combination of atomic orbitals [24]. The hopping element t describes the tunneling between neighbouring atoms. Its high computational efficiency, compared to EPM approaches, stems from two basic assumptions that are commonly made:

1. Representing charge carriers in terms of atomic orbitals relies on electrons that are tightly bound to the atoms in the case that interatomic distances are large. It has nevertheless been found that even for the realistic interatomic distances in a crystal an accurate description of the band structure is possible, based on only a small number of atomic orbitals employed [24]. Existing models are commonly using the sp^3 or $sp^3 s^\star$ approximation, where it is assumed that the valence band is mainly formed by the p-orbitals of the anions and the conduction band from the s-orbitals of the cations [25], but this approximation fails a proper description of the band structure in the outer area of the Brillouin zone, as e.g. an infinite transverse mass at the X-point [26], and it is thus for many semiconductor systems preferable to extend the basis to the $sp^3 d^5 s^\star$ basis, taking the d-orbitals into account [27, 28].

2. The overlap between atomic orbitals decreases fast with the distance between neighbouring atoms. It is thus feasible to include only the nearest neighbours for

each atom, instead of including all atomic orbitals of the whole system. Common approaches limit themselves to second [29] and third [30] nearest neighbours.

To achieve a higher accuracy, the ETBM can be expanded to more atomic orbitals employed as well as to a higher number of nearest neighbour interactions considered. However, these improvements increase the computational effort for a simulation and thus reduce the overall number of atoms that can be described with reasonable computational costs. Additionally, determining the hopping elements to these higher order neighbours becomes likewise complicated. Strain-induced modifications of the electronic structure can be included by adjusting the tight binding hopping elements to strained bond lengths [31] and the required elastic properties can be taken from atomistic models such as the VFF, as well as from continuum approaches.

14.3.1 Electronic Properties of Graphene Quantum Rings in a Magnetic Field

While a number of tight binding based calculations focus on metallic quantum rings [32–34], only few model simulations are available on semiconductor quantum rings, and in particular on self-assembled quantum rings. Early tight binding calculations on self-assembled semiconductor quantum rings [35, 36] undergo further severe simplifications, and do not provide a complete, three-dimensional picture of the system, certainly due to the sheer numerical effort of such a fully three-dimensional description of quantum rings with diameters of up to 200 nm and ring thicknesses of up to 30 nm [35]. As an example of a tight binding based description of a three-dimensional quantum ring, the studies of graphene quantum rings by Potasz and co-workers [37, 38] have therefore been chosen.

14.3.1.1 Model Systems

The study in Ref. [37] is dedicated to hexagonal quantum rings. These can be constructed from six ribbonlike units (each consisting of a number of benzene rings) that are connected to a hexagonal quantum ring, where the parameters W and L denote the width and length of such a ribbon in terms of benzene rings (Fig. 14.5).

In Ref. [38], a triangular graphene quantum ring (TGQR) is investigated, where it is proposed to design such a quantum ring applying reactive ion etching, as used e.g. in Ref. [39] where graphene nanotubes serve as a mask for the TGQR (Fig. 14.6).

In both quantum rings, the carbon atoms are distinguished between type A and type B atoms, which denote equivalent sublattices in a honeycomb lattice. In Figs. 14.5 and 14.6, these two types are marked red and blue. In-plane dangling bonds at the interfaces have been assumed to be passivated with hydrogen.

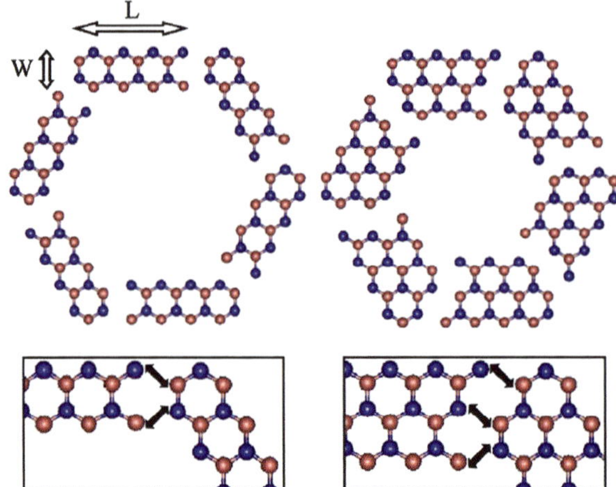

Fig. 14.5 Construction of quantum ring structures from six ribbonlike units. On the left, there are six thinnest possible ribbons, each one benzene ring thick denoted as $W = 1$ arranged in a hexagonal ring structure. The length of each ribbon is given by $L = 4$, the number of one type of atoms in one row. Each ribbon consists of 16 atoms which gives a total of 96 atoms in a ring. On the right, there are six ribbons with width $W = 2$ two benzene ring thick. Each of them consists of 21 atoms giving a total of 126 atoms in a ring. *Small black arrows* in the bottom enlargement indicate bonds and hopping integrals between nearest neighbours in a tight-binding model between neighbouring ribbons. Reprinted with permission from [37]. Copyright 2010, American Physical Society

14.3.1.2 Electronic Properties

The single-particle states of both the triangular and the hexagonal graphene quantum ring have been obtained based on (14.4). Within this model, the Hamiltonian in the nearest-neighbours approximation is denoted as

$$\hat{H} = t \cdot \sum_{\langle i,j \rangle, \sigma} c_{i\sigma}^{\dagger} c_{j\sigma}, \tag{14.5}$$

thus including the spin σ. In Ref. [38], the hopping element $t = -2.5$ eV is chosen. The operators $c_{i\sigma}^{\dagger}$ and $c_{j\sigma}$ denote creation and annihilation operators. $\langle i, j \rangle$ represents a summation over nearest neighbours and $\sigma = \uparrow, \downarrow$ stands for the spin. The single particle energy levels of the TGQR are shown in Fig. 14.7, for a TGQR with a total of 171 atoms (a) and one with a total of 504 atoms (b).

A number of degenerate states are found for the triangular ring, with the total number of degenerate states increasing with the number of atoms in the system. Furthermore, it has been established that these states stem from the orbitals of the outer atoms of the TGQR (depicted red in Fig. 14.6) [40]. Other states close to the Fermi level, however, appear to consist of the orbitals of type A and type B atoms (red and blue in Fig. 14.6), and are mainly localised on the inner edge of the TGQR [38].

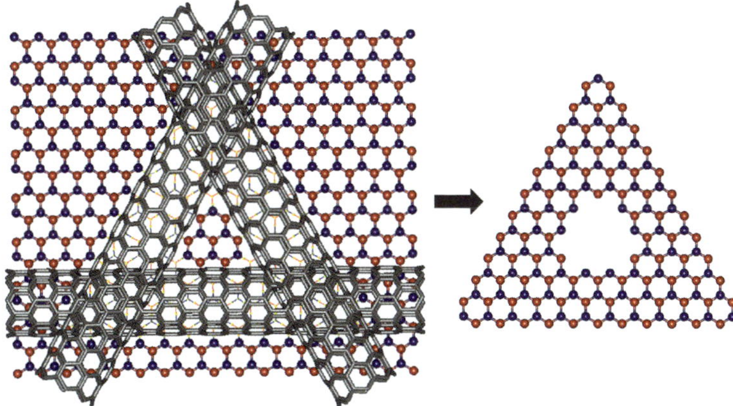

Fig. 14.6 Proposed experimental method for designing a TGQR. Three Carbon nanotubes arranged in an equilateral triangle along zigzag edges play the role of a mask. By using etching methods one can obtain a TGQR with well-defined edges. The circumference of the carbon nanotubes determines the width of the TGQR. *Red (light gray)* and *blue (dark gray)* colors distinguish between two sublattices in the honeycomb graphene lattice. Reprinted with permission from [38]. Copyright 2011, American Physical Society

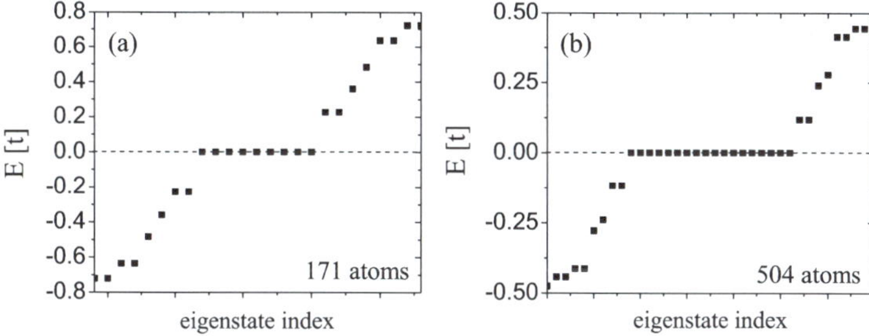

Fig. 14.7 Single particle TB levels for TGQR with (**a**) $N_{\text{width}} = 2$, consisting of 171 atoms and (**b**) $N_{\text{width}} = 5$, consisting of 504 atoms. The degeneracy at the Fermi level (*dashed line*) is a function of the width $N_{\text{zero}} = 3(N_{\text{width}} + 1)$, for (**a**) $N_{\text{zero}} = 9$ and for (**b**) $N_{\text{zero}} = 18$. Reprinted with permission from [38]. Copyright 2011, American Physical Society

For the hexagonal graphene quantum ring, the single-particle spectrum depends likewise on the number of atoms in the system (Fig. 14.8). For the case of thin rings of only one benzene ring thickness ($W = 1$) with a total number of 192 atoms, nearly degenerate shells of energy levels above and below the Fermi level have been observed [37]. The energy splitting between these nearly-degenerate states increases when the thickness of the ring is increased. A shell-like structure of the single-particle states is furthermore only found for ring thicknesses of one benzene ring (Fig. 14.9). Furthermore, a second hopping element t' is introduced, which describes the coupling of the six ribbons, thus allowing to investigate the transformation of six independent graphene ribbons into one graphene quantum ring. With the hopping

Fig. 14.8 Single-particle spectrum near the Fermi level for ring structures with $L = 8$, different widths W and $t' = t$. The shell structure is clearly observed only for the thinnest ring $W = 1$. The *dotted line* indicates the location of the Fermi energy. Reprinted with permission from [37]. Copyright 2010, American Physical Society

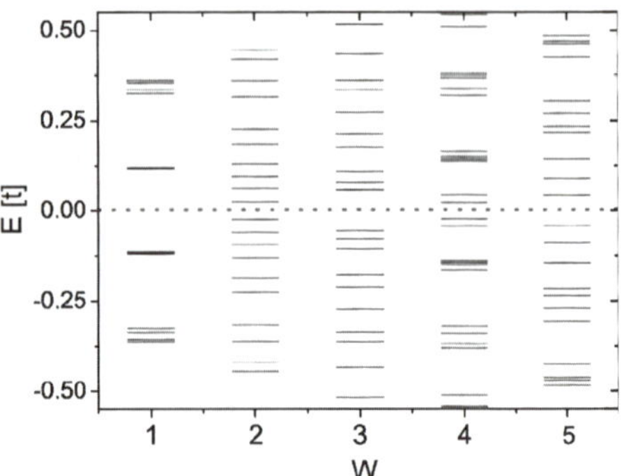

integral $t' = 0$ between the six ribbons, the case of independent ribbons is described, and correspondingly each eigenvalue is at least sixfold degenerate. It can now be seen that for the thin quantum rings ($W = 1$), the single-particle states closest to the Fermi energy remain almost degenerate, when the hopping-integral t' between the six ribbons is increased towards t, thus subsequently introducing a coupling between the ribbons towards a quantum ring (Fig. 14.10). For thicker rings ($W = 2$ and $W = 3$), the sixfold degeneracies are strongly broken (Fig. 14.10).

From the single-particle wave functions and energies, the many-particle properties of the system can be calculated. For this purpose, the many-body Hamiltonian

$$\hat{H}_{\mathrm{mb}} = \sum_{i,\sigma} \varepsilon_i c_{i,\sigma}^{\dagger} c_{i,\sigma} + \frac{1}{2} \sum_{i,j,k,l} \langle ij|V|kl\rangle c_{i,\sigma}^{\dagger} c_{j,\sigma'}^{\dagger}, c_{k,\sigma'} c_{l,\sigma} \qquad (14.6)$$

is diagonalised in the basis of all possible charge carrier configurations of the single-particle states obtained. Here, $\langle ij|V|kl\rangle$ denotes the Coulomb matrix element with i, j, k, l indicating the site of the charge carrier ($i = 1, 2, 3$ indicates on-site, nearest-neighbour and next-nearest neighbour electrons).

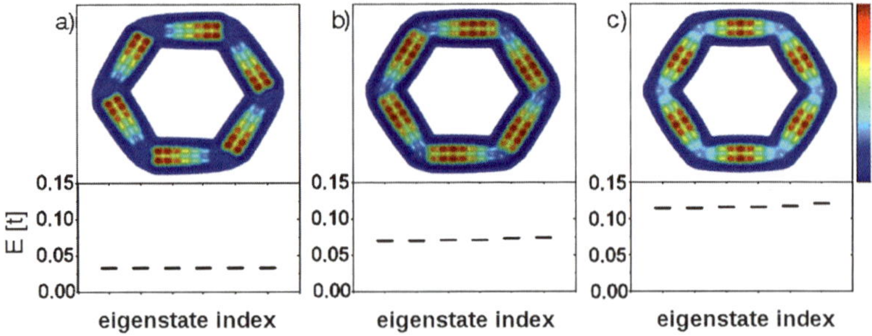

Fig. 14.9 Energy levels and corresponding total electronic densities for the first six states above the Fermi level for the thinnest structure $W = 1$ with $L = 8$ (192 atoms), for (**a**) $t' = 0$, (**b**) $t' = 0.5t$, and (**c**) $t' = t$. Reprinted with permission from [37]. Copyright 2010, American Physical Society

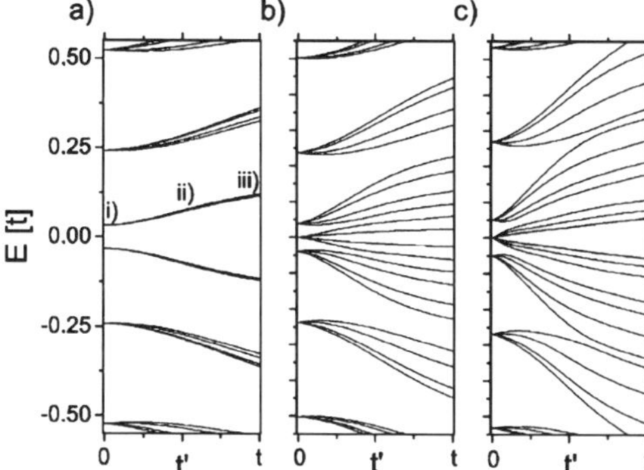

Fig. 14.10 The evolution of the single-particle spectrum from six independent ribbons with $L = 8$ to a hexagonal ring structure spectrum. t' indicate hopping integrals between neighbouring ribbons. (**a**) For the thinnest ring ($W = 1$), the sixfold degeneracy is slightly removed, preserving a shell structure. For thicker structures (**b**) and (**c**), $W = 2$ and $W = 3$, respectively, the sixfold degeneracy is strongly lifted and a shell structure is not observed. Reprinted with permission from [37]. Copyright 2010, American Physical Society

Based on the previously derived single-particle states, a study of the ground and excited electron states is presented as a function of the number of interacting electrons in different quantum rings in Ref. [37].

It is assumed that all states below the Fermi level remain fully occupied, due to the presence of a sufficiently large energy gap around the Fermi level. Additionally, scattering processes between the electron shells have been neglected. All shells then consist of six energy levels, that can be filled each with two electrons (spin up and spin down). The considerations below are all done for thin quantum rings ($W = 1$) with different lengths of each of the six graphene ribbons used to construct the ring ($L = 2$ to $L = 10$, where L denotes the length of each ribbon in terms of benzene molecules). The eigenenergies of the electron states above the Fermi level are then calculated for different total spins. The energy spectra of rings with $L = 4$ (96 atoms) and $L = 8$ (192 atoms) are shown in Fig. 14.11 for half-filled shells (six electrons) and total spins from $S = 0$ to $S = 3$. It is found that for the larger ring ($L = 8$) the ground state energy is minimal for $S = 3$, i.e. for the case that all electrons in the ground state shell above the Fermi level have the same spin. This is also seen in rings with $L > 5$ and half-filled shells. It means that for the larger quantum rings a maximum polarisation is energetically favourable. For the smaller rings, the situation is different and the ground state is found to exhibit $S = 0$ ($L = 3$) or $S = 1$ ($L = 2$ and $L = 4$) in a half-occupied shell. The different behavior of small and large rings can be explained from two competing effects that determine the character of the spin orientations. Occupation of the levels of a shell with small single particle energies favours opposite spin directions thus inducing a small total spin, whereas exchange interactions are maximal in systems with parallel spins, i.e. a large total spin. For larger rings ($L \geq 5$), the splitting between the energy levels inside a shell is small, and thus the behaviour of the ground state is dominated by the exchange effects, resulting in parallel spin orientations. For the smaller rings, the effects of

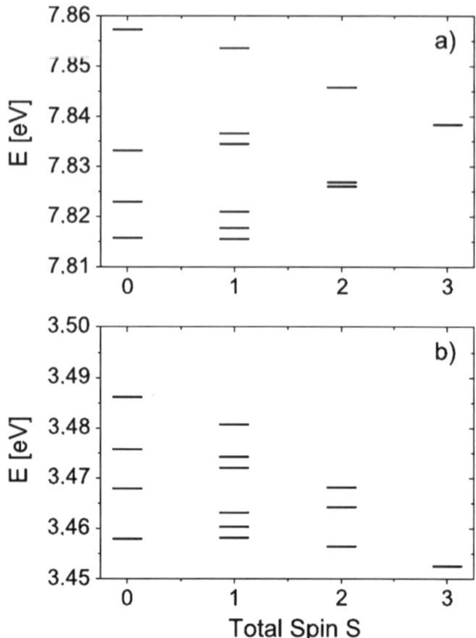

Fig. 14.11 The low-energy spectra for the different total spin S of the half-filled first shell above the Fermi energy for the two thinnest rings $W = 1$ with (**a**) $L = 4$ (96 atoms) and (**b**) $L = 8$ (192 atoms). Reprinted with permission from [37]. Copyright 2010, American Physical Society

single particle energy splittings and exchange interactions are comparable, which leads to alternating total spins $S = 0$ and $S = 1$. This sensitivity of the total spin on the size of the quantum ring will be highly important for the behavior of charge carriers in graphene quantum rings in magnetic fields.

14.4 Current-Spin Density Functional Theory

The Density Functional Theory (DFT) [41] allows for a highly accurate *ab initio* description of periodic systems and small molecules. However, its immense computational effort still limits the accessible system size to approximately 1000 atoms, which is too small for most quantum ring systems, in particular for self-organised quantum rings that can easily exhibit diameters of around 100 nm.

Within the Current-Spin Density Functional Theory (CSDFT) [42], a nanostructured system is described in terms of an electron density which is localised by a confining potential within a continuum picture instead of an atomistic description. This mean-field approach has been designed particularly to provide a good qualitative description of the electronic properties of nanostructures in the presence of a magnetic field, and has been applied to study the electronic structure of quantum dots [43, 44]. Relying on a continuum description of the nanostructure, the CS-DFT does not require the large computational effort arising from an atomistic *ab initio* DFT calculation, and furthermore easily allows for introducing asymmetries and impurities in a system and for studying their influence on the electronic properties.

Within the CSDFT, N electrons are considered in a ring with the radius R by the confining potential:

$$V_c(r) = \frac{m_e}{2}\omega_0^2(r-R)^2 \tag{14.7}$$

where ω_0 is the confinement frequency and m_e is the effective mass of the electron in the ring material [45]. It is convenient, to introduce the *degree of one-dimensionality* C_F instead of using the quantities m_e, ω_0 and R [46]:

$$C_F = \omega_0 \frac{32 m_e r_s^2}{\pi^2 \hbar} \tag{14.8}$$

with $r_s = \pi R/N$ being the one-dimensional average spacing between the electrons confined in the ring. The degree of one-dimensionality is a measure for the confining potential: the larger C_F is, the more narrow is the quantum ring's confining potential.

With the confining potential and the number N of electrons in the system, the Kohn-Sham equations of the electrons in the system can be solved [47]:

$$\left(-\frac{\hbar^2}{2m}\nabla^2 + V_{\text{eff}}(\mathbf{r})\right)\Phi_i(\mathbf{r}) = \varepsilon_i\Phi_i(\mathbf{r}) \tag{14.9}$$

$\Phi_i(\mathbf{r})$ being the Kohn-Sham-Orbitals of the ith electron. The effective potential $V_{\text{eff}}(\mathbf{r})$ is defined as

$$V_{\text{eff}}(\mathbf{r}) = V_c(\mathbf{r}) + e^2\int\frac{\varrho(\mathbf{r}')}{|\mathbf{r}-\mathbf{r}'|}d\mathbf{r}' + V_{\text{xc}}(\mathbf{r}) \tag{14.10}$$

$\varrho(\mathbf{r})$ being the charge density and V_{xc} the exchange-correlation potential:

$$\varrho(\mathbf{r}) = \sum_{i=1}^{N}|\Phi_i(\mathbf{r})|^2 \quad\text{and}\quad V_{\text{xc}}(\mathbf{r}) = \frac{\delta E_{\text{xc}}[\varrho]}{\delta\varrho(\mathbf{r})} \tag{14.11}$$

The Kohn-Sham-equations are then solved in a self-consistent manner to determine the corresponding eigenenergies ε_i.

14.4.1 Electronic Properties of Coupled GaAs/AlGaAs Quantum Rings

A number of studies of semiconductor quantum rings and their electronic and magnetic properties rely on the CSDFT [45, 48, 49]. In Chap. 12, a systematic study of elliptical quantum rings in comparison to ideal rotational symmetric ones based on a CSDFT model is provided in full detail. To illustrate the capabilities of this approach, a study of the electronic structure of coupled few-electron double quantum rings (DQR) [50, 51] shall be presented here.

The DQR in Refs. [50] and [51] consists of two concentric GaAs quantum rings in an $Al_{0.3}Ga_{0.7}As$ matrix. Following experimental data [52], the height of the DQRs is described as a superposition of Gaussian functions as:

$$H(r) = h_{in} \cdot \exp\left[-\frac{r - R_{in}}{\sigma_{in}}\right] + h_{out} \cdot \exp\left[-\frac{r - R_{out}}{\sigma_{out}}\right] \qquad (14.12)$$

with $\sigma^2 = \Delta v_{1/2}^2 / \ln 2$ where $v_{1/2}$ is the half-width of the Gaussians used to model the DQRs. The confining potential is then zero inside the DQR and the conduction band offset V_c between $Al_{0.3}Ga_{0.7}As$ and GaAs outside the DQR. The radius of the inner ring has been fixed to $R_{in} = 22.5$ nm, and the outer radius is varied from 22.5 nm $\leq R_{out} \leq 50$ nm. The half-widths of both rings are $v_{1/2,in} = 12.5$ nm and $v_{1/2,out} = 30$ nm and $h_{in} = h_{out} = 4$ nm. The conduction band offset between GaAs and $Al_{0.3}Ga_{0.7}As$ is $V_c = 262$ meV. Furthermore, an effective electron mass of $m_e = 0.067 m_0$ with m_0 being the electron mass in bulk GaAs has been assumed.

The first two single-particle electron states localised in the DQR ($n = 0$ and $n = 1$) are energetically close and it is seen that the ground state electron is confined in the inner ring for most outer radii and the energy difference between state $n = 1$ that is localised in the outer ring and the ground state $n = 0$ decreases with increasing R_{out} until the states become almost degenerate for an outer radius of $R_{out} = 49$ nm. The reason for this behaviour is that the volume of the inner ring decreases effectively with larger R_{out}, so that for $R_{out} = 50$ nm the outer ring is energetically favourable and the ground state thus localises in the outer ring.

The charge densities for N noninteracting electrons in the DQR are shown in Fig. 14.12 for $N = 2$ to $N = 15$ and for outer ring radii of 45 nm to 50 nm along ϱ. In this calculation, Coulomb interactions between the electrons are neglected. It can be seen, that with increasing R_{out} all charge densities move from the inner ring into the outer ring, and are finally fully confined in the outer ring for $R_{out} = 50$ nm. This results from the fact that the outer ring has a larger volume when the radius R_{out} is increased, whereas the inner ring's volume is decreased at the same time, making the outer ring energetically favourable for the electrons above a certain outer radius.

Once the Coulomb interaction is included, it can be expected that electrons are forced into the outer ring due to the Coulomb repulsion if the inner one is occupied by too many electrons. For the outer ring diameter of $R_{out} = 50$ nm, however, this behaviour might be reversed, such that electrons from the outer ring are pushed into the inner one. In fact, this behaviour can be seen in Fig. 14.13, where the charge densities of N interacting electrons in the DQR are shown for different outer radii and different number of electrons. It can be seen that, in comparison to Fig. 14.12, the electrons are already pushed into the outer ring for the radius $R_{out} = 42.5$ nm, once

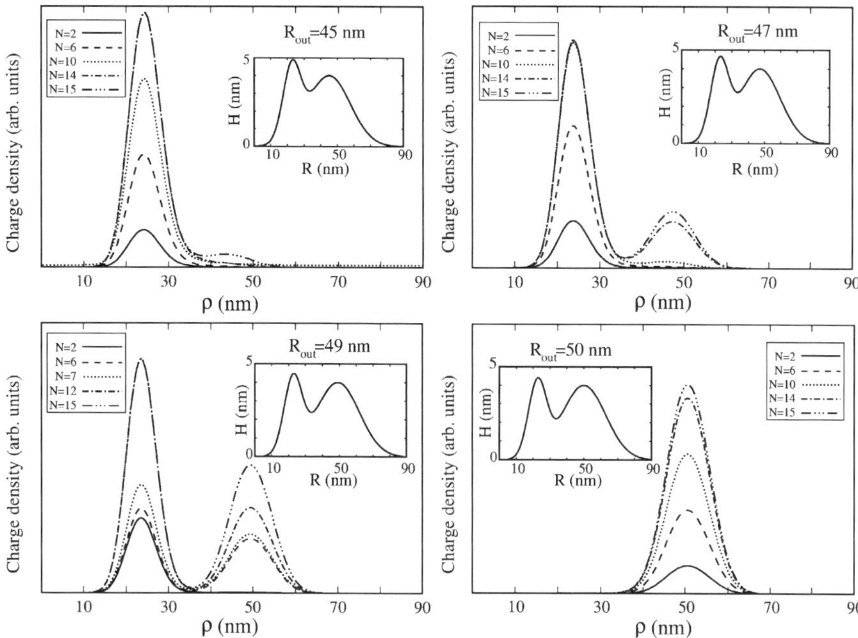

Fig. 14.12 Radial charge density distribution of N noninteracting electrons in DQRs with $R_{in} = 22.5$ nm for different values of R_{out}. The *insets* illustrate the DQR cross-section profile. Reprinted with permission from [51]. Copyright 2006, American Physical Society

a sufficiently large number of electrons is considered ($N = 12$ and $N = 15$). Additionally, for $R_{out} = 50$ nm a charge carrier localisation occurs in the inner ring for $N = 15$, as the Coulomb interaction forces electrons into the inner ring in this case. For an outer radius of $R_{out} = 47.5$ nm, the charge density is found to be extremely sensitive to the number of electrons. $N = 1$ and 2 electrons localise in the inner ring, $N = 3$ to 5 in the outer one and $N = 6$ in the inner one again. For this outer radius, the electrons localised in the inner and the outer ring are very close in energy and such a DQR thus represents a strongly correlated system, where Coulomb-mediated tunneling is highly efficient.

Another effect of the Coulomb interaction is that for $R_{out} = 47.5$ nm the inner ring is found to be completely spin-polarised for $N = 12$ electrons, whereas the outer ring is almost depolarised (see Fig. 14.14) [51]. This is a result of strong energy splittings between the spin-up and the spin-down levels which is much larger in the inner ring than in the outer ring as the Coulomb interaction is much stronger in the inner ring, due a narrower confinement.

A detailed study on the influence of the electronic properties of coupled quantum dot-quantum ring nanostructures based on the effective mass approximation (EMA) can be found in Chap. 18. The EMA will be discussed in detail in the following section.

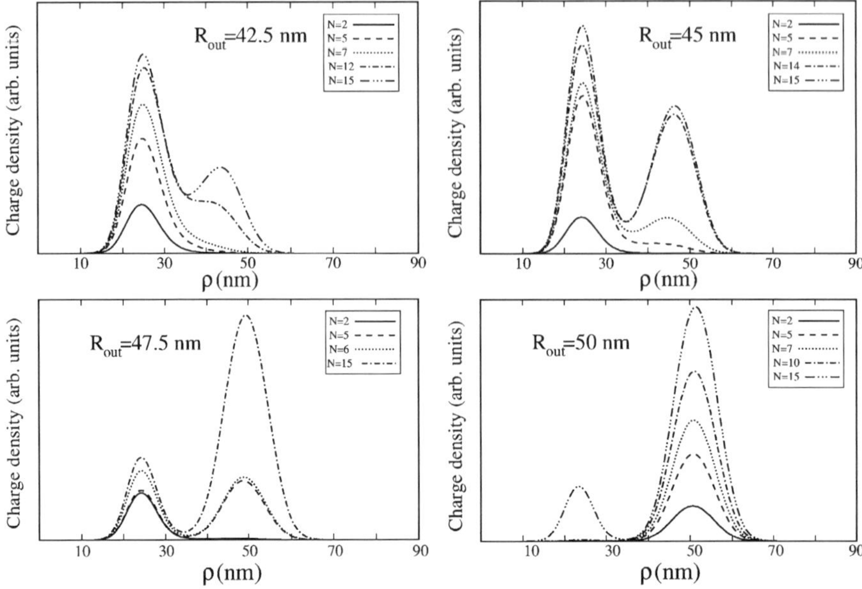

Fig. 14.13 Radial charge density distribution of N interacting electrons in DQRs with $R_{\mathrm{in}} = 22.5$ nm for different values of R_{out}. Reprinted with permission from [51]. Copyright 2006, American Physical Society

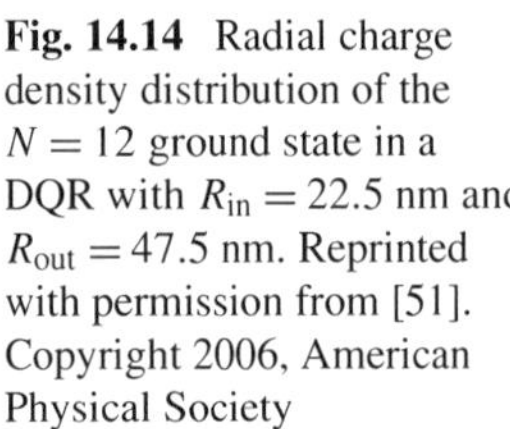

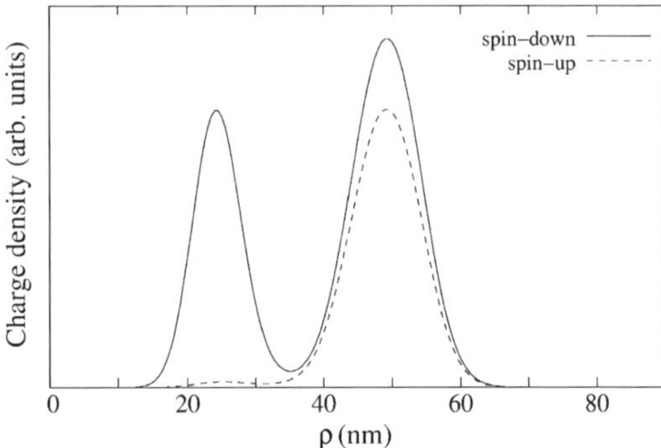

Fig. 14.14 Radial charge density distribution of the $N = 12$ ground state in a DQR with $R_{\mathrm{in}} = 22.5$ nm and $R_{\mathrm{out}} = 47.5$ nm. Reprinted with permission from [51]. Copyright 2006, American Physical Society

14.5 Effective Mass Models

The EMA is a computationally inexpensive approach to the electronic structure of particles in a confining potential. To obtain the single-particle properties of a system, the charge carriers are treated as particles with an effective mass m_e in a finite confining potential, that can have arbitrary shapes in one, two or three dimensions. Correspondingly, a nanostructured system such as a quantum ring, is described in terms of continuous material parameters, thus neglecting the atomistic nature of the system.

Within the EMA, the single-particle Schrödinger equation is solved using the Hamiltonian

$$\hat{H}_{\mathrm{EMA}} = -\frac{\hbar^2}{2m_e(\mathbf{r})}\nabla^2 + V_c(\mathbf{r}) \tag{14.13}$$

where $V_c(\mathbf{r})$ is the confining potential, in the case of the electrons represented by the conduction band offset between the nanostructure and the surrounding matrix material. The effective mass $m_e(\mathbf{r})$ is spatially dependent here, but for some systems constant effective masses are likewise reasonable, if the effective masses do not differ too much and a strong confinement can be assumed. The many-particle properties are then derived by adding a Coulomb interaction term:

$$\hat{H}_{\mathrm{EMA}}^{\mathrm{mp}} = -\frac{\hbar^2}{2m_e(\mathbf{r})}\nabla^2 + V_c(\mathbf{r}) + \sum_{i<j}^{N} \frac{e^2}{4\pi\kappa\kappa_0|\mathbf{r}_i - \mathbf{r}_j|} \tag{14.14}$$

with κ and κ_0 being the relative and the vacuum dielectric constants. Despite the inaccuracy of the EMA neglecting the atomistic nature and assuming a parabolic behaviour of the electronic band, this model has been applied to quite a number of different nanostructured systems [53, 54], and can still provide a first approach to a qualitatively good description of electronic properties of nanostructures at very small computational costs, and it is a powerful model to study the qualitative influences such as material inhomogeneities or deviations from an ideal shape of a nanostructure on its electronic properties, as has been done e.g. by considering realistic shapes for the calculation of optical properties of quantum rings [55] and for studies of the magnetic properties in Chap. 12 of this book.

14.5.1 InGaAs Quantum Rings in a Vertical Electric Field

The EMA has been used to explore the electronic properties of a number of quantum ring systems [56–59] and several studies are dedicated to inhomogeneities in the quantum ring geometry or the influence of electric fields on the electronic properties [60–62]. Chapter 9 provides a more detailed study of the influence of a vertical electric field on Aharonov-Bohm-Oscillations in In(Ga)As quantum rings.

Here, the work of McDonald et al. [62] has been chosen to present the applicability, advantages and disadvantages of the EMA to describe the electronic properties of semiconductor quantum rings.

14.5.1.1 Model System, Strain and Piezoelectric Potential

The model quantum ring is a self-assembled $In_{0.5}Ga_{0.5}As$ quantum ring with a height of $h = 6$ nm, an outer radius of $R_{\mathrm{out}} = 20$ nm and an inner radius of $R_{\mathrm{in}} = 5$ nm, residing on a thin $In_{0.30}Ga_{0.70}As$ wetting layer, and the system is buried in GaAs. The elastic properties have been found using VFF, based on which the

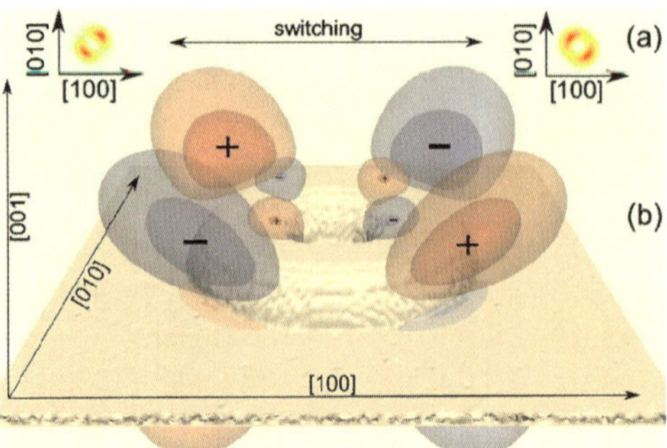

Fig. 14.15 (**a**) Hole density for two different localisations of excitons. (**b**) Illustration of negative and positive piezoelectric potential superimposed upon the structure of the model ring. Each lobe is represented by two isosurfaces of ± 38 and ± 54 meV. Reprinted with permission from [62]. Copyright 2010, American Institute of Physics

piezoelectric potentials have been calculated, including second-order effects as described in Ref. [63].

Strain and polarisation potentials exhibit a twofold symmetry here, which is lower than the rotational symmetry that the quantum ring geometry exhibits. The polarisation potential of the system is shown in Fig. 14.15. However, this potential does not, as claimed in Ref. [62], reduce the actual symmetry of the quantum ring, but only the artificial symmetry seen here that results from the inability of this continuum model to allow for a correct description of the underlying crystal symmetry [64]. The symmetry of nanostructures in the zinc-blende lattice and its treatment in continuum approaches have been matter of detailed discussions, and it has been shown from closer considerations of the atomistic setup of interfaces in the zinc-blende lattice, that quantum dots with a C_{∞}- or C_{4v}-symmetric shape such as cylindric, lens-shaped or pyramidal quantum dots do in fact have only a C_{2v}-symmetry in a more sophisticated atomistic picture [64]. This means, that even without the influences of strain and piezoelectricity, the quantum ring studied here exhibits only a C_{2v}-symmetry. While the artificially increased symmetry in a continuum picture induces only minor modifications to the electronic structure in cases where the charge carriers are localised well in the center of a nanostructure without experiencing much of the interface's influences [65] (e.g. in sufficiently large quantum dots), these effects are of higher significance in quantum rings, where the influence of the interfaces between the quantum ring and the matrix material is more important [12], but a more detailed, systematic investigation of the discrepancy between atomistic and continuum approaches on the electronic properties of semiconductor quantum rings has not been carried out so far.

14.5.1.2 Excitonic Properties

The electron and hole states and eigenenergies have been calculated using an EMA model, where the conduction and valence band energies have been modified by

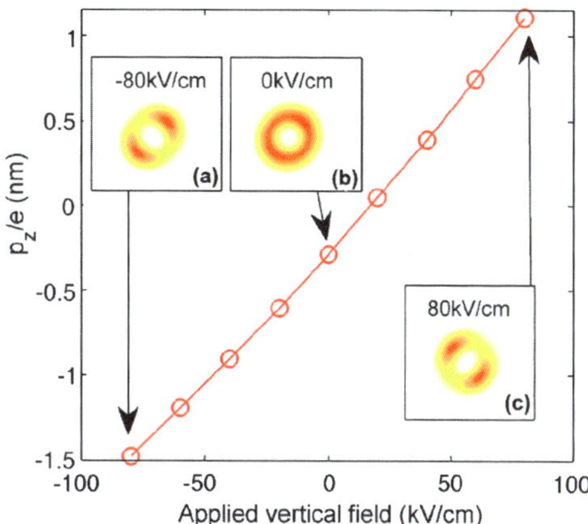

Fig. 14.16 Vertical dipole, p_z, for different electric fields applied vertically, shows changing of sign, and a permanent dipole already at zero field. Insets (**a**), (**b**), and (**c**) show the mean electron probability distribution for −80, 0, and 80 kV/cm and demonstrate switching of the exciton around the ring by changing the sign of the vertical field. Reprinted with permission from [62]. Copyright 2010, American Institute of Physics

the strain profiles obtained from an atomistic VFF. A path integral quantum Monte Carlo simulation [66, 67] has been then employed to compute the excitonic properties of the system in Ref. [62].

An electric field is applied along the [001]-direction, as an additional external potential contribution to the EMA Hamiltonian, and leads to a polarisation of the excitons. It is found that the sign of the external electric field dominates the orientation of charge carriers in the ring, as it pushes the charge carriers into the minima of the polarisation potential, that exhibits a C_{2v}-symmetry and thus the charge carriers are found to be oriented along either the [110]-direction or along the [1$\bar{1}$0]-direction, depending on the sign of the external electric field (Fig. 14.16).

Resulting from the charge carrier localisation, the excitons in such a quantum ring are similarly affected by an external electric field applied vertically, i.e. along the [001]-direction. It is found in Ref. [62], that a positive vertical field applied localises the exciton in the [1$\bar{1}$0]-direction, whereas a negative electric field induces a localisation along [110], which results in an increase of the lateral polarisability tangential to the direction of the confinement.

The fact that already for a zero external electric field a nonzero dipole is observed results from the piezoelectric potentials and the different effective masses of electrons and holes. Due to their larger effective mass, the holes tend to a stronger localisation, and are thus more sensitive to strain-induced modifications of the valence band and the piezoelectric potentials in and around the quantum ring. The excitons then align with the hole state localisation, resulting in a polarisability difference between the [110] and the [1$\bar{1}$0]-polarisation direction even in the absence of an external electric field.

The study in Ref. [62] outlines how a vertical electric field can be employed to modify the polarisability of a self-assembled InGaAs quantum ring, due to the existence of strain and piezoelectric potentials. While the behaviour of excitons in such a quantum ring in the presence of an electric field can be considered well described by this work, the neglected influence of the atomistic nature of the interfaces represents a shortcoming of such a continuum approach, of which the quantitative deviations towards an atomistic modelling remain so far unknown.

14.6 Multiband $\mathbf{k} \cdot \mathbf{p}$ Approaches

Multiband $\mathbf{k} \cdot \mathbf{p}$ approaches are a more sophisticated generalisation of the single-band EMA discussed above. These models describe a band structure of a number of non-perturbatively (i.e. directly) treated bands, where the interaction between these bands allows for a more accurate description of the band structure and hence a more detailed description of the electronic structure of semiconductor nanostructures. Remote bands are considered only perturbatively.

Following Yu and Cardona [68], the element $\hat{H}_{ij}$ of an n-band Hamiltonian matrix $\hat{H}_{n \times n}$, where Γ_e denotes the set of bands directly considered, is calculated as:

$$\hat{H}_{ij} = \varepsilon_i \delta_{ij} + \hat{H}_{ij}^0 + \sum_{l \notin \Gamma_e} \frac{\hat{H}_{il} \hat{H}_{lj}}{\varepsilon_i - \varepsilon_l}$$

$$= \varepsilon_i \delta_{ij} + \langle \Psi_i | \frac{\hbar^2 \mathbf{k}^2}{2m_0} + \frac{\hbar \mathbf{k} \cdot \mathbf{p}}{m_0} | \Psi_j \rangle \tag{14.15}$$

$$+ \sum_{l \notin \Gamma_e} \langle \Psi_i | \frac{\hbar^2 \mathbf{k}^2}{2m_0} + \frac{\hbar \mathbf{k} \cdot \mathbf{p}}{m_0} | \Psi_l \rangle \langle \Psi_l | \frac{\hbar^2 \mathbf{k}^2}{2m_0} + \frac{\hbar \mathbf{k} \cdot \mathbf{p}}{m_0} | \Psi_j \rangle \frac{1}{\varepsilon_i - \varepsilon_l}. \tag{14.16}$$

The influence of the remaining bands l on the directly considered bands of Γ_e is treated perturbatively. By increasing the number of directly considered bands in Γ_e, a more accurate description of the band structure throughout the Brillouin zone can be achieved. The parameters that determine the shape of the band structure, and thus band gaps and effective masses, are typically fitted to ensure a good description of the band structure in the vicinity of the Γ point $\mathbf{k} = (0, 0, 0)$, and with more bands directly considered a larger section of the Brillouin zone can be described accurately. A minimum basis of 15 bands has been found to be required to achieve a reliable description of the dispersion relation throughout the whole Brillouin zone [69], which of course requires a correspondingly large set of material parameters that need to be determined from experiment or *ab initio* calculations.

For the description of direct band gap semiconductor nanostructures, six- [70] and eight-band [71–73] $\mathbf{k} \cdot \mathbf{p}$ models have been highly successful in studies of the electronic structure of nanostructured systems of various shapes, sizes and material compositions [74–77]. The eight-band model includes the three energetically highest valence bands and the lowest conduction band with each spin-up and spin-down components. The six-band model [70] describes only the upper valence bands, thus neglecting the interaction between valence- and conduction bands. A common eight-band $\mathbf{k} \cdot \mathbf{p}$ Hamiltonian for the zinc-blende lattice including strain and piezoelectric contributions can be found in Ref. [78].

The eight-band $\mathbf{k} \cdot \mathbf{p}$ approach allows for a reliable description of the band structure of the energetically lowest conduction band and the three highest valence bands in the vicinity of the Γ point, but becomes inaccurate towards the outer areas of the Brillouin zone, due to the direct interaction with remote bands that are treated only perturbatively in this model. As the small $\mathbf{k}$ vectors around the Γ point are the ones of primary importance for most nanostructures, the eight-band approach still allows for a reliable description of most semiconductor nanostructures and a good agreement with more sophisticated atomistic models is commonly seen (e.g. in Ref. [65]).

While a better accuracy of the dispersion relation of valence- and conduction bands is achieved in the eight-band model in comparison to single-band EMA approaches, the more sophisticated model describes nanostructures also within in a continuum picture and thus neglects all atomistic effects. For the case of the eight-band model, this leads to the similar artificially increased symmetry of zinc-blende nanostructures that is also observed in an EMA, and that is not found in atomistic models [64].

14.6.1 Electronic Properties of CdTe/ZnTe Quantum Rings

A number of studies based on continuum models go beyond the simple EMA, using two- or four-band descriptions for the valence bands [79–81]. Chapter 14 provides a detailed investigation of hole mixing in semiconductor quantum rings, based on a six-band $\mathbf{k} \cdot \mathbf{p}$ approach. The much more sophisticated eight-band approach that includes the important coupling between valence and conduction bands, however has not yet received much attention for the description of electronic properties of semiconductor quantum rings. An eight-band $\mathbf{k} \cdot \mathbf{p}$ approach has been applied to CdTe/ZnTe quantum rings [82], which shall be used as an example of a multiband $\mathbf{k} \cdot \mathbf{p}$ model employed to study the electronic structure of semiconductor quantum rings here.

14.6.2 Model and Formalism

The calculations performed in Ref. [82] rely on the eight-band $\mathbf{k} \cdot \mathbf{p}$ model by Bahder [71], they include strain but neglect the influence of piezoelectric polarisation. However, in this study the Hamiltonian is mainly used to discuss the impact of strain on the interband transition energy, i.e. the energy difference between the conduction band and the heavy-hole subband, rather than the charge carrier localisation of electrons and holes in the systems under consideration. Additionally, a temperature dependence of the CdTe bandgap is assumed:

$$E_g(T) = E_g(0) - \frac{\alpha T^2}{\Theta_D + T} \tag{14.17}$$

where $\alpha = 4.73 \cdot 10^{-4}$ eV/K is the Varshni's proportionality constant [83], $\Theta_D = 160$ K is the Debye temperature and $E_g(0) = 1.923$ eV is the band gap at $T = 0$ K. The elastic properties of the system are calculated on the basis of linear elasticity, where the stress (σ_{ij}) and strain (ε_{ij}) is given as:

$$\sigma_{ij} = C_{ijkl}\left(\varepsilon_{kl} - \varepsilon_0^{kl}\right), \tag{14.18}$$

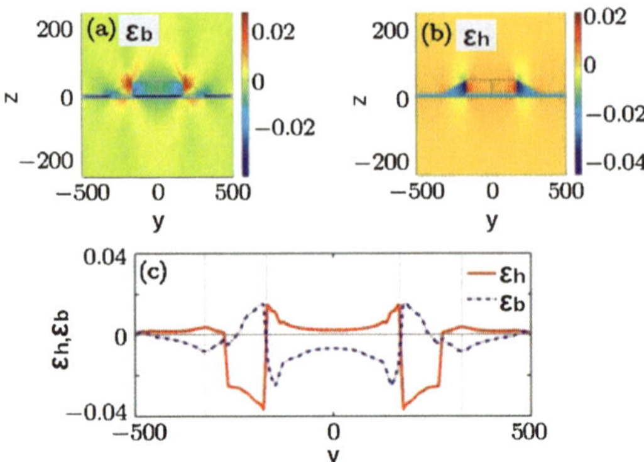

Fig. 14.17 Cross section plots of (**a**) hydrostatic ε_h and (**b**) biaxial ε_b strains in the y-z-plane. (**c**) Plots of hydrostatic and biaxial strains at $z = 0$ along the y axis. Reprinted with permission from [82]. Copyright 2011, Chinese Physics B

with $i, j, k, l = x, y, z$. The initial strain ε_0 arises from the lattice and elastic constants:

$$\varepsilon_0^{xx} = \varepsilon_0^{yy} = \frac{a_m - a(\mathbf{r})}{a(\mathbf{r})} \quad \text{and} \quad \varepsilon_0^{zz} = -\frac{2C_{12}(\mathbf{r})}{C_{11}(\mathbf{r})}\varepsilon_0^{xx} \tag{14.19}$$

Here, $a(\mathbf{r})$ is a spatially dependent lattice constant with $a(\mathbf{r}) = a_m = 0.608$ nm in the ZnTe matrix material and $a(\mathbf{r}) = 0.648$ nm in the CdTe quantum ring. The elastic properties are then determined using a finite-elements method as described in Ref. [84].

The model quantum ring in the study in Ref. [82] has the outer diameter of $R_{\text{out}} = 65$ nm and an inner diameter of $R_{\text{in}} = 32.7$ nm. The height increases from $h = 0$ at the outer radius to the maximum height $h_r = 4.2$ nm at the inner radius, and then abruptly reduces to zero inside the inner radius, thus modelling a volcano-like shape. The ring resides on a wetting layer of a thickness of $w = 0.45$ nm and the whole system is buried in a ZnTe matrix.

14.6.3 Strain and Interband Transition Energy

The hydrostatic and biaxial strains $\varepsilon_h = \varepsilon_{xx} + \varepsilon_{yy} + \varepsilon_{zz}$ and $\varepsilon_b = \varepsilon_{zz} - (\varepsilon_{xx} + \varepsilon_{yy})/2$ are shown in Fig. 14.17 in a cross-section in the y-z-plane, together with a linescan of these properties through the ring's base along the y-axis. It is seen that the hydrostatic strain is mostly compressive inside the ring material, whereas the biaxial strain, which strongly influences the heavy-hole subband is tensile there.

The influence of the quantum ring's height h_r on the biaxial and hydrostatic strain is shown in Fig. 14.18, where ε_b and ε_h are shown for $h_r = 4.2$ nm and $h_r = 10.0$ nm. Only a minor influence of h_r on the hydrostatic strain is found, but the biaxial strain is significantly modified (Fig. 14.18). As a result, a strong impact on the heavy-hole subband can be expected, and thus on the interband transition energy.

The interband transition energies, i.e. the difference between the conduction band minimum and the valence band maximum, are shown in Fig. 14.19 for different

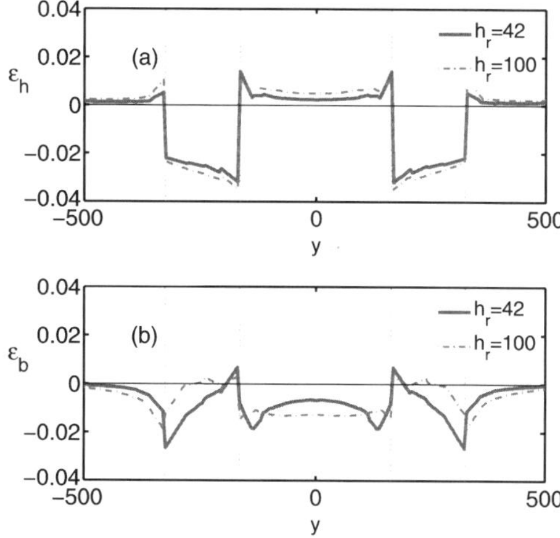

Fig. 14.18 Plots of (**a**) ε_h and (**b**) ε_b strains at $z = 0$ versus the quantum ring height along the y axis. Reprinted with permission from [82]. Copyright 2011, Chinese Physics B

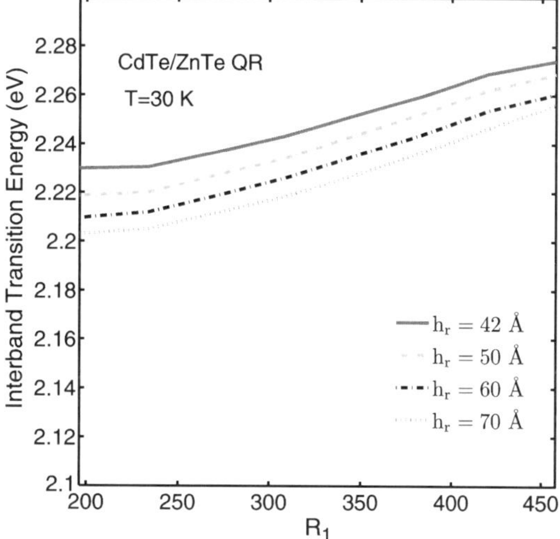

Fig. 14.19 Interband transition energy versus inner quantum ring diameter R_1 and height h_r at $T = 30$ K. The interband transition energy increases with increasing R_1 (or decreasing quantum ring volume), while it shifts downwards with increasing quantum ring height (or increasing quantum ring volume). Reprinted with permission from [82]. Copyright 2011, Chinese Physics B

quantum ring heights as a function of the inner radius for the temperature $T = 30$ K. It can be seen that generally the transition energy increases for both smaller h_r and larger inner radius R_1, i.e. for the case of reducing the quantum ring's volume. This is a result of the fact that the biaxial strain inside the ring is mostly negative, pushing the heavy-hole subband further away from the conduction band, thus increasing the interband transition energy.

Figure 14.20 shows the temperature dependence of the interband transition energy. It can be seen that increasing temperature inside the system leads to a reduction of this energy by approximately 15 meV from 30 K to 100 K for the quantum rings considered. Moreover, a good agreement of the calculations in Ref. [82] with the experiment has been found.

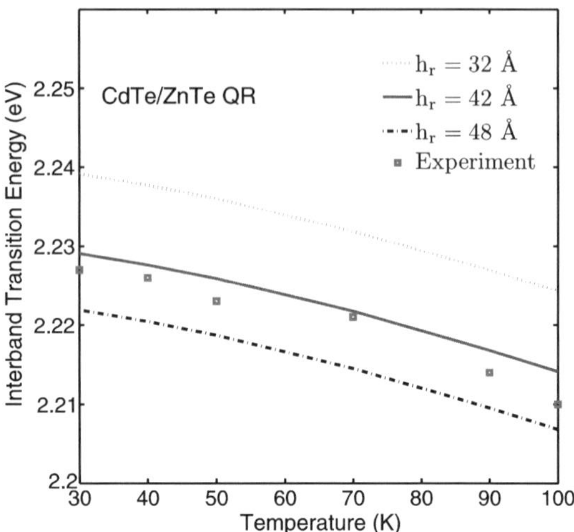

Fig. 14.20 Interband transition energy versus temperature and height of the quantum ring, which is in good agreement with the PL data for the quantum ring height at $h_r = 4.2$ nm [85]. Reprinted with permission from [82]. Copyright 2011, Chinese Physics B

14.6.4 Single-Particle States

The single-particle electron and hole states have not been calculated in Ref. [82], and it can be assumed that generally energy quantisation effects in the system under consideration are quite small as a result of the comparatively large dimensions. This means that the comparison of the interband transition energy with experimentally observed photoluminescence spectra will not suffer from these energy quantisation effects.

However, if an eight-band model is employed to study single-particle states of quantum rings, the symmetry breaking effect of the polarisation potential can be expected to be as important here as it is in single-band EMA calculations. Additionally, the same shortcoming of the single-band EMA applies that the underlying crystal symmetry is not considered, and as this induces already qualitative modifications in quantum dots where the electronic states are commonly localised well in the center of the system [64], this effect can be even more pronounced in quantum rings [12].

It has been shown that this disadvantage of most common multiband $\mathbf{k} \cdot \mathbf{p}$ models can be overcome by taking more than eight bands into account and a minimum of 14 bands has been found to allow for a correct description of the underlying crystal symmetry [86, 87]. It can therefore be assumed that using a more sophisticated multiband $\mathbf{k} \cdot \mathbf{p}$ approach than the well-established eight-band model will allow for a reliable description of the electronic and optical properties of experimentally observed quantum ring systems.

14.7 Conclusions

In this chapter, a variety of theoretical approaches to the electronic and optical properties of semiconductor quantum rings has been presented. In this context, atomistic studies based on the EPM [12] and ETBM [37, 38] calculations as well as continuum-based studies using CSDFT [50, 51], single-band EMA [62] and an eight-band $\mathbf{k} \cdot \mathbf{p}$ model [82] have been reviewed to illustrate and discuss the advantages and disadvantages of each model. Atomistic models do in general allow for a more accurate description of the electronic structure of quantum rings, but as the computational effort of these models increases with the number of atoms in the system, larger quantum rings will be difficult to access with these approaches, and immense computational resources will then be required.

For these systems, continuum models appear to be a more suitable choice. A significant drawback of single-band EMA and the well-established eight-band $\mathbf{k} \cdot \mathbf{p}$ approach, however, is the inability to describe the correct symmetry of the underlying crystal lattice. This problem is well known from studies on semiconductor quantum dots [64], but these effects are expected to be much more pronounced in quantum rings, where the interface regions are of higher importance [12]. As it has been shown that this deficiency can be overcome by including a larger number of directly considered bands in a multiband $\mathbf{k} \cdot \mathbf{p}$ model [86, 87], it is suggested that these more sophisticated continuum approaches should be applied to those quantum ring structures that exceed the dimensions accessible for a reasonable atomistic simulation.

Acknowledgements   The author would like to thank S. Schulz, V.M. Fomin, J.I. Climente, P. Potasz, L. He, and W.P. Hong for their kind support.

References

1. V.M. Fomin, L.F. Chibotaru, J. Nanoelectron. Optoelectron. **4**, 3–19 (2009)
2. V.M. Fomin, J. Nanoelectron. Optoelectron. **6**, 1–3 (2011)
3. A.A. Avetisyan, A.V. Ghazaryan, A.P. Djotyan et al., Acta Phys. Pol. A **116**, 826–828 (2009)
4. J. Even, S. Loualiche, J. Phys. A **37**, L289–L294 (2004)
5. J. Even, C. Cornet, S. Loualiche, Physica E **28**, 514–518 (2005)
6. M. Zarenia, J. Milton Pereira, A. Chaves et al., Phys. Rev. B **81**, 045431 (2010)
7. L.-W. Wang, A. Zunger, Phys. Rev. B **59**, 15806–15818 (1999)
8. A.J. Williamson, L.-W. Wang, A. Zunger, Phys. Rev. B **62**, 12963–12977 (2000)
9. L.-W. Wang, A. Zunger, J. Phys. Chem. **98**, 2158–2165 (1994)
10. R. Singh, G. Bester, Phys. Rev. B **85**, 205405 (2012)
11. L.-W. Wang, M. Califano, A. Zunger et al., Phys. Rev. Lett. **91**, 056404 (2003)
12. W. Zhang, Z. Su, M. Gong et al., Europhys. Lett. **83**, 67004 (2008)
13. W. Zhang, M. Gong, C.-F. Li et al., J. Appl. Phys. **106**, 104314 (2009)
14. P.N. Keating, Phys. Rev. **145**, 637–645 (1966)
15. J.L. Martins, A. Zunger, Phys. Rev. B **30**, R6217–R6220 (1984)
16. A. Franceschetti, H. Fu, L.-W. Wang et al., Phys. Rev. B **60**, 1819–1829 (1999)
17. D.W. Jenkins, J.D. Dow, Phys. Rev. B **39**, 3317–3329 (1989)
18. S. Schulz, G. Czycholl, Phys. Rev. B **72**, 165317 (2005)
19. H. Dierks, G. Czycholl, J. Cryst. Growth **184–185**, 877–881 (1998)
20. T. Saito, Y. Arakawa, Physica E **15**, 169–181 (2002)
21. R. Santoprete, B. Koiller, R.B. Capaz et al., Phys. Rev. B **68**, 235311 (2003)
22. M. Usman, T. Inoue, Y. Harda et al., Phys. Rev. B **84**, 115321 (2011)

23. P. Potasz, A.D. Güçlü, P. Hawrylak, Acta Phys. Pol. A **116**, 832–834 (2006)
24. J.P. Loehr, *Physics of Strained Quantum Well Lasers* (Kluwer, Boston, 1998)
25. J.C. Phillips, *Bonds and Bands in Semiconductors* (Academic, New York, 1973)
26. T.B. Boykin, G. Klimeck, R.C. Bowen et al., Phys. Rev. B **56**, 4102–4107 (1997)
27. J.M. Jancu, R. Scholz, F. Beltram et al., Phys. Rev. B **57**, 6493–6507 (1998)
28. J.G. Díaz, G.W. Bryant, Phys. Rev. B **73**, 075329 (2006)
29. P. Vogl, H.P. Hjalmarson, J.D. Dow, Phys. Chem. Solids **44**, 365–378 (1983)
30. C. Delerue, G. Allan, M. Lannoo, Phys. Rev. B **48**, 11024–11036 (1993)
31. S. Froyen, W.A. Harrison, Phys. Rev. B **20**, 2420–2422 (1979)
32. S.K. Maiti, J. Chowdhury, S.N. Karmakar, J. Phys. Condens. Matter **18**, 5349–5361 (2006)
33. E. Faizabadi, M. Omidi, Phys. Lett. A **373**, 1469–1477 (2009)
34. E. Faizabadi, M. Omidi, Phys. Lett. A **374**, 1762–1768 (2010)
35. M. Lee, C. Bruder, Phys. Rev. B **73**, 085315 (2006)
36. E.R. Hedin, Y.S. Joe, J. Appl. Phys. **110**, 026107 (2011)
37. P. Potasz, A.D. Güçlü, P. Hawrylak, Phys. Rev. B **82**, 075425 (2010)
38. P. Potasz, A.D. Güçlü, O. Voznyy et al., Phys. Rev. B **83**, 174441 (2011)
39. C. Stampfer, J. Güttinger, F. Molitor et al., Appl. Phys. Lett. **92**, 012102 (2008)
40. P. Potasz, A.D. Güçlü, P. Hawrylak, Phys. Rev. B **81**, 033403 (2010)
41. P. Hohenberg, W. Kohn, Phys. Rev. **136**, B864–B871 (1964)
42. G. Vignale, M. Rasolt, Phys. Rev. B **37**, 10685–10696 (1988)
43. M. Ferconi, G. Vignale, Phys. Rev. B **50**, 14722–14725 (1994)
44. M. Pi, M. Barranco, A. Emperador, E. Lippinari, Ll. Serra, Phys. Rev. B **57**, 14783–14792 (1998)
45. S. Viefers, P. Singha Deo, S.M. Reimann et al., Phys. Rev. B **62**, 10668–10673 (2000)
46. S.M. Reimann, M. Koskinen, M. Manninen, Phys. Rev. B **59**, 1613–1616 (1999)
47. W. Kohn, L.J. Sham, Phys. Rev. **140**, A1133–A1138 (1965)
48. J.C. Lin, G.Y. Guo, Phys. Rev. B **65**, 035304 (2002)
49. L.K. Castellano, G.Q. Hai, B. Partoens et al., Phys. Rev. B **74**, 045313 (2006)
50. J. Planelles, J.I. Climente, Eur. Phys. J. B **48**, 65–70 (2001)
51. J.I. Climente, J. Planelles, M. Barranco et al., Phys. Rev. B **73**, 235327 (2006)
52. T. Mano, T. Kuroda, S. Sanguinetti et al., Nano Lett. **5**, 425–428 (2005)
53. A. Wojs, P. Hawrylak, S. Fafard et al., Phys. Rev. B **54**, 5604–5608 (1996)
54. S. Baskoutas, A.F. Terzins, W. Schommers, J. Comput. Theor. Nanosci. **3**, 269–271 (2006)
55. N.A.J.M. Kleemans, I.M.A. Bominaar-Silkens, V.M. Fomin et al., Phys. Rev. Lett. **99**, 146808 (2007)
56. H.F. Cheung, Y. Gefen, E.K. Riedel et al., Phys. Rev. B **37**, 6050–6062 (1988)
57. T. Chakraborty, P. Pietiläinen, Phys. Rev. B **50**, 8460–8468 (1994)
58. S.S. Li, J.B. Xia, J. Appl. Phys. **89**, 3434–3437 (2001)
59. J.A. Barker, R.J. Warburton, E.P. O'Reilly, Phys. Rev. B **69**, 035327 (2004)
60. A. Bruno-Alfonso, A. Latgé, Phys. Rev. B **61**, 15887–15894 (2001)
61. J.M. Llorens, C. Trallero-Giner, A. García-Chrístobal et al., Phys. Rev. B **64**, 035309 (2001)
62. P.G. McDonald, J. Shumway, I. Galbraith, Appl. Phys. Lett. **97**, 173101 (2010)
63. G. Bester, A. Zunger, X. Wu et al., Phys. Rev. B **74**, 081305(R) (2006)
64. G. Bester, A. Zunger, Phys. Rev. B **71**, 045318 (2005)
65. O. Marquardt, D. Mourad, S. Schulz et al., Phys. Rev. B **78**, 235302 (2008)
66. D.M. Ceperley, Rev. Mod. Phys. **67**, 279–355 (1995)
67. M. Harowitz, D. Shin, J. Shumway, J. Low Temp. Phys. **140**, 211–226 (2005)
68. P.Y. Yu, M. Cardona, *Fundamentals of Semiconductors* (Springer, Berlin, 1996)
69. M. Cardona, F.H. Pollack, Phys. Rev. **142**, 530–543 (1966)
70. C. Pryor, Phys. Rev. B **56**, 10404–10411 (1997)
71. T. Bahder, Phys. Rev. B **41**, 11992–12001 (1990)
72. S.L. Chuang, C.S. Chang, Phys. Rev. B **54**, 2491–2504 (1996)
73. E.P. Pokatilov, V.A. Fonoberov, V.M. Fomin et al., Phys. Rev. B **64**, 245328 (2001)
74. O. Stier, D. Bimberg, Phys. Rev. B **55**, 7726–7732 (1997)

75. A.D. Andreev, E.P. O'Reilly, Phys. Rev. B **62**, 15851–15870 (2000)
76. M. Winkelnkemper, S. Schliwa, D. Bimberg, Phys. Rev. B **74**, 155322 (2006)
77. O. Marquardt, T. Hickel, J. Neugebauer, J. Appl. Phys. **106**, 083707 (2009)
78. A. Schliwa, Electronic properties of self-organized quantum dots. PhD Thesis, TU Berlin, Berlin, 2007
79. J. Planelles, W. Jaskólski, J.I. Aliaga, Phys. Rev. B **65**, 033306 (2001)
80. S.S. Li, J.B. Xia, J. Appl. Phys. **91**, 3227–3231 (2002)
81. B. Jia, Z. Yu, Y. Liu, Model. Simul. Mater. Sci. Eng. **17**, 035009 (2009)
82. W.P. Hong, S.H. Park, Chin. Phys. B **20**, 098502 (2011)
83. Y.P. Varshni, Physica **34**, 149–154 (2002)
84. M. Tadić, F.M. Peeters, K.L. Lanssens, J. Appl. Phys. **92**, 5819–5829 (2002)
85. T.W. Kim, E.H. Lee, K.H. Lee et al., Appl. Phys. Lett. **84**, 595–597 (2004)
86. S. Tomić, N. Vukmirović, J. Appl. Phys. **110**, 053710 (2011)
87. O. Marquardt, S. Schulz, C. Freysoldt et al., Opt. Quantum Electron. **44**, 183–188 (2012)

Chapter 15
Coulomb Interaction in Finite-Width Quantum Rings

Benjamin Baxevanis and Daniela Pfannkuche

Abstract Due to the particular confinement and reduced dimensionality of quantum rings, Coulomb interaction plays an important role for their electronic structure. In this chapter, we discuss the dependency of the ground state of an idealized quantum ring contained a small number of interacting electrons on geometric parameters like ring radius, radial confinement and eccentricity. The ring geometry affects both the charge distribution and the spin configuration in a quantum ring. Numerically exact results obtained from path-integral quantum Monte Carlo demonstrate the strong connection between the structure and the total spin of the ground state emerging from the interplay between confinement and Coulomb interaction.

15.1 Introduction

The specific topology of nanostructured quantum rings evokes a variety of features like the occurrence of persistent currents and molecular states [1–4]. The Coulomb interaction between confined electrons plays a vital role in the emergence of these properties [5, 6]. Depending on the process of manufacture the nature of the electrons in a quantum ring can range from predominantly single-particle characteristics to strongly correlated collective features [7–9].

In this chapter, we will show how the geometry of the system and the Coulomb interaction determine the electronic properties of a quantum ring. The simplest model is a one-dimensional ring with electrons moving on circular paths. A finite ring width additionally leads to coupled radial oscillations. Studies have revealed that the energy spectrum and correlation functions have different properties depending on the ring size and ring width [8, 10]. We will review how the adjustment of these parameters can induce a strong localization of the electrons inside a rotating frame. Furthermore, the interaction between the electrons causes the system to develop rotational and vibrational collective modes. In this regime, the energy spectra

B. Baxevanis (✉) · D. Pfannkuche
I. Institute for Theoretical Physics, University of Hamburg, Jungiusstr. 9, 20355 Hamburg, Germany
e-mail: benjamin.baxevanis@physik.uni-hamburg.de

D. Pfannkuche
e-mail: daniela.pfannkuche@physik.uni-hamburg.de

V.M. Fomin (ed.), *Physics of Quantum Rings*, NanoScience and Technology,
DOI 10.1007/978-3-642-39197-2_15, © Springer-Verlag Berlin Heidelberg 2014

can be described by an effective spin Hamiltonian combined with center-of-mass rotation and internal vibrations of the localized electrons [11].

The geometry of the system generates a strong coupling between the orbital degrees of freedom and the combined spin of the electrons. We will discuss how the ground state of the system can be derived based on the permutation symmetry of the particle by a rigorous group-theoretical derivation. The results are compared to exact numerical calculations for small numbers of electrons obtained using path-integral quantum Monte Carlo. The ground state is determined by the parameters of the system which define the crossover from a one-particle regime to a non-trivial many-body state.

Imperfections of the quantum ring structure can perturb the properties of an idealized ring. We consider elliptical deformations of the quantum ring which can arise e.g. from crystal strain. Analysis of the electronic states in deformed rings shows that the orbital degeneracy is lifted and gaps in the energy spectrum are opened depending on the degree of distortion [12, 13]. We demonstrate the effect of elliptical deformation on the few-particle ground state. The distortion of the ring reduces the system's symmetry and destroys the properties of a perfect circular ring especially through transitions of the ground state spin.

15.2 Circular Quantum Ring

A model Hamiltonian for N interacting electrons in a quantum ring can be written as

$$\hat{\mathscr{H}} = \sum_i^N \hat{\mathscr{H}}_i^{(1)} + \sum_{i<j}^N \frac{e^2}{\kappa |\mathbf{r}_i - \mathbf{r}_j|}, \tag{15.1}$$

with κ denoting the dielectric constant of the bulk and the ring is assumed to be perfectly flat in the direction perpendicular to the plain, i.e. the electron motion is strictly two-dimensional. The one-particle Hamiltonian $\hat{\mathscr{H}}_i^{(1)}$ describes an electron confined in radial direction by a displaced harmonic confinement

$$\hat{\mathscr{H}}_i^{(1)} = -\frac{\hbar^2}{2m^*}\left(\frac{\partial^2}{\partial r_i^2} + \frac{1}{r_i}\frac{\partial}{\partial r_i} + \frac{1}{r_i^2}\frac{\partial^2}{\partial\varphi^2}\right) + \frac{1}{2}m^*\omega_0^2(r_i - r_0)^2, \tag{15.2}$$

where r_0 is the radius of the ring, r_i the radial coordinate of the electron and m^* its effective mass. The electrons move freely in angular direction and are bounded in radial direction. We may relate the confinement frequency ω_0 to the width of the ring defined as $W = 2l_0$, where $l_0 = \sqrt{\hbar/m^*\omega_0}$ is the characteristic length of the displaced harmonic oscillator. For instance, experimentally estimated values in small quantum rings are a radius of $r_0 = 14$ nm and a confinement strength of $\hbar\omega_0 = 12$ meV [14].

For narrow quantum rings with $W \ll r_0$ one may assume that only the lowest radial mode is occupied. With the radial motion frozen out the system can be treated

Fig. 15.1 The single-particle energy levels depending on the ring radius ($E \propto 1/r_0^2$). For large radii the lowest energy levels become almost degenerate

as a one-dimensional ring with an electron motion restricted to a circular orbit of radius r_0. In this case, the Schrödinger equation of the azimuthal motion of the electrons reads

$$-\frac{\hbar^2}{2m^*r_0^2}\frac{\partial^2}{\partial\varphi^2}\psi_\ell = E_\ell\psi_\ell. \tag{15.3}$$

The solutions are the eigenfunctions ψ_ℓ and eigenenergies E_ℓ

$$\psi_\ell = \frac{1}{\sqrt{2\pi}}\exp(-i\ell\varphi), \qquad E_\ell = \frac{\hbar^2\ell^2}{2m^*r_0^2}, \tag{15.4}$$

with the boundary condition $\psi_\ell(\varphi) = \psi_\ell(\varphi + 2\pi)$ leading to quantum numbers $\ell = (0, \pm 1, \pm 2, \ldots)$. The total ($z$-component) angular momentum of many particles is given by $L = \sum_i \ell_i$.

In the following, the model of this quasi one-dimensional ring will be utilized to provide an understanding of the results obtained with the more realistic Hamiltonian (15.1). Although the motion of the electrons described by the Hamiltonian (15.1) is not strictly one-dimensional, in many cases the simplified model will serve as a good approximation.

The single-particle energies E_ℓ reveal two important features. First, the kinetic energy of the particles scales with the inverse square of the radius r_0 (Fig. 15.1). All levels except the ground state are doubly degenerate indicating the symmetry between left- and right-moving orbitals. As the radius increases the low-lying levels become almost degenerate.

On the other hand, the kinetic energy has a quadratic dependence on ℓ. It is convenient to introduce a level diagram (Fig. 15.2) in which the different ℓ quantum numbers are marked on the abscissa and the ordinate represents the energy scale. Each bar corresponds to an energy level. At the same time, the bars represent orbitals relevant to the specific energy. An up (down) arrow indicates an electron with spin up (down) occupying the orbital corresponding to the energy. The electron orbitals are labeled s, p and d in analogy to the atomic shell structure. The energy scale in the level diagram may be omitted if only relative energy differences are of interest.

In the diagrams and the following paragraphs energies are given in units of effective Hartree Ha* and lengths in effective Bohr radii a_0^*

$$\mathrm{Ha}^* = \frac{\hbar^2}{m^*a_0^{*2}} = \frac{e^2}{\kappa a_0^*}, \qquad a_0^* = \frac{\kappa\hbar^2}{m^*e^2}. \tag{15.5}$$

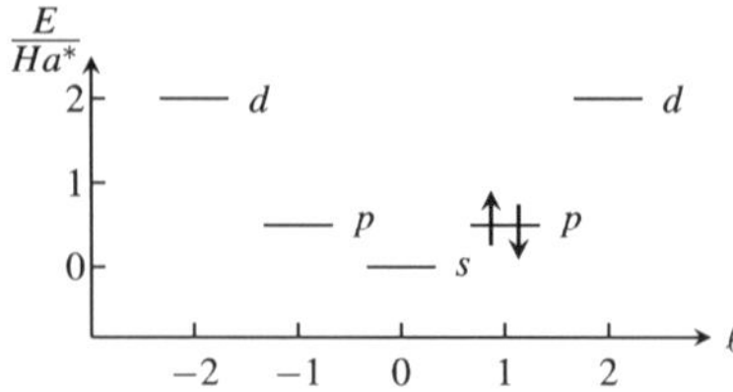

Fig. 15.2 Level diagram of the one-dimensional ring. Each bar symbolizes a single-particle energy level ($E \propto \ell^2$). An *up/down arrow* denotes a spin-up/spin-down electron occupying the orbital with the corresponding energy

An arrangement of electrons in the level diagram represents a spin configuration, i.e. a total-spin eigenfunction constructed by Slater-determinants of single-particle orbitals. Let us consider the influence of the Coulomb interaction on the total energy of a configuration. The Coulomb matrix elements in the basis of the one-dimensional ring read

$$V_{ijkl} = \langle \psi_{\ell_i}, \psi_{\ell_j} | \frac{1}{|\mathbf{r}_1 - \mathbf{r}_2|} | \psi_{\ell_k}, \psi_{\ell_l} \rangle \tag{15.6}$$

$$\approx \int_0^{2\pi} \frac{d\varphi_1}{\sqrt{2\pi}} \int_0^{2\pi} \frac{d\varphi_2}{\sqrt{2\pi}} \frac{e^{i\ell_i\varphi_1} e^{i\ell_j\varphi_2} e^{-i\ell_k\varphi_1} e^{-i\ell_l\varphi_2}}{r_0\sqrt{d^2 + 4\sin^2(\frac{\varphi_1-\varphi_2}{2})}}$$

$$= \delta_{\ell_i+\ell_j, \ell_k+\ell_l} \int_{-\pi}^{\pi} \frac{d\varphi_{\text{rel}}}{\sqrt{2\pi}} \frac{\cos((\ell_i - \ell_k)\varphi_{\text{rel}})}{r_0\sqrt{d^2 + 4\sin^2(\varphi_{\text{rel}})}}, \tag{15.7}$$

using center-of-mass and relative coordinates $\varphi_{\text{cm}} = \frac{\varphi_1+\varphi_2}{2}$ and $\varphi_{\text{rel}} = \frac{\varphi_1-\varphi_2}{2}$, respectively. The 2D-distance between two particles in the one-dimensional ring equals $2r_0|\sin(\frac{\varphi_1-\varphi_2}{2})|$. An auxiliary parameter $d \ll 1$ has been introduced to account for a finite thickness of the ring and to prevent the Coulomb matrix integral from diverging. These matrix elements reveal several important properties of the system. First, the two-particle matrix elements conserve the total angular momentum $\ell_i + \ell_j = \ell_k + \ell_l$, which reflects the circular symmetry of the system. The interaction does not affect the center-of-mass motion.

Furthermore, we can derive the following behavior of the Coulomb interaction. Two electrons occupying orbitals ℓ_i and ℓ_j may form a singlet state ($S = 0$) with a symmetric wave function or a triplet state ($S = 1$) with an antisymmetric wave function. Instead of including the symmetry into the wave function, one may build it into the interaction matrix elements which read as follows

$$E_{\text{coulomb}} = V_{ijij} \pm V_{ijji} = \int_{-\pi}^{\pi} \frac{d\varphi_{\text{rel}}}{\sqrt{2\pi}} \frac{1 \pm \cos((\ell_i - \ell_j)\varphi_{\text{rel}})}{r_0\sqrt{d^2 + 4\sin^2(\varphi_{\text{rel}})}}, \tag{15.8}$$

with the plus sign for the singlet state and the minus sign for the triplet state. The total Coulomb energy is decreased if different orbitals are occupied and the exchange energy is given by the energy difference between singlet and triplet state. The Pauli principle forbids the formation of a triplet state when the same orbital is occupied.

Summarizing the salient points: in a weakly interacting system the lowest orbitals will be occupied by the electrons to minimize the kinetic energy. If the interaction energy dominates, the system will try to reduce the Coulomb interaction. This can be achieved by occupying different orbitals and arranging as in a triplet state to gain exchange energy. An important comment is in order. The Coulomb matrix elements do not only act on a single configuration but couple different ones. Thus, in a strongly interacting system different configurations of the same total angular momentum and total spin are mixed leading to non-trivial correlated states.

15.2.1 Path-Integral Monte Carlo for Fermions in Quantum Rings

The dominating interaction regime in a quantum ring strongly depends on its radius. The kinetic energy of a narrow ring scales with the inverse square of the radius, whereas the interaction energy drops inversely with increasing ring radius

$$E_{\text{kinetic}} \sim \frac{1}{r_0^2} \quad \text{and} \quad E_{\text{coulomb}} \sim \frac{1}{r_0}. \tag{15.9}$$

For small radii, the kinetic energy dominates and the Coulomb interaction can be considered as perturbation. Approximations like Hartree-Fock may be applied, representing the many-body wave functions by a single Slater determinant. Such antisymmetric wave functions obey the Pauli exclusion principle and introduce exchange-type correlations between the electrons.

Increasing the radius of the ring, the Coulomb repulsion determines the energy spectrum and the few-particle wave functions. In this regime, the many-body wave function can not be expressed by only a single Slater-determinant and a Hartree-Fock description fails. Due to the symmetry of the system only many-electron configurations of the same total angular momentum L and total spin quantum number S are mixed by the Coulomb interaction. In this case, the electrons are termed strongly correlated denoting those many-body effects which are not captured by a single-determinant theory.

Methods like exact diagonalization and density functional theory become computational expensional and inadequate respectively in describing the regime governed by strong correlation. Thus, to study the transition from weak to strong correlation path-integral Monte Carlo (PIMC) calculations [15, 16] have been carried out. In the following sections, numerically exact results employing PIMC for the ground state of a system at finite-temperature T are presented. The temperature, however, has been chosen low enough to resemble the $T = 0$ ground state in most cases.

Considering a quantum mechanical system described by the Hamiltonian $\hat{\mathscr{H}}$, the eigenfunctions $|\Phi_i\rangle$, the eigenenergies E_i, then the canonical density matrix of the system in thermal equilibrium is given by

$$\hat{\rho} = \sum_i e^{-\beta E_i} |\Phi_i\rangle\langle\Phi_i| = e^{-\beta \hat{\mathscr{H}}}, \tag{15.10}$$

with inverse temperature $\beta = 1/k_B T$. Here, the canonical partition function is given by the trace over the density matrix $Z = \text{Tr}[\hat{\rho}]$. The expectation value of an operator $\hat{\mathcal{O}}$ is

$$\langle \hat{\mathcal{O}} \rangle = \frac{\text{Tr}[\hat{\mathcal{O}}\hat{\rho}]}{\text{Tr}[\hat{\rho}]} = \frac{1}{Z} \sum_i e^{-\beta E_i} \langle \Phi_i | \hat{\mathcal{O}} | \Phi_i \rangle. \tag{15.11}$$

In principle, one can carry out the above traces in any complete basis, for our continuum PIMC method we choose the position basis [15]. Introducing the set of coordinates for an N-particle system in d dimensions $\mathbf{R} = \{\mathbf{r}_1, \mathbf{r}_2, \ldots, \mathbf{r}_N\}$, where $\mathbf{r}_i$ is the position of the ith particle, the position-space density matrix reads

$$\rho(\mathbf{R}, \mathbf{R}'; \beta) = \langle \mathbf{R} | e^{-\beta \hat{\mathcal{H}}} | \mathbf{R}' \rangle = \sum_i \Phi_i(\mathbf{R}) \Phi_i(\mathbf{R}') e^{-\beta E_i}. \tag{15.12}$$

In order to obtain the total spin of the configuration giving the largest contribution to the thermodynamic ground state, we restrict the density matrix to states of a certain total spin S. Including spin in the density formalism is straight forward. If the Hamiltonian $\hat{\mathcal{H}}$ of the system does not directly operate on the spin, then the single-particle density matrix is diagonal in the spin indices σ,

$$\rho(\mathbf{x}, \mathbf{x}'; \beta) = \rho(\mathbf{r}, \mathbf{r}'; \beta) \delta_{\sigma, \sigma'}, \tag{15.13}$$

with $\mathbf{x} = (\mathbf{r}, \sigma)$ and coordinates $\mathbf{r} = (x, y)$ in two-dimensional space. In exactly the same manner the density matrix of N distinguishable particles reads

$$\rho(\mathbf{x}_1, \ldots, \mathbf{x}_N, \mathbf{x}_1', \ldots, \mathbf{x}_N'; \beta) = \langle \mathbf{R} | e^{-\beta \hat{\mathcal{H}}} | \mathbf{R}' \rangle \delta_{\sigma_1, \sigma_1'} \cdots \delta_{\sigma_N, \sigma_N'}. \tag{15.14}$$

The density matrix of N electrons has to be antisymmetric and we can write down the partition function

$$Z = \frac{1}{N!} \int d\mathbf{R} \sum_{\mathscr{P}} (-1)^P n(\mathscr{P}) \langle \mathbf{R} | e^{-\beta \hat{\mathcal{H}}} | \mathscr{P} \mathbf{R} \rangle, \tag{15.15}$$

with

$$n(\mathscr{P}) = \sum_{\sigma_1 \ldots \sigma_N = \pm \frac{1}{2}} \delta_{\sigma_1, \mathscr{P}(\sigma_1)} \cdots \delta_{\sigma_N, \mathscr{P}(\sigma_N)}. \tag{15.16}$$

In the following, the numerical results shown were obtained by stochastically sampling the partition function (15.15) using the path-integral Monte Carlo method [16]. Quantum states of different S_z and S quantum numbers contribute to the above partition function. In some cases it is useful to constrain the contributions to make a statement about the ground state of the system under study. A common approach is to fix the total spin projection S_z and to decrease the temperature (e.g. [8, 17]). However, this procedure hinders the determination of the total spin S of the ground state. If the energy difference between different S states is smaller than the temperature $k_B T$ the calculated energy for fixed S_z receives contributions from all states with $S \geq S_z$. This is also referred to as spin contamination problem [18]. This procedure turns out to be unsuitable to distinguish states with different total spin lying close in energy. The required low temperatures to resolve such energy differences lead to a disastrous fermion sign problem.

In the following, a different path is adopted by explicitly constructing eigenstates of S^2. Then thermodynamic expectation values receive only contributions from states with defined total spin. Particularly, the free energy will be equal to the energy of the state with spin S if the temperature is lower than the energy of the next higher state with the same spin S. With this approach we are able to determine the spin of the state with the lowest energy which is the ground state. Young operators are employed to properly symmetrize [19].

An instructive example is given to describe our approach. Considering a system of three electrons, the possible spin quantum numbers are $S = 1/2$ and $S = 3/2$. The corresponding Young operators $Y_{S=1/2}$ and $Y_{S=3/2}$ are defined by the normal Young tableaus

$$
Y_{S=1/2} = \begin{array}{|c|c|}\hline 1 & 2 \\\hline 3 \\\cline{1-1}\end{array}, \qquad Y_{S=3/2} = \begin{array}{|c|}\hline 1 \\\hline 2 \\\hline 3 \\\hline\end{array}. \tag{15.17}
$$

Only the normal Young tableaus are needed to treat indistinguishable particles, i.e. the order of the indices has no effect on the properties of expectation values. The Young tableaus represent a combination of permutation operators which act on the particle indices

$$
Y_{S=1/2} = (\mathbb{1} + \mathscr{P}_{12})(\mathbb{1} - \mathscr{P}_{13}) = \mathbb{1} + \mathscr{P}_{12} - \mathscr{P}_{13} - \mathscr{P}_{132}, \tag{15.18}
$$

$$
Y_{S=3/2} = \mathbb{1} - \mathscr{P}_{12} - \mathscr{P}_{23} - \mathscr{P}_{13} + \mathscr{P}_{123} + \mathscr{P}_{132}, \tag{15.19}
$$

where $\mathbb{1}$ is the identity operator. For example, the diagonal part of the $S = 3/2$ density matrix for three electrons reads

$$
\begin{aligned}
\rho_{S=3/2}(\mathbf{R}, \mathbf{R}; \beta) &= \frac{f(N, S)}{N!} \times \langle \mathbf{r}_1 \mathbf{r}_2 \mathbf{r}_3 | e^{-\beta \hat{\mathscr{H}}} \big| Y_{S=3/2}(\mathbf{r}_1 \mathbf{r}_2 \mathbf{r}_3) \rangle \\
&= \frac{1}{6} \times \big(\langle \mathbf{r}_1 \mathbf{r}_2 \mathbf{r}_3 | e^{-\beta \hat{\mathscr{H}}} | \mathbf{r}_1 \mathbf{r}_2 \mathbf{r}_3 \rangle - \langle \mathbf{r}_1 \mathbf{r}_2 \mathbf{r}_3 | e^{-\beta \hat{\mathscr{H}}} | \mathbf{r}_2 \mathbf{r}_1 \mathbf{r}_3 \rangle \\
&\quad - \langle \mathbf{r}_1 \mathbf{r}_2 \mathbf{r}_3 | e^{-\beta \hat{\mathscr{H}}} | \mathbf{r}_1 \mathbf{r}_3 \mathbf{r}_2 \rangle - \langle \mathbf{r}_1 \mathbf{r}_2 \mathbf{r}_3 | e^{-\beta \hat{\mathscr{H}}} | \mathbf{r}_3 \mathbf{r}_2 \mathbf{r}_1 \rangle \\
&\quad + \langle \mathbf{r}_1 \mathbf{r}_2 \mathbf{r}_3 | e^{-\beta \hat{\mathscr{H}}} | \mathbf{r}_2 \mathbf{r}_3 \mathbf{r}_1 \rangle + \langle \mathbf{r}_1 \mathbf{r}_2 \mathbf{r}_3 | e^{-\beta \hat{\mathscr{H}}} | \mathbf{r}_3 \mathbf{r}_2 \mathbf{r}_1 \rangle \big). \tag{15.20}
\end{aligned}
$$

The prefactor $f(N, S)$ accounts for the number of different standard Young tableaus belonging to one normal tableau and depends on the number of particles N and the total spin S

$$
f(N, S) = \frac{(2S + 1)N!}{(N/2 - S)!(N/2 + S + 1)!}. \tag{15.21}
$$

Each summand $\langle \mathbf{R} | e^{-\beta \hat{\mathscr{H}}} | \mathscr{P}_i \mathbf{R} \rangle$ can be expanded in a path integral

$$
\begin{aligned}
&\langle \mathbf{R} | e^{-\beta \hat{\mathscr{H}}} | \mathscr{P}_i \mathbf{R} \rangle \\
&= \int d\mathbf{R}_1 d\mathbf{R}_2 \cdots d\mathbf{R}_{M-1} \rho(\mathbf{R}, \mathbf{R}_1; \tau) \rho(\mathbf{R}_1, \mathbf{R}_2; \tau) \cdots \rho(\mathbf{R}_{M-1}, \mathscr{P}_i \mathbf{R}; \tau),
\end{aligned} \tag{15.22}
$$

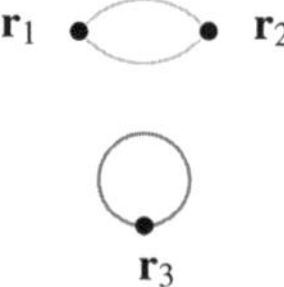

Fig. 15.3 Closed paths in imaginary time for three particles represented by diagrams with enumerated vertices. Two particles, 1 and 2, are involved in a cyclic permutation

where $\mathscr{P}_i$ is one of the permutation operators of the Young operator Y_S. The particle coordinates propagate in imaginary time from $\mathbf{R}$ to $\mathscr{P}_i\mathbf{R}$, i.e. the permutation defines the boundary condition to the imaginary time path (15.22).

A certain configuration of linked paths can be represented by diagrams with enumerated vertices omitting the intermediate points (Fig. 15.3). Each permutation corresponds to an enumerated diagram. Since the path integral is integrated over the whole space, graphs differing only in enumeration of the vertices have the same contribution to the integral. Generally, the properly symmetrized density matrix for spin S can be expressed by a summation of density matrices permuted according to a certain graph G suitably weighted by coefficients $n(G, S)$

$$\rho_S(\mathbf{R}, \mathbf{R}; \beta) = \frac{f(N, S)}{N!}\langle\mathbf{R}|e^{-\beta\hat{\mathscr{H}}}|Y_S\mathbf{R}\rangle = \frac{f(N, S)}{N!}\sum_G n(G, S)\langle\mathbf{R}|e^{-\beta\hat{\mathscr{H}}}|\mathscr{P}_G\mathbf{R}\rangle.$$

(15.23)

The coefficients $n(G, S)$ depend on the type of graph G and the desired spin S. The partition function

$$Z_S = \frac{f(N, S)}{N!}\int d\mathbf{R}\sum_G n(G, S)\langle\mathbf{R}|e^{-\beta\hat{\mathscr{H}}}|\mathscr{P}_G\mathbf{R}\rangle$$

(15.24)

receives only contributions from states with total spin S. Finally, the corresponding free energy for a fixed total spin can be obtained by

$$E_S = -\frac{\partial \ln Z}{\partial \beta}.$$

(15.25)

A related simulation procedure has been proposed by [20] deriving the coefficients $n(G, S)$ by other group-theoretical considerations [21]. Their simulation was carried out by using an effective action $\mathscr{U} = -\ln(\rho_S)$, i.e. they used the full, symmetrized density matrix as a probability function. However, an inefficient sampling of the configuration space was encountered because regions which correspond to different diagrams appeared to be separated by high-potential barriers. Transitions from one region to another occurred extremely rarely.

In the following simulations, we circumvent this problem by sampling the diagrams combined with a multilevel move [15]. By regrowing the path in a new permuted configuration we effectively tunnel through the energy barriers separating different configurations. All calculations were done by taking a sufficiently high number of time slices (~ 200 and more) to obtain a vanishingly small systematical

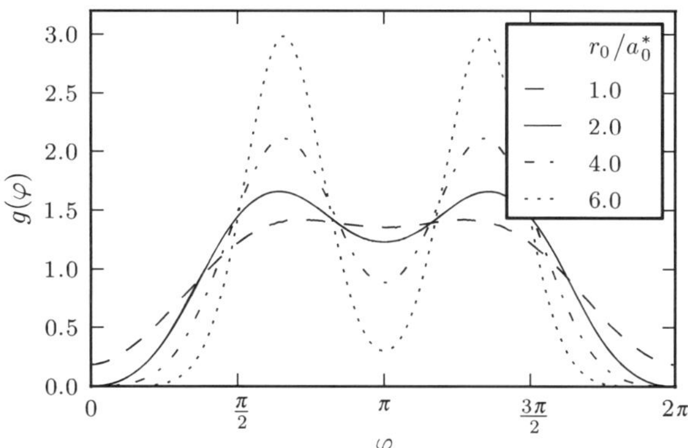

Fig. 15.4 The pair-correlation function $g(\varphi)$ for three-electrons in a ring with confinement $\hbar\omega_0 = 2.0$ Ha* at temperature $k_B T = 0.1$ Ha*. The graph shows g for different ring radii r_0 where the total spin of the ground state for $r_0 = 1.0a_0^*$ is $S = 1/2$ and $S = 3/2$ for all other radii, respectively. All functions are normalized, $\int g(\varphi)d\varphi = 2\pi$

error. All simulation runs were performed until the statistical error reached the order of one per mile of the free energy.

We start to analyze the effect of an increasing ring radius on the three-electron state. The following results were obtained at a temperature sufficiently low to assume that in most cases the observables are determined by the lowest energy few-particle state. The density of the system reveals no internal structure since it only reflects the geometry of the ring. The pair-correlation function, however, gives a direct measurement of the wave function overlap between two electrons. It is defined as

$$g(\mathbf{r}) = \frac{1}{N(N-1)} \left\langle \sum_{i \neq j}^{N} \delta\big(\mathbf{r} - (\mathbf{r}_i - \mathbf{r}_j)\big) \right\rangle, \tag{15.26}$$

where the brackets denote thermodynamic expectation values. The pair-correlation function gives the probability to find two of N electrons at distance $\mathbf{r}$. In general, g depends on the position $\mathbf{r}$, in a radially symmetric system it can be written as a function of the angle φ between $\mathbf{r}$ and the x-axis at a fixed radius.

Figure 15.4 shows the pair-correlation function $g(\varphi)$ for different radii containing three electrons in a ring at temperature $k_B T = 0.1$ Ha* and fixed confinement $\hbar\omega_0 = 2.0$ Ha*. We are able to distinguish three characteristic regimes. For small ring radii, $r_0 \approx 1.0a_0^*$, the Coulomb interaction represents only a weak perturbation. At the origin, however, the function decreases due to the Pauli principle preventing two particles with equal spin projection to coincide. The pair-correlation function, however, does not vanish at the origin because the ground state has total spin $S = 1/2$, thus there is still a finite probability to find two of the three electrons at zero distance.

For $r_0 \approx 2.0a_0^*$ the behavior changes qualitatively. The spin of the ground state switches to $S = 3/2$ and consequently the pair-correlation function vanishes at the origin. These are exchange-type correlations produced by the Pauli exclusion principle. Moreover, the pair-correlation function exhibits two peaks arising due to the avoidance of the two remaining electrons.

Increasing r_0 further to $r_0 \approx 6.0a_0^*$, the pair-correlation function drops to about 10 % of the peak-height in regions between the maxima. The separation of the

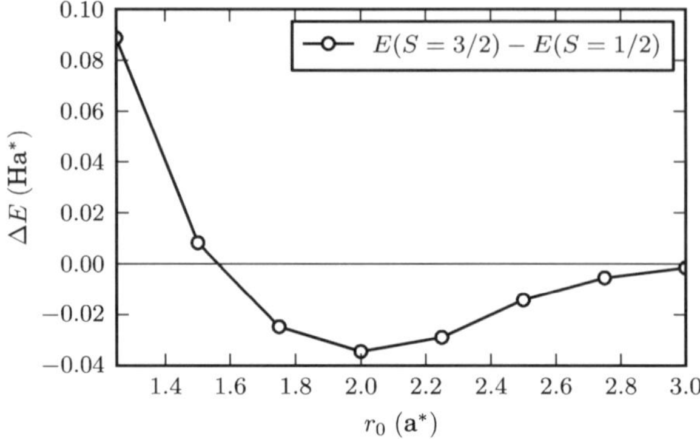

Fig. 15.5 Difference in free energy of the three-electron states as a function of the radius r_0. The system temperature is $k_B T = 0.1$ Ha* and the confinement $\hbar\omega_0 = 2.0$ Ha*. For small radii $r_0 \leq 1.5a_0^*$ the total spin of the ground state is $S = 1/2$ because the electrons arrange such as to minimize the single-particle energy. The energy difference between the states with $S = 3/2$ and $S = 1/2$ decreases as $1/r_0^2$ with increasing radius because the single-particle level splitting reduces. At $r_0 \approx 1.5a_0^*$ the difference in kinetic energy between the $S = 1/2$ and $S = 3/2$ is smaller than the exchange-energy saving associated with the $S = 3/2$ state which becomes the ground state. A further increase in the radius reduces the wave function overlap between the particles and thus to a decreasing exchange energy ($\propto 1/r_0$) until both states $S = 1/2$ and $S = 3/2$ become almost energetically degenerate for $r_0 \approx 3.0a_0^*$

maxima is equal to that of classical point charges arranging at the vertices of an equilateral triangle. The peaks appear because all three electrons try to reduce the Coulomb repulsion which can be accomplished by decreasing the wave function overlap between the particles. In detail, the single-particle energies ($\propto 1/r_0^2$) become almost degenerate and the interaction strongly mixes quasi-degenerate states. As a result, the electrons localize in their intrinsic frame of reference, while the density itself does not show any broken symmetry. It is customary to say a rotating Wigner molecule (RWM) is established [8, 22]. This name refers to the formation of Wigner crystals in electron gases at low density and strong Coulomb interaction [23]. In contrast to a 'crystal', which signifies well localized and distinguishable electrons, the probability density in the rotating reference frame of the electron 'molecule' is strongly diminished, but does not vanish between the two adjacent maxima.

15.2.2 Three-Electron Spin Transition in a Finite-Width Ring

The pair-correlation function reveals different characteristic regimes of the system depending on the ring radius. These are also reflected in the energy difference between spin states of three electrons in the ring during the continuous crossover between the regimes (Fig. 15.5).

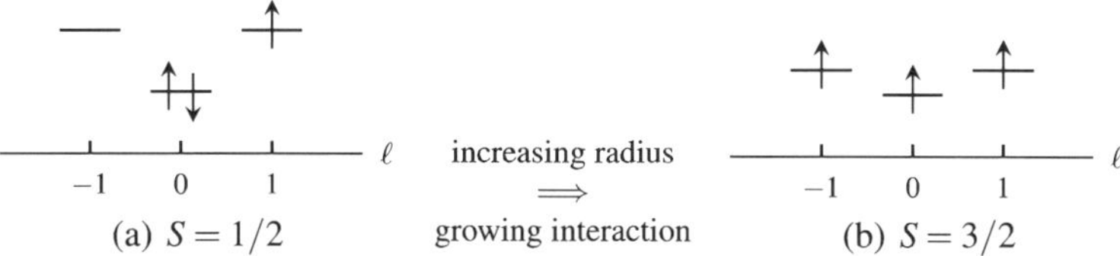

Fig. 15.6 Level diagram for three particles. Weakly interacting particles successively fill the single-particle orbitals (a) to minimize their level energy. The level spacing reduces with increasing radius and the importance of the interaction grows simultaneously. A high-spin configuration is favored due to exchange-energy saving (b)

For a small radius, the state with $S = 1/2$ has the lowest energy. As previously discussed, the electrons interact weakly and thus the Coulomb repulsion can be assumed as perturbation. The ground state resembles the configuration of a non-interacting system. The configuration of electrons occupying the lowest energy levels minimize the total kinetic energy. The resulting $S = 1/2$ ground state corresponds to a configuration with a doubly occupied s-shell and a singly occupied p-orbital shown (Fig. 15.6(a)). The energy difference between the ground state and the excited state with $S = 3/2$ is at least the energy of the first excited single-particle level. The configuration of the state with $S = 3/2$ singly fills each level $\ell = 0$, $\ell = 1$ and $\ell = -1$ because the Pauli principle prevents the double occupation of an orbital by two particles of parallel spin.

As the radius is increased, the single-particle level spacing and thus the energy difference between $S = 1/2$ and $S = 3/2$ decreases as $1/r_0^2$. The importance of the Coulomb interaction grows with respect to the kinetic energy. For strong interaction the configuration with different orbitals occupied minimizes the total energy and is the most important of the three-electron states (Fig. 15.6(b)). The occupation of different orbitals reduces the Coulomb interaction between electrons. Additionally, a parallel spin alignment is energetically favored since the exchange term of the Coulomb interaction reduces the total energy leading to a $S = 3/2$ ground state.

Finally, for a large radius the energy difference between the lowest single-particle levels vanishes. Towards this case, the level energy for both states with $S = 1/2$ and $S = 3/2$ becomes almost equal. The total energy of the many-body states differs almost only by the exchange term of the Coulomb interaction which reduces the energy of the $S = 3/2$ state. With increasing radius, the exchange energy diminishes as $1/r_0$ until the energies of $S = 1/2$ and $S = 3/2$ become almost degenerate at radius $r_0 \approx 3a_0^*$. In conclusion, in a three-electron narrow quantum ring a transition of the ground state with $S = 1/2$ to the $S = 3/2$ ground state occurs with increasing radius.

15.2.3 Rotating Wigner Molecule Interpretation

In the preceding paragraph, we have approached the regime of a rotating Wigner molecule from a weakly interacting system with the concept of single-particle orbitals. This idea fails, however, in strongly interacting systems because many configurations are mixed. Therefore, it is useful to consider a collective concept of the many-body system in the regime of large Coulomb interaction.

In classical mechanics, the particles on a ring are point-charges with defined positions. Their interaction energy can be minimized by forming equilateral polygons. In quantum mechanics, however, the particles are represented by a wave function and the kinetic energy is determined by its curvature, i.e. the second derivative of the wave function with respect to each of the particle coordinates. A rigid configuration of point-charges would mean to have a wave function of delta-distributions which may minimize the Coulomb repulsion but gives a diverging kinetic energy. In a quantum system, the lowest energy state is a compromise between kinetic and interaction energy to minimize the total energy. The wave function should be smoothly distributed in the vicinity of equilibrium points with a maximum at the points of the classical configuration. This could be read off the pair-correlation function in Fig. 15.4.

In this limit of an arising rotating Wigner molecule, Koskinen et al. [24] examined the spectra of a quasi one-dimensional ring with six electrons and identified two classes of elementary excitations: tunneling between different permutational arrangements of the electrons and vibrational excitations of the charge density. Koskinen et al. proposed a simple phenomenological model Hamiltonian describing the many-body spectrum of such an electron 'molecule'

$$\hat{\mathscr{H}}_{\mathrm{eff}} = \frac{\hbar^2 L^2}{2I} + \sum_\alpha \hbar\omega_\alpha n_\alpha + J \sum_{i,j}^{N} \mathbf{S}_i \mathbf{S}_j, \tag{15.27}$$

with the total angular momentum L and a rigid moment of inertia I of the N electrons located at the N vertices of a rotating equilateral polygon. The vibrational excitations are characterized by the vibrational frequencies ω_α and the number of excitation quanta $n_\alpha = (0, 1, 2, \ldots)$. The last term is an anti-ferromagnetic Heisenberg Hamiltonian with $J > 0$ where the summation is taken over the nearest neighbour pairs of electrons located at the corners of a rotating equilateral polygon. The Heisenberg part models the exchange interaction between the particles which at larger distances favors spin-singlet couplings.

The model (15.27) represents an expansion of the system Hamiltonian (15.1) around classical equilibrium positions of the electrons. The fluctuations around the classical points are taken into account by the vibrational states. At first glance, the ground state of the system described by this model Hamiltonian is not trivially evident. The Heisenberg part of the Hamiltonian suggests an anti-ferromagnetic ground state. In the next sections, however, we will see that the symmetry of the wave function introduces a coupling between spin and total angular momentum. Thus, in the case of a ring with a large radius ($J \ll \hbar^2/I$) and strongly interacting electrons, the rotational levels dictate the ground state.

In the following, we will show that the permutation symmetry determines the configuration which minimizes the total energy. More precisely, the state of an equilateral polygon which minimizes the potential energy is accessible to states with total spin S only for certain total angular momenta. This kind of implicit coupling between spin and angular momentum is already known from quantum dots. Reference [25] used generalized Jacobian coordinates to describe three electrons in a quantum dot in a magnetic field and showed that minimum-energy states of different total spin S occur at certain 'magic' angular momenta. Using the Eckart-frame theory, [26] also found these angular momenta and at large angular momenta a molecule-like electronic structure in the quantum dot.

Notably, [27] stated that the non-interacting states of a quantum dot in a magnetic field may be pictured as localized states on rings. This suggests a strong relationship between quantum dots and quantum rings. In the quantum dot, the 'magic' angular momenta determine the ground state in magnetic fields. Here, we demonstrate that these sequences of 'magic' angular momenta also appear in quantum rings. They determine the ground state of the ring with strongly interacting electrons even in the case of absent magnetic fields.

First, let us recall the picture of a rotating Wigner molecule. In this system the Coulomb repulsion is minimized, if the electrons occupy the corners of an equilateral polygon. The symmetry of the system constrains certain angular momenta L to a designated spin S. To show these selection rules, we consider a one-dimensional ring with three electrons forming an equilateral triangle where the eigenfunctions are defined by the quantum numbers (L, S). Then, a cyclic permutation acting on the spatial part of the wave function is equivalent to a rotation by an angle of $2\pi/3$

$$\hat{\mathscr{P}}_{123}|\Phi\rangle = \exp\left(i\frac{2\pi}{3}L\right)|\Phi\rangle. \tag{15.28}$$

We consider the antisymmetric wave function of a $S = 3/2$ state. The spatial part of the wave function is totally antisymmetric because its spin part is symmetric and thus invariant under a cyclic permutation

$$\hat{\mathscr{P}}_{123}|\Phi_{S=3/2}\rangle = |\Phi_{S=3/2}\rangle. \tag{15.29}$$

If $|\Phi_{S=3/2}\rangle \neq 0$, then combining (15.29) and (15.28) gives $\exp(i2\pi L/3) = 1$. This condition leads to a sequence of 'magic' angular momenta associated with total spin $S = 3/2$

$$L = 3k, \quad k = 0, \pm 1, \pm 2, \dots. \tag{15.30}$$

For the $S = 1/2$ states, the wave function can be expanded as

$$|\Psi\rangle = |\Phi^A_{S=1/2}\rangle|\chi^A_{S=1/2}\rangle + |\Phi^B_{S=1/2}\rangle|\chi^B_{S=1/2}\rangle, \tag{15.31}$$

where $|\Phi^A_{S=1/2}\rangle$ and $|\Phi^B_{S=1/2}\rangle$ are the spatial parts of the wave function which transform according to the standard Young tableaus

$$\begin{array}{|c|c|}\hline 1 & 2 \\\hline 3 \\\cline{1-1}\end{array} \quad \text{and} \quad \begin{array}{|c|c|}\hline 1 & 3 \\\hline 2 \\\cline{1-1}\end{array}, \tag{15.32}$$

and $|\chi^A_{S=1/2}\rangle$, $|\chi^B_{S=1/2}\rangle$ are the corresponding spin functions. Under a cyclic permutation the spatial parts of the doublet state transform as

$$\hat{\mathscr{P}}_{123}\left|\Phi^A_{S=1/2}\right\rangle = -\frac{1}{2}\left|\Phi^A_{S=1/2}\right\rangle - \frac{\sqrt{3}}{2}\left|\Phi^B_{S=1/2}\right\rangle, \tag{15.33}$$

$$\hat{\mathscr{P}}_{123}\left|\Phi^B_{S=1/2}\right\rangle = \frac{\sqrt{3}}{2}\left|\Phi^A_{S=1/2}\right\rangle - \frac{1}{2}\left|\Phi^B_{S=1/2}\right\rangle, \tag{15.34}$$

given by Young's orthogonal representation [19]. Equation (15.28) also holds for $\left|\Phi^A_{S=1/2}\right\rangle$ and $\left|\Phi^B_{S=1/2}\right\rangle$ when the electrons form an equilateral triangle and we get the following set of homogeneous linear equations

$$\begin{pmatrix} -1/2 - e^{i\frac{2\pi}{3}L} & \sqrt{3}/2 \\ \sqrt{3}/2 & -1/2 - e^{i\frac{2\pi}{3}L} \end{pmatrix} \begin{pmatrix} \left|\Phi^A_{S=1/2}\right\rangle \\ \left|\Phi^B_{S=1/2}\right\rangle \end{pmatrix} = 0. \tag{15.35}$$

Now, if $\left|\Phi^{A/B}_{S=1/2}\right\rangle \neq 0$, then non-zero solutions exist only if the following equation is fulfilled

$$\left(e^{i\frac{2\pi}{3}} - e^{i\frac{2\pi}{3}L}\right)\left(e^{-i\frac{2\pi}{3}} - e^{i\frac{2\pi}{3}L}\right) = 0. \tag{15.36}$$

The resulting sequences of 'magic' angular momenta associated with total spin $S = 1/2$ are

$$L = 3k + 1 \quad \text{and} \quad L = 3k - 1, \quad k = 0, \pm 1, \pm 2, \ldots. \tag{15.37}$$

Concluding, an equilateral triangle which minimizes the interaction energy is only accessible to the $L = 3k$ ($S = 1$) or $L = 3k \pm 1$ ($S = 0$) states.

Now, let us assume that this system may be described by an electron 'molecule' with rotational and vibrational states (15.27)

$$\hat{\mathscr{H}}_{\text{eff}} = \frac{\hbar^2 L^2}{2I} + \sum_\alpha \hbar\omega_\alpha n_\alpha + J\sum_{i,j}^N \mathbf{S}_i\mathbf{S}_j. \tag{15.38}$$

Since the model is a good approximation to the system for a large ring radius with $J \ll \hbar^2/I$, the ground state of the model Hamiltonian is a state with $L = 0$. Thus, we get a quartet ($S = 3/2$) state as the ground state which is in agreement with our numerical results in the regime of a large ring radius.

The preceding considerations demonstrate that the triplet state not only minimizes the interaction energy but also the kinetic ('rotational') energy of the system. This result is unexpected because in a single-particle picture the level energy of the lowest $S = 3/2$ configuration is higher than the energy of the $S = 1/2$ configuration. Apparently, the single-particle picture breaks down in the strongly correlated regime of a rotating Wigner molecule. This explanation is supported by the results of the numerical calculations (Fig. 15.7). The interaction energy of the $S = 3/2$ state compared to $S = 1/2$ is lowered by the exchange energy. For small radii, the single-particle level energies decrease as $1/r_0^2$. The kinetic energy difference between $S = 3/2$ and $S = 1/2$ becomes small. For $r_0 \geq 1.8a_0^*$, the $S = 3/2$ state is lower in kinetic energy than the $S = 1/2$ state. This many-body effect can not be covered by the single-particle picture.

The trend in the energies (Fig. 15.7) can be interpreted as a consequence of the symmetry constraints on the few-particle wave function. The eigenfunctions of a

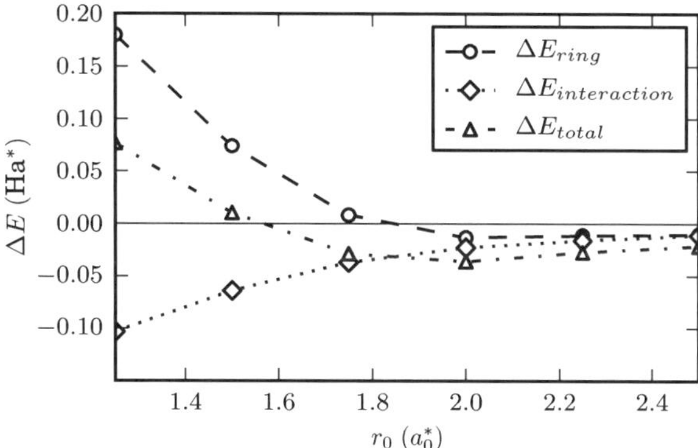

Fig. 15.7 Free energy difference between the $S = 3/2$ state and $S = 1/2$ state of three particles as a function of the ring radius r_0. The confinement is $\hbar\omega_0 = 2.0$ Ha*. The contributions to the total energy are differentiated in one-particle (kinetic/potential) energies E_{ring} and interaction energies $E_{\text{interaction}}$

one-dimensional ring can be classified by the Young tableaus [19] which are employed to construct the spatial part of wave functions associated with a certain spin state S. The three-particle tableaus are

(I) non-physical   (II) doublet ($S = 1/2$)   (III) quartet ($S = 3/2$)

For spin-$\frac{1}{2}$ electrons, only the second and third tableau may be used to construct spatial wave functions associated with the spin states $S = 1/2$ and $S = 3/2$, respectively. The corresponding Young operators symmetrize with respect to particle indices in the same row and antisymmetrize with respect to particle indices in the same column. The symmetry of the three particle wave functions can be deduced from the Young tableaus (Fig. 15.8). For tableau (I) the spatial wave function keeps its sign as two particles exchange (Fig. 15.8a) while the function constructed by tableau (III) changes its sign (Fig. 15.8c). Tableau (II) indicates a mixed symmetry meaning that exchanging the pair $r_1 \leftrightarrow r_2$ keeps the sign of the wave function with $S = 1/2$ while exchanging $r_1 \leftrightarrow r_3$ changes the sign.

The lowest eigenfunction of the Hamiltonian describing the system should be nodeless [28]. Only a totally symmetric function is nodeless which cannot be realized by a fermionic system. In a hypothetical case, however, when the repulsive interaction becomes infinite as two particles approach each other the wave function must vanish at these points (Fig. 15.8a). In this case, the antisymmetric function associated with a quartet ($S = 3/2$) state will also be an eigenfunction with the same energy as the totally symmetric function. The reason is that the totally antisymmetric function is defined as equaling the totally symmetric function for the identity permutation and its negative if two particles are permuted (Fig. 15.8c). The spatial wave function of the doublet ($S = 1/2$) state will necessarily have a higher energy since its wave function possesses an additional node implied by the periodic boundary conditions at a point where two particles are not approaching each other (Fig. 15.8b).

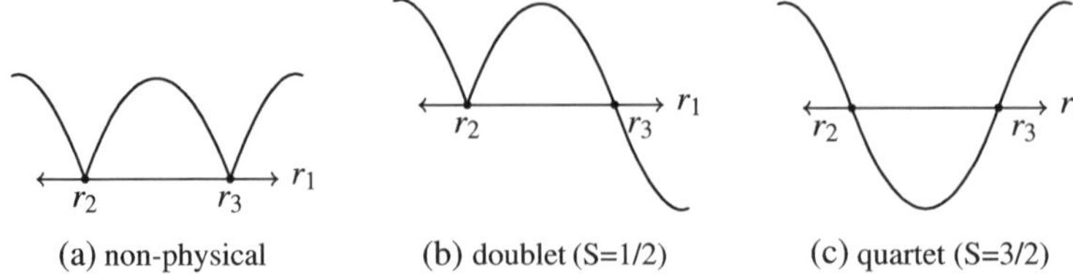

Fig. 15.8 The spatial wave functions associated with different spin states. The particle coordinates r_2 and r_3 are kept fixed while moving the third coordinate r_1 on the ring contour with periodic boundary conditions. A repulsive interaction of the particles is assumed which becomes infinite as two particles approach each other and consequently the wave function vanishes at these meeting points. Figure (**a**) shows a totally symmetric function which is non-physical in the case of spin-$\frac{1}{2}$ fermions. The $S = 3/2$ function (**c**) possesses two nodes at points where two particles approach each other. The $S = 1/2$ function (**b**) has only one node at a point where two particles approach each other. Another node of the $S = 1/2$ function is implied by the periodic boundary conditions and appears at a point where two particles are not passing each other

The number of nodes of the wave functions with $S = 1/2$ and $S = 3/2$ is equal but they are localized at different points in the configuration space. The two roots of the $S = 3/2$ function are at points where the wave function vanishes due to the interaction. However, the $S = 1/2$ wave function possesses three roots and raises the energy.

If we decrease the repulsive interaction from infinity to a large finite value, the energies of the completely symmetric and completely antisymmetric states will split apart. However, for a small splitting the completely antisymmetric $S = 3/2$ state will remain energetically lower than the $S = 1/2$ state.

The preceding considerations have been applied to many particles showing that the ground state of an odd number of electrons in a one-dimensional space with periodic boundary conditions is one of maximum multiplicity [29, 30]. It was argued that the above reasoning also applies for three fermions in three-dimensional space with a one-body potential which confines the particles effectively to the close neighbourhood of a ring [29].

With increasing width, however, the ring becomes two dimensional. The electrons can pass without approaching each other which would lead to an infinite repulsive interaction. Then, the $S = 1/2$ state can reduce its kinetic energy. At a critical width the $S = 1/2$ state will fall in energy below the $S = 3/2$ state. The numerical results support this statement (Fig. 15.9).

15.2.4 Absence of a Spin Transition in a Four-Electron Ring

The energy difference between states in a four electron ring (Fig. 15.10) reveals a $S = 1$ ground state for the whole range of radii. At a large radius $r_0 \approx 3a^*$, however, the states with $S = 0$ and $S = 1$ become degenerate. The energy of the state with $S = 2$ converges towards the other energies and degeneracy of all three states may be expected for radii $r_0 \gg 3.0a^*$.

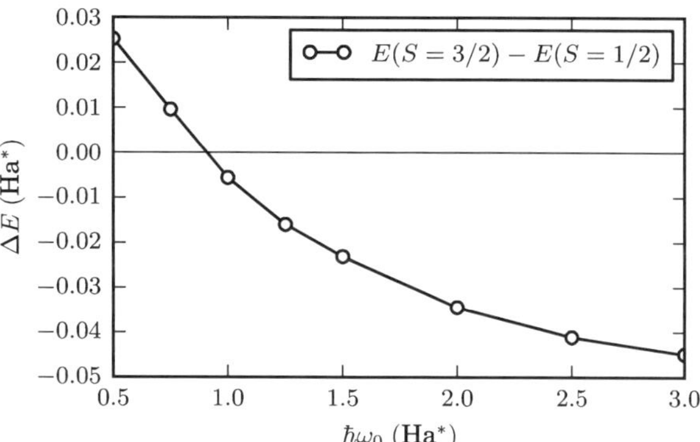

Fig. 15.9 Free energy difference between the $S = 3/2$ state and $S = 1/2$ state of three particles as a function of the confinement $\hbar\omega_0$. The ring radius is $r_0 = 2.0a_0^*$. The ring width is proportional to $1/\sqrt{\hbar\omega_0}$. For a narrow ring ($\hbar\omega_0 \geq 0.8$ Ha*) the total spin of the ground state is $S = 3/2$. A larger ring width ($\hbar\omega_0 \leq 0.8$ Ha*) leads to a $S = 1/2$ ground state

For a small radius, the kinetic energy dominates with respect to the interaction energy. The single-particle orbitals are filled as compactly as possible around $\ell = 0$ respecting the Pauli principle. The resulting configurations are shown in Fig. 15.11. The configuration of the $S = 0$ state is a doubly occupied s-shell and one doubly occupied p-orbital. In the $S = 1$ configuration the s-shell is also doubly filled and both p-orbitals are singly occupied. The level energies of the $S = 0$ and $S = 1$ configurations are equal. The degeneracy is, however, removed by the interaction. The total spin of the Hund's rule ground state is $S = 1$ because the total energy is lowered by the exchange energy compared to the $S = 0$ state. The configuration with $S = 2$ possesses the highest energy because a higher single-particle orbital with $\ell = 2$ is occupied.

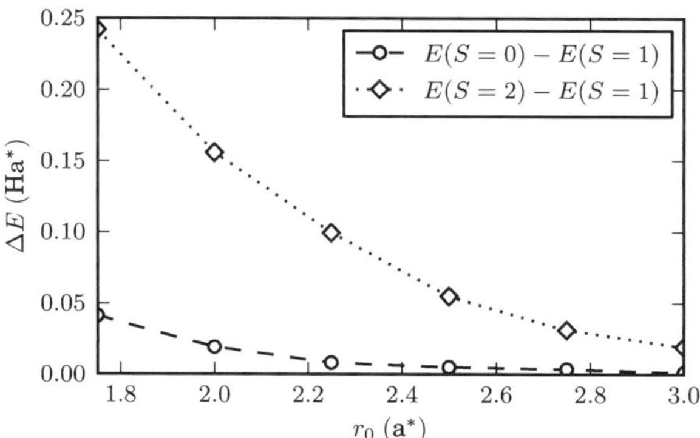

Fig. 15.10 Difference in free energy of the four-electron states as a function of the radius r_0. The system temperature is $k_B T = 0.1$ Ha* and the confinement $\hbar\omega_0 = 2.0$ Ha*. The energy difference between the states $S = 1$ and $S = 0$ reveals that the ground state is a triplet configuration. The singlet and triplet configurations possess equal level energies but the triplet energy is lowered by the exchange energy. As the radius increases, the wave function overlap between the particles reduces and thus the exchange energy decreases as $1/r_0$. At $r_0 \approx 3.0a_0^*$ the exchange energy is small enough that triplet and singlet become degenerate. The orbital energy of the quintet ($S = 2$) is higher than the other configuration and decreases $\propto 1/r_0$. At $r_0 \gg 3.0a_0^*$ the energy spacing between lowest single-particle levels vanishes and the degeneracy of all three states can be expected

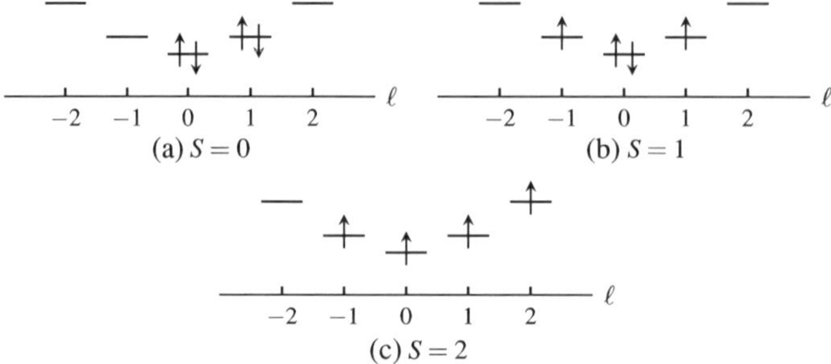

Fig. 15.11 Level diagram for four weakly interacting electrons. The particles successively fill the single-particle orbitals respecting the Pauli principle. The level energy of the states $S = 0$ and $S = 1$ is degenerate. However, interaction removes this degeneracy and reduces the total energy of the $S = 1$ state by the exchange energy

With increasing radius, the single-particle energy levels decrease proportional to ℓ^2/r_0^2. The first three levels become almost degenerate and thus the level energies of the states $S = 0$ and $S = 1$ converge. Due to the small wave function overlap between the particles at large radii, the exchange energy decreases. At $r_0 \approx 3a^*$ both states with $S = 0$ and $S = 1$ become degenerate. The single-particle levels $\ell = \pm 2$ still have a significantly higher energy than the lowest three levels because the level spacing depends quadratically on the quantum number ℓ. Therefore, the spin $S = 2$ state remains energetically separated until more single-particle energy levels become almost degenerated. A spin transition of the ground state is absent in a narrow quantum ring with four electrons when increasing the radius.

Even in the case of four electrons in the quantum ring 'magic' angular momenta can be deduced from the symmetry of the wave function. Carrying out similar calculation as in the three electrons case result in the following sequences of 'magic' angular momenta in four-electron case

$$S = 2 : L = 4k + 2, \tag{15.39}$$

$$S = 1 : L = 4k \quad \text{and} \quad L = 4k \pm 1, \tag{15.40}$$

$$S = 0 : L = 4k \quad \text{and} \quad L = 4k + 2, \tag{15.41}$$

where k is an integer.

For the four-electron ring, the model of a rigid rotator is not complete. It gives us a degenerate $L = 0$ ground state with $S = 0$ and $S = 1$. In this case, we have to take into account the Heisenberg part of the model Hamiltonian (15.27)

$$\hat{\mathcal{H}}_{\text{eff}} = \frac{\hbar^2 L^2}{2I} + \sum_\alpha \hbar \omega_\alpha n_\alpha + J \sum_{i,j}^{N} \mathbf{S}_i \mathbf{S}_j. \tag{15.42}$$

For four electrons, the total spin of the ground state of the Heisenberg Hamiltonian is $S = 0$ [31]. This state, however, does not necessarily belong to the angular momentum $L = 0$. It turns out that the angular momentum of the lowest $S = 0$ eigenstate of the Heisenberg Hamiltonian is $L = 2$ [11]. In contrast, the $S = 1$ state belonging to $L = 0$ has a lower total energy than the next excited state with $S = 0$ and $L = 0$ and thus the total spin of the ground state is $S = 1$. As the radius increases the wave function overlap decreases leading to a small coupling constant J and to the degeneracy between the states with $S = 0$ and $S = 1$. If the radius is increased further, the splitting between the rotational energy levels decreases and all three states $S = 0, 1, 2$ become degenerate. The considerations of the model Hamiltonian are consistent with the numerical results.

15.3 Elliptically Distorted Quantum Rings

Crystal strain and imperfections occurring in fabrication imply that a model of a perfectly circular ring can be insufficient to describe the properties of experimentally realized rings. It was shown that in a magnetic field the electronic spectra of elliptic rings [12] and rings with distortion [32] exhibit anticrossings and that the levels show quenched oscillations as a function of the magnetic field strength. References [3] and [2] found in few-electron curvilinear rings that the presence of distortions opens a gap in the energy spectrum of the ring which depends on the degree of the distortions. References [13] and [33] calculated the electronic states in deformed rings with uniform width and showed that regions of larger curvature are more favourable for the electrons. In this subsection, quantum rings with elliptic shape are investigated to gain a deeper understanding of the effect of distortions on the few-electron ground state. In our model, the coordinates are subject to the transformation

$$x \to x/\sqrt{\delta} \quad \text{and} \quad y \to \sqrt{\delta} \cdot y, \tag{15.43}$$

i.e. one semi-axis is stretched whereas the other semi-axis is shortened. Then, the single-particle potential reads

$$V(r) = \frac{1}{2}m^*\omega_0^2\left(r' - r_0\right)^2 \quad \text{with } r'^2 = \frac{x^2}{\delta} + \delta \cdot y^2. \tag{15.44}$$

With $\delta = 1$ the circular ring with radius r_0 is restored. If $\delta > 1$, then the semi-axes are given by $a = \sqrt{\delta} \cdot r_0$ and $b = r_0/\sqrt{\delta}$. Figure 15.12 depicts the transformation from a circular ring to a distorted ring. This definition of an ellipse does not preserve the circumference but the area while increasing δ.

Before examining the free energies of few electrons in an elliptic ring, we will show the consequences of a distortion on the single-particle levels of a one-dimensional ring. The Hamiltonian of a strictly one-dimensional ring reads

$$\hat{\mathcal{H}}_0 = \frac{\hbar^2}{2m^*r_0^2}\frac{\partial^2}{\partial\varphi^2}, \tag{15.45}$$

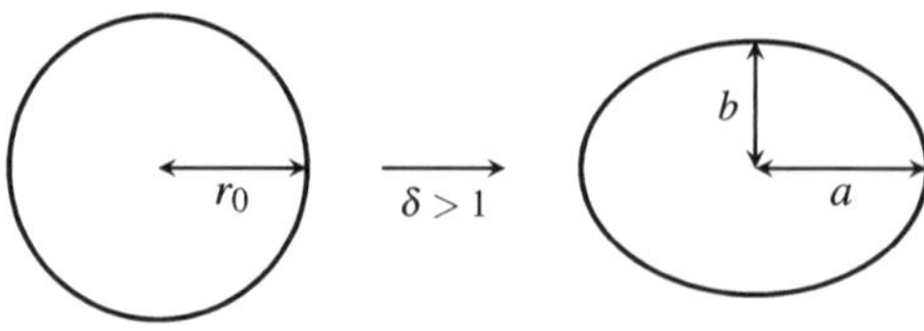

Fig. 15.12 Elliptical deformation of a circular ring with radius r_0 (*left*). The semi-axes of the distorted ring (*right*) are $a = \sqrt{\delta} \cdot r_0$ and $b = r_0/\sqrt{\delta}$

with eigenfunctions $\psi_\ell = \frac{1}{\sqrt{2\pi}} \exp(-i\ell\varphi)$ and eigenenergies $E_\ell = \frac{\hbar^2 \ell^2}{2m^* r_0^2}$. If an electron adopts an one-dimensional elliptic orbit, the Hamiltonian of a distorted ring is given by

$$\hat{\mathcal{H}}' = \frac{\hbar^2}{2m^* r'^2} \frac{\partial^2}{\partial \varphi^2} + \frac{1}{2} m^* \omega_0 (r' - r_0)^2 \quad \text{with } r' = r_0 \sqrt{\frac{\cos^2(\varphi)}{\delta} + \delta \sin^2(\varphi)}.$$

$$(15.46)$$

Introducing a small parameter ε that specifies the deviation of the electron path from the circular orbit gives $\delta \approx 1 + \varepsilon$ and $1/\delta \approx 1 - \varepsilon$. Using this definition and

$$r'^2/r_0^2 = \frac{\cos^2(\varphi)}{\delta} + \delta \sin^2(\varphi) \approx 1 + \varepsilon \left(\sin^2(\varphi) - \cos^2(\varphi) \right) = 1 - \varepsilon \cos(2\varphi),$$

$$(15.47)$$

the first part of the Hamiltonian $\hat{\mathcal{H}}'$ (15.46) can be expanded

$$\hat{\mathcal{T}}' \equiv \frac{\hbar^2}{2m^* r'^2} \frac{\partial^2}{\partial \varphi^2} \approx \left(1 + \varepsilon \cos(2\varphi) \right) \frac{\hbar^2}{2m^* r_0^2} \frac{\partial^2}{\partial \varphi^2}. \tag{15.48}$$

An expression for the second part of the Hamiltonian $\hat{\mathcal{H}}'$ (15.46)

$$\hat{\mathcal{V}}' \equiv \frac{1}{2} m^* \omega_0^2 r_0^2 \left(\sqrt{\frac{\cos^2(\varphi)}{\delta} + \delta \cdot \sin^2(\varphi)} - 1 \right)^2 \tag{15.49}$$

can be derived for small distortion by expanding the square root in (15.49) up to the first-order term

$$\sqrt{\frac{\cos^2(\varphi)}{\delta} + \delta \cdot \sin^2(\varphi)} \approx 1 + \frac{\varepsilon}{2} \left(\sin^2(\varphi) - \cos^2(\varphi) \right) = 1 - \frac{\varepsilon}{2} \cos(2\varphi). \tag{15.50}$$

The approximation for the second part of the Hamiltonian $\hat{\mathcal{H}}'$ reads

$$\hat{\mathcal{V}}' = \varepsilon^2 \frac{m^* \omega_0^2 r_0^2}{8} \cos^2(2\varphi). \tag{15.51}$$

Finally, the approximated Hamiltonian of an electron moving on an elliptic orbit can be written as

$$\hat{\mathcal{H}}' = \hat{\mathcal{T}}' + \hat{\mathcal{V}}' \approx \frac{\hbar^2}{2m^* r_0^2} \frac{\partial^2}{\partial \varphi^2} + \varepsilon \cos(2\varphi) \frac{\hbar^2}{2m^* r_0^2} \frac{\partial^2}{\partial \varphi^2} + \varepsilon^2 \frac{m^* \omega_0^2 r_0^2}{8} \cos^2(2\varphi)$$

$$\equiv \hat{\mathcal{H}}_0 + \hat{\mathcal{H}}_1, \tag{15.52}$$

where $\hat{\mathcal{H}}_0$ is the unperturbed Hamiltonian of a circular ring (15.45) and $\hat{\mathcal{H}}_1$ a perturbation to the system due to distortion

$$\hat{\mathcal{H}}_1 = \varepsilon \cos(2\varphi) \frac{\hbar^2}{2m^* r_0^2} \frac{\partial^2}{\partial \varphi^2} + \varepsilon^2 \frac{m^* \omega_0^2 r_0^2}{8} \cos^2(2\varphi). \tag{15.53}$$

In order to use perturbation theory it is useful to define the following matrix elements of the perturbation Hamiltonian

$$A\left(\ell, \ell'\right) = \langle \psi_\ell | \varepsilon \cos(2\varphi) \frac{\hbar^2}{2m^* r_0^2} \frac{\partial^2}{\partial \varphi^2} | \psi_{\ell'} \rangle = \begin{cases} \frac{\varepsilon}{r_0^2} \frac{\hbar^2 \ell'^2}{4m^*} & \text{if } \ell - \ell' = \pm 2, \\ 0 & \text{else} \end{cases} \tag{15.54}$$

and

$$B\left(\ell, \ell'\right) = \langle \psi_\ell | \varepsilon^2 \frac{m^* \omega_0^2 r_0^2}{8} \cos^2(2\varphi) | \psi_{\ell'} \rangle = \begin{cases} \varepsilon^2 r_0^2 \frac{m^* \omega_0^2}{16} & \text{if } \ell - \ell' = 0 \text{ or } \pm 4, \\ 0 & \text{else.} \end{cases} \tag{15.55}$$

With these expression we examine the energy corrections to the first three levels of a circular ring due to distortion. The first-order energy shift of the unperturbed ground-state with $\ell = 0$ is given by

$$E_0' = \langle \psi_0 | \hat{\mathcal{H}}_1 | \psi_0 \rangle = B(0,0) = \varepsilon^2 \omega_0^2 r_0^2 \frac{m^*}{16}. \tag{15.56}$$

The first two excited states are doubly degenerate and we have to diagonalize the perturbation Hamiltonian in the subspace spanned by these degenerate eigenstates with $\ell = 1$ and $\ell = -1$, leading to the first-order energy shift of the first excited states

$$E_{\pm 1}' = \varepsilon^2 \omega_0^2 r_0^2 \frac{m^*}{16} \pm \frac{\varepsilon}{r_0^2} \frac{\hbar^2}{4m^*}. \tag{15.57}$$

The energy corrections to the first three levels of a circular ring due to small distortion exhibit two important features (Fig. 15.13). The degeneracy of the p-orbitals is removed leading to two levels split by an energy proportional to the distortion and proportional to the inverse square of the radius. Additionally, the energies of all three levels are increased linearly with distortion. This elevation of the energies is proportional to the square of the radius. The ratio between energy shifting and splitting can be tuned by the confinement strength.

With a small distortion the primal states of the p-orbitals hybridize to ψ_1 and ψ_{-1}

$$\psi_{\pm 1}' = \frac{(e^{i\varphi} \pm e^{-i\varphi})}{2\sqrt{\pi}} = \frac{1}{\sqrt{\pi}} \begin{cases} \cos(\varphi) & \text{for } E_{1+}', \\ \sin(\varphi) & \text{for } E_{1-}', \end{cases} \tag{15.58}$$

and the perturbed ground state receives corrections from the states with $|\ell| = 2$ and $|\ell| = 4$

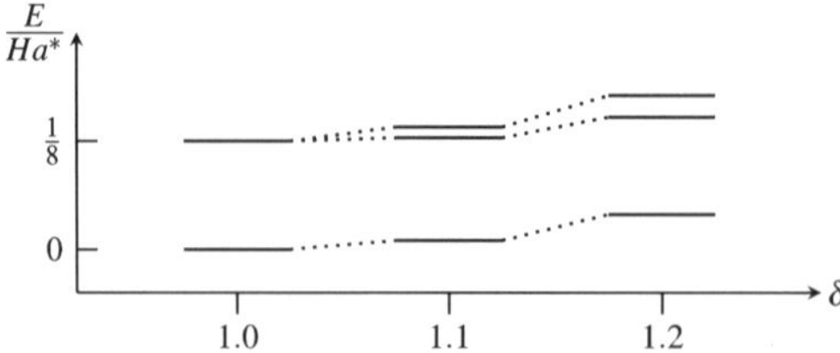

Fig. 15.13 The electronic states of a ring with radius $r_0 = 2.0a_0^*$ and $\hbar\omega_0 = 2.0$ Ha*. The first three energy levels are given as a function of the distortion δ. A small distortion removes the degeneracy of the p-orbitals leading to an energy splitting proportional to the distortion δ and simultaneously elevates the energy of all states ($\propto \delta^2$)

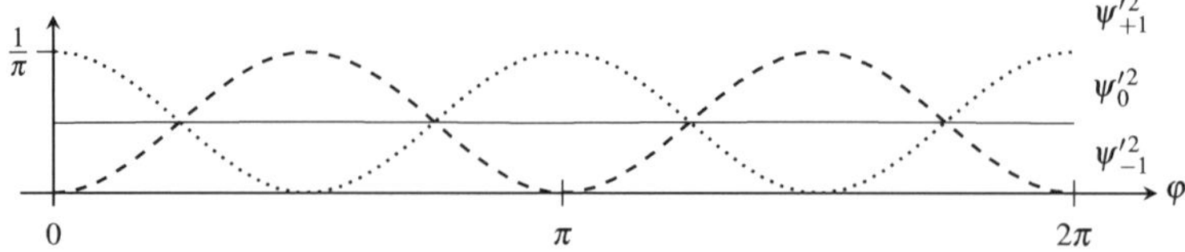

Fig. 15.14 The density of the first three states in an elliptical ring. The densities $\psi_{\pm 1}'^2$ of the hybridized p-orbitals exhibit nodes and anti-nodes at either end of the elliptic ring

$$\psi_0' = \psi_0 + \sum_{k \neq 0} \frac{\langle \psi_k | V | \psi_0 \rangle}{E_0 - E_k} \psi_k = \frac{1}{\sqrt{2\pi}} \left(1 - \frac{\varepsilon}{2r_0^2} \cos(2\varphi) - \frac{\varepsilon^2 r_0^2 m^{*2} \omega_0^2}{128\hbar^2} \cos(4\varphi) \right)$$

$$\approx \psi_0. \tag{15.59}$$

The densities of the perturbed states (Fig. 15.14) reveal that lowest energy state is only weakly perturbed and smeared over the ring. In contrast, two stationary waves originate from the hybridized p-orbitals. One stationary wave possesses nodes and the other wave anti-nodes at either end of the elliptic ring. The distortion breaks the symmetry of the system and removes the orbital degeneracy.

15.3.1 Distorted Three-Electron Ring

In the following, we examine the path-integral Monte Carlo results obtained using a two-dimensional potential (15.44). All simulations are performed at a low temperature $k_B T = 0.1$ Ha*. We first discuss the three electron case. Figure 15.15(a) depicts the free energy of the states with $S = 1/2$ and $S = 3/2$ as a function of distortion. The radius of the undistorted ring is $r_0 = 1.35a_0^*$ with a confinement $\hbar\omega_0 = 2.0$ Ha*.

At zero distortion ($\delta = 1$), the weakly interacting electrons occupy the lowest energy levels of a circular ring to minimize their energy. Consequently, the total spin of the ground state is $S = 1/2$ corresponding to a configuration of a doubly occupied s-orbital and one singly occupied p-orbital (Fig. 15.16a). The configuration of the

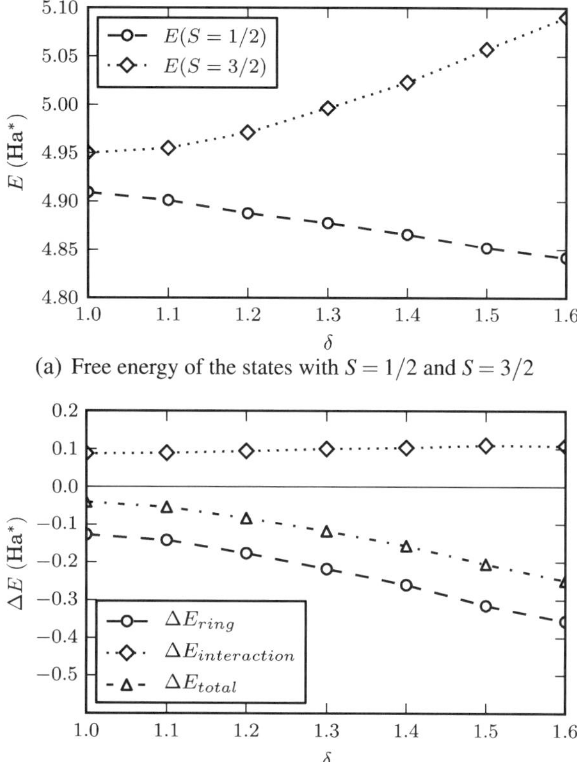

Fig. 15.15 The free energy of the states with $S = 1/2$ and $S = 3/2$ as a function of the distortion (**a**). The ring parameters are $r_0 = 1.35a_0^*$ and $\hbar\omega_0 = 2.0$ Ha*. Figure (**b**) shows the free energy difference between the ground state with $S = 1/2$ and the $S = 3/2$ state. The contributions to the free energy are split into one-particle (kinetic/potential) energies E_{ring} and interaction energies $E_{\text{interaction}}$. The total spin of the ground state is $S = 1/2$ for the whole parameter range. While the ground state energy decreases linearly with distortion, the energy of the $S = 3/2$ state increases quadratically

(a) Free energy of the states with $S = 1/2$ and $S = 3/2$

(b) Contributions to the free energy difference between states with $S = 1/2$ and $S = 3/2$

excited state with $S = 3/2$ is a singly occupied s-orbital and two singly occupied p-orbitals (Fig. 15.16b).

Deforming the circular ring lifts the degeneracy of the p-orbitals and simultaneously pushes all levels to higher energies. For small radii and small distortions, the energy splitting between the hybridized p-orbitals ($\propto \varepsilon/r_0^2$) is larger than the energy shift of the levels ($\propto \varepsilon^2 r_0^2$). In the $S = 1/2$ configuration only one hybridized orbital is occupied. The resulting energy decrease is larger than the increase in energy due to the level shift. Therefore, the hybridization linearly reduces the ground state energy with distortion. In contrast, in the $S = 3/2$ configuration both p-orbitals are

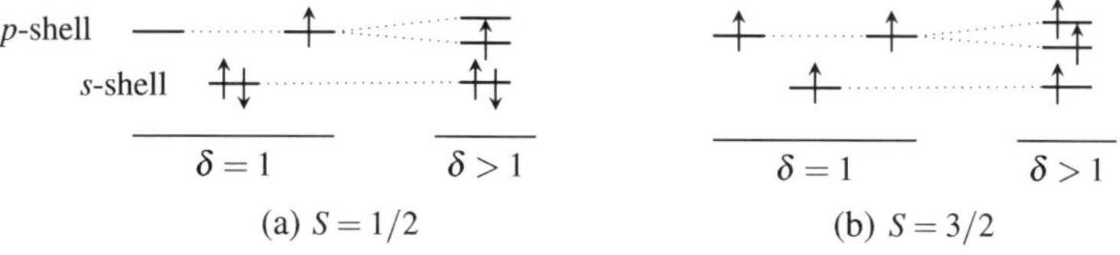

Fig. 15.16 Level diagram for three particles without ($\delta = 1$) and with distortion ($\delta > 1$)

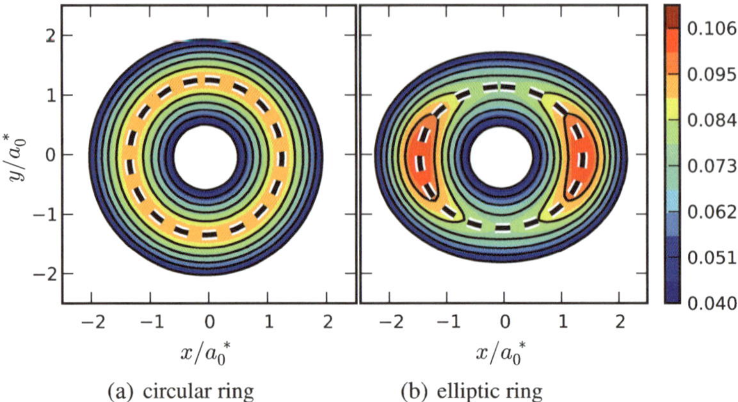

Fig. 15.17 Ground state density $\rho(\mathbf{r})$ of three particles in a *circular ring* (**a**) and an *elliptic ring* with $\delta = 1.2$ (**b**). The two-dimensional density is normalized $\int d\mathbf{r}\rho(\mathbf{r}) = 1$. A *dotted black line* marks the ring potential minimum. While the density of a circular ring is uniform, density peaks appear at both ends of the elliptic ring

occupied and the total energy remains unaffected by the distortion-induced p-shell hybridization. However, the energy shift of the levels increases the energy of the $S = 3/2$ state proportional to the square of the distortion. Figure 15.15(b) indicates that the energy difference between the $S = 1/2$ and $S = 3/2$ states is solely caused by the one-particle energies whereas the interaction energy difference remains almost constant with distortion.

Additionally, the elliptic deformation of the ring changes the electron density due to the reduced symmetry of the system. The preceding treatment of the one-particle problem showed that a two-fold symmetry removes the p-orbital degeneracy producing two stationary waves. The ground state density of three particles (Fig. 15.17) reveals the broken symmetry of the system. While the density of a circular ring is uniform, two peaks appear in the case of an elliptic ring. The appearance of these peaks can be explained with the occupation of the orbitals. As described previously, the configuration of the unperturbed ground state is a doubly occupied s-orbital and a singly occupied p-shell. While the lowest orbital remains almost undisturbed as the ring gets deformed, the singly occupied higher orbital develops a stationary wave which causes the density concentration at the ends of the ring.

In the following, the ring radius is changed from $r_0 = 1.35a_0^*$ to $r_0 = 2.0a_0^*$ which increases the interaction between the electrons. Keeping a confinement of $\hbar\omega_0 = 2.0$ Ha* the ground-state is a quartet with $S = 3/2$. Figure 15.18(a) depicts the free energy of the states with $S = 1/2$ and $S = 3/2$ as a function of the distortion. The state with $S = 1/2$ rises faster in energy with elliptical deformation than the state with $S = 3/2$. Figure 15.18(b) indicates that the difference in interaction energy between $S = 1/2$ and $S = 3/2$ varies slowly with distortion and can be attributed to the changing geometry of the system. The energy of two states mainly differs in the one-particle energy. For the chosen radius the gain in level energy due to the distortion-induced hybridization ($\propto 1/r_0^2$) is not strong enough to fully compensate

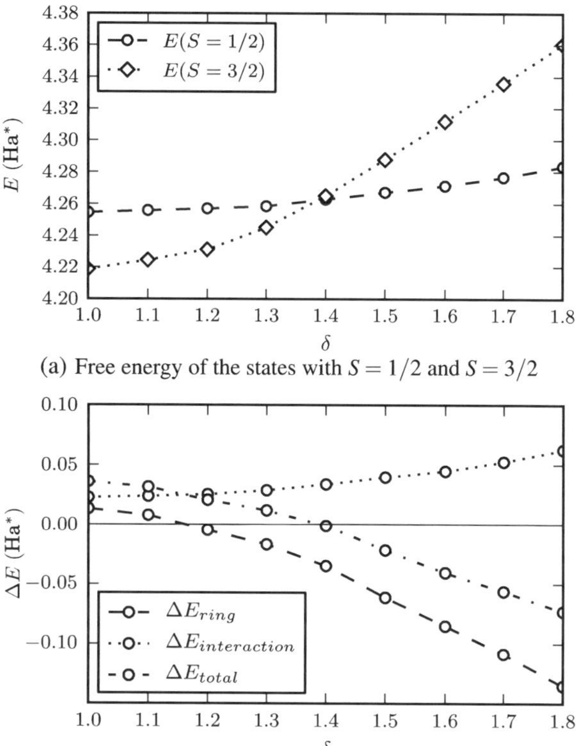

Fig. 15.18 The free energy of the states with $S = 1/2$ and $S = 3/2$ as a function of the distortion (**a**). The ring parameters are $r_0 = 2.0 a_0^*$ and $\hbar\omega_0 = 2.0$ Ha*. Figure (**b**) shows the free energy difference between the $S = 1/2$ state and $S = 3/2$ state. The contributions to the free energy are differentiated in one-particle (kinetic/potential) energies E_{ring} and interaction energies $E_{\mathrm{interaction}}$. The $S = 3/2$ state raises faster in energy with distortion compared to the $S = 1/2$ state leading to a ground state transition form $S = 3/2$ to $S = 1/2$ at $\delta \approx 1.4$

(a) Free energy of the states with $S = 1/2$ and $S = 3/2$

(b) Contributions to the free energy difference between states with $S = 1/2$ and $S = 3/2$

the energy shifting of the levels ($\propto r_0^2$). As a consequence, the energy of the $S = 1/2$ state increases slowly. Both hybridized orbitals are occupied by the $S = 3/2$ state which does not gain energy by the degeneracy splitting. The distortion only shifts the $S = 3/2$ state energy leading to a crossing of the energy levels at $\delta \approx 1.4$. In essence, for this value of deformation the splitting between the hybridized states is sufficient to overcome the exchange-energy saving associated with the $S = 3/2$ state. The total spin of the ground state changes from $S = 3/2$ to $S = 1/2$.

Summarizing, distortion of a three-electron quantum ring leads to a ground state with the lowest possible total spin. Particularly in the case of relatively large radii, a distortion-induced transition of the ground state with $S = 3/2$ to $S = 1/2$ occurs.

15.3.2 Distorted Four-Electron Ring

The three electron case in the previous paragraph revealed that elliptical deformation of the ring raises the energy of a non-degenerate state ($S = 3/2$) compared to a degenerate state ($S = 1/2$) which becomes the ground state. Now, we discuss the

Fig. 15.19 The free energy of the states with $S = 0$ and $S = 1$ as a function of the distortion (**a**). The ring parameters are $r_0 = 2.0a_0^*$ and $\hbar\omega_0 = 2.0$ Ha*. Figure (**b**) shows the free energy difference between the $S = 0$ state and $S = 1$ state. The contributions to the free energy are differentiated in one-particle (kinetic/potential) energies E_{ring} and interaction energies $E_{\text{interaction}}$. The $S = 1$ state raises faster in energy with distortion compared to the $S = 0$ state leading to a ground state transition form $S = 1$ to $S = 0$ at $\delta \approx 1.23$

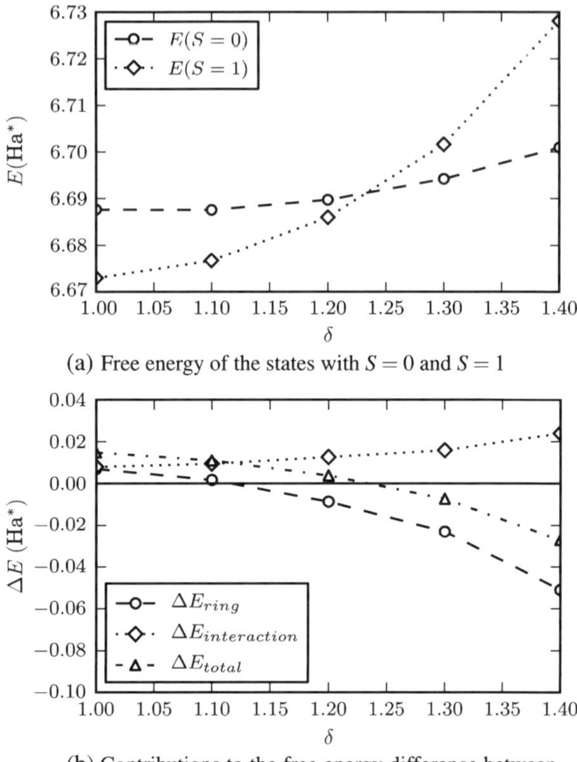

(a) Free energy of the states with $S = 0$ and $S = 1$

(b) Contributions to the free energy difference between states with $S = 0$ and $S = 1$

effect of deformation on the ground state of a four-electron ring. Figure 15.19(a) shows the free energy of the states with $S = 0$ and $S = 1$ for the ring parameters $r_0 = 2.0a_0^*$ and $\hbar\omega_0 = 2.0$ Ha*. For the chosen parameters the energy of the $S = 2$ state is much higher than the energy of the other two states and therefore its discussion will be left out. Figure 15.19(b) shows that again mainly the one-particle energies are affected by distortion.

At zero distortion, the system adopts a Hund's-rule ground state ($S = 1$) with a filled s-shell and two singly occupied p-orbitals (Fig. 15.20b). In contrast, the configuration of the excited $S = 0$ state is a doubly occupied s-orbital and one doubly occupied p-orbital with $\ell = 1$ (Fig. 15.20a). In a circular ring the $S = 0$ state is degenerated with its time-reserved partner which is a configuration of a doubly occupied s-orbital and one doubly occupied p-orbital with $\ell = -1$.

Distortion elevates the energy levels and removes the degeneracy of the p-orbitals. The $S = 1$ state is not degenerate and, as a consequence, raises faster in energy with distortion than the lowest $S = 0$ state which gains energy by the p-shell hybridization. For $\delta \approx 1.23$ the splitting between the hybridized states is sufficient to overcome the exchange-energy savings associated with the $S = 1$ state. The total spin of the ground state can be tuned by elliptical deformation. In the four-electron

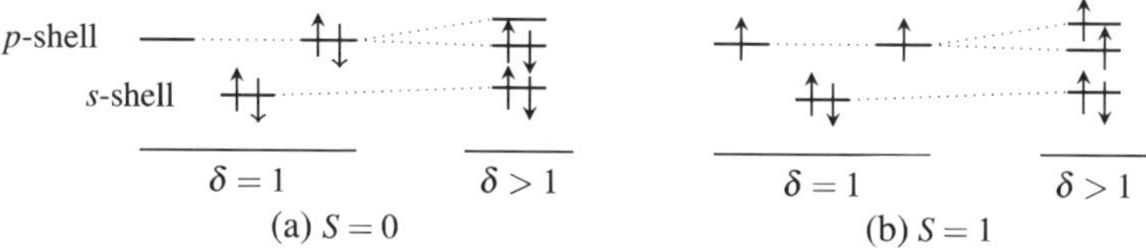

Fig. 15.20 Level diagram for four particles without ($\delta = 1$) and with distortion ($\delta > 1$)

ring the spin transition of the ground state sets in for smaller distortion than in the three-electron case (Fig. 15.18a). The reason is that in the present case one p-orbital is doubly occupied which reduces the total energy of the $S = 0$ state faster than the three-electron $S = 1/2$ state with a singly occupied p-orbital. The question may arise whether the four-electron $S = 2$ state reduces its energy such that there occurs a second spin transition. However, in the $S = 2$ configuration each orbital of the s, p and d-shell is singly occupied. Hybridization does not play any role because both p and d-orbitals are occupied and thus the energy of the $S = 2$ increases with deformation. We have shown that distortion of a four-electron ring also leads to a change of the ground state from $S = 1$ to $S = 0$.

15.4 Conclusions

A path-integral Monte Carlo algorithm suitable to determine spin ground states has been applied to idealized quantum rings of different size and shape. It has been established that the pair-correlation function of three electrons exhibits strong oscillations with increasing radius which can be related to strong correlations between the particles. An analysis of the ground state of three electrons in a ring shows a spin transition to the fully polarized state by increasing the radius. The interpretation as a rotating Wigner molecule and symmetry considerations revealed properties of the system which are beyond the single-particle picture. Analyzing the structure of the wave function gives further insight into the node structure of three strongly interacting electrons in a ring and demonstrates the strong relation between orbital and spin degrees in reduced dimensions. A similar analysis of the four-electron ring revealed that no such spin transitions exist in this case.

Additionally, the effect of an elliptical distortion on the ground state of the ring has been investigated. Elliptically deforming the ring breaks the rotational symmetry of the system and leads to the removal of orbital degeneracy. For strong distortions, the splitting between hybridized states is sufficient to overcome the exchange-energy saving associated with a higher spin state. In this situation the total spin of the ground state always takes the lowest possible value.

It is evident that the spin degree of freedom is strongly linked to the size and shape of a quantum ring. This opens the possibility to prepare high-spin states of confined electrons by choosing the respective dimension of the quantum ring.

References

1. P. Singha Deo, P. Koskinen, M. Koskinen, M. Manninen, Europhys. Lett. **63**(6), 846–852 (2003)
2. M. Szelag, M. Szopa, J. Phys. Conf. Ser. **104**, 012006–012014 (2008)
3. Y.V. Pershin, C. Piermarocchi, Phys. Rev. B **72**(12), 125348 (2005)
4. N. Yang, J.-L. Zhu, Z. Dai, J. Phys. Condens. Matter **20**(29), 295202–295213 (2008)
5. T. Chakraborty, P. Pietiläinen, Phys. Rev. B **52**, 1932–1935 (1995)
6. K. Niemelä, P. Pietiläinen, P. Hyvönen, T. Chakraborty, Europhys. Lett. **36**(7), 533–538 (1996)
7. F. Pederiva, A. Emperador, E. Lipparini, Phys. Rev. B **66**(16), 165314 (2002)
8. P. Borrmann, J. Harting, Phys. Rev. Lett. **86**(14), 3120–3123 (2001)
9. V.M. Fomin (ed.), J. Nanoelectron. Optoelectron. **6**(1) (2011)
10. Y. Saiga, D.S. Hirashima, J. Usukura, Phys. Rev. B **75**(4), 045343–045354 (2007)
11. P. Koskinen, M. Koskinen, M. Manninen, Eur. Phys. J. B **28**(4), 483–489 (2002)
12. L.A. Lavenere-Wanderley, A. Bruno-Alfonso, A. Latge, J. Phys. Condens. Matter **14**(2), 259–270 (2002)
13. D. Gridin, A.T.I. Adamou, R.V. Craster, Phys. Rev. B **69**(15), 155317 (2004)
14. A. Lorke, J.R. Luyken, A.O. Govorov, J.P. Kotthaus, J.M. Garcia, P.M. Petroff, Phys. Rev. Lett. **84**(10), 2223–2226 (2000)
15. D.M. Ceperley, Rev. Mod. Phys. **67**(2), 279–355 (1995)
16. B. Baxevanis, D. Pfannkuche, J. Nanoelectron. Optoelectron. **6**(1), 76–80 (2011)
17. M. Takahashi, M. Imada, J. Phys. Soc. Jpn. **53**(3), 963–974 (1984)
18. R. Egger, W. Häusler, C.H. Mak, H. Grabert, Phys. Rev. Lett. **83**(2), 462 (1999)
19. R. Pauncz, *The Construction of Spin Eigenfunctions: An Exercise Book* (Kluwer Academic/Plenum, New York, 2001)
20. A.P. Lyubartsev, P.N. Vorontsov-Velyaminov, Phys. Rev. A **48**(6), 4075–4083 (1993)
21. M. Hamermesh, *Group Theory and Its Application to Physical Problems* (Dover, New York, 1989)
22. L. Wendler, V.M. Fomin, A.V. Chaplik, A.O. Govorov, Z. Phys. B, Condens. Matter **100**, 211–221 (1996)
23. E. Wigner, Phys. Rev. **46**(11), 1002–1011 (1934)
24. M. Koskinen, M. Manninen, B. Mottelson, S.M. Reimann, Phys. Rev. B **63**(20), 205323 (2001)
25. P. Hawrylak, D. Pfannkuche, Phys. Rev. Lett. **70**(4), 485–488 (1993)
26. P.A. Maksym, Phys. Rev. B **53**(16), 10871–10886 (1996)
27. P.A. Maksym, Physica B, Condens. Matter **184**(1–4), 385–393 (1993)
28. R. Courant, D. Hilbert, *Methods of Mathematical Physics*, vol. I (Interscience, New York, 1953)
29. C. Herring, Phys. Rev. B **11**(5), 2056–2061 (1975)
30. M. Aizenman, E.H. Lieb, Phys. Rev. Lett. **65**(12), 1470–1473 (1990)
31. E.H. Lieb, D. Mattis, J. Math. Phys. **3**(4), 749–751 (1962)
32. J. Planelles, F. Rajadell, J.I. Climente, Nanotechnology **18**(37), 375402–375413 (2007)
33. A. Bruno-Alfonso, A. Latgé, Phys. Rev. B **77**(20), 205303 (2008)

Chapter 16
Differential Geometry Applied to Rings and Möbius Nanostructures

Benny Lassen, Morten Willatzen, and Jens Gravesen

Abstract Nanostructure shape effects have become a topic of increasing interest due to advancements in fabrication technology. In order to pursue novel physics and better devices by tailoring the shape and size of nanostructures, effective analytical and computational tools are indispensable. In this chapter, we present analytical and computational differential geometry methods to examine particle quantum eigenstates and eigenenergies in curved and strained nanostructures. Example studies are carried out for a set of ring structures with different radii and it is shown that eigenstate and eigenenergy changes due to curvature are most significant for the groundstate eventually leading to qualitative and quantitative changes in physical properties. In particular, the groundstate in-plane symmetry characteristics are broken by curvature effects, however, curvature contributions can be discarded at bending radii above 50 nm. In the second part of the chapter, a more complicated topological structure, the Möbius nanostructure, is analyzed and geometry effects for eigenstate properties are discussed including dependencies on the Möbius nanostructure width, length, thickness, and strain.

B. Lassen · M. Willatzen
Mads Clausen Institute, University of Southern Denmark, Alsion 2, 6400 Sønderborg, Denmark

B. Lassen
e-mail: benny@mci.sdu.dk

M. Willatzen
e-mail: willatzen@mci.sdu.dk

M. Willatzen (✉)
Department of Photonics Engineering, Technical University of Denmark, Ørsteds Plads, 2800 Kgs. Lyngby, Denmark
e-mail: morwi@fotonik.dtu.dk

J. Gravesen
Department of Mathematics, Technical University of Denmark, Matematiktorvet, 2800 Kgs. Lyngby, Denmark
e-mail: jgra@dtu.dk

V.M. Fomin (ed.), *Physics of Quantum Rings*, NanoScience and Technology, DOI 10.1007/978-3-642-39197-2_16, © Springer-Verlag Berlin Heidelberg 2014

16.1 Introduction

With the possibility to shape objects in the nano range using novel fabrication technologies it becomes increasingly important to assess experimentally and theoretically the combined influence of shape and size on physical properties of nanostructures. Experimental studies revealing these geometry properties include electronic, magnetic, and optical properties of electrons and holes confined to curved surfaces such as graphene strips and semiconductor nanorings. An exotic nanostructure which has been examined experimentally is the Möbius nanostructure produced by spooling a single crystalline $NbSe_3$ ribbon on a selenium droplet whereby surface tension leads to a twist in the ribbon [1]. Topological insulators having unusual properties and potential interesting applications have been studied experimentally [2, 3] and theoretically for a Möbius graphene strip [4].

Since the first Möbius nanostructure fabrications, several investigations have been carried out on Möbius structures both for classical and nanostructure dimensions [4–13]. Gravesen and Willatzen computed electronic eigenstates and the shape of Möbius nanostructures using differential geometry arguments taking into account bending effects [5, 14]. Heijden and Starostin solved a classical problem in geometry by employing an invariant variational bi-complex formalism to derive the first equilibrium equations for a wide developable strip undergoing large deformations [6, 7]. Ballon et al. showed that classical Möbius ring resonators exhibit fermion-boson rotational symmetry [8]. Yoneya et al. determined theoretically the structure of domain walls in ferromagnetic states on Möbius strips [9]. Guo studied electronic properties of a Möbius graphene strip with a zigzag edge [4]. Optical activity for a Möbius nanostructure was examined by Rockstuhl et al. [10]. Zhao et al. analyzed topological properties of quantum states for a spinless particle hopping in a Möbius ladder [11]. Li and Ram-Mohan [12] theoretically discussed several quantum-mechanical properties of Möbius nanostructures including level splittings, symmetry (in particular the absence of rotational symmetry for eigenstates), the influence of magnetic field and optical transitions. In a recent study, Fomin et al. examined electron localization in inhomogeneous Möbius rings [13].

In order to assess and optimize the influence of geometry on physical properties of nanostructures, it is important to develop effective analytical and computational techniques. This chapter presents analytical techniques to compute quantum-mechanical particle eigenstates confined to complicated geometries. We start out by deriving the governing equations of a conduction electron confined to a curved semiconductor nanoring. It is shown that if the nanoring is characterized by a large aspect ratio, i.e., nanorings where the length is much larger than the cross-sectional dimensions, then the three-dimensional Schrödinger equation can be separated in three ordinary differential equations. Two of these can be solved analytically and lead to sinusoidal solutions, while the wave-function part that depends on the last coordinate parametrizing the length contains curvature contributions. Open and closed nanoring boundary conditions are considered and the influence of boundary conditions on eigenstate symmetries and energies is discussed. The general effect of a curved geometry is to decrease electronic energies compared to the corresponding

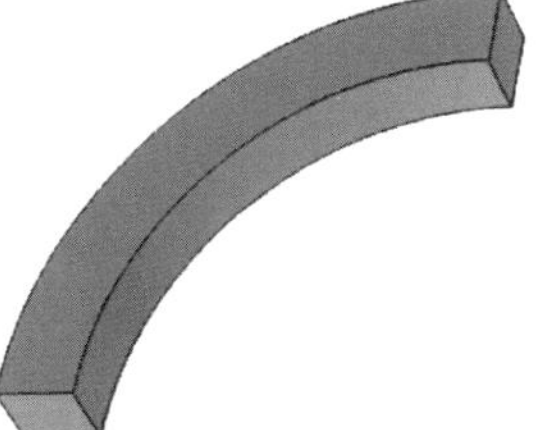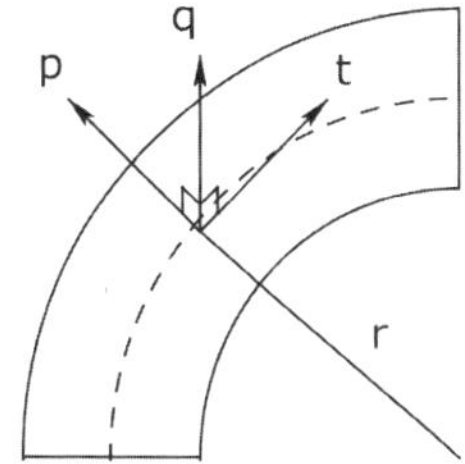

Fig. 16.1 Schematic picture of a curved nanowire structure (*left plot*) and a figure showing the chosen curvilinear coordinates (*right plot*)

straight nanowire of the same volume. The discussion is continued to include contributions from strain through the deformation potential. For typical semiconductors this effect is huge if strain is present and by far the major effect compared to the direct geometry and curvature influence on the Laplacian. In the second part of the chapter, an exotic example of a complicated nanostructure geometry is analyzed: the Möbius structure. By inclusion of the bending energy in determining the shape of a Möbius nanostructure, the median or centerline parametrization is derived and described in terms of the generalized coordinate u^1 and the width and thickness of the Möbius nanostructure are parametrized by coordinates u^2 and u^3. It is shown that for thin Möbius nanostructures, the electronic eigenstates can be written as a semi-separable problem. The thickness-coordinate part of the wavefunction separates out while the median- and width-coordinate parts couple in a non-separable manner. Results are compared with exact finite element calculations for a general Möbius nanostructure. It is verified that the differential geometry formulation presented is accurate whenever the thickness is much smaller than the median length and the width in agreement with the assumptions.

16.1.1 Arc-Length Parametrization

It is convenient for the analysis of complex-shaped nanowires or nanorings to introduce a set of orthogonal vectors where one is the local tangent vector of the nanowire axis and the two others supplement to span the local cross section of the nanowire [15], see Fig. 16.1. The nanowire axis $\mathbf{r}(s)$ is assumed to be parametrized by arc length s, but this is not really a restriction since any curve can be parametrized by arc length. The tangent vector $\mathbf{t}(s) = \mathbf{r}'(s) = d\mathbf{r}/ds$ is a unit vector field along the curve and the two vectors spanning the local cross section of the nanowire are designated $\mathbf{p}(s)$ and $\mathbf{q}(s)$ such that $\mathbf{t}(s), \mathbf{p}(s), \mathbf{q}(s)$ constitute an orthonormal frame at each point $\mathbf{r}(s)$ along the axis. Differentiation of the identities,

$$\mathbf{t} \cdot \mathbf{t} = 1,$$
$$\mathbf{p} \cdot \mathbf{p} = 1,$$
$$\mathbf{q} \cdot \mathbf{q} = 1,$$
$$\mathbf{t} \cdot \mathbf{p} = 0,$$
$$\mathbf{t} \cdot \mathbf{q} = 0,$$
$$\mathbf{p} \cdot \mathbf{q} = 0,$$

$$(16.1)$$

yields

$$2\frac{d\mathbf{t}}{ds} \cdot \mathbf{t} = 0,$$
$$2\frac{d\mathbf{p}}{ds} \cdot \mathbf{p} = 0,$$
$$2\frac{d\mathbf{q}}{ds} \cdot \mathbf{q} = 0, \tag{16.2}$$
$$\frac{d\mathbf{t}}{ds} \cdot \mathbf{p} + \mathbf{t} \cdot \frac{d\mathbf{p}}{ds} = 0,$$
$$\frac{d\mathbf{t}}{ds} \cdot \mathbf{q} + \mathbf{t} \cdot \frac{d\mathbf{q}}{ds} = 0,$$
$$\frac{d\mathbf{p}}{ds} \cdot \mathbf{q} + \mathbf{p} \cdot \frac{d\mathbf{q}}{ds} = 0.$$

If we now let

$$a(s) = \frac{d\mathbf{t}}{ds} \cdot \mathbf{p},$$
$$b(s) = \frac{d\mathbf{t}}{ds} \cdot \mathbf{q}, \tag{16.3}$$
$$c(s) = \frac{d\mathbf{p}}{ds} \cdot \mathbf{q},$$

then we obtain the following equation

$$\frac{d}{ds}\begin{bmatrix}\mathbf{t}\\\mathbf{p}\\\mathbf{q}\end{bmatrix} = \begin{bmatrix}0 & a & b\\-a & 0 & c\\-b & -c & 0\end{bmatrix}\begin{bmatrix}\mathbf{t}\\\mathbf{p}\\\mathbf{q}\end{bmatrix}. \tag{16.4}$$

Observe that the curvature κ of the axis is:

$$\kappa(s) = |\mathbf{r}''(s)| = |\mathbf{t}'(s)| = \sqrt{a^2 + b^2}. \tag{16.5}$$

The above does not uniquely determine $\mathbf{p}, \mathbf{q}$. A typical choice of vector fields $\mathbf{p}, \mathbf{q}$ is to let $\mathbf{p}$ be the principal normal $\mathbf{n} = \mathbf{t}'/\kappa$ and $\mathbf{q}$ is the binormal $\mathbf{b} = \mathbf{t} \times \mathbf{n}$. In this case, (16.4) leads to the Frenet-Serret equations, where $a = \kappa$, $b = 0$, and $c = \tau$ is the torsion of the axis. If further the nanowire axis lies in a plane, the torsion vanishes: $\tau = 0$. It should be noted that the frame chosen here does not always exist in principle; this happens if the curvature vanishes locally. The problem can be circumvented by choosing the minimal rotation frame. In the latter frame $c = 0$ even if the torsion is non-zero.

16.1.2 *Planar Nanowire Axis Curves*

In the following, the analysis is restricted to nanowire structures where the nanowire axis lies in a plane, and we choose $b = 0$ and $c = 0$ which allows for some simplifications in the model setup.

We can now parametrize a curved tube of rectangular cross section, i.e., a tubular neighbourhood in $\mathbb{R}^3$ of the curve $\mathbf{r}(s)$, according to:

$$\mathbf{x}(u^1, u^2, u^3) = \mathbf{r}(u^1) + u^2 \mathbf{p}(u^1) + u^3 \mathbf{q}(u^1),\tag{16.6}$$

where $-\varepsilon_2 \leq u_2 \leq \varepsilon_2$ and $-\varepsilon_3 \leq u_3 \leq \varepsilon_3$ for two constants ε_2 and ε_3. Albeit the formulation is for rectangular-shaped nanowire cross sections it is easy, afterwards, to adjust the theory to curved nanowires with, e.g., a circular cross section.

Using (16.4), simple manipulations give

$$\frac{\partial \mathbf{x}}{\partial u^1} = \left(1 - u^2 a(u^1)\right)\mathbf{t}(u^1),\tag{16.7}$$

$$\frac{\partial \mathbf{x}}{\partial u^2} = \mathbf{p}(u^1),\tag{16.8}$$

$$\frac{\partial \mathbf{x}}{\partial u^3} = \mathbf{q}(u^1).\tag{16.9}$$

The metric tensor $G_{ij} = \frac{\partial \mathbf{x}_i}{\partial u^i} \cdot \frac{\partial \mathbf{x}_j}{\partial u^j}$ of $\mathbb{R}^3$ becomes

$$[G_{ij}] = \begin{bmatrix} (1 - u^2 a(u^1))^2 & 0 & 0 \\ 0 & 1 & 0 \\ 0 & 0 & 1 \end{bmatrix}.\tag{16.10}$$

The determinant is

$$G = \left(1 - u^2 a(u^1)\right)^2,\tag{16.11}$$

and the inverse $[G_{ij}]^{-1} = [G^{ij}]$ reads

$$[G^{ij}] = \begin{bmatrix} \frac{1}{(1 - u^2 a(u^1))^2} & 0 & 0 \\ 0 & 1 & 0 \\ 0 & 0 & 1 \end{bmatrix}.\tag{16.12}$$

If we expand the determinant and the inverse in u^2 and u^3 then we obtain to zeroth order

$$G \approx 1, \qquad \partial_1 G \approx 0,\tag{16.13}$$

$$\partial_2 G \approx -2a(u^1), \qquad \partial_3 G \approx 0,\tag{16.14}$$

$$[G^{ij}] \approx \begin{bmatrix} 1 & 0 & 0 \\ 0 & 1 & 0 \\ 0 & 0 & 1 \end{bmatrix}, \qquad \partial_1[G^{ij}] \approx \begin{bmatrix} 0 & 0 & 0 \\ 0 & 0 & 0 \\ 0 & 0 & 0 \end{bmatrix},\tag{16.15}$$

$$\partial_2\big[G^{ij}\big] \approx \begin{bmatrix} 2a(u^1) & 0 & 0 \\ 0 & 0 & 0 \\ 0 & 0 & 0 \end{bmatrix}, \qquad \partial_3\big[G^{ij}\big] \approx \begin{bmatrix} 0 & 0 & 0 \\ 0 & 0 & 0 \\ 0 & 0 & 0 \end{bmatrix}. \tag{16.16}$$

The Laplace operator $\Delta_{\mathbb{R}^3} = \frac{\partial^2}{\partial x^2} + \frac{\partial^2}{\partial y^2} + \frac{\partial^2}{\partial z^2}$ in $\mathbb{R}^3$ is in the curvilinear coordinates u^1, u^2, u^3 given by [16]

$$\Delta_{\mathbb{R}^3} = G^{ij}\partial_i\partial_j + \left(\frac{G^{ij}}{2}\frac{\partial_j G}{G} + \partial_j G^{ij}\right)\partial_i, \tag{16.17}$$

where $\partial_i = \frac{\partial}{\partial u^i}$ and reads to zeroth order in u^2 and u^3:

$$\Delta_{\mathbb{R}^3} = G^{ij}\partial_i\partial_j + \left(G^{ij}\frac{\partial_j G}{2G} + \partial_j G^{ij}\right)\partial_i \approx \partial_1^2 + \partial_2^2 + \partial_3^2 - a(u^1)\partial_2. \tag{16.18}$$

Next, introducing

$$F = \sqrt{G} = 1 - u^2 a(u^1), \tag{16.19}$$

and letting $\chi = \sqrt{F}\psi$ allow us to write, again to zeroth order in u^2 and u^3,

$$\Delta_{\mathbb{R}^3}\psi = \Delta_{\mathbb{R}^3}\left(\frac{\chi}{\sqrt{F}}\right) \approx \partial_1^2\chi + \partial_2^2\chi + \partial_3^2\chi + \frac{\kappa^2}{4}\chi. \tag{16.20}$$

Hence, the benefit in recasting the Laplacian operator problem in terms of χ instead of ψ is that the right-hand-side of (16.20) is separable in the three coordinates u^1, u^2, u^3 while the right-hand-side of (16.18) is not. When we address the electron one-band heterostructure problem, separability of the scaled wavefunction χ is maintained and considerable computational simplicity is gained.

16.1.3 General Nanowire Axis Parametrization

For most curves, it is difficult to find an explicit arc-length parametrization $\mathbf{r}(s)$. Hence, we need to look for a general parametrization $\mathbf{r}(t)$ with $t = t(s)$ and $|\mathbf{r}'(t)| \neq 1$ as follows:

$$\frac{d}{ds} = \frac{dt}{ds}\frac{d}{dt} = \left(\frac{ds}{dt}\right)^{-1}\frac{d}{dt}, \tag{16.21}$$

$$\frac{d\chi_1}{ds} = \left(\frac{ds}{dt}\right)^{-1}\frac{d\chi_1}{dt}, \tag{16.22}$$

$$\frac{d^2\chi_1}{ds^2} = \left(\frac{ds}{dt}\right)^{-1}\frac{d}{dt}\left(\left(\frac{ds}{dt}\right)^{-1}\frac{d\chi_1}{dt}\right)$$

$$= \left(\frac{ds}{dt}\right)^{-2}\frac{d^2\chi_1}{dt^2} - \left(\frac{ds}{dt}\right)^{-3}\frac{d^2s}{dt^2}\frac{d\chi_1}{dt}. \tag{16.23}$$

Now

$$\frac{\mathrm{d}s}{\mathrm{d}t} = |\mathbf{r}'(t)| = \sqrt{\mathbf{r}' \cdot \mathbf{r}'}, \tag{16.24}$$

$$\frac{\mathrm{d}^2 s}{\mathrm{d}t^2} = \frac{\mathbf{r}' \cdot \mathbf{r}''}{\sqrt{\mathbf{r}' \cdot \mathbf{r}'}}, \tag{16.25}$$

$$\kappa^2 = \frac{|\mathbf{r}' \times \mathbf{r}''|^2}{|\mathbf{r}'|^6} = \frac{(\mathbf{r}' \times \mathbf{r}'') \cdot (\mathbf{r}' \times \mathbf{r}'')}{(\mathbf{r}' \cdot \mathbf{r}')^3} = \frac{|\mathbf{r}'|^2 |\mathbf{r}''|^2 - (\mathbf{r}' \cdot \mathbf{r}')^2}{(\mathbf{r}' \cdot \mathbf{r}'')^3}. \tag{16.26}$$

16.2 Application to the Schrödinger Equation

Let s be nanowire axis arc length. With the above expression for the Laplacian, the Schrödinger equation for an electron with effective mass m and energy E reads in terms of χ and coordinates u^i (applies to zeroth order in s and u_2)

$$\frac{-\hbar^2}{2m}\left(\partial_s^2 \chi + \partial_2^2 \chi + \partial_3^2 \chi + \frac{\kappa^2}{4}\chi\right) + V(u_1, u_2, u_3)\chi = E\chi, \tag{16.27}$$

where $\partial_s \equiv \frac{\partial}{\partial s}$, and the potential V satisfies

$$V(s, u_2, u_3) = 0, \tag{16.28}$$

if (s, u_2, u_3) is a point within the nanowire structure, i.e., the domain: $-\varepsilon_2 \leq u_2 \leq \varepsilon_2$ and $-\varepsilon_3 \leq u_3 \leq \varepsilon_3$. Similarly, the potential V is infinite outside the nanowire structure to mimic the infinite-barrier case.

As the curvature κ is a function of s only it follows that a separable solution $\chi = \chi_1(s)\chi_2(u_2)\chi_3(u_3)$ can be sought. For a general parametrization $t = t(s)$, using the expressions in (16.21)–(16.26) allows us to recast (16.27) as three ordinary differential equations:

$$\chi_1'' - \frac{\mathbf{r}' \cdot \mathbf{r}''}{\mathbf{r}' \cdot \mathbf{r}'}\chi_1' + \left(\frac{(\mathbf{r}' \cdot \mathbf{r}')(\mathbf{r}'' \cdot \mathbf{r}'') - (\mathbf{r}' \cdot \mathbf{r}'')^2}{4(\mathbf{r}' \cdot \mathbf{r}')^2} - (\lambda + \mu)(\mathbf{r}' \cdot \mathbf{r}')\right)\chi_1 = 0, \tag{16.29}$$

$$\partial_2^2 \chi_2 + \nu^2 \chi_2 = 0, \tag{16.30}$$

$$\partial_3^2 \chi_3 + \left(\mu - \nu^2\right)\chi_3 = 0, \tag{16.31}$$

with $\lambda = -\frac{2mE}{\hbar^2}$ and μ and ν separation constants; and a prime (') denotes differentiation with respect to t.

16.2.1 Analytical Solution for χ_2, χ_3

The equations in χ_2 and χ_3 can be solved analytically. The general solution to (16.30) in χ_2 is

$$\chi_2(u^2) = \sin(\nu u^2 + \phi_2),\tag{16.32}$$

where ν and ϕ_2 are constants determined by the hard-wall boundary conditions imposed, i.e.,

$$\chi_2(-\varepsilon^2) = \sin(-\nu\varepsilon^2 + \phi_2) = 0,$$
$$\chi_2(\varepsilon^2) = \sin(\nu\varepsilon^2 + \phi_2) = 0.\tag{16.33}$$

Thus:

$$\nu = \frac{m'\pi}{2\varepsilon^2},\tag{16.34}$$

where m' is an integer different from zero. The other constant, the phase ϕ_2, must be chosen such that

$$\phi_2 = -\nu\varepsilon^2,\tag{16.35}$$

and the boundary conditions in (16.33) are fulfilled. If m' is even (and different from zero), χ_2 becomes

$$\chi_2(u^2) = \sin\left(\frac{m'\pi}{2\varepsilon^2}u^2\right),\tag{16.36}$$

while for m' odd:

$$\chi_2(u^2) = \cos\left(\frac{m'\pi}{2\varepsilon^2}u^2\right).\tag{16.37}$$

In exactly the same way, (16.31) allows the separation constant μ to be determined:

$$\mu - \nu^2 = \left(\frac{n'\pi}{2\varepsilon^3}\right)^2,\tag{16.38}$$

where n' is an integer different from zero. If n' is even and different from zero, the eigenfunction χ_3 is

$$\chi_3(u^3) = \sin\left(\frac{n'\pi}{2\varepsilon^3}u^3\right),\tag{16.39}$$

while for n' odd:

$$\chi_3(u^3) = \cos\left(\frac{n'\pi}{2\varepsilon^3}u^3\right).\tag{16.40}$$

Combining (16.34) and (16.38) yields for μ the result

$$\mu = \left(\frac{m'\pi}{2\varepsilon^2}\right)^2 + \left(\frac{n'\pi}{2\varepsilon^3}\right)^2,\tag{16.41}$$

with $m' = \pm 1, \pm 2, \pm 3$ and $n' = \pm 1, \pm 2, \pm 3$. The possible values of the particle energy E are finally found from the χ_1 eigenvalue equation (16.29) by imposing appropriate boundary conditions given the value of μ.

16.2.2 Case Study: Circular Nanoring

The circular nanoring can be treated analytically. An arc-length parametrization of a circular nanoring is

$$\mathbf{r}(u^1) = \left(R\cos\left(\frac{u^1}{R}\right), R\sin\left(\frac{u^1}{R}\right), 0\right),\tag{16.42}$$

where

$$\left|\mathbf{r}'(u^1)\right| = 1, \qquad \left|\mathbf{r}''(u^1)\right| = \frac{1}{R^2},\tag{16.43}$$

and R is the circular ring radius-of-curvature. Since $\left|\mathbf{r}'(u^1)\right| = 1$, the parametrization is an arc-length parametrization and (16.29) reads

$$\chi_1'' - \left(\lambda + \mu - \frac{1}{4R^2}\right)\chi_1 = 0\tag{16.44}$$

with the general solution

$$\chi_1 = \sin\left(\sqrt{-\lambda - \mu + \frac{1}{4R^2}}\, u^1 + \phi_1\right),\tag{16.45}$$

where ϕ_1 is an arbitrary phase. Further, imposing the boundary conditions:

$$\chi_1(u^1 = 0) = \chi_1(u^1 = L) = 0,\tag{16.46}$$

corresponding to an *open* circular-shaped nanowire structure, gives

$$\chi_1(u^1) = \sin\left(\frac{l\pi}{L}u^1\right), \quad l = \pm 1, \pm 2, \pm 3, \ldots,\tag{16.47}$$

and the associated energy eigenvalue is

$$E = -\frac{\hbar^2\lambda}{2m} = \frac{\hbar^2}{2m}\left[\left(\frac{l\pi}{L}\right)^2 + \left(\frac{m'\pi}{2\varepsilon^2}\right)^2 + \left(\frac{n'\pi}{2\varepsilon^3}\right)^2 - \frac{1}{4R^2}\right].\tag{16.48}$$

A *closed* circular-shaped nanowire structure is subject to less strict boundary conditions

$$\chi_1\left(u^1\right) = \chi_1\left(u^1 + 2\pi R\right), \tag{16.49}$$

corresponding to

$$\sqrt{\frac{1}{4R^2} - \mu - \lambda} \, 2\pi R = 2l\pi, \quad l = 0, \pm 1, \pm 2, \pm 3, \ldots, \tag{16.50}$$

and the following eigenstate solutions (with $L = 2\pi R$):

$$\chi_1\left(u^1\right) = \sin\left(\frac{2l\pi}{L} u^1 + \phi_1\right), \quad l = 0, \pm 1, \pm 2, \pm 3, \ldots. \tag{16.51}$$

The associated energy spectrum is

$$E = -\frac{\hbar^2 \lambda}{2m} = \frac{\hbar^2}{2m}\left[\left(\frac{2l\pi}{L}\right)^2 + \left(\frac{m'\pi}{2\varepsilon^2}\right)^2 + \left(\frac{n'\pi}{2\varepsilon^3}\right)^2 - \frac{1}{4R^2}\right]. \tag{16.52}$$

In particular, note that $l = 0$ is possible since the phase ϕ_1 is arbitrary for the closed circular nanowire axis. This result is the same as found in the case of a cylinder surface of revolution [16] keeping in mind that L in Ref. [16] equals the present $2\varepsilon^2$. It was demonstrated in Ref. [16] that the energy expression (16.52) is an excellent approximation if the cylinder thickness is less than 10 % of the radius R.

16.2.3 Case Study: Elliptic Nanoring

Consider an elliptical-shaped nanoring structure with an axis parametrization:

$$\mathbf{r}\left(u^1\right) = \left(R_1 \cos\left(2\pi \frac{u^1}{L}\right), R_2 \sin\left(2\pi \frac{u^1}{L}\right), 0\right), \tag{16.53}$$

where R_1 and R_2 define the elliptic semi-major and semi-minor axes, and the nanoring corresponds to the range: $0 \le u^1 \le L$.

From (16.53), the following relations are obtained:

$$\mathbf{r}'\left(u^1\right) = \left(-2\pi \frac{R_1}{L} \sin\left(2\pi \frac{u^1}{L}\right), 2\pi \frac{R_2}{L} \cos\left(2\pi \frac{u^1}{L}\right), 0\right), \tag{16.54}$$

$$\mathbf{r}''\left(u^1\right) = \left(-(2\pi)^2 \frac{R_1}{L^2} \cos\left(2\pi \frac{u^1}{L}\right), -(2\pi)^2 \frac{R_2}{L^2} \sin\left(2\pi \frac{u^1}{L}\right), 0\right) \tag{16.55}$$

and

$$\mathbf{r}' \cdot \mathbf{r}' = (2\pi)^2 \frac{R_1^2}{L^2} \sin^2\left(2\pi \frac{u^1}{L}\right) + (2\pi)^2 \frac{R_2^2}{L^2} \cos^2\left(2\pi \frac{u^1}{L}\right), \tag{16.56}$$

Table 16.1 The first three energy levels for an elliptical nanoring. Here m is the InAs electron effective mass equal to $0.022m_0$ [25] where m_0 is the free electron mass

Energy	$E(1) - \frac{\hbar^2}{2m}\mu$ (meV)	$E(2) - \frac{\hbar^2}{2m}\mu$ (meV)	$E(3) - \frac{\hbar^2}{2m}\mu$ (meV)
	-3	23	78

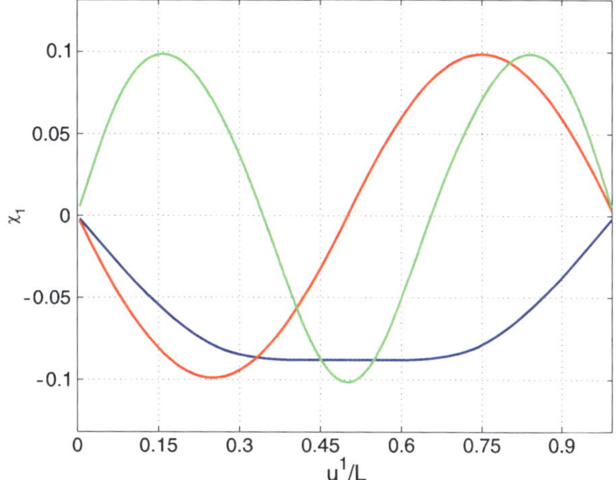

Fig. 16.2 The first three eigenstates for χ_1 for an elliptical nanoring. The *blue, red*, and *green curves* are the groundstate, the first-excited state, and the second-excited state, respectively

$$\mathbf{r}'' \cdot \mathbf{r}'' = (2\pi)^4 \frac{R_1^2}{L^4} \cos^2\left(2\pi\frac{u^1}{L}\right) + (2\pi)^4 \frac{R_2^2}{L^4} \sin^2\left(2\pi\frac{u^1}{L}\right), \qquad (16.57)$$

$$\mathbf{r}' \cdot \mathbf{r}'' = (2\pi)^3 \frac{R_1^2 - R_2^2}{L^3} \sin\left(2\pi\frac{u^1}{L}\right) \cos\left(2\pi\frac{u^1}{L}\right). \qquad (16.58)$$

Using the latter expressions in (16.29) and solving for χ_1 with appropriate boundary conditions, we show in Table 16.1 the first three eigenvalues (relative to $\frac{\hbar^2}{2m}\mu$) for a nanowire with $R_1 = 5$ nm, $R_2 = 8$ nm, and $L = 2\pi \cdot 10$ nm. In our case, we assumed the elliptic nanoring to be cut open such that $\chi_1(u^1 = L) = \chi_1(u^1 = 0) = 0$.

The three associated eigenstates for χ_1 can be seen in Fig. 16.2.

16.3 Strain in Nanorings

In what follows, we consider large aspect-ratio nanowires, i.e., nanowires where the cross-sectional dimensions are much smaller than the nanowire length. The nanoring axis coordinate is u^1 and the cross-sectional coordinates are u^2 and u^3, see Fig. 16.1.

16.3.1 Stress Tensor for a Bent Nanowire

It is assumed that the nanowire is free-standing, i.e., the nanoring structure obeys the boundary stress relations:

$$\sigma_{ik} n_k = 0, \tag{16.59}$$

where σ_{ik} is the stress tensor, n_k is the boundary normal vector components, and summation from 1 to 3 is implied for repeated indices. The indices i and k take on values $1, 2, 3$ corresponding to the u^1, u^2, u^3 coordinate directions, respectively. The normal vector $\mathbf{n}$ always lies in the $u^2 - u^3$ plane, i.e., $n_1 = 0$. Further, locations exist on the cross-sectional boundary where $n_2 = 1$ and $n_3 = 0$ such that (16.59) with $i = 1$ gives $\sigma_{12} = 0$. Similarly, locations exist at the cross-sectional boundary where $n_2 = 0$ and $n_3 = 1$ hence with $i = 1$ we get $\sigma_{13} = 0$. Using similar arguments for $i = 2$ and $i = 3$ we conclude that only σ_{11} can be nonzero at the cross-sectional boundary. Since the cross-sectional dimensions are small, it is reasonable to assume that the stress tensor components are constant over the cross section and we immediately conclude that σ_{11} is the only non-zero stress component *everywhere* in the nanowire.

16.3.2 Strain Tensor Results in the Zincblende Case

Using the stress-strain relations for cubic materials:

$$\sigma_{11} = c_{11}\varepsilon_{11} + c_{12}\varepsilon_{22} + c_{12}\varepsilon_{33}, \tag{16.60}$$

$$\sigma_{22} = c_{12}\varepsilon_{11} + c_{11}\varepsilon_{22} + c_{12}\varepsilon_{33} = 0, \tag{16.61}$$

$$\sigma_{33} = c_{12}\varepsilon_{11} + c_{12}\varepsilon_{22} + c_{11}\varepsilon_{33} = 0, \tag{16.62}$$

$$\sigma_{23} = c_{44}\varepsilon_{23} = 0, \tag{16.63}$$

$$\sigma_{13} = c_{44}\varepsilon_{13} = 0, \tag{16.64}$$

$$\sigma_{12} = c_{44}\varepsilon_{12} = 0, \tag{16.65}$$

where ε_{ij} is the strain tensor and c_{ij} is the stiffness tensor, it is found that

$$\varepsilon_{22} = \varepsilon_{33} = -\frac{c_{12}}{c_{11} + c_{12}}\varepsilon_{11}, \tag{16.66}$$

and all other strain components are zero.

16.3.3 Nonlinear Expression for the Strain Component ε_{11}

With the above relations between strain components, it is possible to find all strain components once we know, say, ε_{11}. The general expression for the strain tensor to second order is [17]

$$\varepsilon_{ik} = \frac{1}{2}\left(\frac{\partial d_i}{\partial x_k} + \frac{\partial d_k}{\partial x_i} + \frac{\partial d_l}{\partial x_i}\frac{\partial d_l}{\partial x_k}\right),\tag{16.67}$$

where d_i is the ith component of the displacement. In the following, we shall restrict our analysis to nanowires bent uniformly, i.e., nanowires with a constant radius of curvature: R along the nanowire length. In this case, if the bending is assumed to be in the u^1-u^2 plane, then we have:

$$\varepsilon_{11} = \frac{1}{2}\left(\frac{\partial d_1}{\partial u^1} + \frac{\partial d_1}{\partial u^1} + \frac{\partial d_1}{\partial u^1}\frac{\partial d_1}{\partial u^1}\right).\tag{16.68}$$

In the presence of bending forces only, a simple geometry analysis to first order in the deformation d_1 shows that:

$$\frac{\partial d_1}{\partial u^1} = \frac{u^2}{R}\tag{16.69}$$

and

$$\varepsilon_{11} = \frac{u^2}{R} + \frac{1}{2}\left(\frac{u^2}{R}\right)^2.\tag{16.70}$$

Note that all other terms (shear second-order components) vanish due to the stress-strain relations given by (16.63)–(16.65) and thus do not contribute to (16.67). It may seem strange to keep the second-order term in the strain expression for ε_{11} since strain values are typically small, i.e., much less than 1. The reason is that the first-order strain term ($\frac{u^2}{R}$), being proportional to the coordinate u^2, corresponds to a parity changing operator. Hence, energy contributions to first order in $\frac{u^2}{R}$ vanish in first-order non-degenerate perturbation theory and the most significant non-zero contribution from strain to electronic eigenstates is of the second order in $\frac{u^2}{R}$. The origin of the second-order terms stems from either second-order perturbation theory in the $\frac{u^2}{R}$ strain term or first-order perturbation theory in the $\frac{1}{2}(\frac{u^2}{R})^2$ strain term. In other words, we must take care in keeping all strain terms to second order when analyzing influence of strain on electronic eigenstates.

With the determination of ε_{11}, the other non-zero diagonal strain components follow from (16.66).

16.3.4 The Strain Hamiltonian Contribution for Conduction Electrons

Having determined the strain tensor, the ingredients needed for the zincblende conduction-band effective-mass problem are assembled. The one-band heterostructure effective-mass equation in the presence of strain reads [20]:

$$-\frac{\hbar^2}{2}\nabla\cdot\left(\frac{1}{m_{\mathrm{eff}}(\mathbf{r})}\nabla\right)\psi(\mathbf{r})+\left[V_{\mathrm{BE}}(\mathbf{r})+D_e(\varepsilon_{11}+\varepsilon_{22}+\varepsilon_{33})\right]\psi(\mathbf{r})=E\psi(\mathbf{r}),$$

$$(16.71)$$

where $\hbar$, m_{eff}, V_{BE}, D_e, E, ψ, $\mathbf{r}$ are Planck's constant divided by 2π, the position-dependent electron effective mass, the position-dependent heterostructure band-edge potential for electrons, the electron hydrostatic deformation potential, the electron energy, electron eigenstate, and position vector, respectively. The above differential equation in the wavefunction ψ can be formulated as a perturbative problem:

$$H\psi=(H_0+H_1)\psi=E\psi,\tag{16.72}$$

$$H_0\psi^0=E^0\psi^0,\tag{16.73}$$

$$H_0=-\frac{\hbar^2}{2}\nabla\cdot\left(\frac{1}{m_{\mathrm{eff}}(\mathbf{r})}\nabla\right)+V_{\mathrm{BE}}(\mathbf{r}),\tag{16.74}$$

$$H_1=D_e(\varepsilon_{11}+\varepsilon_{22}+\varepsilon_{33}),\tag{16.75}$$

where H_0, H_1, E^0, and ψ^0 are the unperturbed Hamiltonian (i.e., in the absence of strain), the strain perturbation, the unperturbed electron energy, and the unperturbed electron energy, respectively. The above splitting of the Hamiltonian into an unperturbed part and a strain Hamiltonian perturbation is only meaningful if the strain energy contribution is substantially smaller than energy differences between any two unperturbed eigenstates. If not so, a Löwdin degenerate perturbation method [19] should be employed.

Let us next rewrite the strain Hamiltonian as:

$$H_1=H_1^A+H_1^B,\tag{16.76}$$

$$H_1^A=D_e\left(1-2\frac{c_{12}}{c_{11}+c_{12}}\right)\frac{u^2}{R},\tag{16.77}$$

$$H_1^B=\frac{1}{2}D_e\left(1-2\frac{c_{12}}{c_{11}+c_{12}}\right)\left(\frac{u^2}{R}\right)^2.\tag{16.78}$$

Evidently, the Hamiltonian part H_1^A contributes to the second order in the perturbation only while H_1^B contributes to the first order.

16.3.5 Computation of Eigenstates for Circular-Bent Nanowires using Differential Geometry

In this section, curvilinear coordinates are used to determine, in a simple way, eigenstates and eigenvalues in the case of a bent nanowire subject to homogeneous strain. Since electrons are completely confined to the nanowire structure and the heterostructure potential is an infinite-barrier potential, the eigenstate problem is a Dirichlet problem, and the effective-mass equation for a nanoring can be recast in terms of the transformed wavefunction χ [refer to (16.20)]:

$$\frac{-\hbar^2}{2m_{\text{eff}}}\left(\partial_1^2 + \partial_2^2 + \partial_3^2 + \frac{\kappa^2}{4} + D_e(\varepsilon_{11} + \varepsilon_{22} + \varepsilon_{33})\right)\chi = \Xi\chi. \tag{16.79}$$

As the curvature $\kappa = \frac{1}{R}$ is a constant for homogeneous bending of a nanoring with a radius of curvature R, it is apparent that a separable solution $\chi = \chi_1(u^1)\chi_2(u^2)\chi_3(u^3)$ can be sought. This follows by noticing that also the strain perturbation (16.76)–(16.78) is a function of u^2 solely. In the general case of inhomogeneous deformations, the strain perturbation will, however, depend on u^2, u^3 and a separable solution cannot be found.

Thus, for the nanoring problem with homogeneous bending strains, insertion of $\chi = \chi_1(u^1)\chi_2(u^2)\chi_3(u^3)$ into (16.79) gives

$$\partial_1^2\chi_1 + \left(\frac{1}{4R^2} - \lambda - \mu\right)\chi_1 = 0, \tag{16.80}$$

$$\partial_2^2\chi_2 - \left[\frac{2m_{\text{eff}}}{\hbar^2}D_e\left(1 - 2\frac{c_{12}}{c_{11}+c_{12}}\right)\left(\frac{u^2}{R} + \frac{1}{2}\left(\frac{u^2}{R}\right)^2\right) - c^2\right]\chi_2 = 0, \tag{16.81}$$

$$\partial_3^2\chi_3 + \left(\mu - c^2\right)\chi_3 = 0, \tag{16.82}$$

with $\lambda = -\frac{2mE}{\hbar^2}$.

In the next section we describe an approach to find exact quasi-analytical solutions to this problem. Following that section we will return to the perturbative approach.

16.4 Results and Discussions

As an example, we consider InAs zincblende nanowires with a rectangular cross section of side lengths: $2\varepsilon_2 = 2\varepsilon_3$ and assume the nanowire centerline length L to be much larger than the cross-sectional dimensions. The zincblende InAs stiffness components are (in units of Pa): $c_{11} = 8.33 \times 10^{10}, c_{12} = 4.53 \times 10^{10}, c_{44} = 3.96 \times 10^{10}$. For InAs [25] the effective mass is $m_{\text{eff}} = 0.022m_0$ where m_0 is the free electron mass and the conduction-band hydrostatic deformation potential $D_e = -5.1$ eV.

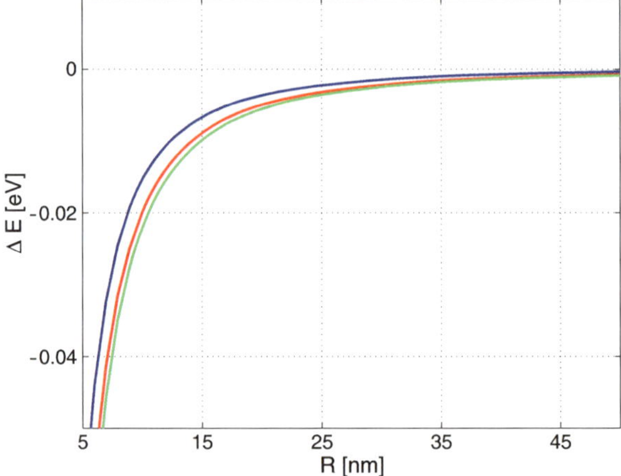

Fig. 16.3 Eigenenergy changes due to strain and geometry-bending as a function of the bending radius. The first three eigenenergy shifts are shown for a InAs zincblende nanowire with $\varepsilon_2 = 2$ nm (refer to the text for details and other parameters used in the calculations). The *blue*, *red*, and *green curves* correspond to the first, second, and third eigenstates, respectively. Both linear and squared contributions in u^2 are included in the strain coefficients

16.4.1 Eigenstate and Eigenenergy Changes due to Circular Bending

In Fig. 16.3, eigenenergy changes due to strain and geometry-bending effects are plotted for a nanowire with cross-sectional side length $\varepsilon_2 = 2$ nm based on (16.80)–(16.82). The first three eigenenergy changes ΔE_β ($\beta = 1, 2, 3$) are shown in the figure as given by the relation:

$$\Delta E_\beta = -\frac{\hbar^2}{2m_{\text{eff}}} \left[\frac{1}{4R^2} + \left(\frac{\beta \pi}{2\varepsilon_2} \right)^2 - c_\beta^2 \right], \tag{16.83}$$

where c_β are the corresponding first three eigenvalues c of (16.81). Note that the first term in the parenthesis of (16.83) is the geometry bending shift present *even* in the absence of strain while the difference between the other two terms is the strain contribution stemming from the nanowire bending.

Evidently, the effect of strain and curvature is rather small for all three states in the range 5 nm $\leq R \leq$ 50 nm. The eigenenergy changes at $R = 15$ nm are approximately -7 meV, -9 meV, and -10 meV for the first, second, and third eigenstates, respectively. Clearly, the dependence on the bending radius is becoming increasingly pronounced as R decreases. This is in qualitative agreement with results obtained using perturbation theory based on the perturbation Hamiltonian H_1 in (16.76). The pure geometry bending effect, i.e., the term $-\frac{\hbar^2}{2m_{\text{eff}}} \frac{1}{4R^2}$ is small only accounting for -2 meV when $R = 15$ nm. Evidently, this term leads to significant contributions in case of large bending only. Quantitative agreement between pertur-

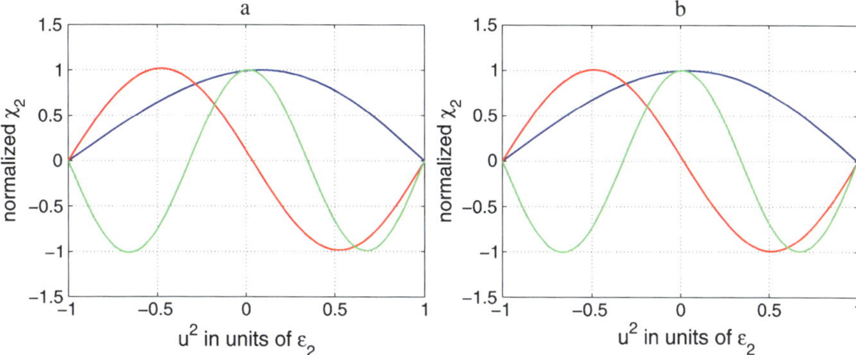

Fig. 16.4 First three χ_2 eigenfunctions as a function of the bending radius for a InAs zincblende nanowire with $\varepsilon_2 = 2$ nm (refer to the text for details and other parameters used in the calculations). The panels **a** and **b** are for a curvature radius equal to 5 nm and 10 nm, respectively. The *blue*, *red*, and *green curves* correspond to the first, second, and third eigenstates, respectively. It is evident that the groundstate is slightly tilted to the right (asymmetric) for a curvature radius equal to 5 nm

bation theory results and the more accurate Frobenius method result is also found but we find no reason to show that here.

Figure 16.4 depicts the first three eigenfunctions χ_2 [see (16.81)] in the case where $\varepsilon_2 = 2$ nm. Clearly, the number of nodes along the u^2 direction is equal to the solution number minus one. As expected, parity is broken due to bending and bending-induced eigenstate asymmetry is visible as the bending radius is decreased to 5 nm. Note also that the state asymmetry is due to strain and not the pure geometry-bending effect since the geometry effect changes the potential by a constant only $(-\frac{1}{4R^2})$. The computed state changes are likely to affect optoelectronic properties only for nanorings with a large bending radius, however, at a bending radius above 50 nm, the asymmetry effect apparently becomes insignificant. A similar calculation for GaAs nanorings shows that significant asymmetries in eigenstates occur at bending radii up to 30 nm.

16.5 How Are the Möbius Strips Constructed?

We consider Möbius strips that are constructed from a planar rectangle by pure bending. Mathematically such a Möbius strip forms a developable surface. It is in particular a ruled surface, i.e., it is of the form $\mathbf{x}(u, v) = \mathbf{r}(u) + v\mathbf{v}(u)$ where $\mathbf{r}$ is some curve on the surface crossing all the rulings. If we let $\mathbf{r}$ be the image of the line down the middle of the rectangle, the *median curve*, then it is a geodesic. Thus the principal normal $\mathbf{n}$ of $\mathbf{r}$ is orthogonal to the tangent plane. We have in particular that $\mathbf{v}$ it is orthogonal to $\mathbf{n}$. As the Frenet-Serret frame $\mathbf{t}, \mathbf{n}, \mathbf{b}$ is an orthonormal basis we can conclude that $\mathbf{v} = \alpha\mathbf{t} + \beta\mathbf{b}$. A ruled surface is developable if and only if $\det(\mathbf{t}, \mathbf{v}, \dot{\mathbf{v}})$ vanishes. As $\dot{\mathbf{v}} = \dot{\alpha}\mathbf{v} + \alpha\kappa\mathbf{n} + \dot{\beta}\mathbf{v} - \beta\tau\mathbf{n}$, where κ and τ is the curvature

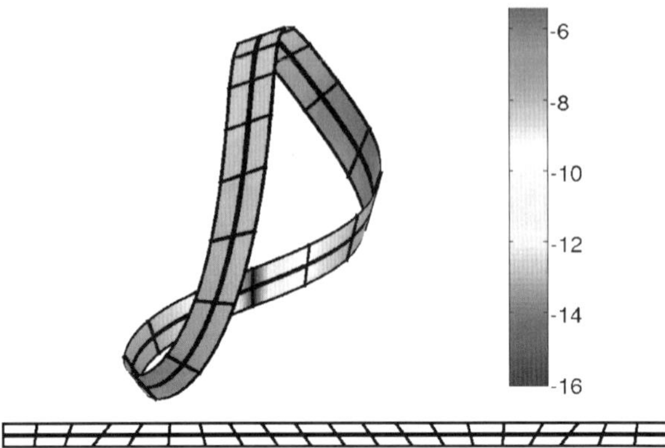

Fig. 16.5 A Möbius strip with length 200 nm and half width $w = 3.333$ nm, coloured according to $\log(M^2)$. The median curve and some of the rulings are shown on both the Möbius strip and the original planar rectangle

and torsion of $\mathbf{r}$ respectively, we see that we have a developable surface if and only if $\alpha\kappa = \beta\tau$. That is, we can parametrize the Möbius strip as

$$\mathbf{x}(u, v) = \mathbf{r}(u) + v\left(\mathbf{b}(u) + \frac{\tau(u)}{\kappa(u)}\mathbf{t}(u)\right), \quad v \in [-w, w], \tag{16.84}$$

where $\mathbf{r}$ is a parametrization of the median curve and w is half the width of the Möbius strip, see Fig. 16.5 and Ref. [5]. Note that the Möbius strip is completely determined by the median curve $\mathbf{r}$. The shape of the Möbius strip is determined by the bending energy which in absence of exterior forces has to be minimized. Locally the strip bends around the rulings and the energy is proportional to the square of the curvature of a section orthogonal to the ruling. One of the principal curvature is zero, $\kappa_1 = 0$, and the corresponding principal direction is in the direction of the rulings. As the principal directions are orthogonal the other principal curvature, κ_2, is exactly the curvature of a section orthogonal to the rulings. So the energy density is proportional to $\kappa_2^2 = (\kappa_1 + \kappa_2)^2 = 4M^2$, where M denotes the mean curvature. We conclude that the energy density is proportional to the square of the mean curvature. To simplify notation, we let

$$\Psi = \frac{\tau}{\kappa}, \quad \text{and} \quad \psi = \frac{d\Psi}{ds}, \tag{16.85}$$

where s denotes arc-length on the median curve. If we assume that the median curve is parametrized by arc-length, then the first and second fundamental forms of the Möbius strip are given by

$$[g_{ij}] = \begin{bmatrix} (1 + v\psi)^2 & \Psi(1 + v\psi) \\ \Psi(1 + v\psi) & 1 + \Psi^2 \end{bmatrix} \quad \text{and} \quad [b_{ij}] = \begin{bmatrix} -\kappa(1 + v\psi) & 0 \\ 0 & 0 \end{bmatrix}, \tag{16.86}$$

respectively. The mean curvature is then

$$M = \frac{\kappa}{2}\frac{1 + \Psi^2}{1 + v\psi}. \tag{16.87}$$

So the bending energy is proportional to

$$E = \frac{1}{2} \int M^2 \, \mathrm{d}A = \frac{1}{8} \int_0^L \int_{-w}^w \frac{\kappa^2 (1 + \Psi^2)^2}{1 + v\psi} \, \mathrm{d}v \, \mathrm{d}s$$

$$= \frac{1}{8} \int_0^L \frac{\kappa^2 (1 + \Psi^2)^2}{\psi} \log\left(\frac{1 + w\psi}{1 - w\psi}\right) \mathrm{d}s, \qquad (16.88)$$

where L is the length of median curve and w still is half the width of the Möbius strip. Finding the exact shape of a Möbius strip is a hard problem with a history going back to [18] where (16.88) was first written down, see [6, 7] and references therein. We do not try to minimize the energy in the space of all curves. Instead, we look at the same three parameter family of curves as in [5]. This family of curves is part of a six-parameter family of median curves of Möbius strips in [21]. The latter family was, in turn, an extension of a single Möbius strip in [22]. Some experiments revealed that the extra three parameters could be set to zero without affecting the final shape of the Möbius strip much. The family of median curves is given as

$$\mathbf{r}(u^1) = \begin{bmatrix} c_1 \sin(u^1) \\ c_2(\sin(u^1) - \tfrac{1}{2}\sin(2u^1)) \\ c_3(\tfrac{5}{3} - \tfrac{5}{2}\cos(u^1) + \cos(2u^1) - \tfrac{1}{6}\cos(3u^1)) \end{bmatrix}. \qquad (16.89)$$

We do not have an arc-length parametrization so we need to change (16.85) to

$$\psi = \frac{\mathrm{d}\Psi}{\mathrm{d}s} = \frac{\mathrm{d}\Psi}{\mathrm{d}u}\frac{\mathrm{d}u}{\mathrm{d}s} = \frac{\dot{\Psi}}{\|\dot{\mathbf{r}}\|}, \qquad (16.90)$$

where $\cdot$ denotes differentiation with respect to u. We want to minimize the energy, but we also want to get a specific length and to obtain a Möbius strip without singularities. We end up with the following constrained optimization problem:

$$\operatorname*{minimize}_{c_1, c_2, c_3} \quad \int_0^{2\pi} \frac{\kappa^2 (1 + \Psi^2)^2}{\psi} \log\left(\frac{1 + h\psi}{1 - h\psi}\right) \|\dot{\mathbf{r}}\| \, \mathrm{d}u, \qquad (16.91a)$$

such that

$$\int_0^{2\pi} \|\dot{\mathbf{r}}\| \, \mathrm{d}u = 200 \text{ nm}, \qquad (16.91b)$$

$$h\psi(u) < 1, \quad u \in [0, 2\pi], \qquad (16.91c)$$

$$h\psi(u) > -1, \quad u \in [0, 2\pi]. \qquad (16.91d)$$

We use the MATLAB function fmincon from the optimization toolbox [23] to solve this problem. All functions are evaluated in 1000 evenly spaced points and the conditions (16.91c) and (16.91d) are only checked in these points. Similarly, the integrals are replaced by a finite sum over these 1000 points. The optimization method

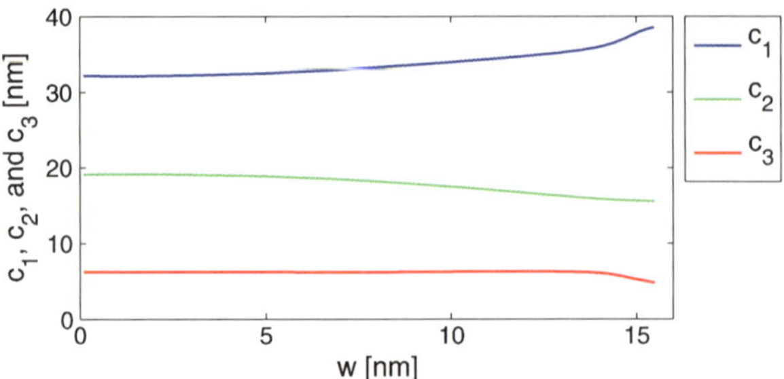

Fig. 16.6 The coefficients c_1, c_2, and c_3 as a function of half width w, with $L = 200$ nm

is gradient driven, so it is necessary to calculate the gradient of both the objective function (16.91a) and the constraints (16.91b)–(16.91d). This is a lengthy but straightforward calculation, which we omit here.

As a result we find the coefficients c_i shown in Fig. 16.6 for a varying set of w values. Values for different lengths of the Möbius strip are obtained by the scaling relations $\tilde{L} = KL$, $\tilde{c}_i = Kc_i$, and $\tilde{w} = Kw$, for a given scaling constant K.

16.6 Curvature Induced Potential

Consider next the one-band envelope-function equation for a conduction electron confined to a semiconductor surface Σ [24]:

$$-\frac{\hbar^2}{2m_e}\left(\Delta_0 + \partial_3^2\right)\chi\left(u^1, u^2\right) + \left[V_S\left(u^1, u^2, u^3\right) + V\left(u^1, u^2, u^3\right)\right]\chi\left(u^1, u^2\right)$$

$$= E\chi\left(u^1, u^2\right), \tag{16.92}$$

where m_e is the effective mass, χ is the envelope eigenfunction, and E its energy. The surface Σ is defined as the center surface corresponding to the third coordinate u^3 being zero, V_S is the deformation potential term proportional to the sum of the diagonal strain components in Cartesian coordinates, and the potential barrier term V due to material inhomogeneity is assumed to be of the infinite-barrier type, i.e.,

$$V\left(u^1, u^2, u^3\right) = \begin{cases} 0, & \text{if } u^3 = 0, \\ \infty, & \text{else.} \end{cases} \tag{16.93}$$

The details of the strain potential V_S in curvilinear coordinates u^1, u^3, u^3 are given in Sect. 16.7. The operator Δ_0 in (16.92) is [16]:

$$\Delta_0 = \Delta_\Sigma + M^2 - K, \tag{16.94}$$

where Δ_Σ is the Laplace-Beltrami operator on Σ, and M and K are the mean and Gaussian curvatures, respectively. In our case, M is non-zero but K is zero.

The effect of M^2 on energies is to shift eigenenergies downwards. Note that since M^2 is much smaller than Δ_Σ), it follows from first-order perturbation theory that

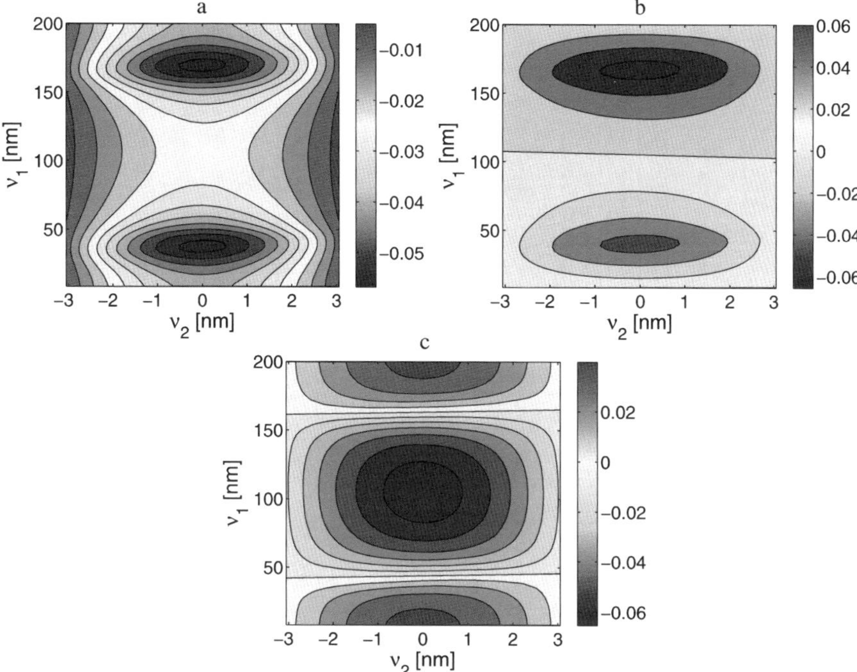

Fig. 16.7 The panels **a**, **b**, and **c** are the groundstate, the first-excited state, and the third-excited state, respectively for the Möbius strip with dimensions $w = 3.333$ nm and $L = 200$ nm. Parameters used in the calculation are given in the main text. See (16.106) for the definition of v_1 and v_2

Möbius structure eigenstates more or less retain the same symmetries (even or odd in the u^1 and u^2 coordinates) as the corresponding eigenstates for the flat cylinder problem even though M^2 is not an even or odd function in u^1 and u^2. This argument is only strictly valid if states are non-degenerate in the absence of the bending term, and if the energy separation between non-degenerate states is larger than the bending energy contribution.

In Fig. 16.7, we plot the first three eigenstates are found by solving (16.79) in the absence of strain effects ($V_S = 0$) using the finite-difference method. The structure considered corresponds to the parameters: $L = 200$ nm, $w = 3.333$ nm.

16.7 Möbius Strip of Finite Thickness

Next, we solve for the eigenstates and associated energies of an electron bound to a Möbius strip of finite thickness. Since it is computationally cumbersome and, for thin Möbius structure, unnecessarily expensive to solve the one-band problem in Cartesian coordinates, it is convenient to formulate the problem in the (u^1, u^2, u^3)

coordinate system, thereby reducing the complexity of the geometry to a simple box at the price of having to solve a more complicated differential equation. In Cartesian coordinates the one-band model is given by

$$-\frac{\hbar^2}{2m_{\text{eff}}}\Delta\chi + [V_S + V]\chi = E\chi. \tag{16.95}$$

The parametrization for a Möbius strip with finite thickness is given by

$$\mathbf{x}(u^1, u^2, u^3) = \mathbf{r}(u^1) + u^2\left(\mathbf{b}(u^1) + \frac{\tau(u^1)}{\kappa(u^1)}\mathbf{t}(u^1)\right) - u^3\mathbf{n}(u^1), \tag{16.96}$$

so that

$$\frac{\partial\mathbf{x}}{\partial u^1} = \left|\mathbf{r}'(u^1)\right|\left((1 + u^2\psi + u^3\kappa)\mathbf{t} - u^3\tau\mathbf{b}\right), \tag{16.97}$$

$$\frac{\partial\mathbf{x}}{\partial u^2} = \mathbf{b} + \Psi\mathbf{t}, \tag{16.98}$$

$$\frac{\partial\mathbf{x}}{\partial u^3} = -\mathbf{n}, \tag{16.99}$$

where Frenet's relations have been used. The metric tensor is now found to be

$$\mathbf{G} = \begin{bmatrix} |\mathbf{r}'|^2((1 + u^2\psi + u^3\kappa)^2 + (u^3\tau)^2) & |\mathbf{r}'|\Psi(1 + \psi) & 0 \\ |\mathbf{r}'|\Psi(1 + \psi) & 1 + \Psi^2 & 0 \\ 0 & 0 & 1 \end{bmatrix}. \tag{16.100}$$

Further on,

$$G = |\mathbf{r}'|^2\left[(1 + u^2\psi + u^3\kappa)^2 + 2(u^3\tau)^2 + 2u^3\kappa\Psi^2(1 + \psi) + \left(\frac{u^3\tau^2}{\kappa}\right)^2\right], \tag{16.101}$$

and

$$\mathbf{G}^{-1} = \frac{1}{G}\begin{bmatrix} 1 + \Psi^2 & -|\mathbf{r}'|\Psi(1 + \psi) & 0 \\ -|\mathbf{r}'|\Psi(1 + \psi) & |\mathbf{r}'|^2((1 + u^2\psi + u^3\kappa)^2 + (u^3\tau)^2) & 0 \\ 0 & 0 & G \end{bmatrix}. \tag{16.102}$$

It is clear from the expression of the metric tensor, that the problem is not separable in any of the three coordinates u^1, u^2, u^3. However, it is possible, for small thicknesses, to carry out a perturbative analysis in terms of the effective-mass equation that couples the Möbius thickness coordinate u^3 to the other two coordinates u^1, u^2. Then, the unperturbed three-dimensional problem is separable in one coordinate and decouples into a two-dimensional problem in u^1, u^2 and a one-dimensional problem in u^3.

As the electron is completely confined to the Möbius strip, Dirichlet conditions $\chi = 0$ are invoked at the boundary:

$$\chi(u^1, u^2 = \pm w, u^3) = \chi(u^1, u^2, u^3 = \pm h) = 0, \tag{16.103}$$

and since the Möbius strip is rotated by $180°$ during one revolution along the median line defined by $u^2 = u^3 = 0$, anti-periodic boundary conditions are imposed at the u^1 end surfaces:

$$\chi\left(u^1 = 0, u^2, u^3\right) = \chi\left(u^1 = 2\pi, -u^2, -u^3\right). \tag{16.104}$$

16.7.1 Inclusion of Strain

As before, this effect is included in its simplest form by an effective potential of the form

$$V_S = a_c \mathrm{Tr}(\varepsilon), \tag{16.105}$$

where a_c is the deformation potential, ε is the strain tensor, and Tr denotes the trace. Since the (u^1, u^2, u^3) coordinate system has been chosen solely based on the convenience of the domain, it is less convenient to use the (u^1, u^2, u^3) coordinate system as a reference configuration in the calculation of the strain. A more useful coordinate system, (v_1, v_2, v_3), is parametrized as follows:

$$v_1 = \int_0^{u^1} \left|\mathbf{r}'(s)\right| \mathrm{d}s + u^2 \Psi, \qquad v_2 = u^2, \quad \text{and} \quad v_3 = u^3. \tag{16.106}$$

In this coordinate system, the Möbius strip is given by the natural domain $[0, L] \times [-w, w] \times [-h, h]$ where L is the length of median curve. This choice of reference configuration is made since the flat nanostrip is neither stretched nor compressed when deformed into the Möbius strip by the deformation $\mathbf{R}(v_1, v_2, v_3) = \mathbf{x}(u^1(v_1, v_2), v_2, v_3)$. Given the deformation $\mathbf{R}$, the strain tensor becomes

$$\varepsilon_{ij} = \frac{1}{2}\left(\delta_{ij} - \frac{\partial \mathbf{R}}{\partial v_i} \cdot \frac{\partial \mathbf{R}}{\partial v_j}\right) = \frac{1}{2}\left(\delta_{ij} - \sum_{k,l=1}^{3} \frac{\partial u_k}{\partial v_i} \frac{\partial \mathbf{x}}{\partial u_k} \cdot \frac{\partial u_l}{\partial v_j} \frac{\partial \mathbf{x}}{\partial u_l}\right). \tag{16.107}$$

16.8 Results

In this section the energies and eigenstates are studied as a function of the width, thickness, and length of the Möbius strip with emphasis on the effect of inclusion of strain in the calculations. The Möbius strip considered here is made of InAs, but similar results would be obtained with other choices of material.

In the inset of Fig. 16.8a the groundstate energy as a function of width is shown for a 200 nm long and 2 nm thick Möbius strip. As expected, the energies decrease as the width is increased, due to electron confinement effects.

In Fig. 16.8a the energy differences relative to the groundstate energy are also shown. The eigenstates are very close in energy. This is due to the relatively large

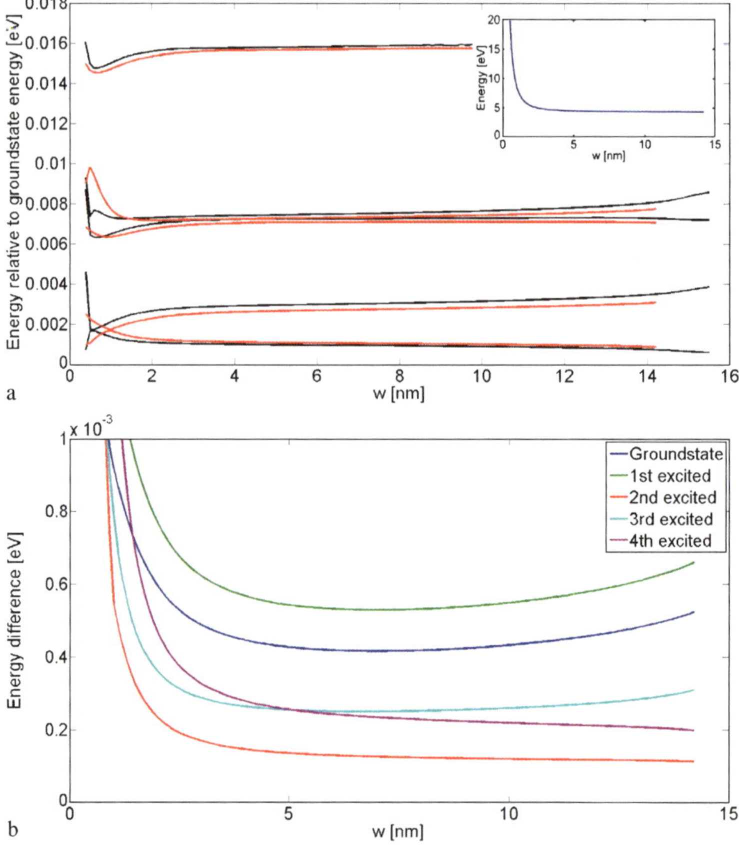

Fig. 16.8 Panel **a** shows the energy difference relative to the ground state energy both for the case including strain (*red lines*) and the case disregarding strain (*black lines*). Panel **b** shows the difference in energy between a model with strain and a model without strain. In both panels the length and thickness of the Möbius strip are 200 nm and 2 nm, respectively

length of the Möbius strip. This fact is most readily illustrated by plotting the energies as a function of length of the Möbius strip, see Fig. 16.9. In this figure it is clearly seen that the level spacing increases as the length of the Möbius strip is decreased. In Fig. 16.8a the energy difference is shown between a model with strain relative to a model without strain. Strain only has a minor influence on the electron energies. This is in agreement with results from paper [14].

In Fig. 16.8b it is furthermore observed that the first and second excited states cross at the width of 2 nm. This can also be seen in the symmetries of the wave-functions shown in Fig. 16.10, where in the left column the first excited state is shown for different widths and in the right column the second excited state is shown. In all plots the wave-function are given as a function of v^1 and v^2 at $v^3 = 0$.

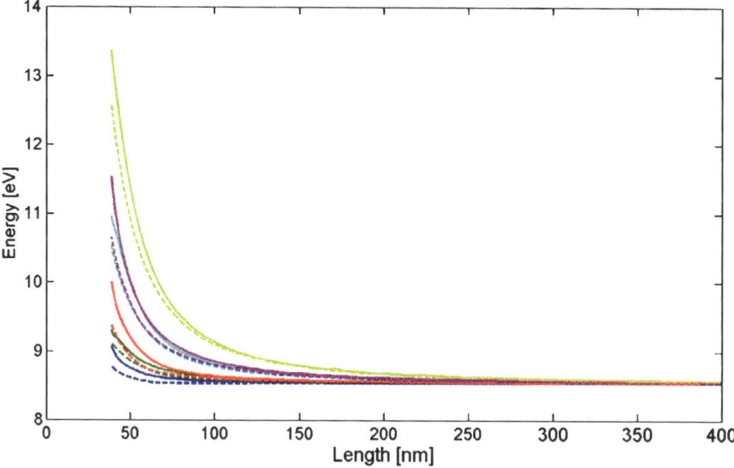

Fig. 16.9 The first 6 eigenenergies for a model with strain (*solid lines*) and a model without strain (*dashed lines*) as a function of the length of the Möbius strip. In the plots the width and thickness of the Möbius strip are 6 nm and 2 nm, respectively

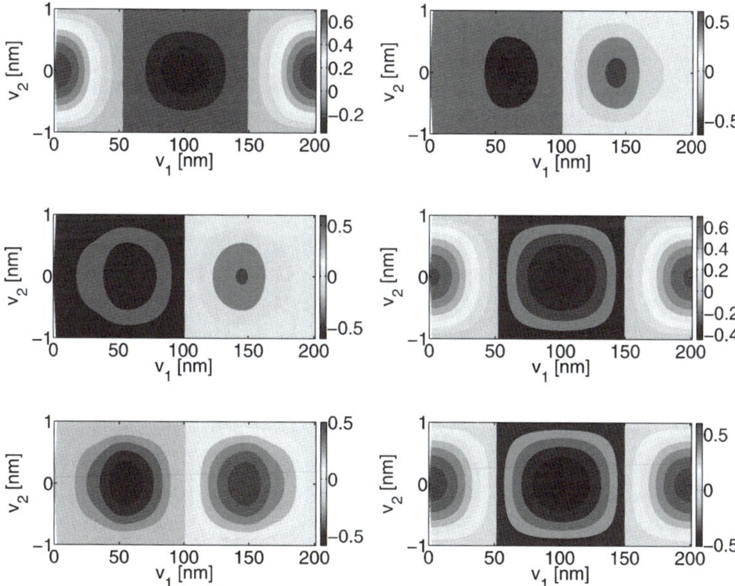

Fig. 16.10 The *left* and *right column* shows the first and second excited state, respectively. The *upper row* wave functions have a width of 1.6 nm, the *middle row* wave functions have a width of 2 nm, and the *lower row* wave functions have a width of 2.4 nm. In all plots the length and thickness of the Möbius strip is 200 nm and 2 nm, respectively, and all plots are given as functions of v^1 and v^2 with $v^3 = 0$

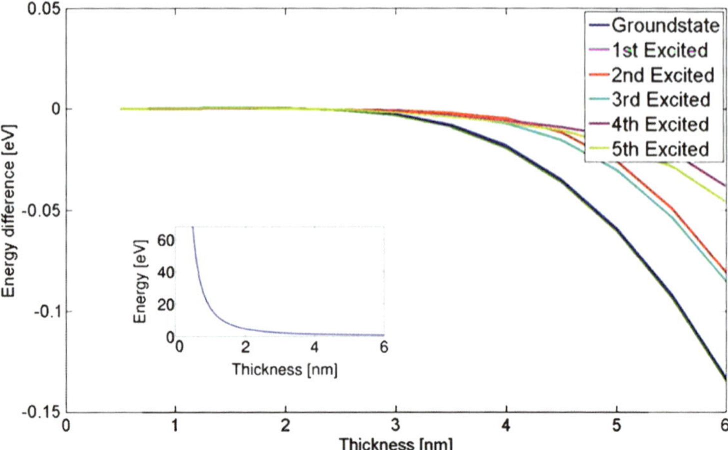

Fig. 16.11 Difference in energies between a model with strain and a model without strain as a function of the thickness of the Möbius strip. The *inset* shows the groundstate energy as a function of thickness using a model with strain. In the plots the length and thickness of the Möbius strip are 200 nm and 2 nm, respectively

Fig. 16.12 The hydrostatic strain distribution as a function of v^1 and v^3 at $v^2 = 0$ for the Möbius strip with length of 200 nm, width of 2 nm, and thickness of 2 nm

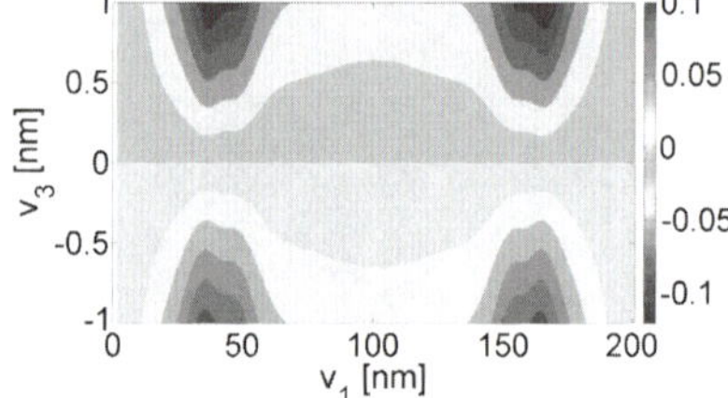

In Fig. 16.11, the influence of the thickness on the electron energies is studied. The energy decreases as the thickness is increased due to a decrease in the confinement of the electron. More interestingly, it is seen that the impact of strain on the eigenenergies increases as the thickness is increased. The reason is that the strain due to bending in the structure increases nearly linearly as a function of the thickness. This is seen in Fig. 16.12, where the hydrostatic strain is shown as a function of v^1 and v^3 at $v^2 = 0$. This can also been seen from the approximate result based on the Euler beam theory, where bending strain is given by ku^3 to the first order in the curvature (similarly to (16.70)).

16.9 Conclusion

Analytical and simple computational differential geometry methods applicable to curved nanostructures are presented and applied to geometries which cannot be solved analytically nor computationally effective using standard coordinate systems.

Test cases of experimental interest are computed for electronic eigenstates of circular and elliptic nanorings as well as Möbius nanostructures, and it is shown that for bending radii of a few nanometers, significant changes in eigenstate symmetry properties and eigenenergy values exist due to curvature and strain effects affecting physical properties. At bending radii above approximately 50 nm, curvature effects are, however, negligible. A detailed study of a complicated geometry structures, the Möbius nanostructure, is discussed in the second part of the chapter. Consequences of curvature, strain, and Möbius nanostructure length, width, and thickness are assessed for electron eigenstates.

References

1. S. Tanda, T. Tsuneta, Y. Okajima, K. Inagaki, K. Yamaya, N. Hatekenaka, Nature (London) **417**, 397 (2002)
2. M. König, S. Wiedmann, C. Brüne, A. Roth, H. Buhmann, L.W. Molenkamp, X.L. Qi, S.C. Zhang, Science **318**, 766 (2008)
3. Y. Ran, Y. Zhang, A. Vishwanath, Nat. Phys. **5**, 298 (2009)
4. Z.L. Guo, Z.R. Gong, H. Dong, C.P. Sun, Phys. Rev. B **80**, 195310 (2009)
5. J. Gravesen, M. Willatzen, Phys. Rev. A **72**, 032108 (2005)
6. E.L. Starostin, G.H.M. van der Heijden, Nat. Mater. **6**, 563 (2007)
7. E.L. Starostin, G.H.M. van der Heijden, Phys. Rev. B **79**, 066602 (2009)
8. D.J. Ballon, H.U. Voss, Phys. Rev. Lett. **101**, 247701 (2008)
9. M. Yoneya, K. Kuboki, M. Hayashi, Phys. Rev. B **78**, 064419 (2008)
10. C. Rockstuhl, C. Menzel, T. Paul, F. Lederer, Phys. Rev. B **79**, 035321 (2009)
11. N. Zhao, H. Dong, S. Yang, C.P. Sun, Phys. Rev. B **79**, 125440 (2009)
12. Z. Li, L.R. Ram-Mohan, Phys. Rev. B **85**, 195438 (2012)
13. V.M. Fomin, S. Kiravittaya, O.G. Schmidt, Phys. Rev. B **86**, 195421 (2012)
14. B. Lassen, M. Willatzen, J. Gravesen, J. Nanoelectron. Optoelectron. **6**, 68 (2011)
15. J. Gravesen, M. Willatzen, Physica B **371**, 112–119 (2006)
16. J. Gravesen, M. Willatzen, L.C. Lew Yan Voon, J. Math. Phys. **46**, 012107 (2005)
17. L.D. Landau, E.M. Lifshitz, *Theory of Elasticity*, 3rd edn. Course of Theoretical Physics, vol. 7 (Butterworth Heinemann, Oxford, 1999)
18. M. Sadowski, in *Verh. 3. Kongr. Techn. Mechanik*, vol. II (1930), pp. 444–451
19. E.O. Kane, J. Phys. Chem. Solids **1**, 249 (1957)
20. P.Y. Yu, M. Cardona, *Fundamentals of Semiconductors*, 4th edn. (Springer, Berlin, 2010)
21. T. Randrup, P. Røgen, Arch. Math. **66**, 511–521 (1996)
22. G. Schwarz, Pac. J. Math. **143**, 195 (1990)
23. The MathWorks Inc., *MATLAB Version 7.8.0* (The MathWorks Inc., Natick, 2009)
24. L.C. Lew Yan Voon, M. Willatzen, *The k · p Method*. Springer Series in Solid State Physics (Springer, Berlin, 2009)
25. I. Vurgaftman, J.R. Meyer, L.R. Ram-Mohan, J. Appl. Phys. **89**, 5815 (2001)

Chapter 17
Hole Mixing in Semiconductor Quantum Rings

Carlos Segarra, Josep Planelles, and Juan I. Climente

Abstract Many applications of semiconductor quantum dots rely on the use of valence band holes. A prominent example is current endeavour to develop quantum information science using the spin of holes rather than that of electrons. Understanding the spin and orbital properties of holes is necessary for further progress. In self-assembled InAs/GaAs quantum dots, the hole ground state is mainly formed by the heavy hole subband. However, there is a finite mixing with the light-hole subband which has been shown to be critical in determining the hole spin properties. A large number of works have then investigated the influence of such coupling in dots. Based on k·p theory, in this chapter we study the influence of hole subband mixing in self-assembled quantum rings. It is shown that the inner cavity of the ring enhances the light hole component of the ground state. As the quasi-1D limit is approached, the light-hole character becomes comparable to that of the heavy hole. In InAs/GaAs quantum rings strain reduces the coupling, but the mixing is still larger than in quantum dots. Strain also gives rise to unusual phenomena, such as partial localization of the heavy holes inside the repulsive core region. Deviations of quantum rings from the perfect axial symmetry are shown to have a minor influence on the hole mixing.

17.1 Hole Mixing in Quantum Dots

An electron excited across the band gap of a semiconductor leaves behind a hole in the otherwise full valence states. This hole behaves like a charged particle, similar to an electron albeit with a few remarkable differences. In zinc-blende semiconductors, these generally include heavier (and significantly anisotropic) effective masses,

C. Segarra · J. Planelles (✉) · J.I. Climente
Departament de Química Física i Analítica, Universitat Jaume I, 12080 Castelló, Spain
e-mail: josep.planelles@uji.es

C. Segarra
e-mail: csegarra@uji.es

J.I. Climente
e-mail: climente@uji.es

V.M. Fomin (ed.), *Physics of Quantum Rings*, NanoScience and Technology,
DOI 10.1007/978-3-642-39197-2_17, © Springer-Verlag Berlin Heidelberg 2014

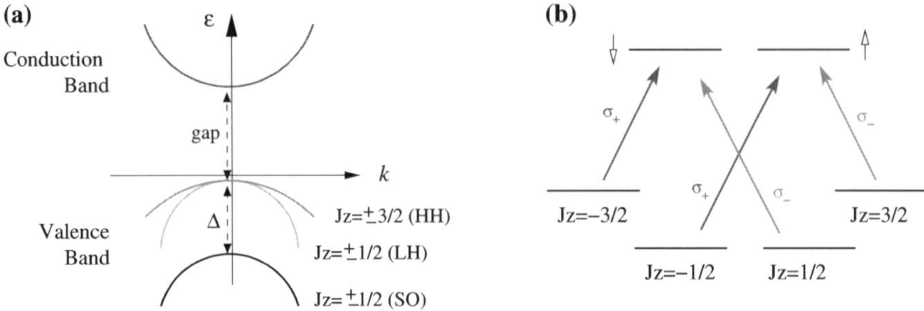

Fig. 17.1 (**a**) Band structure of the zone center in a zinc-blende semiconductor. (**b**) Optical transitions between valence states and conduction states. $\sigma_\pm$ stand for left and right circularly polarized light

stronger spin-orbit interaction (SOI) and multi-band coupling. The reason is that the conduction and valence bands are constructed from different atomic orbitals. While the conduction band is mainly formed from the s orbitals of the crystal atoms, the valence band is formed by p orbitals [1]. As a result, the band structure near the center of the Brillouin zone looks like Fig. 17.1(a). The conduction band (CB) is a single band well isolated from the valence band and higher excited bands. It is doubly degenerate if we consider the spin degree of freedom. Instead, the valence band has a more complex structure. Owing to the spin-orbit interaction, the total microscopic angular momentum of valence states is $\mathbf{J} = 3/2$. The symmetry properties at the zone center (momentum $k = 0$) are like those of p atomic states. Thus, we have a four fold degenerate state ($J = 3/2$, $J_z = +3/2, +1/2 - 1/2, -3/2$) which is separated by an energy distance Δ, the spin-orbit splitting, from a two fold degenerate state ($J = 1/2$, $J_z = +1/2, -1/2$). The $J = 1/2$ state is referred to as the split-off band (SO). The $J = 3/2$ state splits in the presence of a force (i.e. $k \neq 0$) into two subbands, one with $J_z = \pm 3/2$ and another with $J_z = \pm 1/2$. These are referred to as the heavy hole (HH) and light hole (LH) subbands [1].

In bulk semiconductors and quantum wells, hole states are severely affected by the HH-LH mixing, which leads to rather peculiar physics [2, 3]. In quantum dots (QDs), however, low-energy holes can be often described simply as HHs, with HH-LH mixing being but a weak perturbation [4]. The reason is that LHs are generally lighter than HHs (notice in Fig. 17.1(a) that $\frac{1}{m_{HH}^*} = |\frac{\delta^2 \varepsilon_{HH}}{\delta k^2}| < \frac{1}{m_{LH}^*} = |\frac{\delta^2 \varepsilon_{LH}}{\delta k^2}|$). LHs are then more sensitive to the three-dimensional confinement of QDs and show up at higher energies.

The high purity of the hole ground state in QDs has recently sparked great interest in using its spin for quantum information and spintronic applications [5–10]. Indeed, its spin lifetimes are comparable to those of electrons [11] and, owing to the p-like nature of the Bloch function, they suffer much less decoherence than electrons from the hyperfine interaction with lattice nuclei [12–16]. Even if weak, HH-LH mixing in these systems needs to be understood as it sets a limit for the fidelity and execution time of control operations [17, 18]. It is also important for optical initialization and read-out relying on optical orientation [2]. The underlying idea of these processes is summarized in Fig. 17.1(b). The absorption/emission of circularly polarized photons enables selective transitions between HH states ($J_z = \pm 3/2$) and electron states with well-defined spin projection ($\sigma_z = \uparrow, \downarrow$). Photons with the same polarizations but higher energies enable transitions between LH states ($J_z = \pm 1/2$) and electron states with opposite spin projections. Thus, if HH and LH states are significantly coupled the photons give rise to final states with both spin projections and are not spin selective any more. Last, we note that HH-LH mixing is also important to determine other properties of holes in QDs, including the magnetic field dispersion [19, 20] and the tunneling rates in coupled quantum dots [21–27].

Semiconductor quantum rings (QRs) have emerged in the last decade as an alternative to QDs for optoelectronic devices such as lasers and photovoltaics [28–30]. Much of the basic research on these structures has focused on the magnetic response ensuing from their doubly-connected topology, which provides a suitable playground to probe the Aharonov-Bohm effect [31] (see also Chaps. 9, 10 and 12). As a matter of fact, Aharonov-Bohm oscillations of energy and emission intensity have been reported in different experiments [32–36]. These results are well understood from theoretical studies analyzing the influence of the confinement and external fields on the response of electrons, holes and excitons (see e.g. [37–42]). In principle, one can also expect QRs to be suited for quantum information systems, with additional potentialities as compared to QDs owing to the richer magnetic response [43, 44]. Understanding the properties of holes confined in these structures is a necessary step for further development in this direction. Because the strength of the HH-LH mixing is strongly dependent on the details of the quantum confinement [45, 46], the inner cavity of QRs is expected to influence the hole admixture [47].

In this chapter, we analyze the hole ground state properties in InAs/GaAs QRs. Based on a 6-band Burt-Foreman k·p Hamiltonian, we study the effect of quantum confinement on the hole composition. It is shown that the inner cavity of the QR greatly enhances the HH-LH mixing, leading to much higher LH character than in QDs. We also explore the individual role of additional factors such as the elastic strain or ring eccentricity. The accuracy of usual approximations such as position-independent effective masses [48–50] and the axial approximation of the valence band Hamiltonian [51] are assessed. The chapter is organized as follows. In Sect. 17.2 we give details about the theoretical model used to calculate hole states confined in QRs. In Sect. 17.3 we discuss how the different factors influence the HH-LH admixture of the hole. Finally, in Sect. 17.4 we compare the spatial localization of the HH and LH components in QRs subject to strain or structural deformations.

17.2 Theory

An accurate description of holes in InAs/GaAs QRs can be obtained using 6-band k·p Hamiltonians including HH, LH and SO subbands. This requires spanning the Hamiltonian on the basis of periodic Bloch functions $|J, J_z\rangle$:

$$\left|\frac{3}{2}, +\frac{3}{2}\right\rangle = \frac{1}{\sqrt{2}}\left|(X + iY)\uparrow\right\rangle = |\text{HH}_+\rangle,$$

$$\left|\frac{3}{2}, +\frac{1}{2}\right\rangle = \frac{1}{\sqrt{6}}\left|(X + iY)\downarrow\right\rangle - \sqrt{\frac{2}{3}}|Z\uparrow\rangle = |\text{LH}_+\rangle,$$

$$\left|\frac{3}{2}, -\frac{1}{2}\right\rangle = -\frac{1}{\sqrt{6}}\left|(X - iY)\uparrow\right\rangle - \sqrt{\frac{2}{3}}|Z\downarrow\rangle = |\text{LH}_-\rangle,$$

$$\left|\frac{3}{2}, -\frac{3}{2}\right\rangle = \frac{1}{\sqrt{2}}\left|(X - iY)\downarrow\right\rangle = |\text{HH}_-\rangle,$$

$$\left|\frac{1}{2}, +\frac{1}{2}\right\rangle = \frac{1}{\sqrt{3}}\left|(X + iY)\downarrow\right\rangle + \sqrt{\frac{1}{3}}|Z\uparrow\rangle = |\text{SO}_+\rangle,$$

$$\left|\frac{1}{2}, -\frac{1}{2}\right\rangle = -\frac{1}{\sqrt{3}}\left|(X - iY)\uparrow\right\rangle + \sqrt{\frac{1}{3}}|Z\downarrow\rangle = |\text{SO}_-\rangle.$$

The $|3/2, \pm 3/2\rangle$ components correspond to HH, the $|3/2, \pm 1/2\rangle$ to LH and the $|1/2, \pm 1/2\rangle$ to SO. One can see from the explicit $|J, J_z\rangle$ functions above that HH components have pure spin, while LH and SO components contain spin admixture. It then follows that HH-LH mixing has straightforward implications in the spin purity of holes.

Since the Luttinger parameters of InAs and GaAs are quite different, it is convenient to employ position-dependent effective mass parameters. Then, instead of the classical Luttinger Hamiltonian [52, 53] one must use the Burt-Foreman one [48–50]. The full Hamiltonian reads:

$$\mathcal{H}_6 = \mathcal{H}_{\text{bf}} + V(x, y, z)\mathcal{I} + \mathcal{H}_{\text{s}}, \tag{17.1}$$

where $\mathcal{H}_{\text{bf}}$ is the Burt-Foreman Hamiltonian, $V(x, y, z)$ the confining potential, $\mathcal{I}$ the identity matrix and $\mathcal{H}_{\text{s}}$ the strain Hamiltonian. A detailed description of $\mathcal{H}_{\text{bf}}$ can be found in Ref. [54], where the due expression in Cartesian coordinates is given. Because QRs are approximately circular, it is however convenient to use cylindrical coordinates instead. The Burt-Foreman Hamiltonian in atomic units and cylindrical coordinates reads:

$$\mathcal{H}_{\text{bf}} = \frac{1}{2}\mathcal{M}, \tag{17.2}$$

where $\mathcal{M}$ is a rank-6 matrix with the following elements:

$$\mathcal{M}[1, 1] = \frac{\partial}{\partial\rho}(\gamma_1 + \gamma_2)\frac{\partial}{\partial\rho} + \frac{(\gamma_1 + \gamma_2)}{\rho}\frac{\partial}{\partial\rho} + \frac{\partial}{\partial z}(\gamma_1 - 2\gamma_2)\frac{\partial}{\partial z}$$
$$- \frac{(F_z - \frac{3}{2})^2}{\rho^2}(\gamma_1 + \gamma_2) + \frac{(F_z - \frac{3}{2})}{2\rho}\left[\frac{\partial}{\partial\rho}(C_1 + C_2) - (C_1 + C_2)\frac{\partial}{\partial\rho}\right],$$

$$\mathcal{M}[1, 2] = \frac{1}{\sqrt{3}}\left\{\frac{\partial}{\partial\rho}C_1\frac{\partial}{\partial z} - \frac{\partial}{\partial z}C_2\frac{\partial}{\partial\rho} + \frac{(F_z - \frac{1}{2})}{\rho}\left[C_1\frac{\partial}{\partial z} - \frac{\partial}{\partial z}C_2\right]\right\},$$

$$\mathcal{M}[1, 3] = -\sqrt{3}\left\{\frac{\partial}{\partial\rho}\tilde{\gamma}\frac{\partial}{\partial\rho} + \frac{(F_z + \frac{1}{2})}{\rho}\frac{\partial}{\partial\rho}\tilde{\gamma} + \frac{(F_z - \frac{1}{2})}{\rho}\tilde{\gamma}\frac{\partial}{\partial\rho}\right.$$

$$+ \frac{(F_z - \frac{3}{2})(F_z + \frac{1}{2})}{\rho^2} \tilde{\gamma} \bigg\},$$

$$\mathcal{M}[1,4] = 0,$$

$$\mathcal{M}[1,5] = -\frac{1}{\sqrt{6}} \bigg\{ \frac{\partial}{\partial \rho} C_1 \frac{\partial}{\partial z} - \frac{\partial}{\partial z} C_2 \frac{\partial}{\partial \rho} + \frac{(F_z - \frac{1}{2})}{\rho} \bigg[C_1 \frac{\partial}{\partial z} - \frac{\partial}{\partial z} C_2 \bigg] \bigg\},$$

$$\mathcal{M}[1,6] = -\sqrt{6} \bigg\{ \frac{\partial}{\partial \rho} \tilde{\gamma} \frac{\partial}{\partial \rho} + \frac{(F_z + \frac{1}{2})}{\rho} \frac{\partial}{\partial \rho} \tilde{\gamma} + \frac{(F_z - \frac{1}{2})}{\rho} \tilde{\gamma} \frac{\partial}{\partial \rho}$$

$$+ \frac{(F_z - \frac{3}{2})(F_z + \frac{1}{2})}{\rho^2} \tilde{\gamma} \bigg\},$$

$$\mathcal{M}[2,1] = \frac{1}{\sqrt{3}} \bigg\{ \frac{\partial}{\partial z} C_1 \frac{\partial}{\partial \rho} - \frac{\partial}{\partial \rho} C_2 \frac{\partial}{\partial z} + \frac{(F_z - \frac{3}{2})}{\rho} \bigg[C_2 \frac{\partial}{\partial z} - \frac{\partial}{\partial z} C_1 \bigg] \bigg\},$$

$$\mathcal{M}[2,2] = \frac{\partial}{\partial \rho} (\gamma_1 - \gamma_2) \frac{\partial}{\partial \rho} + \frac{(\gamma_1 - \gamma_2)}{\rho} \frac{\partial}{\partial \rho} + \frac{\partial}{\partial z} (\gamma_1 + 2\gamma_2) \frac{\partial}{\partial z}$$

$$- \frac{(F_z - \frac{1}{2})^2}{\rho^2} (\gamma_1 - \gamma_2) + \frac{(F_z - \frac{1}{2})}{6\rho} \bigg[\frac{\partial}{\partial \rho} (C_1 + C_2) - (C_1 + C_2) \frac{\partial}{\partial \rho} \bigg],$$

$$\mathcal{M}[2,3] = \frac{1}{3} \bigg\{ \frac{\partial}{\partial \rho} (C_1 + C_2) \frac{\partial}{\partial z} - \frac{\partial}{\partial z} (C_1 + C_2) \frac{\partial}{\partial \rho}$$

$$+ \frac{(F_z + \frac{1}{2})}{\rho} \bigg[(C_1 + C_2) \frac{\partial}{\partial z} - \frac{\partial}{\partial z} (C_1 + C_2) \bigg] \bigg\},$$

$$\mathcal{M}[2,4] = \sqrt{3} \bigg\{ \frac{\partial}{\partial \rho} \tilde{\gamma} \frac{\partial}{\partial \rho} + \frac{(F_z + \frac{3}{2})}{\rho} \frac{\partial}{\partial \rho} \tilde{\gamma} + \frac{(F_z + \frac{1}{2})}{\rho} \tilde{\gamma} \frac{\partial}{\partial \rho}$$

$$+ \frac{(F_z + \frac{3}{2})(F_z - \frac{1}{2})}{\rho^2} \tilde{\gamma} \bigg\},$$

$$\mathcal{M}[2,5] = \sqrt{2} \bigg\{ \frac{\partial}{\partial \rho} \gamma_2 \frac{\partial}{\partial \rho} - 2 \frac{\partial}{\partial z} \gamma_2 \frac{\partial}{\partial z} + \frac{\gamma_2}{\rho} \frac{\partial}{\partial \rho} - \frac{(F_z - \frac{1}{2})^2}{\rho^2} \gamma_2$$

$$+ \frac{(F_z - \frac{1}{2})}{6\rho} \bigg[\frac{\partial}{\partial \rho} (C_1 + C_2) - (C_1 + C_2) \frac{\partial}{\partial \rho} \bigg] \bigg\},$$

$$\mathcal{M}[2,6] = -\frac{1}{3\sqrt{2}} \bigg\{ \frac{\partial}{\partial \rho} (C_1 - 2C_2) \frac{\partial}{\partial z} + \frac{\partial}{\partial z} (2C_1 - C_2) \frac{\partial}{\partial \rho}$$

$$+ \frac{(F_z + \frac{1}{2})}{\rho} \bigg[(C_1 - 2C_2) \frac{\partial}{\partial z} + \frac{\partial}{\partial z} (2C_1 - C_2) \bigg] \bigg\},$$

$$\mathcal{M}[3,1] = -\sqrt{3} \bigg\{ \frac{\partial}{\partial \rho} \tilde{\gamma} \frac{\partial}{\partial \rho} - \frac{(F_z - \frac{3}{2})}{\rho} \frac{\partial}{\partial \rho} \tilde{\gamma} - \frac{(F_z - \frac{1}{2})}{\rho} \tilde{\gamma} \frac{\partial}{\partial \rho}$$

$$+ \frac{(F_z - \frac{3}{2})(F_z + \frac{1}{2})}{\rho^2} \tilde{\gamma} \bigg\},$$

$$\mathcal{M}[3,2] = \frac{1}{3}\left\{ \frac{\partial}{\partial z}(C_1+C_2)\frac{\partial}{\partial \rho} - \frac{\partial}{\partial \rho}(C_1+C_2)\frac{\partial}{\partial z} \right.$$
$$\left. + \frac{(F_z - \frac{1}{2})}{\rho}\left[(C_1+C_2)\frac{\partial}{\partial z} - \frac{\partial}{\partial z}(C_1+C_2) \right] \right\},$$

$$\mathcal{M}[3,3] = \frac{\partial}{\partial \rho}(\gamma_1-\gamma_2)\frac{\partial}{\partial \rho} + \frac{(\gamma_1-\gamma_2)}{\rho}\frac{\partial}{\partial \rho} + \frac{\partial}{\partial z}(\gamma_1+2\gamma_2)\frac{\partial}{\partial z}$$
$$- \frac{(F_z+\frac{1}{2})^2}{\rho^2}(\gamma_1-\gamma_2) - \frac{(F_z+\frac{1}{2})}{6\rho}\left[\frac{\partial}{\partial \rho}(C_1+C_2) - (C_1+C_2)\frac{\partial}{\partial \rho} \right],$$

$$\mathcal{M}[3,4] = \frac{1}{\sqrt{3}}\left\{ \frac{\partial}{\partial z}C_1\frac{\partial}{\partial \rho} - \frac{\partial}{\partial \rho}C_2\frac{\partial}{\partial z} - \frac{(F_z+\frac{3}{2})}{\rho}\left[C_2\frac{\partial}{\partial z} - \frac{\partial}{\partial z}C_1 \right] \right\},$$

$$\mathcal{M}[3,5] = \frac{1}{3\sqrt{2}}\left\{ \frac{\partial}{\partial \rho}(C_1-2C_2)\frac{\partial}{\partial z} + \frac{\partial}{\partial z}(2C_1-C_2)\frac{\partial}{\partial \rho} \right.$$
$$\left. - \frac{(F_z-\frac{1}{2})}{\rho}\left[(C_1-2C_2)\frac{\partial}{\partial z} + \frac{\partial}{\partial z}(2C_1-C_2) \right] \right\},$$

$$\mathcal{M}[3,6] = \sqrt{2}\left\{ \frac{\partial}{\partial \rho}\gamma_2\frac{\partial}{\partial \rho} - 2\frac{\partial}{\partial z}\gamma_2\frac{\partial}{\partial z} + \frac{\gamma_2}{\rho}\frac{\partial}{\partial \rho} - \frac{(F_z+\frac{1}{2})^2}{\rho^2}\gamma_2 \right.$$
$$\left. - \frac{(F_z+\frac{1}{2})}{6\rho}\left[\frac{\partial}{\partial \rho}(C_1+C_2) - (C_1+C_2)\frac{\partial}{\partial \rho} \right] \right\},$$

$$\mathcal{M}[4,1] = 0,$$

$$\mathcal{M}[4,2] = \sqrt{3}\left\{ \frac{\partial}{\partial \rho}\tilde{\gamma}\frac{\partial}{\partial \rho} - \frac{(F_z-\frac{1}{2})}{\rho}\frac{\partial}{\partial \rho}\tilde{\gamma} - \frac{(F_z+\frac{1}{2})}{\rho}\tilde{\gamma}\frac{\partial}{\partial \rho} \right.$$
$$\left. + \frac{(F_z+\frac{3}{2})(F_z-\frac{1}{2})}{\rho^2}\tilde{\gamma} \right\},$$

$$\mathcal{M}[4,3] = \frac{1}{\sqrt{3}}\left\{ \frac{\partial}{\partial \rho}C_1\frac{\partial}{\partial z} - \frac{\partial}{\partial z}C_2\frac{\partial}{\partial \rho} - \frac{(F_z+\frac{1}{2})}{\rho}\left[C_1\frac{\partial}{\partial z} - \frac{\partial}{\partial z}C_2 \right] \right\},$$

$$\mathcal{M}[4,4] = \frac{\partial}{\partial \rho}(\gamma_1+\gamma_2)\frac{\partial}{\partial \rho} + \frac{(\gamma_1+\gamma_2)}{\rho}\frac{\partial}{\partial \rho} + \frac{\partial}{\partial z}(\gamma_1-2\gamma_2)\frac{\partial}{\partial z}$$
$$- \frac{(F_z+\frac{3}{2})^2}{\rho^2}(\gamma_1+\gamma_2) - \frac{(F_z+\frac{3}{2})}{2\rho}\left[\frac{\partial}{\partial \rho}(C_1+C_2) - (C_1+C_2)\frac{\partial}{\partial \rho} \right],$$

$$\mathcal{M}[4,5] = \sqrt{6}\left\{ \frac{\partial}{\partial \rho}\tilde{\gamma}\frac{\partial}{\partial \rho} - \frac{(F_z-\frac{1}{2})}{\rho}\frac{\partial}{\partial \rho}\tilde{\gamma} - \frac{(F_z+\frac{1}{2})}{\rho}\tilde{\gamma}\frac{\partial}{\partial \rho} \right.$$
$$\left. + \frac{(F_z+\frac{3}{2})(F_z-\frac{1}{2})}{\rho^2}\tilde{\gamma} \right\},$$

$$\mathcal{M}[4,6] = -\frac{1}{\sqrt{6}}\left\{ \frac{\partial}{\partial \rho}C_1\frac{\partial}{\partial z} - \frac{\partial}{\partial z}C_2\frac{\partial}{\partial \rho} - \frac{(F_z+\frac{1}{2})}{\rho}\left[C_1\frac{\partial}{\partial z} - \frac{\partial}{\partial z}C_2 \right] \right\},$$

$$\mathcal{M}[5,1] = -\frac{1}{\sqrt{6}}\left\{\frac{\partial}{\partial z}C_1\frac{\partial}{\partial \rho} - \frac{\partial}{\partial \rho}C_2\frac{\partial}{\partial z} + \frac{(F_z - \frac{3}{2})}{\rho}\left[C_2\frac{\partial}{\partial z} - \frac{\partial}{\partial z}C_1\right]\right\},$$

$$\mathcal{M}[5,2] = \sqrt{2}\left\{\frac{\partial}{\partial \rho}\gamma_2\frac{\partial}{\partial \rho} - 2\frac{\partial}{\partial z}\gamma_2\frac{\partial}{\partial z} + \frac{\gamma_2}{\rho}\frac{\partial}{\partial \rho} - \frac{(F_z - \frac{1}{2})^2}{\rho^2}\gamma_2\right.$$
$$\left. + \frac{(F_z - \frac{1}{2})}{6\rho}\left[\frac{\partial}{\partial \rho}(C_1 + C_2) - (C_1 + C_2)\frac{\partial}{\partial \rho}\right]\right\},$$

$$\mathcal{M}[5,3] = \frac{1}{3\sqrt{2}}\left\{\frac{\partial}{\partial \rho}(2C_1 - C_2)\frac{\partial}{\partial z} + \frac{\partial}{\partial z}(C_1 - 2C_2)\frac{\partial}{\partial \rho}\right.$$
$$\left. + \frac{(F_z + \frac{1}{2})}{\rho}\left[(2C_1 - C_2)\frac{\partial}{\partial z} + \frac{\partial}{\partial z}(C_1 - 2C_2)\right]\right\},$$

$$\mathcal{M}[5,4] = \sqrt{6}\left\{\frac{\partial}{\partial \rho}\tilde{\gamma}\frac{\partial}{\partial \rho} + \frac{(F_z + \frac{3}{2})}{\rho}\frac{\partial}{\partial \rho}\tilde{\gamma} + \frac{(F_z + \frac{1}{2})}{\rho}\tilde{\gamma}\frac{\partial}{\partial \rho}\right.$$
$$\left. + \frac{(F_z + \frac{3}{2})(F_z - \frac{1}{2})}{\rho^2}\tilde{\gamma}\right\},$$

$$\mathcal{M}[5,5] = \frac{\partial}{\partial \rho}\gamma_1\frac{\partial}{\partial \rho} + \frac{\partial}{\partial z}\gamma_1\frac{\partial}{\partial z} + \frac{\gamma_1}{\rho}\frac{\partial}{\partial \rho} - \frac{(F_z - \frac{1}{2})^2}{\rho^2}\gamma_1$$
$$+ \frac{(F_z - \frac{1}{2})}{3\rho}\left[\frac{\partial}{\partial \rho}(C_1 + C_2) - (C_1 + C_2)\frac{\partial}{\partial \rho}\right] - 2\Delta(\rho, z),$$

$$\mathcal{M}[5,6] = -\frac{1}{3}\left\{\frac{\partial}{\partial \rho}(C_1 + C_2)\frac{\partial}{\partial z} - \frac{\partial}{\partial z}(C_1 + C_2)\frac{\partial}{\partial \rho}\right.$$
$$\left. + \frac{(F_z + \frac{1}{2})}{\rho}\left[(C_1 + C_2)\frac{\partial}{\partial z} - \frac{\partial}{\partial z}(C_1 + C_2)\right]\right\},$$

$$\mathcal{M}[6,1] = -\sqrt{6}\left\{\frac{\partial}{\partial \rho}\tilde{\gamma}\frac{\partial}{\partial \rho} - \frac{(F_z - \frac{3}{2})}{\rho}\frac{\partial}{\partial \rho}\tilde{\gamma} - \frac{(F_z - \frac{1}{2})}{\rho}\tilde{\gamma}\frac{\partial}{\partial \rho}\right.$$
$$\left. + \frac{(F_z - \frac{3}{2})(F_z + \frac{1}{2})}{\rho^2}\tilde{\gamma}\right\},$$

$$\mathcal{M}[6,2] = -\frac{1}{3\sqrt{2}}\left\{\frac{\partial}{\partial \rho}(2C_1 - C_2)\frac{\partial}{\partial z} + \frac{\partial}{\partial z}(C_1 - 2C_2)\frac{\partial}{\partial \rho}\right.$$
$$\left. - \frac{(F_z - \frac{1}{2})}{\rho}\left[(2C_1 - C_2)\frac{\partial}{\partial z} + \frac{\partial}{\partial z}(C_1 - 2C_2)\right]\right\},$$

$$\mathcal{M}[6,3] = \sqrt{2}\left\{\frac{\partial}{\partial \rho}\gamma_2\frac{\partial}{\partial \rho} - 2\frac{\partial}{\partial z}\gamma_2\frac{\partial}{\partial z} + \frac{\gamma_2}{\rho}\frac{\partial}{\partial \rho} - \frac{(F_z + \frac{1}{2})^2}{\rho^2}\gamma_2\right.$$
$$\left. - \frac{(F_z + \frac{1}{2})}{6\rho}\left[\frac{\partial}{\partial \rho}(C_1 + C_2) - (C_1 + C_2)\frac{\partial}{\partial \rho}\right]\right\},$$

$$\mathcal{M}[6,4] = -\frac{1}{\sqrt{6}}\left\{\frac{\partial}{\partial z}C_1\frac{\partial}{\partial \rho} - \frac{\partial}{\partial \rho}C_2\frac{\partial}{\partial z} - \frac{(F_z + \frac{3}{2})}{\rho}\left[C_2\frac{\partial}{\partial z} - \frac{\partial}{\partial z}C_1\right]\right\},$$

$$
\mathcal{M}[6,5] = -\frac{1}{3}\left\{ \frac{\partial}{\partial z}(C_1+C_2)\frac{\partial}{\partial \rho} - \frac{\partial}{\partial \rho}(C_1+C_2)\frac{\partial}{\partial z} \right.
$$

$$
\left. + \frac{(F_z - \frac{1}{2})}{\rho}\left[(C_1+C_2)\frac{\partial}{\partial z} - \frac{\partial}{\partial z}(C_1+C_2) \right] \right\},
$$

$$
\mathcal{M}[6,6] = \frac{\partial}{\partial \rho}\gamma_1\frac{\partial}{\partial \rho} + \frac{\partial}{\partial z}\gamma_1\frac{\partial}{\partial z} + \frac{\gamma_1}{\rho}\frac{\partial}{\partial \rho} - \frac{(F_z + \frac{1}{2})^2}{\rho^2}\gamma_1
$$

$$
- \frac{(F_z + \frac{1}{2})}{3\rho}\left[\frac{\partial}{\partial \rho}(C_1+C_2) - (C_1+C_2)\frac{\partial}{\partial \rho} \right] - 2\Delta(\rho, z).
$$

Here γ_i are the position-dependent Luttinger parameters, $\tilde{\gamma} = (\gamma_2 + \gamma_3)/2$ (axial approximation [51]), $C_1 = 1 + \gamma_1 - 2\gamma_2 - 6\gamma_3$ and $C_2 = 1 + \gamma_1 - 2\gamma_2$, $\Delta(\rho, z)$ is the spin-orbit splitting and $F_z = m + J_z$ is the total angular momentum z-projection, which is the sum of the envelope angular momentum m and the Bloch angular momentum J_z.

The strain terms are given by:

$$
\mathcal{H}_s = \begin{pmatrix}
p+q & -s & r & 0 & \frac{s}{\sqrt{2}} & \sqrt{2}r \\
-s^* & p-q & 0 & -r & \sqrt{2}q & \sqrt{\frac{3}{2}}s \\
r^* & 0 & p-q & -s & -\sqrt{\frac{3}{2}}s^* & \sqrt{2}q \\
0 & -r^* & -s^* & p+q & -\sqrt{2}r^* & \frac{s^*}{\sqrt{2}} \\
\frac{s^*}{\sqrt{2}} & \sqrt{2}q & -\sqrt{\frac{3}{2}}s & -\sqrt{2}r & p & 0 \\
\sqrt{2}r^* & \sqrt{\frac{3}{2}}s^* & \sqrt{2}q & \frac{s}{\sqrt{2}} & 0 & p
\end{pmatrix}, \tag{17.3}
$$

where

$$
p = a\,\mathrm{Tr}(\varepsilon), \tag{17.4}
$$

$$
q = b\left(\frac{\varepsilon_{xx}}{2} + \frac{\varepsilon_{yy}}{2} - \varepsilon_{zz} \right), \tag{17.5}
$$

$$
s = d(\varepsilon_{xz} - i\varepsilon_{yz}), \tag{17.6}
$$

$$
r = -\frac{\sqrt{3}}{2}b(\varepsilon_{xx} - \varepsilon_{yy}) + id\varepsilon_{xy}. \tag{17.7}
$$

Here ε is the strain tensor, which is calculated by minimizing the elastic energy [55], and a, b, d are the valence band deformation potentials.

The eigenstates of Hamiltonian (17.1) are six-component spinorial vectors, with each component composed of an envelope and a Bloch part. For the axially symmetric structures, the spinors can be classified by F_z and the main quantum number k.

$$|F_z, k\rangle = \begin{pmatrix} f^{(1)}_{F_z-3/2}(\rho, z)|\text{HH}_+\rangle \\ f^{(2)}_{F_z-1/2}(\rho, z)|\text{LH}_+\rangle \\ f^{(3)}_{F_z+1/2}(\rho, z)|\text{LH}_-\rangle \\ f^{(4)}_{F_z+3/2}(\rho, z)|\text{HH}_-\rangle \\ f^{(5)}_{F_z-1/2}(\rho, z)|\text{SO}_+\rangle \\ f^{(6)}_{F_z+1/2}(\rho, z)|\text{SO}_-\rangle \end{pmatrix} \tag{17.8}$$

where $f^{(i)}_m(\rho, z)$ is the envelope function of the i-th component. In QRs with strong vertical confinement the ground state is formed by $|F_z = +3/2, k = 0\rangle$ or $|F_z = -3/2, k = 0\rangle$. These two states are Kramers-degenerate in the absence of magnetic fields. For our discussion it is convenient to display one of these states, e.g. $|F_z = +3/2, k\rangle$ (an analogous discussion would follow for $|F_z = -3/2, k\rangle$):

$$|3/2, 0\rangle = \begin{pmatrix} f^{(1)}_0(\rho, z)|\text{HH}_+\rangle \\ f^{(2)}_1(\rho, z)|\text{LH}_+\rangle \\ f^{(3)}_2(\rho, z)|\text{LH}_-\rangle \\ f^{(4)}_3(\rho, z)|\text{HH}_-\rangle \\ f^{(5)}_1(\rho, z)|\text{SO}_+\rangle \\ f^{(6)}_2(\rho, z)|\text{SO}_-\rangle \end{pmatrix}. \tag{17.9}$$

Owing to the low angular momentum, the HH ($f^{(1)}_0$) component is dominant, as expected from a single-band approach.

For numerical simulations, we shall consider self-assembled InAs QDs with lens shape (spherical casket) embedded in a GaAs matrix. The height at the apex is $H = 3$ nm and the radius is $R = 10$ nm. The QR is formed by introducing a repulsive core in the center with radius R_{in} (see Fig. 17.2(a)). This is an idealization of the realistic volcano shape of InAs/GaAs QRs [56] (see also Chap. 4), which captures the fact that the potential governing the electron motion in the plane is asymmetric in the radial direction, with a profound minimum under the ring apex and a smooth (abrupt) increase on the outer (inner) side, see Fig. 1b in Ref. [39]. Material parameters of InAs and GaAs (Luttinger parameters, Δ, lattice constants) are taken from Ref. [57]. The deformation potentials are taken from Ref. [58] and the valence band offset is set to $V_c = 0.265$ eV [59]. Hamiltonian (17.1) is solved numerically using COMSOL Multiphysics.

17.3 Hole Mixing

In this section we investigate the composition of the hole ground state as a function of the QR geometry. The composition is given in terms of the weight of each com-

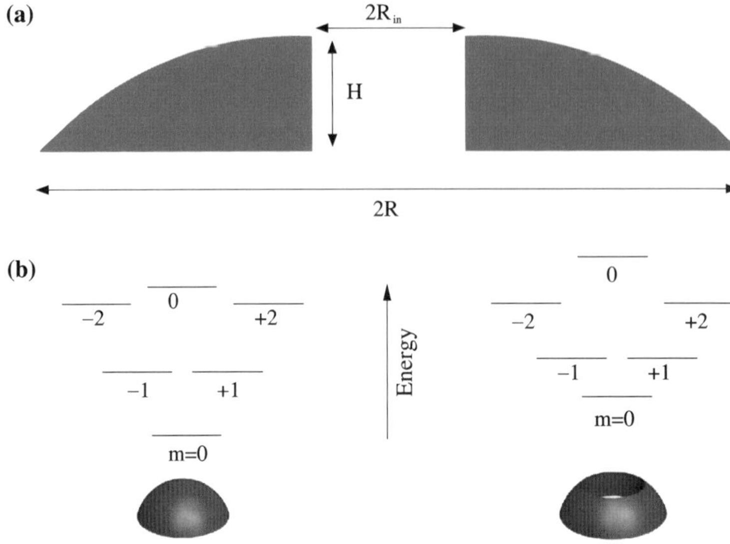

Fig. 17.2 (**a**) Schematic representation of the QR cross-section. (**b**) Energy levels of a QD (*left*) and a QR (*right*). Note that the ring cavity unstabilizes states with small angular momentum $|m|$

ponent within the spinor (17.8). For example, the weight of the $|HH_+\rangle$ component is:

$$c_{HH_+} = \frac{\langle f^{(1)}|f^{(1)}\rangle}{\sum_i \langle f^{(i)}|f^{(i)}\rangle}. \tag{17.10}$$

We start by considering the effect of quantum confinement alone. Constant (InAs) mass parameter is taken and strain effects are disregarded. Figure 17.3 shows the weight of the different spinor components as a function of R_{in}. Panel (a) shows the total HH, LH and SO weights, and panel (b) shows the J_z resolved components. While the SO components remain negligible in all the range under study, one can see that the LH character of the ground state rapidly increases as we depart from the QD limit ($R_{in} = 0$ nm), mainly due to an increase of the $|LH_-\rangle$ component. For $R_{in} = 5$ nm, the LH character is as large as $\sim$20 %, over three times larger than in QDs [47]. The LH character will be even stronger in narrower QRs. The increase of the LH character in QRs can be explained from two factors: (i) the distinct QR energy structure and (ii) the enhanced lateral confinement.

Factor (i) can be understood by comparing the energy level diagrams of lens-shaped QDs and QRs, which is sketched in Fig. 17.2(b)—a single-band model has been used for simplicity. The ground state of the QD has angular momentum $m = 0$, and the excited states follow the energy diagram shown on the figure. The $m = 0$ state has maximum charge density in the center of the QD, while states with increasing $|m|$ are gradually offcentered by centrifugal terms. Switching from a QD to a QR means including a repulsive core in the center. Because of the charge density

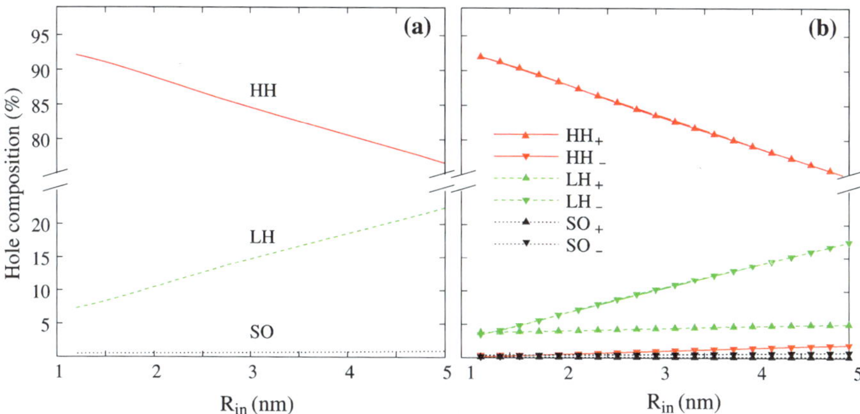

Fig. 17.3 Composition of the hole ground state in a QR with increasing inner radius R_{in}. (**a**) Comparison of HH, LH and SO character. *Solid line*: $c_{HH_+} + c_{HH_-}$, *dashed line*: $c_{LH_+} + c_{LH_-}$, *dotted line*: $c_{SO_+} + c_{SO_-}$. (**b**) J_z resolved components. Strain is disregarded and constant (InAs) effective mass is used

distribution, the effect of the core is stronger on states with small $|m|$. Thus, in QRs the $m = 0$ states approach the $|m| = 1$ and $|m| = 2$ ones. Since the HH-LH coupling terms mix states with different angular momenta, it follows that the repulsive core of QRs favors HH-LH coupling. Mixing of $|m| = 0$ with $|m| = 2$ is more important than with $|m| = 1$ (compare $|LH_-\rangle$ with $|LH_+\rangle$ in Fig. 17.3(b)) because both components have the same chirality [24, 47]. Factor (ii) is related to the anisotropy of hole masses. In InAs structures grown along the [001] direction, $m^{\perp}_{LH} > m^{\perp}_{HH}$. Thus, with increasing lateral confinement the kinetic energy of HH approaches that of LH and the HH-LH mixing becomes stronger. If the formation of QRs does not involve enhanced lateral confinement (e.g. the material of the cavity is pushed towards the ring edges [60]), the LH character still increases due to factor (i), but to a lesser extent than shown here [47].

We next analyze how the previous result is modified by the inclusion of other factors present in realistic self-assembled QRs. First, we consider the fact that the effective mass is different inside and outside the QR. Using the Burt-Foreman Hamiltonian with position-dependent effective masses, one obtains the ground state LH character shown in Fig. 17.4 (short-dashed lines). It can be seen that the LH admixture is essentially the same as that reported above. This is because most of the wave function is localized inside the InAs ring, with only weak leakage into the GaAs matrix. Interestingly, the influence of the position-dependent effective mass is weak even in the case of small R_{in}, where the hole tunnels across the inner GaAs core (see Sect. 17.4).

Second, we consider the influence of artificially imposing axial symmetry in the valence band Hamiltonian. The 6-band Luttinger Hamiltonian (Burt-Foreman

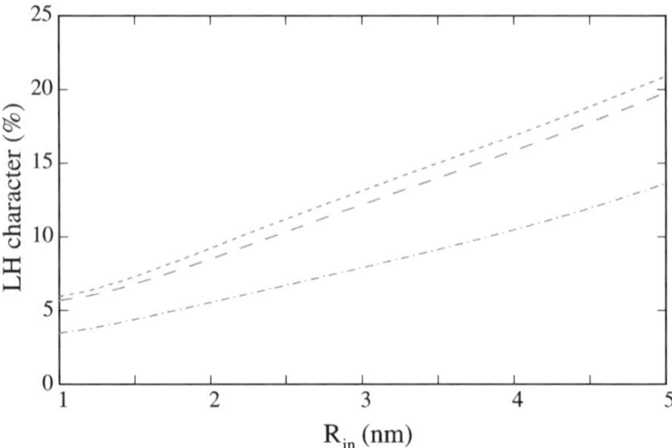

Fig. 17.4 Weight of the LH components in the hole ground state in a QR with increasing inner radius R_{in}. Position-dependent effective mass is used. *Short-dashed lines*: unstrained rings with axial approximation. *Long-dashed lines*: unstrained rings without axial approximation. *Dashed-dotted lines*: strained rings without axial approximation

Hamiltonian with constant mass) in Cartesian coordinates reads:

$$\mathcal{H}_{\text{LK}} = \begin{pmatrix} P+Q & -S & R & 0 & \frac{S}{\sqrt{2}} & \sqrt{2}R \\ -S^* & P-Q & 0 & -R & \sqrt{2}Q & \sqrt{\frac{3}{2}}S \\ R^* & 0 & P-Q & -S & -\sqrt{\frac{3}{2}}S^* & \sqrt{2}Q \\ 0 & -R^* & -S^* & P+Q & -\sqrt{2}R^* & \frac{S^*}{\sqrt{2}} \\ \frac{S^*}{\sqrt{2}} & \sqrt{2}Q & -\sqrt{\frac{3}{2}}S & -\sqrt{2}R & P & 0 \\ \sqrt{2}R^* & \sqrt{\frac{3}{2}}S^* & \sqrt{2}Q & \frac{S}{\sqrt{2}} & 0 & P \end{pmatrix}, \tag{17.11}$$

where

$$P = \frac{\gamma_1}{2}p^2, \tag{17.12}$$

$$Q = \frac{\gamma_2}{2}\left(p_x^2 + p_y^2 - 2p_z^2\right), \tag{17.13}$$

$$S = \sqrt{3}\gamma_3 p_z p_-, \tag{17.14}$$

$$R = -\frac{\sqrt{3}}{2}\gamma_2\left(p_x^2 - p_y^2\right) + i\sqrt{3}\gamma_3 p_x p_y, \tag{17.15}$$

and p_j is the j-projection of the momentum and $p_\pm = (p_x \pm ip_y)$. As can be seen, in (17.11) the term R lacks axial symmetry. To simplify the study of axially symmetric nanostructures, Sercel and Vahala proposed replacing it by [54]:

$$R = -\frac{\sqrt{3}}{2}\gamma p_-^2, \tag{17.16}$$

where $\gamma = (\gamma_2 + \gamma_3)/2$. This is known as the "axial approximation", for it enables analytical integration of the angular coordinate. Because self-assembled QDs and

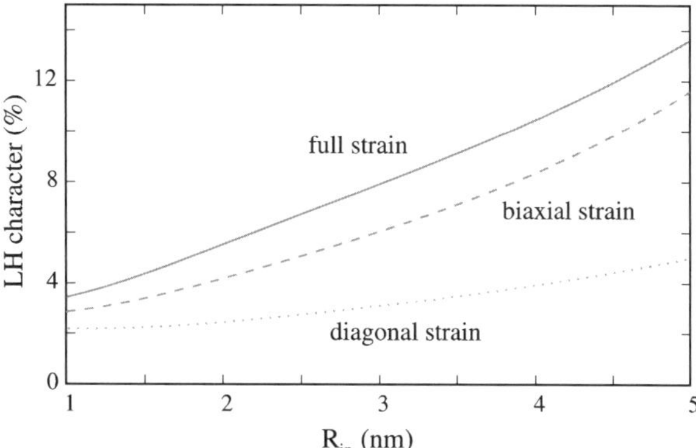

Fig. 17.5 Weight of the light hole components in the hole ground state of a QR with increasing inner radius R_{in}. Three approximations for strain are compared: diagonal strain (*dotted line*), biaxial strain (*dashed line*) and full strain (*solid lines*)

QRs are roughly axially symmetric, this approximation is employed in most multiband descriptions of the valence band structure (including Hamiltonian $\mathcal{H}_{\mathrm{bf}}$). However, in the current context where high purity hole states is desirable for applications, it is worth assessing to which extent the approximation provides accurate estimates of the hole mixing. The long-dashed line in Fig. 17.4 shows the ground state LH character without the axial approximation. The weight of the LH component is found to increase by 1–2 %, although the qualitative trend as a function of inner radii remains the same.

Next, we consider the effect of strain forces, which are known to have a deep impact on the electronic structure of valence band holes in InAs/GaAs QDs [58, 61, 62] and QRs [63] (see also Chap. 13). The results, dashed-dotted lines in Fig. 17.4, show a moderate decrease of the LH character of the QRs as compared to the unstrained case. Still, the same trend as observed before holds. Namely, the inner cavity of the ring systematically enhances the LH character.

The fact that strain-free simulations provide a reasonable reference is somewhat surprising, as biaxial strain is known to split HH and LH states energetically [61, 62]. One may then expect strained QRs to display much smaller LH character [47]. For further insight into this issue, in Fig. 17.5 the role of strain is analyzed in more detail. The LH character of the ground state is compared for different degrees of approximation to $\mathcal{H}_{\mathrm{s}}$: diagonal strain (dotted lines), biaxial strain [64] (dashed lines) and full strain (solid lines). One can see that diagonal strain predicts a strong decrease of the LH character as compared to the unstrained case of Fig. 17.4. This is because the $p+q$ and $p-q$ terms in (17.3) split HH and LH energetically, thus weakening the HH-LH mixing. By contrast, the biaxial approximation ((17.3) but setting the shear terms to zero, $d = 0$) shows larger LH character, close to the full-strain value. This indicates that the weakening of the HH-LH mixing due to the diagonal terms is largely compensated by the strong off-diagonal terms of the Hamiltonian. As a result, the fully strained system has LH component of the same order as the strain-free system.

Next, we take into account that self-assembled InAs/GaAs QRs often deviate from the exact circular geometry assumed so far. This is because the anisotropic redistribution of the QD material during the capping and annealing processes re-

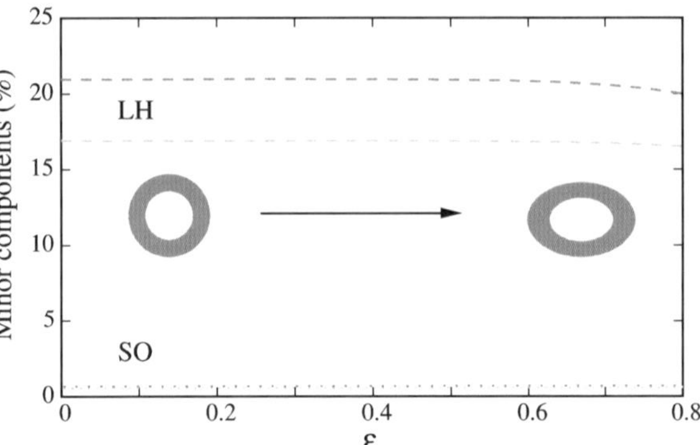

Fig. 17.6 Effect of the eccentricity on the composition of the hole ground state (minor components shown only). The QR has $R = 10$ nm. *Light* and *dark lines* correspond to $R_{in} = 4$ nm and $R_{in} = 5$ nm, respectively

sults in elongated ring-shaped islands [56]. These deviations are known to affect the magnetic response of the QRs [34] and are required for quantitative interpretation of experimental observations [39, 65]. In Fig. 17.6 we study the hole mixing in QRs subject to an increasing degree of ellipticity. We start from circular QDs and let the eccentricity ε increase while keeping the basis area constant. The semi-major (semi-minor) axis R_a (R_b) of the elliptical QR is then:

$$R_a = R/\left(1 - \varepsilon^2\right)^{1/4}, \tag{17.17}$$

$$R_b = R^2/R_a. \tag{17.18}$$

The hole states are calculated using the strain-free Hamiltonian in Cartesian coordinates for a QR with $R_{in} = 4$ nm. The result is shown in Fig. 17.6 (light lines). As can be seen, the weight of both LH and SO components is barely affected by the eccentricity. This is in spite of the fact that for $R_{in} = 4$ nm and $\varepsilon = 0.8$, the ring width, $W = R - R_{in}$, changes from 6 nm all over the ring to 8.9 nm and 4 nm along the major and minor axis, respectively. The same behavior is observed for different inner radii (see e.g. $R_{in} = 5$ nm, dark lines in the figure).

17.4 Hole Localization

It has been suggested that strain plays an important role to determine the localization of electrons and holes in QRs, eventually leading to spatially separated (type-II-like) carriers [63]. In this section we investigate how strain influences the localization of the different hole components. In Fig. 17.7 we plot the potential energy originating in the diagonal strain terms for a QR with a small inner cavity ($R_{in} = 1$ nm, left panels) and a large inner cavity ($R_{in} = 5$ nm, right panels). The upper panel corresponds to HH potential and the lower one to LH potential. Dark (light) colors stand for strain-induced potential well (barrier). As expected, Fig. 17.7 shows that strain stabilizes (unstabilizes) the HH (LH). It is worth noting that the small R_{in} geometry is just a small departure from the lens-shaped QD. Yet, the inner cavity brings about significant differences. For HH, the strain in the cavity yields a potential minimum,

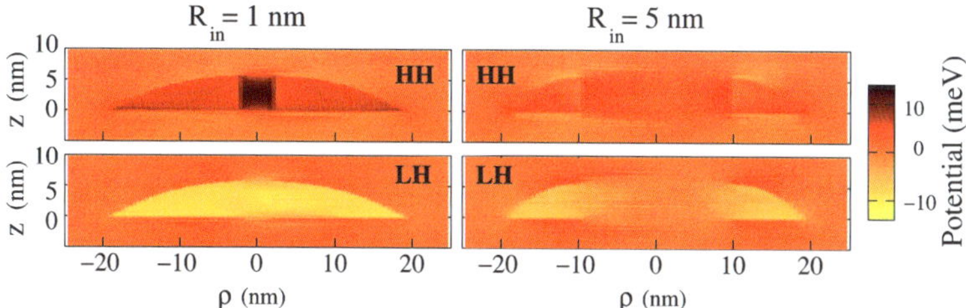

Fig. 17.7 Potential induced by strain on HH (*top panels*) and LH (*bottom panels*) confined in a QR. *Left* and *right panels* are for $R_{in} = 1$ nm and $R_{in} = 5$ nm, respectively. Note that positive potential is attractive for holes

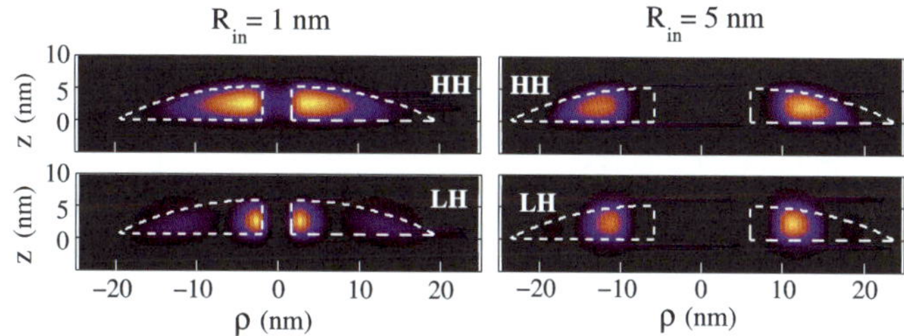

Fig. 17.8 Wave function of the HH (*top panels*) and LH (*bottom panels*) confined in a strained QR. *Dashed lines* show the edges of the QR structure. *Left* and *right panels* are for $R_{in} = 1$ nm and $R_{in} = 5$ nm, respectively. *Dashed lines* show the edges of the QR structure

which is more attractive than the InAs region itself. For LH, it is just the opposite. The inner cavity is strongly repulsive and strain favors localization around it, including the GaAs regions above and below the ring [47].

The actual hole localization, considering both strain and confinement potentials, is illustrated in Fig. 17.8 for the $m = 0$ (HH) and $m = 2$ (LH) components of the ground state. For large inner cavities ($R_{in} = 5$ nm, right panels), both HH and LH share localization inside the QR. By contrast, for small cavities ($R_{in} = 1$ nm, left panels) the HH shows a sizable density inside the cavity, while the LH stays away from it. Because the LH component has a node, half of it localizes around the core and the other half on the distant side, with significant leaking above the QR. This behavior is similar to that of holes in vertically coupled QDs with thin interdot barriers [58, 66], except that here the localization in the GaAs region takes place for HH instead of LH. A more dramatic localization of HH in the inner cavity has been predicted by considering diagonal strain terms only [47]. This indicates that shear strain reduces this phenomenon through the enhancement of HH-LH coupling.

In typical self-assembled InAs/GaAs QRs $R_{in} \gg 1$ nm [56]. In such a case, the strain in the core is weaker and cannot compete with the repulsive confinement potential (i.e., we are in the case illustrated by the right panels in Figs. 17.7 and 17.8). Yet, the results of this section suggest that using strained materials with small band-offset a spatial decoupling of HH and LH may be engineered.

17.5 Conclusions

Motivated by current interest in using confined hole spins for quantum information and spintronic devices, we have investigated the hole mixing in InAs/GaAs QRs. We have shown that the hole ground state has a fairly pure HH character in the QD limit, but it rapidly gains LH character as the inner cavity grows and the quasi-1D (wire-like) limit is approached. Because LH components have mixed spin projections, this suggests that narrow QRs are less suited than QDs for applications relying on hole spin. On the other hand, the large LH component may be of interest for applications requiring enhanced tunneling rates [23, 24], enhanced spin-orbit mediated control [27, 67] and optoelectronic devices emitting z-polarized light [46]. Experiments probing the polarization of emitted light, like the ones in Ref. [68], would confirm the different strength of LH coupling in QRs and QDs.

Deviations of the QR from perfect axial symmetry due to eccentricity or the asymmetric nature of the zinc-blende Luttinger-Kohn Hamiltonian (beyond the axial approximation [51]) have been found to have a minor influence on the HH-LH admixture. The same holds for position-dependent effective mass. This validates the use of simplified models for qualitative estimates.

The role of strain has been investigated in detail. As in QDs, it reduces the LH character of the ground state by introducing a potential which splits HH and LH states energetically [61, 62]. However, this is partially compensated by the enhancement of the off-diagonal terms coupling HH and LH through biaxial and shear strain. As a result, the actual degree of HH-LH mixing is of the same order as that expected in strain-free systems. Besides, the qualitative trends set by quantum confinement are not altered. Last, we have also shown that strain can be used to engineer the wavefunction localization in QRs with narrow inner cavities, where the strong strain potential may compete against the band-offset potential.

Acknowledgements　Support from MICINN project CTQ2011-27324 and UJI-Bancaixa project P1-1A2009-03 is acknowledged.

References

1. P.Y. Yu, M. Cardona, *Fundamentals of Semiconductors* (Springer, Berlin, 1999)
2. M. Dyakonov (ed.), *Spin Physics in Semiconductors* (Springer, Berlin, 2010)
3. R. Winkler, *Spin-Orbit Coupling Effects in Two-Dimensional Electron and Hole Systems* (Springer, Berlin, 2003)
4. L. Jacak, P. Hawrylak, A. Wójs, *Quantum Dots* (Springer, Berlin, 1998)
5. D. Brunner, B.D. Gerardot, P.A. Dalgarno, G. Wüst, K. Karrai, N.G. Stoltz, P.M. Petroff, R.J. Warburton, Science **325**, 70 (2009)
6. A. Greilich, S.G. Carter, D. Kim, A.S. Bracker, D. Gammon, Nat. Photonics **5**, 702 (2011)
7. T.M. Godden, J.H. Quilter, A.J. Ramsay, Y. Wu, P. Brereton, S.J. Boyle, I.J. Luxmoore, J. Puebla-Nunez, A.M. Fox, M.S. Skolnick, Phys. Rev. Lett. **108**, 017402 (2012)
8. K. De Greve, P.L. McMahon, D. Press, T.D. Ladd, D. Bisping, C. Schneider, M. Kamp, L. Worschech, S. Höfling, A. Forchel, Y. Yamamoto, Nat. Phys. **7**, 827 (2011)

9. K. Müller, A. Bechtold, C. Ruppert, C. Hautmann, J.S. Wildmann, T. Kaldewey, M. Bichler, H.J. Krenner, G. Abstreiter, M. Betz, J.J. Finley, Phys. Rev. B **85**, 241306(R) (2012)
10. E.A. Stinaff, M. Scheibner, A.S. Bracker, I.V. Ponomarev, V.L. Korenev, M.E. Ware, M.F. Doty, T.L. Reinecke, D. Gammon, Science **311**, 636 (2006)
11. D. Heiss, S. Schaeck, H. Huebl, M. Bichler, G. Abstreiter, J.J. Finley, D.V. Bulaev, D. Loss, Phys. Rev. B **76**, 241306(R) (2007)
12. B.D. Gerardot, D. Brunner, P.A. Dalgarno, P. Öhberg, S. Seidl, M. Kroner, K. Karrai, N.G. Stoltz, P.M. Petroff, R.J. Warburton, Nature **451**, 441 (2008)
13. J. Fischer, W.A. Coish, D.V. Bulaev, D. Loss, Phys. Rev. B **78**, 155329 (2008)
14. B. Eble, C. Testelin, P. Desfonds, F. Bernardot, A. Balocchi, T. Amand, A. Miard, A. Lemaitre, X. Marie, M. Chamarro, Phys. Rev. Lett. **102**, 146601 (2009)
15. P. Fallahi, S.T. Yilmaz, A. Imamoglu, Phys. Rev. Lett. **105**, 257402 (2010)
16. E.A. Chekhovich, A.B. Krysa, M.S. Skolnick, A.I. Tartakovskii, Phys. Rev. Lett. **106**, 027402 (2011)
17. C.-Y. Lu, Y. Zhao, A.N. Vamivakas, C. Matthiesen, S. Fält, A. Badolato, M. Atatüre, Phys. Rev. B **81**, 035332 (2010)
18. M.F. Doty, J.I. Climente, A. Greilich, M. Yakes, A.S. Bracker, D. Gammon, Phys. Rev. B **81**, 035308 (2010)
19. J.H. Blokland, F.J.P. Wijnen, P.C.M. Christiansen, U. Zeitler, J.C. Maan, Phys. Rev. B **75**, 23305 (2007)
20. J.I. Climente, J. Planelles, M. Pi, F. Malet, Phys. Rev. B **72**, 233305 (2005)
21. W. Jaskólski, M. Zielinski, G.W. Bryant, Phys. Rev. B **74**, 195339 (2006)
22. G. Bester, A. Zunger, Phys. Rev. B **72**, 165334 (2005)
23. M.F. Doty, J.I. Climente, M. Korkusinski, M. Scheibner, A.S. Bracker, P. Hawrylak, D. Gammon, Phys. Rev. Lett. **102**, 047401 (2009)
24. J.I. Climente, M. Korkusinski, G. Goldoni, P. Hawrylak, Phys. Rev. B **78**, 115323 (2008)
25. T. Chwiej, B. Szafran, Phys. Rev. B **81**, 075302 (2010)
26. A.I. Yakimov, A.A. Bloshkin, A.V. Dvurechenskii, Semicond. Sci. Technol. **24**, 095002 (2009)
27. G. Katsaros, V.N. Golovach, P. Spathis, N. Ares, M. Stoffel, F. Fournel, O.G. Schmidt, L.I. Glazman, S. De Franceschi, Phys. Rev. Lett. **107**, 246601 (2011)
28. F. Suarez, D. Granados, M. Luisa Dotor, J.M. Garcia, Nanotechnology **15**, S126 (2004)
29. J. Wu, Z.M. Wang, V.G. Dorogan, S. Li, Z. Zhou, H. Li, J. Lee, E.S. Kim, Y.I. Mazur, G.J. Salamo, Appl. Phys. Lett. **101**, 043904 (2012)
30. O. Tangmettajittakul, P. Boonpeng, P. Changmoung, S. Thainoi, S. Rattanathammaphan, S. Panyakeow, in *Photovoltaics Specialists Conference 37th IEEE* (2011), 002665
31. Y. Aharonov, D. Bohm, Phys. Rev. **115**, 485 (1959)
32. A. Lorke, J. Luyken, A.O. Govorov, J.P. Kotthaus, Phys. Rev. Lett. **84**, 2223 (2000)
33. M. Bayer, M. Korkusinski, P. Hawrylak, T. Gutbrod, M. Michel, A. Forchel, Phys. Rev. Lett. **90**, 186801 (2003)
34. N.A.J.M. Kleemans, I.M.A. Bominaar-Silkens, V.M. Fomin, V.N. Gladilin, D. Granados, A.G. Taboada, J.M. Garcia, P. Offermans, U. Zeitler, J.C. Christianen Maan, J.T. Devreese, P.M. Koenraad, Phys. Rev. Lett. **99**, 146808 (2007)
35. F. Ding, B. Li, N. Akopian, U. Perinetti, Y.H. Chen, F.M. Peeters, A. Rastelli, V. Zwiller, O.G. Schmidt, J. Nanoelectron. Optoelectron. **6**, 51 (2011)
36. F. Ding, N. Akopian, B. Li, U. Perinetti, A. Govorov, F.M. Peeters, C.C. Bof Bufon, C. Deneke, Y.H. Chen, A. Rastelli, O.G. Schmidt, V. Zwiller, Phys. Rev. B **82**, 075309 (2010)
37. J.I. Climente, J. Planelles, W. Jaskolski, Phys. Rev. B **68**, 075307 (2003)
38. L.G.G.V. Dias da Silva, S.E. Ulloa, A.O. Govorov, Phys. Rev. B **70**, 155318 (2004)
39. V.M. Fomin, V.N. Gladilin, S.N. Klimin, J.T. Devreese, Phys. Rev. B **76**, 235320 (2007)
40. N. Cukaric, M. Tadic, F.M. Peeters, Superlattices Microstruct. **48**, 491 (2010)
41. A.O. Govorov, S.E. Ulloa, K. Karrai, R.J. Warburton, Phys. Rev. B **66**, 081309(R) (2002)
42. B. Li, F.M. Peeters, Phys. Rev. B **83**, 115448 (2011)
43. E. Waltersson, E. Lindroth, I. Pilskog, J.P. Hansen, Phys. Rev. B **79**, 115318 (2009)

44. M. Szopa, E. Zipper, J. Phys. Conf. Ser. **213**, 012006 (2006)
45. C. Segarra, J.I. Climente, J. Planelles, J. Phys. Condens. Matter **24**, 115801 (2012)
46. J. Planelles, F. Rajadell, J.I. Climente, J. Phys. Chem. C **114**, 8337 (2010)
47. J.I. Climente, J. Planelles, J. Nanoelectron. Optoelectron. **6**, 81 (2011)
48. M.G. Burt, J. Phys. Condens. Matter **4**, 6651 (1992)
49. M.G. Burt, J. Phys. Condens. Matter **11**, R53 (1999)
50. B.A. Foreman, Phys. Rev. B **48**, 4964 (1993)
51. P.C. Sercel, K.J. Vahala, Phys. Rev. B **42**, 3690 (1990)
52. J.M. Luttinger, W. Kohn, Phys. Rev. **97**, 869 (1955)
53. J.M. Luttinger, Phys. Rev. **102**, 1030 (1956)
54. L. Voon, M. Willatzen, *The k·p Method: Electronic Properties of Semiconductors* (Springer, Berlin, 2009)
55. F. Rajadell, M. Royo, J. Planelles, J. Appl. Phys. **111**, 014303 (2012)
56. P. Offermans, P.M. Koenraad, J.H. Wolter, D. Granados, J.M. Garcia, Appl. Phys. Lett. **87**, 131902 (2005)
57. I. Vurgaftman, J.R. Meyer, L.R. Ram-Mohan, J. Appl. Phys. **89**, 5815 (2001)
58. M. Tadic, F.M. Peeters, K.L. Janssens, M. Korkusiński, P. Hawrylak, J. Appl. Phys. **92**, 5819 (2002)
59. J.A. Barker, R.J. Warburton, E.P. O'Reilly, Phys. Rev. B **69**, 035327 (2004)
60. R. Blossey, A. Lorke, Phys. Rev. E **65**, 021603 (2002)
61. C. Pryor, Phys. Rev. B **57**, 7190 (1998)
62. W. Sheng, J.P. Leburton, Phys. Status Solidi **237**, 394 (2003)
63. M. Tadic, N. Cukaric, V. Asoski, F.M. Peeters, Phys. Rev. B **84**, 125307 (2011)
64. C.Y.P. Chao, S.L. Chuang, Phys. Rev. B **46**, 4110 (1992)
65. N.A.J.M. Kleemans, J.H. Blokland, A.G. Taboada, H.C.M. van Genuchten, M. Bozkurt, V.M. Fomin, V.N. Gladilin, D. Granados, J.M. García, P.C.M. Christianen, J.C. Maan, J.T. Devreese, P.M. Koenraad, Phys. Rev. B **80**, 155318 (2009)
66. C. Pryor, Phys. Rev. Lett. **80**, 3579 (1998)
67. S.E. Economou, J.I. Climente, A. Badolato, A.S. Bracker, D. Gammon, M.F. Doty, Phys. Rev. B **86**, 085319 (2012)
68. T. Flissikowski, I.A. Akimov, A. Hundt, F. Henneberger, Phys. Rev. B **68**, 0161309(R) (2003)

Chapter 18
Engineering of Electron States and Spin Relaxation in Quantum Rings and Quantum Dot-Ring Nanostructures

Marcin Kurpas, Elżbieta Zipper, and Maciej M. Maśka

Abstract Quantum nanostructures are frequently referred to as artificial atoms. Like the natural atoms they show a discrete spectrum of energy levels but at the same time they exhibit new physics which has no analogue in real atoms. Electrons in an atom are attracted to the nucleus by a potential that diminishes inversely proportional to the distance from the center of the atom. This feature, together with the Coulomb interactions between electrons, determines properties of natural atoms. On the other hand, in quantum nanostructures one can (almost) freely design the shape of the confinement potential. As a result, a variety of properties of quantum nanostructure can be modified according to the designer's will.

The aim of this chapter is to demonstrate in a detailed way how these properties depend on the geometry of the confinement potentials. Quantum rings can be narrow, quasi-one-dimensional objects with periodic boundary conditions. But they also can be wide, like a quantum dot with a small hole in the center that leads to a nontrivial topology. The electronic properties, determined by the energy spectrum and the distribution of the wave functions, are very different in these limiting cases. They are also different from the properties of a quantum dot. In this chapter we demonstrate how selected properties are modified when the confinement changes in such a way that the nanostructure evolves from a quantum dot to a wide quantum ring and than to a narrow, quasi-one-dimensional nanoring. Besides single quantum rings we describe properties of two coupled nanorings and of systems composed of a quantum ring coupled to a quantum dot. We are mainly interested in properties that are connected to possible applications of quantum nanostructures. There is a common belief that such systems are among the most promising candidates for realization of qubits in quantum computing. However, physical implementation requires, among others, relatively long decoherence time, much longer than the gate operation times. Assuming the spin-orbit-mediated electron-phonon interaction as the dominant relaxation mechanism for spin qubits, we show how the relaxation time depends on the details of the confinement potential. We first compare the re-

M. Kurpas · E. Zipper · M.M. Maśka (✉)
Department of Theoretical Physics, University of Silesia, Uniwersytecka 4, 40-007 Katowice, Poland
e-mail: maciej.maska@us.edu.pl

V.M. Fomin (ed.), *Physics of Quantum Rings*, NanoScience and Technology,
DOI 10.1007/978-3-642-39197-2_18, © Springer-Verlag Berlin Heidelberg 2014

laxation times calculated for quantum dots and quantum rings of different shapes and sizes. However, it seems that complex structures composed of a quantum dot surrounded by a quantum ring can have far more interesting properties due to the high controllability of the spatial distribution of the electronic wave function. The results indicate that the main factor that determines the relaxation time is the so-called overlap factor, i.e., the overlap of the radial parts of the wave functions of the ground and first two excited states. With this knowledge, one can try to optimize the confinement potential with respect to the relaxation time. By tuning the relative positions of the bottoms of the ring and dot confinement potentials one can control the overlap factor, what in turn allows to control the relaxation time. The same effect can be achieved by modifying the height of the potential barrier between the ring and the dot. A high controllability is also expected in a similar system where the central quantum dot is replaced by a small ring, i.e., in a system of two coupled concentric quantum rings.

The wave function engineering allows one to control not only the relaxation time. We demonstrate that the same overlap factor determines also optical properties of quantum nanostructures. Using realistic parameters we show that changing the shape of the confinement potential it is possible to modify the microwave and infrared absorption cross sections of the dot-ring nanostructure. That way, the nanostructures can be moved over from highly absorbing to almost transparent. The last property analyzed in this chapter is the conductivity of a system composed of many dot-ring nanostructures. Apart from unique properties of a single nanostructure, interesting behavior emerges when such structures are combined into a two-dimensional array. If they are located sufficiently close to each other, electrons can tunnel between them, making a system that resembles a narrow band crystal. Since the tunneling rate depends on the overlap of the electron wave functions on adjoining structures, the transport properties would be dependent on the shape of the confinement potential. As a result, a metal-insulator transition can be easily induced in the array. We demonstrate a way how to control the confinement potential globally for the whole array.

18.1 Introduction

Ring-like nanostructures have long fascinated physicists [1–7]. These nanometer-size rings which are the nanoscopic analogues of benzene have many intriguing features. The magnetic properties of such non-simply connected quantum systems are related to the possibility of trapping flux in their interior at accessible magnetic fields.

The high degree of symmetry of quantum rings (QRs) imply the existence of persistent currents (PC) [1] due to the conservation of electron angular momentum. Since the first experimental verification by Levy et al. [2] PC has been found in metallic [3, 4, 8, 9] and semiconducting rings [5]. Nowadays technology allows the preparation and characterization of very small, high-mobility semiconductor structures of ring geometry with very good resolution [6, 7, 10–12]. The ability to fill semiconductor QRs with one or a few electrons allows even to detect persistent current carried by single electrons [5] and the magnetoinduced change of the ground state [10, 11]. Also, the possibility of creating self-sustaining orbital currents in mesoscopic systems of ring geometry has been predicted [13] but no experimental confirmation of this effect has been reported.

The ring geometry which supports two opposite currents was used to propose flux qubit built on the orbital PC states [14] in some analogy to flux qubit on superconducting ring with Josephson junctions [15]. However the orbital degrees of freedom have much shorter relaxation times than devices based on the electron spin degrees of freedom [16]. It is thus of interest to discuss the possibility of building a spin qubit or spin memory device on a defect free QR as the solid state devices seem to be the most promising because of their scalability, tunability and relatively long coherence times.

The most frequently used in this context are semiconductor quantum dots (QDs) [17] which indicate the great potential for device applications in spintronics, optoelectronics, and quantum computing. The spin of a single electron in a semiconductor quantum dot (QD) placed in magnetic field is a natural two-level system suitable for use as a qubit or spin memory device in spintronics [18, 19]. However, quantum confinement properties of QDs can be deeply mutated to crater-like nanostructures called hereafter quantum rings [5–7, 10, 11, 20–22]. Just like QDs, QRs possess atom-like properties making them potential candidates for both basic research and future applications, e.g., in quantum information processing and spintronics.

The long term promise of such devices depends crucially on the relaxation and decoherence times which can be different in QRs and QDs due to different dispersion relation and different confinement properties. It is thus important to relate the devices build on QRs to those constructed on QDs. In this context it was shown that QRs are also attractive for the realization of spin qubits as the relaxation and decoherence processes take place in the time scale that is sufficiently long for spin manipulations and readout [23]. The problem of spin relaxation in QRs will be discussed in the first part of this chapter. The presented considerations are general and apply both for electrostatically defined QRs (EQRs) [6, 7, 24] which can be primarily controlled electrically and self-assembled QRs (SQRs) [5, 10, 11, 21, 22] where spin manipulations are done using optical techniques.

Nowadays nanotechnology enables a precise control of structural parameters of quantum nanostructures both at the fabrication stage [25–28] as well as dynamically while operating the device, e.g., by electrostatic potential [24]. It enables to produce complex systems where different building blocks such as QRs [28, 29] and QDs [30, 31] are combined together into a single structure [25–27]. Such composed systems have rich energy spectrum and exceptionally high degree of tunability as the coupling between the components is an additional parameter that can be controlled. They are highly relevant to new technologies in which the control and manipulations of electron spin and wave functions play an important role and the potential use of such devices requires deep theoretical analysis. Thus in the second part of this chapter an investigation of two-dimensional complex nanostructures: one in the form of double concentric quantum rings and second consisting of a quantum ring with quantum dot inside named afterwords a dot-ring nanostructure (DRN) is presented. In particular the combined dot-ring structure seems to be very interesting because, as it will be shown, by changing the confinement potential, i.e., the parameters of the potential barrier $V_0(r)$ separating the dot from the ring and/or the potential well offset V_{QD}-V_{QR} (see Fig. 18.4b), one can considerably alter coherent, optical and conducting properties of the DRN. It will be demonstrated that by such manipulations one can change the spin relaxation time T_1 of DRNs, used as spin qubits or spin memory devices, by orders of magnitude. These features are mostly determined by the so called overlap factor (OF) which reflects the distribution of the wave functions in the system and can be largely modified by the form of the confinement potential.

It will be also briefly pointed out that confinement-based wave function engineering may be used to control

(a) the cross section for optical absorption at microwave and infra-red range from strong to negligible,
(b) the conducting properties of an array of DRNs from highly conducting to insulating.

The more comprehensive discussion of the last two problems which are beyond the scope of this chapter can be found in Ref. [32].

18.2 Quantum Confinement in Semiconductor Quantum Ring and Formation of Spin Qubits

Consider a 2D semiconductor QR of finite thickness and radius r_0 set by a confinement potential $V(r)$. The ring contains a *single or a few electrons* and is placed in a static magnetic field **B** parallel to the ring plane. The in-plane orientation is favourable because it does not disturb the distance between orbital levels [33] (for perpendicular magnetic field see [23]).

The Hamiltonian of such a ring with spin-orbit and electron-phonon interaction can be written in the form

$$H = H_R + \frac{e\hbar}{2m^*}\hat{\boldsymbol{\sigma}} \cdot \mathbf{B} + H_{SO} + H_{\text{e-ph}}, \tag{18.1}$$

where

$$H_R = \frac{1}{2m^*}\mathbf{p}^2 + V(r), \tag{18.2}$$

with m^* representing the effective electron mass,

$$H_{SO} = \beta(-\hat{\sigma}_x p_x + \hat{\sigma}_y p_y) \tag{18.3}$$

is the Dresselhaus SO interaction due to the bulk inversion asymmetry. For GaAs heterostructures the parameter β takes the values from 10^5 to 3×10^5 cm/s. $H_{\text{e-ph}}$

is the Hamiltonian of the electron-phonon coupling and its particular form will be given later in the text.

The energy spectrum of H_R consists of a set of discrete levels E_{nl} due to radial motion with radial quantum numbers $n = 0, 1, 2, \ldots$, and rotational motion with angular momentum quantum numbers $l = 0, \pm 1, \pm 2 \ldots$.

The single particle wave function is of the form

$$\Psi_{nl} = R_{nl}(r) \exp(il\phi)\chi_\sigma, \tag{18.4}$$

with the radial part $R_{nl}(r)$ and the spin part χ_σ. For 1D QRs the energy spectra and the wave functions can be calculated analytically, for finite-width QRs both E_{nl} and Ψ_{nl} need numerical evaluation. The application of magnetic field **B** splits the orbital energy levels by Zeeman energy

$$\Delta_Z = g_s \mu_B B, \tag{18.5}$$

where g_s is the Landé factor and μ_B is the Bohr magneton. Another important energy gap is the distance from the highest occupied orbital state (nl) to the excited orbital state $(n'l')$

$$\Delta_{nl,n'l'} = E_{n'l'} - E_{nl}. \tag{18.6}$$

If the following relation holds

$$k_B T \ll \Delta_Z \ll \Delta_{nl,nl\pm 1} \tag{18.7}$$

the two Zeeman sublevels of the orbital nl are well separated from the others and the ring can be well approximated as a two-state system—a qubit. At this point let us assume that the orbital l, called afterwards 'operating orbital' is occupied by a single electron, i.e., for $l = 0$ the number of electrons is $N_e = 1$, for $|l| = 1$, $N_e = 3$, etc.

In order to get better understanding of processes governing the relaxation different quantum rings will be considered. The radii and confining potentials of three of them (A, B, C) are chosen to roughly reproduce the energy spectra of the recently grown InGaAs rings described in Refs. [10, 11], and [5], respectively. As the experiments measuring the spin relaxation times are performed mainly at $B \geq 1$ T [24, 34] in the following the magnitude of the magnetic field will be set to $B = 2$ T which gives, for the Landé factor $|g_s| = 0.8$ [34], the electron spin Zeeman splitting $\Delta_Z = 0.092$ meV. The confining potential used to model rings A, B and C is assumed to be of the following form,

$$V_1(r) = \frac{1}{2}m^*\omega_0^2(r - r_0)^2, \tag{18.8}$$

where the parameters are collected in Table 18.1. For the remaining rings *the same radius as for ring* C has been assumed, but the potential takes on different shapes.

Table 18.1 The parameters and relaxation times of three modeled InGaAs quantum rings corresponding (in alphabetical order) to the experimental rings described in Refs. [10, 11], and [5]. The ring geometry has been reached by the confining potential $V_1(r)$ (18.8); $\hbar\omega_0$ is the potential strength. $B = 2$ T has been assumed

Ring	r_0 [nm]	$\hbar\omega_0$ [meV]	T_1^0 [ms]	T_1^1 [ms]	T_1^2 [ms]	T_1^3 [ms]
A	20	15	0.12	0.4	0.5	0.55
B	14	12	2	1.5	1	0.86
C	11.5	25	10	15	13	11

In order to be able to compare results for QRs and QDs the potential is parametrized in such a way that it can reproduce both harmonic potential of a QD as well as a δ-like potential of a quasi one-dimensional (1D) QR. It is given by

$$V_2(r) = \frac{1}{2}m^*\omega_0^2\left[(1-k)r^2 + \frac{k}{1-k}(r-r_0)^2\right]$$

$$= \frac{1}{2}m^*\omega_{QD}^2 r^2 + \frac{1}{2}m^*\omega_{QR}^2(r-r_0)^2. \tag{18.9}$$

It is a superposition of QD and QR potentials, where the confinement (a measure of radial localization of the electron wave function) is given by $\hbar\omega_{QD}$ and $\hbar\omega_{QR}$, respectively. For $k = 0$ the second term vanishes and the potential describes harmonic QD. On the other hand, in the $k \to 1$ limit it describes a 1D QR. Therefore, changing k from 0 to 1 one can observe how the properties of a quantum system evolves while moving from QD to QR. The radius of QR is defined by r_0 in (18.8), i.e., it is the distance from the center of the ring to the minimum of the confining potential. The definition of the radius of harmonic QD is not so unambiguous—r_0 is defined by the shape of the ground state wave function $\Psi(r, \phi) \propto \exp(-r/r_0)$. In order to ensure that the size of the system does not depend on k, and therefore that its properties depend only on the shape of the potential, ω_0 in (18.9) is given by the relation

$$\omega_0 = \frac{2\hbar}{m^* r_0^2}, \tag{18.10}$$

what gives the radius of the QD equal to r_0. Figure 18.1 shows the radial part of the wave function for ground state ($l = 0$) and the first excited state ($l = 1$) for different values of k: For $k = 0$ the potential models QD (Fig. 18.1a), for $0 < k < 1$ one gets QRs of decreasing thickness (Fig. 18.1b, c), reaching at $k = 0.99$ a quasi 1D ring (Fig. 18.1d). The corresponding shapes of the confining potential are shown in the insets.

Note that the model calculations are for circularly symmetric nanostructures, whereas some of the experimentally fabricated rings have slightly different symmetry and therefore slightly different energy spectrum. Moreover, other possible factors which may modify the solutions of the Schrödinger equation, like the polarization effects [35, 36], are neglected. They are, however, connected to a particular

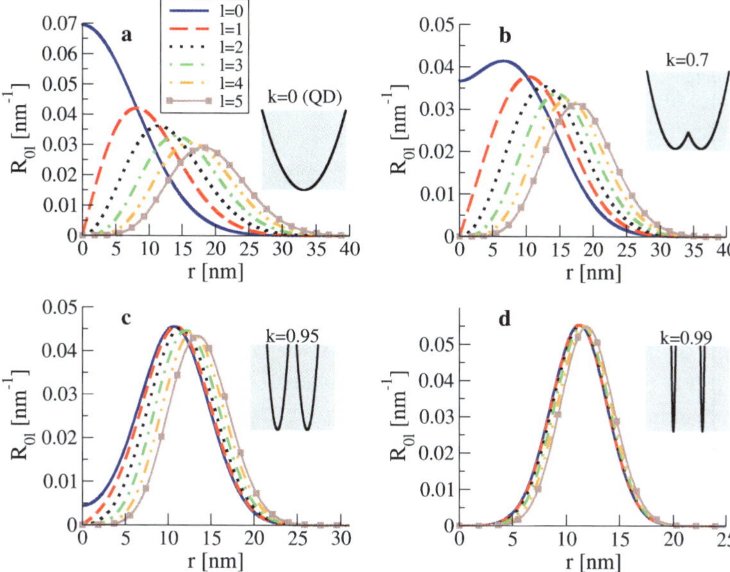

Fig. 18.1 The radial part $R_{0l}(r)$ of the electron wave function (18.4) plotted as a function of radius r for orbital quantum number l equal 0 and 1. Panels (**a**)–(**d**) include results for different shapes of the confining potential (18.9) (shown in the *inset plots*). In all cases $r_0 = 11.5$ nm, $B = 1$ T have been assumed

nanostructure fabrication method, whereas the calculations are as general as possible.

In order to use quantum rings in quantum computation it is necessary to establish a way to perform single qubit operations and to implement efficient quantum logic gates on pairs of qubits. During the past few years a big progress has been made towards full control of quantum states of single and coupled spins in QDs [33, 37, 38]. Going carefully through all this one finds that most of those features are shared also by QRs.

The QR qubit can be initialized by, e.g., thermal equilibration followed by optical pumping, coherently manipulated (through magnetic resonance technique or by faster electrical and optical gates) and read out using both electrical and optical techniques [21, 22, 33]. Coherent coupling of EQRs leading to the formation of, e.g., the CNOT gate can be obtained in an analogous way as for QDs [18, 19], by assembling a system of two coplanar QRs with the possibility of tuning their exchange coupling J by gating the barrier between them. Such coupling can be switched on and off by electrical impulses. Quantum gates for SQRs can be accomplished by electric or photonic connections [21, 22, 38, 39]. Recently a scheme for coherent coupling of spin qubits placed microcavities by entanglement swapping [40] has been proposed. Such long-distance entanglement is a crucial ingredient for quantum communication.

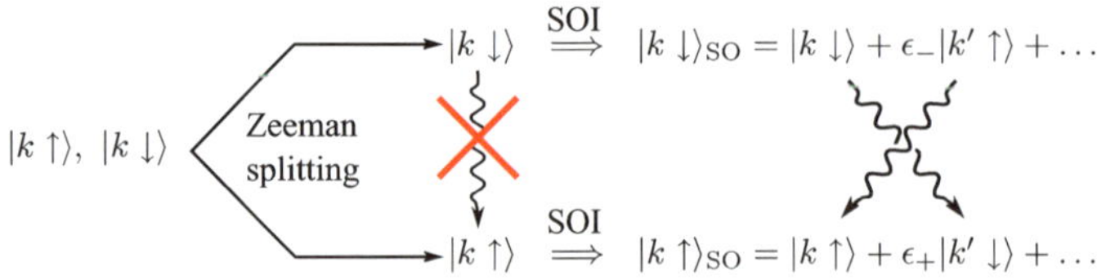

Fig. 18.2 Direct phonon transitions between pure Zeeman split states is impossible. However, the SOI gives a small admixture of opposite spin state, that enables transitions [24]

Single qubit rotations together with the CNOT gate form a universal set of quantum logic gates. Remarkably these operations are very fast, on the order of pico to nanoseconds [41]. Thus very many coherent operations can be performed until the qubit's state is destroyed by decoherence.

18.3 Spin Relaxation and Decoherence

The main difficulty in development of the spin based devices is to keep them in the quantum regime for a sufficiently long time. The ideal situation would be to cut off the interaction with the environment that is the main source of destruction of a quantum state. This is, however, an impossible task as this interaction is equally needed (for steering, measurement etc.) as unwanted (decoherence, relaxation).

Electron spin decoherence in semiconductors is caused primarily by spin-lattice relaxation via phonon scattering and spin-orbit (SOI) interaction and by hyperfine (HFI) interaction with nuclear spins [16, 42, 43].

At magnetic fields $B < 0.1$ T the dominant relaxation mechanism is the HFI but for larger fields this mechanism is suppressed by the mismatch between the nuclear and electron Zeeman energies. At 0.5 T $< B < 10$ T the SOI causes spin relaxation by mixing the spin and orbital states (see Fig. 18.2) and providing the mechanism for coupling of spins to (mainly) piezoelectric phonons. H_{SO} (18.3) gives a small admixture of the state of the opposite spin to each of the orbital states with spin up or down. This enables the phonon transition between the two states and is responsible for spin relaxation. The prolongation of spin relaxation times for small nanostructures stems from a drastic reduction in spin-phonon coupling mediated by the combination of electron-phonon and SOI, due to strong confinement. It has been shown [42] that in a magnetic field of a few Tesla the relaxation is dominated by single-phonon processes mediated by SOI; it has been positively verified in several experiments [24, 33, 34, 44].

The comprehensive analysis of relaxation due to phonons in QDs has been given in Refs. [16, 33, 42, 45, 46] and [47]. As the developed formalism [42] is valid for all circularly symmetric III–V systems it can be adopted also for semiconducting QRs. Below the main ideas of this derivation for a qubit composed of Zeeman split states of an electron at the energy level E_{nl} (defined in Sect. 18.2) are sketched.

In polar crystals, such as GaAs, the main relaxation mechanism is due to piezo-electric interaction with acoustic phonons (with linear dispersion relation). The direct phonon coupling to the spin states is prohibited, however such coupling is made possible by the SOI which mixes the Zeeman-split ground orbital state with excited orbital states of opposite spin. The perturbed eigenstates are [33]

$$|nl \uparrow\rangle^{(1)} = |nl \uparrow\rangle + \sum_{n'l' \neq nl} \frac{\langle n'l' \downarrow |H_{SO}|nl \uparrow\rangle}{E_{nl} - E_{n'l'} - \Delta_Z}|n'l' \downarrow\rangle, \qquad (18.11)$$

$$|nl \downarrow\rangle^{(1)} = |nl \downarrow\rangle + \sum_{n'l' \neq nl} \frac{\langle n'l' \uparrow |H_{SO}|nl \downarrow\rangle}{E_{nl} - E_{n'l'} + \Delta_Z}|n'l' \uparrow\rangle, \qquad (18.12)$$

where the spin-orbit Hamiltonian H_{SO} is given by (18.3) [42]. The relaxation rate between these dressed states can be calculated using the Fermi's golden rule:

$$\frac{1}{T_1^l} \equiv \Gamma^l = \frac{2\pi}{\hbar}\left|{}^{(1)}\langle nl \uparrow |H_{\text{e-ph}}|nl \downarrow\rangle^{(1)}\right|^2 D(\Delta_Z), \qquad (18.13)$$

where $D(\Delta_Z)$ is the phonon density of states at the Zeeman energy, which for bulk acoustic phonons increases quadratically with this energy. The electron-phonon coupling is described by

$$H_{\text{e-ph}}^{\mathbf{q}j} = M_{\mathbf{q}j}e^{i\mathbf{q}\mathbf{r}}\left(b_{\mathbf{q}j}^{\dagger} + b_{\mathbf{q}j}\right), \qquad (18.14)$$

where $M_{\mathbf{q}j}$ is a measure of the electric-field strength of a phonon with wave vector $\mathbf{q}$ and phonon branch j, $\mathbf{r}$ is the position vector of the electron and $b_{\mathbf{q}j}^{\dagger}$ ($b_{\mathbf{q}j}$) are the phonon creation (annihilation) operators.

The electric field and thus $M_{\mathbf{q}j}$ scales as $1/\sqrt{q}$ for piezoelectric phonons and as $\sqrt{q}$ for deformational potential phonons. Thus if both types of phonons are present the first mechanism dominates at small energies. For small Zeeman energies the matrix element of $e^{i\mathbf{q}\mathbf{r}}$ can be calculated in the long wave length limit (the so called dipole approximation):

$$\langle nl|e^{i\mathbf{q}\mathbf{r}}|n'l'\rangle \approx \mathbf{q}\langle nl|\mathbf{r}|n'l'\rangle. \qquad (18.15)$$

First, we will estimate the dependence of Γ^l governed by phonons on the energy Δ_Z. The spin-flip matrix element $|H_{\text{e-ph}}^{nl\uparrow\downarrow}|^2 \propto M_{\mathbf{q}j}^2 q^2 \Delta_Z^2$. Since the phonons carry away the spin-flip energy, one gets $E_{ph} \propto \Delta_Z$, and since $E_{ph} \propto q$ the matrix element is proportional to $M_{\mathbf{q}j}^2 \Delta_Z^4$. Thus, its final dependence on Δ_Z is determined by $M_{\mathbf{q}j}$. For piezoelectric phonons $M_{\mathbf{q}j} \propto 1/\sqrt{q}$ and

$$\left|H_{\text{e-ph}}^{nl\uparrow\downarrow}\right|^2 \propto \Delta_Z^3. \qquad (18.16)$$

Inserting (18.16) it into (18.13) and taking into account that $D(\Delta_Z) \propto \Delta_Z^2$ one finds that

$$\Gamma^l \propto \Delta_Z^5. \qquad (18.17)$$

This strong dependence has been experimentally observed in many experiments on QDs [24, 33, 34, 44].

Because T_1^l given by (18.13) is due to SOI admixture of the higher orbital states it depends on the distance $\Delta_{nl,n'l'}$ between the relevant energy states and on the dipole matrix elements (18.15) called afterwards overlap factors $\Xi_{nl,n'l'}$.

Assuming that the electrons participating in relaxation occupy merely the states with $n = 0$, after some algebra the final formula for the relaxation rate from spin down to spin up state for an electron in the E_{0l} state reads

$$\frac{1}{T_1^l} = \frac{\Delta_Z^5}{\eta} \left(\sum_{n'l'} \frac{\Xi_{0l,n'l'}^2}{\Delta_{0l,n'l'}} \right)^2, \tag{18.18}$$

where $\Delta_{0l,n'l'} \equiv E_{n'l'} - E_{0l}$ and the overlap factor $\Xi_{0l,n'l'}$ is given by

$$\Xi_{0l,n'l'} = \int_0^\infty R_{n'l'}^* R_{0l} r^2 dr \tag{18.19}$$

with $l' = l \pm 1$. The parameter η is expressed as

$$\eta = \frac{\hbar^5}{\Lambda_p (2\pi)^4 (m^*)^2}, \tag{18.20}$$

where Λ_p is the dimensionless constant depending on the strength of the effective spin-piezoelectric phonon coupling and the magnitude of SOI, $\Lambda_p = 0.007$ for GaAs type systems [34, 42].

The above formulas apply for very low temperatures (e.g., when thermal phonons are negligible) (see (18.7)). At finite temperature (18.13) should be multiplied by a factor $N_{\Delta_Z} + 1$ for the transition with emission of a phonon and by N_{Δ_Z} in the case of phonon absorption, $N_{\Delta_Z} = [\exp(\Delta_Z/kT) - 1]^{-1}$. The temperature dependence of spin relaxation is out of the scope of the chapter and will not be discussed in details.

All other mechanisms of spin relaxation have been estimated [42] to be much less efficient than the mechanism via admixing of spin and orbitals by the spin-orbit interaction.

Because the relaxation rate for an electron in QRs (and in QDs) is determined by the admixture to the first excited orbital state. It is because the energy distance $\Delta_{0,l,n'l'}$ to the higher lying states is too big to make those states important for relaxation—it will be different for complex dot-ring structures discussed in Sect. 18.4.1. The formula (18.18) for a single electron in the highest occupied state l takes on the following form:

$$T_1^l = \frac{\eta}{\Delta_Z^5} \frac{\Delta_l^2}{\Xi_l^4}, \tag{18.21}$$

where

$$\varXi_l = \int_0^\infty R_{0l}^* R_{0l'} r^2 dr, \qquad \Delta_l = E_{0l'} - E_{0l} \quad \left(l' = l \pm 1\right). \tag{18.22}$$

Since in this case Δ_l and $\varXi_l$ depend only on l all other indices have been dropped. It follows from (18.21) that T_1^l depends on Δ_l, i.e., on the number of electrons N_e and on $\varXi_l$ which depends, in turn, on the wave functions of the neighbouring l states (see (18.22) and Fig. 18.1a–d). Notice, that in contrast to QDs where $\Delta_l = \hbar\omega_0$, for QRs the energy gaps between neighbouring l states are l-dependent and increase with increasing l (faster for a thinner ring), tending to

$$\Delta_l^{1D} = \frac{\hbar^2}{2m^* r_0^2}(2l + 1) \tag{18.23}$$

for a 1D ring (see Fig. 18.3a).

Let us first discuss the case of a single electron ($N_e = 1$, $l = 0$, solid blue lines in Fig. 18.3a–c). For harmonic QDs ($k = 0$) the formula (18.22) simplifies to

$$\left(\varXi_0^{\text{dot}}\right)^2 = \frac{\hbar^2}{4\pi^2 m^* \hbar\omega_0}. \tag{18.24}$$

Replacing $\varXi_l$ in (18.21) by $\varXi_0^{\text{dot}}$ gives

$$T_1^{0,\text{dot}} = \Lambda_p^{-1} \frac{\hbar(\hbar\omega_0)^4}{\Delta_Z^5}, \tag{18.25}$$

i.e., the relaxation time for QDs obtained in Ref. [42].

It is interesting to compare relaxation times for QRs and QDs—it is shown below that $T_1^{0,\text{dot}}$ is a higher limit of T_1^0. From (18.21) and (18.25) one can find the formula relating them:

$$T_1^0 = T_1^{0,\text{dot}} \left(\frac{\Delta_0}{\hbar\omega_0}\right)^2 \left(\frac{\varXi_0^{\text{dot}}}{\varXi_0}\right)^4. \tag{18.26}$$

It follows from Figs. 18.3a and 18.3b that for a singly occupied QR of arbitrary thickness

$$\left.\begin{aligned} \hbar\omega_0 &> \Delta_0, \\ \varXi_0^{\text{dot}} &< \varXi_0 \end{aligned}\right\} \quad \Rightarrow \quad T_1^0 < T_1^{0,\text{dot}}, \tag{18.27}$$

i.e., assuming the same size, QD is the structure having the longest relaxation time. To understand the first of the inequalities let us rewrite (18.23) in terms of $\hbar\omega_0 = 2\hbar^2/m^* r_0^2$

$$\Delta_l^{1D} = \frac{1}{4}\hbar\omega_0(2l + 1) \xrightarrow{l=0} \frac{\hbar\omega_0}{4}. \tag{18.28}$$

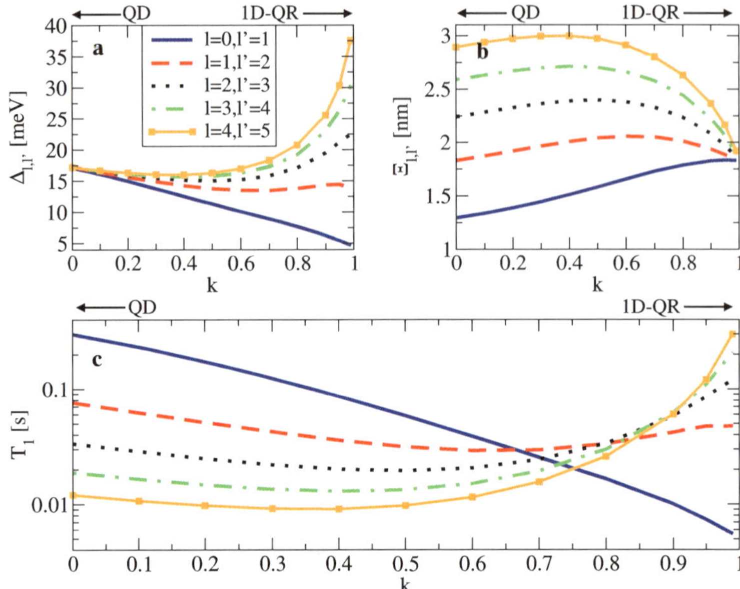

Fig. 18.3 (**a**) The orbital energy gap Δ_l as a function of the potential parameter k, for different values of l (corresponding to different occupation N_e). For $k = 0$ the potential (18.9) models QD and Δ_l is l-independent. Increasing k we reach (for $k \to 1$) the 1D-QR limit with Δ_l defined by (18.23); (**b**) The overlap factor Ξ_l; (**c**) the relaxation time T_1 plotted as a function of k for different orbital states l. $r_0 = 11.5$ nm has been assumed

It is clear that for 1D QR with one electron Δ_0 never reaches $\hbar\omega_0$. For rings of finite thickness Δ_0 changes smoothly between Δ_0 and $\hbar\omega_0$.

The inequality between the overlap factors Ξ_0 and Ξ_0^{dot}, follows from the difference in shape and distribution of the radial parts R_{0l} (Fig. 18.1a–d). For QD the radial functions are concentrated closer to $r = 0$ than for QRs where they stay mostly at larger r. Additionally, for QD the difference between the $R_{00}(r)$ and $R_{01}(r)$ is much more significant than for QR. Both these properties result in a smaller value of Ξ_0 for QD than for QR, leading to the relations given by (18.27).

The situation, however, changes if the number of electrons is larger than one. It was shown in Refs. [6, 7, 33] that in such cases, the electron-electron interaction is well described by the constant interaction model—it shifts the single electron spectra by a multiple of the charging energy, $(N_e - 1)E_C$. Assuming this, the values of Δ_l in Fig. 18.3 correspond to the energies after the charging energy has been subtracted. One can see from Fig. 18.1a that for QDs with $l > 0$ the maxima of the wave functions move to larger r leading to an increase of Ξ_l^{dot} (Fig. 18.3b) and subsequent decrease of T_1^{dot} (Fig. 18.3c). For QRs of the large thickness ($0 < k < 0.8$) the situation is similar as for QDs, but for thinner rings with $k > 0.8$ both the decrease of Ξ_l (Fig. 18.3b) and the simultaneous increase of Δ_l (Fig. 18.3a) lead to a substantial increase of T_1^l (see Fig. 18.3c and Table 18.2). Such relatively thin rings with relaxation times exceeding milliseconds for $B = 2$ T are within reach for

Table 18.2 The relaxation time T_1^l for different values of the orbital number l (equivalently, the number of electrons N_e) and different shapes of the potential $V_2(r)$. $r_0 = 11.5$ nm and $B_\parallel = 2$ T have been assumed

Relaxation time T_1^l [s]

k	$l = 0$	$l = 1$	$l = 2$	$l = 3$	$l = 4$
0	0.30	0.076	0.034	0.019	0.0129
0.2	0.17	0.051	0.025	0.0158	0.01
0.4	0.086	0.036	0.020	0.013	0.009
0.6	0.039	0.029	0.021	0.015	0.011
0.8	0.017	0.033	0.034	0.030	0.026
0.95	0.007	0.047	0.086	0.11	0.12
0.99	0.006	0.047	0.12	0.21	0.30

nowadays nanotechnology. In Table 18.1 the relaxation times for the rings A–C are presented. It can be seen that T_1 increase considerably with decreasing radius of the rings reaching the value $T_1^0 = 10$ ms already for singly occupied ring C. However, because the rings A–C are relatively thick we do not get the essential increase of $T_1^l > 10$ ms.

The above model considerations apply for InGaAs/GaAs rings but the underlying physics is similar in other systems with somehow different set of parameters. In GaAs and GaAs/AlGaAs nanosystems the spin g_s factor changes in a range $g_s \sim$ 0.2–0.4 [33]. Assuming that material properties entering (18.21) are roughly the same as for InGaAs/GaAs and $N_e = 1$ one obtains, e.g., for the ring B made out of material with $g_s = 0.4$, $T_1 \sim 64$ ms and for the ring C (with $g_s = 0.4$), $T_1 \sim$ 0.32 s. However, one has to stress again that these very long relaxation times have been obtained taking into account only SO mediated interaction with piezoelectric phonons. Considering also other mechanisms of relaxation, (e.g., due to fluctuations of the electric and magnetic field, deformational phonons, multiphonon processes, and circuit noise) which were neglected in the above model calculations, can further limit the relaxation time.

The spin decoherence time T_2 for nanosystems made out of III–V semiconductors is limited by HFI as it was shown [16] that SOI does not lead to pure dephasing. It was estimated as $T_2 \sim 10$–100 μs [48, 49] for the considered magnetic field. Besides, several strategies have been proposed to decrease the randomness in the nuclear-spin system which can be useful also for QRs. Polarization of nuclei [50] and putting the nuclear spins in a particular quantum state [51] are very promising. An alternative approach is to use a quantum ring with holes instead of electrons. For a hole the hyperfine coupling is expected to be much weaker than for an electron because of the p-symmetry of the valence band [52]. Recent experiments have shown that hole spins remain coherent an order of magnitude longer than electron spins [53].

Because of the detrimental effect of nuclear spins one can use different material. If QRs were made not of III–V semiconductors (with non-zero nuclear spin) but of the group IV isotopes with zero nuclear spins, the coherence times should be longer because of the absence or very small (in isotopically not purified) hyperfine interaction. As a result one could then get $T_2 \sim 2T_1$, which is a relatively long time.

Besides '*natural*' semiconductors there exists another material having amazing capabilities for electronics. Carbon nanotubes constitute a new class of ballistic low dimensional quantum systems which also can be used for the implementation of a qubit [54, 55]. They are attractive because the zero nuclear spin of the dominant isotope ^{12}C yields a strongly reduced hyperfine interaction.

Because of ubiquitous nature of Si in modern electronics the estimations for Si rings [56] are important. It is known that the magnitude of the SOI in Si is ten times smaller than in GaAs and thus the relaxation times should be hundred times longer [57]. However, for Si and SiGe systems $g_s \sim 2$ and these two factors make T_1 of the same order as for GaInAs rings. On the other hand, stronger g_s allows spin control at smaller magnetic fields.

Si, the best semiconducting material for charge based electronics also seems to be a promising choice for spintronics and for quantum computing [37]. Thus the decoherence times of electron spins in material with few or no nuclear spins as well as decoherence times for hole spins are expected to be much longer than for the group III–V semiconductors [57–62]. One should stress that in all materials considered above the decoherence times are much longer than the time for initialization, qubit operations and measurement allowing for quantum error correction scheme to be efficient.

Let us briefly summarize the first part of this chapter. Spin relaxation in QRs due to SOI induced admixture mechanism turned out to have different characteristics than in QDs due to different geometry (confinement) properties. It is highly determined by the detailed shape and distribution of the radial parts of the wave functions reflected in the *overlap factor* Ξ. It follows that for singly occupied structures QDs have always longer relaxation time than QRs (assuming the same material parameters) but for relatively thin rings with higher occupation the relaxation time can exceed those for QDs (due to the increased size of the energy gaps).

The estimated relaxation times (at $B = 2$ T) for experimentally fabricated rings (see Table 18.1) are in the range of the milliseconds and for $B = 1$ T from a few milliseconds to a few seconds. Thus QRs, like QDs, are attractive for quantum information processing.

18.4 Complex Ring-Ring and Dot-Ring Nanostructures

In the preceding chapter is was shown that spin relaxation strongly depends on the overlap factor Ξ_l and on the energy gaps Δ_l between the ground and excited orbital states. From this can infer that to increase T_1 one should construct a structure in which the OF will be smaller, i.e., the wave functions of the ground and excited orbital states will be spatially separated. It can be achieved e.g., by making two coupled concentric rings [28] or by dividing the ring into two concentric rings by

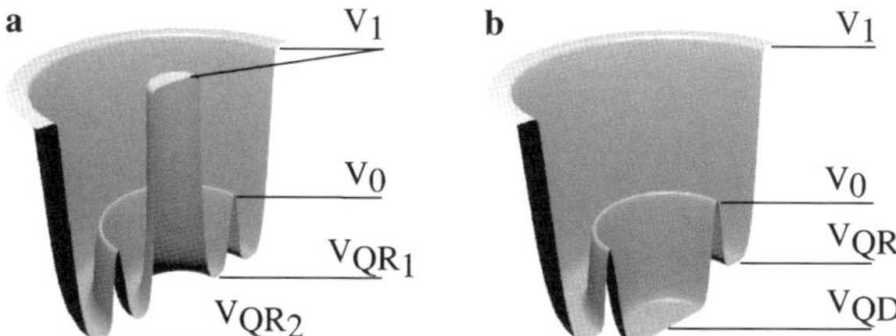

Fig. 18.4 (**a**) Cross section of the confining potential of a double ring with marked bottom of the QRs potential V_{QR_i}, $i = 1, 2$, top of the barrier potential (V_0) and the value of the potential outside (V_1). (**b**) Similar cross section of the confining potential of a QR with a QD inside (DRN)

introducing a circular barrier. Then, one can try to adjust the confining potential in such a way that the ground state wave function is localized in one of the rings, whereas the excited states are localized in the other one. In order to introduce the barrier the ring has to be sufficiently thick. Figure 18.4a shows a cross-section of the confining potential.

To be specific, the data presented below has been obtained for the ring ($V_0 = 0$) of radius $r_0 = 30$ nm and width ~ 40 nm, filled with a *single electron*. The circular barrier splitting the structure into two concentric rings is of the height $V_0 = 30$ meV. In this and the following sections the confinement potential defining the structure is assumed to have a Gaussian form [63, 64] and the height $V_1 = 80$ meV. The in-plane magnetic field is $B = 2$ T (the Zeeman splitting $\Delta_Z = 0.092$ meV), the structures are assumed to be made of InGaAs.

In Fig. 18.5 the distribution of the wave functions of the three lowest energy states involved in the spin relaxation for different values of the width of the inner ring are shown. The aim was to separate the wave functions of the ground and excited orbital states in order to increase T_1 which for the original single thick ring is $T_1 = 1$ ms. For narrow inner rings the wave functions go together, so the OF is large (Fig. 18.6b). However they start to split when the inner ring gets thicker. The most favourable situation is when the inner ring is transformed into QD where the ground state wave function is situated mainly in the center of the nanostructure (Fig. 18.5d). The resulting relaxation times are shown in Fig. 18.6c. It follows that it is of interest to investigate in more detail the properties of QR with a QD inside called afterwards a DRN.

18.4.1 Spin Relaxation in Dot-Ring Nanostructures

A cross section of the confinement potential defining a DRN used throughout the text is presented in Fig. 18.4b. Such a potential, which conserves the circular symmetry, can be obtained in many ways, e.g., using atomic force microscope to locally oxidize the surface of a sample [6, 7], by self-assembly techniques (in particular by pulsed droplet epitaxy (PDE)) [21, 22, 25, 26] by the split gates [65] or by lithog-

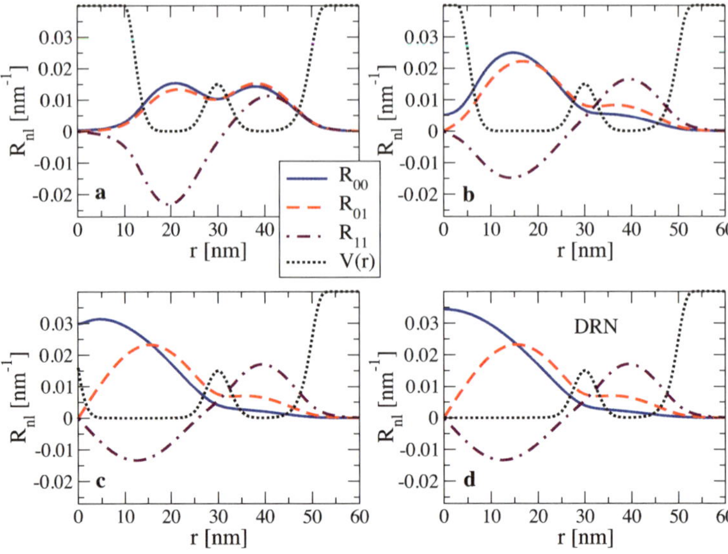

Fig. 18.5 The distribution of the wave functions of the three lowest energy states for different values of the inner ring width $V_0 = 30$ meV, $V_1 = 80$ meV

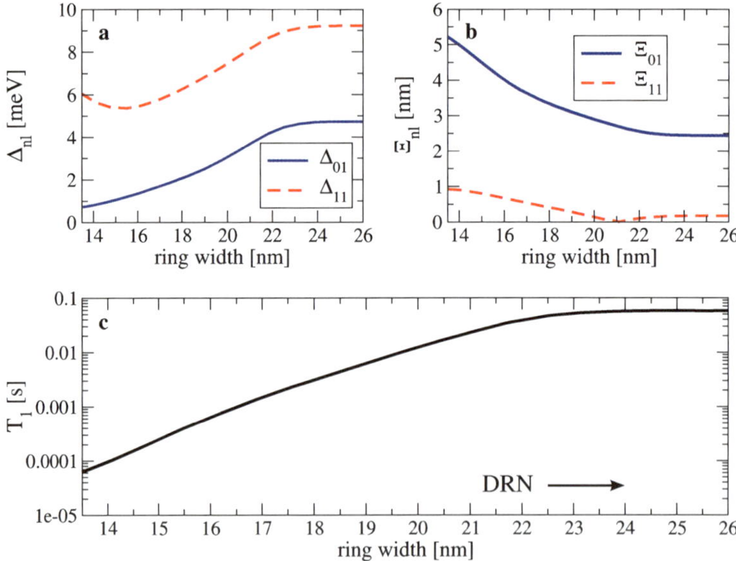

Fig. 18.6 The dependence of the orbital gaps Δ_{nl} (**a**), overlap factors Ξ_{nl} (**b**), relaxation times (**c**) on the inner ring width. $V_0 = 30$ meV, $V_{QR_1} = V_{QR_2} = 0$ meV, $V_1 = 80$ meV

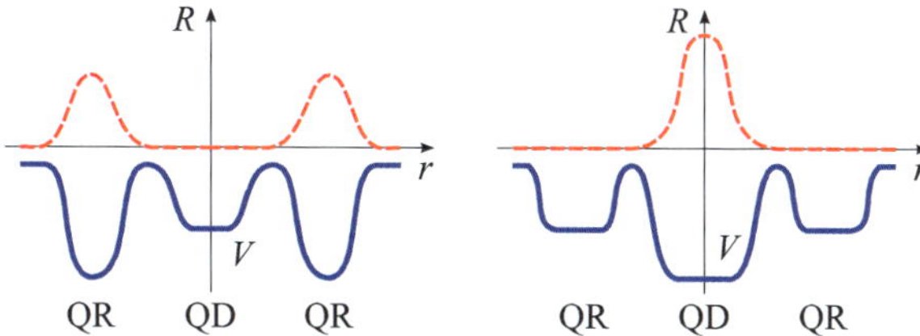

Fig. 18.7 Schematic illustration of the radial part of the ground state wave function (*dashed line*) in the case when the QD potential is deeper than the QR potential (*left panel*) or vice versa (*right panel*). The cross section of the confining potential is represented by the *solid line*

raphy. The influence of magnetic field on such single and few electron systems has been investigated in Ref. [66].

It will be shown below that by changing the parameters of the potential barrier $V_0(r)$ separating the dot from the ring and/or the potential well offset V_{QD}-V_{QR} (see Fig. 18.4b), one can considerably alter its coherent, optical and conducting properties. For example, depending on the relative positions of the bottoms of QR and QD parts the ground state electron occupies mainly the lower part of the structure which is reflected in the respective shape of the wave function (see Fig. 18.7). Moreover, by fine-tuning the confinement potential it is possible to have, e.g., the ground state located in the QD, whereas the lowest excited state in the QR (or *vice versa*). This way one can easily control the OF and all the properties which depend on it. In the following it will be shown how this feature can be exploited to modify the relaxation time. Also, a brief discussion on how such manipulations influence the optical absorption at microwave and infrared frequencies and conducting properties of a set of DRNs will be presented (for investigations of these two problems see [32]).

For concreteness, one has to assume some values of the DRN parameters. The height of the potential V_1 defining the structure can be different for different fabrication methods [67]. It can be rather shallow for modulated barrier structures ($V_1 \sim 25$–100 meV) and large for deep-etched structures ($V_1 > 10^3$ meV). In the following $V_1 = 80$ meV is assumed. Additionally, the radius of the DRN is assumed $r_0 = 50$ nm, the barrier is located at $r_0^{\text{barrier}} = 27$ nm and the zero potential energy is set at the level of V_{QR} (i.e., the potential well offset is then equal V_{QD}). It is also assumed that DRN is occupied by a *single* electron which couples to photonic (optical absorption) or phononic (spin relaxation) degrees of freedom. For such processes the selection rules allow the electron transitions to and from the ground state to the states with the orbital number $l = \pm 1$ only. Then the relevant OFs are $\Xi_{00,nl}$ and the relevant energy gaps are $\Delta_{00,nl}$ (see (18.19)) with $l = \pm 1$. For brevity these quantities are called Ξ_{nl} and Δ_{nl} respectively.

Equation (18.18) allows one to discuss the change of T_1 governed by the change of the confinement potential that defines the investigated DRN structure. It turns out that for a single electron occupying the $l = 0$ state the relaxation time is determined by the SOI to (at most) two lowest (allowed by the selection rule) excited orbital

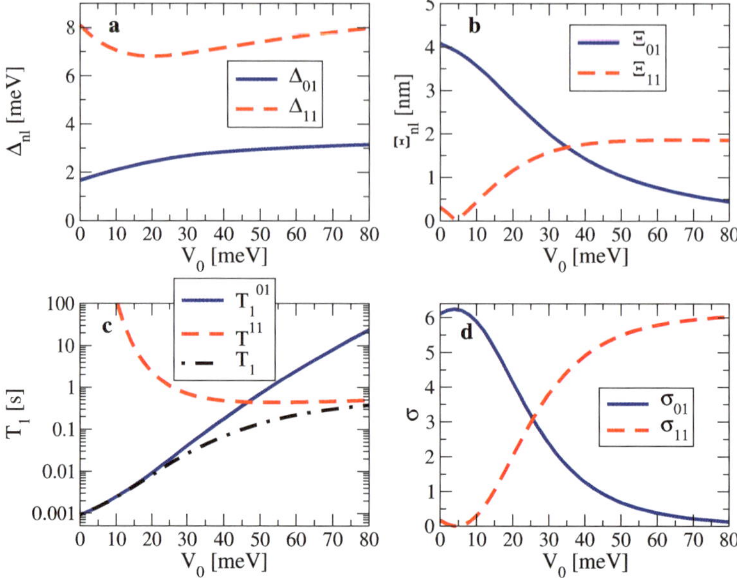

Fig. 18.8 Orbital energy gaps (**a**), overlap factor (**b**) and relaxation time (**c**) for a DRN as a function of the separating barrier height V_0. The barrier is located at $r_0^{\text{barrier}} = 27$ nm. The line T_1 shows the overall relaxation time of the nanostructure, whereas lines labeled T_1^{01} and T_1^{11} show individual relaxation times for phonon coupling to the R_{01} and R_{11} states respectively. Note that for $V_0 = 0$ the nanostructure is a big quantum dot of $r_0 \sim 50$ nm. (**d**) Cross-section for photon absorption

levels, thus

$$\frac{1}{T_1} = \frac{4\Delta_z^5}{\eta}\left(\Gamma^{01} + \Gamma^{11}\right)^2, \tag{18.29}$$

where

$$\Gamma^{01} = \frac{\Xi_{01}^2}{\Delta_{01}}, \qquad \Gamma^{11} = \frac{\Xi_{11}^2}{\Delta_{11}}.$$

The quantities entering T_1 depend on the potential confining the electrons which determines the orbital energy spectrum, the shape of the orbital wave functions and therefore the OF. In Figs. 18.8a–c Δ_{nl}, Ξ_{nl} and T_1 are plotted as functions of V_0; the numerical values are given in Table 18.3.

Comparing these figures one can see that, in contrast to Ξ_{nl}, Δ_{nl} hardly depends on V_0 and it is Ξ_{nl} which will determine T_1. Indeed, when Ξ_{nl} increases then T_1^{nl} decreases and *vice versa*. For small V_0 the dominant contribution to relaxation is given by the state E_{01} (Fig. 18.8c). Increasing V_0, the wave function of E_{01} moves over to QR which results in a decrease of Ξ_{01} (increase of T_1^{01}) with simultaneous increase of Ξ_{11} (decrease of T_1^{11}). For $V_0 \approx 45$ meV the contributions to T_1 from E_{01} and E_{11} are equal and by further increasing the height of the barrier it becomes the *higher* excited state (E_{11}) that determines T_1—a rather unusual situation. Similar

Table 18.3 The values of Δ_{nl}, Ξ_{nl}, T_1 and $\sigma_{i,f}$ as a function of the height of the barrier V_0. The parameters are: $r_0 = 50$ nm, $r_0^{\text{barrier}} = 27$ nm, $V_{\text{QD}} = V_{\text{QR}} = 0$ meV, $V_1 = 80$ meV. The cross-sections for photon absorption presented in the last two columns are discussed in Sect. 18.4.2

V_0	Δ_{01} [meV]	Δ_{11} [meV]	Ξ_{01} [nm]	Ξ_{11} [nm]	T_1 [s]	σ_{01}	σ_{11}
0	1.67	8.09	4.08	0.31	9.17E–04	6.12	0.17
10	2.10	7.06	3.57	0.44	2.45E–03	5.89	0.3
20	2.45	6.81	2.77	1.16	8.17E–03	4.15	2.02
40	2.84	7.16	1.43	1.76	6.84E–02	1.28	4.91
60	3.03	7.62	0.76	1.86	0.22	0.39	5.78
80	3.15	7.97	0.44	1.85	0.37	0.13	6.02

phenomenon has been obtained experimentally by changing the shape of the QD by electrical gating [24].

Similar considerations can be done for DRNs by changing, instead of the barrier height, the potential well offset V_{QD}. Such manipulations can be easily done experimentally by the application of the gate potential below the QD. Changing the depth of the QD one can move individual wave functions between the QD and QR. By changing considerably the V_{QD} it is possible to change the effective geometry of a nanostructure from QD through DRN to QR as was schematically shown in Fig. 18.7. The possible applications of such features are given in Sect. 18.4.3. The results of the calculations of relevant Δ's and Ξ's are presented in Fig. 18.9 and the corresponding relaxation times are given in Table 18.4. The above studies show that by changing the confinement parameters one can change T_1 three orders of magnitude.

One should stress that the important feature of such studies is not only the value of T_1 itself but the possibility to change it by external conditions which can be steered by electric fields.

18.4.2 Optical Absorption of Dot-Ring Nanostructures

The discussed above wave function engineering allows to control not only the relaxation time. Another example of a parameter that can be changed is the intraband absorption of microwave and infrared radiation. The cross section for photon absorption due to electron transition from the ith bound state $E_i(n_i, l_i)$ to the fth bound state $E_f(n_f, l_f)$ in the dipole approximation is given by the formula [68–70]

$$\sigma_{i,f} = \frac{16\pi^2 \beta \hbar \omega \Xi_{i,f}^2}{n_2} \delta(E_f - E_i - \hbar\omega)\big[f(E_i) - f(E_f)\big], \qquad (18.30)$$

where $\beta = 1/137$ is the fine structure constant, $l_f = l_i \pm 1$, n_2 is the refractive index and $f(E)$ is the Fermi-Dirac distribution function. At low temperature and with the

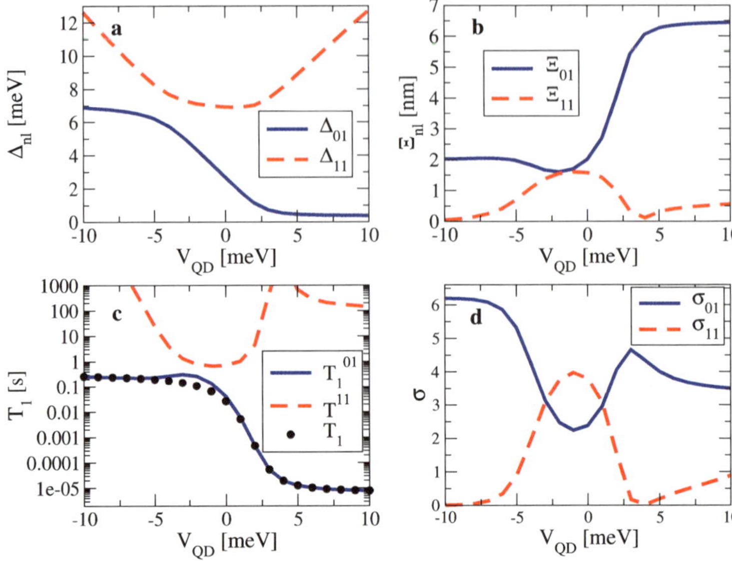

Fig. 18.9 Orbital energy gaps (**a**), overlap factor (**b**) and relaxation time (**c**) for a DRN as a function of the potential offset V_{QD}. *Closed dots* (T_1) show the overall relaxation time of the nanostructure, whereas *solid line* (T_1^{01}) and *dashed line* (T_1^{11}) show individual relaxation times for phonon coupling to states R_{01} and R_{11}, respectively. (**d**) Cross-section for photon absorption

Table 18.4 The values of Δ_{nl}, Ξ_{nl}, T_1 and $\sigma_{i,f}$ as a function of the height of the potential well offset V_{QD}. The parameters are: $V_0 = 30$ meV, $V_{QR} = 0$ meV, $V_1 = 80$ meV

V_{QD}	Δ_{01} [meV]	Δ_{11} [meV]	Ξ_{01} [nm]	Ξ_{11} [nm]	T_1[s]	σ_{01}	σ_{11}
-10	6.89	12.63	2.02	4.66E–02	0.26	6.20	6.0E–03
-8	6.74	10.8	2.04	0.15	0.24	6.16	5.04E–02
-4	5.8	7.72	1.83	1.08	0.17	4.26	1.97
0	2.67	6.9	2.0	1.58	2.64E–02	2.38	3.80
4	0.54	8.10	6.06	0.12	1.94E–05	4.34	2.54E–02
6	0.43	9.65	6.35	0.41	1.03E–05	3.80	0.36
10	0.38	12.72	6.44	0.56	27.8E–06	3.50	0.89

delta function replaced by the Lorentzian function with half-width Γ the maximum cross-section at the resonance frequency is given by

$$\sigma_{i,f}^{m} = \frac{16\pi^2 \beta\, \Xi_{i,f}^2}{n_2 \Gamma} \Delta_{i,f}. \tag{18.31}$$

Then, by manipulating the confinement potential one can control $\Xi_{i,f} \equiv \Xi_{nl,n'l'}$ and $\Delta_{i,f} \equiv \Delta_{nl,n'l'}$ which enter the above equation. This way, one is able to engineer the DRNs according to their applications:

Fig. 18.10 Electron wave functions in a one-dimensional array of DRNs in the case when electron are located in the QDs (*left panel*, electron hopping is not possible) and in the QRs (*right panel*, hopping is possible)

(a) one can design DRNs to get most effective absorption required for efficient infrared and microwave photodetectors, or

(b) one can design DRNs with negligible absorption at given frequencies, i.e., structures which will be transparent for the respective photons.

It was demonstrated in Ref. [32] that by changing V_{QD} one can smoothly move over from highly absorbing to almost transparent DRNs. The absorbed photon energy can be changed to large extent by changing the radius of DRN, the barrier height and the material (e.g., for the structure with $m^* = 0.04m_e$ it changes from microwaves to far infrared). The last two columns in Tables 18.3 and 18.4 as well as Figs. 18.8d and 18.9d present the cross-sections for different values of the parameters of the confining potential.

18.4.3 Conducting Properties of Arrays of Dot-Ring Nanostructures

Apart from unique properties of a single DRN, interesting behaviour emerges when such structures are combined into a two-dimensional array. If they are located sufficiently close to each other, electrons can tunnel from one DRN to another one, making a system that resembles a narrow band crystal. The tunneling rate depends on the overlap of the electron wave functions on adjoining structures. And since it is possible to control the shape of the wave functions, one can control the overlap, and thereby manipulate the transport properties of the crystal-like structure. If the electron wave functions are located in the QDs, the overlap is effectively zero and the system behaves like an insulator. On the other hand, when the wave functions are located in the QRs, the overlap is much larger which results in metallic character. These two situations for a one-dimensional arrays are illustrated in Fig. 18.10.

Controlling individual DRNs in an array may be involved—especially in two-dimensional arrays—since it requires to supply voltage to every gate. However, if one is interested only in global properties of such array, there is no need to control individual DRN's. In order to be able to control all the DRNs in the same way, i.e., to force electrons to occupy QDs or QRs in all nanostructures, one can place the QDs in one layer and the QRs in another one, located above (or below) the first layer. Then, the difference between the energy of states localized in QDs and states localized in QRs is proportional to the strength of electric field applied to the whole array perpendicularly to the layers. Details of realization of this idea and possible applications are given in Ref. [32]. Note that in order to have a true metal-insulator phase transition in an array of DRN's Coulomb correlations have to be taken into account. This problem has been studied, e.g., in Ref. [71], where the Hubbard model has been used to describe "solid-state" physics in an array of quantum dots [72, 73].

18.5 Summary

Among the most significant nanoscience advances a relevant place is taken by quantum confinement effects that take place in semiconductor nanostructures. The unprecedented level of control over individual electrons will enable exploration of new regime and pave the way for tests of simple quantum protocols. In particular fascinating phenomena based on a carrier confinement in ring shaped nanostructures have intrigued physicists for many years. It was found that nanorings with $R < 20$ nm can be considered as almost ideal quantum systems [39] and thus can be, besides QDs, excellent systems for spin studies.

The investigations presented in the first part of the chapter concerned quantum rings with a single or a few electrons. The results have shown that such rings can be treated as quantum bits or spin memory devices. The crucial point for quantum information processing is the necessity to keep coherence on sufficiently long time. The estimated relaxation times (at $B = 2$ T) for investigated QRs are in the range of a few milliseconds. However, for the realization of quantum-nanostructure-based devices the electronic properties should be freely accessible for engineering. For the first time it is possible to fabricate nanostructures which are designed on demand for a specific device functions with optimized properties.

Modern nanotechnology enables to produce complex systems where different building blocks such as QDs [30, 31] and QRs [28, 29] are combined together within a single structure [25–28]. It is possible to build a complex system composed of coupled elements where the coupling constant is an additional parameter that can be controlled. It was shown that a quantum ring with a quantum dot inside is a structure in which one can perform an effective wave function engineering which influences several measurable properties.

Systematic studies of the influence of the shape and the height of the confinement potential on relaxation times have been presented. It has been shown that changes of the confinement potential, e.g., by electrostatic gating can modify the relaxation times by orders of magnitude. Also, a brief discussion [32] of the influence of the confinement potential on optical absorption and on conducting properties of an array of such nanostructures (transition of metal-insulator type) has been presented.

The presented model calculations show that the above mentioned macroscopic variables can be modified by changing on demand the microscopic features of the nanosystem such as the shape and distribution of the wave functions. The wavelength range of the absorption spectra may be largely expanded from microwaves to infra-red by utilizing QNs of different size. Combined quantum structures are highly relevant to new technologies in which the control and manipulations of electron spin and wave functions play an important role. The results should serve as a hint for experimentalists in order to fabricate quantum nanostructures with, depending on destination, the best properties.

Acknowledgements M.M.M. and M.K. acknowledges support from the Foundation for Polish Science under the "TEAM" program for the years 2011–2014.

References

1. M. Büttiker, Y. Imry, R. Landauer, Phys. Lett. A **96**, 365 (1969)
2. L.P. Lévy, G. Dolan, J. Dunsmuir, H. Bouchiat, Phys. Rev. Lett. **64**, 2074 (1990)
3. D. Mailly, C. Chapelier, A. Benoit, Phys. Rev. Lett. **70**, 2020 (1993)
4. H. Bluhm, N.C. Koschnick, J.A. Bert, M.E. Huber, K.A. Moler, Phys. Rev. Lett. **102**, 136802 (2009)
5. N.A.J.M. Kleemans, I.M.A. Bominaar-Silkens, V.M. Fomin, V.N. Gladilin, D. Granados, A.G. Taboada, J.M. García, P. Offermans, U. Zeitler, P.C.M. Christianen, J.C. Maan, J.T. Devreese, P.M. Koenraad, Phys. Rev. Lett. **99**, 146808 (2009)
6. A. Fuhrer, S.L. Scher, T. Ihn, T. Henzel, K. Ensslin, W. Wegscheider, M. Bichler, Nature **413**, 822 (2001)
7. T. Ihn, A. Fuhrer, K. Ensslin, W. Wegscheider, M. Bichler, Physica E **26**, 225 (2005)
8. V. Chandrasekhar, R.A. Webb, M.J. Brady, M.B. Ketchen, W.J. Gallagher, A. Kleinsasser, Phys. Rev. Lett. **67**, 3578 (1991)
9. A.C. Bleszynski-Jayich, W.E. Shanks, B. Peaudecerf, E. Ginossar, F. von Oppen, L. Glazman, J.G.E. Harris, Science **326**, 5950 (2009)
10. W. Lei, C. Notthoff, A. Lorke, D. Reuter, A.D. Wieck, Appl. Phys. Lett. **96**, 33111 (2010)
11. A. Lorke, R.J. Luyken, A.O. Govorov, J.P. Kotthaus, Phys. Rev. Lett. **84**, 2223 (2000)
12. P. Offermans, P.M. Koenraad, J.H. Wolter, D. Granados, J.M. García, M. Fomin, V.N. Gladilin, J.T. Devreese, Appl. Phys. Lett. **87**, 131902 (2005)
13. D. Wohlleben, M. Esser, P. Freche, E. Zipper, M. Szopa, Phys. Rev. Lett. **66**, 3191 (1991)
14. E. Zipper, M. Kurpas, M. Szelag, J. Dajka, M. Szopa, Phys. Rev. B **74**, 125426 (2006)
15. J.E. Mooij, T.P. Orlando, L.S. Levitov, L. Tian, C.H. van der Wal, S. Lloyd, Science **285**, 1036 (1999)
16. V.N. Golovach, A.V. Khaetskii, D. Loss, Phys. Rev. Lett. **93**, 016601 (2004)
17. M. Nakahara, T. Ohmi, *Quantum Computing: From Linear Algebra to Physical Realizations* (CRC Press, Boca Raton, 2008)
18. D. Loss, D.P. DiVincenzo, Phys. Rev. A **57**, 120 (1998)
19. L.M.K. Vandersypen, R. Hanson, L.H. Willems van Beveren, J.M. Elzerman, J.S. Greidanus, S. De Franceschi, L.P. Kouwenhoven, in *Quantum Computing and Quantum Bits in Mesoscopic Systems*, ed. by A. Leggett, B. Ruggiero, P. Silvestrini (Kluwer Academic Plenum, Dordrecht, 2004)
20. V. Baranwal, G. Biasiol, S. Heun, A. Locatelli, T.O. Mentes, M. Niño Orti, Phys. Rev. B **80**, 155328 (2009)
21. T. Mano, T. Kuroda, K. Mitsuishi, M. Yamagiwa, X.-J. Guo, K. Furuya, K. Sakoda, N. Koguchi, J. Cryst. Growth **301**, 740 (2007)
22. T. Kuroda, T. Mano, T. Ochiai, S. Sanguinetti, K. Sakoda, G. Kido, N. Koguchi, Phys. Rev. B **72**, 20530 (2005)
23. E. Zipper, M. Kurpas, J. Sadowski, M.M. Maśka, J. Phys. Condens. Matter **23**, 115302 (2011)
24. S. Amasha, K. MacLean, P. Iuliana, D.M. Zumbühl, M.A. Kastner, M.P. Hanson, A.C. Gossard, Phys. Rev. Lett. **100**, 046803 (2008)

25. C. Somaschini, S. Bietti, N. Koguchi, S. Sanguinetti, Nanotechnology **22**, 185602 (2011)
26. S. Sanguinetti, C. Somaschini, S. Bietti, N. Koguchi, Nanomater. Nanotech. **1**, 14 (2011)
27. I. Shorubalko, A. Pfund, R. Leturcq, M.T. Borgstörm, F. Gramm, E. Müller, E. Gini, K. Ensslin, Nanotechnology **18**, 044014 (2007)
28. C. Somaschini, S. Bietti, N. Koguchi, S. Sanguinetti, Appl. Phys. Lett. **97**, 203109 (2010)
29. L.G.G.V.D. da Silva, J.M. Villas-Bôas, S.E. Ulloa, Phys. Rev. B **76**, 155306 (2007)
30. Y.Y. Wang, M.W. Wu, Phys. Rev. B **74**, 165312 (2006)
31. M. Raith, P. Stano, J. Fabian, Phys. Rev. B **83**, 195318 (2011)
32. E. Zipper, M. Kurpas, M. Maśka, New J. Phys. **14**, 093029 (2012)
33. R. Hanson, L.P. Kouwenhoven, J.R. Petta, S. Tarucha, L.M.K. Vandersypen, Rev. Mod. Phys. **79**, 1217 (2007)
34. M. Kroutvar, Y. Ducommun, D. Heiss, M. Bichler, D. Schuh, G. Abstreiter, J.J. Finley, Nature **432**, 81 (2004)
35. K.S. Virk, D.R. Reichman, M.S. Hybertsen, Phys. Rev. B **86**, 165332 (2012)
36. T.H. Stievater, X. Li, T. Cubel, D.G. Steel, D. Gammon, D.S. Katzer, D. Park, Appl. Phys. Lett. **81**, 4251 (2002)
37. R. Hanson, D.D. Awschalom, Nature **453**, 1043 (2008)
38. T.D. Ladd, F. Jelezko, R. Laflamme, Y. Nakamura, C. Monroe, J.L. O'Brienl, Nature **464**, 45 (2010)
39. M. Abbarchi, C.A. Mastrandrea, A. Vinattieri, S. Sanguinetti, T. Mano, T. Kuroda, N. Koguchi, K. Sakoda, M. Gurioli, Phys. Rev. B **79**, 085308 (2009)
40. M. Kurpas, E. Zipper, Eur. Phys. J. D **50**, 201 (2008)
41. R.A. Żak, B. Röthlisberger, S. Chesi, D. Loss, Riv. Nuovo Cimento **33**, 7 (2010)
42. A.V. Khaetskii, Y.V. Nazarov, Phys. Rev. B **64**, 125316 (2001)
43. P. Stano, J. Fabian, Phys. Rev. B **72**, 155410 (2005)
44. J. Dreiser, M. Atatüre, C. Galland, T. Müller, A. Badolato, A. Imamoglu, Phys. Rev. B **77**, 075317 (2008)
45. P. Stano, J. Fabian, Phys. Rev. B **74**, 045320 (2006)
46. P. Stano, J. Fabian, Phys. Rev. Lett. **96**, 186602 (2006)
47. L.M. Woods, T.L. Reinecke, Y. Lyanda-Geller, Phys. Rev. B **66**, 161318(R) (2002)
48. W. Yao, R.-B. Liu, L.J. Scham, Phys. Rev. B **74**, 195301 (2006)
49. H. Bluhm, S. Foletti, I. Neder, M. Rudner, D. Mahalu, V. Umansky, A. Yacoby. arXiv:1005.2995
50. X. Xu, W. Yao, B. Sun, D.G. Steel, A.S. Bracker, D. Gammon, L.J. Sham, Nature **459**, 1105 (2009)
51. G. Giedke, J.M. Taylor, D. D'Alessandro, M.D. Lukin, A. Imamoglu, Phys. Rev. A **74**, 032316 (2006)
52. E.I. Gryncharova, V.I. Perel, Sov. Phys. Semicond. **11**, 997 (1977)
53. D. Brunner, B.D. Geradot, P.A. Dalgarno, G. Wuest, K. Karrai, N.G. Stolz, P.M. Petroff, R.J. Warburton, Science **325**, 70 (2009)
54. F. Kuemmeth, S. Ilani, D.C. Ralph, P.L. McEuen, Nature **452**, 448 (2008)
55. B. Trauzettel, D.V. Bulayev, D. Loss, G. Burkhard, Nat. Phys. **3**, 192 (2007)
56. C.-H. Lee, C.W. Liu, H.-T. Chang, S.W. Lee, J. Appl. Phys. **107**, 056103 (2010)
57. A.M. Tyryshkin, S.A. Lyon, A.V. Astashkin, A.M. Raitsimring, Phys. Rev. B **68**, 193207 (2003)
58. R. de Sousa, S. Das Sarma, Phys. Rev. B **68**, 115322 (2003)
59. W. Jantsch, Z. Wilamowski, N. Sanderfeld, M. Muhlberger, F. Schaffler, Physica E **13**, 504 (2002)
60. F. Jelezko, T. Gaebel, I. Popa, A. Gruber, J. Wrachtrup, Phys. Rev. Lett. **92**, 076401 (2004)
61. G. Balasubramanian et al., Nat. Mater. **8**, 383 (2009)
62. T.D. Ladd, D. Maryenko, Y. Yamamoto, E. Abe, K.M. Itoh, Phys. Rev. B **71**, 014401 (2005)
63. M. Ciurla, J. Adamowski, B. Szafran, S. Bednarek, Physica E **15**, 261 (2002)
64. K. Lis, S. Bednarek, B. Szafran, J. Adamowski, Physica E **17**, 494 (2003)
65. N.B. Zhitenev, M. Brodsky, R.C. Ashoori, L.N. Pfeiffer, K.W. West, Science **285**, 715 (1999)

66. B. Szafran, F.M. Peeters, S. Bednarek, Phys. Rev. B **70**, 125310 (2004)
67. M. Bayer, S.N. Walck, T.L. Reinecke, A. Forchel, Phys. Rev. B **57**, 6584 (1998)
68. V. Halonen, P. Pietiläinen, T. Chakraborty, Europhys. Lett. **33**, 377 (1996)
69. V. Milanovic, Z. Ikonic, Phys. Rev. B **39**, 7982 (1989)
70. V. Bondarenko, Y. Zhao, J. Phys. Condens. Matter **15**, 1377 (2003)
71. C.A. StaEord, S. Das Sarma, Phys. Rev. Lett. **72**, 3590 (1994)
72. L.P. Kouwenhoven et al., Phys. Rev. Lett. **65**, 361 (1990)
73. R.J. Hsug, J.M. Hong, K.Y. Lee, Surf. Sci. **263**, 415 (1992)

Index

V.M. Fomin (ed.), *Physics of Quantum Rings*, NanoScience and Technology,
DOI 10.1007/978-3-642-39197-2, © Springer-Verlag Berlin Heidelberg 2014

Printed by Printforce, the Netherlands